COURSE OF THEORETICAL

VOLUME 6

FLUID MECHANICS

Second Edition

Other Titles in the Series

LANDAU and LIFSHITZ: COURSE OF THEORETICAL PHYSICS

Volume 1 Mechanics, 3rd Edition

Volume 2 The Classical Theory of Fields, 4th Edition

Volume 3 Quantum Mechanics (Non-relativistic Theory), 3rd Edition

Volume 4 Quantum Electrodynamics, 2nd Edition

Volume 5 Statistical Physics, Part 1, 3rd Edition

Volume 7 Theory of Elasticity, 3rd Edition

Volume 8 Electrodynamics of Continuous Media, 2nd Edition

Volume 9 Statistical Physics, Part 2

Volume 10 Physical Kinetics

Also by L. D. Landau and E. M. Lifshitz (based on the Course of Theoretical Physics)

A SHORTER COURSE OF THEORETICAL PHYSICS

Volume 1: Mechanics and Electrodynamics

Volume 2: Quantum Mechanics

FLUID MECHANICS

Second Edition

by

L. D. LANDAU and E. M. LIFSHITZ

Institute of Physical Problems, U.S.S.R. Academy of Sciences

Volume 6 of Course of Theoretical Physics

Second English Edition, Revised

Translated from the Russian by

J. B. SYKES and W. H. REID

ELSEVIER

BUTTERWORTH
HEINEMANN

AMSTERDAM • BOSTON • HEIDELBERG • LONDON • NEW YORK • OXFORD
PARIS • SAN DIEGO • SAN FRANCISCO • SINGAPORE • SYDNEY • TOKYO

Butterworth-Heinemann is an imprint of Elsevier
The Boulevard, Langford Lane, Kidlington, Oxford, OX5 1GB
30 Corporate Drive, Suite 400, Burlington, MA 01803, USA

First edition 1959
Reprinted 1963, 1966, 1975, 1978, 1982
Second edition 1987
Reprinted with corrections 1989, 1998, 1999, 2000, 2003, 2004, 2005, 2006, 2007, 2008, 2009

British Library Cataloguing in Publication Data
Landau, L. D.
Fluid mechanics, -2nd ed. – (Course
Of theoretical physics; v. 6)
1. Fluid mechanics
I. Title II. Lifshitz, E. M. III. Series
532 QC145.2

Library of Congress Cataloging-in-Publication Data
Landau, L. D. (Lev Davidovich), 1908–1968.
Fluid mechanics.
(Volume 6 of Course of theoretical physics)
Includes index.
Translation of: Mekhanika sploshnykh sred.
1. Fluid mechanics. I. Lifshitz, E. M.
(Evgenii Mikhailovich) II. Title. III. Series:
Teoreticheskaia fizika. English; v. 6.
QA901. L283 1987 532 86-30498

ISBN: 978-0-7506-2767-2

Printed and bound in the United Kingdom

Transferred to Digital Printing, 2010

CONTENTS

IV. BOUNDARY LAYERS

V. THERMAL CONDUCTION IN FLUIDS

VI. DIFFUSION

VII. SURFACE PHENOMENA

VIII. SOUND

PREFACE TO THE SECOND ENGLISH EDITION

The content and treatment in this edition remain in accordance with what was said in the preface to the first edition (see below). My chief care in revising and augmenting has been to comply with this principle.

Despite the lapse of thirty years, the previous edition has, with very slight exceptions, not gone out of date. Its material has been only fairly slightly supplemented and modified. About ten new sections have been added.

In recent decades, fluid mechanics has undergone extremely rapid development, and there has accordingly been a great increase in the literature of the subject. The development has been mainly in applications, however, and in an increasing complexity of the problems accessible to theoretical calculation (with or without computers). These include, in particular, various problems of instability and its development, including non-linear regimes. All such topics are beyond the scope of our book; in particular, stability problems are discussed, as previously, mainly in terms of results.

There is also no treatment of non-linear waves in dispersive media, which is by now a significant branch of mathematical physics. The purely hydrodynamic subject of this theory consists in waves with large amplitude on the surface of a liquid. Its principal physical applications are in plasma physics, non-linear optics, various problems of electrodynamics, and so on, and in that respect they belong in other volumes of the *Course*.

There have been important changes in our understanding of the mechanism whereby turbulence occurs. Although a consistent theory of turbulence is still a thing of the future, there is reason to suppose that the right path has finally been found. The basic ideas now available and the results obtained are discussed in three sections (§§30–32) written jointly with M. I. Rabinovich, to whom I am deeply grateful for this valuable assistance. A new area in continuum mechanics over the last few decades is that of liquid crystals. This combines features of the mechanics of liquid and elastic media. Its principles are discussed in the new edition of *Theory of Elasticity*.

This book has a special place among those I had occasion to write jointly with L. D. Landau. He gave it a part of his soul. That branch of theoretical physics, new to him at the time, caught his fancy, and in a very typical way he set about thinking through it *ab initio* and deriving its basic results. This led to a number of original papers which appeared in various journals, but several of his conclusions or ideas were not published elsewhere than in the book, and in some instances even his priority was not established till later. In the new edition, I have added an appropriate reference to his authorship in all such cases that are known to me.

In the revision of this book, as in other volumes of the *Course*, I have had the help and advice of many friends and colleagues. I should like to mention in particular numerous discussions with G. I. Barenblatt, L. P. Pitaevskiĭ, Ya. G. Sinaĭ, and Ya. B. Zel'dovich. Several useful comments came from A. A. Andronov, S. I. Anisimov, V. A. Belokon', A. L. Fabrikant, V. P. Kraĭnov, A. G. Kulikovskiĭ, M. A. Liberman, R. V. Polovin, and A. V. Timofeev. To all of them I express my sincere gratitude.

Institute of Physical Problems E. M. LIFSHITZ
August 1984

PREFACE TO THE FIRST ENGLISH EDITION

The present book deals with fluid mechanics, i.e. the theory of the motion of liquids and gases.

The nature of the book is largely determined by the fact that it describes fluid mechanics as a branch of theoretical physics, and it is therefore markedly different from other textbooks on the same subject. We have tried to develop as fully as possible all matters of physical interest, and to do so in such a way as to give the clearest possible picture of the phenomena and their interrelation. Accordingly, we discuss neither approximate methods of calculation in fluid mechanics, nor empirical theories devoid of physical significance. On the other hand, accounts are given of some topics not usually found in textbooks on the subject: the theory of heat transfer and diffusion in fluids; acoustics; the theory of combustion; the dynamics of superfluids; and relativistic fluid dynamics.

In a field which has been so extensively studied as fluid mechanics it was inevitable that important new results should have appeared during the several years since the last Russian edition was published. Unfortunately, our preoccupation with other matters has prevented us from including these results in the English edition. We have merely added one further chapter, on the general theory of fluctuations in fluid dynamics.

We should like to express our sincere thanks to Dr Sykes and Dr Reid for their excellent translation of the book, and to Pergamon Press for their ready agreement to our wishes in various matters relating to its publication.

Moscow 1958

L. D. LANDAU
E. M. LIFSHITZ

EVGENIĬ MIKHAĬLOVICH LIFSHITZ (1915–1985)†

Soviet physics suffered a heavy loss on 29 October 1985 with the death of the outstanding theoretical physicist Academician Evgeniĭ Mikhaĭlovich Lifshitz.

Lifshitz was born on 21 February 1915 in Khar'kov. In 1933 he graduated from the Khar'kov Polytechnic Institute. He worked at the Khar'kov Physicotechnical Institute from 1933 to 1938 and at the Institute of Physical Problems of the USSR Academy of Sciences in Moscow from 1939 until his death. He was elected an associate member of the USSR Academy of Sciences in 1966 and a full member in 1979.

Lifshitz's scientific activity began very early. He was among L. D. Landau's first students and at 19 he co-authored with him a paper on the theory of pair production in collisions. This paper, which has not lost its significance to this day, outlined many methodological features of modern relativistically invariant techniques of quantum field theory. It includes, in particular, a consistent allowance for retardation.

Modern ferromagnetism theory is based on the "Landau-Lifshitz" equation, which describes the dynamics of the magnetic moment in a ferromagnet. A 1935 article on this subject is one of the best known papers on the physics of magnetic phenomena. The derivation of the equation is accompanied by development of a theory of ferromagnetic resonance and of the domain structure of ferromagnets.

In a 1937 paper on the Boltzmann kinetic equation for electrons in a magnetic field, E. M. Lifshitz developed a drift approximation extensively used much later, in the 50s, in plasma theory.

A paper published in 1939 on deuteron dissociation in collisions remains a brilliant example of the use of quasi-classical methods in quantum mechanics.

A most important step towards the development of a theory of second-order phase transitions, following the work by L. D. Landau, was a paper by Lifshitz dealing with the change of the symmetry of a crystal, of its space group, in transitions of this type (1941). Many years later the results of this paper came into extensive use, and the terms "Lifshitz criterion" and "Lifshitz point," coined on its basis have become indispensable components of modern statistical physics.

A decisive role in the detection of an important physical phenomenon, second sound in superfluid helium, was played by a 1944 paper by E. M. Lifshitz. It is shown in it that second sound is effectively excited by a heater having an alternating temperature. This was precisely the method used to observe second sound in experiment two years later.

A new approach to the theory of molecular-interaction forces between condensed bodies was developed by Lifshitz in 1954–1959. It is based on the profound physical idea that these forces are manifestations of stresses due to quantum and thermal fluctuations of an electromagnetic field in a medium. This idea was pursued to develop a very elegant and general theory in which the interaction forces are expressed in terms of electrodynamic material properties such as the complex dielectric permittivity. This theory of E. M.

---

† By A. F. Andreev, A. S. Borovik-Romanov, V. L. Ginzburg, L. P. Gor'kov, I. E. Dzyaloshinskiĭ, Ya. B. Zel'dovich, M. I. Kaganov, L. P. Pitaevskiĭ, E. L. Feĭnberg, and I. M. Khalatnikov; published in Russian in *Uspekhi fizicheskikh nauk* **148**, 549–550, 1986. This translation is by J. G. Adashko (first published in *Soviet Physics Uspekhi* **29**, 294–295, 1986), and is reprinted by kind permission of the American Institute of Physics.

Lifshitz stimulated many studies and was confirmed by experiment. It gained him the M. V. Lomonosov Prize in 1958.

E. M. Lifshitz made a fundamental contribution in one of the most important branches of modern physics, the theory of gravitation. His research into this field started with a classical 1946 paper on the stability of cosmological solutions of Einstein's theory of gravitation. The perturbations were divided into distinctive classes—scalar, with variation of density, vector, describing vortical motion, and finally tensor, describing gravitational waves. This classification is still of decisive significance in the analysis of the origin of the universe. From there, E. M. Lifshitz tackled the exceedingly difficult question of the general character of the singularities of this theory. Many years of labor led in 1972 to a complete solution of this problem in papers written jointly with V. A. Belinskiĭ and I. M. Khalatnikov, which earned their authors the 1974 L. D. Landau Prize. The singularity was found to have a complicated oscillatory character and could be illustratively represented as contraction of space in two directions with simultaneous expansion in the third. The contraction and expansion alternate in time according to a definite law. These results elicited a tremendous response from specialists, altered radically our ideas concerning relativistic collapse, and raised a host of physical and mathematical problems that still await solution.

His life-long occupation was the famous Landau and Lifshitz *Course of Theoretical Physics*, to which he devoted about 50 years. (The first edition of *Statistical Physics* was written in 1937. A new edition of *Theory of Elasticity* went to press shortly before his last illness.) The greater part of the *Course* was written by Lifshitz together with his teacher and friend L. D. Landau. After the automobile accident that made Landau unable to work, Lifshitz completed the edition jointly with Landau's students. He later continued to revise the previously written volumes in the light of the latest advances in science. Even in the hospital, he discussed with visiting friends the topics that should be subsequently included in the *Course*.

The *Course of Theoretical Physics* became world famous. It was translated in its entirety into six languages. Individual volumes were published in 10 more languages. In 1972 L. D. Landau and E. M. Lifshitz were awarded the Lenin Prize for the volumes published by then.

The *Course of Theoretical Physics* remains a monument to E. M. Lifshitz as a scientist and a pedagogue. It has educated many generations of physicists, is being studied, and will continue to teach students in future generations.

A versatile physicist, E. M. Lifshitz dealt also with applications. He was awarded the USSR State Prize in 1954.

A tremendous amount of E.M. Lifshitz's labor and energy was devoted to Soviet scientific periodicals. From 1946 to 1949 and from 1955 to his death he was deputy editor-in-chief of the *Journal of Experimental and Theoretical Physics*. His extreme devotion to science, adherence to principles, and meticulousness greatly helped to make this journal one of the best scientific periodicals in the world.

E. M. Lifshitz accomplished much in his life. He will remain in our memory as a remarkable physicist and human being. His name will live forever in the history of Soviet physics.

NOTATION

ρ density
p pressure
T temperature
s entropy per unit mass
ε internal energy per unit mass
$w = \varepsilon + p/\rho$ heat function (enthalpy)
$\gamma = c_p/c_v$ ratio of specific heats at constant pressure and constant volume
η dynamic viscosity
$\nu = \eta/\rho$ kinematic viscosity
κ thermal conductivity
$\chi = \kappa/\rho c_p$ thermometric conductivity
R Reynolds number
c velocity of sound
M ratio of fluid velocity to velocity of sound (Mach number)

Vector and tensor (three-dimensional) suffixes are denoted by Latin letters $i, k, l, \ldots$. Summation over repeated ("dummy") suffixes is everywhere implied. The unit tensor is δ_{ik}:

References to other volumes in the *Course of Theoretical Physics:*

Fields = Vol. 2 (*The Classical Theory of Fields*, fourth English edition, 1975).
QM = Vol. 3 (*Quantum Mechanics*, third English edition, 1977).
SP 1 = Vol. 5 (*Statistical Physics*, Part 1, third English edition, 1980).
ECM = Vol. 8 (*Electrodynamics of Continuous Media*, second English edition, 1984).
SP 2 = Vol. 9 (*Statistical Physics*, Part 2, English edition, 1980).
PK = Vol. 10 (*Physical Kinetics*, English edition, 1981).

All are published by Pergamon Press.

CHAPTER I

IDEAL FLUIDS

§1. The equation of continuity

Fluid dynamics concerns itself with the study of the motion of fluids (liquids and gases). Since the phenomena considered in fluid dynamics are macroscopic, a fluid is regarded as a continuous medium. This means that any small volume element in the fluid is always supposed so large that it still contains a very great number of molecules. Accordingly, when we speak of infinitely small elements of volume, we shall always mean those which are "physically" infinitely small, i.e. very small compared with the volume of the body under consideration, but large compared with the distances between the molecules. The expressions *fluid particle* and *point in a fluid* are to be understood in a similar sense. If, for example, we speak of the displacement of some fluid particle, we mean not the displacement of an individual molecule, but that of a volume element containing many molecules, though still regarded as a point.

The mathematical description of the state of a moving fluid is effected by means of functions which give the distribution of the fluid velocity $\mathbf{v} = \mathbf{v}(x, y, z, t)$ and of any two thermodynamic quantities pertaining to the fluid, for instance the pressure $p(x, y, z, t)$ and the density $\rho(x, y, z, t)$. All the thermodynamic quantities are determined by the values of any two of them, together with the equation of state; hence, if we are given five quantities, namely the three components of the velocity $\mathbf{v}$, the pressure p and the density ρ, the state of the moving fluid is completely determined.

All these quantities are, in general, functions of the coordinates x, y, z and of the time t. We emphasize that $\mathbf{v}(x, y, z, t)$ is the velocity of the fluid at a given point (x, y, z) in space and at a given time t, i.e. it refers to fixed points in space and not to specific particles of the fluid; in the course of time, the latter move about in space. The same remarks apply to ρ and p.

We shall now derive the fundamental equations of fluid dynamics. Let us begin with the equation which expresses the conservation of matter. We consider some volume V_0 of space. The mass of fluid in this volume is $\int \rho \, dV$, where ρ is the fluid density, and the integration is taken over the volume V_0. The mass of fluid flowing in unit time through an element $\mathbf{df}$ of the surface bounding this volume is $\rho\mathbf{v}\cdot\mathbf{df}$; the magnitude of the vector $\mathbf{df}$ is equal to the area of the surface element, and its direction is along the normal. By convention, we take $\mathbf{df}$ along the outward normal. Then $\rho\mathbf{v}\cdot\mathbf{df}$ is positive if the fluid is flowing out of the volume, and negative if the flow is into the volume. The total mass of fluid flowing out of the volume V_0 in unit time is therefore

$$\oint \rho \mathbf{v}\cdot\mathbf{df},$$

where the integration is taken over the whole of the closed surface surrounding the volume in question.

Next, the decrease per unit time in the mass of fluid in the volume V_0 can be written

$$-\frac{\partial}{\partial t}\int \rho\, \mathrm{d}V.$$

Equating the two expressions, we have

$$\frac{\partial}{\partial t}\int \rho\, \mathrm{d}V = -\oint \rho\, \mathbf{v}\cdot \mathbf{df}. \tag{1.1}$$

The surface integral can be transformed by Green's formula to a volume integral:

$$\oint \rho\, \mathbf{v}\cdot \mathbf{df} = \int \operatorname{div}(\rho\mathbf{v})\, \mathrm{d}V.$$

Thus

$$\int \left[\frac{\partial \rho}{\partial t} + \operatorname{div}(\rho\mathbf{v})\right] \mathrm{d}V = 0.$$

Since this equation must hold for any volume, the integrand must vanish, i.e.

$$\partial\rho/\partial t + \operatorname{div}(\rho\mathbf{v}) = 0. \tag{1.2}$$

This is the *equation of continuity*. Expanding the expression div $(\rho\mathbf{v})$, we can also write (1.2) as

$$\partial\rho/\partial t + \rho \operatorname{div}\mathbf{v} + \mathbf{v}\cdot \mathbf{grad}\, \rho = 0. \tag{1.3}$$

The vectoı

$$\mathbf{j} = \rho\mathbf{v} \tag{1.4}$$

is called the *mass flux density*. Its direction is that of the motion of the fluid, while ıts magnitude equals the mass of fluid flowing in unit time through unit area perpendicular to the velocity.

§2. Euler's equation

Let us consider some volume in the fluid. The total force acting on this volume is equal to the integral

$$-\oint p\, \mathbf{df}$$

of the pressure, taken over the surface bounding the volume. Transforming it to a volume integral, we have

$$-\oint p\, \mathbf{df} = -\int \mathbf{grad}\, p\, \mathrm{d}V.$$

Hence we see that the fluid surrounding any volume element $\mathrm{d}V$ exerts on that element a force $-\mathrm{d}V\, \mathbf{grad}\, p$. In other words, we can say that a force $-\mathbf{grad}\, p$ acts on unit volume of the fluid.

We can now write down the equation of motion of a volume element in the fluid by equating the force $-\mathbf{grad}\, p$ to the product of the mass per unit volume (ρ) and the acceleration $\mathrm{d}\mathbf{v}/\mathrm{d}t$:

$$\rho\, \mathrm{d}\mathbf{v}/\mathrm{d}t = -\mathbf{grad}\, p. \tag{2.1}$$

The derivative $\mathrm{d}\mathbf{v}/\mathrm{d}t$ which appears here denotes not the rate of change of the fluid velocity at a fixed point in space, but the rate of change of the velocity of a given fluid particle as it moves about in space. This derivative has to be expressed in terms of quantities referring to points fixed in space. To do so, we notice that the change $\mathrm{d}\mathbf{v}$ in the velocity of the given fluid particle during the time $\mathrm{d}t$ is composed of two parts, namely the change during $\mathrm{d}t$ in the velocity at a point fixed in space, and the difference between the velocities (at the same instant) at two points $\mathrm{d}\mathbf{r}$ apart, where $\mathrm{d}\mathbf{r}$ is the distance moved by the given fluid particle during the time $\mathrm{d}t$. The first part is $(\partial\mathbf{v}/\partial t)\mathrm{d}t$, where the derivative $\partial\mathbf{v}/\partial t$ is taken for constant x, y, z, i.e. at the given point in space. The second part is

$$\mathrm{d}x\frac{\partial\mathbf{v}}{\partial x}+\mathrm{d}y\frac{\partial\mathbf{v}}{\partial y}+\mathrm{d}z\frac{\partial\mathbf{v}}{\partial z} = (\mathrm{d}\mathbf{r}\cdot\mathbf{grad})\mathbf{v}.$$

Thus

$$\mathrm{d}\mathbf{v} = (\partial\mathbf{v}/\partial t)\mathrm{d}t + (\mathrm{d}\mathbf{r}\cdot\mathbf{grad})\mathbf{v},$$

or, dividing both sides by $\mathrm{d}t$,†

$$\frac{\mathrm{d}\mathbf{v}}{\mathrm{d}t} = \frac{\partial\mathbf{v}}{\partial t} + (\mathbf{v}\cdot\mathbf{grad})\mathbf{v}. \tag{2.2}$$

Substituting this in (2.1), we find

$$\frac{\partial\mathbf{v}}{\partial t} + (\mathbf{v}\cdot\mathbf{grad})\mathbf{v} = -\frac{1}{\rho}\mathbf{grad}\,p. \tag{2.3}$$

This is the required equation of motion of the fluid; it was first obtained by L. Euler in 1755. It is called *Euler's equation* and is one of the fundamental equations of fluid dynamics.

If the fluid is in a gravitational field, an additional force $\rho\mathbf{g}$, where $\mathbf{g}$ is the acceleration due to gravity, acts on any unit volume. This force must be added to the right-hand side of equation (2.1), so that equation (2.3) takes the form

$$\frac{\partial\mathbf{v}}{\partial t} + (\mathbf{v}\cdot\mathbf{grad})\mathbf{v} = -\frac{\mathbf{grad}\,p}{\rho} + \mathbf{g}. \tag{2.4}$$

In deriving the equations of motion we have taken no account of processes of energy dissipation, which may occur in a moving fluid in consequence of internal friction (viscosity) in the fluid and heat exchange between different parts of it. The whole of the discussion in this and subsequent sections of this chapter therefore holds good only for motions of fluids in which thermal conductivity and viscosity are unimportant; such fluids are said to be *ideal*.

The absence of heat exchange between different parts of the fluid (and also, of course, between the fluid and bodies adjoining it) means that the motion is adiabatic throughout the fluid. Thus the motion of an ideal fluid must necessarily be supposed adiabatic.

In adiabatic motion the entropy of any particle of fluid remains constant as that particle moves about in space. Denoting by s the entropy per unit mass, we can express the condition for adiabatic motion as

$$\mathrm{d}s/\mathrm{d}t = 0, \tag{2.5}$$

† The derivative $\mathrm{d}/\mathrm{d}t$ thus defined is called the *substantial* time derivative, to emphasize its connection with the moving substance.

where the total derivative with respect to time denotes, as in (2.1), the rate of change of entropy for a given fluid particle as it moves about. This condition can also be written

$$\partial s/\partial t + \mathbf{v}\cdot\mathbf{grad}\, s = 0. \tag{2.6}$$

This is the general equation describing adiabatic motion of an ideal fluid. Using (1.2), we can write it as an "equation of continuity" for entropy:

$$\partial(\rho s)/\partial t + \operatorname{div}(\rho s\mathbf{v}) = 0. \tag{2.7}$$

The product $\rho s\mathbf{v}$ is the *entropy flux density*.

The adiabatic equation usually takes a much simpler form. If, as usually happens, the entropy is constant throughout the volume of the fluid at some initial instant, it retains everywhere the same constant value at all times and for any subsequent motion of the fluid. In this case we can write the adiabatic equation simply as

$$s = \text{constant}, \tag{2.8}$$

and we shall usually do so in what follows. Such a motion is said to be *isentropic*.

We may use the fact that the motion is isentropic to put the equation of motion (2.3) in a somewhat different form. To do so, we employ the familiar thermodynamic relation

$$\mathrm{d}w = T\,\mathrm{d}s + V\,\mathrm{d}p,$$

where w is the heat function per unit mass of fluid (enthalpy), $V = 1/\rho$ is the specific volume, and T is the temperature. Since s = constant, we have simply

$$\mathrm{d}w = V\,\mathrm{d}p = \mathrm{d}p/\rho,$$

and so $(\mathbf{grad}\, p)/\rho = \mathbf{grad}\, w$. Equation (2.3) can therefore be written in the form

$$\partial\mathbf{v}/\partial t + (\mathbf{v}\cdot\mathbf{grad})\mathbf{v} = -\mathbf{grad}\, w. \tag{2.9}$$

It is useful to notice one further form of Euler's equation, in which it involves only the velocity. Using a formula well known in vector analysis,

$$\tfrac{1}{2}\mathbf{grad}\, v^2 = \mathbf{v}\times\mathbf{curl}\,\mathbf{v} + (\mathbf{v}\cdot\mathbf{grad})\mathbf{v},$$

we can write (2.9) in the form

$$\partial\mathbf{v}/\partial t - \mathbf{v}\times\mathbf{curl}\,\mathbf{v} = -\mathbf{grad}\,(w + \tfrac{1}{2}v^2). \tag{2.10}$$

If we take the curl of both sides of this equation, we obtain

$$\frac{\partial}{\partial t}(\mathbf{curl}\,\mathbf{v}) = \mathbf{curl}\,(\mathbf{v}\times\mathbf{curl}\,\mathbf{v}), \tag{2.11}$$

which involves only the velocity.

The equations of motion have to be supplemented by the boundary conditions that must be satisfied at the surfaces bounding the fluid. For an ideal fluid, the boundary condition is simply that the fluid cannot penetrate a solid surface. This means that the component of the fluid velocity normal to the bounding surface must vanish if that surface is at rest:

$$v_n = 0. \tag{2.12}$$

In the general case of a moving surface, v_n must be equal to the corresponding component of the velocity of the surface.

At a boundary between two immiscible fluids, the condition is that the pressure and the velocity component normal to the surface of separation must be the same for the two fluids, and each of these velocity components must be equal to the corresponding component of the velocity of the surface.

As has been said at the beginning of §1, the state of a moving fluid is determined by five quantities: the three components of the velocity **v** and, for example, the pressure p and the density ρ. Accordingly, a complete system of equations of fluid dynamics should be five in number. For an ideal fluid these are Euler's equations, the equation of continuity, and the adiabatic equation.

PROBLEM

Write down the equations for one-dimensional motion of an ideal fluid in terms of the variables a, t, where a (called a *Lagrangian variable*†) is the x coordinate of a fluid particle at some instant $t = t_0$.

SOLUTION. In these variables the coordinate x of any fluid particle at any instant is regarded as a function of t and its coordinate a at the initial instant: $x = x(a, t)$. The condition of conservation of mass during the motion of a fluid element (the equation of continuity) is accordingly written $\rho\, dx = \rho_0\, da$, or

$$\rho\left(\frac{\partial x}{\partial a}\right)_t = \rho_0,$$

where $\rho_0(a)$ is a given initial density distribution. The velocity of a fluid particle is, by definition, $v = (\partial x/\partial t)_a$, and the derivative $(\partial v/\partial t)_a$ gives the rate of change of the velocity of the particle during its motion. Euler's equation becomes

$$\left(\frac{\partial v}{\partial t}\right)_a = -\frac{1}{\rho_0}\left(\frac{\partial p}{\partial a}\right)_t,$$

and the adiabatic equation is

$$(\partial s/\partial t)_a = 0.$$

§3. Hydrostatics

For a fluid at rest in a uniform gravitational field, Euler's equation (2.4) takes the form

$$\mathbf{grad}\, p = \rho \mathbf{g}. \tag{3.1}$$

This equation describes the mechanical equilibrium of the fluid. (If there is no external force, the equation of equilibrium is simply $\mathbf{grad}\, p = 0$, i.e. $p =$ constant; the pressure is the same at every point in the fluid.)

Equation (3.1) can be integrated immediately if the density of the fluid may be supposed constant throughout its volume, i.e. if there is no significant compression of the fluid under the action of the external force. Taking the z-axis vertically upward, we have

$$\partial p/\partial x = \partial p/\partial y = 0, \qquad \partial p/\partial z = -\rho g.$$

Hence

$$p = -\rho g z + \text{constant}.$$

If the fluid at rest has a free surface at height h, to which an external pressure p_0, the same at every point, is applied, this surface must be the horizontal plane $z = h$. From the condition $p = p_0$ for $z = h$, we find that the constant is $p_0 + \rho g h$, so that

$$p = p_0 + \rho g(h - z). \tag{3.2}$$

† Although such variables are usually called Lagrangian, the equations of motion in these coordinates were first obtained by Euler, at the same time as equations (2.3).

For large masses of liquid, and for a gas, the density ρ cannot in general be supposed constant; this applies especially to gases (for example, the atmosphere). Let us suppose that the fluid is not only in mechanical equilibrium but also in thermal equilibrium. Then the temperature is the same at every point, and equation (3.1) may be integrated as follows. We use the familiar thermodynamic relation

$$d\Phi = -s\,dT + V\,dp,$$

where Φ is the thermodynamic potential (Gibbs free energy) per unit mass. For constant temperature

$$d\Phi = V\,dp = dp/\rho.$$

Hence we see that the expression $(\mathbf{grad}\, p)/\rho$ can be written in this case as $\mathbf{grad}\,\Phi$, so that the equation of equilibrium (3.1) takes the form

$$\mathbf{grad}\,\Phi = \mathbf{g}.$$

For a constant vector $\mathbf{g}$ directed along the negative z-axis we have

$$\mathbf{g} \equiv -\mathbf{grad}\,(gz).$$

Thus

$$\mathbf{grad}\,(\Phi + gz) = 0,$$

whence we find that throughout the fluid

$$\Phi + gz = \text{constant}; \tag{3.3}$$

gz is the potential energy of unit mass of fluid in the gravitational field. The condition (3.3) is known from statistical physics to be the condition for thermodynamic equilibrium of a system in an external field.

We may mention here another simple consequence of equation (3.1). If a fluid (such as the atmosphere) is in mechanical equilibrium in a gravitational field, the pressure in it can be a function only of the altitude z (since, if the pressure were different at different points with the same altitude, motion would result). It then follows from (3.1) that the density

$$\rho = -\frac{1}{g}\frac{dp}{dz} \tag{3.4}$$

is also a function of z only. The pressure and density together determine the temperature, which is therefore again a function of z only. Thus, in mechanical equilibrium in a gravitational field, the pressure, density and temperature distributions depend only on the altitude. If, for example, the temperature is different at different points with the same altitude, then mechanical equilibrium is impossible.

Finally, let us derive the equation of equilibrium for a very large mass of fluid, whose separate parts are held together by gravitational attraction—a star. Let ϕ be the Newtonian gravitational potential of the field due to the fluid. It satisfies the differential equation

$$\triangle\phi = 4\pi G\rho, \tag{3.5}$$

where G is the Newtonian constant of gravitation. The gravitational acceleration is $-\mathbf{grad}\,\phi$, and the force on a mass ρ is $-\rho\,\mathbf{grad}\,\phi$. The condition of equilibrium is therefore

$$\mathbf{grad}\,p = -\rho\,\mathbf{grad}\,\phi.$$

Dividing both sides by ρ, taking the divergence of both sides, and using equation (3.5), we obtain

$$\operatorname{div}\left(\frac{1}{\rho}\mathbf{grad}\,p\right) = -4\pi G\rho. \tag{3.6}$$

It must be emphasized that the present discussion concerns only mechanical equilibrium; equation (3.6) does not presuppose the existence of complete thermal equilibrium.

If the body is not rotating, it will be spherical when in equilibrium, and the density and pressure distributions will be spherically symmetrical. Equation (3.6) in spherical polar coordinates then takes the form

$$\frac{1}{r^2}\frac{\mathrm{d}}{\mathrm{d}r}\left(\frac{r^2}{\rho}\frac{\mathrm{d}p}{\mathrm{d}r}\right) = -4\pi G\rho. \tag{3.7}$$

§4. The condition that convection be absent

A fluid can be in mechanical equilibrium (i.e. exhibit no macroscopic motion) without being in thermal equilibrium. Equation (3.1), the condition for mechanical equilibrium, can be satisfied even if the temperature is not constant throughout the fluid. However, the question then arises of the stability of such an equilibrium. It is found that the equilibrium is stable only when a certain condition is fulfilled. Otherwise, the equilibrium is unstable, and this leads to the appearance in the fluid of currents which tend to mix the fluid in such a way as to equalize the temperature. This motion is called *convection*. Thus the condition for a mechanical equilibrium to be stable is the condition that convection be absent. It can be derived as follows.

Let us consider a fluid element at height z, having a specific volume $V(p, s)$, where p and s are the equilibrium pressure and entropy at height z. Suppose that this fluid element undergoes an adiabatic upward displacement through a small interval ξ; its specific volume then becomes $V(p', s)$, where p' is the pressure at height $z+\xi$. For the equilibrium to be stable, it is necessary (though not in general sufficient) that the resulting force on the element should tend to return it to its original position. This means that the element must be heavier than the fluid which it "displaces" in its new position. The specific volume of the latter is $V(p', s')$, where s' is the equilibrium entropy at height $z+\xi$. Thus we have the stability condition

$$V(p', s') - V(p', s) > 0.$$

Expanding this difference in powers of $s' - s = \xi\,\mathrm{d}s/\mathrm{d}z$, we obtain

$$\left(\frac{\partial V}{\partial s}\right)_p \frac{\mathrm{d}s}{\mathrm{d}z} > 0. \tag{4.1}$$

The formulae of thermodynamics give

$$\left(\frac{\partial V}{\partial s}\right)_p = \frac{T}{c_p}\left(\frac{\partial V}{\partial T}\right)_p,$$

where c_p is the specific heat at constant pressure. Both c_p and T are positive, so that we can write (4.1) as

$$\left(\frac{\partial V}{\partial T}\right)_p \frac{\mathrm{d}s}{\mathrm{d}z} > 0. \tag{4.2}$$

The majority of substances expand on heating, i.e. $(\partial V/\partial T)_p > 0$. The condition that convection be absent then becomes

$$ds/dz > 0, \tag{4.3}$$

i.e. the entropy must increase with height.

From this we easily find the condition that must be satisfied by the temperature gradient dT/dz. Expanding the derivative ds/dz, we have

$$\frac{ds}{dz} = \left(\frac{\partial s}{\partial T}\right)_p \frac{dT}{dz} + \left(\frac{\partial s}{\partial p}\right)_T \frac{dp}{dz} = \frac{c_p}{T}\frac{dT}{dz} - \left(\frac{\partial V}{\partial T}\right)_p \frac{dp}{dz} > 0.$$

Finally, substituting from (3.4) $dp/dz = -g/V$, we obtain

$$-dT/dz < g\beta T/c_p, \tag{4.4}$$

where $\beta = (1/V)(\partial V/\partial T)_p$ is the thermal expansion coefficient. For a column of gas in equilibrium which can be taken as a thermodynamically perfect gas, $\beta T = 1$ and (4.4) becomes

$$-dT/dz < g/c_p. \tag{4.5}$$

Convection occurs if these conditions are not satisfied, i.e. if the temperature decreases upwards with a gradient whose magnitude exceeds the value given by (4.4) and (4.5).†

§5. Bernoulli's equation

The equations of fluid dynamics are much simplified in the case of steady flow. By *steady flow* we mean one in which the velocity is constant in time at any point occupied by fluid. In other words, **v** is a function of the coordinates only, so that $\partial \mathbf{v}/\partial t = 0$. Equation (2.10) then reduces to

$$\tfrac{1}{2}\,\mathbf{grad}\, v^2 - \mathbf{v}\times\mathbf{curl}\,\mathbf{v} = -\mathbf{grad}\, w. \tag{5.1}$$

We now introduce the concept of *streamlines*. These are lines such that the tangent to a streamline at any point gives the direction of the velocity at that point; they are determined by the following system of differential equations:

$$\frac{dx}{v_x} = \frac{dy}{v_y} = \frac{dz}{v_z}. \tag{5.2}$$

In steady flow the streamlines do not vary with time, and coincide with the paths of the fluid particles. In non-steady flow this coincidence no longer occurs: the tangents to the streamlines give the directions of the velocities of fluid particles at various points in space at a given instant, whereas the tangents to the paths give the directions of the velocities of given fluid particles at various times.

We form the scalar product of equation (5.1) with the unit vector tangent to the streamline at each point; this unit vector is denoted by **l**. The projection of the gradient on any direction is, as we know, the derivative in that direction. Hence the projection of **grad** w is $\partial w/\partial l$. The vector $\mathbf{v}\times\mathbf{curl}\,\mathbf{v}$ is perpendicular to **v**, and its projection on the direction of **l** is therefore zero.

† For water at 20°C, the right-hand side of (4.4) is about one degree per 6.7 km; for air, the right-hand side of (4.5) is about one degree per 100 m.

Thus we obtain from equation (5.1)

$$\frac{\partial}{\partial l}(\tfrac{1}{2}v^2 + w) = 0.$$

It follows from this that $\frac{1}{2}v^2 + w$ is constant along a streamline:

$$\tfrac{1}{2}v^2 + w = \text{constant}. \tag{5.3}$$

In general the constant takes different values for different streamlines. Equation (5.3) is called *Bernoulli's equation.*†

If the flow takes place in a gravitational field, the acceleration **g** due to gravity must be added to the right-hand side of equation (5.1). Let us take the direction of gravity as the z-axis, with z increasing upwards. Then the cosine of the angle between the directions of **g** and **l** is equal to the derivative $-\mathrm{d}z/\mathrm{d}l$, so that the projection of **g** on **l** is

$$-g\,\mathrm{d}z/\mathrm{d}l.$$

Accordingly, we now have

$$\frac{\partial}{\partial l}(\tfrac{1}{2}v^2 + w + gz) = 0.$$

Thus Bernoulli's equation states that along a streamline

$$\tfrac{1}{2}v^2 + w + gz = \text{constant}. \tag{5.4}$$

§6. The energy flux

Let us choose some volume element fixed in space, and find how the energy of the fluid contained in this volume element varies with time. The energy of unit volume of fluid is

$$\tfrac{1}{2}\rho v^2 + \rho\varepsilon,$$

where the first term is the kinetic energy and the second the internal energy, ε being the internal energy per unit mass. The change in this energy is given by the partial derivative

$$\frac{\partial}{\partial t}(\tfrac{1}{2}\rho v^2 + \rho\varepsilon).$$

To calculate this quantity, we write

$$\frac{\partial}{\partial t}(\tfrac{1}{2}\rho v^2) = \tfrac{1}{2}v^2\frac{\partial\rho}{\partial t} + \rho\mathbf{v}\cdot\frac{\partial\mathbf{v}}{\partial t},$$

or, using the equation of continuity (1.2) and the equation of motion (2.3),

$$\frac{\partial}{\partial t}(\tfrac{1}{2}\rho v^2) = -\tfrac{1}{2}v^2\,\mathrm{div}(\rho\mathbf{v}) - \mathbf{v}\cdot\mathbf{grad}\,p - \rho\mathbf{v}\cdot(\mathbf{v}\cdot\mathbf{grad})\mathbf{v}.$$

In the last term we replace $\mathbf{v}\cdot(\mathbf{v}\cdot\mathbf{grad})\mathbf{v}$ by $\frac{1}{2}\mathbf{v}\cdot\mathbf{grad}\,v^2$, and $\mathbf{grad}\,p$ by $\rho\,\mathbf{grad}\,w - \rho T\,\mathbf{grad}\,s$ (using the thermodynamic relation $\mathrm{d}w = T\mathrm{d}s + (1/\rho)\mathrm{d}p$), obtaining

$$\frac{\partial}{\partial t}(\tfrac{1}{2}\rho v^2) = -\tfrac{1}{2}v^2\,\mathrm{div}\,(\rho\mathbf{v}) - \rho\mathbf{v}\cdot\mathbf{grad}\,(\tfrac{1}{2}v^2 + w) + \rho T\mathbf{v}\cdot\mathbf{grad}\,s.$$

† It was derived for an incompressible fluid (§10) by D. Bernoulli in 1738.

In order to transform the derivative $\partial(\rho\varepsilon)/\partial t$, we use the thermodynamic relation

$$d\varepsilon = T ds - p dV = T ds + (p/\rho^2) d\rho.$$

Since $\varepsilon + p/\rho = \varepsilon + pV$ is simply the heat function w per unit mass, we find

$$d(\rho\varepsilon) = \varepsilon d\rho + \rho d\varepsilon = w d\rho + \rho T ds,$$

and so

$$\frac{\partial(\rho\varepsilon)}{\partial t} = w\frac{\partial \rho}{\partial t} + \rho T\frac{\partial s}{\partial t} = -w \operatorname{div}(\rho \mathbf{v}) - \rho T \mathbf{v} \cdot \mathbf{grad}\, s.$$

Here we have also used the general adiabatic equation (2.6).

Combining the above results, we find the change in the energy to be

$$\frac{\partial}{\partial t}(\tfrac{1}{2}\rho v^2 + \rho\varepsilon) = -(\tfrac{1}{2}v^2 + w)\operatorname{div}(\rho \mathbf{v}) - \rho \mathbf{v} \cdot \mathbf{grad}(\tfrac{1}{2}v^2 + w),$$

or, finally,

$$\frac{\partial}{\partial t}(\tfrac{1}{2}\rho v^2 + \rho\varepsilon) = -\operatorname{div}[\rho \mathbf{v}(\tfrac{1}{2}v^2 + w)]. \tag{6.1}$$

In order to see the meaning of this equation, let us integrate it over some volume:

$$\frac{\partial}{\partial t}\int(\tfrac{1}{2}\rho v^2 + \rho\varepsilon) dV = -\int \operatorname{div}[\rho \mathbf{v}(\tfrac{1}{2}v^2 + w)] dV,$$

or, converting the volume integral on the right into a surface integral,

$$\frac{\partial}{\partial t}\int(\tfrac{1}{2}\rho v^2 + \rho\varepsilon) dV = -\oint \rho \mathbf{v}(\tfrac{1}{2}v^2 + w) \cdot d\mathbf{f}. \tag{6.2}$$

The left-hand side is the rate of change of the energy of the fluid in some given volume. The right-hand side is therefore the amount of energy flowing out of this volume in unit time. Hence we see that the expression

$$\rho \mathbf{v}(\tfrac{1}{2}v^2 + w) \tag{6.3}$$

may be called the *energy flux density* vector. Its magnitude is the amount of energy passing in unit time through unit area perpendicular to the direction of the velocity.

The expression (6.3) shows that any unit mass of fluid carries with it during its motion an amount of energy $w + \frac{1}{2}v^2$. The fact that the heat function w appears here, and not the internal energy ε, has a simple physical significance. Putting $w = \varepsilon + p/\rho$, we can write the flux of energy through a closed surface in the form

$$-\oint \rho \mathbf{v}(\tfrac{1}{2}v^2 + \varepsilon) \cdot d\mathbf{f} - \oint p \mathbf{v} \cdot d\mathbf{f}.$$

The first term is the energy (kinetic and internal) transported through the surface in unit time by the mass of fluid. The second term is the work done by pressure forces on the fluid within the surface.

§7. The momentum flux

We shall now give a similar series of arguments for the momentum of the fluid. The momentum of unit volume is $\rho\mathbf{v}$. Let us determine its rate of change, $\partial(\rho\mathbf{v})/\partial t$. We shall use tensor notation. We have

$$\frac{\partial}{\partial t}(\rho v_i) = \rho\frac{\partial v_i}{\partial t} + \frac{\partial \rho}{\partial t} v_i.$$

Using the equation of continuity (1.2) in the form

$$\frac{\partial \rho}{\partial t} = -\frac{\partial(\rho v_k)}{\partial x_k},$$

and Euler's equation (2.3) in the form

$$\frac{\partial v_i}{\partial t} = -v_k\frac{\partial v_i}{\partial x_k} - \frac{1}{\rho}\frac{\partial p}{\partial x_i},$$

we obtain

$$\frac{\partial}{\partial t}(\rho v_i) = -\rho v_k \frac{\partial v_i}{\partial x_k} - \frac{\partial p}{\partial x_i} - v_i\frac{\partial(\rho v_k)}{\partial x_k}$$

$$= -\frac{\partial p}{\partial x_i} - \frac{\partial}{\partial x_k}(\rho v_i v_k).$$

We write the first term on the right in the form

$$\frac{\partial p}{\partial x_i} = \delta_{ik}\frac{\partial p}{\partial x_k},$$

and finally obtain

$$\frac{\partial}{\partial t}(\rho v_i) = -\frac{\partial \Pi_{ik}}{\partial x_k}, \tag{7.1}$$

where the tensor Π_{ik} is defined as

$$\Pi_{ik} = p\delta_{ik} + \rho v_i v_k. \tag{7.2}$$

This tensor is clearly symmetrical.

To see the meaning of the tensor Π_{ik}, we integrate equation (7.1) over some volume:

$$\frac{\partial}{\partial t}\int \rho v_i \mathrm{d}V = -\int \frac{\partial \Pi_{ik}}{\partial x_k}\mathrm{d}V.$$

The integral on the right is transformed into a surface integral by Green's formula:†

$$\frac{\partial}{\partial t}\int \rho v_i \mathrm{d}V = -\oint \Pi_{ik}\mathrm{d}f_k. \tag{7.3}$$

The left-hand side is the rate of change of the *i*th component of the momentum contained in the volume considered. The surface integral on the right is therefore the

† The rule for transforming an integral over a closed surface into one over the volume bounded by that surface can be formulated as follows: the surface element $\mathrm{d}f_i$ must be replaced by the operator $\mathrm{d}V\cdot\partial/\partial x_i$, which is to be applied to the whole of the integrand.

amount of momentum flowing out through the bounding surface in unit time. Consequently, $\Pi_{ik}\mathrm{d}f_k$ is the ith component of the momentum flowing through the surface element $\mathrm{d}\mathbf{f}$. If we write $\mathrm{d}f_k$ in the form $n_k\,\mathrm{d}f$, where $\mathrm{d}f$ is the area of the surface element, and $\mathbf{n}$ is a unit vector along the outward normal, we find that $\Pi_{ik}n_k$ is the flux of the ith component of momentum through unit surface area. We may notice that, according to (7.2), $\Pi_{ik}n_k = pn_i + \rho v_i v_k n_k$. This expression can be written in vector form

$$p\mathbf{n} + \rho\mathbf{v}(\mathbf{v}\cdot\mathbf{n}). \tag{7.4}$$

Thus Π_{ik} is the ith component of the amount of momentum flowing in unit time through unit area perpendicular to the x_k-axis. The tensor Π_{ik} is called the *momentum flux density tensor*. The energy flux is determined by a vector, energy being a scalar; the momentum flux, however, is determined by a tensor of rank two, the momentum itself being a vector.

The vector (7.4) gives the momentum flux in the direction of $\mathbf{n}$, i.e. through a surface perpendicular to $\mathbf{n}$. In particular, taking the unit vector $\mathbf{n}$ to be directed parallel to the fluid velocity, we find that only the longitudinal component of momentum is transported in this direction, and its flux density is $p + \rho v^2$. In a direction perpendicular to the velocity, only the transverse component (relative to $\mathbf{v}$) of momentum is transported, its flux density being just p.

§8. The conservation of circulation

The integral

$$\Gamma = \oint \mathbf{v}\cdot\mathrm{d}\mathbf{l},$$

taken along some closed contour, is called the *velocity circulation* round that contour.

Let us consider a closed contour drawn in the fluid at some instant. We suppose it to be a "fluid contour", i.e. composed of the fluid particles that lie on it. In the course of time these particles move about, and the contour moves with them. Let us investigate what happens to the velocity circulation. In other words, let us calculate the time derivative

$$\frac{\mathrm{d}}{\mathrm{d}t}\oint \mathbf{v}\cdot\mathrm{d}\mathbf{l}.$$

We have written here the total derivative with respect to time, since we are seeking the change in the circulation round a "fluid contour" as it moves about, and not round a contour fixed in space.

To avoid confusion, we shall temporarily denote differentiation with respect to the coordinates by the symbol δ, retaining the symbol d for differentiation with respect to time. Next, we notice that an element $\mathrm{d}\mathbf{l}$ of the length of the contour can be written as the difference $\delta\mathbf{r}$ between the position vectors $\mathbf{r}$ of the points at the ends of the element. Thus we write the velocity circulation as $\oint \mathbf{v}\cdot\delta\mathbf{r}$. In differentiating this integral with respect to time, it must be borne in mind that not only the velocity but also the contour itself (i.e. its shape) changes. Hence, on taking the time differentiation under the integral sign, we must differentiate not only $\mathbf{v}$ but also $\delta\mathbf{r}$:

$$\frac{\mathrm{d}}{\mathrm{d}t}\oint \mathbf{v}\cdot\delta\mathbf{r} = \oint \frac{\mathrm{d}\mathbf{v}}{\mathrm{d}t}\cdot\delta\mathbf{r} + \oint \mathbf{v}\cdot\frac{\mathrm{d}\delta\mathbf{r}}{\mathrm{d}t}.$$

Since the velocity $\mathbf{v}$ is just the time derivative of the position vector $\mathbf{r}$, we have

$$\mathbf{v}\cdot\frac{\mathrm{d}\delta\mathbf{r}}{\mathrm{d}t} = \mathbf{v}\cdot\delta\frac{\mathrm{d}\mathbf{r}}{\mathrm{d}t} = \mathbf{v}\cdot\delta\mathbf{v} = \delta(\tfrac{1}{2}v^2).$$

The integral of a total differential along a closed contour, however, is zero. The second integral therefore vanishes, leaving

$$\frac{\mathrm{d}}{\mathrm{d}t}\oint\mathbf{v}\cdot\delta\mathbf{r} = \oint\frac{\mathrm{d}\mathbf{v}}{\mathrm{d}t}\cdot\delta\mathbf{r}.$$

It now remains to substitute for the acceleration $\mathrm{d}\mathbf{v}/\mathrm{d}t$ its expression from (2.9):

$$\mathrm{d}\mathbf{v}/\mathrm{d}t = -\,\mathbf{grad}\,w.$$

Using Stokes' formula, we then have

$$\oint\frac{\mathrm{d}\mathbf{v}}{\mathrm{d}t}\cdot\delta\mathbf{r} = \oint\mathbf{curl}\left(\frac{\mathrm{d}\mathbf{v}}{\mathrm{d}t}\right)\cdot\delta\mathbf{f} = 0,$$

since $\mathbf{curl}\,\mathbf{grad}\,w \equiv 0$. Thus, going back to our previous notation, we find†

$$\frac{\mathrm{d}}{\mathrm{d}t}\oint\mathbf{v}\cdot\mathrm{d}\mathbf{l} = 0,$$

or

$$\oint\mathbf{v}\cdot\mathrm{d}\mathbf{l} = \text{constant}. \tag{8.1}$$

We have therefore reached the conclusion that, in an ideal fluid, the velocity circulation round a closed "fluid" contour is constant in time (*Kelvin's theorem* (1869) or the *law of conservation of circulation*).

It should be emphasized that this result has been obtained by using Euler's equation in the form (2.9), and therefore involves the assumption that the flow is isentropic. The theorem does not hold for flows which are not isentropic.‡

By applying Kelvin's theorem to an infinitesimal closed contour δC and transforming the integral according to Stokes' theorem, we get

$$\oint\mathbf{v}\cdot\mathrm{d}\mathbf{l} = \int\mathbf{curl}\,\mathbf{v}\cdot\mathrm{d}\mathbf{f} \cong \delta\mathbf{f}\cdot\mathbf{curl}\,\mathbf{v} = \text{constant}, \tag{8.2}$$

where $\mathrm{d}\mathbf{f}$ is a fluid surface element spanning the contour δC. The vector $\mathbf{curl}\,\mathbf{v}$ is often called the *vorticity* of the fluid flow at a given point. The constancy of the product (8.2) can be intuitively interpreted as meaning that the vorticity moves with the fluid.

PROBLEM

Show that, in flow which is not isentropic, any moving particle carries with it a constant value of the product $(1/\rho)\,\mathbf{grad}\,s\cdot\mathbf{curl}\,\mathbf{v}$ (H. Ertel 1942).

---

† This result remains valid in a uniform gravitational field, since in that case $\mathbf{curl}\,\mathbf{g} \equiv 0$.

‡ Mathematically, it is necessary that there should be a one-to-one relation between p and ρ (which for isentropic flow is $s(p, \rho) = $ constant); then $-(1/\rho)\,\mathbf{grad}\,p$ can be written as the gradient of some function, a result which is needed in deriving Kelvin's theorem.

SOLUTION. When the flow is not isentropic, the right-hand side of Euler's equation (2.3) cannot be replaced by $-\mathbf{grad}\, w$, and (2.11) becomes

$$\partial\boldsymbol{\omega}/\partial t = \mathbf{curl}\,(\mathbf{v}\times\boldsymbol{\omega}) + (1/\rho^2)\,\mathbf{grad}\,\rho\times\mathbf{grad}\,p,$$

where for brevity $\boldsymbol{\omega} = \mathbf{curl}\,\mathbf{v}$. We multiply scalarly by $\mathbf{grad}\,s$; since $s = s(p, \rho)$, $\mathbf{grad}\,s$ is a linear function of $\mathbf{grad}\,p$ and $\mathbf{grad}\,\rho$, and $\mathbf{grad}\,s\cdot(\mathbf{grad}\,\rho\times\mathbf{grad}\,p) = 0$. The expression on the right-hand side can then be transformed as follows:

$$\begin{aligned}\mathbf{grad}\,s\cdot\partial\boldsymbol{\omega}/\partial t &= \mathbf{grad}\,s\cdot\mathbf{curl}\,(\mathbf{v}\times\boldsymbol{\omega})\\ &= -\mathrm{div}\,[\mathbf{grad}\,s\times(\mathbf{v}\times\boldsymbol{\omega})]\\ &= -\mathrm{div}\,[\mathbf{v}(\boldsymbol{\omega}\cdot\mathbf{grad}\,s)] + \mathrm{div}\,[\boldsymbol{\omega}(\mathbf{v}\cdot\mathbf{grad}\,s)]\\ &= -(\boldsymbol{\omega}\cdot\mathbf{grad}\,s)\,\mathrm{div}\,\mathbf{v} - \mathbf{v}\cdot\mathbf{grad}\,(\boldsymbol{\omega}\cdot\mathbf{grad}\,s) + \boldsymbol{\omega}\cdot\mathbf{grad}\,(\mathbf{v}\cdot\mathbf{grad}\,s).\end{aligned}$$

From (2.6), $\mathbf{v}\cdot\mathbf{grad}\,s = -\partial s/\partial t$, and therefore

$$\frac{\partial}{\partial t}(\boldsymbol{\omega}\cdot\mathbf{grad}\,s) + \mathbf{v}\cdot\mathbf{grad}\,(\boldsymbol{\omega}\cdot\mathbf{grad}\,s) + (\boldsymbol{\omega}\cdot\mathbf{grad}\,s)\,\mathrm{div}\,\mathbf{v} = 0.$$

The first two terms can be combined as $\mathrm{d}(\boldsymbol{\omega}\cdot\mathbf{grad}\,s)/\mathrm{d}t$, where $\mathrm{d}/\mathrm{d}t = \partial/\partial t + \mathbf{v}\cdot\mathbf{grad}$; in the last term, we put from (1.3) $\rho\,\mathrm{div}\,\mathbf{v} = -\mathrm{d}\rho/\mathrm{d}t$. The result is

$$\frac{\mathrm{d}}{\mathrm{d}t}\left(\frac{\boldsymbol{\omega}\cdot\mathbf{grad}\,s}{\rho}\right) = 0,$$

which gives the required conservation law.

§9. Potential flow

From the law of conservation of circulation we can derive an important result. Let us at first suppose that the flow is steady, and consider a streamline of which we know that **curl v** is zero at some point. We draw an arbitrary infinitely small closed contour to encircle the streamline at that point. In the course of time, this contour moves with the fluid, but always encircles the same streamline. Since the product (8.2) must remain constant, it follows that **curl v** must be zero at every point on the streamline.

Thus we reach the conclusion that, if at any point on a streamline the vorticity is zero, the same is true at all other points on that streamline. If the flow is not steady, the same result holds, except that instead of a streamline we must consider the path described in the course of time by some particular fluid particle;† we recall that in non-steady flow these paths do not in general coincide with the streamlines.

At first sight it might seem possible to base on this result the following argument. Let us consider steady flow past some body. Let the incident flow be uniform at infinity; its velocity **v** is a constant, so that $\mathbf{curl}\,\mathbf{v} \equiv 0$ on all streamlines. Hence we conclude that **curl v** is zero along the whole of every streamline, i.e. in all space.

A flow for which $\mathbf{curl}\,\mathbf{v} = 0$ in all space is called a *potential flow* or *irrotational flow*, as opposed to *rotational flow*, in which the curl of the velocity is not everywhere zero. Thus we should conclude that steady flow past any body, with a uniform incident flow at infinity, must be potential flow.

Similarly, from the law of conservation of circulation, we might argue as follows. Let us suppose that at some instant we have potential flow throughout the volume of the fluid. Then the velocity circulation round any closed contour in the fluid is zero.‡ By Kelvin's

† To avoid misunderstanding, we may mention here that this result has no meaning in turbulent flow. We may also remark that a non-zero vorticity may occur on a streamline after the passage of a shock wave. We shall see that this is because the flow is no longer isentropic (§114).

‡ Here we suppose for simplicity that the fluid occupies a simply-connected region of space. The same final result would be obtained for a multiply-connected region, but restrictions on the choice of contours would have to be made in the derivation.

theorem, we could then conclude that this will hold at any future instant, i.e. we should find that, if there is potential flow at some instant, then there is potential flow at all subsequent instants (in particular, any flow for which the fluid is initially at rest must be a potential flow). This is in accordance with the fact that, if $\mathbf{curl\,v} = 0$, equation (2.11) is satisfied identically.

In fact, however, all these conclusions are of only very limited validity. The reason is that the proof given above that $\mathbf{curl\,v} = 0$ all along a streamline is, strictly speaking, invalid for a line which lies in the surface of a solid body past which the flow takes place, since the presence of this surface makes it impossible to draw a closed contour in the fluid encircling such a streamline. The equations of motion of an ideal fluid therefore admit solutions for which *separation* occurs at the surface of the body: the streamlines, having followed the surface for some distance, become separated from it at some point and continue into the fluid. The resulting flow pattern is characterized by the presence of a "surface of tangential discontinuity" proceeding from the body; on this surface the fluid velocity, which is everywhere tangential to the surface, has a discontinuity. In other words, at this surface one layer of fluid "slides" on another. Figure 1 shows a surface of discontinuity which separates moving fluid from a region of stationary fluid behind the body. From a mathematical point of view, the discontinuity in the tangential velocity component corresponds to a surface on which the curl of the velocity is non-zero.

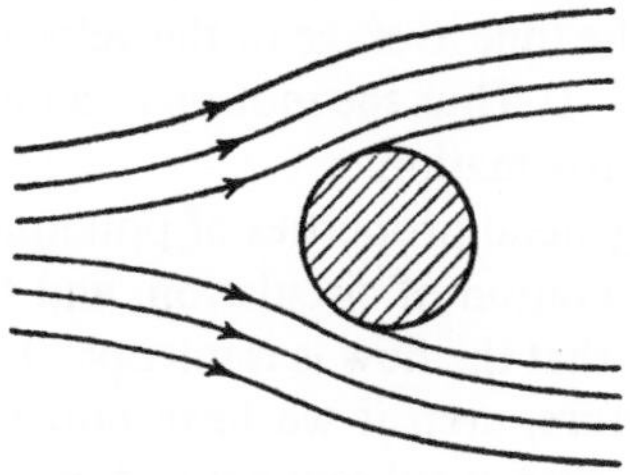

FIG. 1

When such discontinuous flows are included, the solution of the equations of motion for an ideal fluid is not unique: besides continuous flow, they admit also an infinite number of solutions possessing surfaces of tangential discontinuity starting from any prescribed line on the surface of the body past which the flow takes place. It should be emphasized, however, that none of these discontinuous solutions is physically significant, since tangential discontinuities are absolutely unstable, and therefore the flow would in fact become turbulent (see Chapter III).

The actual physical problem of flow past a given body has, of course, a unique solution. The reason is that ideal fluids do not really exist; any actual fluid has a certain viscosity, however small. This viscosity may have practically no effect on the motion of most of the fluid, but, no matter how small it is, it will be important in a thin layer of fluid adjoining the body. The properties of the flow in this *boundary layer* decide the choice of one out of the infinity of solutions of the equations of motion for an ideal fluid. It is found that, in the general case of flow past bodies of arbitrary form, solutions with separation must be taken, which in turn will result in turbulence.

In spite of what we have said above, the study of the solutions of the equations of motion for continuous steady potential flow past bodies is in some cases meaningful. Although, in the general case of flow past bodies of arbitrary form, the actual flow pattern bears almost no relation to the pattern of potential flow, for bodies of certain special ("streamlined"—§46) shapes the flow may differ very little from potential flow; more precisely, it will be potential flow except in a thin layer of fluid at the surface of the body and in a relatively narrow "wake" behind the body.

Another important case of potential flow occurs for small oscillations of a body immersed in fluid. It is easy to show that, if the amplitude a of the oscillations is small compared with the linear dimension l of the body ($a \ll l$), the flow past the body will be potential flow. To show this, we estimate the order of magnitude of the various terms in Euler's equation

$$\partial \mathbf{v}/\partial t + (\mathbf{v}\cdot\mathbf{grad})\mathbf{v} = -\mathbf{grad}\, w.$$

The velocity $\mathbf{v}$ changes markedly (by an amount of the same order as the velocity $\mathbf{u}$ of the oscillating body) over a distance of the order of the dimension l of the body. Hence the derivatives of $\mathbf{v}$ with respect to the coordinates are of the order of u/l. The order of magnitude of $\mathbf{v}$ itself (at fairly small distances from the body) is determined by the magnitude of $\mathbf{u}$. Thus we have $(\mathbf{v}\cdot\mathbf{grad})\mathbf{v} \sim u^2/l$. The derivative $\partial \mathbf{v}/\partial t$ is of the order of ωu, where ω is the frequency of the oscillations. Since $\omega \sim u/a$, we have $\partial \mathbf{v}/\partial t = u^2/a$. It now follows from the inequality $a \ll l$ that the term $(\mathbf{v}\cdot\mathbf{grad})\mathbf{v}$ is small compared with $\partial \mathbf{v}/\partial t$ and can be neglected, so that the equation of motion of the fluid becomes $\partial \mathbf{v}/\partial t = -\mathbf{grad}\, w$. Taking the curl of both sides, we obtain $\partial(\mathbf{curl}\,\mathbf{v})/\partial t = 0$, whence $\mathbf{curl}\,\mathbf{v} = \text{constant}$. In oscillatory motion, however, the time average of the velocity is zero, and therefore $\mathbf{curl}\,\mathbf{v} = \text{constant}$ implies that $\mathbf{curl}\,\mathbf{v} = 0$. Thus the motion of a fluid executing small oscillations is potential flow to a first approximation.

We shall now obtain some general properties of potential flow. We first recall that the derivation of the law of conservation of circulation, and therefore all its consequences, were based on the assumption that the flow is isentropic. If the flow is not isentropic, the law does not hold, and therefore, even if we have potential flow at some instant, the vorticity will in general be non-zero at subsequent instants. Thus only isentropic flow can in fact be potential flow.

In potential flow, the velocity circulation along any closed contour is zero:

$$\oint \mathbf{v}\cdot d\mathbf{l} = \int \mathbf{curl}\,\mathbf{v}\cdot d\mathbf{f} = 0. \tag{9.1}$$

It follows from this that, in particular, closed streamlines cannot exist in potential flow.† For, since the direction of a streamline is at every point the direction of the velocity, the circulation along such a line can never be zero.

In rotational flow the velocity circulation is not in general zero. In this case there may be closed streamlines, but it must be emphasized that the presence of closed streamlines is not a necessary property of rotational flow.

Like any vector field having zero curl, the velocity in potential flow can be expressed as the gradient of some scalar. This scalar is called the *velocity potential*; we shall denote it by ϕ:

$$\mathbf{v} = \mathbf{grad}\,\phi. \tag{9.2}$$

† This result, like (9.1), may not be valid for motion in a multiply-connected region of space. In potential flow in such a region, the velocity circulation may be non-zero if the closed contour round which it is taken cannot be contracted to a point without crossing the boundaries of the region.

Writing Euler's equation in the form (2.10)

$$\partial \mathbf{v}/\partial t + \tfrac{1}{2}\,\mathbf{grad}\,v^2 - \mathbf{v}\times\mathbf{curl}\,\mathbf{v} = -\mathbf{grad}\,w$$

and substituting $\mathbf{v} = \mathbf{grad}\,\phi$, we have

$$\mathbf{grad}\left(\frac{\partial\phi}{\partial t} + \tfrac{1}{2}v^2 + w\right) = 0,$$

whence

$$\partial\phi/\partial t + \tfrac{1}{2}v^2 + w = f(t), \tag{9.3}$$

where $f(t)$ is an arbitrary function of time. This equation is a first integral of the equations of potential flow. The function $f(t)$ in equation (9.3) can be put equal to zero without loss of generality, because the potential is not uniquely defined: since the velocity is the space derivative of ϕ, we can add to ϕ any function of the time.

For steady flow we have (taking the potential ϕ to be independent of time) $\partial\phi/\partial t = 0$, $f(t) = \text{constant}$, and (9.3) becomes Bernoulli's equation:

$$\tfrac{1}{2}v^2 + w = \text{constant}. \tag{9.4}$$

It must be emphasized here that there is an important difference between the Bernoulli's equation for potential flow and that for other flows. In the general case, the "constant" on the right-hand side is a constant along any given streamline, but is different for different streamlines. In potential flow, however, it is constant throughout the fluid. This enhances the importance of Bernoulli's equation in the study of potential flow.

§10. Incompressible fluids

In a great many cases of the flow of liquids (and also of gases), their density may be supposed invariable, i.e. constant throughout the volume of the fluid and throughout its motion. In other words, there is no noticeable compression or expansion of the fluid in such cases. We then speak of *incompressible flow*.

The general equations of fluid dynamics are much simplified for an incompressible fluid. Euler's equation, it is true, is unchanged if we put $\rho = \text{constant}$, except that ρ can be taken under the gradient operator in equation (2.4):

$$\frac{\partial\mathbf{v}}{\partial t} + (\mathbf{v}\cdot\mathbf{grad})\mathbf{v} = -\mathbf{grad}\left(\frac{p}{\rho}\right) + \mathbf{g}. \tag{10.1}$$

The equation of continuity, on the other hand, takes for constant ρ the simple form

$$\operatorname{div}\mathbf{v} = 0. \tag{10.2}$$

Since the density is no longer an unknown function as it was in the general case, the fundamental system of equations in fluid dynamics for an incompressible fluid can be taken to be equations involving the velocity only. These may be the equation of continuity (10.2) and equation (2.11):

$$\frac{\partial}{\partial t}(\mathbf{curl}\,\mathbf{v}) = \mathbf{curl}(\mathbf{v}\times\mathbf{curl}\,\mathbf{v}). \tag{10.3}$$

Bernoulli's equation too can be written in a simpler form for an incompressible fluid. Equation (10.1) differs from the general Euler's equation (2.9) in that it has $\mathbf{grad}\,(p/\rho)$ in place of **grad** w. Hence we can write down Bernoulli's equation immediately by simply replacing the heat function in (5.4) by p/ρ:

$$\tfrac{1}{2}v^2 + p/\rho + gz = \text{constant}. \tag{10.4}$$

For an incompressible fluid, we can also write p/ρ in place of w in the expression (6.3) for the energy flux, which then becomes

$$\rho \mathbf{v}\left(\tfrac{1}{2}v^2 + \frac{p}{\rho}\right). \tag{10.5}$$

For we have, from a well-known thermodynamic relation, the expression $d\varepsilon = Tds - pdV$ for the change in internal energy; for $s = \text{constant}$ and $V = 1/\rho = \text{constant}$, $d\varepsilon = 0$, i.e. $\varepsilon = \text{constant}$. Since constant terms in the energy do not matter, we can omit ε in $w = \varepsilon + p/\rho$.

The equations are particularly simple for potential flow of an incompressible fluid. Equation (10.3) is satisfied identically if $\mathbf{curl}\,\mathbf{v} = 0$. Equation (10.2), with the substitution $\mathbf{v} = \mathbf{grad}\,\phi$, becomes

$$\triangle\phi = 0, \tag{10.6}$$

i.e. Laplace's equation† for the potential ϕ. This equation must be supplemented by boundary conditions at the surfaces where the fluid meets solid bodies. At fixed solid surfaces, the fluid velocity component v_n normal to the surface must be zero, whilst for moving surfaces it must be equal to the normal component of the velocity of the surface (a given function of time). The velocity v_n, however, is equal to the normal derivative of the potential ϕ: $v_n = \partial\phi/\partial n$. Thus the general boundary conditions are that $\partial\phi/\partial n$ is a given function of coordinates and time at the boundaries.

For potential flow, the velocity is related to the pressure by equation (9.3). In an incompressible fluid, we can replace w in this equation by p/ρ:

$$\partial\phi/\partial t + \tfrac{1}{2}v^2 + p/\rho = f(t). \tag{10.7}$$

We may notice here the following important property of potential flow of an incompressible fluid. Suppose that some solid body is moving through the fluid. If the result is potential flow, it depends at any instant only on the velocity of the moving body at that instant, and not, for example, on its acceleration. For equation (10.6) does not explicitly contain the time, which enters the solution only through the boundary conditions, and these contain only the velocity of the moving body.

From Bernoulli's equation, $\tfrac{1}{2}v^2 + p/\rho = \text{constant}$, we see that, in steady flow of an incompressible fluid (not in a gravitational field), the greatest pressure occurs at points where the velocity is zero. Such a point usually occurs on the surface of a body past which the fluid is moving (at the point O in Fig. 2), and is called a *stagnation point*. If $\mathbf{u}$ is the velocity of the incident current (i.e. the fluid velocity at infinity), and p_0 the pressure at infinity, the pressure at the stagnation point is

$$p_{\max} = p_0 + \tfrac{1}{2}\rho u^2. \tag{10.8}$$

If the velocity distribution in a moving fluid depends on only two coordinates (x and y, say), and the velocity is everywhere parallel to the xy-plane, the flow is said to be *two-*

† The velocity potential was first introduced by Euler, who obtained an equation of the form (10.6) for it; this form later became known as Laplace's equation.

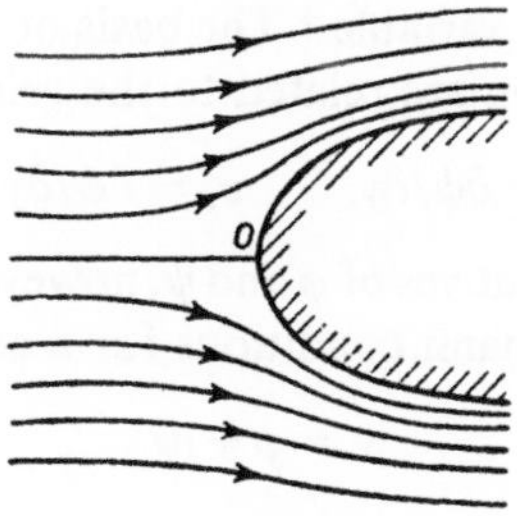

FIG. 2

dimensional or *plane flow*. To solve problems of two-dimensional flow of an incompressible fluid, it is sometimes convenient to express the velocity in terms of what is called the *stream function*. From the equation of continuity $\operatorname{div} \mathbf{v} \equiv \partial v_x/\partial x + \partial v_y/\partial y = 0$ we see that the velocity components can be written as the derivatives

$$v_x = \partial\psi/\partial y, \qquad v_y = -\partial\psi/\partial x \tag{10.9}$$

of some function $\psi(x, y)$, called the stream function. The equation of continuity is then satisfied automatically. The equation that must be satisfied by the stream function is obtained by substituting (10.9) in equation (10.3). We then obtain

$$\frac{\partial}{\partial t}\triangle\psi - \frac{\partial\psi}{\partial x}\frac{\partial}{\partial y}\triangle\psi + \frac{\partial\psi}{\partial y}\frac{\partial}{\partial x}\triangle\psi = 0. \tag{10.10}$$

If we know the stream function we can immediately determine the form of the streamlines for steady flow. For the differential equation of the streamlines (in two-dimensional flow) is $dx/v_x = dy/v_y$ or $v_y\,dx - v_x\,dy = 0$; it expresses the fact that the direction of the tangent to a streamline is the direction of the velocity. Substituting (10.9), we have

$$\frac{\partial\psi}{\partial x}dx + \frac{\partial\psi}{\partial y}dy = d\psi = 0,$$

whence ψ = constant. Thus the streamlines are the family of curves obtained by putting the stream function $\psi(x, y)$ equal to an arbitrary constant.

If we draw a curve between two points A and B in the xy-plane, the mass flux Q across this curve is given by the difference in the values of the stream function at these two points, regardless of the shape of the curve. For, if v_n is the component of the velocity normal to the curve at any point, we have

$$Q = \rho\oint_A^B v_n\,dl = \rho\oint_A^B(-v_y\,dx + v_x\,dy) = \rho\int_A^B d\psi,$$

or

$$Q = \rho(\psi_B - \psi_A). \tag{10.11}$$

There are powerful methods of solving problems of two-dimensional potential flow of an incompressible fluid past bodies of various profiles, involving the application of the theory of functions of a complex variable.† The basis of these methods is as follows. The potential and the stream function are related to the velocity components by‡

$$v_x = \partial\phi/\partial x = \partial\psi/\partial y, \qquad v_y = \partial\phi/\partial y = -\partial\psi/\partial x.$$

These relations between the derivatives of ϕ and ψ, however, are the same, mathematically, as the well-known Cauchy–Riemann conditions for a complex expression

$$w = \phi + i\psi \tag{10.12}$$

to be an analytic function of the complex argument $z = x + iy$. This means that the function $w(z)$ has at every point a well-defined derivative

$$\frac{\mathrm{d}w}{\mathrm{d}z} = \frac{\partial\phi}{\partial x} + i\frac{\partial\psi}{\partial x} = v_x - iv_y. \tag{10.13}$$

The function w is called the *complex potential*, and $\mathrm{d}w/\mathrm{d}z$ the *complex velocity*. The modulus and argument of the latter give the magnitude v of the velocity and the angle θ between the direction of the velocity and that of the x-axis:

$$\mathrm{d}w/\mathrm{d}z = ve^{-i\theta}. \tag{10.14}$$

At a solid surface past which the flow takes place, the velocity must be along the tangent. That is, the profile contour of the surface must be a streamline, i.e. ψ = constant along it; the constant may be taken as zero, and then the problem of flow past a given contour reduces to the determination of an analytic function $w(z)$ which takes real values on the contour. The statement of the problem is more involved when the fluid has a free surface; an example is found in Problem 9.

The integral of an analytic function round any closed contour C is well known to be equal to $2\pi i$ times the sum of the residues of the function at its simple poles inside C; hence

$$\oint w'\,\mathrm{d}z = 2\pi i \sum_k A_k,$$

where A_k are the residues of the complex velocity. We also have

$$\oint w'\,\mathrm{d}z = \oint (v_x - iv_y)(\mathrm{d}x + i\mathrm{d}y)$$

$$= \oint (v_x\mathrm{d}x + v_y\mathrm{d}y) + i\oint (v_x\mathrm{d}y - v_y\mathrm{d}x).$$

The real part of this expression is just the velocity circulation Γ round the contour C. The imaginary part, multiplied by ρ, is the mass flux across C; if there are no sources of fluid within the contour, this flux is zero and we then have simply

$$\Gamma = 2\pi i \sum_k A_k; \tag{10.15}$$

all the residues A_k are in this case purely imaginary.

† A more detailed account of these methods and their numerous applications may be found in many books which treat fluid dynamics from a more mathematical standpoint. Here, we shall describe only the basic idea.

‡ The existence of the stream function depends, however, only on the flow's being two-dimensional, not necessarily a potential flow.

Finally, let us consider the conditions under which the fluid may be regarded as incompressible. When the pressure changes adiabatically by Δp, the density changes by $\Delta\rho = (\partial\rho/\partial p)_s\Delta p$. According to Bernoulli's equation, however, Δp is of the order of ρv^2 in steady flow. We shall show in §64 that the derivative $(\partial p/\partial\rho)_s$ is the square of the velocity c of sound in the fluid, so that $\Delta\rho \sim \rho v^2/c^2$. The fluid may be regarded as incompressible if $\Delta\rho/\rho \ll 1$. We see that a necessary condition for this is that the fluid velocity be small compared with that of sound:

$$v \ll c. \qquad (10.16)$$

However, this condition is sufficient only in steady flow. In non-steady flow, a further condition must be fulfilled. Let τ and l be a time and a length of the order of the times and distances over which the fluid velocity undergoes significant changes. If the terms $\partial\mathbf{v}/\partial t$ and $(1/\rho)\,\mathbf{grad}\,p$ in Euler's equation are comparable, we find, in order of magnitude, $v/\tau \sim \Delta p/l\rho$ or $\Delta p \sim l\rho v/\tau$, and the corresponding change in ρ is $\Delta\rho \sim l\rho v/\tau c^2$. Now comparing the terms $\partial\rho/\partial t$ and $\rho\,\mathrm{div}\,\mathbf{v}$ in the equation of continuity, we find that the derivative $\partial\rho/\partial t$ may be neglected (i.e. we may suppose ρ constant) if $\Delta\rho/\tau \ll \rho v/l$, or

$$\tau \gg l/c. \qquad (10.17)$$

If the conditions (10.16) and (10.17) are both fulfilled, the fluid may be regarded as incompressible. The condition (10.17) has an obvious meaning: the time l/c taken by a sound signal to traverse the distance l must be small compared with the time τ during which the flow changes appreciably, so that the propagation of interactions in the fluid may be regarded as instantaneous.

PROBLEMS

PROBLEM 1. Determine the shape of the surface of an incompressible fluid subject to a gravitational field, contained in a cylindrical vessel which rotates about its (vertical) axis with a constant angular velocity Ω.

SOLUTION. Let us take the axis of the cylinder as the z-axis. Then $v_x = -y\Omega$, $v_y = x\Omega$, $v_z = 0$. The equation of continuity is satisfied identically, and Euler's equation (10.1) gives

$$x\Omega^2 = \frac{1}{\rho}\frac{\partial p}{\partial x}, \qquad y\Omega^2 = \frac{1}{\rho}\frac{\partial p}{\partial y}, \qquad \frac{1}{\rho}\frac{\partial p}{\partial z} + g = 0.$$

The general integral of these equations is

$$p/\rho = \tfrac{1}{2}\Omega^2(x^2+y^2) - gz + \text{constant}.$$

At the free surface $p = $ constant, so that the surface is a paraboloid:

$$z = \tfrac{1}{2}\Omega^2(x^2+y^2)/g,$$

the origin being taken at the lowest point of the surface.

PROBLEM 2. A sphere, with radius R, moves with velocity $\mathbf{u}$ in an incompressible ideal fluid. Determine the potential flow of the fluid past the sphere.

SOLUTION. The fluid velocity must vanish at infinity. The solutions of Laplace's equation $\triangle\phi = 0$ which vanish at infinity are well known to be $1/r$ and the derivatives, of various orders, of $1/r$ with respect to the coordinates (the origin is taken at the centre of the sphere). On account of the complete symmetry of the sphere, only one constant vector, the velocity $\mathbf{u}$, can appear in the solution, and, on account of the linearity of both Laplace's equation and the boundary condition, ϕ must involve $\mathbf{u}$ linearly. The only scalar which can be formed from $\mathbf{u}$ and the derivatives of $1/r$ is the scalar product $\mathbf{u}\cdot\mathbf{grad}(1/r)$. We therefore seek ϕ in the form

$$\phi = \mathbf{A}\cdot\mathbf{grad}(1/r) = -(\mathbf{A}\cdot\mathbf{n})/r^2,$$

where $\mathbf{n}$ is a unit vector in the direction of $\mathbf{r}$. The constant $\mathbf{A}$ is determined from the condition that the normal components of the velocities **v** and **u** must be equal at the surface at the sphere, i.e. $\mathbf{v}\cdot\mathbf{n} = \mathbf{u}\cdot\mathbf{n}$ for $r = R$. This condition gives $\mathbf{A} = \frac{1}{2}\mathbf{u}R^3$, so that

$$\phi = -\frac{R^3}{2r^2}\mathbf{u}\cdot\mathbf{n}, \qquad \mathbf{v} = \frac{R^3}{2r^3}[3\mathbf{n}(\mathbf{u}\cdot\mathbf{n}) - \mathbf{u}].$$

The pressure distribution is given by equation (10.7):

$$p = p_0 - \tfrac{1}{2}\rho v^2 - \rho\partial\phi/\partial t,$$

where p_0 is the pressure at infinity. To calculate the derivative $\partial\phi/\partial t$, we must bear in mind that the origin (which we have taken at the centre of the sphere) moves with velocity **u**. Hence

$$\partial\phi/\partial t = (\partial\phi/\partial\mathbf{u})\cdot\dot{\mathbf{u}} - \mathbf{u}\cdot\mathbf{grad}\,\phi.$$

The pressure distribution over the surface of the sphere is given by the formula

$$p = p_0 + \tfrac{1}{8}\rho u^2(9\cos^2\theta - 5) + \tfrac{1}{2}\rho R\mathbf{n}\cdot\mathrm{d}\mathbf{u}/\mathrm{d}t,$$

where θ is the angle between **n** and **u**.

PROBLEM 3. The same as Problem 2, but for an infinite cylinder moving perpendicular to its axis.†

SOLUTION. The flow is independent of the axial coordinate, so that we have to solve Laplace's equation in two dimensions. The solutions which vanish at infinity are the first and higher derivatives of $\log r$ with respect to the coordinates, where **r** is the radius vector perpendicular to the axis of the cylinder. We seek a solution in the form

$$\phi = \mathbf{A}\cdot\mathbf{grad}\log r = \mathbf{A}\cdot\mathbf{n}/r,$$

and from the boundary conditions we obtain $\mathbf{A} = -R^2\mathbf{u}$, so that

$$\phi = -\frac{R^2}{r}\mathbf{u}\cdot\mathbf{n}, \qquad \mathbf{v} = \frac{R^2}{r^2}[2\mathbf{n}(\mathbf{u}\cdot\mathbf{n}) - \mathbf{u}].$$

The pressure at the surface of the cylinder is given by

$$p = p_0 + \tfrac{1}{2}\rho u^2(4\cos^2\theta - 3) + \rho R\mathbf{n}\cdot\mathrm{d}\mathbf{u}/\mathrm{d}t.$$

PROBLEM 4. Determine the potential flow of an incompressible ideal fluid in an ellipsoidal vessel rotating about a principal axis with angular velocity Ω, and determine the total angular momentum of the fluid.

SOLUTION. We take Cartesian coordinates x, y, z along the axes of the ellipsoid at a given instant, the z-axis being the axis of rotation. The velocity of points in the vessel wall is

$$\mathbf{u} = \mathbf{\Omega}\times\mathbf{r},$$

so that the boundary condition $v_n = \partial\phi/\partial n = u_n$ is

$$\partial\phi/\partial n = \Omega(xn_y - yn_x),$$

or, using the equation of the ellipsoid $x^2/a^2 + y^2/b^2 + z^2/c^2 = 1$,

$$\frac{x}{a^2}\frac{\partial\phi}{\partial x} + \frac{y}{b^2}\frac{\partial\phi}{\partial y} + \frac{z}{c^2}\frac{\partial\phi}{\partial z} = xy\Omega\left(\frac{1}{b^2} - \frac{1}{a^2}\right)$$

The solution of Laplace's equation which satisfies this boundary condition is

$$\phi = \Omega\frac{a^2 - b^2}{a^2 + b^2}xy. \tag{1}$$

The angular momentum of the fluid in the vessel is

$$M = \rho\int(xv_y - yv_x)\,\mathrm{d}V.$$

---

† The solution of the more general problems of potential flow past an ellipsoid and an elliptical cylinder may be found in: N. E. Kochin, I. A. Kibel' and N. V. Roze, *Theoretical Hydromechanics* (*Teoreticheskaya gidromekhanika*), Part 1, chapter VII, Moscow 1963; H. Lamb, *Hydrodynamics*, 6th ed., §§103–116, Cambridge 1932.

Integrating over the volume V of the ellipsoid, we have

$$M = \frac{\Omega\rho V}{5}\frac{(a^2-b^2)^2}{a^2+b^2}.$$

Formula (1) gives the absolute motion of the fluid relative to the instantaneous position of the axes x, y, z which are fixed to the rotating vessel. The motion relative to the vessel (i.e. relative to a rotating system of coordinates x, y, z) is found by subtracting the velocity $\mathbf{\Omega}\times\mathbf{r}$ from the absolute velocity; denoting the relative velocity of the fluid by $\mathbf{v}'$, we have

$$v'_x = \frac{\partial\phi}{\partial x} + y\Omega = \frac{2\Omega a^2}{a^2+b^2}y, \qquad v'_y = -\frac{2\Omega b^2}{a^2+b^2}x, \qquad v'_z = 0.$$

The paths of the relative motion are found by integrating the equations $\dot{x} = v'_x$, $\dot{y} = v'_y$, and are the ellipses $x^2/a^2 + y^2/b^2 = $ constant, which are similar to the boundary ellipse.

PROBLEM 5. Determine the flow near a stagnation point (Fig. 2).

SOLUTION. A small part of the surface of the body near the stagnation point may be regarded as plane. Let us take it as the xy-plane. Expanding ϕ for x, y, z small, we have as far as the second-order terms

$$\phi = ax + by + cz + Ax^2 + By^2 + Cz^2 + Dxy + Eyz + Fzx;$$

a constant term in ϕ is immaterial. The constant coefficients are determined so that ϕ satisfies the equation $\triangle\phi = 0$ and the boundary conditions $v_z = \partial\phi/\partial z = 0$ for $z = 0$ and all x, y, $\partial\phi/\partial x = \partial\phi/\partial y = 0$ for $x = y = z = 0$ (the stagnation point). This gives $a = b = c = 0$; $C = -A - B$, $E = F = 0$. The term Dxy can always be removed by an appropriate rotation of the x and y axes. We then have

$$\phi = Ax^2 + By^2 - (A+B)z^2. \tag{1}$$

If the flow is axially symmetrical about the z-axis (symmetrical flow past a solid of revolution), we must have $A = B$, so that

$$\phi = A(x^2 + y^2 - 2z^2).$$

The velocity components are $v_x = 2Ax$, $v_y = 2Ay$, $v_z = -4Az$. The streamlines are given by equations (5.2), from which we find $x^2z = c_1$, $y^2z = c_2$, i.e. the streamlines are cubical hyperbolae.

If the flow is uniform in the y-direction (e.g. flow in the z-direction past a cylinder with its axis in the y-direction), we must have $B = 0$ in (1), so that

$$\phi = A(x^2 - z^2).$$

The streamlines are the hyperbolae $xz = $ constant.

PROBLEM 6. Determine the potential flow near an angle formed by two intersecting planes.

SOLUTION. Let us take polar coordinates r, θ in the cross-sectional plane (perpendicular to the line of intersection), with the origin at the vertex of the angle; θ is measured from one of the arms of the angle. Let the angle be α radians; for $\alpha < \pi$ the flow takes place within the angle, for $\alpha > \pi$ outside it. The boundary condition that the normal velocity component vanish means that $\partial\phi/\partial\theta = 0$ for $\theta = 0$ and $\theta = \alpha$. The solution of Laplace's equation satisfying these conditions can be written†

$$\phi = Ar^n \cos n\theta, \qquad n = \pi/\alpha,$$

so that

$$v_r = nAr^{n-1}\cos n\theta, \qquad v_\theta = -nAr^{n-1}\sin n\theta.$$

For $n < 1$ (flow outside an angle; Fig. 3), v_r becomes infinite as $1/r^{1-n}$ at the origin. For $n > 1$ (flow inside an angle; Fig. 4), v becomes zero for $r = 0$.

The stream function, which gives the form of the streamlines, is $\psi = Ar^n \sin n\theta$. The expressions obtained for ϕ and ψ are the real and imaginary parts of the complex potential $w = Az^n$.‡

PROBLEM 7. A spherical hole with radius a is suddenly formed in an incompressible fluid filling all space. Determine the time taken for the hole to be filled with fluid (Besant 1859; Rayleigh 1917).

---

† We take the solution which involves the lowest positive power of r, since r is small.

‡ If the boundary planes are supposed infinite, Problems 5 and 6 involve degeneracy, in that the values of the constants A and B in the solutions are indeterminate. In actual cases of flow past finite bodies, they are determined by the general conditions of the problem.

SOLUTION. The flow after the formation of the hole will be spherically symmetrical, the velocity at every point being directed to the centre of the hole. For the radial velocity $v_r \equiv v < 0$ we have Euler's equation in spherical polar coordinates:

$$\frac{\partial v}{\partial t} + v\frac{\partial v}{\partial r} = -\frac{1}{\rho}\frac{\partial p}{\partial r}. \tag{1}$$

The equation of continuity gives

$$r^2 v = F(t), \tag{2}$$

where $F(t)$ is an arbitrary function of time; this equation expresses the fact that, since the fluid is incompressible, the volume flowing through any spherical surface is independent of the radius of that surface.

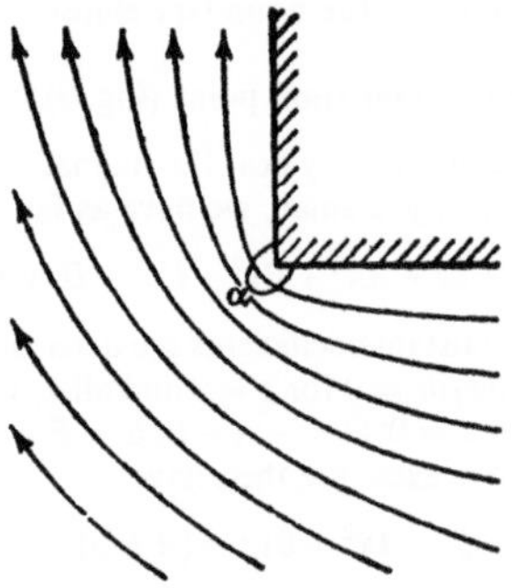

FIG. 3

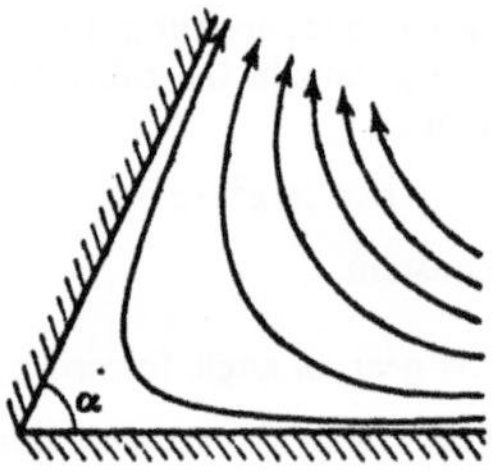

FIG. 4

Substituting v from (2) in (1), we have

$$\frac{F'(t)}{r^2} + v\frac{\partial v}{\partial r} = -\frac{1}{\rho}\frac{\partial p}{\partial r}.$$

Integrating this equation over r from the instantaneous radius $R = R(t) \leqslant a$ of the hole to infinity, we obtain

$$-\frac{F'(t)}{R} + \tfrac{1}{2}V^2 = \frac{p_0}{\rho} \tag{3}$$

where $V = \mathrm{d}R(t)/\mathrm{d}t$ is the rate of change of the radius of the hole, and p_0 is the pressure at infinity; the fluid velocity at infinity is zero, and so is the pressure at the surface of the hole. From equation (2) for points on the surface of the hole we find

$$F(t) = R^2(t)\,V(t),$$

and, substituting this expression for $F(t)$ in (3), we obtain the equation

$$-\frac{3V^2}{2} - \tfrac{1}{2}R\frac{\mathrm{d}V^2}{\mathrm{d}R} = \frac{p_0}{\rho}. \tag{4}$$

The variables are separable; integrating with the boundary condition $V = 0$ for $R = a$ (the fluid being initially at rest), we have

$$V \equiv \frac{dR}{dt} = -\sqrt{\left[\frac{2p_0}{3\rho}\left(\frac{a^3}{R^3}-1\right)\right]}.$$

Hence we have for the required total time for the hole to be filled

$$\tau = \sqrt{\frac{3\rho}{2p_0}}\int_0^a \frac{dR}{\sqrt{[(a/R)^3-1]}}.$$

This integral reduces to a beta function, and we have finally

$$\tau = \sqrt{\frac{3a^2\rho\pi}{2p_0}\frac{\Gamma(5/6)}{\Gamma(1/3)}} = 0{\cdot}915a\sqrt{\frac{\rho}{p_0}}.$$

Problem 8. A sphere immersed in an incompressible fluid expands according to a given law $R = R(t)$. Determine the fluid pressure at the surface of the sphere.

Solution. Let the required pressure be $P(t)$. Calculations exactly similar to those of Problem 7, except that the pressure at $r = R$ is $P(t)$ and not zero, give instead of (3) the equation

$$-\frac{F'(t)}{R(t)}+\tfrac{1}{2}V^2 = \frac{p_0}{\rho}-\frac{P(t)}{\rho}$$

and accordingly instead of (4) the equation

$$\frac{p_0-P(t)}{\rho} = -\frac{3V^2}{2}-RV\frac{dV}{dR}.$$

Bearing in mind the fact that $V = dR/dt$, we can write the expression for $P(t)$ in the form

$$P(t) = p_0+\tfrac{1}{2}\rho\left[\frac{d^2(R^2)}{dt^2}+\left(\frac{dR}{dt}\right)^2\right].$$

Problem 9. Determine the form of a jet emerging from an infinitely long slit in a plane wall.

Solution. Let the wall be along the x-axis in the xy-plane, and the aperture be the segment $-\frac{1}{2}a \leqslant x \leqslant \frac{1}{2}a$ of that axis, the fluid occupying the half-plane $y > 0$. Far from the wall ($y \to \infty$) the fluid velocity is zero, and the pressure is p_0, say.

At the free surface of the jet (BC and $B'C'$ in Fig. 5a) the pressure $p = 0$, while the velocity takes the constant value $v_1 = \sqrt{(2p_0/\rho)}$, by Bernoulli's equation. The wall lines are streamlines, and continue into the free boundary of the jet. Let ψ be zero on the line ABC; then, on the line $A'B'C'$, $\psi = -Q/\rho$, where $Q = \rho a_1 v_1$ is the rate at which the fluid emerges in the jet (a_1, v_1 being the jet width and velocity at infinity). The potential ϕ varies from $-\infty$ to $+\infty$ both along ABC and along $A'B'C'$; let ϕ be zero at B and B'. Then, in the plane of the complex variable w, the region of flow is an infinite strip of width Q/ρ (Fig. 5b). (The points in Fig. 5b, c, d are named to correspond with those in Fig. 5a.)

We introduce a new complex variable, the logarithm of the complex velocity:

$$\zeta = -\log\left[\frac{1}{v_1 e^{\frac{1}{2}i\pi}}\frac{dw}{dz}\right] = \log\frac{v_1}{v}+i(\tfrac{1}{2}\pi+\theta); \qquad (1)$$

here $v_1 e^{\frac{1}{2}i\pi}$ is the complex velocity of the jet at infinity. On $A'B'$ we have $\theta = 0$; on AB, $\theta = -\pi$; on BC and $B'C'$, $v = v_1$, while at infinity in the jet $\theta = -\frac{1}{2}\pi$. In the plane of the complex variable ζ, therefore, the region of flow is a semi-infinite strip of width π in the right half-plane (Fig. 5c). If we can now find a conformal transformation which carries the strip in the w-plane into the half-strip in the ζ-plane (with the points corresponding as in Fig. 5), we shall have determined w as a function of dw/dz, and w can then be found by a simple quadrature.

In order to find the desired transformation, we introduce one further auxiliary complex variable, u, such that the region of flow in the u-plane is the upper half-plane, the points B and B' corresponding to $u = \pm 1$, the points C and C' to $u = 0$, and the infinitely distant points A and A' to $u = \pm\infty$ (Fig. 5d). The dependence of w on this auxiliary variable is given by the conformal transformation which carries the upper half of the u-plane into the strip in the w-plane. With the above correspondence of points, this transformation is

$$w = -\frac{Q}{\rho\pi}\log u. \qquad (2)$$

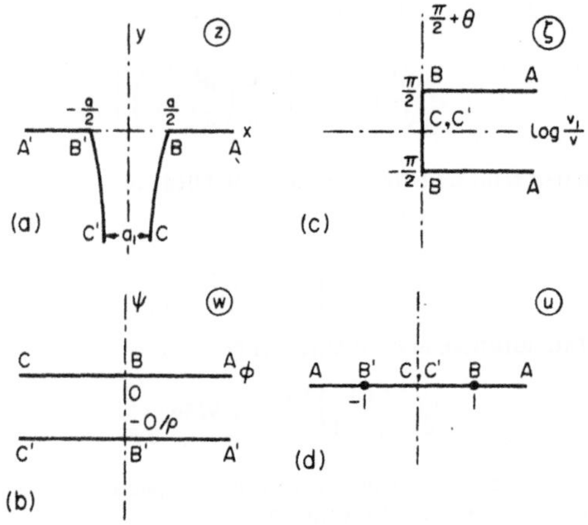

FIG. 5

In order to find the dependence of ζ on u, we have to find a conformal transformation of the half-strip in the ζ-plane into the upper half of the u-plane. Regarding this half-strip as a triangle with one vertex at infinity, we can find the desired transformation by means of the well-known Schwarz–Christoffel formula; it is

$$\zeta = -i\sin^{-1}u. \tag{3}$$

Formulae (2) and (3) give the solution of the problem, since they furnish the dependence of $\mathrm{d}w/\mathrm{d}z$ on w in parametric form.

Let us now determine the form of the jet. On BC we have $w = \phi$, $\zeta = i(\frac{1}{2}\pi+\theta)$, while u varies from 1 to 0. From (2) and (3) we obtain

$$\phi = -\frac{Q}{\rho\pi}\log(-\cos\theta), \tag{4}$$

and from (1) we have

$$\mathrm{d}\phi/\mathrm{d}z = v_1 e^{-i\theta},$$

or

$$\mathrm{d}z \equiv \mathrm{d}x + i\,\mathrm{d}y = \frac{1}{v_1}e^{i\theta}\,\mathrm{d}\phi = \frac{a_1}{\pi}e^{i\theta}\tan\theta\,\mathrm{d}\theta,$$

whence we find, by integration with the conditions $y = 0$, $x = \frac{1}{2}a$ for $\theta = -\pi$, the form of the jet, expressed parametrically. In particular, the compression of the jet is $a_1/a = \pi/(2+\pi) = 0{\cdot}61$.

§11. The drag force in potential flow past a body

Let us consider the problem of potential flow of an incompressible ideal fluid past some solid body. This problem is, of course, completely equivalent to that of the motion of a fluid when the same body moves through it. To obtain the latter case from the former, we need only change to a system of coordinates in which the fluid is at rest at infinity. We shall, in fact, say in what follows that the body is moving through the fluid.

Let us determine the nature of the fluid velocity distribution at great distances from the moving body. The potential flow of an incompressible fluid satisfies Laplace's equation, $\triangle\phi = 0$. We have to consider solutions of this equation which vanish at infinity, since the fluid is at rest there. We take the origin somewhere inside the moving body; the coordinate system moves with the body, but we shall consider the fluid velocity distribution at a particular instant. As we know, Laplace's equation has a solution $1/r$, where r is the distance from the origin. The gradient and higher space derivatives of $1/r$ are also solutions. All these solutions, and any linear combination of them, vanish at infinity. Hence the general form of the required solution of Laplace's equation at great distances from the body is

$$\phi = -\frac{a}{r} + \mathbf{A}\cdot\mathbf{grad}\frac{1}{r} + \dots,$$

where a and $\mathbf{A}$ are independent of the coordinates; the omitted terms contain higher-order derivatives of $1/r$. It is easy to see that the constant a must be zero. For the potential $\phi = -a/r$ gives a velocity

$$\mathbf{v} = -\,\mathbf{grad}\,(a/r) = a\mathbf{r}/r^3.$$

Let us calculate the corresponding mass flux through some closed surface, say a sphere with radius R. On this surface the velocity is constant and equal to a/R^2; the total flux through it is therefore $\rho(a/R^2)4\pi R^2 = 4\pi\rho a$. But the flux of an incompressible fluid through any closed surface must, of course, be zero. Hence we conclude that $a = 0$.

Thus ϕ contains terms of order $1/r^2$ and higher. Since we are seeking the velocity at large distances, the terms of higher order may be neglected, and we have

$$\phi = \mathbf{A}\cdot\mathbf{grad}(1/r) = -\mathbf{A}\cdot\mathbf{n}/r^2, \tag{11.1}$$

and the velocity $\mathbf{v} = \mathbf{grad}\,\phi$ is

$$\mathbf{v} = (\mathbf{A}\cdot\mathbf{grad})\,\mathbf{grad}\frac{1}{r} = \frac{3(\mathbf{A}\cdot\mathbf{n})\mathbf{n}-\mathbf{A}}{r^3}, \tag{11.2}$$

where $\mathbf{n}$ is a unit vector in the direction of $\mathbf{r}$. We see that at large distances the velocity diminishes as $1/r^3$. The vector $\mathbf{A}$ depends on the actual shape and velocity of the body, and can be determined only by solving completely the equation $\triangle\phi = 0$ at all distances, taking into account the appropriate boundary conditions at the surface of the moving body.

The vector $\mathbf{A}$ which appears in (11.2) is related in a definite manner to the total momentum and energy of the fluid in its motion past the body. The total kinetic energy of the fluid (the internal energy of an incompressible fluid is constant) is $E = \frac{1}{2}\int \rho v^2\,\mathrm{d}V$, where the integration is taken over all space outside the body. We take a region of space V bounded by a sphere with large radius R, whose centre is at the origin, and first integrate only over V, later letting R tend to infinity. We have identically

$$\int v^2\,\mathrm{d}V = \int u^2\,\mathrm{d}V + \int(\mathbf{v}+\mathbf{u})\cdot(\mathbf{v}-\mathbf{u})\,\mathrm{d}V,$$

where $\mathbf{u}$ is the velocity of the body. Since $\mathbf{u}$ is independent of the coordinates, the first integral on the right is simply $u^2(V - V_0)$, where V_0 is the volume of the body. In the second integral, we write the sum $\mathbf{v}+\mathbf{u}$ as $\mathbf{grad}\,(\phi+\mathbf{u}\cdot\mathbf{r})$; using the facts that $\operatorname{div}\mathbf{v} = 0$ (equation of continuity) and $\operatorname{div}\mathbf{u} \equiv 0$, we have

$$\int v^2\,\mathrm{d}V = u^2(V - V_0) + \int \operatorname{div}\left[(\phi+\mathbf{u}\cdot\mathbf{r})(\mathbf{v}-\mathbf{u})\right]\mathrm{d}V.$$

The second integral is now transformed into an integral over the surface S of the sphere and the surface S_0 of the body:

$$\int v^2\,\mathrm{d}V = u^2(V - V_0) + \oint_{S+S_0} (\phi + \mathbf{u}\cdot\mathbf{r})(\mathbf{v}-\mathbf{u})\cdot\mathrm{d}\mathbf{f}.$$

On the surface of the body, the normal components of $\mathbf{v}$ and $\mathbf{u}$ are equal by virtue of the boundary conditions; since the vector $\mathrm{d}\mathbf{f}$ is along the normal to the surface, it is clear that the integral over S_0 vanishes identically. On the remote surface S we substitute the expressions (11.1), (11.2) for ϕ and $\mathbf{v}$, and neglect terms which vanish as $R \to \infty$. Writing the surface element on the sphere S in the form $\mathrm{d}\mathbf{f} = \mathbf{n}R^2\,\mathrm{d}o$, where $\mathrm{d}o$ is an element of solid angle, we obtain

$$\int v^2\,\mathrm{d}V = u^2(\tfrac{4}{3}\pi R^3 - V_0) + \int [3(\mathbf{A}\cdot\mathbf{n})(\mathbf{u}\cdot\mathbf{n}) - (\mathbf{u}\cdot\mathbf{n})^2 R^3]\,\mathrm{d}o.$$

Finally, effecting the integration† and multiplying by $\frac{1}{2}\rho$, we obtain the following expression for the total energy of the fluid:

$$E = \tfrac{1}{2}\rho(4\pi\mathbf{A}\cdot\mathbf{u} - V_0 u^2). \tag{11.3}$$

As has been mentioned already, the exact calculation of the vector $\mathbf{A}$ requires a complete solution of the equation $\triangle\phi = 0$, taking into account the particular boundary conditions at the surface of the body. However, the general nature of the dependence of $\mathbf{A}$ on the velocity $\mathbf{u}$ of the body can be found directly from the facts that the equation is linear in ϕ, and the boundary conditions are linear in both ϕ and $\mathbf{u}$. It follows from this that $\mathbf{A}$ must be a linear function of the components of $\mathbf{u}$. The energy E given by formula (11.3) is therefore a quadratic function of the components of $\mathbf{u}$, and can be written in the form

$$E = \tfrac{1}{2}m_{ik}u_i u_k, \tag{11.4}$$

where m_{ik} is some constant symmetrical tensor, whose components can be calculated from those of $\mathbf{A}$; it is called the *induced-mass tensor*.

Knowing the energy E, we can obtain an expression for the total momentum $\mathbf{P}$ of the fluid. To do so, we notice that infinitesimal changes in E and $\mathbf{P}$ are related by‡ $\mathrm{d}E = \mathbf{u}\cdot\mathrm{d}\mathbf{P}$;

† The integration over o is equivalent to averaging the integrand over all directions of the vector $\mathbf{n}$ and multiplying by 4π. To average expressions of the type $(\mathbf{A}\cdot\mathbf{n})(\mathbf{B}\cdot\mathbf{n}) \equiv A_i n_i B_k n_k$, where $\mathbf{A}$, $\mathbf{B}$ are constant vectors, we notice that

$$\overline{(\mathbf{A}\cdot\mathbf{n})(\mathbf{B}\cdot\mathbf{n})} = A_i B_k \overline{n_i n_k} = \tfrac{1}{3}\delta_{ik}A_i B_k = \tfrac{1}{3}\mathbf{A}\cdot\mathbf{B}.$$

‡ For, let the body be accelerated by some external force $\mathbf{F}$. The momentum of the fluid will thereby be increased; let it increase by $\mathrm{d}\mathbf{P}$ during a time $\mathrm{d}t$. This increase is related to the force by $\mathrm{d}\mathbf{P} = \mathbf{F}\,\mathrm{d}t$, and on scalar multiplication by the velocity $\mathbf{u}$ we have $\mathbf{u}\cdot\mathrm{d}\mathbf{P} = \mathbf{F}\cdot\mathbf{u}\,\mathrm{d}t$, i.e. the work done by the force $\mathbf{F}$ acting through the distance $\mathbf{u}\,\mathrm{d}t$, which in turn must be equal to the increase $\mathrm{d}E$ in the energy of the fluid.

It should be noticed that it would not be possible to calculate the momentum directly as the integral $\int\rho\mathbf{v}\,\mathrm{d}V$ over the whole volume of the fluid. The reason is that this integral, with the velocity $\mathbf{v}$ distributed in accordance with (11.2), diverges, in the sense that the result of the integration, though finite, depends on how the integral is taken: on effecting the integration over a large region, whose dimensions subsequently tend to infinity, we obtain a value depending on the shape of the region (sphere, cylinder, etc.). The method of calculating the momentum which we use here, starting from the relation $\mathbf{u}\cdot\mathrm{d}\mathbf{P} = \mathrm{d}E$, leads to a completely definite final result, given by formula (11.6), which certainly satisfies the physical relation between the rate of change of the momentum and the forces acting on the body.

it follows from this that, if E is expressed in the form (11.4), the components of **P** must be

$$P_i = m_{ik} u_k. \tag{11.5}$$

Finally, a comparison of formulae (11.3), (11.4) and (11.5) shows that **P** is given in terms of **A** by

$$\mathbf{P} = 4\pi\rho\mathbf{A} - \rho V_0 \mathbf{u}. \tag{11.6}$$

It must be noticed that the total momentum of the fluid is a perfectly definite finite quantity.

The momentum transmitted to the fluid by the body in unit time is $d\mathbf{P}/dt$. With the opposite sign it evidently gives the reaction **F** of the fluid, i.e. the force acting on the body:

$$\mathbf{F} = -d\mathbf{P}/dt. \tag{11.7}$$

The component of **F** parallel to the velocity of the body is called the *drag force*, and the perpendicular component is called the *lift force.*

If it were possible to have potential flow past a body moving uniformly in an ideal fluid, we should have **P** = constant, since **u** = constant, and so **F** = 0. That is, there would be no drag and no lift; the pressure forces exerted on the body by the fluid would balance out (a result known as *d'Alembert's paradox*). The origin of this paradox is most clearly seen by considering the drag. The presence of a drag force in uniform motion of a body would mean that, to maintain the motion, work must be continually done by some external force, this work being either dissipated in the fluid or converted into kinetic energy of the fluid, and the result being a continual flow of energy to infinity in the fluid. There is, however, by definition no dissipation of energy in an ideal fluid, and the velocity of the fluid set in motion by the body diminishes so rapidly with increasing distance from the body that there can be no flow of energy to infinity.

However, it must be emphasized that all these arguments relate only to the motion of a body in an infinite volume of fluid. If, for example, the fluid has a free surface, a body moving uniformly parallel to this surface will experience a drag. The appearance of this force (called *wave drag*) is due to the occurrence of a system of waves propagated on the free surface, which continually remove energy to infinity.

Suppose that a body is executing an oscillatory motion under the action of an external force **f**. When the conditions discussed in §10 are fulfilled, the fluid surrounding the body moves in a potential flow, and we can use the relations previously obtained to derive the equations of motion of the body. The force **f** must be equal to the time derivative of the total momentum of the system, and the total momentum is the sum of the momentum $M\mathbf{u}$ of the body (M being the mass of the body) and the momentum **P** of the fluid:

$$M\,d\mathbf{u}/dt + d\mathbf{P}/dt = \mathbf{f}.$$

Using (11.5), we then obtain

$$M\,du_i/dt + m_{ik}\,du_k/dt = f_i,$$

which can also be written

$$\frac{du_k}{dt}(M\delta_{ik} + m_{ik}) = f_i. \tag{11.8}$$

This is the equation of motion of a body immersed in an ideal fluid.

Let us now consider what is in some ways the converse problem. Suppose that the fluid executes some oscillatory motion on account of some cause external to the body. This motion will set the body in motion also.† We shall derive the equation of motion of the body.

We assume that the velocity of the fluid varies only slightly over distances of the order of the dimension of the body. Let **v** be what the fluid velocity at the position of the body would be if the body were absent; that is, **v** is the velocity of the unperturbed flow. According to the above assumption, **v** may be supposed constant throughout the volume occupied by the body. We denote the velocity of the body by **u** as before.

The force which acts on the body and sets it in motion can be determined as follows. If the body were wholly carried along with the fluid (i.e. if $\mathbf{v} = \mathbf{u}$), the force acting on it would be the same as the force which would act on the liquid in the same volume if the body were absent. The momentum of this volume of fluid is $\rho V_0 \mathbf{v}$, and therefore the force on it is $\rho V_0\, \mathrm{d}\mathbf{v}/\mathrm{d}t$. In reality, however, the body is not wholly carried along with the fluid; there is a motion of the body relative to the fluid, in consequence of which the fluid itself acquires some additional motion. The resulting additional momentum of the fluid is $m_{ik}(u_k - v_k)$, since in (11.5) we must now replace **u** by the velocity $\mathbf{u} - \mathbf{v}$ of the body relative to the fluid. The change in this momentum with time results in the appearance of an additional reaction force on the body of $-m_{ik}\,\mathrm{d}(u_k - v_k)/\mathrm{d}t$. Thus the total force on the body is

$$\rho V_0 \frac{\mathrm{d}v_i}{\mathrm{d}t} - m_{ik}\frac{\mathrm{d}}{\mathrm{d}t}(u_k - v_k).$$

This force is to be equated to the time derivative of the body momentum. Thus we obtain the following equation of motion:

$$\frac{\mathrm{d}}{\mathrm{d}t}(Mu_i) = \rho V_0 \frac{\mathrm{d}v_i}{\mathrm{d}t} - m_{ik}\frac{\mathrm{d}}{\mathrm{d}t}(u_k - v_k).$$

Integrating both sides with respect to time, we have

$$(M\delta_{ik} + m_{ik})u_k = (m_{ik} + \rho V_0 \delta_{ik})v_k. \qquad (11.9)$$

We put the constant of integration equal to zero, since the velocity **u** of the body in its motion caused by the fluid must vanish when **v** vanishes. The relation obtained determines the velocity of the body from that of the fluid. If the density of the body is equal to that of the fluid ($M = \rho V_0$), we have $\mathbf{u} = \mathbf{v}$, as we should expect.

PROBLEMS

PROBLEM 1. Obtain the equation of motion for a sphere executing an oscillatory motion in an ideal fluid, and for a sphere set in motion by an oscillating fluid.

SOLUTION. Comparing (11.1) with the expression for ϕ for flow past a sphere obtained in §10, Problem 2, we see that

$$\mathbf{A} = \tfrac{1}{2}R^3\mathbf{u},$$

where R is the radius of the sphere. The total momentum transmitted to the fluid by the sphere is, according to (11.6), $\mathbf{P} = \tfrac{2}{3}\pi\rho R^3\mathbf{u}$, so that the tensor m_{ik} is

$$m_{ik} = \tfrac{2}{3}\pi\rho R^3\delta_{ik}.$$

† For example, we may be considering the motion of a body in a fluid through which a sound wave is propagated, the wavelength being large compared with the dimension of the body.

The drag on the moving sphere is

$$\mathbf{F} = -\tfrac{2}{3}\pi\rho R^3 d\mathbf{u}/dt,$$

and the equation of motion of the sphere oscillating in the fluid is

$$\tfrac{4}{3}\pi R^3(\rho_0 + \tfrac{1}{2}\rho)\frac{d\mathbf{u}}{dt} = \mathbf{f},$$

where ρ_0 is the density of the sphere. The coefficient of $d\mathbf{u}/dt$ is the *virtual mass* of the sphere; it consists of the actual mass of the sphere and the induced mass, which in this case is half the mass of the fluid displaced by the sphere.

If the sphere is set in motion by the fluid, we have for its velocity, from (11.9),

$$\mathbf{u} = \frac{3\rho}{\rho + 2\rho_0}\mathbf{v}.$$

If the density of the sphere exceeds that of the fluid ($\rho_0 > \rho$), $u < v$, i.e. the sphere "lags behind" the fluid; if $\rho_0 < \rho$, on the other hand, the sphere "goes ahead".

PROBLEM 2. Express the moment of the forces acting on a body moving in a fluid in terms of the vector **A**.

SOLUTION. As we know from mechanics, the moment **M** of the forces acting on a body is determined from its Lagrangian function (in this case, the energy E) by the relation $\delta E = \mathbf{M}\cdot\delta\boldsymbol{\theta}$, where $\delta\boldsymbol{\theta}$ is the vector of an infinitesimal rotation of the body, and δE is the resulting change in E. Instead of rotating the body through an angle $\delta\theta$ (and correspondingly changing the components m_{ik}), we may rotate the fluid through an angle $-\delta\theta$ relative to the body (and correspondingly change the velocity **u**). We have $\delta\mathbf{u} = -\delta\boldsymbol{\theta}\times\mathbf{u}$, so that

$$\delta E = \mathbf{P}\cdot\delta\mathbf{u} = -\delta\boldsymbol{\theta}\cdot\mathbf{u}\times\mathbf{P}.$$

Using the expression (11.6) for **P**, we then obtain the required formula:

$$\mathbf{M} = -\mathbf{u}\times\mathbf{P} = 4\pi\rho\mathbf{A}\times\mathbf{u}.$$

§12. Gravity waves

The free surface of a liquid in equilibrium in a gravitational field is a plane. If, under the action of some external perturbation, the surface is moved from its equilibrium position at some point, motion will occur in the liquid. This motion will be propagated over the whole surface in the form of waves, which are called *gravity waves*, since they are due to the action of the gravitational field. Gravity waves appear mainly on the surface of the liquid; they affect the interior also, but less and less at greater and greater depths.

We shall here consider gravity waves in which the velocity of the moving fluid particles is so small that we may neglect the term $(\mathbf{v}\cdot\mathbf{grad})\mathbf{v}$ in comparison with $\partial\mathbf{v}/\partial t$ in Euler's equation. The physical significance of this is easily seen. During a time interval of the order of the period τ of the oscillations of the fluid particles in the wave, these particles travel a distance of the order of the amplitude a of the wave. Their velocity v is therefore of the order of a/τ. It varies noticeably over time intervals of the order of τ and distances of the order of λ in the direction of propagation (where λ is the wavelength). Hence the time derivative of the velocity is of the order of v/τ, and the space derivatives are of the order of v/λ. Thus the condition $(\mathbf{v}\cdot\mathbf{grad})\mathbf{v} \ll \partial\mathbf{v}/\partial t$ is equivalent to

$$\frac{1}{\lambda}\left(\frac{a}{\tau}\right)^2 \ll \frac{a}{\tau}\cdot\frac{1}{\tau},$$

or

$$a \ll \lambda, \tag{12.1}$$

i.e. the amplitude of the oscillations in the wave must be small compared with the wavelength. We have seen in §9 that, if the term $(\mathbf{v}\cdot\mathbf{grad})\mathbf{v}$ in the equation of motion may be neglected, we have potential flow. Assuming the fluid incompressible, we can therefore use equations (10.6) and (10.7). The term $\frac{1}{2}v^2$ in the latter equation may be neglected, since it contains the square of the velocity; putting $f(t) = 0$ and including a term ρgz on account of the gravitational field, we obtain

$$p = -\rho gz - \rho\partial\phi/\partial t. \tag{12.2}$$

We take the z-axis vertically upwards, as usual, and the xy-plane in the equilibrium surface of the liquid.

Let us denote by ζ the z coordinate of a point on the surface; ζ is a function of x, y and t. In equilibrium $\zeta = 0$, so that ζ gives the vertical displacement of the surface in its oscillations. Let a constant pressure p_0 act on the surface. Then we have at the surface, by (12.2),

$$p_0 = -\rho g\zeta - \rho\partial\phi/\partial t.$$

The constant p_0 can be eliminated by redefining the potential ϕ, adding to it a quantity $p_0 t/\rho$ independent of the coordinates. We then obtain the condition at the surface as

$$g\zeta + (\partial\phi/\partial t)_{z=\zeta} = 0. \tag{12.3}$$

Since the amplitude of the wave oscillations is small, the displacement ζ is small. Hence we can suppose, to the same degree of approximation, that the vertical component of the velocity of points on the surface is simply the time derivative of ζ:

$$v_z = \partial\zeta/\partial t.$$

But $v_z = \partial\phi/\partial z$, so that

$$(\partial\phi/\partial z)_{z=\zeta} = \partial\zeta/\partial t = -\left(\frac{1}{g}\frac{\partial^2\phi}{\partial t^2}\right)_{z=\zeta}.$$

Since the oscillations are small, we can take the value of the derivatives at $z = 0$ instead of $z = \zeta$. Thus we have finally the following system of equations to determine the motion in a gravitational field:

$$\triangle\phi = 0, \tag{12.4}$$

$$\left(\frac{\partial\phi}{\partial z} + \frac{1}{g}\frac{\partial^2\phi}{\partial t^2}\right)_{z=0} = 0. \tag{12.5}$$

We shall here consider waves on the surface of a liquid whose area is unlimited, and we shall also suppose that the wavelength is small in comparison with the depth of the liquid; we can then regard the liquid as infinitely deep. We shall therefore omit the boundary conditions at the sides and bottom.

Let us consider a gravity wave propagated along the x-axis and uniform in the y-direction; in such a wave, all quantities are independent of y. We shall seek a solution which is a simple periodic function of time and of the coordinate x, i.e. we put

$$\phi = f(z)\cos(kx - \omega t).$$

Here ω is what is called the *circular frequency* (we shall say simply the *frequency*) of the wave; k is called the *wave number*; $\lambda = 2\pi/k$ is the *wavelength*.

Substituting in the equation $\triangle\phi = 0$, we have

$$\mathrm{d}^2f/\mathrm{d}z^2 - k^2 f = 0.$$

The solution which decreases as we go into the interior of the liquid (i.e. as $z \to -\infty$) is

$$\phi = Ae^{kz}\cos(kx - \omega t). \tag{12.6}$$

We have also to satisfy the boundary condition (12.5). Substituting (12.6), we obtain

$$\omega^2 = kg \tag{12.7}$$

as the relation between the wave number and the frequency of a gravity wave (the *dispersion relation*).

The velocity distribution in the moving liquid is found by simply taking the space derivatives of ϕ:

$$v_x = -Ake^{kz}\sin(kx - \omega t), \qquad v_z = Ake^{kz}\cos(kx - \omega t). \tag{12.8}$$

We see that the velocity diminishes exponentially as we go into the liquid. At any given point in space (i.e. for given x, z) the velocity vector rotates uniformly in the xz-plane, its magnitude remaining constant.

Let us also determine the paths of fluid particles in the wave. We temporarily denote by x, z the coordinates of a moving fluid particle (and not of a point fixed in space), and by x_0, z_0 the values of x and z at the equilibrium position of the particle. Then $v_x = dx/dt$, $v_z = dz/dt$, and on the right-hand side of (12.8) we may approximate by writing x_0, z_0 in place of x, z, since the oscillations are small. An integration with respect to time then gives

$$\left.\begin{aligned} x - x_0 &= -A\frac{k}{\omega}e^{kz_0}\cos(kx_0 - \omega t), \\ z - z_0 &= -A\frac{k}{\omega}e^{kz_0}\sin(kx_0 - \omega t). \end{aligned}\right\} \tag{12.9}$$

Thus the fluid particles describe circles about the points (x_0, z_0) with a radius which diminishes exponentially with increasing depth.

The velocity of propagation U of the wave is, as we shall show in §67, $U = \partial\omega/\partial k$. Substituting here $\omega = \sqrt{(kg)}$, we find that the velocity of propagation of gravity waves on an unbounded surface of infinitely deep liquid is

$$U = \tfrac{1}{2}\sqrt{(g/k)} = \tfrac{1}{2}\sqrt{(g\lambda/2\pi)}. \tag{12.10}$$

It increases with the wavelength.

Long gravity waves

Having considered gravity waves whose length is small compared with the depth of the liquid, let us now discuss the opposite limiting case of waves whose length is large compared with the depth. These are called *long* waves.

Let us examine first the propagation of long waves in a channel. The channel is supposed to be along the x-axis, and of infinite length. The cross-section of the channel may have any shape, and may vary along its length. We denote the cross-sectional area of the liquid in the channel by $S = S(x, t)$. The depth and width of the channel are supposed small in comparison with the wavelength.

We shall here consider longitudinal waves, in which the liquid moves along the channel. In such waves the velocity component v_x along the channel is large compared with the components v_y, v_z.

We denote v_x by v simply, and omit small terms. The x-component of Euler's equation can then be written in the form

$$\frac{\partial v}{\partial t} = -\frac{1}{\rho}\frac{\partial p}{\partial x},$$

and the z-component in the form

$$\frac{1}{\rho}\frac{\partial p}{\partial z} = -g;$$

we omit terms quadratic in the velocity, since the amplitude of the wave is again supposed small. From the second equation we have, since the pressure at the free surface ($z = \zeta$) must be p_0,

$$p = p_0 + g\rho(\zeta - z).$$

Substituting this expression in the first equation, we obtain

$$\partial v/\partial t = -g\partial\zeta/\partial x. \tag{12.11}$$

The second equation needed to determine the two unknowns v and ζ can be derived similarly to the equation of continuity; it is essentially the equation of continuity for the case in question. Let us consider a volume of liquid bounded by two plane cross-sections of the channel at a distance $\mathrm{d}x$ apart. In unit time a volume $(Sv)_x$ of liquid flows through one plane, and a volume $(Sv)_{x+\mathrm{d}x}$ through the other. Hence the volume of liquid between the two planes changes by

$$(Sv)_{x+\mathrm{d}x} - (Sv)_x = \frac{\partial(Sv)}{\partial x}\mathrm{d}x.$$

Since the liquid is incompressible, however, this change must be due simply to the change in the level of the liquid. The change per unit time in the volume of liquid between the two planes considered is $(\partial S/\partial t)\mathrm{d}x$. We can therefore write

$$\frac{\partial S}{\partial t}\mathrm{d}x = -\frac{\partial(Sv)}{\partial x}\mathrm{d}x,$$

or

$$\frac{\partial S}{\partial t} + \frac{\partial(Sv)}{\partial x} = 0. \tag{12.12}$$

This is the required equation of continuity.

Let S_0 be the equilibrium cross-sectional area of the liquid in the channel. Then $S = S_0 + S'$, where S' is the change in the cross-sectional area caused by the wave. Since the change in the liquid level is small, we can write S' in the form $b\zeta$, where b is the width of the channel at the surface of the liquid. Equation (12.12) then becomes

$$b\frac{\partial\zeta}{\partial t} + \frac{\partial(S_0 v)}{\partial x} = 0. \tag{12.13}$$

Differentiating (12.13) with respect to t and substituting $\partial v/\partial t$ from (12.11), we obtain

$$\frac{\partial^2\zeta}{\partial t^2} - \frac{g}{b}\frac{\partial}{\partial x}\left(S_0\frac{\partial\zeta}{\partial x}\right) = 0. \tag{12.14}$$

If the channel cross-section is the same at all points, then S_0 = constant and

$$\frac{\partial^2\zeta}{\partial t^2} - \frac{gS_0}{b}\frac{\partial^2\zeta}{\partial x^2} = 0. \tag{12.15}$$

This is called a *wave equation*: as we shall show in §64, it corresponds to the propagation of waves with a velocity U which is independent of frequency and is the square root of the coefficient of $\partial^2\zeta/\partial x^2$. Thus the velocity of propagation of long gravity waves in channels is

$$U = \sqrt{(gS_0/b)}. \tag{12.16}$$

In an entirely similar manner, we can consider long waves in a large tank, which we suppose infinite in two directions (those of x and y). The depth of liquid in the tank is denoted by h. The component v_z of the velocity is now small. Euler's equations take a form similar to (12.11):

$$\frac{\partial v_x}{\partial t} + g\frac{\partial\zeta}{\partial x} = 0, \qquad \frac{\partial v_y}{\partial t} + g\frac{\partial\zeta}{\partial y} = 0. \tag{12.17}$$

The equation of continuity is derived in the same way as (12.12) and is

$$\frac{\partial h}{\partial t} + \frac{\partial(hv_x)}{\partial x} + \frac{\partial(hv_y)}{\partial y} = 0.$$

We write the depth h as $h_0 + \zeta$, where h_0 is the equilibrium depth. Then

$$\frac{\partial\zeta}{\partial t} + \frac{\partial(h_0 v_x)}{\partial x} + \frac{\partial(h_0 v_y)}{\partial y} = 0. \tag{12.18}$$

Let us assume that the tank has a horizontal bottom (h_0 = constant). Differentiating (12.18) with respect to t and substituting (12.17), we obtain

$$\frac{\partial^2\zeta}{\partial t^2} - gh_0\left(\frac{\partial^2\zeta}{\partial x^2} + \frac{\partial^2\zeta}{\partial y^2}\right) = 0. \tag{12.19}$$

This is again a (two-dimensional) wave equation; it corresponds to waves propagated with a velocity

$$U = \sqrt{(gh_0)}. \tag{12.20}$$

PROBLEMS

Problem 1. Determine the velocity of propagation of gravity waves on an unbounded surface of liquid with depth h.

Solution. At the bottom of the liquid, the normal velocity component must be zero, i.e. $v_z = \partial\phi/\partial z = 0$ for $z = -h$. From this condition we find the ratio of the constants A and B in the general solution

$$\phi = [Ae^{kz} + Be^{-kz}]\cos(kx - \omega t).$$

The result is

$$\phi = A\cos(kx - \omega t)\cosh k(z+h).$$

From the boundary condition (12.5) we find the relation between k and ω to be

$$\omega^2 = gk\tanh kh.$$

The velocity of propagation of the wave is

$$U = \frac{1}{2}\sqrt{\frac{g}{k\tanh kh}}\left[\tanh kh + \frac{kh}{\cosh^2 kh}\right].$$

For $kh \gg 1$ we have the result (12.10), and for $kh \ll 1$ the result (12.20).

PROBLEM 2. Determine the relation between frequency and wavelength for gravity waves on the surface separating two liquids, the upper liquid being bounded above by a fixed horizontal plane, and the lower liquid being similarly bounded below. The density and depth of the lower liquid are ρ and h, those of the upper liquid are ρ' and h', and $\rho > \rho'$.

SOLUTION. We take the xy-plane as the equilibrium plane of separation of the two liquids. Let us seek a solution having in the two liquids the forms

$$\left.\begin{aligned} \phi &= A\cosh k(z+h)\cos(kx-\omega t), \\ \phi' &= B\cosh k(z-h')\cos(kx-\omega t), \end{aligned}\right\} \quad (1)$$

so that the conditions at the upper and lower boundaries are satisfied; see the solution to Problem 1. At the surface of separation, the pressure must be continuous; by (12.2), this gives the condition

$$\rho g\zeta + \rho\frac{\partial\phi}{\partial t} = \rho' g\zeta + \rho'\frac{\partial\phi'}{\partial t} \quad \text{for} \quad z = \zeta,$$

or

$$\zeta = \frac{1}{g(\rho-\rho')}\left(\rho'\frac{\partial\phi'}{\partial t} - \rho\frac{\partial\phi}{\partial t}\right). \quad (2)$$

Moreover, the velocity component v_z must be the same for each liquid at the surface of separation. This gives the condition

$$\partial\phi/\partial z = \partial\phi'/\partial z \quad \text{for} \quad z = 0. \quad (3)$$

Now $v_z = \partial\phi/\partial z = \partial\zeta/\partial t$ and, substituting (2), we have

$$g(\rho-\rho')\frac{\partial\phi}{\partial z} = \rho'\frac{\partial^2\phi'}{\partial t^2} - \rho\frac{\partial^2\phi}{\partial t^2}. \quad (4)$$

Substituting (1) in (3) and (4) gives two homogeneous linear equations for A and B, and the condition of compatibility gives

$$\omega^2 = \frac{kg(\rho-\rho')}{\rho\coth kh + \rho'\coth kh'}.$$

For $kh \gg 1$, $kh' \gg 1$ (both liquids very deep),

$$\omega^2 = kg\frac{\rho-\rho'}{\rho+\rho'},$$

while for $kh \ll 1$, $kh' \ll 1$ (long waves),

$$\omega = k\sqrt{\frac{g(\rho-\rho')hh'}{\rho h' + \rho' h}}.$$

Lastly, if $kh \gtrsim 1$ and $kh' \ll 1$,

$$\omega^2 = k^2 gh'(\rho-\rho')/\rho.$$

PROBLEM 3. Determine the relation between frequency and wavelength for gravity waves propagated simultaneously on the surface of separation and on the upper surface of two liquid layers, the lower (density ρ) being infinitely deep, and the upper (density ρ') having depth h' and a free upper surface.

SOLUTION. We take the xy-plane as the equilibrium plane of separation of the two liquids. Let us seek a solution having in the two liquids the forms

$$\left.\begin{aligned} \phi &= Ae^{kz}\cos(kx-\omega t), \\ \phi' &= [Be^{-kz} + Ce^{kz}]\cos(kx-\omega t). \end{aligned}\right\} \quad (1)$$

At the surface of separation, i.e. for $z = 0$, we have the conditions (see Problem 2)

$$\frac{\partial\phi}{\partial z} = \frac{\partial\phi'}{\partial z}, \quad g(\rho-\rho')\frac{\partial\phi}{\partial z} = \rho'\frac{\partial^2\phi'}{\partial t^2} - \rho\frac{\partial^2\phi}{\partial t^2}, \tag{2}$$

and at the upper surface, i.e. for $z = h'$, the condition

$$\frac{\partial\phi'}{\partial z} + \frac{1}{g}\frac{\partial^2\phi'}{\partial t^2} = 0. \tag{3}$$

The first equation (2), on substitution of (1), gives $A = C - B$, and the remaining two conditions then give two equations for B and C; from the condition of compatibility we obtain a quadratic equation for ω^2, whose roots are

$$\omega^2 = kg\frac{(\rho-\rho')(1-e^{-2kh})}{\rho+\rho'+(\rho-\rho')e^{-2kh'}}, \qquad \omega^2 = kg.$$

For $h' \to \infty$ these roots correspond to waves propagated independently on the surface of separation and on the upper surface.

PROBLEM 4. Determine the characteristic frequencies of oscillation (see §69) of a liquid with depth h in a rectangular tank with width a and length b.

SOLUTION. We take the x and y axes along two sides of the tank. Let us seek a solution in the form of a stationary wave:

$$\phi = f(x, y)\cosh k(z+h)\cos\omega t.$$

We obtain for f the equation

$$\frac{\partial^2 f}{\partial x^2} + \frac{\partial^2 f}{\partial y^2} + k^2 f = 0,$$

and the condition at the free surface gives, as in Problem 1, the relation

$$\omega^2 = gk\tanh kh.$$

We take the solution of the equation for f in the form

$$f = \cos px\cos qy, \qquad p^2 + q^2 = k^2.$$

At the sides of the tank we must have the conditions

$$v_x = \partial\phi/\partial x = 0 \quad \text{for} \quad x = 0, a;$$

$$v_y = \partial\phi/\partial y = 0 \quad \text{for} \quad y = 0, b.$$

Hence we find $p = m\pi/a$, $q = n\pi/b$, where m, n are integers. The possible values of k^2 are therefore

$$k^2 = \pi^2\left(\frac{m^2}{a^2} + \frac{n^2}{b^2}\right).$$

§13. Internal waves in an incompressible fluid

There is a kind of gravity wave which can be propagated inside an incompressible fluid. Such waves are due to an inhomogeneity of the fluid caused by the gravitational field. The pressure (and therefore the entropy s) necessarily varies with height; hence any displacement of a fluid particle in height destroys the mechanical equilibrium, and consequently causes an oscillatory motion. For, since the motion is adiabatic, the particle carries with it to its new position its old entropy s, which is not the same as the equilibrium value at the new position.

We shall suppose below that the wavelength is small in comparison with distances over which the gravitational field causes a marked change in density†; and we shall regard the fluid itself as incompressible. This means that we can neglect the change in its density caused by the pressure change in the wave. The change in density caused by thermal expansion cannot be neglected, since it is this that causes the phenomenon in question.

Let us write down a system of hydrodynamic equations for this motion. We shall use a suffix 0 to distinguish the values of quantities in mechanical equilibrium, and a prime to mark small deviations from those values. Then the equation of conservation of the entropy $s = s_0 + s'$ can be written, to the first order of smallness,

$$\partial s'/\partial t + \mathbf{v}\cdot\mathbf{grad}\, s_0 = 0, \tag{13.1}$$

where s_0, like the equilibrium values of other quantities, is a given function of the vertical coordinate z.

Next, in Euler's equation we again neglect the term $(\mathbf{v}\cdot\mathbf{grad})\mathbf{v}$ (since the oscillations are small); taking into account also the fact that the equilibrium pressure distribution is given by $\mathbf{grad}\, p_0 = \rho_0\mathbf{g}$, we have to the same accuracy

$$\frac{\partial \mathbf{v}}{\partial t} = -\frac{\mathbf{grad}\, p}{\rho} + \mathbf{g} = -\frac{\mathbf{grad}\, p'}{\rho_0} + \frac{\mathbf{grad}\, p_0}{\rho_0^{\,2}}\rho'.$$

Since, from what has been said above, the change in density is due only to the change in entropy, and not to the change in pressure, we can put

$$\rho' = \left(\frac{\partial \rho_0}{\partial s_0}\right)_p s',$$

and we then obtain Euler's equation in the form

$$\frac{\partial \mathbf{v}}{\partial t} = \frac{\mathbf{g}}{\rho_0}\left(\frac{\partial \rho_0}{\partial s_0}\right)_p s' - \mathbf{grad}\frac{p'}{\rho_0}. \tag{13.2}$$

We can take ρ_0 under the gradient operator, since, as stated above, we always neglect the change in the equilibrium density over distances of the order of a wavelength. The density may likewise be supposed constant in the equation of continuity, which then becomes

$$\operatorname{div}\mathbf{v} = 0. \tag{13.3}$$

We shall seek a solution of equations (13.1)–(13.3) in the form of a plane wave:

$$\mathbf{v} = \text{constant} \times e^{i(\mathbf{k}\cdot\mathbf{r}-\omega t)},$$

and similarly for s' and p'. Substitution in the equation of continuity (13.3) gives

$$\mathbf{v}\cdot\mathbf{k} = 0, \tag{13.4}$$

† The density and pressure gradients are related by

$$\mathbf{grad}\, p = (\partial p/\partial \rho)_s\, \mathbf{grad}\, \rho = c^2\, \mathbf{grad}\, \rho,$$

where c is the speed of sound in the fluid. The hydrostatic equation $\mathbf{grad}\, p = \rho g$ thus gives $\mathbf{grad}\, \rho = (\rho/c^2)\mathbf{g}$. The density in the gravitational field therefore varies considerably over distances $l \cong c^2/g$. For air and water, $l \cong 10$ km and 200 km respectively.

i.e. the fluid velocity is everywhere perpendicular to the *wave vector* **k** (a transverse wave). Equations (13.1) and (13.2) give

$$i\omega s' = \mathbf{v}\cdot\mathbf{grad}\, s_0, \qquad -i\omega\mathbf{v} = \frac{1}{\rho_0}\left(\frac{\partial\rho_0}{\partial s_0}\right)_p s'\mathbf{g} - \frac{i\mathbf{k}}{\rho_0}p'.$$

The condition $\mathbf{v}\cdot\mathbf{k} = 0$ gives with the second of these equations

$$ik^2 p' = \left(\frac{\partial\rho_0}{\partial s_0}\right)_p s'\mathbf{g}\cdot\mathbf{k},$$

and, eliminating **v** and s' from the two equations, we obtain the desired dispersion relation,

$$\omega^2 = {\omega_0}^2 \sin^2\theta, \tag{13.5}$$

where

$${\omega_0}^2 = -\frac{g}{\rho}\left(\frac{\partial\rho}{\partial s}\right)_p \frac{\mathrm{d}s}{\mathrm{d}z}. \tag{13.6}$$

Here and henceforward we omit the suffix zero to the equilibrium values of thermodynamic quantities; the z-axis is vertically upwards, and θ is the angle between this axis and the direction of **k**. If the expression on the right of (13.6) is positive, the condition for the stability of the equilibrium distribution $s(z)$ (the condition that convection be absent—see §4) is fulfilled.

We see that the frequency depends only on the direction of the wave vector, and not on its magnitude. For $\theta = 0$ we have $\omega = 0$; this means that waves of the type considered, with the wave vector vertical, cannot exist.

If the fluid is in both mechanical equilibrium and complete thermodynamic equilibrium, its temperature is constant and we can write

$$\frac{\mathrm{d}s}{\mathrm{d}z} = \left(\frac{\partial s}{\partial p}\right)_T \frac{\mathrm{d}p}{\mathrm{d}z} = -\rho g\left(\frac{\partial s}{\partial p}\right)_T.$$

Finally, using the well-known thermodynamic relations

$$\left(\frac{\partial s}{\partial p}\right)_T = \frac{1}{\rho^2}\left(\frac{\partial\rho}{\partial T}\right)_p, \qquad \left(\frac{\partial\rho}{\partial s}\right)_p = \frac{T}{c_p}\left(\frac{\partial\rho}{\partial T}\right)_p,$$

where c_p is the specific heat per unit mass, we find

$$\omega_0 = \sqrt{\frac{Tg}{c_p\rho}}\left|\left(\frac{\partial\rho}{\partial T}\right)_p\right|. \tag{13.7}$$

In particular, for a perfect gas,

$$\omega_0 = \frac{g}{\sqrt{(c_p T)}}. \tag{13.8}$$

The dependence of the frequency on the direction of the wave vector has the result that the wave propagation velocity $\mathbf{U} = \partial\omega/\partial\mathbf{k}$ is not parallel to **k**. Representing $\omega(\mathbf{k})$ in the form

$$\omega = \omega_0\sqrt{[1 - (\mathbf{k}\cdot\boldsymbol{\nu}/k)^2]},$$

where $\boldsymbol{\nu}$ is a unit vector in the vertically upward direction, and differentiating, we find

$$\mathbf{U} = -(\omega_0^2/\omega k)(\mathbf{n}\cdot\boldsymbol{\nu})[\boldsymbol{\nu}-(\mathbf{n}\cdot\boldsymbol{\nu})\mathbf{n}] \tag{13.9}$$

(where $\mathbf{n} = \mathbf{k}/k$). This is perpendicular to $\mathbf{k}$, and its magnitude is

$$U = (\omega_0/k)\cos\theta.$$

Its vertical component is

$$\mathbf{U}\cdot\boldsymbol{\nu} = -(\omega_0/k)\cos\theta\sin\theta.$$

§14. Waves in a rotating fluid

Another kind of internal wave can be propagated in an incompressible fluid uniformly rotating as a whole. These waves are due to the Coriolis forces which occur in rotation.

We shall consider the fluid in coordinates rotating with it. With this treatment, the mechanical equations of motion must include additional (centrifugal and Coriolis) terms. Correspondingly, forces (per unit mass of fluid) must be added on the right of Euler's equation. The centrifugal force can be written as $\mathbf{grad}\,\frac{1}{2}(\boldsymbol{\Omega}\times\mathbf{r})^2$, where $\boldsymbol{\Omega}$ is the angular velocity vector of the fluid rotation. This term can be combined with the force $-(1/\rho)\,\mathbf{grad}\,p$ by using an effective pressure

$$P = p - \tfrac{1}{2}\rho(\boldsymbol{\Omega}\times\mathbf{r})^2. \tag{14.1}$$

The Coriolis force is $2\mathbf{v}\times\boldsymbol{\Omega}$, and occurs only when the fluid has a motion relative to the rotating coordinates, $\mathbf{v}$ being the velocity in those coordinates. We can transfer this term to the left-hand side of Euler's equation, writing the equation as

$$\partial\mathbf{v}/\partial t + (\mathbf{v}\cdot\mathbf{grad})\mathbf{v} + 2\boldsymbol{\Omega}\times\mathbf{v} = -(1/\rho)\mathbf{grad}\,P. \tag{14.2}$$

The equation of continuity is unchanged; for an incompressible fluid, it is simply $\operatorname{div}\mathbf{v} = 0$.

We shall again assume the wave amplitude to be small, and neglect the term quadratic in the velocity in (14.2), which becomes

$$\partial\mathbf{v}/\partial t + 2\boldsymbol{\Omega}\times\mathbf{v} = -(1/\rho)\,\mathbf{grad}\,p', \tag{14.3}$$

where p' is the variable part of the pressure in the wave, and ρ is a constant. The pressure can be eliminated by taking the curl of both sides. The right-hand side gives zero, and on the left-hand side, since the fluid is incompressible,

$$\begin{aligned}\mathbf{curl}\,(\boldsymbol{\Omega}\times\mathbf{v}) &= \boldsymbol{\Omega}\operatorname{div}\mathbf{v} - (\boldsymbol{\Omega}\cdot\mathbf{grad})\mathbf{v}\\ &= -(\boldsymbol{\Omega}\cdot\mathbf{grad})\mathbf{v}.\end{aligned}$$

Taking the direction of $\boldsymbol{\Omega}$ as the z-axis, we write the resulting equation as

$$\frac{\partial}{\partial t}\mathbf{curl}\,\mathbf{v} = 2\Omega\frac{\partial\mathbf{v}}{\partial z}. \tag{14.4}$$

We seek the solution as a plane wave

$$\mathbf{v} = \mathbf{A}e^{i(\mathbf{k}\cdot\mathbf{r}-\omega t)}, \tag{14.5}$$

which, since $\operatorname{div}\mathbf{v} = 0$, satisfies the transversality condition

$$\mathbf{k}\cdot\mathbf{A} = 0. \tag{14.6}$$

Substitution of (14.5) in (14.4) gives

$$\omega \mathbf{k}\times\mathbf{v} = 2i\Omega k_z \mathbf{v}. \tag{14.7}$$

The dispersion relation for these waves is found by eliminating $\mathbf{v}$ from this vector equation. Vector multiplication on both sides by $\mathbf{k}$ gives

$$-\omega k^2 \mathbf{v} = 2i\Omega k_z \mathbf{k}\times\mathbf{v}$$

and a comparison of the two equations yields the dependence of ω on $\mathbf{k}$:

$$\omega = 2\Omega k_z/k = 2\Omega\cos\theta, \tag{14.8}$$

where θ is the angle between $\mathbf{k}$ and $\mathbf{\Omega}$.

With (14.4), (14.7) takes the form

$$\mathbf{n}\times\mathbf{v} = i\mathbf{v},$$

where $n = \mathbf{k}/k$. If we use the complex wave amplitude in the form $\mathbf{A} = \mathbf{a} + i\mathbf{b}$ with real vectors $\mathbf{a}$ and $\mathbf{b}$, it follows that $\mathbf{n}\times\mathbf{b} = \mathbf{a}$: the vectors $\mathbf{a}$ and $\mathbf{b}$ (both lying in the plane perpendicular to $\mathbf{k}$) are at right angles and equal in magnitude. By taking their directions as the x and y axes, and separating real and imaginary parts in (14.5), we find

$$v_x = a\cos(\omega t - \mathbf{k}\cdot\mathbf{r}), \qquad v_y = -a\sin(\omega t - \mathbf{k}\cdot\mathbf{r}).$$

The wave is thus circularly polarized: at each point in space, the vector $\mathbf{v}$ rotates in the course of time, remaining constant in magnitude.†

The wave propagation velocity is

$$\mathbf{U} = \partial\omega/\partial\mathbf{k} = (2\Omega/k)[\boldsymbol{\nu} - \mathbf{n}(\mathbf{n}\cdot\boldsymbol{\nu})], \tag{14.9}$$

where $\boldsymbol{\nu}$ is a unit vector along $\mathbf{\Omega}$; as with internal gravity waves, it is perpendicular to the wave vector. Its magnitude and its component along $\mathbf{\Omega}$ are

$$U = (2\Omega/k)\sin\theta, \qquad \mathbf{U}\cdot\boldsymbol{\nu} = (2\Omega/k)\sin^2\theta = U\sin\theta.$$

These are called *inertial waves*. Since the Coriolis forces do no work on the moving fluid, the energy in the waves is entirely kinetic energy.

One particular form of axially symmetrical (not plane) inertial waves can be propagated along the axis of rotation of the fluid; see Problem 1.

There is one more comment to be made, regarding steady motions in a rotating fluid rather than wave propagation in it.

Let l be a characteristic length for such motion, and u a characteristic velocity. In order of magnitude, the term $(\mathbf{v}\cdot\mathbf{grad})\mathbf{v}$ in (14.2) is u^2/l, and $2\mathbf{\Omega}\times\mathbf{v}$ is Ωu. The former can be neglected in comparison with the latter if $u/l\Omega \ll 1$, and the equation of steady motion then reduces to

$$2\mathbf{\Omega}\times\mathbf{v} = -(1/\rho)\,\mathbf{grad}\,P \tag{14.10}$$

or

$$2\Omega v_y = (1/\rho)\partial P/\partial x, \qquad 2\Omega v_x = -(1/\rho)\partial P/\partial y, \qquad \partial P/\partial z = 0,$$

† This motion is relative to rotating coordinates. For fixed coordinates, it is combined with the rotation of the whole fluid.

where x and y are Cartesian coordinates in the plane perpendicular to the axis of rotation. Hence we see that P, and therefore v_x and v_y, are independent of the longitudinal coordinate z. Next, eliminating P from the first two equations, we get

$$\frac{\partial v_x}{\partial x}+\frac{\partial v_y}{\partial y}=0,$$

and the equation div $\mathbf{v}=0$ then shows that $\partial v_z/\partial z=0$. Thus steady motion (in rotating coordinates) in a rapidly rotating fluid is a superposition of two independent motions: two-dimensional flow in the transverse plane and axial flow independent of z (J. Proudman 1916).

PROBLEMS

PROBLEM 1. Determine the motion in an axially symmetrical wave propagated along the axis of an incompressible fluid rotating as a whole (W. Thomson 1880).

SOLUTION. We take cylindrical polar coordinates r, ϕ, z, with the z-axis parallel to $\mathbf{\Omega}$. In an axially symmetrical wave, all quantities are independent of the angle variable ϕ. The dependence on time and on the coordinate z is given by a factor $\exp[i(kz-\omega t)]$. Taking components in (14.3), we get

$$-i\omega v_r-2\Omega v_\phi=-(1/\rho)\partial p'/\partial r, \tag{1}$$

$$-i\omega v_\phi+2\Omega v_r=0, \qquad -i\omega v_z=-(ik/\rho)p'. \tag{2}$$

These are to be combined with the equation of continuity

$$\frac{1}{r}\frac{\partial}{\partial r}(rv_r)+ikv_z=0. \tag{3}$$

Expressing v_ϕ and p' in terms of v_r by means of (2) and (3) and substituting in (1), we find the equation

$$\frac{\mathrm{d}^2F}{\mathrm{d}r^2}+\frac{1}{r}\frac{\mathrm{d}F}{\mathrm{d}r}+\left[\frac{4\Omega^2k^2}{\omega^2}-k^2-\frac{1}{r^2}\right]F=0 \tag{4}$$

for the function $F(r)$ which determines the radial dependence of v_r:

$$v_r=F(r)e^{i(\omega t-kz)}.$$

The solution that vanishes for $r=0$ is

$$F=\text{constant}\times J_1[kr\sqrt{\{(4\Omega^2/\omega^2)-1\}}], \tag{5}$$

where J_1 is a Bessel function of order 1.

The motion comprises regions between coaxial cylinders with radius r_n such that

$$kr_n\sqrt{\{(4\Omega^2/\omega^2)-1\}}=x_n,$$

where $x_1, x_2, \ldots$ are the successive zeros of $J_1(x)$. On these cylindrical surfaces $v_r=0$, and the fluid therefore does not cross them.

For these waves in an infinite fluid, ω is independent of k. The possible values of the frequency are, however, restricted by the condition $\omega<2\Omega$; if this is not satisfied, (4) has no solution satisfying the necessary conditions of finiteness.

If the rotating fluid is bounded by a cylindrical wall with radius R, we have to use the condition $v_r=0$ at the wall. This gives the relation

$$ka\sqrt{\{(4\Omega^2/\omega^2)-1\}}=x_n$$

between ω and k for a wave with a given n (the number of coaxial regions in it).

PROBLEM 2. Derive an equation describing an arbitrary small perturbation of the pressure in a rotating fluid.

SOLUTION. Equation (14.3) in components is

$$\frac{\partial v_x}{\partial t}-2\Omega v_y=-\frac{1}{\rho}\frac{\partial p'}{\partial x}, \qquad \frac{\partial v_y}{\partial t}+2\Omega v_x=-\frac{1}{\rho}\frac{\partial p'}{\partial y}, \qquad \frac{\partial v_z}{\partial t}=-\frac{1}{\rho}\frac{\partial p'}{\partial z}. \tag{1}$$

Differentiating these with respect to x, y, and z, adding, and using div $\mathbf{v} = 0$, we find

$$\frac{1}{\rho}\triangle p' = 2\Omega\left(\frac{\partial v_y}{\partial x} - \frac{\partial v_x}{\partial y}\right).$$

Differentiation with respect to t, again using equations (1), gives

$$\frac{1}{\rho}\frac{\partial}{\partial t}\triangle p' = 4\Omega^2\frac{\partial v_z}{\partial z},$$

and by a further differentiation with respect to t we arrive at the final equation

$$\frac{\partial^2}{\partial t^2}\triangle p' + 4\Omega^2\frac{\partial^2 p'}{\partial z^2} = 0. \tag{2}$$

For periodic perturbations with frequency ω, this becomes

$$\frac{\partial^2 p'}{\partial x^2} + \frac{\partial^2 p'}{\partial y^2} + \left(1 - \frac{4\Omega^2}{\omega^2}\right)\frac{\partial^2 p'}{\partial z^2} = 0. \tag{3}$$

For waves having the form (14.5), this of course gives the known dispersion relation (14.8), with $\omega < 2\Omega$ and a negative coefficient of $\partial^2 p'/\partial z^2$ in (3). Perturbations from a point source are propagated along generators of a cone whose axis is along $\mathbf{\Omega}$ and whose vertical angle is 2θ, where $\sin\theta = \omega/2\Omega$.

When $\omega > 2\Omega$, the coefficient of $\partial^2 p'/\partial z^2$ in (3) is positive, and this equation becomes Laplace's equation by an obvious change in the z scale. In this case, a point source of perturbation affects the whole volume of the fluid, to an extent that decreases away from the source according to a power law.

CHAPTER II

VISCOUS FLUIDS

§15. The equations of motion of a viscous fluid

Let us now study the effect of energy dissipation, occurring during the motion of a fluid, on that motion itself. This process is the result of the thermodynamic irreversibility of the motion. This irreversibility always occurs to some extent, and is due to internal friction (viscosity) and thermal conduction.

In order to obtain the equations describing the motion of a viscous fluid, we have to include some additional terms in the equation of motion of an ideal fluid. The equation of continuity, as we see from its derivation, is equally valid for any fluid, whether viscous or not. Euler's equation, on the other hand, requires modification.

We have seen in §7 that Euler's equation can be written in the form

$$\frac{\partial}{\partial t}(\rho v_i) = -\frac{\partial \Pi_{ik}}{\partial x_k},$$

where Π_{ik} is the momentum flux density tensor. The momentum flux given by formula (7.2) represents a completely reversible transfer of momentum, due simply to the mechanical transport of the different particles of fluid from place to place and to the pressure forces acting in the fluid. The viscosity (internal friction) causes another, irreversible, transfer of momentum from points where the velocity is large to those where it is small.

The equation of motion of a viscous fluid may therefore be obtained by adding to the "ideal" momentum flux (7.2) a term $-\sigma'_{ik}$ which gives the irreversible "viscous" transfer of momentum in the fluid. Thus we write the momentum flux density tensor in a viscous fluid in the form

$$\Pi_{ik} = p\delta_{ik} + \rho v_i v_k - \sigma'_{ik} = -\sigma_{ik} + \rho v_i v_k. \qquad (15.1)$$

The tensor

$$\sigma_{ik} = -p\delta_{ik} + \sigma'_{ik} \qquad (15.2)$$

is called the *stress tensor*, and σ'_{ik} the *viscous stress tensor*. σ_{ik} gives the part of the momentum flux that is not due to the direct transfer of momentum with the mass of moving fluid.†

The general form of the tensor σ'_{ik} can be established as follows. Processes of internal friction occur in a fluid only when different fluid particles move with different velocities, so that there is a relative motion between various parts of the fluid. Hence σ'_{ik} must depend on the space derivatives of the velocity. If the velocity gradients are small, we may suppose

† We shall see below that σ'_{ik} contains a term proportional to δ_{ik}, i.e. of the same form as the term $p\delta_{ik}$. When the momentum flux tensor is put in such a form, therefore, we should specify what is meant by the pressure p; see the end of §49.

that the momentum transfer due to viscosity depends only on the first derivatives of the velocity. To the same approximation, σ'_{ik} may be supposed a linear function of the derivatives $\partial v_i/\partial x_k$. There can be no terms in σ'_{ik} independent of $\partial v_i/\partial x_k$, since σ'_{ik} must vanish for $\mathbf{v}$ = constant. Next, we notice that σ'_{ik} must also vanish when the whole fluid is in uniform rotation, since it is clear that in such a motion no internal friction occurs in the fluid. In uniform rotation with angular velocity $\mathbf{\Omega}$, the velocity $\mathbf{v}$ is equal to the vector product $\mathbf{\Omega}\times\mathbf{r}$. The sums

$$\frac{\partial v_i}{\partial x_k}+\frac{\partial v_k}{\partial x_i}$$

are linear combinations of the derivatives $\partial v_i/\partial x_k$, and vanish when $\mathbf{v} = \mathbf{\Omega}\times\mathbf{r}$. Hence σ'_{ik} must contain just these symmetrical combinations of the derivatives $\partial v_i/\partial x_k$.

The most general tensor of rank two satisfying the above conditions is

$$\sigma'_{ik} = \eta\left(\frac{\partial v_i}{\partial x_k}+\frac{\partial v_k}{\partial x_i}-\tfrac{2}{3}\delta_{ik}\frac{\partial v_l}{\partial x_l}\right)+\zeta\delta_{ik}\frac{\partial v_l}{\partial x_l}, \tag{15.3}$$

with coefficients η and ζ independent of the velocity. In making this statement we use the fact that the fluid is isotropic, as a result of which its properties must be described by scalar quantities only (in this case, η and ζ). The terms in (15.3) are arranged so that the expression in parentheses has the property of vanishing on contraction with respect to i and k.† The constants η and ζ are called *coefficients of viscosity*, and ζ often the *second viscosity*. As we shall show in §§16 and 49, they are both positive:

$$\eta > 0, \qquad \zeta > 0. \tag{15.4}$$

The equations of motion of a viscous fluid can now be obtained by simply adding the expressions $\partial\sigma'_{ik}/\partial x_k$ to the right-hand side of Euler's equation

$$\rho\left(\frac{\partial v_i}{\partial t}+v_k\frac{\partial v_i}{\partial x_k}\right) = -\frac{\partial p}{\partial x_i}.$$

Thus we have

$$\rho\left(\frac{\partial v_i}{\partial t}+v_k\frac{\partial v_i}{\partial x_k}\right) = -\frac{\partial p}{\partial x_i}+\frac{\partial}{\partial x_k}\left\{\eta\left(\frac{\partial v_i}{\partial x_k}+\frac{\partial v_k}{\partial x_i}-\tfrac{2}{3}\delta_{ik}\frac{\partial v_l}{\partial x_l}\right)\right\}+\frac{\partial}{\partial x_i}\left(\zeta\frac{\partial v_l}{\partial x_l}\right). \tag{15.5}$$

This is the most general form of the equations of motion of a viscous fluid. The quantities η and ζ are functions of pressure and temperature. In general, p and T, and therefore η and ζ, are not constant throughout the fluid, so that η and ζ cannot be taken outside the gradient operator.

In most cases, however, the viscosity coefficients do not change noticeably in the fluid, and they may be regarded as constant. We then have equations (15.5), in vector form, as

$$\rho\left[\frac{\partial \mathbf{v}}{\partial t}+(\mathbf{v}\cdot\mathbf{grad})\mathbf{v}\right] = -\mathbf{grad}\, p+\eta\triangle\mathbf{v}+(\zeta+\tfrac{1}{3}\eta)\mathbf{grad}\,\mathrm{div}\,\mathbf{v}. \tag{15.6}$$

This is called the *Navier–Stokes equation*. It becomes considerably simpler if the fluid may be regarded as incompressible, so that div $\mathbf{v} = 0$, and the last term on the right of (15.6)

† That is, on taking the sum of the components with $i = k$.

is zero. In discussing viscous fluids, we shall almost always regard them as incompressible, and accordingly use the equation of motion in the form†

$$\frac{\partial \mathbf{v}}{\partial t} + (\mathbf{v} \cdot \mathbf{grad})\mathbf{v} = -\frac{1}{\rho}\,\mathbf{grad}\, p + \frac{\eta}{\rho}\triangle \mathbf{v}. \tag{15.7}$$

The stress tensor in an incompressible fluid takes the simple form

$$\sigma_{ik} = -p\delta_{ik} + \eta\left(\frac{\partial v_i}{\partial x_k} + \frac{\partial v_k}{\partial x_i}\right). \tag{15.8}$$

We see that the viscosity of an incompressible fluid is determined by only one coefficient. Since most fluids may be regarded as practically incompressible, it is this viscosity coefficient η which is generally of importance. The ratio

$$\nu = \eta/\rho \tag{15.9}$$

is called the *kinematic viscosity* (while η itself is called the *dynamic viscosity*). We give below the values of η and ν for various fluids, at a temperature of 20° C:

	η (g/cm sec)	ν (cm²/sec)
Water	0·010	0·010
Air	0·00018	0·150
Alcohol	0·018	0·022
Glycerine	8·5	6·8
Mercury	0·0156	0·0012

It may be mentioned that the dynamic viscosity of a gas at a given temperature is independent of the pressure. The kinematic viscosity, however, is inversely proportional to the pressure.

The pressure can be eliminated from equation (15.7) in the same way as from Euler's equation. Taking the curl of both sides, we obtain, instead of equation (2.11) as for an ideal fluid,

$$\frac{\partial}{\partial t}(\mathbf{curl}\,\mathbf{v}) = \mathbf{curl}\,(\mathbf{v} \times \mathbf{curl}\,\mathbf{v}) + \nu\triangle\,(\mathbf{curl}\,\mathbf{v})$$

Since the fluid is incompressible, the equation can be transformed by expanding the product in the first term on the right and using the equation div $\mathbf{v} = 0$:

$$\frac{\partial}{\partial t}(\mathbf{curl}\cdot\mathbf{v}) + (\mathbf{v}\cdot\mathbf{grad})\,\mathbf{curl}\,\mathbf{v} - (\mathbf{curl}\,\mathbf{v}\cdot\mathbf{grad})\,\mathbf{v}$$
$$= \nu\triangle\,\mathbf{curl}\,\mathbf{v}. \tag{15.10}$$

† Equation (15.7) was first stated as a result of studies on models by C. L. Navier (1827). A derivation, similar to the modern one, for equations (15.6) (without the ζ term) and (15.7) was given by G. G. Stokes (1845).

When the velocity distribution is known, the pressure distribution in the fluid can be found by solving the Poisson-type equation

$$\triangle p = -\rho \frac{\partial v_i}{\partial x_k}\frac{\partial v_k}{\partial x_i} = -\rho \frac{\partial^2 v_i v_k}{\partial x_k \partial x_i}, \tag{15.11}$$

which is obtained by taking the divergence of (15.7).

We may also give the equation satisfied by the stream function $\psi(x, y)$ in two-dimensional flow of an incompressible viscous fluid. It is derived by substituting (10.9) in (15.10):

$$\frac{\partial}{\partial t}\triangle\psi - \frac{\partial\psi}{\partial x}\frac{\partial\triangle\psi}{\partial y} + \frac{\partial\psi}{\partial y}\frac{\partial\triangle\psi}{\partial x} - \nu\triangle\triangle\psi = 0. \tag{15.12}$$

We must also write down the boundary conditions on the equations of motion of a viscous fluid. There are always forces of molecular attraction between a viscous fluid and the surface of a solid body, and these forces have the result that the layer of fluid immediately adjacent to the surface is brought completely to rest, and "adheres" to the surface. Accordingly, the boundary conditions on the equations of motion of a viscous fluid require that the fluid velocity should vanish at fixed solid surfaces:

$$\mathbf{v} = 0. \tag{15.13}$$

It should be emphasized that both the normal and the tangential velocity component must vanish, whereas for an ideal fluid the boundary conditions require only the vanishing of v_n.†

In the general case of a moving surface, the velocity **v** must be equal to the velocity of the surface.

It is easy to write down an expression for the force acting on a solid surface bounding the fluid. The force acting on an element of the surface is just the momentum flux through this element. The momentum flux through the surface element **df** is

$$\Pi_{ik}\mathrm{d}f_k = (\rho v_i v_k - \sigma_{ik})\mathrm{d}f_k.$$

Writing $\mathrm{d}f_k$ in the form $\mathrm{d}f_k = n_k \mathrm{d}f$, where **n** is a unit vector along the normal, and recalling that $\mathbf{v} = 0$ at a solid surface,‡ we find that the force **P** acting on unit surface area is

$$P_i = -\sigma_{ik} n_k = p n_i - \sigma'_{ik} n_k. \tag{15.14}$$

The first term is the ordinary pressure of the fluid, while the second is the force of friction, due to the viscosity, acting on the surface. We must emphasize that **n** in (15.14) is a unit vector along the outward normal to the fluid, i.e. along the inward normal to the solid surface.

† We may note that, in general, Euler's equations cannot be satisfied with the extra boundary condition (in comparison with the case of an ideal fluid) that the tangential velocity be zero. Mathematically, this occurs becuase the equation is first-order in the derivatives with respect to the coordinates, whereas the Navier–Stokes equation is second-order.

‡ In determining the force acting on the surface, each surface element must be considered in a frame of reference in which it is at rest. The force is equal to the momentum flux only when the surface is fixed.

If we have a surface of separation between two immiscible fluids, the conditions at the surface are that the velocities of the fluids must be equal and the forces which they exert on each other must be equal and opposite. The latter condition is written

$$n_{1,k}\sigma_{1,ik} + n_{2,k}\sigma_{2,ik} = 0,$$

where the suffixes 1 and 2 refer to the two fluids. The normal vectors $\mathbf{n}_1$ and $\mathbf{n}_2$ are in opposite directions, i.e. $\mathbf{n}_1 = -\mathbf{n}_2 \equiv \mathbf{n}$, so that we can write

$$n_i\sigma_{1,ik} = n_i\sigma_{2,ik}. \tag{15.15}$$

At a free surface of the fluid the condition

$$\sigma_{ik}n_k \equiv \sigma'_{ik}n_k - pn_i = 0 \tag{15.16}$$

must hold.

EQUATIONS OF MOTION IN CURVILINEAR COORDINATES

We give below, for reference, the equations of motion for a viscous incompressible fluid in frequently used curvilinear coordinates. In cylindrical polar coordinates r, ϕ, z the components of the stress tensor are

$$\sigma_{rr} = -p + 2\eta\frac{\partial v_r}{\partial r}, \qquad \sigma_{r\phi} = \eta\left(\frac{1}{r}\frac{\partial v_r}{\partial \phi} + \frac{\partial v_\phi}{\partial r} - \frac{v_\phi}{r}\right),$$

$$\sigma_{\phi\phi} = -p + 2\eta\left(\frac{1}{r}\frac{\partial v_\phi}{\partial \phi} + \frac{v_r}{r}\right), \qquad \sigma_{\phi z} = \eta\left(\frac{\partial v_\phi}{\partial z} + \frac{1}{r}\frac{\partial v_z}{\partial \phi}\right),$$

$$\sigma_{zz} = -p + 2\eta\frac{\partial v_z}{\partial z}, \qquad \sigma_{zr} = \eta\left(\frac{\partial v_z}{\partial r} + \frac{\partial v_r}{\partial z}\right). \tag{15.17}$$

The three components of the Navier–Stokes equation are

$$\frac{\partial v_r}{\partial t} + (\mathbf{v}\cdot\mathbf{grad})v_r - \frac{v_\phi^2}{r} = -\frac{1}{\rho}\frac{\partial p}{\partial r} + \nu\left(\triangle v_r - \frac{2}{r^2}\frac{\partial v_\phi}{\partial \phi} - \frac{v_r}{r^2}\right),$$

$$\frac{\partial v_\phi}{\partial t} + (\mathbf{v}\cdot\mathbf{grad})v_\phi + \frac{v_r v_\phi}{r} = -\frac{1}{\rho r}\frac{\partial p}{\partial \phi} + \nu\left(\triangle v_\phi + \frac{2}{r^2}\frac{\partial v_r}{\partial \phi} - \frac{v_\phi}{r^2}\right),$$

$$\frac{\partial v_z}{\partial t} + (\mathbf{v}\cdot\mathbf{grad})v_z = -\frac{1}{\rho}\frac{\partial p}{\partial z} + \nu\triangle v_z, \tag{15.18}$$

where

$$(\mathbf{v}\cdot\mathbf{grad})f = v_r\frac{\partial f}{\partial r} + \frac{v_\phi}{r}\frac{\partial f}{\partial \phi} + v_z\frac{\partial f}{\partial z},$$

$$\triangle f = \frac{1}{r}\frac{\partial}{\partial r}\left(r\frac{\partial f}{\partial r}\right) + \frac{1}{r^2}\frac{\partial^2 f}{\partial \phi^2} + \frac{\partial^2 f}{\partial z^2}.$$

The equation of continuity is

$$\frac{1}{r}\frac{\partial(rv_r)}{\partial r} + \frac{1}{r}\frac{\partial v_\phi}{\partial \phi} + \frac{\partial v_z}{\partial z} = 0. \tag{15.19}$$

In spherical polar coordinates r, ϕ, θ we have for the stress tensor

$$\begin{aligned}
\sigma_{rr} &= -p+2\eta\frac{\partial v_r}{\partial r},\\
\sigma_{\phi\phi} &= -p+2\eta\left(\frac{1}{r\sin\theta}\frac{\partial v_\phi}{\partial\phi}+\frac{v_r}{r}+\frac{v_\theta\cot\theta}{r}\right),\\
\sigma_{\theta\theta} &= -p+2\eta\left(\frac{1}{r}\frac{\partial v_\theta}{\partial\theta}+\frac{v_r}{r}\right),\\
\sigma_{r\theta} &= \eta\left(\frac{1}{r}\frac{\partial v_r}{\partial\theta}+\frac{\partial v_\theta}{\partial r}-\frac{v_\theta}{r}\right),\\
\sigma_{\theta\phi} &= \eta\left(\frac{1}{r\sin\theta}\frac{\partial v_\theta}{\partial\phi}+\frac{1}{r}\frac{\partial v_\phi}{\partial\theta}-\frac{v_\phi\cot\theta}{r}\right),\\
\sigma_{\phi r} &= \eta\left(\frac{\partial v_\phi}{\partial r}+\frac{1}{r\sin\theta}\frac{\partial v_r}{\partial\phi}-\frac{v_\phi}{r}\right),
\end{aligned}\tag{15.20}$$

while the Navier–Stokes equations are

$$\begin{aligned}
&\frac{\partial v_r}{\partial t}+(\mathbf{v}\cdot\mathbf{grad})v_r-\frac{v_\theta{}^2+v_\phi{}^2}{r}\\
&\qquad=-\frac{1}{\rho}\frac{\partial p}{\partial r}+\nu\left[\triangle v_r-\frac{2}{r^2\sin^2\theta}\frac{\partial(v_\theta\sin\theta)}{\partial\theta}-\frac{2}{r^2\sin\theta}\frac{\partial v_\phi}{\partial\phi}-\frac{2v_r}{r^2}\right],\\
&\frac{\partial v_\theta}{\partial t}+(\mathbf{v}\cdot\mathbf{grad})v_\theta+\frac{v_r v_\theta}{r}-\frac{v_\phi{}^2\cot\theta}{r}\\
&\qquad=-\frac{1}{\rho r}\frac{\partial p}{\partial\theta}+\nu\left[\triangle v_\theta-\frac{2\cos\theta}{r^2\sin^2\theta}\frac{\partial v_\phi}{\partial\phi}+\frac{2}{r^2}\frac{\partial v_r}{\partial\theta}-\frac{v_\theta}{r^2\sin^2\theta}\right],\\
&\frac{\partial v_\phi}{\partial t}+(\mathbf{v}\cdot\mathbf{grad})v_\phi+\frac{v_r v_\phi}{r}+\frac{v_\theta v_\phi\cot\theta}{r}\\
&\qquad=-\frac{1}{\rho r\sin\theta}\frac{\partial p}{\partial\phi}+\nu\left[\triangle v_\phi+\frac{2}{r^2\sin\theta}\frac{\partial v_r}{\partial\phi}+\frac{2\cos\theta}{r^2\sin^2\theta}\frac{\partial v_\theta}{\partial\phi}-\frac{v_\phi}{r^2\sin^2\theta}\right],
\end{aligned}\tag{15.21}$$

where

$$(\mathbf{v}\cdot\mathbf{grad})f=v_r\frac{\partial f}{\partial r}+\frac{v_\theta}{r}\frac{\partial f}{\partial\theta}+\frac{v_\phi}{r\sin\theta}\frac{\partial f}{\partial\phi},$$

$$\triangle f=\frac{1}{r^2}\frac{\partial}{\partial r}\left(r^2\frac{\partial f}{\partial r}\right)+\frac{1}{r^2\sin\theta}\frac{\partial}{\partial\theta}\left(\sin\theta\frac{\partial f}{\partial\theta}\right)+\frac{1}{r^2\sin^2\theta}\frac{\partial^2 f}{\partial\phi^2}=0.$$

The equation of continuity is

$$\frac{1}{r^2}\frac{\partial(r^2v_r)}{\partial r}+\frac{1}{r\sin\theta}\frac{\partial(v_\theta\sin\theta)}{\partial\theta}+\frac{1}{r\sin\theta}\frac{\partial v_\phi}{\partial\phi}=0.\tag{15.22}$$

§16. Energy dissipation in an incompressible fluid

The presence of viscosity results in the dissipation of energy, which is finally transformed into heat. The calculation of the energy dissipation is especially simple for an incompressible fluid.

The total kinetic energy of an incompressible fluid is

$$E_{\text{kin}} = \tfrac{1}{2}\rho \int v^2 \mathrm{d}V.$$

We take the time derivative of this energy, writing $\partial(\frac{1}{2}\rho v^2)/\partial t = \rho v_i \partial v_i/\partial t$ and substituting for $\partial v_i/\partial t$ the expression for it given by the Navier–Stokes equation:

$$\frac{\partial v_i}{\partial t} = -v_k \frac{\partial v_i}{\partial x_k} - \frac{1}{\rho}\frac{\partial p}{\partial x_i} + \frac{1}{\rho}\frac{\partial \sigma'_{ik}}{\partial x_k}.$$

The result is

$$\frac{\partial}{\partial t}(\tfrac{1}{2}\rho v^2) = -\rho \mathbf{v}\cdot(\mathbf{v}\cdot\mathbf{grad})\mathbf{v} - \mathbf{v}\cdot\mathbf{grad}\, p + v_i \frac{\partial \sigma'_{ik}}{\partial x_k}$$

$$= -\rho(\mathbf{v}\cdot\mathbf{grad})\left(\tfrac{1}{2}v^2 + \frac{p}{\rho}\right) + \operatorname{div}(\mathbf{v}\cdot\boldsymbol{\sigma}') - \sigma'_{ik}\frac{\partial v_i}{\partial x_k}.$$

Here $\mathbf{v}\cdot\boldsymbol{\sigma}'$ denotes the vector whose components are $v_i\sigma'_{ik}$. Since $\operatorname{div}\mathbf{v} = 0$ for an incompressible fluid, we can write the first term on the right as a divergence:

$$\frac{\partial}{\partial t}(\tfrac{1}{2}\rho v^2) = -\operatorname{div}\left[\rho\mathbf{v}\left(\tfrac{1}{2}v^2 + \frac{p}{\rho}\right) - \mathbf{v}\cdot\boldsymbol{\sigma}'\right] - \sigma'_{ik}\frac{\partial v_i}{\partial x_k}. \qquad (16.1)$$

The expression in brackets is just the energy flux density in the fluid: the term $\rho\mathbf{v}(\frac{1}{2}v^2 + p/\rho)$ is the energy flux due to the actual transfer of fluid mass, and is the same as the energy flux in an ideal fluid (see (10.5)). The second term, $\mathbf{v}\cdot\boldsymbol{\sigma}'$, is the energy flux due to processes of internal friction. For the presence of viscosity results in a momentum flux σ'_{ik}; a transfer of momentum, however, always involves a transfer of energy, and the energy flux is clearly equal to the scalar product of the momentum flux and the velocity.

If we integrate (16.1) over some volume V, we obtain

$$\frac{\partial}{\partial t}\int \tfrac{1}{2}\rho v^2 \mathrm{d}V = -\oint\left[\rho\mathbf{v}\left(\tfrac{1}{2}v^2 + \frac{p}{\rho}\right) - \mathbf{v}\cdot\boldsymbol{\sigma}'\right]\cdot\mathbf{df} - \int \sigma'_{ik}\frac{\partial v_i}{\partial x_k}\mathrm{d}V. \qquad (16.2)$$

The first term on the right gives the rate of change of the kinetic energy of the fluid in V owing to the energy flux through the surface bounding V. The integral in the second term is consequently the decrease per unit time in the kinetic energy owing to dissipation.

If the integration is extended to the whole volume of the fluid, the surface integral vanishes (since the velocity vanishes at infinity†), and we find the energy dissipated per unit time in the whole fluid to be

$$\dot{E}_{\text{kin}} = -\int \sigma'_{ik}\frac{\partial v_i}{\partial x_k}\mathrm{d}V = -\tfrac{1}{2}\int \sigma'_{ik}\left(\frac{\partial v_i}{\partial x_k} + \frac{\partial v_k}{\partial x_i}\right)\mathrm{d}V,$$

† We are considering the motion of the fluid in a system of coordinates such that the fluid is at rest at infinity. Here, and in similar cases, we speak, for the sake of definiteness, of an infinite volume of fluid, but this implies no loss of generality. For a fluid enclosed in a finite volume, the surface integral again vanishes, because the velocity at the surface vanishes.

since the tensor σ'_{ik} is symmetrical. In incompressible fluids, the tensor σ'_{ik} is given by (15.8), so that we have finally for the energy dissipation in an incompressible fluid

$$\dot{E}_{\text{kin}} = -\tfrac{1}{2}\eta \int \left(\frac{\partial v_i}{\partial x_k} + \frac{\partial v_k}{\partial x_i}\right)^2 \mathrm{d}V. \tag{16.3}$$

The dissipation leads to a decrease in the mechanical energy, i.e. we must have $\dot{E}_{\text{kin}} < 0$. The integral in (16.3), however, is always positive. We therefore conclude that the viscosity coefficient η is always positive.

PROBLEM

Transform the integral (16.3) for potential flow into an integral over the surface bounding the region of flow.

SOLUTION. Putting $\partial v_i/\partial x_k = \partial v_k/\partial x_i$ and integrating once by parts, we find

$$\dot{E}_{\text{kin}} = -2\eta \int \left(\frac{\partial v_i}{\partial x_k}\right)^2 \mathrm{d}V = -2\eta \int v_i \frac{\partial v_i}{\partial x_k} \mathrm{d}f_k,$$

or

$$\dot{E}_{\text{kin}} = -\eta \int \mathbf{grad}\, v^2 \cdot \mathrm{d}\mathbf{f}.$$

§17. Flow in a pipe

We shall now consider some simple problems of motion of an incompressible viscous fluid.

Let the fluid be enclosed between two parallel planes moving with a constant relative velocity $\mathbf{u}$. We take one of these planes as the xz-plane, with the x-axis in the direction of $\mathbf{u}$. It is clear that all quantities depend only on y, and that the fluid velocity is everywhere in the x-direction. We have from (15.7) for steady flow

$$\mathrm{d}p/\mathrm{d}y = 0, \quad \mathrm{d}^2v/\mathrm{d}y^2 = 0.$$

(The equation of continuity is satisfied identically.) Hence p = constant, $v = ay + b$. For $y = 0$ and $y = h$ (h being the distance between the planes) we must have respectively $v = 0$ and $v = u$. Thus

$$v = yu/h. \tag{17.1}$$

The fluid velocity distribution is therefore linear. The mean fluid velocity is

$$\bar{v} = \frac{1}{h}\int_0^h v\,\mathrm{d}y = \tfrac{1}{2}u. \tag{17.2}$$

From (15.14) we find that the normal component of the force on either plane is just p, as it should be, while the tangential friction force on the plane $y = 0$ is

$$\sigma_{xy} = \eta\,\mathrm{d}v/\mathrm{d}y = \eta u/h; \tag{17.3}$$

the force on the plane $y = h$ is $-\eta u/h$.

Next, let us consider steady flow between two fixed parallel planes in the presence of a pressure gradient. We choose the coordinates as before; the x-axis is in the direction of motion of the fluid. The Navier–Stokes equations give, since the velocity clearly depends only on y,

$$\frac{\partial^2 v}{\partial y^2} = \frac{1}{\eta}\frac{\partial p}{\partial x}, \qquad \frac{\partial p}{\partial y} = 0.$$

The second equation shows that the pressure is independent of y, i.e. it is constant across the depth of the fluid between the planes. The right-hand side of the first equation is therefore a function of x only, while the left-hand side is a function of y only; this can be true only if both sides are constant. Thus $\mathrm{d}p/\mathrm{d}x$ = constant, i.e. the pressure is a linear function of the coordinate x along the direction of flow. For the velocity we now obtain

$$v = \frac{1}{2\eta}\frac{\mathrm{d}p}{\mathrm{d}x}y^2 + ay + b.$$

The constants a and b are determined from the boundary conditions, $v = 0$ for $y = 0$ and $y = h$. The result is

$$v = -\frac{1}{2\eta}\frac{\mathrm{d}p}{\mathrm{d}x}y(y-h). \tag{17.4}$$

Thus the velocity varies parabolically across the fluid, reaching its maximum value in the middle. The mean fluid velocity (averaged over the depth of the fluid) is

$$\bar{v} = -\frac{h^2}{12\eta}\frac{\mathrm{d}p}{\mathrm{d}x}. \tag{17.5}$$

The frictional force acting on one of the fixed planes is

$$\sigma_{xy} = \eta(\partial v/\partial y)_{y=0} = -\tfrac{1}{2}h\,\mathrm{d}p/\mathrm{d}x. \tag{17.6}$$

Finally, let us consider steady flow in a pipe with arbitrary cross-section (the same along the whole length of the pipe, however). We take the axis of the pipe as the x-axis. The fluid velocity is evidently along the x-axis at all points, and is a function of y and z only. The equation of continuity is satisfied identically, while the y and z components of the Navier–Stokes equation again give $\partial p/\partial y = \partial p/dz = 0$, i.e. the pressure is constant over the cross-section of the pipe. The x-component of equation (15.7) gives

$$\frac{\partial^2 v}{\partial y^2} + \frac{\partial^2 v}{\partial z^2} = \frac{1}{\eta}\frac{\mathrm{d}p}{\mathrm{d}x}. \tag{17.7}$$

Hence we again conclude that $\mathrm{d}p/\mathrm{d}x$ = constant; the pressure gradient may therefore be written $-\Delta p/l$, where Δp is the pressure difference between the ends of the pipe and l is its length.

Thus the velocity distribution for flow in a pipe is determined by a two-dimensional equation of the form $\triangle v$ = constant. This equation has to be solved with the boundary condition $v = 0$ at the circumference of the cross-section of the pipe. We shall solve the equation for a pipe with circular cross-section. Taking the origin at the centre of the circle and using polar coordinates, we have by symmetry $v = v(r)$. Using the expression for the Laplacian in polar coordinates, we have

$$\frac{1}{r}\frac{\mathrm{d}}{\mathrm{d}r}\left(r\frac{\mathrm{d}v}{\mathrm{d}r}\right) = -\frac{\Delta p}{\eta l}.$$

Integration gives

$$v = -\frac{\Delta p}{4\eta l}r^2 + a\log r + b. \tag{17.8}$$

The constant a must be put equal to zero, since the velocity must remain finite at the centre of the pipe. The constant b is determined from the requirement that $v = 0$ for $r = R$, where R is the radius of the pipe. We then find

$$v = \frac{\Delta p}{4\eta l}(R^2 - r^2). \tag{17.9}$$

Thus the velocity distribution across the pipe is parabolic.

It is easy to determine the mass Q of fluid passing per unit time through any cross-section of the pipe (called the *discharge*). A mass $\rho \cdot 2\pi r v\, dr$ passes per unit time through an annular element $2\pi r\, dr$ of the cross-sectional area. Hence

$$Q = 2\pi\rho \int_0^R rv\, dr.$$

Using (17.9), we obtain

$$Q = \frac{\pi\Delta p}{8\nu l} R^4. \tag{17.10}$$

The mass of fluid is thus proportional to the fourth power of the radius of the pipe.†

PROBLEMS

PROBLEM 1. Determine the flow in a pipe of annular cross-section, the internal and external radii being R_1, R_2.

SOLUTION. Determining the constants a and b in the general solution (17.8) from the conditions that $v = 0$ for $r = R_1$ and $r = R_2$, we find

$$v = \frac{\Delta p}{4\eta l}\left[R_2^{\,2} - r^2 + \frac{R_2^{\,2} - R_1^{\,2}}{\log(R_2/R_1)}\log\frac{r}{R_2}\right].$$

The discharge is

$$Q = \frac{\pi\Delta p}{8\nu l}\left[R_2^{\,4} - R_1^{\,4} - \frac{(R_2^{\,2} - R_1^{\,2})^2}{\log(R_2/R_1)}\right].$$

PROBLEM 2. The same as Problem 1, but for a pipe of elliptical cross-section.

SOLUTION. We seek a solution of equation (17.7) in the form $v = Ay^2 + Bz^2 + C$. The constants A, B, C are determined from the requirement that this expression must satisfy the boundary condition $v = 0$ on the circumference of the ellipse (i.e. $Ay^2 + Bz^2 + C = 0$ must be the same as the equation $y^2/a^2 + z^2/b^2 = 1$, where a and b are the semi-axes of the ellipse). The result is

$$v = \frac{\Delta p}{2\eta l}\frac{a^2 b^2}{a^2 + b^2}\left(1 - \frac{y^2}{a^2} - \frac{z^2}{b^2}\right).$$

† The dependence of Q on Δp and R given by this formula was established empirically by G. Hagen (1839) and J. L. M. Poiseuille (1840) and theoretically justified by G. G. Stokes (1845).

Parallel viscous flow between fixed walls is often called *Poiseuille flow* in the literature; equation (17.4) relates to two-dimensional Poiseuille flow.

The discharge is

$$Q = \frac{\pi \Delta p}{4\nu l} \frac{a^3 b^3}{a^2 + b^2}.$$

PROBLEM 3. The same as Problem 1, but for a pipe whose cross-section is an equilateral triangle with side a.

SOLUTION. The solution of equation (17.7) which vanishes on the bounding triangle is

$$v = \frac{\Delta p}{l} \frac{2}{\sqrt{3} a \eta} h_1 h_2 h_3,$$

where h_1, h_2, h_3 are the lengths of the perpendiculars from a given point in the triangle to its three sides. For each of the expressions $\triangle h_1, \triangle h_2, \triangle h_3$ (where $\triangle = \partial^2/\partial z^2 + \partial^2/\partial y^2$) is zero; this is seen at once from the fact that each of the perpendiculars h_1, h_2, h_3 may be taken as the axis of y or z, and the result of applying the Laplacian to a coordinate is zero. We therefore have

$$\triangle (h_1 h_2 h_3) = 2(h_1 \,\mathbf{grad}\, h_2 \cdot \mathbf{grad}\, h_3 + h_2 \,\mathbf{grad}\, h_3 \cdot \mathbf{grad}\, h_1 + h_3 \,\mathbf{grad}\, h_1 \cdot \mathbf{grad}\, h_2)$$

But $\mathbf{grad}\, h_1 = \mathbf{n}_1$, $\mathbf{grad}\, h_2 = \mathbf{n}_2$, $\mathbf{grad}\, h_3 = \mathbf{n}_3$, where $\mathbf{n}_1, \mathbf{n}_2, \mathbf{n}_3$ are unit vectors along the perpendiculars h_1, h_2, h_3. Any two of $\mathbf{n}_1, \mathbf{n}_2, \mathbf{n}_3$ are at an angle $2\pi/3$, so that $\mathbf{grad}\, h_1 \cdot \mathbf{grad}\, h_2 = \mathbf{n}_1 \cdot \mathbf{n}_2 = \cos(2\pi/3) = -\frac{1}{2}$, and so on. We thus obtain the relation

$$\triangle (h_1 h_2 h_3) = -(h_1 + h_2 + h_3) = -\tfrac{1}{2}\sqrt{3} a,$$

and we see that equation (17.7) is satisfied. The discharge is

$$Q = \frac{\sqrt{3} a^4 \Delta p}{320 \nu l}.$$

PROBLEM 4. A cylinder with radius R_1 moves parallel to its axis with velocity u inside a coaxial cylinder with radius R_2. Determine the motion of a fluid occupying the space between the cylinders.

SOLUTION. We take cylindrical polar coordinates, with the z-axis along the axis of the cylinders. The velocity is everywhere along the z-axis and depends only on r (as does the pressure): $v_z = v(r)$. We obtain for v the equation

$$\triangle v = \frac{1}{r} \frac{\mathrm{d}}{\mathrm{d}r}\left(r \frac{\mathrm{d}v}{\mathrm{d}r}\right) = 0;$$

the term $(\mathbf{v} \cdot \mathbf{grad})\mathbf{v} = v\, \partial v/\partial z$ vanishes identically. Using the boundary conditions $v = u$ for $r = R_1$ and $v = 0$ for $r = R_2$, we find

$$v = u \frac{\log(r/R_2)}{\log(R_1/R_2)}.$$

The frictional force per unit length of either cylinder is $2\pi\eta u/\log(R_2/R_1)$.

PROBLEM 5. A layer of fluid with thickness h is bounded above by a free surface and below by a fixed plane inclined at an angle α to the horizontal. Determine the flow due to gravity.

SOLUTION. We take the fixed plane as the xy-plane, with the x-axis in the direction of flow (Fig. 6). We seek a solution depending only on z. The Navier–Stokes equations with $v_x = v(z)$ in a gravitational field are

$$\eta \frac{\mathrm{d}^2 v}{\mathrm{d}z^2} + \rho g \sin\alpha = 0, \qquad \frac{\mathrm{d}p}{\mathrm{d}z} + \rho g \cos\alpha = 0.$$

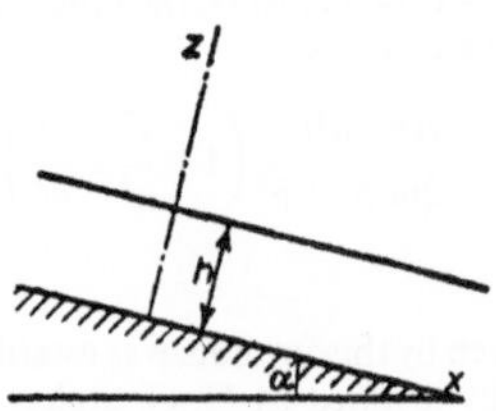

FIG. 6

At the free surface ($z = h$) we must have $\sigma_{xz} = \eta\, dv/dz = 0$, $\sigma_{zz} = -p = -p_0$ (p_0 being the atmospheric pressure). For $z = 0$ we must have $v = 0$. The solution satisfying these conditions is

$$p = p_0 + \rho g(h - z)\cos\alpha, \qquad v = \frac{\rho g \sin\alpha}{2\eta} z(2h - z).$$

The discharge, per unit length in the y-direction, is

$$Q = \rho \int_0^h v\, dz = \frac{\rho g h^3 \sin\alpha}{3\nu}.$$

PROBLEM 6. Determine the way in which the pressure falls along a tube of circular cross-section in which a viscous perfect gas is flowing isothermally (bearing in mind that the dynamic viscosity η of a perfect gas is independent of the pressure).

SOLUTION. Over any short section of the pipe the gas may be supposed incompressible, provided that the pressure gradient is not too great, and we can therefore use formula (17.10), according to which

$$-\frac{dp}{dx} = \frac{8\eta Q}{\pi\rho R^4}.$$

Over greater distances, however, ρ varies, and the pressure is not a linear function of x. According to the equation of state, the gas density $\rho = mp/T$, where m is the mass of a molecule, so that

$$-\frac{dp}{dx} = \frac{8\eta Q T}{\pi m R^4} \cdot \frac{1}{p}.$$

(The discharge Q of the gas through the tube is obviously the same, whether or not the gas is incompressible.) From this we find

$$p_2^{\,2} - p_1^{\,2} = \frac{16\eta Q T}{\pi m R^4} l,$$

where p_2, p_1 are the pressures at the ends of a section of the tube with length l.

§18. Flow between rotating cylinders

Let us now consider the motion of a fluid between two infinite coaxial cylinders with radii R_1, R_2 ($R_2 > R_1$), rotating about their axis with angular velocities Ω_1, Ω_2.† We take cylindrical polar coordinates r, ϕ, z, with the z-axis along the axis of the cylinders. It is evident from symmetry that

$$v_z = v_r = 0, \qquad v_\phi = v(r), \qquad p = p(r).$$

The Navier–Stokes equation in cylindrical polar coordinates gives in this case two equations:

$$dp/dr = \rho v^2/r, \tag{18.1}$$

$$\frac{d^2 v}{dr^2} + \frac{1}{r}\frac{dv}{dr} - \frac{v}{r^2} = 0. \tag{18.2}$$

The latter equation has solutions of the form r^n; substitution gives $n = \pm 1$, so that

$$v = ar + \frac{b}{r}.$$

The constants a and b are found from the boundary conditions, according to which the fluid velocity at the inner and outer cylindrical surfaces must be equal to that of the

† Flow between rotating cylinders is often called *Couette flow* in the literature (M. Couette 1890). In the limit $R_1 \to R_2$, it becomes the flow (17.1) between moving parallel planes, referred to as two-dimensional Couette flow.

corresponding cylinder: $v = R_1\Omega_1$ for $r = R_1$, $v = R_2\Omega_2$ for $r = R_2$. As a result we find the velocity distribution to be

$$v = \frac{\Omega_2 R_2^2 - \Omega_1 R_1^2}{R_2^2 - R_1^2} r + \frac{(\Omega_1 - \Omega_2) R_1^2 R_2^2}{R_2^2 - R_1^2} \frac{1}{r}. \tag{18.3}$$

The pressure distribution is then found from (18.1) by straightforward integration.

For $\Omega_1 = \Omega_2 = \Omega$ we have simply $v = \Omega r$, i.e. the fluid rotates rigidly with the cylinders. When the outer cylinder is absent ($\Omega_2 = 0$, $R_2 = \infty$) we have $v = \Omega_1 R_1^2/r$.

Let us also determine the moment of the frictional forces acting on the cylinders. The frictional force acting on unit area of the inner cylinder is along the tangent to the surface and, from (15.14), is equal to the component $\sigma'_{r\phi}$ of the stress tensor. Using formulae (15.17), we find

$$[\sigma'_{r\phi}]_{r=R_1} = \eta\left[\left(\frac{\partial v}{\partial r} - \frac{v}{r}\right)\right]_{r=R_1}$$
$$= -2\eta\frac{(\Omega_1 - \Omega_2)R_2^2}{R_2^2 - R_1^2}.$$

The moment of this force is found by multiplying by R_1, and the total moment M_1 acting on unit length of the cylinder by multiplying the result by $2\pi R_1$. We thus have

$$M_1 = -\frac{4\pi\eta(\Omega_1 - \Omega_2)R_1^2 R_2^2}{R_2^2 - R_1^2}. \tag{18.4}$$

The moment of the forces acting on the outer cylinder is $M_2 = -M_1$. When $\Omega_2 = 0$ and the gap between the cylinders is small ($\delta \equiv R_2 - R_1 \ll R_2$), (18.4) becomes

$$M_2 = \eta RSu/\delta, \tag{18.5}$$

where $S \cong 2\pi R$ is the surface area of the cylinder per unit length, and $u = \Omega_1 R$ is its peripheral velocity.†

The following general remark may be made concerning the solutions of the equations of motion of a viscous fluid which we have obtained in §§17 and 18. In all these cases the non-linear term $(\mathbf{v}\cdot\mathbf{grad})\mathbf{v}$ in the equations which determine the velocity distribution is identically zero, so that we are actually solving linear equations, a fact which very much simplifies the problem. For this reason all the solutions also satisfy the equations of motion for an incompressible ideal fluid, say in the form (10.2) and (10.3). This is why formulae (17.1) and (18.3) do not contain the viscosity coefficient at all. This coefficient appears only in formulae, such as (17.9), which relate the velocity to the pressure gradient in the fluid, since the presence of a pressure gradient is due to the viscosity; an ideal fluid could flow in a pipe even if there were no pressure gradient.

§19. The law of similarity

In studying the motion of viscous fluids we can obtain a number of important results from simple arguments concerning the dimensions of various physical quantities. Let us

† The solution of the more complex problem of the motion of a viscous fluid in a narrow space between cylinders whose axes are parallel but not coincident may be found in: N. E. Kochin, I. A. Kibel' and N. V. Roze. *Theoretical Hydromechanics* (*Teoreticheskaya gidromekhanika*), Part 2, p. 534, Moscow 1963; A. Sommerfeld, *Mechanics of Deformable Bodies*, §36, New York 1950.

consider any particular type of motion, for instance the motion of a body of some definite shape through a fluid. If the body is not a sphere, its direction of motion must also be specified: e.g. the motion of an ellipsoid in the direction of its greatest or least axis. Alternatively, we may be considering flow in a region with boundaries having a definite form (a pipe with given cross-section, etc.).

In such a case we say that bodies of the same shape are *geometrically similar*; they can be obtained from one another by changing all linear dimensions in the same ratio. Hence, if the shape of the body is given, it suffices to specify any one of its linear dimensions (the radius of a sphere or of a cylindrical pipe, one semi-axis of a spheroid with given eccentricity, and so on) in order to determine its dimensions completely.

We shall at present consider steady flow. If, for example, we are discussing flow past a solid body (which case we shall take below, for definiteness), the velocity of the main stream must therefore be constant. We shall suppose the fluid incompressible.

Of the parameters which characterize the fluid itself, only the kinematic viscosity $\nu = \eta/\rho$ appears in the equations of hydrodynamics (the Navier–Stokes equations); the unknown functions which have to be determined by solving the equations are the velocity $\mathbf{v}$ and the ratio p/ρ of the pressure p to the constant density ρ. Moreover, the flow depends, through the boundary conditions, on the shape and dimensions of the body moving through the fluid and on its velocity. Since the shape of the body is supposed given, its geometrical properties are determined by one linear dimension, which we denote by l. Let the velocity of the main stream be u. Then any flow is specified by three parameters, ν, u and l. These quantities have the following dimensions:

$$\nu = \text{cm}^2/\text{sec}, \qquad l = \text{cm}, \qquad u = \text{cm}/\text{sec}.$$

It is easy to verify that only one dimensionless quantity can be formed from the above three, namely ul/ν. This combination is called the *Reynolds number* and is denoted by R:

$$\mathrm{R} = \rho ul/\eta = ul/\nu. \tag{19.1}$$

Any other dimensionless parameter can be written as a function of R.

We shall now measure lengths in terms of l, and velocities in terms of u, i.e. we introduce the dimensionless quantities $\mathbf{r}/l$, $\mathbf{v}/u$. Since the only dimensionless parameter is the Reynolds number, it is evident that the velocity distribution obtained by solving the equations of incompressible flow is given by a function having the form

$$\mathbf{v} = u\mathbf{f}(\mathbf{r}/l, \mathrm{R}). \tag{19.2}$$

It is seen from this expression that, in two different flows of the same type (for example, flow past spheres with different radii by fluids with different viscosities), the velocities $\mathbf{v}/u$ are the same functions of the ratio $\mathbf{r}/l$ if the Reynolds number is the same for each flow. Flows which can be obtained from one another by simply changing the unit of measurement of coordinates and velocities are said to be *similar*. Thus flows of the same type with the same Reynolds number are similar. This is called the *law of similarity* (O. Reynolds 1883).

A formula similar to (19.2) can be written for the pressure distribution in the fluid. To do so, we must construct from the parameters ν, l, u some quantity with the dimensions of pressure divided by density; this quantity can be u^2, for example. Then we can say that $p/\rho u^2$ is a function of the dimensionless variable $\mathbf{r}/l$ and the dimensionless parameter R. Thus

$$p = \rho u^2 f(\mathbf{r}/l, \mathrm{R}). \tag{19.3}$$

Finally, similar considerations can also be applied to quantities which characterize the flow but are not functions of the coordinates. Such a quantity is, for instance, the drag force F acting on the body. We can say that the dimensionless ratio of F to some quantity formed from ν, u, l, ρ and having the dimensions of force must be a function of the Reynolds number alone. Such a combination of ν, u, l, ρ can be $\rho u^2 l^2$, for example. Then

$$F = \rho u^2 l^2 f(\mathrm{R}). \tag{19.4}$$

If the force of gravity has an important effect on the flow, then the latter is determined not by three but by four parameters, l, u, ν and the acceleration g due to gravity. From these parameters we can construct not one but two independent dimensionless quantities. These can be, for instance, the Reynolds number and the *Froude number*, which is

$$\mathrm{F} = u^2/lg. \tag{19.5}$$

In formulae (19.2)–(19.4) the function f will now depend on not one but two parameters (R and F), and two flows will be similar only if both these numbers have the same values.

Finally, we may say a little regarding non-steady flows. A non-steady flow of a given type is characterized not only by the quantities ν, u, l but also by some time interval τ characteristic of the flow, which determines the rate of change of the flow. For instance, in oscillations, according to a given law, of a solid body, of a given shape, immersed in a fluid, τ may be the period of oscillation. From the four quantities ν, u, l, τ we can again construct two independent dimensionless quantities, which may be the Reynolds number and the number

$$\mathrm{S} = u\tau/l, \tag{19.6}$$

sometimes called the *Strouhal number*. Similar motion takes place in these cases only if both these numbers have the same values.

If the oscillations of the fluid occur spontaneously (and not under the action of a given external exciting force), then for motion of a given type S will be a definite function of R:

$$\mathrm{S} = f(\mathrm{R}).$$

§20. **Flow with small Reynolds numbers**

The Navier–Stokes equation is considerably simplified in the case of flow with small Reynolds numbers. For steady flow of an incompressible fluid, this equation is

$$(\mathbf{v}\cdot\mathbf{grad})\mathbf{v} = -(1/\rho)\mathbf{grad}\,p + (\eta/\rho)\triangle\mathbf{v}.$$

The term $(\mathbf{v}\cdot\mathbf{grad})\mathbf{v}$ is of the order of magnitude of u^2/l, u and l having the same meaning as in §19. The quantity $(\eta/\rho)\triangle\mathbf{v}$ is of the order of magnitude of $\eta u/\rho l^2$. The ratio of the two is just the Reynolds number. Hence the term $(\mathbf{v}\cdot\mathbf{grad})\mathbf{v}$ may be neglected if the Reynolds number is small, and the equation of motion reduces to a linear equation

$$\eta\triangle\mathbf{v} - \mathbf{grad}\,p = 0. \tag{20.1}$$

Together with the equation of continuity

$$\operatorname{div}\mathbf{v} = 0 \tag{20.2}$$

it completely determines the motion. It is useful to note also the equation

$$\triangle\,\mathbf{curl}\,\mathbf{v} = 0, \tag{20.3}$$

which is obtained by taking the curl of equation (20.1).

As an example, let us consider rectilinear and uniform motion of a sphere in a viscous fluid (G. G. Stokes 1851). The problem of the motion of a sphere, it is clear, is exactly equivalent to that of flow past a fixed sphere, the fluid having a given velocity **u** at infinity. The velocity distribution in the first problem is obtained from that in the second problem by simply subtracting the velocity **u**; the fluid is then at rest at infinity, while the sphere moves with velocity $-\mathbf{u}$. If we regard the flow as steady, we must, of course, speak of the flow past a fixed sphere, since, when the sphere moves, the velocity of the fluid at any point in space varies with time.

Since $\operatorname{div}(\mathbf{v}-\mathbf{u}) = \operatorname{div}\mathbf{v} = 0$, $\mathbf{v}-\mathbf{u}$ can be expressed as the curl of some vector **A**:

$$\mathbf{v}-\mathbf{u} = \mathbf{curl}\,\mathbf{A},$$

with **curl A** equal to zero at infinity. The vector **A** must be axial, in order for its curl to be polar, like the velocity. In flow past a sphere, a completely symmetrical body, there is no preferred direction other than that of **u**. This parameter **u** must appear linearly in **A**, because the equation of motion and its boundary conditions are linear. The general form of a vector function $\mathbf{A}(\mathbf{r})$ satisfying all these requirements is $\mathbf{A} = f'(r)\mathbf{n}\times\mathbf{u}$, where **n** is a unit vector parallel to the position vector **r** (the origin being taken at the centre of the sphere), and $f'(r)$ is a scalar function of r. The product $f'(r)\mathbf{n}$ can be represented as the gradient of another function $f(r)$. We shall thus look for the velocity in the form

$$\mathbf{v} = \mathbf{u} + \mathbf{curl}\,(\mathbf{grad}\,f\times\mathbf{u}) = \mathbf{u} + \mathbf{curl}\,\mathbf{curl}\,(f\mathbf{u}); \tag{20.4}$$

the last expression is obtained by noting that **u** is constant.

To determine the function f, we use equation (20.3). Since

$$\mathbf{curl}\,\mathbf{v} = \mathbf{curl}\,\mathbf{curl}\,\mathbf{curl}(f\mathbf{u}) = (\mathbf{grad}\,\operatorname{div} - \triangle)\,\mathbf{curl}(f\mathbf{u})$$

$$= -\triangle\,\mathbf{curl}(f\mathbf{u}),$$

(20.3) takes the form $\triangle^2\,\mathbf{curl}(f\mathbf{u}) = \triangle^2(\mathbf{grad}\,f\times\mathbf{u}) = (\triangle^2\mathbf{grad}\,f)\times\mathbf{u} = 0$. It follows from this that

$$\triangle^2\,\mathbf{grad}\,f = 0. \tag{20.5}$$

A first integration gives

$$\triangle^2 f = \text{constant}.$$

It is easy to see that the constant must be zero, since the velocity difference $\mathbf{v}-\mathbf{u}$ must vanish at infinity, and so must its derivatives. The expression $\triangle^2 f$ contains fourth derivatives of f, whilst the velocity is given in terms of the second derivatives of f. Thus we have

$$\triangle^2 f \equiv \frac{1}{r^2}\frac{\mathrm{d}}{\mathrm{d}r}\left(r^2\frac{\mathrm{d}}{\mathrm{d}r}\right)\triangle f = 0.$$

Hence

$$\triangle f = 2a/r + c.$$

The constant c must be zero if the velocity $v-u$ is to vanish at infinity. From $\triangle f = 2a/r$ we obtain

$$f = ar + b/r. \tag{20.6}$$

The additive constant is omitted, since it is immaterial (the velocity being given by derivatives of f).

Substituting in (20.4), we have after a simple calculation

$$\mathbf{v} = \mathbf{u} - a\frac{\mathbf{u}+\mathbf{n}(\mathbf{u}\cdot\mathbf{n})}{r} + b\frac{3\mathbf{n}(\mathbf{u}\cdot\mathbf{n})-\mathbf{u}}{r^3}. \tag{20.7}$$

The constants a and b have to be determined from the boundary conditions: at the surface of the sphere ($r = R$), $\mathbf{v} = 0$, i.e.

$$-\mathbf{u}\left(\frac{a}{R}+\frac{b}{R^3}-1\right)+\mathbf{n}(\mathbf{u}\cdot\mathbf{n})\left(-\frac{a}{R}+\frac{3b}{R^3}\right) = 0.$$

Since this equation must hold for all $\mathbf{n}$, the coefficients of $\mathbf{u}$ and $\mathbf{n}(\mathbf{u}\cdot\mathbf{n})$ must each vanish. Hence $a = \frac{3}{4}R$, $b = \frac{1}{4}R^3$. Thus we have finally

$$f = \tfrac{3}{4}Rr + \tfrac{1}{4}R^3/r, \tag{20.8}$$

$$\mathbf{v} = -\tfrac{3}{4}R\frac{\mathbf{u}+\mathbf{n}(\mathbf{u}\cdot\mathbf{n})}{r} - \tfrac{1}{4}R^3\frac{\mathbf{u}-3\mathbf{n}(\mathbf{u}\cdot\mathbf{n})}{r^3} + \mathbf{u}, \tag{20.9}$$

or, in spherical polar components with the axis parallel to $\mathbf{u}$,

$$\left.\begin{aligned} v_r &= u\cos\theta\left[1-\frac{3R}{2r}+\frac{R^3}{2r^3}\right],\\ v_\theta &= -u\sin\theta\left[1-\frac{3R}{4r}-\frac{R^3}{4r^3}\right]. \end{aligned}\right\} \tag{20.10}$$

This gives the velocity distribution about the moving sphere. To determine the pressure, we substitute (20.4) in (20.1):

$$\begin{aligned} \mathbf{grad}\, p = \eta\triangle\mathbf{v} &= \eta\triangle\,\mathbf{curl}\,\mathbf{curl}\,(f\mathbf{u})\\ &= \eta\triangle\,(\mathbf{grad}\,\mathrm{div}\,(f\mathbf{u})-\mathbf{u}\triangle f). \end{aligned}$$

But $\triangle^2 f = 0$, and so

$$\mathbf{grad}\, p = \mathbf{grad}[\eta\triangle\,\mathrm{div}(f\mathbf{u})] = \mathbf{grad}(\eta\mathbf{u}\cdot\mathbf{grad}\,\triangle f).$$

Hence

$$p = \eta\mathbf{u}\cdot\mathbf{grad}\,\triangle f + p_0, \tag{20.11}$$

where p_0 is the fluid pressure at infinity. Substitution for f leads to the final expression

$$p = p_0 - \tfrac{3}{2}\eta\frac{\mathbf{u}\cdot\mathbf{n}}{r^2}R. \tag{20.12}$$

Using the above formulae, we can calculate the force $\mathbf{F}$ exerted on the sphere by the moving fluid (or, what is the same thing, the drag on the sphere as it moves through the fluid). To do so, we take spherical polar coordinates with the axis parallel to $\mathbf{u}$; by symmetry, all quantities are functions only of r and of the polar angle θ. The force $\mathbf{F}$ is evidently parallel to the velocity $\mathbf{u}$. The magnitude of this force can be determined from (15.14). Taking from this formula the components, normal and tangential to the surface, of the force on an element of the surface of the sphere, and projecting these components on the direction of $\mathbf{u}$, we find

$$F = \oint(-p\cos\theta + \sigma'_{rr}\cos\theta - \sigma'_{r\theta}\sin\theta)\mathrm{d}f, \tag{20.13}$$

where the integration is taken over the whole surface of the sphere.

Substituting the expressions (20.10) in the formulae

$$\sigma'_{rr} = 2\eta\frac{\partial v_r}{\partial r}, \qquad \sigma'_{r\theta} = \eta\left(\frac{1}{r}\frac{\partial v_r}{\partial \theta} + \frac{\partial v_\theta}{\partial r} - \frac{v_\theta}{r}\right)$$

(see (15.20)), we find that at the surface of the sphere

$$\sigma'_{rr} = 0, \qquad \sigma'_{r\theta} = -(3\eta/2R)u\sin\theta,$$

while the pressure (20.12) is $p = p_0 - (3\eta/2R)u\cos\theta$. Hence the integral (20.13) reduces to $F = (3\eta u/2R)\oint df$. In this way we finally arrive at *Stokes' formula* for the drag on a sphere moving slowly in a fluid:†

$$F = 6\pi\eta R u. \tag{20.14}$$

The drag is proportional to the velocity and linear size of the body. This could have been foreseen from dimensional arguments: the fluid density ρ does not appear in the approximate equations (20.1), (20.2), and so the force F which they give must be expressed only in terms of η, u and R; from these, only one combination with the dimensions of force can be formed, namely the product $\eta R u$.

A similar dependence occurs for slowly moving bodies with other shapes. The direction of the drag on a body of arbitrary shape is not the same as that of the velocity; the general form of the dependence of $\mathbf{F}$ on $\mathbf{u}$ can be written

$$F_i = \eta a_{ik} u_k, \tag{20.15}$$

where a_{ik} is a tensor of rank two, independent of the velocity. It is important to note that this tensor is symmetrical, a result which holds in the linear approximation with respect to the velocity, and is a particular case of a general law valid for slow motion accompanied by dissipative processes (see *SP*1, §121).

Refinement of Stokes' formula

The above solution of the problem of flow past a sphere is not valid at large distances, even if the Reynolds number is small. To see this, let us estimate the term $(\mathbf{v}\cdot\mathbf{grad})\mathbf{v}$ neglected in (20.1). At large distances, $\mathbf{v} \cong \mathbf{u}$; the velocity derivatives there are of the order of uR/r^2, as is seen from (20.9). Hence $(\mathbf{v}\cdot\mathbf{grad})\mathbf{v} \sim u^2R/r^2$. The terms retained in (20.1) are of the order of $\eta R u/\rho r^3$, as can be seen from the same expression (20.9) for the velocity or (20.12) for the pressure. The condition $\eta R u/\rho r^3 \gg u^2 R/r^2$ is satisfied only for distances such that

$$r \ll \nu/u. \tag{20.16}$$

At greater distances, the terms neglected are not negligible, and the velocity distribution so found is incorrect.

---

† With a view to later applications, it may be mentioned that calculations with (20.7) and the constants a and b undetermined give

$$F = 8\pi\eta a u. \tag{20.14a}$$

The drag can also be calculated for a slowly moving ellipsoid with any shape. The relevant formulae are given by H. Lamb, *Hydrodynamics*, 6th ed., §339, Cambridge 1932. Here we shall give the limiting expressions for a plane circular disk with radius R moving perpendicular to its plane:

$$F = 16\eta R u,$$

and for a similar disk moving in its plane:

$$F = 32\eta R u/3.$$

To find the velocity distribution at large distances from the body, we have to include the term $(\mathbf{v}\cdot\mathbf{grad})\mathbf{v}$ omitted from (20.1). Since at these distances $\mathbf{v}$ is almost the same as $\mathbf{u}$, we can approximately replace $\mathbf{v}\cdot\mathbf{grad}$ by $\mathbf{u}\cdot\mathbf{grad}$. We then find for the velocity at large distances the linear equation

$$(\mathbf{u}\cdot\mathbf{grad})\mathbf{v} = -(1/\rho)\ \mathbf{grad}\ p + \nu\Delta\mathbf{v} \tag{20.17}$$

(C. W. Oseen 1910). We shall not pause to give here the procedure for solving this equation for flow past a sphere,† but merely mention that the velocity distribution thus obtained can be used to derive a more accurate formula for the drag on the sphere, which includes the next term in the expansion of the drag in powers of the Reynolds number $\mathrm{R} = uR/\nu$:

$$F = 6\pi\eta uR(1+3uR/8\nu). \tag{20.18}$$

In solving the problem of flow past an infinite cylinder at right angles to its axis, Oseen's equation has to be used from the start; the equation (20.1) has in this case no solution satisfying the boundary conditions at the surface of the cylinder and also at infinity. The drag per unit length of the cylinder is found to be

$$F = \frac{4n\eta u}{\frac{1}{2} - C - \log(uR/4\nu)} = \frac{4\pi\eta u}{\log(3{\cdot}70\nu/uR)}, \tag{20.19}$$

where $C = 0{\cdot}577\ldots$ is Euler's constant (H. Lamb 1911).‡

Another comment should be made regarding the problem of flow past a sphere. The replacement of $\mathbf{v}$ by $\mathbf{u}$ in the non-linear term in (20.17) is valid at large distances from the sphere, $r \gg R$. It is therefore natural that Oseen's equation, while correctly refining the picture of flow at large distances, does not do the same at short distances. This is evident from the fact that the solution of (20.17) which satisfies the necessary conditions at infinity does not satisfy the exact condition that the velocity be zero on the surface of the sphere, which is met only by the zero-order term in the expansion of the velocity in powers of the Reynolds number and not even by the first-order term.

It might therefore seem at first sight that the solution of Oseen's equation cannot be used for a valid calculation of the correction term in the drag. This is not so, however, for the following reason. The contribution to $\mathbf{F}$ from the motion of the fluid at short distances (for which $u \ll \nu/r$) has to be expandable in powers of $\mathbf{u}$. The first non-zero correction term in the vector $\mathbf{F}$ arising from this contribution therefore has to be proportional to $\mathbf{u}u^2$, and gives a second-order correction relative to the Reynolds number; it thus does not affect the first-order correction in (20.18).

Further corrrections to Stokes' formula and a valid refinement of the flow pattern at short distances can not be obtained by a direct solution of (20.17). Although these refinements themselves are not very important, there is considerable methodological interest in deriving and analysing a consistent perturbation theory for solving problems of viscous flow at small Reynolds numbers (S. Kaplun and P. A. Lagerstrom 1957; I.

† It is given by N. E. Kochin, I. A. Kibel' and N. V. Roze, *Theoretical Hydromechanics* (*Teoreticheskaya gidromekhanika*), Part 2, chapter II, §§25–26, Moscow 1963; H. Lamb, *Hydrodynamics*, 6th ed., §§342–3, Cambridge 1932.

‡ The impossibility of calculating the drag in the cylinder problem by means of (20.1) is evident from dimensional arguments. As already mentioned, the result would have to be expressed in terms of η, u and R, but in this case we are concerned with the force per unit length of the cylinder, and the only quantity having the right dimensions would be ηu, which is independent of the size of the body and therefore does not vanish as $R \to 0$; this is physically absurd.

Proudman and J. R. A. Pearson 1957). We shall describe the existing situation and give all expressions needed to illustrate it, without going through the calculations in detail.†

To show explicitly the small parameter R, the Reynolds number, we use the dimensionless velocity and position vector $\mathbf{v}' = \mathbf{v}/u$, $\mathbf{r}' = \mathbf{r}/R$, and in the rest of this section denote them by **v** and **r** without the primes. The exact solution of the equation of motion (which we take in the form (15.10) with the pressure eliminated) is then

$$\mathrm{R}\,\mathbf{curl}\,(\mathbf{v}\times\mathbf{curl}\,\mathbf{v}) + \triangle\,\mathbf{curl}\,\mathbf{v} = 0. \tag{20.20}$$

We distinguish two regions of space around the sphere: the near region with $r \ll 1/\mathrm{R}$, and the far region with $r \gg 1$. These together cover all space, overlapping in the intermediate range

$$1/\mathrm{R} \gg r \gg 1. \tag{20.21}$$

In a consistent perturbation theory, the initial approximation in the near region is the Stokes approximation, i.e. the solution of the equation $\triangle\,\mathbf{curl}\,\mathbf{v} = 0$ obtained from (20.20) by neglecting the term which contains the factor R. This solution is given by formulae (20.10); in dimensionless variables, it is

$$v_r^{(1)} = \cos\theta\left(1 - \frac{3}{2r} + \frac{1}{2r^3}\right), \qquad v_\theta^{(1)} = -\sin\theta\left(1 - \frac{3}{4r} - \frac{1}{4r^3}\right),$$

$$r \ll 1/\mathrm{R}, \tag{20.22}$$

the superscript (1) denoting the first approximation.

The first approximation in the far region is simply the constant $\mathbf{v}^{(1)} = \boldsymbol{\nu}$ corresponding to the unperturbed uniform incoming flow ($\boldsymbol{\nu}$ being a unit vector in the direction of the flow). Substitution of $\mathbf{v} = \boldsymbol{\nu} + \mathbf{v}^{(2)}$ in (20.20) gives for $\mathbf{v}^{(2)}$ Oseen's equation

$$\mathrm{R}\,\mathbf{curl}\,(\boldsymbol{\nu}\times\mathbf{curl}\,\mathbf{v}^{(2)}) + \triangle\,\mathbf{curl}\,\mathbf{v}^{(2)} = 0. \tag{20.23}$$

The solution must satisfy the condition that the velocity $\mathbf{v}^{(2)}$ be zero at infinity and the condition for joining to the solution (20.22) in the intermediate range. The latter excludes, in particular, solutions that increase too rapidly with decreasing r.‡ The appropriate solution is

$$v_r^{(1)} + v_r^{(2)} = \cos\theta + \frac{3}{2r^2\mathrm{R}}\{1 - [1 + \tfrac{1}{2}r\mathrm{R}(1+\cos\theta)]e^{-\frac{1}{2}r\mathrm{R}(1-\cos\theta)}\},$$

$$v_\theta^{(1)} + v_\theta^{(2)} = -\sin\theta + \frac{3}{4r}\sin\theta\, e^{-\frac{1}{2}r\mathrm{R}(1-\cos\theta)},$$

$$r \gg 1. \tag{20.24}$$

---

† These may be found in M. Van Dyke, *Perturbation Methods in Fluid Mechanics*, New York 1964. The calculations there are given not in terms of the velocity $\mathbf{v}(\mathbf{r})$ but in the more compact, less visualizable, terminology of the stream function. For axially symmetrical flow, including flow past a sphere, the stream function $\psi(r,\theta)$ in spherical polar coordinates is defined by

$$v_r = \frac{1}{r^2\sin\theta}\frac{\partial\psi}{\partial\theta},$$

$$v_\theta = -\frac{1}{r\sin\theta}\frac{\partial\psi}{\partial r}, \quad v_\phi = 0.$$

These satisfy identically the continuity equation (15.22).

‡ To determine the numerical coefficients in the solution, we have also to take account of the condition that the total amount of fluid passing through any closed surface around the sphere must be zero.

Note that the variable for the far region is really the product $\rho = r\mathrm{R}$, not the radial coordinate r itself. When this variable is used, R disappears from (20.20), in accordance with the fact that when $r \gtrsim 1/\mathrm{R}$ the viscous and inertia terms in the equation become comparable in order of magnitude. The number R occurs in the solution only through the boundary condition for joining to that in the near region. The expansion of $\mathbf{v}(\mathbf{r})$ in the far region is therefore an expansion in powers of R for given values of $\rho = r\mathrm{R}$, since the second terms in (20.24), when expressed in terms of ρ, contain R as a factor.

To test the correctness of joining for the solutions (20.22) and (20.24), we observe that in the intermediate range (20.21) $r\mathrm{R} \ll 1$ and the expressions (20.24) can be expanded in powers of this variable. As far as the first two terms (apart from the uniform flow), we have

$$\left.\begin{aligned} v_r &= \cos\theta\left(1-\frac{3}{2r}\right)+\frac{3\mathrm{R}}{16}(1-\cos\theta)(1+3\cos\theta), \\ v_\theta &= -\sin\theta\left(1-\frac{3}{4r}\right)-\frac{3\mathrm{R}}{8}\sin\theta\,(1-\cos\theta). \end{aligned}\right\} \tag{20.25}$$

In the same range, on the other hand, $r \gg 1$ and therefore we can omit the terms in $1/r^3$ in (20.22); the remaining terms are the same as the first terms in (20.25), and the second terms there will be made use of later.

On going to the next approximation in the near region, we write $\mathbf{v} = \mathbf{v}^{(1)} + \mathbf{v}^{(2)}$ and obtain from (20.20) an equation for the correction in the second approximation:

$$\triangle\, \mathbf{curl}\, \mathbf{v}^{(2)} = -\mathrm{R}\, \mathbf{curl}\, (\mathbf{v}^{(1)} \times \mathbf{curl}\, \mathbf{v}^{(1)}). \tag{20.26}$$

The solution of this equation must satisfy the condition of vanishing on the surface of the sphere and that of joining to the solution in the far region; the latter means that the leading terms in the function $\mathbf{v}^{(2)}(\mathbf{r})$ when $r \gg 1$ must agree with the second terms in (20.25). The appropriate solution is

$$\begin{aligned} v_r^{(2)} &= \frac{3\mathrm{R}}{8}v_r^{(1)}+\frac{3\mathrm{R}}{32}\left(1-\frac{1}{r}\right)^2\left(2+\frac{1}{r}+\frac{1}{r^2}\right)(1-3\cos^2\theta), \\ v_\theta^{(2)} &= \frac{3\mathrm{R}}{8}v_\theta^{(1)}+\frac{3\mathrm{R}}{32}\left(1-\frac{1}{r}\right)\left(4+\frac{1}{r}+\frac{1}{r^2}+\frac{2}{r^3}\right)\sin\theta\cos\theta, \\ & r \ll 1/\mathrm{R}. \end{aligned} \tag{20.27}$$

In the intermediate region, only the terms without a factor $1/r$ remain in these expressions, and they do in fact agree with the second terms in (20.25).

From the velocity distribution (20.27), we can calculate the correction to Stokes' formula for the drag. The second terms in (20.27), because of their angular dependence, do not contribute to the drag; the first terms give the correction $3\mathrm{R}/8$ shown in (20.18). According to the above discussion, the exact velocity distribution near the sphere leads in this approximation to the same result for the drag as the solution of Oseen's equation.

The next approximation can be obtained by continuing the procedure described. It involves logarithmic terms in the velocity distribution; in the expression (20.18) for the drag, the brackets are replaced by

$$1+\frac{3}{8}\mathrm{R}-\frac{9}{40}\mathrm{R}^2\log(1/\mathrm{R}),$$

the logarithm being assumed large.†

† See I. Proudman and J. R. A. Pearson, *Journal of Fluid Mechanics* **2**, 237, 1957.

PROBLEMS

PROBLEM 1. Determine the motion of a fluid occupying the space between two concentric spheres with radii R_1, R_2 ($R_2 > R_1$), rotating uniformly about different diameters with angular velocities Ω_1, Ω_2; the Reynolds numbers $\Omega_1 R_1^2/\nu$, $\Omega_2 R_2^2/\nu$ are small compared with unity.

SOLUTION. On account of the linearity of the equations, the motion between two rotating spheres may be regarded as a superposition of the two motions obtained when one sphere is at rest and the other rotates. We first put $\Omega_2 = 0$, i.e. only the inner sphere is rotating. It is reasonable to suppose that the fluid velocity at every point is along the tangent to a circle in a plane perpendicular to the axis of rotation with its centre on the axis. On account of the axial symmetry, the pressure gradient in this direction is zero. Hence the equation of motion (20.1) becomes $\triangle \mathbf{v} = 0$. The angular velocity vector $\mathbf{\Omega}_1$ is an axial vector. Arguments similar to those given previously show that the velocity can be written as

$$\mathbf{v} = \mathbf{curl}[f(r)\mathbf{\Omega}_1] = \mathbf{grad} f \times \mathbf{\Omega}_1.$$

The equation of motion then gives $\mathbf{grad} \triangle f \times \mathbf{\Omega}_1 = 0$. Since the vector $\mathbf{grad} \triangle f$ is parallel to the position vector, and the vector product $\mathbf{r} \times \mathbf{\Omega}_1$ cannot be zero for given $\mathbf{\Omega}_1$ and arbitrary $\mathbf{r}$, we must have $\mathbf{grad} \triangle f = 0$, so that

$$\triangle f = \text{constant}.$$

Integrating, we find

$$f = ar^2 + \frac{b}{r}, \qquad \mathbf{v} = \left(\frac{b}{r^3} - 2a\right)\mathbf{\Omega}_1 \times \mathbf{r}.$$

The constants a and b are found from the conditions that $\mathbf{v} = 0$ for $r = R_2$ and $\mathbf{v} = \mathbf{u}$ for $r = R_1$, where $\mathbf{u} = \mathbf{\Omega}_1 \times \mathbf{r}$ is the velocity of points on the rotating sphere. The result is

$$\mathbf{v} = \frac{R_1^3 R_2^3}{R_2^3 - R_1^3}\left(\frac{1}{r^3} - \frac{1}{R_2^3}\right)\mathbf{\Omega}_1 \times \mathbf{r}.$$

The fluid pressure is constant ($p = p_0$). Similarly, we have for the case where the outer sphere rotates and the inner one is at rest ($\Omega_1 = 0$)

$$\mathbf{v} = \frac{R_1^3 R_2^3}{R_2^3 - R_1^3}\left(\frac{1}{R_1^3} - \frac{1}{r^3}\right)\mathbf{\Omega}_2 \times \mathbf{r}.$$

In the general case where both spheres rotate, we have

$$\mathbf{v} = \frac{R_1^3 R_2^3}{R_2^3 - R_1^3}\left\{\left(\frac{1}{r^3} - \frac{1}{R_2^3}\right)\mathbf{\Omega}_1 \times \mathbf{r} + \left(\frac{1}{R_1^3} - \frac{1}{r^3}\right)\mathbf{\Omega}_2 \times \mathbf{r}\right\}.$$

If the outer sphere is absent ($R_2 = \infty$, $\Omega_2 = 0$), i.e. we have simply a sphere with radius R rotating in an infinite fluid, then

$$\mathbf{v} = (R^3/r^3)\mathbf{\Omega} \times \mathbf{r}.$$

Let us calculate the moment of the frictional forces acting on the sphere in this case. If we take spherical polar coordinates with the polar axis parallel to $\mathbf{\Omega}$, we have $v_r = v_\theta = 0$, $v_\phi = v = (R^2\Omega/r^2)\sin\theta$. The frictional force on unit area of the sphere is

$$\sigma'_{r\phi} = \eta\left(\frac{\partial v}{\partial r} - \frac{v}{r}\right)_{r=R} = -3\eta\Omega\sin\theta.$$

The total moment on the sphere is

$$M = \int_0^\pi \sigma'_{r\phi} R \sin\theta . 2\pi R^2 \sin\theta \, d\theta,$$

whence we find

$$M = -8\pi\eta R^3 \Omega.$$

If the inner sphere is absent, $\mathbf{v} = \mathbf{\Omega}_2 \times \mathbf{r}$, i.e. the fluid simply rotates rigidly with the sphere surrounding it.

PROBLEM 2. Determine the velocity of a spherical drop of fluid (with viscosity η') moving under gravity in a fluid with viscosity η (W. Rybczyński 1911).

SOLUTION. We use a system of coordinates in which the drop is at rest. For the fluid outside the drop we again seek a solution of equation (20.5) in the form (20.6), so that the velocity has the form (20.7). For the fluid inside the drop, we have to find a solution which does not have a singularity at $r = 0$ (and the second derivatives of f, which determine the velocity, must also remain finite). This solution is

$$f = \tfrac{1}{4}Ar^2 + \tfrac{1}{8}Br^4,$$

and the corresponding velocity is

$$\mathbf{v} = -A\mathbf{u} + Br^2[\mathbf{n}(\mathbf{u}\cdot\mathbf{n}) - 2\mathbf{u}].$$

At the surface of the sphere† the following conditions must be satisfied. The normal velocity components outside ($\mathbf{v}_e$) and inside ($\mathbf{v}_i$) the drop must be zero:

$$v_{i,r} = v_{e,r} = 0.$$

The tangential velocity component must be continuous:

$$v_{i,\theta} = v_{e,\theta},$$

as must be the component $\sigma_{r\theta}$ of the stress tensor:

$$\sigma_{i,r\theta} = \sigma_{e,r\theta}.$$

The condition that the stress tensor components σ_{rr} be equal need not be written down; it would determine the required velocity u, which is more simply found in the manner shown below. From the above four conditions we obtain four equations for the constants a, b, A, B, whose solutions are

$$a = R\frac{2\eta + 3\eta'}{4(\eta + \eta')}, \qquad b = R^3\frac{\eta'}{4(\eta + \eta')}, \qquad A = -BR^2 = \frac{\eta}{2(\eta + \eta')}.$$

By (20.14a), we have for the drag

$$F = 2\pi u\eta R(2\eta + 3\eta')/(\eta + \eta').$$

As $\eta' \to \infty$ (corresponding to a solid sphere) this formula becomes Stokes' formula. In the limit $\eta' \to 0$ (corresponding to a gas bubble) we have $F = 4\pi u\eta R$, i.e the drag is two-thirds of that on a solid sphere.

Equating F to the force of gravity on the drop, $\frac{4}{3}\pi R^3(\rho - \rho')g$, we find

$$u = \frac{2R^2 g(\rho - \rho')(\eta + \eta')}{3\eta(2\eta + 3\eta')}.$$

PROBLEM 3. Two parallel plane circular disks (with radius R) lie one above the other a small distance apart; the space between them is filled with fluid. The disks approach at a constant velocity u, displacing the fluid. Determine the resistance to their motion (O. Reynolds).

SOLUTION. We take cylindrical polar coordinates, with the origin at the centre of the lower disk, which we suppose fixed. The flow is axially symmetric and, since the fluid layer is thin, predominantly radial: $v_z \ll v_r$, and also $\partial v_r/\partial r \ll \partial v_r/\partial z$. Hence the equations of motion become

$$\eta\frac{\partial^2 v_r}{\partial z^2} = \frac{\partial p}{\partial r}, \qquad \frac{\partial p}{\partial z} = 0, \tag{1}$$

$$\frac{1}{r}\frac{\partial(rv_r)}{\partial r} + \frac{\partial v_z}{\partial z} = 0, \tag{2}$$

with the boundary conditions

$$\text{at } z = 0: \quad v_r = v_z = 0;$$
$$\text{at } z = h: \quad v_r = 0, \quad v_z = -u;$$
$$\text{at } r = R: \quad p = p_0,$$

† We may neglect the change of shape of the drop in its motion, since this change is of a higher order of smallness. However, it must be borne in mind that, in order that the moving drop should in fact be spherical, the forces due to surface tension at its boundary must exceed the forces due to pressure differences, which tend to make the drop non-spherical. This means that we must have $\eta u/R \ll \alpha/R$, where α is the surface-tension coefficient, or, substituting $u \sim R^2 g\rho/\eta$,

$$R \ll \sqrt{(\alpha/\rho g)}.$$

where h is the distance between the disks, and p_0 the external pressure. From equations (1) we find

$$v_r = \frac{1}{2\eta}\frac{dp}{dr}z(z-h).$$

Integrating equation (2) with respect to z, we obtain

$$u = -\frac{1}{r}\frac{d}{dr}\int_0^h rv_r dz = -\frac{h^3}{12\eta r}\frac{d}{dr}\left(r\frac{dp}{dr}\right),$$

whence

$$p = p_0 + \frac{3\eta u}{h^3}(R^2 - r^2).$$

The total resistance to the moving disk is

$$F = 3\pi\eta u R^4/2h^3.$$

§21. The laminar wake

In steady flow of a viscous fluid past a solid body, the flow at great distances behind the body has certain characteristics which can be investigated independently of the particular shape of the body.

Let us denote by $\mathbf{U}$ the constant velocity of the incident current; we take the direction of $\mathbf{U}$ as the x-axis, with the origin somewhere inside the body. The actual fluid velocity at any point may be written $\mathbf{U}+\mathbf{v}$; $\mathbf{v}$ vanishes at infinity.

It is found that, at great distances behind the body, the velocity $\mathbf{v}$ is noticeably different from zero only in a relatively narrow region near the x-axis. This region, called the *laminar wake*,† is reached by fluid particles which move along streamlines passing fairly close to the body. Hence the flow in the wake is essentially rotational. The reason is that rotational flow of a viscous fluid past a solid body is due to the surface of the body.‡ This is easily seen if we recall that, in the pattern of potential flow for an ideal fluid, only the normal velocity component is zero on the surface of the body, not the tangential component $\mathbf{v}_t$. The boundary condition of adhesion for a real fluid makes $\mathbf{v}_t$ also zero, however. If the pattern of potential flow were maintained, this would cause a non-zero discontinuity of $\mathbf{v}_t$, i.e. the occurrence of a surface vorticity. The viscosity smooths out the discontinuity, and the rotational state penetrates into the fluid, from which it passes by convection into the wake region.

On the other hand, the viscosity has almost no effect at any point on streamlines that do not pass near the body, and the vorticity, which is zero in the incident current, remains practically zero on these streamlines, as it would in an ideal fluid. Thus the flow at great distances from the body may be regarded as potential flow everywhere except in the wake.

We shall now derive formulae relating the properties of the flow in the wake to the forces acting on the body. The total momentum transported by the fluid through any closed surface surrounding the body is equal to the integral of the momentum flux density tensor over that surface, $\oint \Pi_{ik} df_k$. The components of the tensor Π_{ik} are

$$\Pi_{ik} = p\delta_{ik} + \rho(U_i + v_i)(U_k + v_k).$$

† In contradistinction to the turbulent wake; see §37.

‡ The fact that the relation $\mathbf{curl}\ \mathbf{v} = 0$ does not remain valid along a streamline which passes over a solid surface has already been noted (§9).

We write the pressure in the form $p = p_0 + p'$, where p_0 is the pressure at infinity. The integration of the constant term $p_0\delta_{ik} + \rho U_i U_k$ gives zero, since the vector integral $\oint \mathbf{df}$ over a closed surface is zero. The integral $\oint \rho v_k \mathrm{d}f_k$ also vanishes: since the total mass of fluid in the volume considered is constant, the total mass flux through the surface surrounding the volume must be zero. Finally, the velocity $\mathbf{v}$ far from the body is small compared with $\mathbf{U}$. Hence, if the surface in question is sufficiently far from the body, we can neglect the term $\rho v_i v_k$ in Π_{ik} as compared with $\rho U_k v_i$. Thus the total momentum flux is

$$\oint (p'\delta_{ik} + \rho U_k v_i)\mathrm{d}f_k.$$

Let us now take the fluid volume concerned to be the volume between two infinite planes x = constant, one of them far in front of the body and the other far behind it. The integral over the infinitely distant "lateral" surface vanishes (since $p' = \mathbf{v} = 0$ at infinity), and it is therefore sufficient to integrate only over the two planes. The momentum flux thus obtained is evidently the difference between the total momentum flux entering through the forward plane and that leaving through the backward plane. This difference, however, is just the quantity of momentum transmitted to the body by the fluid per unit time, i.e. the force $\mathbf{F}$ exerted on the body.

Thus the components of the force $\mathbf{F}$ are

$$F_x = \left(\iint\limits_{x=x_2} - \iint\limits_{x=x_1}\right)(p' + \rho U v_x)\mathrm{d}y\mathrm{d}z,$$

$$F_y = \left(\iint\limits_{x=x_2} - \iint\limits_{x=x_1}\right)\rho U v_y \mathrm{d}y\mathrm{d}z,$$

$$F_z = \left(\iint\limits_{x=x_2} - \iint\limits_{x=x_1}\right)\rho U v_z \mathrm{d}y\mathrm{d}z,$$

where the integration is taken over the infinite planes $x = x_1$ (far behind the body) and $x = x_2$ (far in front of it). Let us first consider the expression for F_x.

Outside the wake we have potential flow, and therefore Bernoulli's equation

$$p + \tfrac{1}{2}\rho(\mathbf{U}+\mathbf{v})^2 = \text{constant} \equiv p_0 + \tfrac{1}{2}\rho U^2$$

holds, or, neglecting the term $\frac{1}{2}\rho v^2$ in comparison with $\rho\mathbf{U}\cdot\mathbf{v}$,

$$p' = -\rho U v_x.$$

We see that in this approximation the integrand in F_x vanishes everywhere outside the wake. In other words, the integral over the plane $x = x_2$ (which lies in front of the body and does not intersect the wake) is zero, and the integral over the plane $x = x_1$ need be taken only over the area covered by the cross-section of the wake. Inside the wake, however, the pressure change p' is of the order of ρv^2, i.e. small compared with $\rho U v_x$. Thus we reach the result that the drag on the body is

$$F_x = -\rho U \iint v_x \mathrm{d}y\mathrm{d}z, \tag{21.1}$$

where the integration is taken over the cross-sectional area of the wake far behind the body. The velocity v_x in the wake is, of course, negative: the fluid moves more slowly than it would if the body were absent. Attention is called to the fact that the integral in (21.1) gives the amount by which the discharge through the wake falls short of its value in the absence of the body.

Let us now consider the force (whose components are F_y, F_z) which tends to move the body transversely. This force is called the *lift*. Outside the wake, where we have potential flow, we can write $v_y = \partial\phi/\partial y$, $v_z = \partial\phi/\partial z$; the integral over the plane $x = x_2$, which does not meet the wake, is zero:

$$\iint v_y \, dy\, dz = \iint \frac{\partial \phi}{\partial y} dy\, dz = 0, \qquad \iint \frac{\partial \phi}{\partial z} dy\, dz = 0,$$

since $\phi = 0$ at infinity. We therefore find for the lift

$$F_y = -\rho U \iint v_y \, dy\, dz, \qquad F_z = -\rho U \iint v_z \, dy\, dz. \tag{21.2}$$

The integration in these formulae is again taken only over the cross-sectional area of the wake. If the body has an axis of symmetry (not necessarily complete axial symmetry), and the flow is parallel to this axis, then the flow past the body has an axis of symmetry also. In this case the lift is, of course, zero.

Let us return to the flow in the wake. An estimate of the magnitudes of various terms in the Navier–Stokes equation shows that the term $\nu \triangle \mathbf{v}$ can in general be neglected at distances r from the body such that $rU/\nu \gg 1$ (cf. the derivation of the opposite condition (20.16)); these are the distances at which the flow outside the wake may be regarded as potential flow. It is not possible to neglect that term inside the wake even at these distances, however, since the transverse derivatives $\partial^2\mathbf{v}/\partial y^2$, $\partial^2\mathbf{v}/\partial z^2$ are large compared with $\partial^2\mathbf{v}/\partial x^2$.

Let Y be of the order of magnitude of the width of the wake, i.e. the distances from the x-axis at which the velocity $\mathbf{v}$ falls off markedly. The order of magnitude of the terms in the Navier–Stokes equation is then

$$(\mathbf{v}\cdot\mathbf{grad})\mathbf{v} \sim U\,\partial v/\partial x \sim Uv/x, \quad \nu\triangle\mathbf{v} \sim \nu\,\partial^2 v/\partial y^2 \sim \nu v/Y^2.$$

If these two magnitudes are comparable, we find

$$Y = \sqrt{(\nu x/U)}. \tag{21.3}$$

This quantity is in fact small compared with x, by the assumed condition $Ux/\nu \gg 1$. Thus the width of the laminar wake increases as the square root of the distance from the body.

In order to determine how the velocity decreases with increasing x in the wake, we return to formula (21.1). The region of integration has an area of the order of Y^2. Hence the integral can be estimated as $F_x \sim \rho U v Y^2$, and by using the relation (21.3) we obtain

$$v \sim F_x/\rho\nu x. \tag{21.4}$$

Having thus elucidated the qualitative features of laminar flow far from the body, we will now derive some quantitative formulae describing the flow pattern inside and outside the wake.

Flow inside the wake

In the Navier–Stokes equation for steady flow,

$$(\mathbf{v}\cdot\mathbf{grad})\,\mathbf{v} = -\mathbf{grad}\,(p/\rho) + \nu\triangle\mathbf{v}, \tag{21.5}$$

we use far from the body Oseen's approximation, replacing the term $(\mathbf{v}\cdot\mathbf{grad})\mathbf{v}$ by $(\mathbf{U}\cdot\mathbf{grad})\mathbf{v}$; cf. (20.17). Furthermore, inside the wake the derivative with respect to the longitudinal coordinate x in $\triangle \mathbf{v}$ can be neglected in comparison with the transverse derivatives. We thus start from the equation

$$U\frac{\partial \mathbf{v}}{\partial x} = -\mathbf{grad}\,(p/\rho) + \nu\left(\frac{\partial^2 \mathbf{v}}{\partial y^2} + \frac{\partial^2 \mathbf{v}}{\partial z^2}\right). \tag{21.6}$$

We seek the solution of this in the form $\mathbf{v} = \mathbf{v}_1 + \mathbf{v}_2$, where $\mathbf{v}_1$ is the solution of

$$U\frac{\partial \mathbf{v}_1}{\partial x} = \nu\left(\frac{\partial^2 \mathbf{v}_1}{\partial y^2} + \frac{\partial^2 \mathbf{v}_1}{\partial z^2}\right). \tag{21.7}$$

The quantity $\mathbf{v}_2$ arising from the term $-\mathbf{grad}\,(p/\rho)$ in the initial equation (21.6) may be sought as the gradient of a scalar Φ.† Since, far from the body, the derivatives with respect to x are small in comparison with those with respect to y and z, in the approximation considered we may neglect the term $\partial\Phi/\partial x$, i.e. take $v_x = v_{1x}$. We thus have for v_x the equation

$$U\frac{\partial v_x}{\partial x} = \nu\left(\frac{\partial^2 v_x}{\partial y^2} + \frac{\partial^2 v_x}{\partial z^2}\right). \tag{21.8}$$

This is formally the same as the two-dimensional equation of heat conduction, with x/U in place of the time, and the viscosity ν in place of the thermometric conductivity. The solution which decreases with increasing y and z (for fixed x) and gives an infinitely narrow wake as $x \to 0$ (in this approximation the dimensions of the body are regarded as small) is (cf. §51)

$$v_x = -\frac{F_x}{4\pi\rho\nu x}\exp\{-U(y^2+z^2)/4\nu x\}. \tag{21.9}$$

The constant coefficient in this formula is expressed in terms of the drag by means of formula (21.1), in which the integration may be extended over the whole yz-plane because of the rapid convergence. If the Cartesian coordinates are replaced by spherical polar coordinates r, θ, ϕ with the polar axis along the x-axis, then the region of the wake, $\sqrt{(y^2+z^2)} \ll x$, corresponds to $\theta \ll 1$. In these coordinates, formula (21.9) becomes

$$v_x = -\frac{F_x}{4\pi\rho\nu r}\exp\{-Ur\theta^2/4\nu\}. \tag{21.10}$$

The term in $\partial\Phi/\partial x$ (with Φ given by formula (21.12) below), which we have omitted, would give a term in v_x which contains an additional small factor θ.

The form of v_{1y} and v_{1z} must be the same as (21.9) but with different coefficients. We take the direction of the lift as the y-axis (so that $F_z = 0$). According to (21.2) we have, since $\Phi = 0$ at infinity,

$$\iint v_y\,dy\,dz = \iint (v_{1y} + \partial\Phi/\partial y)\,dy\,dz$$

$$= \iint v_{1y}\,dy\,dz = -F_y/\rho U$$

$$\iint v_{1z}\,dy\,dz = 0.$$

† The velocity potential will be denoted in the rest of this section by Φ, so as to distinguish it from the azimuthal angle ϕ in spherical polar coordinates.

It is therefore clear that v_{1y} diffffers from (21.9) in that F_x is replaced by F_y, and $v_{1z} = 0$. Thus we find

$$v_y = -\frac{F_x}{4\pi\rho v x}\exp\{-U(y^2+z^2)/4vx\} + \partial\Phi/\partial y, \quad v_z = \partial\Phi/\partial z. \tag{21.11}$$

To determine the function Φ, we proceed as follows. We write the equation of continuity, neglecting the longitudinal derivative:

$$\operatorname{div}\mathbf{v} \cong \frac{\partial v_y}{\partial y} + \frac{\partial v_z}{\partial z} = \left(\frac{\partial^2}{\partial y^2} + \frac{\partial^2}{\partial z^2}\right)\Phi + \frac{\partial v_{1y}}{\partial y} = 0.$$

Differentiating this equation with respect to x and using equation (21.7) for v_{1y}, we obtain

$$\left(\frac{\partial^2}{\partial y^2} + \frac{\partial^2}{\partial z^2}\right)\frac{\partial\Phi}{\partial x} = -\frac{\partial}{\partial y}\left(\frac{\partial v_{1y}}{\partial x}\right)$$

$$= -\frac{v}{U}\left(\frac{\partial^2}{\partial y^2} + \frac{\partial^2}{\partial z^2}\right)\frac{\partial v_{1y}}{\partial y}.$$

Hence

$$\partial\Phi/\partial x = -(v/U)\partial v_{1y}/\partial y.$$

Finally, substituting the expression for v_{1y} (the first term in (21.11)) and integrating with respect to x, we have

$$\Phi = -\frac{F_y}{2\pi\rho U}\frac{y}{y^2+z^2}\{\exp[-U(y^2+z^2)/4vx] - 1\}; \tag{21.12}$$

the constant of integration is chosen so that Φ remains finite when $y = z = 0$. In spherical polar coordinates (with the azimuthal angle ϕ measured from the xy-plane)

$$\Phi = -\frac{F_y}{2\pi\rho U}\frac{\cos\phi}{r\theta}\{\exp[-Ur\theta^2/4v] - 1\}. \tag{21.13}$$

It is seen from (21.11)–(21.13) that v_y and v_z, unlike v_x, contain terms which decrease only as $1/\theta^2$ when we move away from the axis of the wake, as well as those which decrease exponentially with increasing θ (for a given r).

If there is no lift, the flow in the wake is axially symmetrical, and $\Phi \equiv 0$.‡

FLOW OUTSIDE THE WAKE

Outside the wake, potential flow may be assumed. Since we are interested only in the terms in the potential Φ which decrease least rapidly at large distances, we seek a solution of Laplace's equation

$$\triangle\Phi = \frac{1}{r^2}\frac{\partial}{\partial r}\left(r^2\frac{\partial\Phi}{\partial r}\right) + \frac{1}{r^2\sin\theta}\frac{\partial}{\partial\theta}\left(\sin\theta\frac{\partial\Phi}{\partial\theta}\right) + \frac{1}{r^2\sin^2\theta}\frac{\partial^2\Phi}{\partial\phi^2} = 0$$

‡ This is true, in particular, for the wake behind a sphere. In this connection it may be noted that the formulae obtained, like (21.16) below, are in agreement with the velocity distribution (20.24) for flow at very low Reynolds numbers. In this case, the whole of the flow pattern described is moved to very large distances $r \gg l/R$, where l is the size of the body.

as a sum of two terms:

$$\Phi = \frac{a}{r} + \frac{\cos\phi}{r} f(\theta), \tag{21.14}$$

of which the first is spherically symmetrical and belongs to the force F_x, while the second is symmetrical about the xy-plane and belongs to the force F_y.

We obtain for the function $f(\theta)$ the equation

$$\frac{\mathrm{d}}{\mathrm{d}\theta}\left(\sin\theta \frac{\mathrm{d}f}{\mathrm{d}\theta}\right) - \frac{f}{\sin\theta} = 0.$$

The solution of this equation finite as $\theta \to \pi$ is

$$f = b \cot\tfrac{1}{2}\theta. \tag{21.15}$$

The coefficient b must be determined from the condition for joining the solution to that inside the wake. The reason is that (21.13) relates to the angle range $\theta \ll 1$, and (21.14) to $\theta \gg \sqrt{(\nu/Ur)}$. These ranges overlap when $\sqrt{(\nu/Ur)} \ll \theta \ll 1$, and (21.13) then becomes

$$\Phi = \frac{F_y}{2\pi\rho U} \frac{\cos\phi}{r\theta},$$

and the second term in (21.14) is $(2b/r\theta)\cos\phi$. Comparison of these expressions shows that we must take $b = F_y/4\pi\rho U$.

To determine the coefficient a in (21.14), we notice that the total mass flux through a sphere S with large radius r equals zero, as for any closed surface. The rate of inflow through the part S_0 of S intercepted by the wake is

$$-\iint_{S_0} v_x \,\mathrm{d}y\,\mathrm{d}z = F_x/\rho U.$$

Hence the same quantity must flow out through the rest of the surface of the sphere, i.e. we must have

$$\oint_{S-S_0} \mathbf{v}\cdot\mathbf{df} = F_x/\rho U.$$

Since S_0 is small compared with S, we can put

$$\oint_S \mathbf{v}\cdot\mathbf{df} = \int_S \mathbf{grad}\,\Phi\cdot\mathbf{df} = -4\pi a = F_x/\rho U, \tag{21.16}$$

whence $a = -F_x/4\pi\rho U$.

The complete expression for the velocity potential is thus

$$\Phi = \frac{1}{4\pi\rho U r}(-F_x + F_y \cos\phi \cot\tfrac{1}{2}\theta), \tag{21.17}$$

which gives the flow everywhere outside the wake far from the body. The potential diminishes with increasing distance as $1/r$; the velocity accordingly decreases as $1/r^2$. If there is no lift, the flow outside the wake is axially symmetrical.

§22. The viscosity of suspensions

A fluid in which numerous fine solid particles are suspended (forming a *suspension*) may be regarded as a homogeneous medium if we are concerned with phenomena whose characteristic lengths are large compared with the dimensions of the particles. Such a medium has an effective viscosity η which is different from the viscosity η_0 of the original fluid. The value of η can be calculated for the case where the concentration of the suspended particles is small (i.e. their total volume is small in comparison with that of the fluid). The calculations are relatively simple for the case of spherical particles (A. Einstein 1906).

It is necessary to consider first the effect of a single solid globule, immersed in a fluid, on flow having a constant velocity gradient. Let the unperturbed flow be described by a linear velocity distribution

$$v_{0i} = \alpha_{ik} x_k, \tag{22.1}$$

where α_{ik} is a constant symmetrical tensor. The fluid pressure is constant:

$$p_0 = \text{constant},$$

and in future we shall take p_0 to be zero, i.e. measure only the deviation from this constant value. If the fluid is incompressible (div $\mathbf{v}_0 = 0$), the sum of the diagonal elements, or trace, of the tensor α_{ik} must be zero:

$$\alpha_{ii} = 0. \tag{22.2}$$

Now let a small sphere with radius R be placed at the origin. We denote the altered fluid velocity by $\mathbf{v} = \mathbf{v}_0 + \mathbf{v}_1$; $\mathbf{v}_1$ must vanish at infinity, but near the sphere $\mathbf{v}_1$ is not small compared with $\mathbf{v}_0$. It is clear from the symmetry of the flow that the sphere remains at rest, so that the boundary condition is $\mathbf{v} = 0$ for $r = R$.

The required solution of the equations of motion (20.1) to (20.3) may be obtained at once from the solution (20.4), with the function f given by (20.6), if we notice that the space derivatives of this solution are themselves solutions. In the present case we desire a solution depending on the components of the tensor α_{ik} as parameters (and not on the vector $\mathbf{u}$ as in §20). Such a solution is

$$\mathbf{v}_1 = \mathbf{curl}\,\mathbf{curl}\,[(\boldsymbol{\alpha}\cdot\mathbf{grad})f], \qquad p = \eta_0 \alpha_{ik} \partial^2 \triangle f/\partial x_i \partial x_k,$$

where $(\boldsymbol{\alpha}\cdot\mathbf{grad})f$ denotes a vector whose components are $\alpha_{ik}\partial f/\partial x_k$. Expanding these expressions and determining the constants a and b in the function $f = ar + b/r$ so as to satisfy the boundary conditions at the surface of the sphere, we obtain the following formulae for the velocity and pressure:

$$v_{1i} = \frac{5}{2}\left(\frac{R^5}{r^4} - \frac{R^3}{r^2}\right)\alpha_{kl} n_i n_k n_l - \frac{R^5}{r^4}\alpha_{ik} n_k, \tag{22.3}$$

$$p = -5\eta_0 \frac{R^3}{r^3}\alpha_{ik} n_i n_k, \tag{22.4}$$

where $\mathbf{n}$ is a unit vector in the direction of the position vector.

Returning now to the problem of determining the effective viscosity of a suspension, we calculate the mean value (over the volume) of the momentum flux density tensor Π_{ik}, which, in the linear approximation with respect to the velocity, is the same as the stress tensor $-\sigma_{ik}$:

$$\bar{\sigma}_{ik} = (1/V)\int \sigma_{ik}\,\mathrm{d}V.$$

The integration here may be taken over the volume V of a sphere with large radius, which is then extended to infinity.

First of all, we have the identity

$$\bar{\sigma}_{ik} = \eta_0\left(\overline{\frac{\partial v_i}{\partial x_k}} + \overline{\frac{\partial v_k}{\partial x_i}}\right) - \bar{p}\delta_{ik} + \frac{1}{V}\int\left\{\sigma_{ik} - \eta_0\left(\frac{\partial v_i}{\partial x_k} + \frac{\partial v_k}{\partial x_i}\right) + p\,\delta_{ik}\right\}\mathrm{d}V. \qquad (22.5)$$

The integrand on the right is zero except within the solid spheres; since the concentration of the suspension is supposed small, the integral may be calculated for a single sphere as if the others were absent, and then multiplied by the concentration n of the suspension (the number of spheres per unit volume). The direct calculation of this integral would require an investigation of internal stresses in the spheres. We can circumvent this difficulty, however, by transforming the volume integral into a surface integral over an infinitely distant sphere, which lies entirely in the fluid. To do so, we note that the equation of motion $\partial\sigma_{il}/\partial x_l = 0$ leads to the identity

$$\sigma_{ik} = \partial(\sigma_{il}x_k)/\partial x_l;$$

hence the transformation of the volume integral into a surface integral gives

$$\bar{\sigma}_{ik} = \eta_0\left(\overline{\frac{\partial v_i}{\partial x_k}} + \overline{\frac{\partial v_k}{\partial x_i}}\right) + n\oint\{\sigma_{il}x_k\,\mathrm{d}f_l - \eta_0(v_i\,\mathrm{d}f_k + v_k\,\mathrm{d}f_i)\}.$$

We have omitted the term in $\bar{p}$, since the mean pressure is necessarily zero; $\bar{p}$ is a scalar, which must be given by a linear combination of the components α_{ik}, and the only such scalar is $\alpha_{ii} = 0$.

In calculating the integral over a sphere with very large radius, only the terms of order $1/r^2$ need be retained in the expression (22.3) for the velocity. A simple calculation gives the value of the integral as

$$n\eta_0 \cdot 20\pi R^3\{5\alpha_{lm}\overline{n_i n_k n_l n_m} - \alpha_{il}\overline{n_k n_l}\},$$

where the bar denotes an average with respect to directions of the unit vector $\mathbf{n}$. Effecting the averaging,† we finally have

$$\bar{\sigma}_{ik} = \eta_0\left(\overline{\frac{\partial v_i}{\partial x_k}} + \overline{\frac{\partial v_k}{\partial x_i}}\right) + 5\eta_0\alpha_{ik}\cdot\tfrac{4}{3}\pi R^3 n. \qquad (22.6)$$

The first term in (22.6), on substitution of v_0 from (22.1), gives $2\eta_0\alpha_{ik}$; the first-order small component is identically zero after averaging with respect to the directions of $\mathbf{n}$, as it should be, since the effect resides entirely in the integral separated in (22.5). Hence the required relative correction to the effective viscosity η of the suspension is determined by the ratio of the second and first terms in (22.6). We thus obtain

$$\eta = \eta_0(1 + \tfrac{5}{2}\phi), \quad \phi = 4\pi R^3 n/3, \qquad (22.7)$$

† The required mean values of products of components of the unit vector are symmetrical tensors, which can be formed only from the unit tensor δ_{ik}. We then easily find

$$\overline{n_i n_k} = \tfrac{1}{3}\delta_{ik},$$

$$\overline{n_i n_k n_l n_m} = \tfrac{1}{15}(\delta_{ik}\delta_{lm} + \delta_{il}\delta_{km} + \delta_{im}\delta_{kl}).$$

where ϕ is the small ratio of the total volume of the spheres to the total volume of the suspension.

The corresponding calculations and results become very lengthy even for a suspension of spheroidal particles.† As an illustration, we give the numerical values of the correction factor A in the formula

$$\eta = \eta_0(1 + A\phi), \qquad \phi = 4\pi ab^2 n/3,$$

for various values of a/b, where a and $b = c$ are the semi-axes of the spheroids:

a/b	0·1	0·2	0·5	1·0	2	5	10
A	8·04	4·71	2·85	2·5	2·91	5·81	13·6

The correction increases on either side of the value $a/b = 1$ which corresponds to spherical particles.

§23. Exact solutions of the equations of motion for a viscous fluid

If the non-linear terms in the equations of motion of a viscous fluid do not vanish identically, the solving of these equations offers great difficulties, and exact solutions can be obtained only in a very small number of cases. Such solutions are of considerable methodological interest, if not always of physical interest (because in practice turbulence occurs when the Reynolds number is sufficiently large).

We give below examples of exact solutions of the equations of motion for a viscous fluid.

Entrainment of fluid by a rotating disk

An infinite plane disk immersed in a viscous fluid rotates uniformly about its axis. Determine the motion of the fluid caused by this motion of the disk (T. von Kármán 1921).

We take cylindrical polar coordinates, with the plane of the disk as the plane $z = 0$. Let the disk rotate about the z-axis with angular velocity Ω. We consider the unbounded volume of fluid on the side $z > 0$. The boundary conditions are

$$v_r = 0, \quad v_\phi = \Omega r, \quad v_z = 0 \quad \text{for} \quad z = 0,$$
$$v_r = 0, \quad v_\phi = 0 \quad \text{for} \quad z = \infty.$$

The axial velocity v_z does not vanish as $z \to \infty$, but tends to a constant negative value determined by the equations of motion. The reason is that, since the fluid moves radially away from the axis of rotation, especially near the disk, there must be a constant vertical flow from infinity in order to satisfy the equation of continuity. We seek a solution of the equations of motion in the form

$$\left.\begin{aligned} v_r = r\Omega F(z_1); \quad v_\phi = r\Omega G(z_1); \quad v_z = \sqrt{(\nu\Omega)}H(z_1); \\ p = -\rho\nu\Omega P(z_1), \quad \text{where} \quad z_1 = \sqrt{(\Omega/\nu)}z. \end{aligned}\right\} \tag{23.1}$$

In this velocity distribution, the radial and azimuthal velocities are proportional to the distance from the axis of rotation, while v_z is constant on each horizontal plane.

† In the flow of a suspension of non-spherical particles, the presence of velocity gradients has an orienting effect on them. The simultaneous action of orienting hydrodynamic forces and disorienting rotary Brownian motion gives rise to an anisotropic distribution of the particles as regards their orientation in space. This, however, need not be considered when calculating the correction to the viscosity η: the anisotropy of the orientation distribution is itself dependent on the velocity gradients (linearly in the first approximation), and including it would give stress tensor terms non-linear in the gradients.

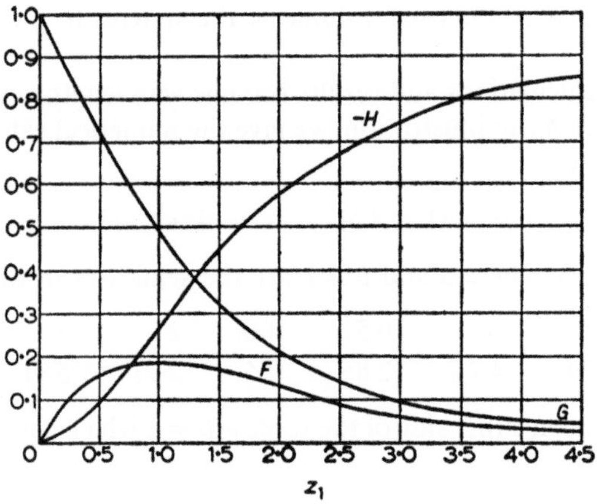

FIG. 7

Substituting in the Navier–Stokes equation and in the equation of continuity, we obtain the following equations for the functions F, G, H and P:

$$\left.\begin{aligned} F^2-G^2+F'H &= F'', \qquad 2FG+G'H = G'', \\ HH' &= P'+H'', \qquad 2F+H' = 0; \end{aligned}\right\} \tag{23.2}$$

the prime denotes differentiation with respect to z_1. The boundary conditions are

$$\left.\begin{aligned} &F=0, \quad G=1, \quad H=0 \quad \text{for} \quad z_1=0. \\ &F=0, \quad G=0 \qquad\qquad \text{for} \quad z_1=\infty. \end{aligned}\right\} \tag{23.3}$$

We have therefore reduced the solution of the problem to the integration of a system of ordinary differential equations in one variable; this can be achieved numerically. Figure 7 shows the functions F, G and $-H$ thus obtained. The limiting value of H as $z_1 \to \infty$ is $-0{\cdot}886$; in other words, the fluid velocity at infinity is $v_z(\infty) = -0{\cdot}886\sqrt{(\nu\Omega)}$.

The frictional force acting on unit area of the disk perpendicularly to the radius is $\sigma_{z\phi} = \eta(\partial v_\phi/\partial z)_{z=0}$. Neglecting edge effects, we may write the moment of the frictional forces acting on a disk with large but finite radius R as

$$M = 2\int_0^R 2\pi r^2 \sigma_{z\phi}\, dr = \pi R^4 \rho \sqrt{(\nu\Omega^3)}G'(0).$$

The factor 2 in front of the integral appears because the disk has two sides exposed to the fluid. A numerical calculation of the function G leads to the formula

$$M = -1{\cdot}94\, R^4 \rho \sqrt{(\nu\Omega^3)}. \tag{23.4}$$

FLOW IN DIVERGING AND CONVERGING CHANNELS

Determine the steady flow between two plane walls meeting at an angle α (Fig. 8 shows a cross-section of the two planes); the fluid flows out from the line of intersection of the planes (G. Hamel 1917).

We take cylindrical polar coordinates r, z, ϕ, with the z-axis along the line of intersection of the planes (the point O in Fig. 8), and the angle ϕ measured as shown in Fig. 8. The flow is

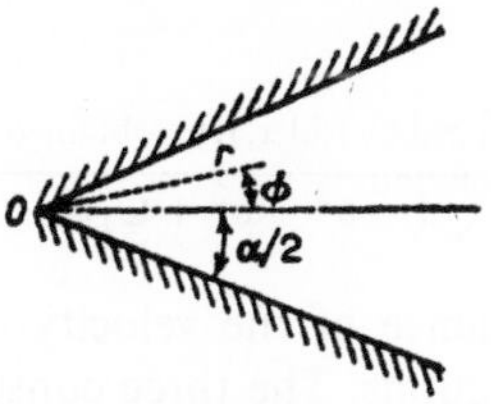

FIG. 8

uniform in the z-direction, and we naturally assume it to be entirely radial, i.e.

$$v_\phi = v_z = 0, \qquad v_r = v(r, \phi).$$

The equations (15.18) give

$$v\frac{\partial v}{\partial r} = -\frac{1}{\rho}\frac{\partial p}{\partial r} + \nu\left(\frac{\partial^2 v}{\partial r^2} + \frac{1}{r^2}\frac{\partial^2 v}{\partial \phi^2} + \frac{1}{r}\frac{\partial v}{\partial r} - \frac{v}{r^2}\right), \tag{23.5}$$

$$-\frac{1}{\rho r}\frac{\partial p}{\partial \phi} + \frac{2\nu}{r^2}\frac{\partial v}{\partial \phi} = 0, \tag{23.6}$$

$$\partial(rv)/\partial r = 0.$$

It is seen from the last of these that rv is a function of ϕ only. Introducing the function

$$u(\phi) = rv/6\nu, \tag{23.7}$$

we obtain from (23.6)

$$\frac{1}{\rho}\frac{\partial p}{\partial \phi} = \frac{12\nu^2}{r^2}\frac{du}{d\phi},$$

whence

$$\frac{p}{\rho} = \frac{12\nu^2}{r^2}u(\phi) + f(r).$$

Substituting this expression in (23.5), we have

$$\frac{d^2 u}{d\phi^2} + 4u + 6u^2 = \frac{1}{6\nu^2}r^3 f'(r),$$

from which we see that, since the left-hand side depends only on ϕ and the right-hand side only on r, each must be a constant, which we denote by $2C_1$. Thus $f'(r) = 12\nu^2 C_1/r^3$, whence $f(r) = -6\nu^2 C_1/r^2 + \text{constant}$, and we have for the pressure

$$\frac{p}{\rho} = \frac{6\nu^2}{r^2}(2u - C_1) + \text{constant}. \tag{23.8}$$

For $u(\phi)$ we have the equation

$$u'' + 4u + 6u^2 = 2C_1,$$

which, on multiplication by u' and one integration, gives

$$\tfrac{1}{2}u'^2 + 2u^2 + 2u^3 - 2C_1 u - 2C_2 = 0.$$

Hence we have

$$2\phi = \pm \int \frac{du}{\sqrt{(-u^3 - u^2 + C_1 u + C_2)}} + C_3, \tag{23.9}$$

which gives the required dependence of the velocity on ϕ; the function $u(\phi)$ can be expressed in terms of elliptic functions. The three constants C_1, C_2, C_3 are determined from the boundary conditions at the walls

$$u(\pm\tfrac{1}{2}\alpha) = 0 \tag{23.10}$$

and from the condition that the same mass Q of fluid passes in unit time through any cross-section r = constant:

$$Q = \rho \int_{-\alpha/2}^{\alpha/2} vr\, d\phi = 6\nu\rho \int_{-\alpha/2}^{\alpha/2} u\, d\phi. \tag{23.11}$$

Q may be either positive or negative. If $Q > 0$, the line of intersection of the planes is a source, i.e. the fluid emerges from the vertex of the angle: this is called *flow in a diverging channel*. If $Q < 0$, the line of intersection is a sink, and we have *flow in a converging channel*. The ratio $|Q|/\nu\rho$ is dimensionless and plays the part of the Reynolds number in the problem considered.

Let us first discuss converging flow ($Q < 0$). To investigate the solution (23.9)–(23.11) we make the assumptions, which will be justified later, that the flow is symmetrical about the plane $\phi = 0$ (i.e. $u(\phi) = u(-\phi)$), and that the function $u(\phi)$ is everywhere negative (i.e. the velocity is everywhere towards the vertex) and decreases monotonically from $u = 0$ at $\phi = \pm\frac{1}{2}\alpha$ to $u = -u_0 < 0$ at $\phi = 0$, so that u_0 is the maximum value of $|u|$. Then for $u = -u_0$ we must have $du/d\phi = 0$, whence it follows that $u = -u_0$ is a zero of the cubic expression under the radical in the integrand of (23.9). We can therefore write

$$-u^3 - u^2 + C_1 u + C_2 = (u + u_0)\{-u^2 - (1 - u_0)u + q\},$$

where q is another constant. Thus

$$2\phi = \pm \int_{-u_0}^{u} \frac{du}{\sqrt{[(u + u_0)\{-u^2 - (1 - u_0)u + q\}]}}, \tag{23.12}$$

the constants u_0 and q being determined from the conditions

$$\left.\begin{aligned} \alpha &= \int_{-u_0}^{0} \frac{du}{\sqrt{[(u + u_0)\{-u^2 - (1 - u_0)u + q\}]}}, \\ \tfrac{1}{6}\mathrm{R} &= \int_{-u_0}^{0} \frac{u\, du}{\sqrt{[(u + u_0)\{-u^2 - (1 - u_0)u + q\}]}} \end{aligned}\right\} \tag{23.13}$$

($\mathrm{R} = |Q|/\nu\rho$); the constant q must be positive, since otherwise these integrals would be complex. The two equations just given may be shown to have solutions u_0 and q for any R and $\alpha < \pi$. In other words, convergent symmetrical flow (Fig. 9) is possible for any aperture

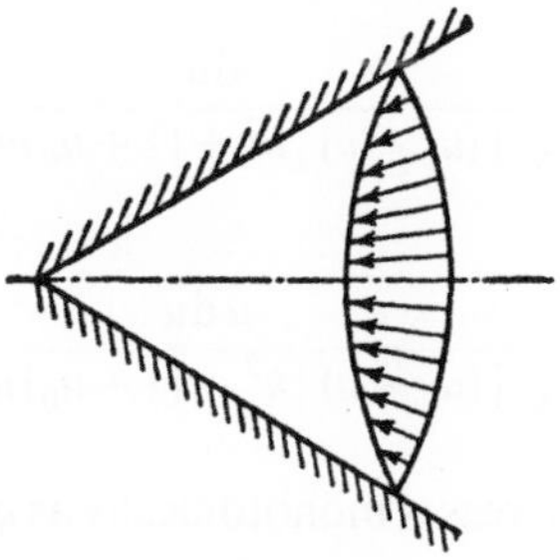

FIG. 9

angle $\alpha < \pi$ and any Reynolds number. Let us consider in more detail the flow for very large R. This corresponds to large u_0. Writing (23.12) (for $\phi > 0$) as

$$2(\tfrac{1}{2}\alpha - \phi) = \int_u^0 \frac{du}{\sqrt{[(u+u_0)\{-u^2-(1-u_0)u+q\}]}},$$

we see that the integrand is small throughout the range of integration if $|u|$ is not close to u_0. This means that $|u|$ can differ appreciably from u_0 only for ϕ close to $\pm\frac{1}{2}\alpha$, i.e. in the immediate neighbourhood of the walls.† In other words, we have $u \cong$ constant $= -u_0$ for almost all angles ϕ, and in addition $u_0 = R/6\alpha$, as we see from equations (23.13). The velocity v itself is $|Q|/\rho\alpha r$, giving a non-viscous potential flow with velocity independent of angle and inversely proportional to r. Thus, for large Reynolds numbers, the flow in a converging channel differs very little from potential flow of an ideal fluid. The effect of the viscosity appears only in a very narrow layer near the walls, where the velocity falls rapidly to zero from the value corresponding to the potential flow (Fig. 10).

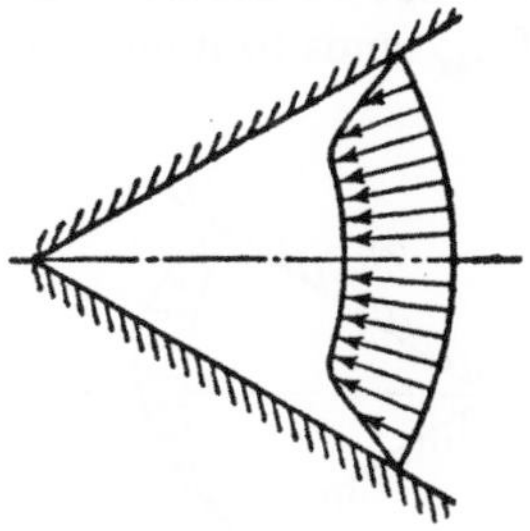

FIG. 10

Now let $Q > 0$, so that we have divergent flow. At first we again suppose that the flow is symmetrical about the plane $\phi = 0$, and that $u(\phi)$ (where now $u > 0$) varies monotonically from zero at $\phi = \pm\frac{1}{2}\alpha$ to $u_0 > 0$ at $\phi = 0$. Instead of (23.13) we now have

† The question may be asked how the integral can cease to be small, even if $u \cong -u_0$. The answer is that, for u_0 very large, one of the roots of $-u^2-(1-u_0)u+q = 0$ is close to $-u_0$, so that the radicand has two almost coincident zeros, the whole integral therefore being "almost divergent" at $u = -u_0$.

$$\left.\begin{aligned} \alpha &= \int_0^{u_0} \frac{du}{\sqrt{[(u_0-u)\{u^2+(1+u_0)u+q\}]}}, \\ \tfrac{1}{6}\mathrm{R} &= \int_0^{u_0} \frac{u\,du}{\sqrt{[(u_0-u)\{u^2+(1+u_0)u+q\}]}}. \end{aligned}\right\} \tag{23.14}$$

If we regard u_0 as given, then α increases monotonically as q decreases, and takes its greatest value for $q = 0$:

$$\alpha_{\max} = \int_0^{u_0} \frac{du}{\sqrt{[u(u_0-u)(u+u_0+1)]}}.$$

It is easy to see that for given q, on the other hand, α is a monotonically decreasing function of u_0. Hence it follows that u_0 is a monotonically decreasing function of q for given α, so that its greatest value is for $q = 0$ and is given by the above equation. The maximum $\mathrm{R} = \mathrm{R}_{\max}$ corresponds to the maximum u_0. Using the substitutions $k^2 = u_0/(1+2u_0)$, $u = u_0 \cos^2 x$, we can write the dependence of $\mathrm{R}_{\max}$ on α in the parametric form

$$\left.\begin{aligned} \alpha &= 2\sqrt{(1-2k^2)} \int_0^{\pi/2} \frac{dx}{\sqrt{(1-k^2\sin^2 x)}}, \\ \mathrm{R}_{\max} &= -6\alpha\frac{1-k^2}{1-2k^2} + \frac{12}{\sqrt{(1-2k^2)}} \int_0^{\pi/2} \sqrt{(1-k^2\sin^2 x)}\,dx. \end{aligned}\right\} \tag{23.15}$$

Thus symmetrical flow, everywhere divergent (Fig. 11a), is possible for a given aperture angle only for Reynolds numbers not exceeding a definite value. As $\alpha \to \pi$ $(k \to 0)$, $\mathrm{R}_{\max} \to 0$; as $\alpha \to 0$ $(k \to 1/\sqrt{2})$, $\mathrm{R}_{\max}$ tends to infinity as $18{\cdot}8/\alpha$.

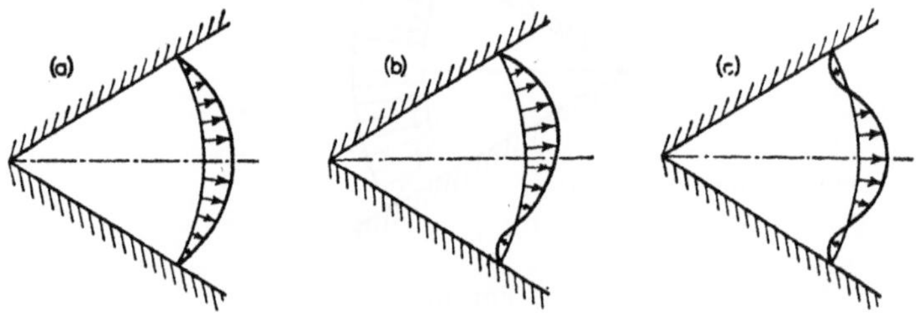

FIG. 11

For $\mathrm{R} > \mathrm{R}_{\max}$ the assumption of symmetrical flow, everywhere divergent, is unjustified, since the conditions (23.14) cannot be satisfied. In the range of angles $-\frac{1}{2}\alpha \leqslant \phi \leqslant \frac{1}{2}\alpha$ the function $u(\phi)$ must now have maxima or minima. The values of $u(\phi)$ corresponding to these extrema must again be zeros of the polynomial under the radical sign. It is therefore clear that the trinomial $u^2 + (1+u_0)u + q$ (with $u_0 > 0$, $q > 0$) must have two real negative roots in the range mentioned, so that the radicand can be written

$(u_0 - u)\ (u + u_0')\ (u + u_0'')$, where $u_0 > 0$, $u_0' > 0$, $u_0'' > 0$; we suppose $u_0' < u_0''$. The function $u(\phi)$ can evidently vary in the range $u_0 \geqslant u \geqslant -u_0'$, $u = u_0$ corresponding to a positive maximum of $u(\phi)$, and $u = -u_0'$ to a negative minimum. Without pausing to make a detailed investigation of the solutions obtained in this way, we may mention that for $R > R_{max}$ a solution appears in which the velocity has one maximum and one minimum, the flow being asymmetric about the plane $\phi = 0$ (Fig. 11b). When R increases further, a symmetrical solution with one maximum and two minima appears (Fig. 11c), and so on. In all these solutions, therefore, there are regions of both outward and inward flow (though of course the total discharge Q is positive). As $R \to \infty$ the number of alternating minima and maxima increases without limit, so that there is no definite limiting solution. We may emphasize that in divergent flow as $R \to \infty$ the solution does not, therefore, tend to the solution of Euler's equations as it does for convergent flow. Finally, it may be mentioned that, as R increases, the steady divergent flow of the kind described becomes unstable soon after R exceeds R_{max}, and in practice a non-steady or *turbulent* flow occurs (Chapter III).

SUBMERGED JET

Determine the flow in a jet emerging from the end of a narrow tube into an infinite space filled with the fluid—the *submerged jet* (L. Landau 1943).

We take spherical polar coordinates r, θ, ϕ, with the polar axis in the direction of the jet at its point of emergence, and with this point as origin. The flow is symmetrical about the polar axis, so that $v_\phi = 0$ and v_θ, v_r are functions of r and θ only. The same total momentum flux (the "momentum of the jet") must pass through any closed surface surrounding the origin (in particular, through an infinitely distant surface). For this to be so, the velocity must be inversely proportional to r, so that

$$v_r = F(\theta)/r, \qquad v_\theta = f(\theta)/r, \tag{23.16}$$

where F and f are some functions of θ only. The equation of continuity is

$$\frac{1}{r^2}\frac{\partial(r^2 v_r)}{\partial r} + \frac{1}{r\sin\theta}\frac{\partial}{\partial\theta}(v_\theta \sin\theta) = 0.$$

Hence we find that

$$F(\theta) = -\mathrm{d}f/\mathrm{d}\theta - f\cot\theta. \tag{23.17}$$

The components $\Pi_{r\phi}$, $\Pi_{\theta\phi}$ of the momentum flux density tensor in the jet vanish identically by symmetry. We assume that the components $\Pi_{\theta\theta}$ and $\Pi_{\phi\phi}$ also vanish; this assumption is justified when we obtain a solution satisfying all the necessary conditions. Using the expressions (15.20) for the components of the tensor σ_{ik}, and formulae (23.16), (23.17), we easily see that the relation

$$\sin^2\theta\,\Pi_{r\theta} = \frac{1}{2}\frac{\partial}{\partial\theta}[\sin^2\theta(\Pi_{\phi\phi} - \Pi_{\theta\theta})]$$

holds between the components of the momentum flux density tensor in the jet. Hence it follows that $\Pi_{r\theta} = 0$. Thus only the component Π_{rr} is non-zero, and it varies as $1/r^2$. It is easy to see that the equations of motion $\partial\Pi_{ik}/\partial x_k = 0$ are automatically satisfied.

Next, we write

$$(\Pi_{\theta\theta} - \Pi_{\phi\phi})/\rho = (f^2 + 2\nu f\cot\theta - 2\nu f')/r^2 = 0,$$

or

$$d(1/f)/d\theta + (1/f)\cot\theta + 1/2\nu = 0.$$

The solution of this equation is

$$f = -2\nu \sin\theta/(A - \cos\theta), \tag{23.18}$$

and then we have from (23.17)

$$F = 2\nu\left\{\frac{A^2-1}{(A-\cos\theta)^2} - 1\right\}. \tag{23.19}$$

The pressure distribution is found from the equation

$$\Pi_{\theta\theta}/\rho = p/\rho + f(f + 2\nu\cot\theta)/r^2 = 0,$$

which gives

$$p - p_0 = -\frac{4\rho\nu^2(A\cos\theta - 1)}{r^2(A-\cos\theta)^2}, \tag{23.20}$$

with p_0 the pressure at infinity. The constant A can be found in terms of the momentum of the jet, i.e. the total momentum flux in it. This flux is equal to the integral over the surface of a sphere

$$P = \oint \Pi_{rr}\cos\theta\, df = 2\pi\int_0^\pi r^2\Pi_{rr}\cos\theta\sin\theta\, d\theta.$$

The value of Π_{rr} is given by

$$\frac{1}{\rho}\Pi_{rr} = \frac{4\nu^2}{r^2}\left\{\frac{(A^2-1)^2}{(A-\cos\theta)^4} - \frac{A}{A-\cos\theta}\right\},$$

and a calculation of the integral gives

$$P = 16\pi\nu^2\rho A\left\{1 + \frac{4}{3(A^2-1)} - \tfrac{1}{2}A\log\frac{A+1}{A-1}\right\}. \tag{23.21}$$

Formulae (23.16)–(23.21) give the solution of the problem. When A varies from 1 to ∞, the jet momentum P takes all values between ∞ and 0.

The streamlines are determined by the equation $dr/v_r = r\,d\theta/v_\theta$, integration of which gives

$$\frac{r\sin^2\theta}{A-\cos\theta} = \text{constant}. \tag{23.22}$$

Figure 12 shows the characteristic form of the streamlines. The flow is a jet which comes from the origin and sucks in the surrounding fluid. If we arbitrarily regard as the boundary of the jet the surface where the streamlines have the least distance ($r\sin\theta$) from the axis, it is a cone with angle $2\theta_0$, where $\cos\theta_0 = 1/A$.

In the limiting case of a weak jet (small P, corresponding to large A), we have from (23.21)

$$P = 16\pi\nu^2\rho/A.$$

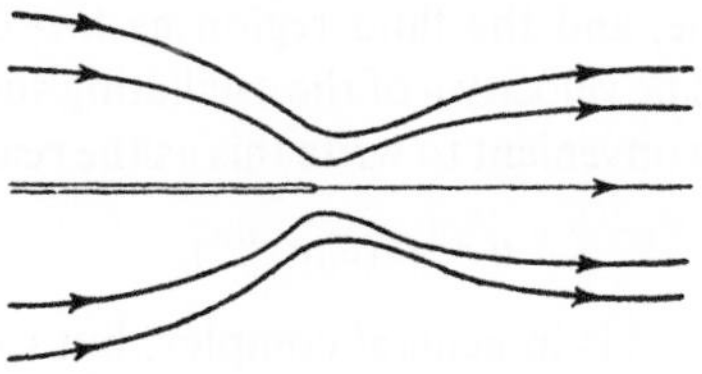

FIG. 12

In this case, the velocity is

$$v_\theta = -\frac{P}{8\pi\nu\rho}\frac{\sin\theta}{r}, \qquad v_r = \frac{P}{4\pi\nu\rho}\frac{\cos\theta}{r}. \tag{23.23}$$

In the opposite limit of a strong jet (large P, corresponding to $A \to 1$)†, we have

$$A = 1 + \tfrac{1}{2}\theta_0^2, \qquad \theta_0^2 = 64\pi\nu^2\rho/3P.$$

For large angles ($\theta \cong 1$), the velocity distribution is given by

$$v_\theta = -(2\nu/r)\cot\tfrac{1}{2}\theta, \qquad v_r = -2\nu/r; \tag{23.24}$$

for small angles ($\theta \cong \theta_0$),

$$v_\theta = -\frac{4\nu\theta}{(\theta_0^2+\theta^2)r}, \qquad v_r = \frac{8\nu\theta_0^2}{(\theta_0^2+\theta^2)^2 r}. \tag{23.25}$$

The solution here obtained is exact for a jet regarded as emerging from a point source. If the finite dimensions of the tube mouth are taken into account, the solution becomes the first term of an expansion in powers of the ratio of these dimensions to the distance r from the mouth of the tube. This is why, if we calculate from the above solution the total mass flux through a closed surface surrounding the origin, the result is zero. A non-zero total mass flux is obtained when further terms in the above-mentioned expansion are considered.‡

§24. Oscillatory motion in a viscous fluid

When a solid body immersed in a viscous fluid oscillates, the flow thereby set up has a number of characteristic properties. In order to study these, it is convenient to begin with a simple but typical example (G. G. Stokes 1851). Let us suppose that an incompressible fluid is bounded by an infinite plane surface which executes a simple harmonic oscillation in its own plane, with frequency ω. We require the resulting motion of the fluid. We take the

† However, the flow in a sufficiently strong jet is actually turbulent (§36). The Reynolds number for the jet considered is represented by the dimensionless parameter $\sqrt{(P/\rho\nu^2)}$.

‡ See Yu. B. Rumer, *Prikladnaya matematika i mekhanika* **16**, 255, 1952.

The submerged laminar jet with a non-zero angular momentum has been discussed by L. G. Loĭtsyanskiĭ (*ibid.* **17**, 3, 1953).

The hydrodynamic equations for any steady axially symmetrical flow of an incompressible viscous fluid with the velocity decreasing as $1/r$ can be reduced to a single second-order ordinary linear differential equation; see N. A. Slezkin, *Uchenye zapiski Moskovskogo gosudarstvennogo universiteta*, No. 2, 1934; *Prikladnaya matematika i mekhanika* **18**, 764, 1954.

solid surface as the yz-plane, and the fluid region as $x > 0$; the y-axis is taken in the direction of the oscillation. The velocity u of the oscillating surface is a function of time, of the form $A\cos(\omega t + \alpha)$. It is convenient to write this as the real part of a complex quantity:

$$u = \mathrm{re}(u_0 e^{-i\omega t}),$$

where the constant $u_0 = Ae^{-i\alpha}$ is in general complex, but can always be made real by a proper choice of the origin of time.

So long as the calculations involve only linear operations on the velocity u, we may omit the sign re and proceed as if u were complex, taking the real part of the final result. Thus we write

$$u_y = u = u_0 e^{-i\omega t}. \tag{24.1}$$

The fluid velocity must satisfy the boundary condition $\mathbf{v} = \mathbf{u}$ for $x = 0$, i.e. $v_x = v_z = 0$, $v_y = u$.

It is evident from symmetry that all quantities will depend only on the coordinate x and the time t. From the equation of continuity div $\mathbf{v} = 0$ we therefore have $\partial v_x/\partial x = 0$, whence v_x = constant = zero, from the boundary condition. Since all quantities are independent of the coordinates y and z, and since v_x is zero, it follows that $(\mathbf{v}\cdot\mathbf{grad})\mathbf{v} = 0$ identically. The equation of motion (15.7) becomes

$$\partial\mathbf{v}/\partial t = -(1/\rho)\,\mathbf{grad}\,p + \nu\triangle\mathbf{v}. \tag{24.2}$$

This is a linear equation. Its x-component is $\partial p/\partial x = 0$, i.e. p = constant.

It is further evident from symmetry that the velocity $\mathbf{v}$ is everywhere in the y-direction. For $v_y = v$ we have by (24.2)

$$\partial v/\partial t = \nu\partial^2 v/\partial x^2, \tag{24.3}$$

that is, a (one-dimensional) heat conduction equation. We shall look for a solution of this equation which is periodic in x and t, of the form

$$v = u_0 e^{i(kx - \omega t)},$$

so that $v = u$ for $x = 0$. Substituting in (24.3), we find

$$i\omega = \nu k^2, \qquad k = (1+i)/\delta, \qquad \delta = \sqrt{(2\nu/\omega)}, \tag{24.4}$$

so that the velocity is

$$v = u_0 e^{-x/\delta} e^{i(x/\delta - \omega t)}; \tag{24.5}$$

the choice of the sign of $\sqrt{i}$ in (24.4) is determined by the need for the velocity to decrease into the fluid.

Thus transverse waves can occur in a viscous fluid, with the velocity $v_y = v$ perpendicular to the direction of propagation. They are, however, rapidly damped as we move away from the solid surface whose motion generates the waves. The amplitude damping is exponential, the *depth of penetration* being δ.† This depth decreases with increasing frequency of the wave, but increases with the kinematic viscosity of the fluid.

The frictional force on the solid surface is evidently in the y-direction. The force per unit area is

$$\sigma_{xy} = \eta(\partial v_y/\partial x)_{x=0} = \sqrt{(\tfrac{1}{2}\omega\eta\rho)}\,(i-1)u. \tag{24.6}$$

† Over a distance δ, the wave amplitude decreases by a factor of e; over one wavelength, it decreases by a factor of $e^{2\pi} \cong 540$.

Supposing u_0 real and taking the real part of (24.6), we have

$$\sigma_{xy} = -\sqrt{(\omega\eta\rho)}u_0 \cos(\omega t + \tfrac{1}{4}\pi).$$

The velocity of the oscillating surface, however, is $u = u_0 \cos \omega t$. There is therefore a phase difference between the velocity and the frictional force.†

It is easy to calculate also the (time) average of the energy dissipation in the above problem. This may be done by means of the general formula (16.3); in this particular case, however, it is simpler to calculate the required dissipation directly as the work done by the frictional forces. The energy dissipated per unit time per unit area of the oscillating plane is equal to the mean value of the product of the force σ_{xy} and the velocity $u_y = u$:

$$-\overline{\sigma_{xy}u} = \tfrac{1}{2}u_0^2\sqrt{(\tfrac{1}{2}\omega\eta\rho)}. \tag{24.7}$$

It is proportional to the square root of the frequency of the oscillations, and to the square root of the viscosity.

An explicit solution can also be given of the problem of a fluid set in motion by a plane surface moving in its plane according to any law $u = u(t)$. We shall not pause to give the corresponding calculations here, since the required solution of equation (24.3) is formally identical with that of an analogous problem in the theory of thermal conduction, which we shall discuss in §52 (the solution is formula (52.15)). In particular, the frictional force on unit area of the surface is given by

$$\sigma_{xy} = -\sqrt{\frac{\eta\rho}{\pi}} \int_{-\infty}^{t} \frac{\mathrm{d}u(\tau)}{\mathrm{d}\tau} \frac{\mathrm{d}\tau}{\sqrt{(t-\tau)}}; \tag{24.8}$$

cf. (52.14).

Let us now consider the general case of an oscillating body with any shape. In the case of an oscillating plane considered above, the term $(\mathbf{v}\cdot\mathbf{grad})\mathbf{v}$ in the equation of motion of the fluid was identically zero. This does not happen, of course, for a surface with arbitrary shape. We shall assume, however, that this term is small in comparison with the other terms, so that it may be neglected. The conditions necessary for this procedure to be valid will be examined below.

We shall therefore begin, as before, from the linear equation (24.2). We take the curl of both sides; the term **curl grad** p vanishes identically, giving

$$\partial(\mathbf{curl}\,\mathbf{v})/\partial t = \nu \triangle\, \mathbf{curl}\,\mathbf{v}, \tag{24.9}$$

i.e. **curl v** satisfies a heat conduction equation. We have seen above, however, that such an equation gives an exponential decrease of the quantity which satisfies it. We can therefore say that the vorticity decreases towards the interior of the fluid. In other words, the motion of the fluid caused by the oscillations of the body is rotational in a certain layer round the

† For oscillations of a half-plane (parallel to its edge) there is an additional frictional force due to edge effects. The problem of the motion of a viscous fluid caused by oscillations of a half-plane, and also the more general problem of the oscillations of a wedge with any angle, can be solved by a class of solutions of the equation $\triangle f + k^2 f = 0$, used in the theory of diffraction by a wedge. We give here, for reference, only one result: the increase in the frictional force on a half-plane, arising from the edge effect, can be regarded as the result of increasing the area of the half-plane by moving the edge a distance $\tfrac{1}{2}\delta$, with δ as in (24.4) (L. D. Landau 1947).

body, while at larger distances it rapidly changes to potential flow. The depth of penetration of the rotational flow is of the order of δ.

Two important limiting cases are possible here: the quantity δ may be either large or small compared with the dimension of the oscillating body. Let l be the order of magnitude of this dimension. We first consider the case $\delta \gg l$; this implies that $l^2\omega \ll \nu$. Besides this condition, we shall also suppose that the Reynolds number is small. If a is the amplitude of the oscillations, the velocity of the body is of the order of $a\omega$. The Reynolds number for the flow in question is therefore $\omega al/\nu$. We thus suppose that

$$l^2\omega \ll \nu, \qquad \omega al/\nu \ll 1. \tag{24.10}$$

This is the case of low frequencies of oscillation, which in turn means that the velocity varies only slowly with time, and therefore that we can neglect the derivative $\partial\mathbf{v}/\partial t$ in the general equation of motion $\partial\mathbf{v}/\partial t + (\mathbf{v}\cdot\mathbf{grad})\mathbf{v} = -(1/\rho)\mathbf{grad}\,p + \nu\Delta\mathbf{v}$. The term $(\mathbf{v}\cdot\mathbf{grad})\mathbf{v}$, on the other hand, can be neglected because the Reynolds number is small.

The absence of the term $\partial\mathbf{v}/\partial t$ from the equation of motion means that the flow is steady. Thus, for $\delta \gg l$, the flow can be regarded as steady at any given instant. This means that the flow at any given instant is what it would be if the body were moving uniformly with its instantaneous velocity. If, for example, we are considering the oscillations of a sphere immersed in the fluid, with a frequency satisfying the inequalities (24.10) (l being now the radius of the sphere), then we can say that the drag on the sphere will be that given by Stokes' formula (20.14) for uniform motion of the sphere at small Reynolds numbers.

Let us now consider the opposite case, where $l \gg \delta$. In order that the term $(\mathbf{v}\cdot\mathbf{grad})\mathbf{v}$ should again be negligible, it is necessary that the amplitude of the oscillations should be small in comparison with the dimensions of the body:

$$l^2\omega \gg \nu, \qquad a \ll l; \tag{24.11}$$

in this case, it should be noticed, the Reynolds number need not be small. The above inequality is obtained by estimating the magnitude of $(\mathbf{v}\cdot\mathbf{grad})\mathbf{v}$. The operator $(\mathbf{v}\cdot\mathbf{grad})$ denotes differentiation in the direction of the velocity. Near the surface of the body, however, the velocity is nearly tangential. In the tangential direction the velocity changes appreciably only over distances of the order of the dimension of the body. Hence

$$(\mathbf{v}\cdot\mathbf{grad})\mathbf{v} \sim v^2/l \sim a^2\omega^2/l,$$

since the velocity itself is of the order of $a\omega$. The derivative $\partial\mathbf{v}/\partial t$, however, is of the order of $v\omega \sim a\omega^2$. Comparing these, we see that

$$(\mathbf{v}\cdot\mathbf{grad})\mathbf{v} \ll \partial\mathbf{v}/\partial t$$

if $a \ll l$. The terms $\partial\mathbf{v}/\partial t$ and $\nu\triangle\mathbf{v}$ are then easily seen to be of the same order.

We may now discuss the nature of the flow round an oscillating body when the conditions (24.11) hold. In a thin layer near the surface of the body the flow is rotational, but in the rest of the fluid we have potential flow.† Hence the flow everywhere except in the layer adjoining the body is given by the equations

$$\mathbf{curl}\,\mathbf{v} = 0, \qquad \mathrm{div}\,\mathbf{v} = 0. \tag{24.12}$$

† For oscillations of a plane surface not only $\mathbf{curl}\,\mathbf{v}$ but also $\mathbf{v}$ itself decreases exponentially with characteristic distance δ. This is because the oscillating plane does not displace the fluid, and therefore the fluid remote from it remains at rest. For oscillations of bodies with other shapes the fluid is displaced, and therefore executes a motion where the velocity decreases appreciably only over distances of the order of the dimension of the body.

Hence it follows that $\triangle \mathbf{v} = 0$, and the Navier–Stokes equation reduces to Euler's equation. The flow is therefore ideal everywhere except in the surface layer. Since this layer is thin, in solving equations (24.12) to determine the flow of the rest of the fluid we should take as boundary condtions those which must be satisfied at the surface of the body, i.e. that the fluid velocity be equal to that of the body. The solutions of the equations of motion for an ideal fluid cannot satisfy these conditions, however. We can require only the fulfilment of the corresponding condition for the fluid velocity component normal to the surface.

Although equations (24.12) are inapplicable in the surface layer of fluid, the velocity distribution obtained by solving them satisfies the necessary boundary condition for the normal velocity component, and the actual variation of this component near the surface therefore has no significant properties. The tangential component would be found, by solving the equations (24.12), to have some value different from the corresponding velocity component of the body, whereas these velocity components should be equal also. Hence the tangential velocity component must change rapidly in the surface layer. The nature of this variation is easily determined. Let us consider any portion of the surface of the body, with dimension large compared with δ, but small compared with the dimension of the body. Such a portion may be regarded as approximately plane, and therefore we can use the results obtained above for a plane surface. Let the x-axis be directed along the normal to the portion considered, and the y-axis parallel to the tangential velocity component of the surface there. We denote by v_y the tangential component of the fluid velocity relative to the body; v_y must vanish on the surface. Lastly, let $v_0 e^{-i\omega t}$ be the value of v_y found by solving equations (24.12). From the results obtained at the beginning of this section, we can say that in the surface layer the quantiy v_y will fall off towards the surface according to the law†

$$v_y = v_0 e^{-i\omega t}[1 - e^{-(1-i)x\sqrt{(\omega/2\nu)}}]. \tag{24.13}$$

Finally, the total amount of energy dissipated in unit time will be given by the integral

$$\overline{\dot{E}_{\text{kin}}} = -\tfrac{1}{2}\sqrt{(\tfrac{1}{2}\omega\eta\rho)}\oint |v_0|^2 \, df \tag{24.14}$$

taken over the surface of the oscillating body.

In the Problems at the end of this section we calculate the drag on various bodies oscillating in a viscous fluid. Here we shall make the following general remark regarding these forces. Writing the velocity of the body in the complex form $u = u_0 e^{-i\omega t}$, we obtain a drag F proportional to the velocity u, and also complex: $F = \beta u$, where $\beta = \beta_1 + i\beta_2$ is a complex constant. This expression can be written as the sum of two terms with real coefficients:

$$F = (\beta_1 + i\beta_2)u = \beta_1 u - \beta_2 \dot{u}/\omega, \tag{24.15}$$

one proportional to the velocity u and the other to the acceleration $\dot{u}$.

The (time) average of the energy dissipation is given by the mean product of the drag and the velocity, where of course we must first take the real parts of the expressions given above, i.e. $u = \tfrac{1}{2}(u_0 e^{-i\omega t} + u_0{}^* e^{i\omega t})$, $F = \tfrac{1}{2}(u_0 \beta e^{-i\omega t} + u_0{}^* \beta^* e^{i\omega t})$. Noticing that the mean values of $e^{\pm 2i\omega t}$ are zero, we have

$$\overline{\dot{E}_{\text{kin}}} = \overline{Fu} = \tfrac{1}{4}(\beta + \beta^*)|u_0|^2 = \tfrac{1}{2}\beta_1 |u_0|^2. \tag{24.16}$$

† The velocity distribution (24.13) is written in a frame where the solid body is at rest ($v_y = 0$ when $x = 0$). Hence v_0 must be taken as the solution of the problem of potential flow past a body at rest.

Thus we see that the energy dissipation arises only from the real part of β; the corresponding part of the drag (24.15), proportional to the velocity, may be called the *dissipative part*. The other part of the drag, proportional to the acceleration and determined by the imaginary part of β, does not involve the dissipation of energy and may be called the *inertial part*.

Similar considerations hold for the moment of the forces on a body executing rotary oscillations in a viscous fluid.

PROBLEMS

PROBLEM 1. Determine the frictional force on each of two parallel solid planes, between which is a layer of viscous fluid, when one of the planes oscillates in its own plane.

SOLUTION. We seek a solution of equation (24.3) in the form†

$$v = (A \sin kx + B \cos kx)e^{-i\omega t},$$

and determine A and B from the conditions $v = u = u_0 e^{-i\omega t}$ for $x = 0$ and $v = 0$ for $x = h$, where h is the distance between the planes. The result is

$$v = u\frac{\sin k(h-x)}{\sin kh}.$$

The frictional force per unit area on the moving plane is

$$P_{1y} = \eta(\partial v/\partial x)_{x=0} = -\eta ku \cot kh,$$

while that on the fixed plane is

$$P_{2y} = -\eta(\partial v/\partial x)_{x=h} = \eta ku \operatorname{cosec} kh,$$

the real parts of all quantities being understood.

PROBLEM 2. Determine the frictional force on an oscillating plane covered by a layer of fluid with thickness h, the upper surface being free.

SOLUTION. The boundary condition at the solid plane is $v = u$ for $x = 0$, and that at the free surface is $\sigma_{xy} = \eta\partial v/\partial x = 0$ for $x = h$. We find the velocity

$$v = u\frac{\cos k(h-x)}{\cos kh}.$$

The frictional force is

$$P_y = \eta(\partial v/\partial x)_{x=0} = \eta ku \tan kh.$$

PROBLEM 3. A plane disk with large radius R executes rotary oscillations with small amplitude about its axis, the angle of rotation being $\theta = \theta_0 \cos \omega t$, where $\theta_0 \ll 1$. Determine the moment of the frictional forces acting on the disk.

SOLUTION. For oscillations with small amplitude the term $(\mathbf{v} \cdot \mathbf{grad})\mathbf{v}$ in the equation of motion is always small compared with $\partial\mathbf{v}/\partial t$, whatever the frequency ω. If $R \gg \delta$, the disk may be regarded as infinite in determining the velocity distribution. We take cylindrical polar coordinates, with the z-axis along the axis of rotation, and seek a solution such that $v_r = v_z = 0$, $v_\phi = v = r\Omega(z, t)$. For the angular velocity $\Omega(z, t)$ of the fluid we obtain the equation

$$\partial\Omega/\partial t = \nu\partial^2\Omega/\partial z^2.$$

The solution of this equation which is $-\omega\theta_0 \sin \omega t$ for $z = 0$ and zero for $z = \infty$ is

$$\Omega = -\omega\theta_0 e^{-z/\delta} \sin(\omega t - z/\delta).$$

† In all the Problems to this section k and δ are defined as in (24.4).

The moment of the frictional forces on both sides of the disk is

$$M = 2\int_0^R r\cdot 2\pi r\eta(\partial v/\partial z)_{z=0}\,\mathrm{d}r = \omega\theta_0\pi\sqrt{(\omega\rho\eta)}R^4\cos(\omega t - \tfrac{1}{4}\pi).$$

PROBLEM 4. Determine the flow between two parallel planes when there is a pressure gradient which varies harmonically with time.

SOLUTION. We take the xz-plane half-way between the two planes, with the x-axis parallel to the pressure gradient, which we write in the form

$$-(1/\rho)\partial p/\partial x = ae^{-i\omega t}.$$

The velocity is everywhere in the x-direction, and is determined by the equation

$$\partial v/\partial t = ae^{-i\omega t} + \nu\partial^2 v/\partial y^2.$$

The solution of this equation which satisfies the conditions $v = 0$ for $y = \pm\frac{1}{2}h$ is

$$v = \frac{ia}{\omega}e^{-i\omega t}\left[1 - \frac{\cos ky}{\cos\frac{1}{2}kh}\right].$$

The mean value of the velocity over a cross-section is

$$\bar{v} = \frac{ia}{\omega}e^{-i\omega t}\left(1 - \frac{2}{kh}\tan\tfrac{1}{2}kh\right).$$

For $h/\delta \ll 1$ this becomes

$$\bar{v} \cong ae^{-i\omega t}h^2/12\nu,$$

in agreement with (17.5), while for $h/\delta \gg 1$ we have

$$\bar{v} \cong (ia/\omega)e^{-i\omega t},$$

in accordance with the fact that in this case the velocity must be almost constant over the cross-section, varying only in a thin surface layer.

PROBLEM 5. Determine the drag on a sphere with radius R which executes translatory oscillations in a fluid.

SOLUTION. We write the velocity of the sphere in the form $\mathbf{u} = \mathbf{u}_0 e^{-i\omega t}$. As in §20, we seek the fluid velocity in the form $\mathbf{v} = e^{-i\omega t}\,\mathbf{curl\,curl}\,f\mathbf{u}_0$, where f is a function of r only (the origin is taken at the instantaneous position of the centre of the sphere). Substituting in (24.9) and effecting transformations similar to those in §20, we obtain the equation

$$\triangle^2 f + (i\omega/\nu)\triangle f = 0$$

(instead of the equation $\triangle^2 f = 0$ in §20). Hence we have

$$\triangle f = \text{constant}\times e^{ikr}/r,$$

the solution being chosen which decreases exponentially with r. Integrating, we have

$$\mathrm{d}f/\mathrm{d}r = [ae^{ikr}(r - 1/ik) + b]/r^2; \tag{1}$$

the function f itself is not needed, since only the derivatives f' and f'' appear in the velocity. The constants a and b are determined from the condition that $\mathbf{v} = \mathbf{u}$ for $r = R$, and are found to be

$$a = -\frac{3R}{2ik}e^{-ikR}, \qquad b = -\tfrac{1}{2}R^3\left(1 - \frac{3}{ikR} - \frac{3}{k^2R^2}\right). \tag{2}$$

It may be pointed out that, at high frequencies ($R \gg \delta$), $a \to 0$ and $b \to -\frac{1}{2}R^3$, the values for potential flow obtained in §10, Problem 2; this is in accordance with what was said in §24.

The drag is calculated from formula (20.13), in which the integration is over the surface of the sphere. The result is

$$F = 6\pi\eta R\left(1 + \frac{R}{\delta}\right)u + 3\pi R^2\sqrt{(2\eta\rho/\omega)}\left(1 + \frac{2R}{9\delta}\right)\frac{\mathrm{d}u}{\mathrm{d}t}. \tag{3}$$

For $\omega = 0$ this becomes Stokes' formula, while for large frequencies we have

$$F = \tfrac{2}{3}\pi\rho R^3\frac{\mathrm{d}u}{\mathrm{d}t} + 3\pi R^2\sqrt{(2\eta\rho\omega)}u.$$

The first term in this expression corresponds to the inertial force in potential flow past a sphere (see §11, Problem 1), while the second gives the limit of the dissipative force. This second term could also have been found by calculating the energy dissipation according to (24.14); see Problem 6.

PROBLEM 6. Find the expression, in the limit of high frequencies ($\delta \ll R$), for the dissipative drag on an infinite cylinder with radius R oscillating at right angles to its axis.

SOLUTION. The velocity distribution round a cylinder at rest in a transverse flow is

$$v = (R^2/r^2)[2\mathbf{n}(\mathbf{u}\cdot\mathbf{n}) - \mathbf{u}] - \mathbf{u};$$

see §10, Problem 3. From this, we find as the tangential velocity at the surface of the cylinder

$$v_0 = -2u\sin\phi,$$

where r and ϕ are polar coordinates in the transverse plane, with ϕ measured from the direction of $\mathbf{u}$. From (24.14) we find the energy dissipated per unit length of the cylinder:

$$\overline{\dot{E}}_{\text{kin}} = \pi u^2 R\sqrt{(2\eta\rho\omega)}.$$

Comparison with (24.15) and (24.16) gives the result

$$F_{\text{dis}} = 2\pi u R\sqrt{(2\eta\rho\omega)}.$$

PROBLEM 7. Determine the drag on a sphere moving in an arbitrary manner, the velocity being given by a function $u(t)$.

SOLUTION. We represent $u(t)$ as a Fourier integral:

$$u(t) = \frac{1}{2\pi}\int_{-\infty}^{\infty} u_\omega e^{-i\omega t}\,d\omega, \qquad u_\omega = \int_{-\infty}^{\infty} u(\tau)e^{i\omega\tau}\,d\tau.$$

Since the equations are linear, the total drag may be written as the integral of the drag forces for velocities which are the separate Fourier components $u_\omega e^{-i\omega t}$; these forces are given by (3) of Problem 5, and are

$$\pi\rho R^3 u_\omega e^{-i\omega t}\left\{\frac{6\nu}{R^2} - \frac{2i\omega}{3} + \frac{3\sqrt{(2\nu)}}{R}(1-i)\sqrt{\omega}\right\}.$$

Noticing that $(du/dt)_\omega = -i\omega u_\omega$, we can rewrite this as

$$\pi\rho R^3 e^{-i\omega t}\left\{\frac{6\nu}{R^2}u_\omega + \tfrac{2}{3}(\dot{u})_\omega + \frac{3\sqrt{(2\nu)}}{R}(\dot{u})_\omega\frac{1+i}{\sqrt{\omega}}\right\}.$$

On integration over $\omega/2\pi$, the first and second terms give respectively $u(t)$ and $\dot{u}(t)$. To integrate the third term, we notice first of all that for negative ω this term must be written in the complex conjugate form, $(1+i)\sqrt{\omega}$ being replaced by $(1-i)/\sqrt{|\omega|}$; this is because formula (3) of Problem 5 was derived for a velocity $u = u_0 e^{-i\omega t}$ with $\omega > 0$, and for a velocity $u_0 e^{i\omega t}$ we should obtain the complex conjugate. Instead of an integral over ω from $-\infty$ to $+\infty$, we can therefore take twice the real part of the integral from 0 to ∞. We write

$$\frac{1}{\pi}\text{re}\left\{(1+i)\int_0^\infty \frac{(\dot{u})_\omega e^{-i\omega t}}{\sqrt{\omega}}d\omega\right\} = \frac{1}{\pi}\text{re}\left\{(1+i)\int_{-\infty}^{\infty}\int_0^\infty \frac{\dot{u}(\tau)e^{i\omega(\tau-t)}}{\sqrt{\omega}}d\omega\,d\tau\right\}$$

$$= \frac{1}{\pi}\text{re}\left\{(1+i)\int_{-\infty}^{t}\int_0^\infty \frac{\dot{u}(\tau)e^{-i\omega(t-\tau)}}{\sqrt{\omega}}d\omega\,d\tau + (1+i)\int_t^\infty\int_0^\infty \frac{\dot{u}(\tau)e^{i\omega(\tau-t)}}{\sqrt{\omega}}d\omega\,d\tau\right\}$$

$$= \sqrt{\frac{2}{\pi}}\text{re}\left\{\int_{-\infty}^{t}\frac{\dot{u}(\tau)}{\sqrt{(t-\tau)}}d\tau + i\int_t^\infty \frac{\dot{u}(\tau)}{\sqrt{(\tau-t)}}d\tau\right\}$$

$$= \sqrt{\frac{2}{\pi}}\int_{-\infty}^{t}\frac{\dot{u}(\tau)}{\sqrt{(t-\tau)}}d\tau.$$

Thus we have finally for the drag

$$F = 2\pi\rho R^3 \left\{ \frac{1}{3}\frac{du}{dt} + \frac{3\nu u}{R^2} + \frac{3}{R}\sqrt{\frac{\nu}{\pi}} \int_{-\infty}^{t} \frac{du}{d\tau} \frac{d\tau}{\sqrt{(t-\tau)}} \right\}. \tag{4}$$

PROBLEM 8. Determine the drag on a sphere which at time $t = 0$ begins to move with a uniform acceleration, $u = \alpha t$.

SOLUTION. Putting, in formula (4) of Problem 7, $u = 0$ for $t < 0$ and $u = \alpha t$ for $t > 0$ we have for $t > 0$

$$F = 2\pi\rho R^3 \alpha \left[\frac{1}{3} + \frac{3\nu t}{R^2} + \frac{6}{R}\sqrt{\frac{t\nu}{\pi}} \right].$$

PROBLEM 9. The same as Problem 8, but for a sphere brought instantaneously into uniform motion.

SOLUTION. We have $u = 0$ for $t < 0$ and $u = u_0$ for $t > 0$. The derivative du/dt is zero except at the instant $t = 0$, when it is infinite, but the time integral of du/dt is finite, and equals u_0. As a result, we have for all $t > 0$

$$F = 6\pi\rho\nu R u_0 \left[1 + \frac{R}{\sqrt{(\pi\nu t)}} \right] + \tfrac{2}{3}\pi\rho R^3 u_0 \delta(t),$$

where $\delta(t)$ is the delta function. For $t \to \infty$ this expression tends asymptotically to the value given by Stokes' formula. The impulsive drag on the sphere at $t = 0$ is obtained by integrating the last term and is $\frac{2}{3}\pi\rho R^3 u_0$.

PROBLEM 10. Determine the moment of the forces on a sphere executing rotary oscillations about a diameter in a viscous fluid.

SOLUTION. For the same reasons as in §20, Problem 1, the pressure-gradient term can be omitted from the equation of motion, so that we have $\partial \mathbf{v}/\partial t = \nu \triangle \mathbf{v}$. We seek a solution in the form $\mathbf{v} = \mathbf{curl}\, f \mathbf{\Omega}_0 e^{-i\omega t}$, where $\mathbf{\Omega} = \mathbf{\Omega}_0 e^{-i\omega t}$ is the angular velocity of rotation of the sphere. We then obtain for f, instead of the equation $\triangle f = \text{constant}$,

$$\triangle f + k^2 f = \text{constant}.$$

Omitting an unimportant constant term in the solution of this equation, we find $f = ae^{ikr}/r$, taking the solution which vanishes at infinity. The constant a is determined from the boundary condition that $\mathbf{v} = \mathbf{\Omega} \times \mathbf{r}$ at the surface of the sphere. The result is

$$f = \frac{R^3}{r(1 - ikR)} e^{ik(r-R)}, \qquad \mathbf{v} = (\mathbf{\Omega} \times \mathbf{r}) \left(\frac{R}{r} \right)^3 \frac{1 - ikr}{1 - ikR} e^{ik(r-R)},$$

where R is the radius of the sphere. A calculation like that in §20, Problem 1, gives the following expression for the moment of the forces exerted on the sphere by the fluid:

$$M = -\frac{8\pi}{3} \eta R^3 \Omega \frac{3 + 6R/\delta + 6(R/\delta)^2 + 2(R/\delta)^3 - 2i(R/\delta)^2(1 + R/\delta)}{1 + 2R/\delta + 2(R/\delta)^2}.$$

For $\omega \to 0$ (i.e. $\delta \to \infty$), we obtain $M = -8\pi\eta R^3 \Omega$, corresponding to uniform rotation of the sphere (see §20, Problem 1). In the opposite limiting case $R/\delta \gg 1$, we find

$$M = \frac{4\sqrt{2}}{3} \pi R^4 \sqrt{(\eta\rho\omega)}\,(i - 1)\Omega.$$

This expression can also be obtained directly: for $\delta \ll R$ each element of the surface of the sphere may be regarded as plane, and the frictional force acting on it is found by substituting $u = \Omega R \sin\theta$ in formula (24.6).

PROBLEM 11. Determine the moment of the forces on a hollow sphere filled with viscous fluid and executing rotary oscillations about a diameter.

SOLUTION. We seek the velocity in the same form as in Problem 10. For f we take the solution $(a/r)\sin kr$, which is finite everywhere within the sphere, including the centre. Determining a from the boundary condition, we have

$$\mathbf{v} = (\mathbf{\Omega} \times \mathbf{r}) \left(\frac{R}{r} \right)^3 \frac{kr \cos kr - \sin kr}{kR \cos kR - \sin kR}.$$

A calculation of the moment of the frictional forces gives the expression

$$M = \tfrac{8}{3}\pi\eta R^3\Omega \frac{k^2R^2 \sin kR + 3kR\cos kR - 3\sin kR}{kR\cos kR - \sin kR}.$$

The limiting value for $R/\delta \gg 1$ is of course the same as in the preceding problem. If $R/\delta \ll 1$ we have

$$M = \tfrac{8}{15}\pi\rho\omega R^5\Omega\left(i - \frac{R^2\omega}{35\nu}\right).$$

The first term corresponds to the inertial forces occurring in the rigid rotation of the whole fluid.

§25. **Damping of gravity waves**

Arguments similar to those given above can be advanced concerning the velocity distribution near the free surface of a fluid. Let us consider oscillatory motion occurring near the surface (for example, gravity waves). We suppose that the conditions (24.11) hold, the dimension l being now replaced by the wavelength λ:

$$\lambda^2\omega \gg \nu, \qquad a \ll \lambda; \tag{25.1}$$

a is the amplitude of the wave, and ω its frequency. Then we can say that the flow is rotational only in a thin surface layer, while throughout the rest of the fluid we have potential flow, just as we should for an ideal fluid.

The motion of a viscous fluid must satisfy the boundary conditions (15.16) at the free surface; these require that certain combinations of the space derivatives of the velocity should vanish. The flow obtained by solving the equations of ideal-fluid dynamics does not satisfy these conditions, however. As in the discussion of v_y in the previous section, we may conclude that the corresponding velocity derivatives decrease rapidly in a thin surface layer. It is important to notice that this does not imply a large velocity gradient as it does near a solid surface.

Let us calculate the energy dissipation in a gravity wave. Here we must consider the dissipation, not of the kinetic energy alone, but of the mechanical energy E_{mech}, which includes both the kinetic energy and the potential energy in the gravitational field. It is clear, however, that the presence or absence of a gravitational field cannot affect the energy dissipation due to processes of internal friction in the fluid. Hence $\dot{E}_{\text{mech}}$ is given by the same formula (16.3):

$$\dot{E}_{\text{mech}} = -\tfrac{1}{2}\eta \int \left(\frac{\partial v_i}{\partial x_k} + \frac{\partial v_k}{\partial x_i}\right)^2 \mathrm{d}V.$$

In calculating this integral for a gravity wave, it is to be noticed that, since the volume of the surface region of rotational flow is small, while the velocity gradient there is not large, the existence of this region may be ignored, unlike what was possible for oscillations of a solid surface. In other words, the integration is to be taken over the whole volume of fluid, which, as we have seen, moves as if it were an ideal fluid.

The flow in a gravity wave for an ideal fluid, however, has already been determined in §12. Since we have potential flow,

$$\partial v_i/\partial x_k = \partial^2\phi/\partial x_k \partial x_i = \partial v_k/\partial x_i,$$

so that

$$\dot{E}_{\text{mech}} = -2\eta \int \left(\frac{\partial^2 \phi}{\partial x_i \partial x_k}\right)^2 \mathrm{d}V.$$

The potential ϕ has the form

$$\phi = \phi_0 \cos(kx - \omega t + \alpha) e^{kz}.$$

We are interested, of course, not in the instantaneous value of the energy dissipation, but in its mean value with respect to time. Noticing that the mean values of the squared sine and cosine are the same, we find

$$\bar{\dot{E}}_{\text{mech}} = -8\eta k^4 \int \overline{\phi^2}\, \mathrm{d}V. \tag{25.2}$$

The energy E_{mech} itself may be calculated for a gravity wave by using a theorem of mechanics that, in any system executing small oscillations (with small amplitude, that is), the mean kinetic and potential energies are equal. We can therefore write $\bar{E}_{\text{mech}}$ simply as twice the kinetic energy:

$$\bar{E}_{\text{mech}} = \rho \int \overline{v^2}\, \mathrm{d}V = \rho \int \overline{(\partial\phi/\partial x_i)^2}\, \mathrm{d}V,$$

whence

$$\bar{E}_{\text{mech}} = 2\rho k^2 \int \overline{\phi^2}\, \mathrm{d}V. \tag{25.3}$$

The damping of the waves is conveniently characterized by the *damping coefficient* γ, defined as

$$\gamma = |\bar{\dot{E}}_{\text{mech}}| / 2\bar{E}_{\text{mech}}. \tag{25.4}$$

In the course of time, the energy of the wave decreases according to the law $\bar{E}_{\text{mech}} = \text{constant} \times e^{-2\gamma t}$; since the energy is proportional to the square of the amplitude, the latter decreases with time as $e^{-\gamma t}$.

Using (25.2), (25.3), we find

$$\gamma = 2\nu k^2. \tag{25.5}$$

Substituting here (12.7), we obtain the damping coefficient for gravity waves in the form

$$\gamma = 2\nu\omega^4/g^2. \tag{25.6}$$

PROBLEMS

PROBLEM 1. Determine the damping coefficient for long gravity waves propagated in a channel with constant cross-section; the frequency is supposed so large that $\sqrt{(\nu/\omega)}$ is small compared with the depth of the fluid in the channel and the width of the channel.

SOLUTION. The principal dissipation of energy occurs in the surface layer of fluid, where the velocity changes from zero at the boundary to the value $v = v_0 e^{-i\omega t}$ which it has in the wave. The mean energy dissipation per unit length of the channel is by (24.14) $l|v_0|^2\sqrt{(\eta\rho\omega/8)}$, where l is the perimeter of the part of the channel cross-section occupied by the fluid. The mean energy of the fluid (again per unit length) is $S\rho\overline{v^2} = \frac{1}{2}S\rho|v_0|^2$, where S is the cross-sectional area of the fluid in the channel. The damping coefficient is $\gamma = l\sqrt{(\nu\omega/8S^2)}$. For a channel with rectangular section, therefore,

$$\gamma = \frac{2h + a}{2\sqrt{2ah}}\sqrt{(\nu\omega)},$$

where a is the width and h the depth of the fluid.

PROBLEM 2. Determine the flow in a gravity wave on a very viscous fluid ($\nu \gtrsim \omega\lambda^2$).

SOLUTION. The calculation of the damping coefficient as shown above is valid only when this coefficient is small ($\gamma \ll \omega$), so that the motion may be regarded as that of an ideal fluid to a first approximation. For arbitrary viscosity we seek a solution of the equations of motion

$$\frac{\partial v_x}{\partial t} = \nu\left(\frac{\partial^2 v_x}{\partial x^2} + \frac{\partial^2 v_x}{\partial z^2}\right) - \frac{1}{\rho}\frac{\partial p}{\partial x},$$

$$\frac{\partial v_z}{\partial t} = \nu\left(\frac{\partial^2 v_z}{\partial x^2} + \frac{\partial^2 v_z}{\partial z^2}\right) - \frac{1}{\rho}\frac{\partial p}{\partial z} - g,$$

$$\frac{\partial v_x}{\partial x} + \frac{\partial v_z}{\partial z} = 0$$

which depends on t and x as $e^{-i\omega t + ikx}$, and diminishes in the interior of the fluid ($z > 0$). We find

$$v_x = e^{-i\omega t + ikx}(Ae^{kz} + Be^{mz}), \qquad v_z = e^{-i\omega t + ikx}\left(-iAe^{kz} - \frac{ik}{m}Be^{mz}\right),$$

$$p/\rho = e^{-i\omega t + ikx}\,\omega Ae^{kz}/k - gz, \quad \text{where} \quad m = \sqrt{(k^2 - i\omega/\nu)}.$$

The boundary conditions at the fluid surface are

$$\sigma_{zz} = -p + 2\eta\,\partial v_z/\partial z = 0, \qquad \sigma_{xz} = \eta\left(\frac{\partial v_x}{\partial z} + \frac{\partial v_z}{\partial x}\right) = 0 \text{ for } z = \zeta.$$

In the second condition we can immediately put $z = 0$ instead of $z = \zeta$. The first condition, however, should be differentiated with respect to t, after which we replace $g\partial\zeta/\partial t$ by gv_z and then put $z = 0$. The condition that the resulting two homogeneous equations for A and B be compatible gives

$$\left(2 - \frac{i\omega}{\nu k^2}\right)^2 + \frac{g}{\nu^2 k^3} = 4\sqrt{\left(1 - \frac{i\omega}{\nu k^2}\right)}. \tag{1}$$

This equation gives ω as a function of the wave number k; ω is complex, its real part giving the frequency of the oscillations and its imaginary part the damping coefficient. The solutions of equation (1) that have a physical meaning are those whose imaginary parts are negative (corresponding to damping of the wave); only two roots of (1) meet this requirement. If $\nu k^2 \ll \sqrt{(gk)}$ (the condition (25.1)), then the damping coefficient is small, and (1) gives approximately $\omega = \pm\sqrt{(gk)} - i.2\nu k^2$, a result which we already know. In the opposite limiting case $\nu k^2 \gg \sqrt{(gk)}$, equation (1) has two purely imaginary roots, corresponding to damped aperiodic flow. One root is $\omega = -ig/2\nu k$, while the other is much larger (of order νk^2), and therefore of no interest, since the corresponding motion is strongly damped.

CHAPTER III

TURBULENCE

§26. Stability of steady flow

For any problem of viscous flow under given steady conditions there must in principle exist an exact steady solution of the equations of fluid dynamics. These solutions formally exist for all Reynolds numbers. Yet not every solution of the equations of motion, even if it is exact, can actually occur in Nature. Those which do must not only obey the equations of fluid dynamics, but also be stable. Any small perturbations which arise must decrease in the course of time. If, on the contrary, the small perturbations which inevitably occur in the flow tend to increase with time, the flow is unstable and cannot actually exist.†

The mathematical investigation of the stability of a given flow with respect to infinitely small perturbations will proceed as follows. On the steady solution concerned (whose velocity distribution is $\mathbf{v}_0(\mathbf{r})$, say), we superpose a non-steady small perturbation $\mathbf{v}_1(\mathbf{r}, t)$, which must be such that the resulting velocity $\mathbf{v} = \mathbf{v}_0 + \mathbf{v}_1$ satisfies the equations of motion. The equation for $\mathbf{v}_1$ is obtained by substituting in the equations

$$\frac{\partial \mathbf{v}}{\partial t} + (\mathbf{v}\cdot\mathbf{grad})\mathbf{v} = -\frac{\mathbf{grad}\, p}{\rho} + \nu\triangle\mathbf{v}, \qquad \operatorname{div}\mathbf{v} = 0 \tag{26.1}$$

the velocity and pressure

$$\mathbf{v} = \mathbf{v}_0 + \mathbf{v}_1, \qquad p = p_0 + p_1, \tag{26.2}$$

where the known functions $\mathbf{v}_0$ and p_0 satisfy the unperturbed equations

$$(\mathbf{v}_0\cdot\mathbf{grad})\mathbf{v}_0 = -\frac{\mathbf{grad}\, p_0}{\rho} + \nu\triangle\mathbf{v}_0, \qquad \operatorname{div}\mathbf{v}_0 = 0. \tag{26.3}$$

Omitting terms above the first order in $\mathbf{v}_1$, we obtain

$$\frac{\partial \mathbf{v}_1}{\partial t} + (\mathbf{v}_0\cdot\mathbf{grad})\mathbf{v}_1 + (\mathbf{v}_1\cdot\mathbf{grad})\mathbf{v}_0 = -\frac{\mathbf{grad}\, p_1}{\rho} + \nu\triangle\mathbf{v}_1, \qquad \operatorname{div}\mathbf{v}_1 = 0. \tag{26.4}$$

The boundary condition is that $\mathbf{v}_1$ vanish on fixed solid surfaces.

Thus $\mathbf{v}_1$ satisfies a system of homogeneous linear differential equations, with coefficients that are functions of the coordinates only, and not of the time. The general solution of such equations can be represented as a sum of particular solutions in which $\mathbf{v}_1$ depends on time

† In the previous edition, instability with respect to infinitesimal perturbations was called *absolute instability*. This adjective will not now be used in the present context, but will serve (in accordance with more customary terminology) as a contrast to *convected* (§28).

as $e^{-i\omega t}$. The frequencies ω of the perturbations are not arbitrary, but are determined by solving the equations (26.4) with the appropriate boundary conditions. The frequencies are in general complex. If there are ω whose imaginary parts are positive, $e^{-i\omega t}$ will increase indefinitely with time. In other words, such perturbations, once having arisen, will increase, i.e. the flow is unstable with respect to such perturbations. For the flow to be stable it is necessary that the imaginary part of any possible frequency ω be negative. The perturbations that arise will then decrease exponentially with time.

Such a mathematical investigation of stability is extremely complicated, however. The theoretical problem of the stability of steady flow past bodies with finite dimensions has not yet been solved. It is certain that steady flow is stable for sufficiently small Reynolds numbers. The experimental data seem to indicate that, when R increases, it eventually reaches a value R_{cr} (the *critical Reynolds number*) beyond which the flow is unstable with respect to infinitesimal disturbances. For sufficiently large Reynolds numbers ($R > R_{cr}$), steady flow past solid bodies is therefore impossible. The critical Reynolds number is not, of course, a universal constant, but takes a different value for each type of flow. These values appear to be of the order of 10 to 100; for example, in flow across a cylinder undamped non-steady flow has been observed for $R = ud/\nu \cong 30$, d being the diameter of the cylinder.

Let us now consider the nature of the non-steady flow which is established as a result of the instability of steady flow at large Reynolds numbers (L. D. Landau 1944). We begin by examining the properties of this flow at Reynolds numbers only slightly greater than R_{cr}. For $R < R_{cr}$ the imaginary parts of the complex frequencies $\omega = \omega_1 + i\gamma_1$ for all possible small perturbations are negative ($\gamma_1 < 0$). For $R = R_{cr}$ there is one frequency whose imaginary part is zero. For $R > R_{cr}$ the imaginary part of this frequency is positive, but, when R is close to R_{cr}, γ_1 is small in comparison with the real part ω_1.† The function $\mathbf{v}_1$ corresponding to this frequency is of the form

$$\mathbf{v}_1 = A(t)\mathbf{f}(x, y, z), \tag{26.5}$$

where $\mathbf{f}$ is some complex function of the coordinates, and the complex amplitude $A(t)$ is‡

$$A(t) = \text{constant} \times e^{\gamma_1 t} e^{-i\omega_1 t}. \tag{26.6}$$

This expression for $A(t)$ is actually valid, however, only during a short interval of time after the disruption of the steady flow; the factor $e^{\gamma_1 t}$ increases rapidly with time, whereas the method of determining $\mathbf{v}_1$ given above, which leads to expressions like (26.5) and (26.6), applies only when $|\mathbf{v}_1|$ is small. In reality, of course, the modulus $|A|$ of the amplitude of the non-steady flow does not increase without limit, but tends to a finite value. For R close to R_{cr} (we always mean, of course, $R > R_{cr}$), this finite value is small, and can be determined as follows.

Let us find the time derivative of the squared amplitude $|A|^2$. For very small values of t, when (26.6) is still valid, we have $\mathrm{d}|A|^2/\mathrm{d}t = 2\gamma_1|A|^2$. This expression is really just the first term in an expansion in series of powers of A and A^*. As the modulus $|A|$ increases (still remaining small), subsequent terms in this expansion must be taken into account. The

† The set (or *spectrum*) of all possible perturbation frequencies for a given type of flow includes both separate isolated values (the *discrete spectrum*) and the whole of various frequency ranges (the *continuous spectrum*). It seems that for flow past finite bodies the frequencies with $\gamma_1 > 0$ can occur only in the discrete spectrum. The reason is that the perturbations corresponding to the frequencies in the continuous spectrum are in general not zero at infinity, but the unperturbed flow there is certainly a stable homogeneous plane-parallel flow.

‡ As usual, we understand the real part of (26.6).

next terms are those of the third order in A. However, we are not interested in the exact value of the derivative $d\,|A|^2/dt$, but in its time average, taken over times large compared with the period $2\pi/\omega_1$ of the factor $e^{-i\omega_1 t}$; we recall that, since $\omega_1 \gg \gamma_1$, this period is small compared with the time $1/\gamma_1$ required for the amplitude modulus $|A|$ to change appreciably. The third-order terms, however, must contain the periodic factor, and therefore vanish on averaging.† The fourth-order terms include one which is proportional to $A^2 A^{*2} = |A|^4$ and which does not vanish on averaging. Thus we have as far as fourth-order terms

$$\overline{d\,|A|^2/dt} = 2\gamma_1 |A|^2 - \alpha |A|^4, \tag{26.7}$$

where α (the *Landau constant*) may be either positive or negative.

We are interested in the case where an infinitesimal perturbation (superimposed on the original flow) first becomes unstable for $R > R_{cr}$. This corresponds to $\alpha > 0$. We have not put bars above $|A|^2$ and $|A|^4$ in (26.7), since the averaging is only over time intervals short compared with $1/\gamma_1$. For the same reason, in solving the equation we proceed as if the bar were omitted above the derivative also. The solution of equation (26.7) is

$$1/|A|^2 = \alpha/2\gamma_1 + \text{constant} \times e^{-2\gamma_1 t}.$$

Hence it is clear that $|A|^2$ tends asymptotically to a finite limit:

$$|A|^2_{max} = 2\gamma_1/\alpha. \tag{26.8}$$

The quantity γ_1 is some function of the Reynolds number. Near R_{cr} it can be expanded as a series of powers of $R - R_{cr}$. But $\gamma_1(R_{cr}) = 0$, by the definition of the critical Reynolds number. Hence we have to the first order

$$\gamma_1 = \text{constant} \times (R - R_{cr}). \tag{26.9}$$

Substituting this in (26.8), we see that the modulus $|A|$ of the amplitude is proportional to the square root of $R - R_{cr}$:

$$|A|_{max} \propto \sqrt{(R - R_{cr})}. \tag{26.10}$$

Let us now briefly discuss the case where $\alpha < 0$ in (26.7). The two terms in that expansion are then insufficient to determine the limiting amplitude of the perturbation, and we have to include a negative term of higher order; let this be $-\beta|A|^6$ with $\beta > 0$, which gives

$$|A|^2_{max} = \frac{|\alpha|}{2\beta} \pm \sqrt{\left(\frac{\alpha^2}{4\beta^2} + \frac{2|\alpha|}{\beta}\gamma_1\right)}, \tag{26.11}$$

with γ_1 as in (26.9). The dependence is shown in Fig. 13b; Fig. 13a corresponds to $\alpha > 0$, (26.10). When $R > R_{cr}$, there can be no steady flow; when $R = R_{cr}$, the perturbation discontinuously reaches a non-zero amplitude, though this is still assumed so small that the expansion in powers of $|A|^2$ is valid.‡ In the range $R_{cr}' < R < R_{cr}$, the unperturbed flow is *metastable*, being stable with respect to infinitesimal perturbations but unstable with respect to those with finite amplitude (the continuous curve; the broken curve shows the unstable branch).

† Strictly speaking, the third-order terms give, on averaging, not zero, but fourth-order terms, which we suppose included among the fourth-order terms in the expansion.

‡ Such systems are said to have *hard* self-excitation, in contrast to those with *soft* self-excitation, which are unstable with respect to infinitesimal perturbations.

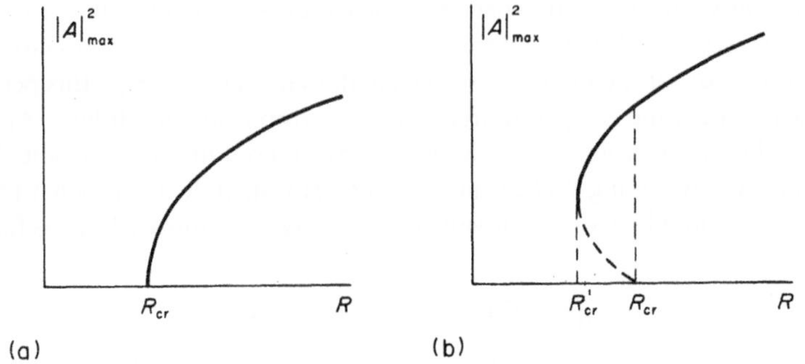

FIG. 13

Let us now return to the non-steady flow which occurs when $R > R_{cr}$, as a result of the instability with respect to small perturbations. For R close to R_{cr} the latter flow can be represented by superposing on the steady flow $\mathbf{v}_0(\mathbf{r})$ a periodic flow $\mathbf{v}_1(\mathbf{r}, t)$, with a small but finite amplitude which increases with R as in (26.10). The velocity distribution in this flow is of the form

$$\mathbf{v}_1 = \mathbf{f}(\mathbf{r})e^{-i(\omega_1 t+\beta_1)}, \tag{26.12}$$

where $\mathbf{f}$ is a complex function of the coordinates, and β_1 is some initial phase. For large $R - R_{cr}$, the separation of the velocity into $\mathbf{v}_0$ and $\mathbf{v}_1$ is no longer meaningful. We then have simply some periodic flow with frequency ω_1. If, instead of the time, we use as an independent variable the phase $\phi_1 \equiv \omega_1 t + \beta_1$, then we can say that the function $\mathbf{v}(\mathbf{r}, \phi_1)$ is a periodic function of ϕ_1, with period 2π. This function, however, is no longer a simple trigonometrical function. Its expansion in Fourier series

$$\mathbf{v} = \sum_p \mathbf{A}_p(\mathbf{r})e^{-i\phi_1 p} \tag{26.13}$$

(where the summation is over all integers p, positive and negative) includes not only terms with the fundamental frequency ω_1, but also terms whose frequencies are integral multiples of ω_1.

Equation (26.7) determines only the modulus of the time factor $A(t)$, and not its phase ϕ_1, which remains essentially indeterminate, and depends on the particular initial conditions which happen to occur at the instant when the flow begins. The initial phase β_1 can have any value, depending on these conditions. Thus the periodic flow under consideration is not uniquely determined by the given steady external conditions in which the flow takes place. One quantity—the initial phase of the velocity—remains arbitrary. We may say that the flow has one degree of freedom, whereas steady flow, which is entirely determined by the external conditions, has no degrees of freedom.

PROBLEM

Derive the equation for the energy balance between the unperturbed flow and a superimposed perturbation, without assuming that the latter is weak.

SOLUTION. Substituting (26.2) in (26.1), but not omitting the term of the second order in $\mathbf{v}_1$, we have

$$\partial\mathbf{v}_1/\partial t + (\mathbf{v}_0\cdot\mathbf{grad})\mathbf{v}_1 + (\mathbf{v}_1\cdot\mathbf{grad})\mathbf{v}_0 + (\mathbf{v}_1\cdot\mathbf{grad})\mathbf{v}_1 = -\mathbf{grad}\, p_1 + (1/R)\triangle\mathbf{v}_1; \tag{1}$$

all quantities are assumed to be brought to dimensionless form, as described in §19. Taking the scalar product of this equation with $\mathbf{v}_1$ and using the equations div $\mathbf{v}_0 = 0$, div $\mathbf{v}_1 = 0$, we obtain

$$\frac{\partial}{\partial t}(\tfrac{1}{2}v_1^2) = -v_{1i}v_{1k}\frac{\partial v_{0i}}{\partial x_k} - \frac{1}{\mathrm{R}}\frac{\partial v_{1i}}{\partial x_k}\frac{\partial v_{1i}}{\partial x_k} + \frac{\partial}{\partial x_k}\left\{-\tfrac{1}{2}v_1^2(v_{0k}+v_{1k}) - p_1 v_{1k} + \frac{1}{\mathrm{R}}v_{1i}\frac{\partial v_{1i}}{\partial x_k}\right\}.$$

The last term on the right gives zero on integration over the whole region of the flow, since $\mathbf{v}_0 = \mathbf{v}_1 = 0$ on the boundary surfaces of the region or at infinity. This gives as the required relation

$$\dot{E}_1 = T - D/\mathrm{R}, \tag{2}$$

$$E_1 = \int \tfrac{1}{2}v_1^2 \,\mathrm{d}V, \qquad T = -\int v_{1i}v_{1k}\frac{\partial v_{0i}}{\partial x_k}\mathrm{d}V, \qquad D = \int\left(\frac{\partial v_{1i}}{\partial x_k}\right)^2 \mathrm{d}V. \tag{3}$$

The functional T represents the energy exchange between the unperturbed flow and the perturbation, and may have either sign. The functional D is the dissipative energy loss, and $D > 0$ always. Note that the term in (1) non-linear in $\mathbf{v}_1$ does not contribute to the relation (2).

The relation (2) provides a lower limit of R_{cr} (O. Reynolds 1894; W. M'F. Orr 1907): the derivative $\mathrm{d}E_1/\mathrm{d}t$ must be negative, i.e. the perturbation decreases with time, if $\mathrm{R} < \mathrm{R}_E$, where

$$\mathrm{R}_E = \min(D/T), \tag{4}$$

the minimum of the functional being taken with respect to functions $\mathbf{v}_1(\mathbf{r})$ which satisfy the boundary conditions and the equation div $\mathbf{v}_1 = 0$. The existence of a finite minimum arises mathematically from the fact that T and D are both second-order homogeneous functionals. This proves the existence of a lower limit of R for metastability, below which the unperturbed flow is stable with respect to any perturbations. The "energy estimate" given by (4) is, however, much too low in the majority of cases.

§27. Stability of rotary flow

To investigate the stability of steady flow between two rotating cylinders (§18) in the limit of very large Reynolds numbers, we can use a simple method like that used in §4 to derive the condition for mechanical stability of a fluid at rest in a gravitational field (Rayleigh 1916). The principle of the method is to consider any small element of the fluid and to suppose that this element is displaced from the path which it follows in the flow concerned. As a result of this displacement, forces appear which act on the displaced element. If the original flow is stable, these forces must tend to return the element to its original position.

Each fluid element in the unperturbed flow moves in a circle r = constant about the axis of the cylinders. Let $\mu(r) = mr^2\dot{\phi}$ be the angular momentum of an element with mass m, $\dot{\phi}$ being the angular velocity. The centrifugal force acting on it is μ^2/mr^3; this force is balanced by the radial pressure gradient in the rotating fluid. Let us now suppose that a fluid element at a distance r_0 from the axis is slightly displaced from its path, being moved to a distance $r > r_0$ from the axis. The angular momentum of the element remains equal to its original value $\mu_0 = \mu(r_0)$. The centrifugal force acting on the element in its new position is therefore μ_0^2/mr^3. In order that the element should tend to return to its initial position, this force must be less than the equilibrium value μ^2/mr^3 which is balanced by the pressure gradient at the distance r. Thus the necessary condition for stability is $\mu^2 - \mu_0^2 > 0$. Expanding $\mu(r)$ in powers of the positive difference $r - r_0$, we can write this condition in the form

$$\mu\,\mathrm{d}\mu/\mathrm{d}r > 0. \tag{27.1}$$

According to formula (18.3), the angular velocity $\dot{\phi}$ of the moving fluid particles is

$$\dot{\phi} = \frac{\Omega_2 R_2^2 - \Omega_1 R_1^2}{R_2^2 - R_1^2} + \frac{(\Omega_1 - \Omega_2)R_1^2 R_2^2}{R_2^2 - R_1^2}\frac{1}{r^2}.$$

Calculating $\mu = mr^2\dot{\phi}$ and omitting factors which are certainly positive, we can write the condition (27.1) as

$$(\Omega_2 R_2^{\,2} - \Omega_1 R_1^{\,2})\dot{\phi} > 0. \tag{27.2}$$

The angular velocity $\dot{\phi}$ varies monotonically from Ω_1 on the inner cylinder to Ω_2 on the outer cylinder. If the two cylinders rotate in opposite directions, i.e. if Ω_1 and Ω_2 have opposite signs, the function $\dot{\phi}$ changes sign between the cylinders, and its product with the constant number $\Omega_2 R_2^{\,2} - \Omega_1 R_1^{\,2}$ cannot be everywhere positive. Thus in this case (27.2) does not hold at all points in the fluid, and the flow is unstable.

Now let the two cylinders be rotating in the same direction; taking this direction of rotation as positive, we have $\Omega_1 > 0$, $\Omega_2 > 0$. Then $\dot{\phi}$ is everywhere positive, and for the condition (27.2) to be fulfilled it is necessary that

$$\Omega_2 R_2^{\,2} > \Omega_1 R_1^{\,2}. \tag{27.3}$$

If $\Omega_2 R_2^{\,2} < \Omega_1 R_1^{\,2}$ the flow is unstable. For example, if the outer cylinder is at rest ($\Omega_2 = 0$), while the inner one rotates, then the flow is unstable. If, on the other hand, the inner cylinder is at rest ($\Omega_1 = 0$), the flow is stable.

It must be emphasized that no account has been taken, in the above arguments, of the effect of the viscous forces when the fluid element is displaced. The method is therefore applicable only for small viscosities, i.e. for large R.

To investigate the stability of the flow for any R, it is necessary to follow the general method, starting from equations (26.4); for flow between rotating cylinders, this was first done by G. I. Taylor (1924). In the present case the unperturbed velocity distribution $\mathbf{v}_0$ depends only on the (cylindrical) radial coordinate r, and not on the angle ϕ or the axial coordinate z. The complete set of independent solutions of equations (26.4) may therefore be sought in the form

$$\mathbf{v}_1(r, \phi, z) = e^{i(n\phi + kz - \omega t)}\mathbf{f}(r), \tag{27.4}$$

the direction of the vector $\mathbf{f}(r)$ being arbitrary. The wave number k, which takes a continuous range of values, determines the periodicity of the perturbation in the z-direction. The number n takes only integral values 0, 1, 2, . . . , as follows from the condition for the function to be single-valued with respect to the variable ϕ; the value $n = 0$ corresponds to axially symmetrical perturbations. The permissible values of the frequency ω are found by solving the equations with the necessary boundary conditions ($\mathbf{v}_1 = 0$ for $r = R_1$ and $r = R_2$). The problem thus formulated yields in general, for given n and k, a discrete series of eigenfrequencies $\omega = \omega_n^{(j)}(k)$, where j labels the branches of the function $\omega_n(k)$; these frequencies are in general complex.

The rolе of the Reynolds number in this case may be taken by $\Omega_1 R_1^{\,2}/\nu$ or $\Omega_2 R_2^{\,2}/\nu$ for given values of the ratios R_1/R_2 and Ω_1/Ω_2 which determine the type of flow. Let us follow the change of some eigenfrequency $\omega = \omega_n^{(j)}(k)$ as the Reynolds number gradually increases. The point where instability appears (for a particular form of perturbation) is determined by the value of R for which the function $\gamma(k) = \mathrm{im}\,\omega$ first becomes zero for some k. For $\mathrm{R} < \mathrm{R}_{\mathrm{cr}}$, the function $\gamma(k)$ is always negative, but for $\mathrm{R} > \mathrm{R}_{\mathrm{cr}}$ we have $\gamma > 0$ in some range of k. Let k_{cr} be the value of k for which $\gamma(k) = 0$ when $\mathrm{R} = \mathrm{R}_{\mathrm{cr}}$. The corresponding function (27.4) gives the nature of the flow which occurs (superimposed on the original flow) in the fluid at the instant when the original flow ceases to be stable; it is periodic along the axis of the cylinders, with period $2\pi/k_{\mathrm{cr}}$. The actual limit of stability is, of course, determined by the form of the perturbation, i.e. the function $\omega_n^{(j)}(k)$, for which R_{cr}

is least, and it is these "most dangerous" perturbations that are of interest here. As a rule (see below) they are axially symmetrical. Because of the great complexity of the calculation, a fairly complete study of them has been made only in the case where the space between the cylinders is narrow: $h \equiv R_2 - R_1 \ll R = \frac{1}{2}(R_1 + R_2)$. The results are as follows.†

It is found that a purely imaginary function $\omega(k)$ corresponds to the solution which gives the smallest R_{cr}. Hence, when $k = k_{cr}$, not only im ω but ω itself is zero. This means that the first instability of steady rotary flow leads to the appearance of another flow which is also steady.‡ It consists of toroidal *Taylor vortices* arranged in a regular manner along the cylinders. For the case where the two cylinders rotate in the same direction, Fig. 14 shows schematically the projections of the streamlines of these vortices on the meridional cross-

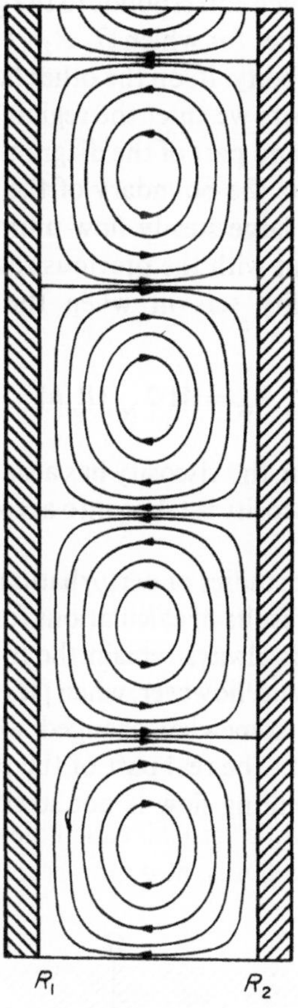

FIG. 14

† A detailed treatment is given by N. E. Kochin, I. A. Kibel' and N. V. Roze, *Theoretical Hydromechanics* (*Teoreticheskaya gidromekhanika*), Part 2, Moscow 1963; S. Chandrasekhar, *Hydrodynamic and Hydromagnetic Stability*, Oxford 1961; P. G. Drazin and W. H. Reid, *Hydrodynamic Stability*, Cambridge 1981.

‡ In such cases there is said to be *exchange of stabilities*. The experimental and numerical results for several particular cases suggest that this property is a general one for the flow considered and does not depend on h being small.

section plane of the cylinders; the velocity $\mathbf{v}_1$ actually has an azimuthal component also. The length $2\pi/k_{cr}$ of each period contains two vortices with opposite directions of rotation.

For R slightly greater than R_{cr} there is not one value of k but a whole range, for which im $\omega > 0$. However, it should not be thought that the resulting flow will be a superposition of flows with various periodicities. In reality, for each R a flow with a definite periodicity occurs which stabilizes the total flow. This periodicity, however, cannot be determined from the linearized equation (26.4).

Figure 15 shows the approximate form of the curve separating the regions of unstable (shaded) and stable flow for a given value of R_1/R_2. The right-hand branch of the curve, corresponding to rotation of the two cylinders in the same direction, is asymptotic to the line $\Omega_2 R_2^2 = \Omega_1 R_1^2$; this property is in fact a general one, not dependent on the smallness of h. When the Reynolds number increases, for a given type of flow, we move upwards along a line through the origin which corresponds to the given value of Ω_1/Ω_2. In the right-hand part of the diagram, such lines for which $\Omega_2 R_2^2/\Omega_1 R_1^2 > 1$ do not meet the curve which bounds the region of instability. If, on the other hand, $\Omega_2 R_2^2/\Omega_1 R_1^2 < 1$, then for sufficiently large Reynolds numbers we enter the region of instability, in accordance with the condition (27.3). In the left-hand part of the diagram (Ω_1 and Ω_2 with opposite signs), any line through the origin meets the boundary of the shaded region; that is, when the Reynolds number is sufficiently large steady flow ultimately becomes unstable for any ratio $|\Omega_2/\Omega_1|$, again in agreement with the previous results. For $\Omega_2 = 0$ (when only the inner cylinder rotates), instability sets in when the Reynolds number, defined as $R = h\Omega_1 R_1/\nu$, is

$$R_{cr} = 41{\cdot}2\sqrt{(R/h)}. \tag{27.5}$$

In the flow under consideration, the viscosity has a stabilizing effect: a flow stable when $\nu = 0$ remains stable when the viscosity is taken into account, and one that is unstable may become stable for a viscous fluid.

There have been no systematic studies of perturbations without axial symmetry in flow between rotating cylinders. The results of calculations for particular cases suggest that the axially symmetrical perturbations always remain the most dangerous on the right-hand side of Fig. 15. On the left-hand side, however, when $|\Omega_2/\Omega_1|$ is sufficiently large, the form of the boundary curve may be somewhat changed when perturbations without axial symmetry are taken into account. The real part of the perturbation frequency then does not tend to zero, and so the resulting flow is not steady, which considerably alters the nature of the instability.

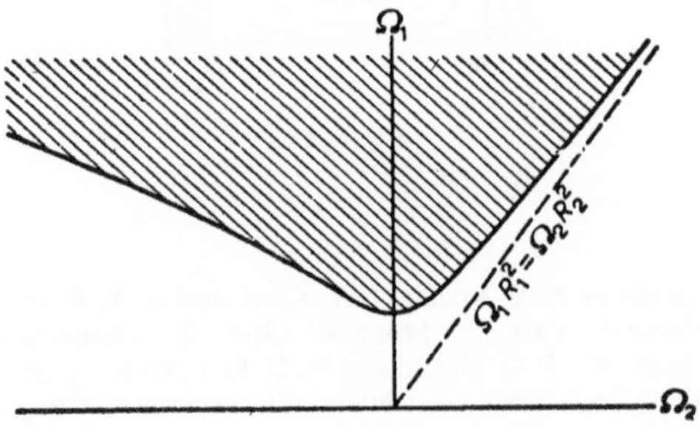

FIG. 15

The limiting case (as $h \to 0$) of flow between rotating cylinders is flow between two parallel planes in relative motion (see §17). This flow is stable with respect to infinitely small perturbations for any value of $\mathrm{R} = uh/\nu$, where u is the relative velocity of the planes.

§28. Stability of flow in a pipe

The steady flow in a pipe discussed in §17 loses its stability in an unusual manner. Since the flow is uniform in the x-direction (along the pipe), the unperturbed velocity distribution $\mathbf{v}_0$ is independent of x. Similarly to the procedure in §27, we can therefore seek solutions of equations (26.4) in the form

$$\mathbf{v}_1 = e^{i(kx-\omega t)}\mathbf{f}(y, z). \tag{28.1}$$

Here also there is a value $\mathrm{R} = \mathrm{R}_{cr}$ for which $\gamma = \mathrm{im}\,\omega$ first becomes zero for some value of k. It is of importance, however, that the real part of the function $\omega(k)$ is not now zero.

For values of R only slightly exceeding R_{cr}, the range of values of k for which $\gamma(k) > 0$ is small and lies near the point for which $\gamma(k)$ is a maximum, i.e. $\mathrm{d}\gamma/\mathrm{d}k = 0$ (as seen from Fig. 16). Let a slight perturbation occur in some part of the flow; it is a wave packet obtained by superposing a series of components with the form (28.1). In the course of time, the components for which $\gamma(k) > 0$ will be amplified, while the remainder will be damped. The amplified wave packet thus formed will also be carried downstream with a velocity equal to the group velocity $\mathrm{d}\omega/\mathrm{d}k$ of the packet (§67); since we are now considering waves whose wave numbers lie in a small range near the point where $\mathrm{d}\gamma/\mathrm{d}k = 0$, the quantity

$$\mathrm{d}\omega/\mathrm{d}k \cong \mathrm{d}(\mathrm{re}\,\omega)/\mathrm{d}k \tag{28.2}$$

is real, and is therefore the actual velocity of propagation of the packet.

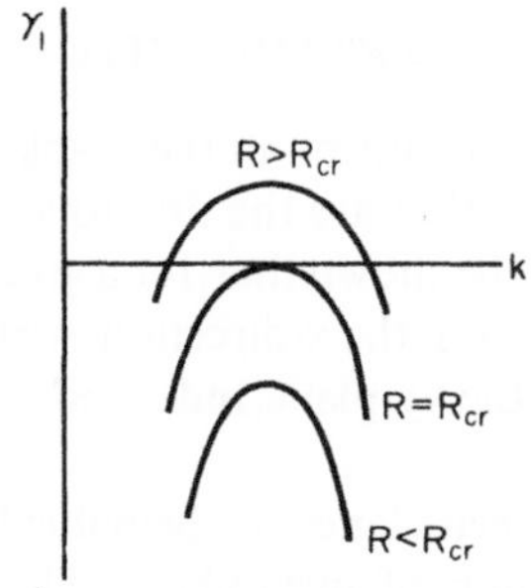

FIG. 16

This downstream displacement of the perturbations is very important, and causes the loss of stability to be totally different from that described in §27.

Since the positiveness of $\mathrm{im}\,\omega$ now implies only an amplification of the perturbation as it moves downstream, there are two possibilities. In one case, despite the movement of the wave packet, the perturbation increases without limit in the course of time at any point fixed in space; this kind of instability with respect to any infinitesimal perturbations will be called *absolute instability*. In the other case, the packet is carried away so swiftly that at any point fixed in space the perturbation tends to zero as $t \to \infty$; this kind will be called *convected instability*.† For Poiseuille flow, it appears that the second kind occurs; see the next footnote but four.

The difference between the two cases is a relative one, in the sense that it depends on the choice of the frame of reference with respect to which the instability is considered: an instability convected in one frame becomes absolute in another frame moving with the packet, and an absolute instability becomes convected in a frame that moves away from the packet with sufficient speed. In the present case, however, the physical significance of the difference is given by the existence of a preferred frame of reference in which the instability should be regarded, namely that in which the pipe walls are at rest. Moreover, since actual pipes have a large but finite length, a perturbation arising anywhere may in principle be carried out of the pipe before it actually disrupts the laminar flow.

Since the perturbations increase with the coordinate x (downstream), and not with time at a given point, it is reasonable to investigate this type of instability as follows. Let us suppose that, at a given point, a continuously acting perturbation with a given frequency ω is applied to the flow, and examine what will happen to this perturbation as it is carried downstream. Inverting the function $\omega(k)$, we find what wave number k corresponds to the given (real) frequency ω. If im $k < 0$, the factor e^{ikx} increases with x, i.e. the perturbation is amplified downstream. The curve in the ωR-plane given by the equation im $k(\omega, \mathrm{R}) = 0$, called the *neutral stability curve* or *neutral curve*, defines the region of stability, and separates, for each R, the frequencies of perturbations which are amplified and damped downstream.

The actual calculations are extremely complicated. A complete analytical investigation has been made only for plane Poiseuille flow (between two parallel planes; C. C. Lin 1945). We shall give the results here.‡

The (unperturbed) flow between the planes is uniform not only in the direction of flow (along the x-axis) but throughout the xz-plane (the y-axis being perpendicular to the planes). We can therefore seek solutions of equations (26.4) in the form

$$\mathbf{v}_1 = e^{i(k_x x + k_z z - \omega t)} \mathbf{f}(y) \tag{28.3}$$

with the wave vector $\mathbf{k}$ having any direction in the xz-plane. We are interested, however, only in the growing perturbations that are the first to appear as R increases, since these govern the limit of stability. It can be shown that, for a given value of the wave number, the first perturbation not damped has $\mathbf{k}$ in the x-direction, with $f_z = 0$. It is therefore sufficient to consider only perturbations in the xy-plane, independent of z and two-dimensional (like the unperturbed flow).††

The neutral curve for flow between planes is schematically shown in Fig. 17. The shaded area within the curve is the region of instability.§ The smallest value of R at which

---

† The general method of establishing the type of instability is described in *PK*, §62.

‡ See C. C. Lin, *The Theory of Hydrodynamic Stability*, Cambridge 1955. A discussion of these and later studies of the topic is to be found in the book by Drazin and Reid mentioned in a previous footnote.

†† The proof of this statement (H. B. Squire 1933) is that the equations (26.4) with a perturbation having the form (28.3) can be brought to a form in which they differ from the equations for two-dimensional perturbations only in that R is replaced by R $\cos\phi$, ϕ being the angle between $\mathbf{k}$ and $\mathbf{v}_0$ in the xz-plane. The critical number $\tilde{\mathrm{R}}_{cr}$ for three-dimensional perturbations with a given k is therefore $\tilde{\mathrm{R}}_{cr} = \mathrm{R}_{cr} \sec\phi > \mathrm{R}_{cr}$, where R_{cr} is calculated for two-dimensional perturbations.

§ The neutral curve in the kR-plane has a similar form. Since both ω and k are real on the neutral curve, the curves in the two planes represent the same dependence expressed in terms of different variables.

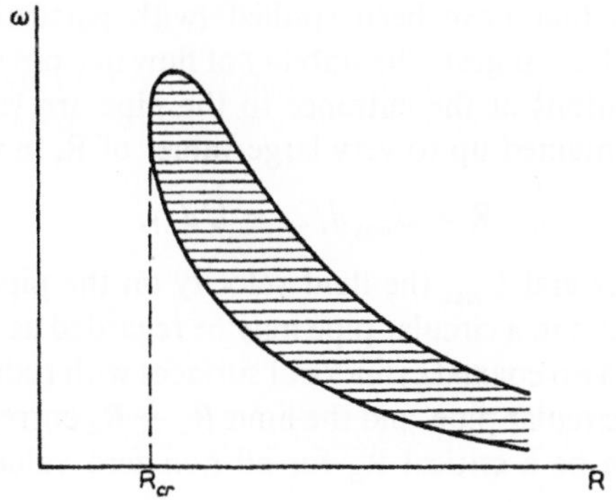

FIG. 17

undamped perturbations are possible is found to be $R_{cr} = 5772$ according to later and more accurate calculations by S. A. Orszag (1971); the Reynolds number is here defined as

$$R = U_{max}\, h/2\nu, \tag{28.4}$$

where U_{max} is the maximum flow velocity and $\frac{1}{2}h$ is half the distance between the planes, i.e. the distance over which the velocity increases from zero to its maximum value.† The value $R = R_{cr}$ corresponds to a perturbation wave number $k_{cr} = 2{\cdot}04/h$. As $R \to \infty$, the two branches of the neutral curve approach the R-axis asymptotically, with $\omega h/U_{max} \cong R^{-3/11}$ and $R^{-3/7}$ for the upper and lower branches respectively; on each branch, ω and k are related by $\omega h/U \cong (kh)^3$.

Thus, for any non-zero frequency ω that does not exceed a certain maximum value ($\sim U/h$), there is a finite range of R values in which the perturbations are amplified.‡ It is noteworthy that in this case a small but finite viscosity of the fluid has, in a sense, a destabilizing effect in comparison with the situation for a strictly ideal fluid.†† For, when $R \to \infty$, perturbations with any finite frequency are damped, but when a finite viscosity is introduced we eventually reach a region of instability; a further increase in the viscosity (decrease in R) finally brings us out of this region.

For flow in a pipe with circular cross-section, no complete theoretical study of the stability has yet been made, but the available results give good reason to suppose that the flow has stability (both absolute and convected) with respect to infinitesimal perturbations at any Reynolds number. When the unperturbed flow is axially symmetrical, the perturbations may be sought in the form

$$\mathbf{v}_1 = e^{i(n\phi + kz - \omega t)}\mathbf{f}(r), \tag{28.5}$$

as in (27.4). It may be regarded as proved that axially symmetrical perturbations ($n = 0$) are always damped. No undamped perturbations have been found, either, among those

† Another definition of R for two-dimensional Poiseuille flow is also used in the literature: $R = \bar{U}h/\nu$, where $\bar{U}$ is the fluid velocity averaged over the cross-section. Since $\bar{U} = \frac{2}{3}U_{max}$, we have $\bar{U}h/\nu = 4R/3$ when R is defined according to (28.4).

‡ The proof that the instability of two-dimensional Poiseuille flow is convected has been given by S. V. Iordanskiĭ and A. G. Kulikovskiĭ, *Soviet Physics JETP* **22**, 915, 1966. The proof relates, however, only to the range of very large R, where the two branches of the neutral curve are close to the abscissa axis; that is, $kh \ll 1$ on each branch. The problem remains unresolved for R values such that $kh \sim 1$ on the neutral curve.

†† This property was discovered by W. Heisenberg (1924).

without axial symmetry that have been studied (with particular values of n and in particular Reynolds number ranges). The stability of flow in a pipe is also suggested by the fact that, when perturbations at the entrance to the pipe are very carefully prevented, laminar flow can be maintained up to very large values of R, in practice up to $\mathrm{R} \cong 10^5$, where

$$\mathrm{R} = U_{\max} d/2\nu = \bar{U} d/\nu, \qquad (28.6)$$

d being the pipe diameter and $U_{\max}$ the fluid velocity on the pipe axis.

Flow between planes and in a circular pipe may be regarded as limiting cases of flow in an annular pipe between two coaxial cylindrical surfaces with radii R_1 and R_2 ($R_2 > R_1$). When $R_1 = 0$ we have a circular pipe, and the limit $R_1 \to R_2$ corresponds to flow between planes. There appears to be a critical R_{cr} for all non-zero values of $R_1/R_2 < 1$; when $R_1/R_2 \to 0$, $\mathrm{R}_{cr} \to \infty$.

For each of these Poiseuille flows there is also a critical number R_{cr}' which determines the limit of stability with respect to perturbations with finite amplitude. When $\mathrm{R} < \mathrm{R}_{cr}'$, undamped non-steady flow in the pipe is impossible. If turbulent flow occurs in any section of the pipe, then for $\mathrm{R} < \mathrm{R}_{cr}'$ the turbulent region will be carried downstream and will diminish in size until it disappears completely; if, on the other hand, $\mathrm{R} > \mathrm{R}_{cr}'$, the turbulent region will enlarge in the course of time to include more and more of the flow. If perturbations of the flow occur continually at the entrance to the pipe, then for $\mathrm{R} < \mathrm{R}_{cr}'$ they will be damped out at some distance down the pipe, no matter how strong they are initially. If, on the other hand, $\mathrm{R} > \mathrm{R}_{cr}'$, the flow becomes turbulent throughout the pipe, and this can be achieved by perturbations that are weaker, if R is greater. In the range between R_{cr}' and R_{cr}, laminar flow is metastable. For a pipe with circular cross-section, undamped turbulence has been observed for $\mathrm{R} \cong 1800$, and for flow between parallel planes for $\mathrm{R} \cong 1000$ and upwards.

Since the disruption of laminar flow in a pipe is "hard", it is accompanied by a discontinuous change in the drag force. For flow in a pipe with $\mathrm{R} > \mathrm{R}_{cr}'$ there are essentially two different dependences of the drag on R, one for laminar and the other for turbulent flow (see §43). The drag has a discontinuity, whatever the value of R at which the change from one to the other occurs.

One further remark may be made, to complete this section. The limit of stability (neutral curve) obtained for flow in an infinitely long pipe has also another significance. Let us consider flow in a pipe whose length is very great (in comparison with its width) but finite. Let certain boundary conditions be imposed at each end, by specifying the velocity profile (for example, we can imagine the ends of the pipe to be closed with porous seals which create a uniform profile); everywhere except near the ends of the pipe, the unperturbed velocity profile may be taken to have the Poiseuille form independent of x. For a finite system thus defined, we can propose the problem of stability with respect to infinitesimal perturbations; the general procedure for establishing the condition for such *global stability* is described in *PK*, §65. It can be shown that the above-mentioned neutral curve for an infinite pipe is also the limit of global stability in a finite pipe, whatever the specific boundary conditions at its ends.†

§29. **Instability of tangential discontinuities**

Flows in which two layers of incompressible fluid move relative to each other, one "sliding" on the other, are unstable if the fluid is ideal; the surface of separation between

† See A. G. Kulikovskiĭ, *Journal of Applied Mathematics and Mechanics* **32**, 100, 1968.

these two fluid layers would be a *surface of tangential discontinuity*, on which the fluid velocity tangential to the surface is discontinuous (H. Helmholtz 1868, W. Kelvin 1871). We shall see below (§35) what is the actual nature of the flow resulting from this instability; here we shall prove the above statement.

If we consider a small portion of the surface of discontinuity and the flow near it, we may regard this portion as plane, and the fluid velocities $\mathbf{v}_1$ and $\mathbf{v}_2$ on each side of it as constants. Without loss of generality we can suppose that one of these velocities is zero; this can always be achieved by a suitable choice of the coordinate system. Let $\mathbf{v}_2 = 0$, and $\mathbf{v}_1$ be denoted by $\mathbf{v}$ simply; we take the direction of $\mathbf{v}$ as the x-axis, and the z-axis along the normal to the surface.

Let the surface of discontinuity receive a slight perturbation, in which all quantities—the coordinates of points on the surface, the pressure, and the fluid velocity—are periodic functions, proportional to $e^{i(kx-\omega t)}$. We consider the fluid on the side where its velocity is $\mathbf{v}$, and denote by $\mathbf{v}'$ the small change in the velocity due to the perturbation. According to the equations (26.4) (with constant $\mathbf{v}_0 = \mathbf{v}$ and $\nu = 0$), we have the following system of equations for the perturbation $\mathbf{v}'$:

$$\operatorname{div}\mathbf{v}' = 0, \qquad \frac{\partial \mathbf{v}'}{\partial t} + (\mathbf{v}\cdot\mathbf{grad})\mathbf{v}' = -\frac{\mathbf{grad}\, p'}{\rho}.$$

Since $\mathbf{v}$ is along the x-axis, the second equation can be rewritten as

$$\frac{\partial \mathbf{v}'}{\partial t} + v\frac{\partial \mathbf{v}'}{\partial x} = -\frac{\mathbf{grad}\, p'}{\rho}. \tag{29.1}$$

If we take the divergence of both sides, then the left-hand side gives zero by virtue of $\operatorname{div}\mathbf{v}' = 0$, so that p' must satisfy Laplace's equation:

$$\triangle p' = 0. \tag{29.2}$$

Let $\zeta = \zeta(x, t)$ be the displacement in the z-direction of points on the surface of discontinuity, due to the perturbation. The derivative $\partial\zeta/\partial t$ is the rate of change of the surface coordinate ζ for a given value of x. Since the fluid velocity component normal to the surface of discontinuity is equal to the rate of displacement of the surface itself, we have to the necessary approximation

$$\partial\zeta/\partial t = v'_z - v\partial\zeta/\partial x, \tag{29.3}$$

where, of course, the value of v'_z on the surface must be taken.

We seek p' in the form $p' = f(z)\, e^{i(kx-\omega t)}$. Substituting in (29.2), we have for $f(z)$ the equation $\mathrm{d}^2 f/\mathrm{d}z^2 - k^2 f = 0$, whence $f = \text{constant} \times e^{\pm kz}$. Suppose that the space on the side under consideration (side 1) corresponds to positive values of z. Then we must take $f = \text{constant} \times e^{-kz}$, so that

$$p'_1 = \text{constant} \times e^{i(kx-\omega t)}\, e^{-kz}. \tag{29.4}$$

Substituting this expression in the z-component of equation (29.1), we find†

$$v'_z = kp'_1/i\rho_1(kv - \omega). \tag{29.5}$$

† The case $kv = \omega$, though possible in principle, is not of interest here, since instability can arise only from complex frequencies ω, not from real ω.

The displacement ζ may also be sought in a form proportional to the same exponential factor $e^{i(kx-\omega t)}$, and we obtain from (29.3) $v'_z = i\zeta(kv-\omega)$. This gives, instead of (29.5),

$$p'_1 = -\zeta\rho_1(kv-\omega)^2/k. \tag{29.6}$$

The pressure p'_2 on the other side of the surface is given by a similar formula, where now $v = 0$ and the sign is changed (since in this region $z < 0$, and all quantities must be proportional to e^{kz}, not e^{-kz}). Thus

$$p'_2 = \zeta\rho_2\omega^2/k. \tag{29.7}$$

We have written different densities ρ_1 and ρ_2 in order to include the case where we have a boundary separating two different immiscible fluids.

Finally, from the condition that the pressures p'_1 and p'_2 be equal on the surface of discontinuity, we obtain $\rho_1(kv-\omega)^2 = -\rho_2\omega^2$, from which the desired relation between ω and k is found to be

$$\omega = kv\frac{\rho_1 \pm i\sqrt{(\rho_1\rho_2)}}{\rho_1+\rho_2}. \tag{29.8}$$

We see that ω is complex, and there are always ω having a positive imaginary part. Thus tangential discontinuities are unstable, even with respect to infinitely small perturbations.† In this form, the result is true for very small viscosities. In that case, it is meaningless to distinguish convected and absolute instability, since as k increases the imaginary part of ω increases without limit, and hence the amplification coefficient of the perturbation as it is carried along may be as large as we please.

When finite viscosity is taken into account, the tangential discontinuity is no longer sharp; the velocity changes from one value to another across a layer with finite thickness. The problem of the stability of such a flow is mathematically entirely similar to that of the stability of flow in a laminar boundary layer with a point of inflexion in the velocity profile (§41). The experimental and numerical results indicate that instability sets in very soon, and perhaps is always present.‡

§30. Quasi-periodic flow and frequency locking††

In the following discussion (§§30–32) it will be convenient to use certain geometrical representations. To do so, we define the mathematical concept of the *space of states* for the fluid, each point in which corresponds to a particular velocity distribution or velocity field in the fluid. States at adjacent instants then correspond to adjacent points.§

A steady flow is represented by a point, and a periodic flow by a closed curve in the space of states; these are called respectively a *limit point* or *critical point*, and a *limit cycle*. If the

† If the direction of the wave vector **k** (in the xy-plane) is not the same as that of **v** but is at an angle ϕ to it, v in (29.8) is replaced by $v\cos\phi$, as is clear from the fact that the unperturbed velocity occurs in the initial linearized Euler's equation only in the combination **v**·**grad**. Such perturbations also are evidently unstable.

‡ Numerical calculations of the stability have been made for plane-parallel flows whose velocities vary between $\pm v_0$ according to a law such as $v = v_0 \tanh(z/h)$; the Reynolds number is then $R = v_0 h/\nu$. The neutral curve in the kR-plane starts from the origin, so that for each R value there is a range of k values (increasing with R) for which the flow is stable.

††§§ 30–32 were written jointly with M. I. Rabinovich.

§ In the mathematical literature, this functional space with an infinity of dimensions (or the spaces with a finite number of dimensions which may replace it in some cases; see below) is often called *phase space*. We shall avoid this term here, in order to prevent confusion with its more specific usual meaning in physics.

flows are stable, then adjacent curves representing the establishment of the flow tend to a limit point or cycle as $t \to \infty$.

A limit cycle (or point) has in the space of states a certain *domain of attraction*, and paths which begin in that region will eventually reach the limit cycle. In this connection, the limit cycle is called an *attractor*. It should be emphasized that for flow in a given volume with given boundary conditions (and a given value of R) there may be more than one attractor. Cases can occur where the space of states contains various attractors, each with its own domain of attraction. That is, when $R > R_{cr}$ there may be more than one stable flow regime, and the different regimes occur in accordance with the way in which the R value is reached. It should be emphasized that these various stable regimes are solutions of a *non-linear* set of equations of motion.†

Let us now consider the phenomena which occur when the Reynolds number is further increased beyond the critical value at which the periodic flow discussed in §26 is established. As R increases, a point is eventually reached where this flow in its turn becomes unstable. The instability should in principle be examined similarly to the procedure in §26 for determining the instability of the original steady flow. The unperturbed flow is now the periodic flow $\mathbf{v}_0(\mathbf{r}, t)$ with frequency ω_1, and in the equations of motion we substitute $\mathbf{v} = \mathbf{v}_0 + \mathbf{v}_2$, where $\mathbf{v}_2$ is a small correction. For $\mathbf{v}_2$ we again obtain a linear equation, but the coefficients are now functions of time as well as of the coordinates, and are periodic functions of time, with period $T_1 = 2\pi/\omega_1$. The solution of such an equation is to be sought in the form

$$\mathbf{v}_2 = \Pi(\mathbf{r}, t)\, e^{-i\omega t}, \tag{30.1}$$

where $\Pi(\mathbf{r}, t)$ is a periodic function of time, with the same period T_1. The instability again occurs when there is a frequency $\omega = \omega_2 + i\gamma_2$ whose imaginary part $\gamma_2 > 0$; the real part ω_2 gives the new frequency which appears.

During the period T_1, the perturbation (30.1) changes by a factor $\mu \equiv e^{-i\omega T_1}$. This factor is called the *multiplier* of the periodic flow, and is a convenient characteristic of the amplification or damping of perturbations in that flow. A periodic flow of a continuous medium (a fluid) corresponds to an infinity of multipliers and an infinity of possible independent perturbations. It ceases to be stable at the value $R_{cr,2}$ for which one or more multipliers reach unit modulus, i.e. μ crosses the unit circle in the complex plane. Since the equations are real, the multipliers must cross this circle in complex conjugate pairs, or singly with real values $+1$ or -1. The loss of stability of the periodic flow is accompanied by a particular qualitative change in the path pattern in the space of states near the now unstable limit cycle; this change is called a local *bifurcation*. The nature of the bifurcation is largely determined by the points at which the multipliers cross the unit circle.‡

Let us consider the bifurcation when the unit circle is crossed by a pair of complex conjugate multipliers having the form $\mu = \exp(\mp 2\pi\alpha i)$ where α is irrational. This causes the occurrence of a secondary flow with a new independent frequency $\omega_2 = \alpha\omega_1$, leading to a quasi-periodic flow with two incommensurate frequencies. The counterpart of this flow in the space of states is a path in the form of an open winding on a two-dimensional

---

† This is the situation, for example, when Couette flow ceases to be stable; the new flow pattern that is established depends in fact on the history of the process whereby the cylinders are caused to rotate with particular angular velocities.

‡ A multiplier cannot be zero, since a perturbation cannot disappear in a finite time (one period T_1).

torus†, the now unstable limit cycle being the generator of the torus; the frequency ω_1 corresponds to rotation round the generator, and ω_2 to rotation round the torus (Fig. 18). Just as, when the first periodic flow appeared, there was one degree of freedom, we now have two arbitrary quantities (phases), so that the flow has two degrees of freedom. The loss of stability of a periodic motion, accompanied by the creation of a two-dimensional torus, is a typical phenomenon in fluid dynamics.

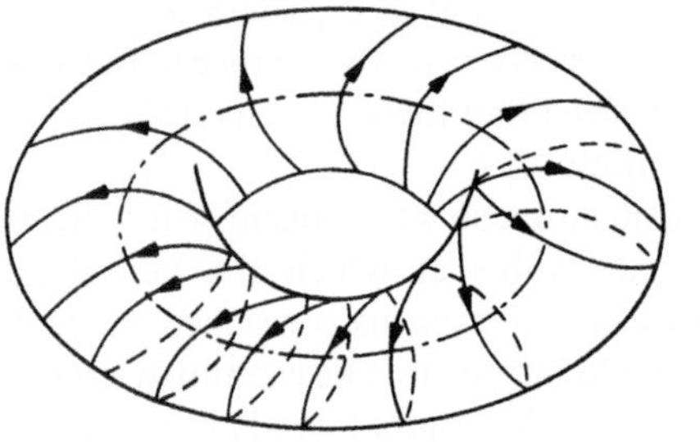

FIG. 18

Let us consider a hypothetical complication of the flow resulting from such a bifurcation, when the Reynolds number increases further ($R > R_{cr,2}$). It would be reasonable to suppose that, as R goes on increasing, new periods will successively appear. In terms of geometrical representations, this would signify loss of stability of the two-dimensional torus and the formation near it of a three-dimensional one, followed by a further bifurcation and its replacement by a four-dimensional one, and so on. The intervals between the Reynolds numbers corresponding to the successive appearance of new frequencies rapidly become shorter, and the flows are on smaller and smaller scales. The flow thus rapidly acquires a complicated and confused form, and is said to be *turbulent*, in contrast to the regular *laminar* flow, in which the fluid moves, as it were, in layers having different velocities.

Assuming now that this way or *scenario* of development of turbulence is in fact possible,‡ we write the general form of the function $\mathbf{v}(\mathbf{r}, t)$, whose time dependence is governed by some number N of different frequencies ω_i. It may be regarded as a function of N different phases $\phi_i = \omega_i t + \beta_i$ (and of the coordinates), periodic in each with period 2π. Such a function may be expressed as a series

$$\mathbf{v}(\mathbf{r}, t) = \sum \mathbf{A}_{p_1 p_2 \ldots p_N}(\mathbf{r}) \exp\left\{-i \sum_{i=1}^{n} p_i \phi_i\right\}, \tag{30.2}$$

which is a generalization of (26.13), the summation being over all integers $p_1, p_2, \ldots, p_N$. The flow described by this formula involves N arbitrary initial phases β_i and has N degrees of freedom.††

† We use the mathematical terminology, in which *torus* denotes a surface without the enclosed volume. Thus a two-dimensional torus is the two-dimensional surface of a three-dimensional "doughnut".

‡ It was proposed by L. D. Landau (1944) and independently by E. Hopf (1948).

†† If we take the phases ϕ_i as coordinates representing the path on an N-dimensional torus, the corresponding velocities are constants $\dot{\phi}_i = \omega_i$. For this reason, quasi-periodic flow can be described as movement on a torus with constant velocity.

States whose phases differ only by an integral multiple of 2π are physically identical. Thus the essentially different values of each phase lie in the range $0 \leqslant \phi_i \leqslant 2\pi$. Let us consider a pair of phases, $\phi_1 = \omega_1 t + \beta_1$ and $\phi_2 = \omega_2 t + \beta_2$. At some instant, let $\phi_1 = \alpha$. Then ϕ_1 will have the "same" value as α at every time

$$t = \frac{\alpha - \beta_1}{\omega_1} + 2\pi s \frac{1}{\omega_1},$$

where s is any integer. At these times,

$$\phi_2 = \beta_2 + (\omega_2/\omega_1)\,(\alpha - \beta_1 + 2\pi s).$$

The different frequencies are incommensurate, and therefore ω_2/ω_1 is irrational. If we reduce each value of ϕ_2 to a value in the range from 0 to 2π by subtracting an appropriate integral multiple of 2π, we therefore find that, when s varies from 0 to ∞, ϕ_2 takes values indefinitely close to any given number in that range. That is, in the course of a sufficiently long time ϕ_1 and ϕ_2 simultaneously take values indefinitely close to any specified pair. The same is true of every phase. In this turbulence model, therefore, in the course of a sufficiently long time, the fluid passes through states indefinitely close to any specified state defined by any possible set of simultaneous values of the phases ϕ_i. The time to do so, however, increases very rapidly with N and becomes so great that in practice no trace of any periodicity remains.†

It should be emphasized here that the path of turbulence development discussed above is essentially based on linear treatments. It has in fact been assumed that, when new periodic solutions appear through the evolution of secondary instabilities, the already existing periodic solutions do not disappear, but on the contrary remain almost unchanged. In this model, turbulent flow is just a superposition of a large number of such unchanged solutions. In general, however, the nature of the solutions changes when the Reynolds number increases and they cease to be stable. The perturbations interact, and this may either simplify or complicate the flow. Here is an illustration of the first possibility.

Let us take a simple case by supposing that the perturbed solution contains only two independent frequencies. As already mentioned, the geometrical representation of such a flow is an open winding on a two-dimensional torus. A perturbation with frequency ω_1 arising at $R = R_{cr,1}$ may naturally be assumed to be stronger near $R = R_{cr,2}$ (where the perturbation with frequency ω_2 arises) and therefore taken as unchanged for relatively small changes in R in that neighbourhood. Then, to describe the evolution of the perturbation with frequency ω_2 against the background of the periodic flow with frequency ω_1, we use a new variable

$$a_2(t) = |a_2(t)| e^{-i\phi_2(t)}; \tag{30.3}$$

$|a_2|$ is the shortest distance to the torus generator (the now unstable limit cycle for frequency ω_1), i.e. the relative amplitude of the secondary periodic flow, and ϕ_2 is the phase of the latter. Let us consider the behaviour of $a_2(t)$ at discrete instants that are multiples of

† In established turbulent flow of this type, the probability for the system (fluid) to be in a given small volume near a chosen point in the space of phases $\phi_1, \phi_2, \ldots, \phi_N$ is the ratio of this volume $(\delta\phi)^N$ to the total volume $(2\pi)^N$ We can therefore say that in the course of a sufficiently long time the system will be in the neighbourhood of a given point only for a fraction $e^{-\kappa N}$ of the time, where $\kappa = \log(2\pi/\delta\phi)$.

the period $T_1 = 2\pi/\omega_1$. During one period, the perturbation with frequency ω_2 changes by a factor μ, where

$$\mu = |\mu| \exp(-2\pi i \omega_2/\omega_1)$$

is its multiplier; after an integral number τ of such periods, a_2 is multiplied by μ^τ. We assume that $R - R_{cr}$ is small; the growth factor of the perturbation is then also small, and $|\mu| - 1$ is positive but small, so that a_2 changes only slightly during the period T_1; the phase ϕ_2 varies simply in proportion to τ. We can thus treat the discrete variable τ as if it were continuous and represent the variation of $a_2(\tau)$ by a differential equation in τ.

The concept of the multiplier relates to very short time intervals after the onset of instability, when the perturbation is still describable by linear equations. In this range, $a_2(\tau)$ varies as μ^τ according to the above discussion, and

$$\mathrm{d}a_2/\mathrm{d}\tau = a_2(\tau) \log \mu;$$

just above the critical Reynolds number,

$$\begin{aligned} \log \mu &= \log |\mu| - 2\pi i \omega_2/\omega_1 \\ &\cong |\mu| - 1 - 2\pi i \omega_2/\omega_1. \end{aligned} \tag{30.4}$$

This is the first term in an expansion of $\mathrm{d}a_2/\mathrm{d}\tau$ in powers of a_2 and a_2^*, and when $|a_2|$ increases (still remaining small) the next term has to be taken into account. The term containing the same oscillatory factor is the third-order one $\propto a_2|a_2|^2$. We thus have

$$\mathrm{d}a_2/\mathrm{d}\tau = a_2 \log \mu - \beta_2 a_2 |a_2|^2, \tag{30.5}$$

where β_2, like μ, is a complex parameter depending on R, with re $\beta_2 > 0$; compare the corresponding discussion relating to (26.7). The real part of this equation gives immediately the steady value of the modulus:

$$|a_2^{(0)}|^2 = (|\mu| - 1)/\mathrm{re}\, \beta_2.$$

The imaginary part gives an equation for the phase $\phi_2(\tau)$; with the above steady value of the modulus, it is

$$\mathrm{d}\phi_2/\mathrm{d}\tau = 2\pi\omega_2/\omega_1 + |a_2^{(0)}|^2 \,\mathrm{im}\, \beta_2. \tag{30.6}$$

According to this, ϕ_2 rotates at a constant rate, a property which is, however, valid only in the approximation considered: as $R - R_{cr}$ increases, the rotation is no longer uniform, and the rate of rotation on the torus is itself a function of ϕ_2. To take account of this, we add on the right-hand side of (30.6) a small perturbation $\Phi(\phi_2)$; since all the physically different values of ϕ_2 lie in the range from 0 to 2π, $\Phi(\phi_2)$ is periodic with period 2π. Next, we approximate the irrational ratio ω_2/ω_1 by a rational fraction (which can be done with any desired degree of accuracy): $\omega_2/\omega_1 = m_2/m_1 + \Delta/2\pi$, where m_1 and m_2 are integers. The equation then becomes

$$\mathrm{d}\phi_2/\mathrm{d}\tau = 2\pi m_2/m_1 + \Delta + |a_2^{(0)}|^2 \,\mathrm{im}\, \beta_2 + \Phi(\phi_2). \tag{30.7}$$

We shall now consider phase values only at times that are a multiple of $m_1 T_1$, i.e. for values of $\tau = m_1 \bar{\tau}$, where $\bar{\tau}$ is an integer. The first term on the right of (30.7) causes in a time $m_1 T_1$ a change in phase by $2\pi m_2$, that is, by an integral multiple of 2π, which can simply be omitted. The whole right-hand side is then a small quantity, so that the change in the function $\phi_2(\bar{\tau})$ can be described by a differential equation in the continuous variable $\bar{\tau}$:

$$\frac{1}{m_1}\frac{d\phi_2}{d\bar{\tau}} = \Delta + |a_2^{(0)}|^2 \operatorname{im}\beta + \Phi(\phi_2); \tag{30.8}$$

in one step of the discrete variable $\bar{\tau}$, ϕ_2/m_1 changes only slightly.

In the general case, (30.8) has steady solutions $\phi_2 = \phi_2^{(0)}$ for which the right-hand side of the equation is zero. The fact that ϕ_2 is constant for times that are multiples of $m_1 T_1$ means that there is a limit cycle on the torus: the path is closed after m_1 turns. Since $\Phi(\phi_2)$ is periodic, such solutions occur in pairs (one pair in the simplest case): one on the ascending and one on the descending part of $\Phi(\phi_2)$. Of these two, only the latter is stable, for which (30.8) has near $\phi_2 = \phi_2^{(0)}$ the form

$$d\phi_2/d\bar{\tau} = -\text{constant} \times (\phi_2 - \phi_2^{(0)})$$

with the constant positive, and there is in fact a solution tending to $\phi_2 = \phi_2^{(0)}$; the second solution is unstable, and the constant is negative.

The formation of a stable limit cycle on the torus is equivalent to *frequency locking* – the disappearance of the quasi-periodic flow and the establishment of a new periodic one. This phenomenon, which in a system with many degrees of freedom can occur in many ways, prevents the occurrence of a flow that is a superposition of flows having a large number of incommensurate frequencies. In this sense, we can say that the probability of the actual occurrence of the Landau–Hopf scenario is very small; this, of course, does not mean that in particular cases several incommensurate frequencies may not appear before locking occurs.

§31. Strange attractors

There is as yet no complete theory of the origin of turbulence in various types of hydrodynamic flow. Various scenarios have, however, been proposed for the process whereby the flow becomes disordered, based mainly on computer studies of model systems of differential equations, partly supported by experiments. The purpose of the discussion in §§31 and 32 will be merely to give some account of these ideas, without going into the relevant results of such studies. It should only be noted that the experimental results relate to hydrodynamic flows in restricted volumes, and these are the flows to be considered in what follows.†

First of all, the following important general remark is to be made. In the analysis of the stability of periodic flow, only those multipliers are of interest whose moduli are close to 1 and which can cross the unit circle when R changes slightly. In viscous flow, the number of these "dangerous" multipliers is always finite, for the following reason. The various types (modes) of perturbation allowed by the equations of motion have different spatial scales, i.e. distances over which $\mathbf{v}_2$ varies significantly. As the scale of the motion decreases, the velocity gradients in it increase and it is retarded to a greater extent by the viscosity. If the allowed modes are arranged in order of decreasing scale, only a finite number at the

† We shall in fact be concerned with thermal convection in restricted volumes, and with Couette flow between coaxial cylinders with finite length. The theoretical ideas on the mechanism of turbulence formation in the boundary layer and in the wake in flow past finite bodies have not so far been much developed, despite the existence of a considerable quantity of experimental results.

beginning can be dangerous; those sufficiently far along the sequence are certain to be strongly damped and correspond to multipliers with small modulus. This enables us to suppose that the possible types of instability of periodic viscous flow can be analysed in essentially the same way as for a dissipative discrete mechanical system described by a finite number of variables; hydrodynamically, these may be, for example, the amplitudes of the Fourier components of the velocity field with respect to the coordinates. The space of states correspondingly has a finite number of dimensions.

Mathematically, we have to consider the time variation of a system that is represented by equations having the form

$$\dot{\mathbf{x}}(t) = \mathbf{F}(\mathbf{x}), \tag{31.1}$$

where $\mathbf{x}(t)$ is a vector in the space of n quantities $x^{(1)}, x^{(2)}, \ldots, x^{(n)}$, which describe the system; the function $\mathbf{F}$ depends on a parameter whose variation may alter the nature of the flow.† For a dissipative system, the divergence of $\dot{\mathbf{x}}$ in $\mathbf{x}$-space is negative; this expresses the contraction of the volumes in that space during the motion:‡

$$\operatorname{div} \dot{\mathbf{x}} = \operatorname{div} \mathbf{F} \equiv \partial F^{(i)}/\partial x^{(i)} < 0. \tag{31.2}$$

Let us now return to the possible results of interaction between different periodic flows. Frequency locking simplifies the flow, but the interaction may also eliminate the quasi-periodicity in such a way as to complicate the picture significantly. So far, it has been tacitly assumed that when the periodic flow becomes unstable an additional periodic flow occurs. This is not logically necessary, however. If the velocity fluctuation amplitudes are limited, this means only that there is a limited volume in the space of states which contains the paths corresponding to steady viscous flow, but we cannot say in advance what the pattern of paths in that volume will be. They may tend to a limit cycle or to an open winding on the torus (corresponding to periodic and quasi-periodic flow), or they may behave quite differently, taking a complicated and confused form. This possibility is extremely important for our understanding of the mathematical nature of turbulence formation and the elucidation of its mechanism.

One can get an idea of the complicated and confused form of the paths within the limited volume containing them, by assuming that all the paths in the volume are unstable. They may include not only unstable cycles but also open paths which wind indefinitely through the limited region, without leaving it. The instability signifies that two points very close together in the space of states will move far apart as they continue along their respective paths; points initially close together may also belong to the same path, since the volume is limited and an open path can pass indefinitely close to itself. This complicated and irregular behaviour of the paths is associated with turbulent flow.

This picture has a further feature: the sensitivity of the flow to small changes in the initial conditions. If the flow is stable, a slight uncertainty in specifying these conditions causes only a similar uncertainty in the determination of the final state. If the flow is unstable, the initial uncertainty increases with time and the ultimate state of the system cannot be predicted (N. S. Krylov 1944; M. Born 1952).

---

† In mathematical terms, $\mathbf{F}$ is the *vector field* of the system. If it does not depend explicitly on the time, as in (31.1), the system is said to be *autonomous*.

‡ For a Hamiltonian mechanical system, the divergence is zero by Liouville's theorem; the components of $\mathbf{x}$ are in that case the generalized coordinates q and momenta p of the system.

An attracting set of unstable paths in the space of states of a dissipative system can in fact exist (E. N. Lorenz 1963), and it is usually called a *stochastic attractor* or *strange attractor*.†

At first sight, the requirement that all paths belonging to the attractor be unstable appears incompatible with the requirement that all adjacent paths tend to it as $t \to \infty$, since the instability implies that the paths move apart. The apparent contradiction is eliminated if we note that the paths can be unstable in some directions in the space of states and stable (that is, attractive) in other directions. In an n-dimensional space of states, the paths belonging to a strange attractor cannot be unstable in all $n-1$ directions (one direction being along the path), since this would mean a continuous increase in the initial volume in the space of states, which is not possible for a dissipative system. Consequently, adjacent paths tend towards the attractor paths in some directions and away from them in other (unstable) directions; see Fig. 19. These are called *saddle paths*, and it is the set of saddle paths that forms the strange attractor.

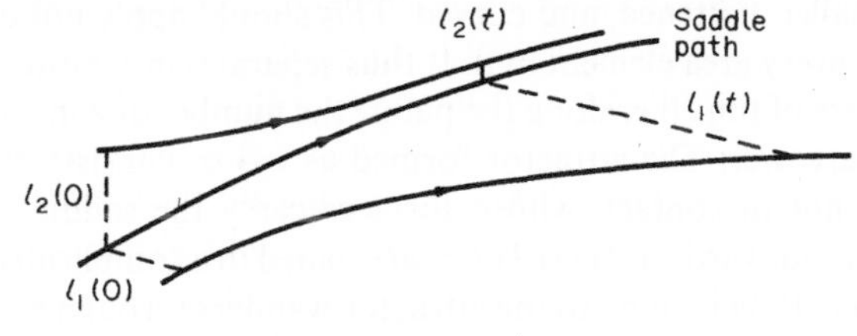

FIG. 19

The strange attractor may appear after only a few bifurcations forming new periods: even an infinitesimal non-linearity can eliminate a quasi-periodic regime (an open winding on the torus) and form a strange attractor on the torus (D. Ruelle and F. Takens 1971). This cannot occur, however, at the second bifurcation (from the end of the steady regime). Here, an open winding on the two-dimensional torus is formed. When the small non-linearity is taken into account, the torus continues to exist, so that the strange attractor could be accommodated on it. But a two-dimensional surface cannot carry an attracting set of unstable paths. The reason is that paths in the space of states cannot intersect one another (or themselves), since this would contradict the causality principle in the behaviour of classical systems, whereby the state of the system at any instant uniquely determines its behaviour at subsequent instants. On a two-dimensional surface, the impossibility of intersections makes the paths so orderly that they cannot become sufficiently random.

Even at the third bifurcation, however, a strange attractor can (but need not) be formed. This attractor, which replaces the three-frequency quasi-periodic regime, lies on a three-dimensional torus (S. Newhouse, D. Ruelle and F. Takens 1978).

The complicated and confused paths in a strange attractors lie in a limited volume in the space of states. There is not yet a known classification of the possible types of strange attractor that can occur in actual problems of fluid dynamics, nor even a set of criteria on

† In contrast to ordinary attractors (stable limit cycles, limit points, and so on); the word "strange" reflects the complexity of its structure, to be discussed later. In the physics literature, "strange attractor" also denotes more complicated attracting manifolds containing stable as well as unstable paths, but having such small domains of attraction as to be undetectable in either physical or numerical experiments.

which such a classification should be based. The available information as to the structure of strange attractors is derived essentially only from a study of instances arising in the computer solution of model systems of ordinary differential equations, which are quite different from the actual equations of fluid dynamics. It is, however, possible to draw some general conclusions about the structure of strange attractors from the saddle-type instability of the paths and the dissipative property of the system.

For clarity, we will refer to a three-dimensional space of states and imagine the attractor inside a two-dimensional torus. Let us consider a set of paths on the way to the attractor, which describe transient flow regimes leading to the establishment of "steady" turbulence. In a transverse cross-section the paths, or rather their traces, occupy a certain area; let us see how this area varies in size and shape along the paths. We note that the volume element near a saddle path expands in one transverse direction and contracts in the other; since the system is dissipative, the latter effect is the stronger, and volumes must decrease. These directions must vary along the paths, since otherwise the latter would get too far away and there would be too great a change in the fluid velocity. The net result is that the cross-section becomes smaller, flattened, and curved. This should apply not only to the whole cross-section but to every area element in it. It thus separates into nested zones separated by voids. In the course of time (i.e. along the paths) the number of zones rapidly increases, and they become narrower. The attractor formed as $t \to \infty$ consists of an uncountable manifold of layers not in contact, whose surfaces carry the saddle paths (with their attracting directions "outwards"). These layers are joined in a complicated manner at their sides and ends; each path belonging to the attractor wanders through all the layers and in the course of a sufficiently long time passes indefinitely close to any point of the attractor – the *ergodic* property. The total volume of the layers and their total cross-sectional area are zero.

In mathematical language, such manifolds in one direction are Cantor sets. The Cantorian structure is the most characteristic property of the attractor and more generally of an n-dimensional ($n > 3$) space of states.

The volume of the strange attractor in its space of states is always zero. It may, however, be non-zero in another space with fewer dimensions. The latter is found as follows. We divide the whole of n-dimensional space into small cubes with edge ε and volume ε^n. Let $N(\varepsilon)$ be the least number of cubes which completely cover the attractor. We define the attractor dimension D as the limit†

$$D = \lim_{\varepsilon \to 0} \frac{\log N(\varepsilon)}{\log (1/\varepsilon)}. \tag{31.3}$$

The existence of this limit signifies that the volume of the attractor in D-dimensional space is finite: when ε is small, $N(\varepsilon) \cong V\varepsilon^{-D}$ (where V is a constant), and $N(\varepsilon)$ may therefore be regarded as the number of D-dimensional cubes covering the volume V in D-dimensional space. When defined in accordance with (31.3), the dimension evidently cannot exceed the total dimension n of the space of states, but may be less, and unlike the ordinary dimension it may be non-integral, as happens for Cantor sets.‡

† This is known in mathematics as the limiting capacity of the manifold. Its definition is similar to that of Hausdorff or fractal dimensions.

‡ The n-dimensional cubes covering the set may be "almost empty", and for this reason we can have $D < n$. For ordinary sets, the definition (31.3) gives obvious results. For example, with a set of N isolated points, $N(\varepsilon) = N$ and $D = 0$; for a line segment with length L, $N(\varepsilon) = L/\varepsilon$ and $D = 1$; for a two-dimensional surface area A, $N(\varepsilon) = S/\varepsilon^2$ and $D = 2$; and so on.

The following point is important. If turbulent flow is already established (the strange attractor has been reached), then the flow in a dissipative system (a viscous fluid) is the same in principle as stochastic flow of a non-dissipative system with a space of states having fewer dimensions. This is because, for steady flow, the viscous dissipation of energy is compensated on the average over a long time by the energy coming from the average flow (or from some other source of disequilibrium). Consequently, if we trace the development in time of a "volume" element belonging to the attractor (in some space whose dimension is determined by that of the attractor), it will be conserved on average, the compression in some directions being compensated by the extension due to the divergence of adjacent paths in other directions. This property can be used to obtain a different estimate of the attractor dimension.

Because the motion on the strange attractor is ergodic, as mentioned above, its average properties can be established by analysing the motion along one unstable path belonging to the attractor in the space of states. That is, we assume that an individual path reproduces the properties of the attractor if the motion along it lasts for a sufficient time.

Let $\mathbf{x} = \mathbf{x}_0(t)$ be the equation of such a path, a solution of (31.1). Let us consider the deformation of a "spherical" volume element as it moves along this path. The deformation is given by the equations (31.1) linearized with respect to the difference $\xi = \mathbf{x} - \mathbf{x}_0(t)$, i.e. the deviation of paths adjacent to the one considered. These equations, written in components, are

$$\dot{\xi}^{(i)} = A_{ik}(t)\,\xi^{(k)}, \qquad A_{ik}(t) = [\partial F^{(i)}/\partial x^{(k)}]_{x = x_0(t)}. \tag{31.4}$$

In the movement along the path, the volume element is compressed in some directions and stretched in others, the sphere becoming an ellipsoid. Both the directions and the lengths of the semi-axes vary; let the latter be $l_s(t)$, where s labels the directions. The *Lyapunov characteristic indices* are

$$L_s = \lim_{t \to \infty} \frac{1}{t} \log \frac{l_s(t)}{l(0)}, \tag{31.5}$$

where $l(0)$ is the radius of the original sphere, at a time arbitrarily chosen as $t = 0$. The quantities thus determined are real, and equal in number to the dimension n of the space. One of them (corresponding to the direction along the path) is zero.†

The sum of the Lyapunov indices gives the mean change, along the path, in the volume element in the space of states. The relative local change in volume at any point on the path is given by the divergence $\operatorname{div} \mathbf{x} = \operatorname{div} \xi = A_{ii}(t)$. It can be shown that the mean divergence along the path is‡

$$\lim_{t \to \infty} \frac{1}{t} \int_0^t \operatorname{div} \xi \, dt = \sum_{s=1}^{n} L_s. \tag{31.6}$$

For a dissipative system, this sum is negative; volumes in an n-dimensional space of states are compressed. The dimension of the strange attractor is defined so that volumes are

† Of course, the solution of (31.4), with specified initial conditions at $t = 0$, actually represents an adjacent path only if all the distances $l_s(t)$ remain small. This, however, does not make meaningless the definition (31.5), which involves indefinitely long times: for any large t, we can choose $l(0)$ so small that the linearized equations remain valid throughout the time concerned.

‡ See V. I. Oseledets, *Transactions of the Moscow Mathematical Society* **19**, 197, 1969.

conserved on average in "its" space. To do so, we arrange the Lyapunov indices in the order $L_1 \geqslant L_2 \geqslant \ldots \geqslant L_n$, and take account of as many stable directions as is necessary to compensate the stretching, by means of compression. The attractor dimension D_L thus defined is between m and $m+1$, where m is the number of indices, in the sequence, whose sum is still positive but becomes negative when L_{m+1} is included.† The fractional part of $D_L = m+d$ $(d < 1)$ is found from

$$\sum_{s=1}^{m} L_s + L_{m+1} d = 0 \qquad (31.7)$$

(F. Ledrappier 1981). Since, in calculating d, we take into account only the least stable directions (omitting the negative L_s that are largest in modulus, at the end of the sequence), the estimate D_L of the dimension is in general too high. This estimate offers in principle a way of determining the dimension of the attractor from measurements of the time dependence of the velocity fluctuations in the turbulent flow.

§32. Transition to turbulence by period doubling

Let us now consider the loss of stability of a periodic flow when the multiplier passes through -1 or $+1$.

In an n-dimensional space of states, $n-1$ multipliers determine the behaviour of the paths in $n-1$ different directions near the periodic path considered (which are not the same as the direction of the tangent at each point of that path). Let a multiplier near ± 1 correspond to the lth direction, say. The other $n-2$ multipliers are small in modulus, and therefore all the paths in the corresponding $n-2$ directions will in the course of time come close to a two-dimensional surface Σ containing the lth direction and the direction of the tangents. One can say that near the limit cycle the space of states is almost two-dimensional as $t \to \infty$ (it cannot be strictly two-dimensional, since the paths can lie on either side of Σ and go from one side to the other). Let the flux of paths near Σ be cut by a surface σ. Each path, on repeatedly passing through σ, determines in accordance with the initial point of intersection $\mathbf{x}_j$ the next point of intersection $\mathbf{x}_{j+1}$. The relation $\mathbf{x}_{j+1} = f(\mathbf{x}_j; \mathrm{R})$ is called a *Poincaré mapping* or *sequence mapping*; it depends on R, in this case the Reynolds number,‡ whose value determines the closeness to the bifurcation where the periodic flow ceases to be stable. Since all paths are close to Σ, the set of points where they meet σ is almost one-dimensional and can be approximated by a line; the Poincaré mapping becomes the one-dimensional transformation

$$x_{j+1} = f(x_j; \mathrm{R}), \qquad (32.1)$$

with x simply a coordinate along the line.†† The discrete variable j acts as the time measured in units of the period.

The mapping (32.1) affords an alternative method of determining the nature of the flow near the bifurcation. The periodic flow itself corresponds to a *fixed point* of the transformation (32.1) – the value $x_j = x_*$ which is unchanged by the mapping, i.e. for which $x_{j+1} = x_j$. The multiplier is the derivative $\mu = \mathrm{d}x_{j+1}/\mathrm{d}x_j$ taken at $x_j = x_*$. The points $x_j = x_* + \xi$ near x_* are mapped into $x_{j+1} \cong x_* + \mu\xi$. The fixed point is stable (and

† Including the zero Lyapunov index adds one to D_L, corresponding to the dimension along the path.
‡ Or the Rayleigh number in the case of thermal convection (§56).
†† In this section x has of course nothing to do with the coordinate in physical space.

is an attractor of the mapping) if $|\mu| < 1$: by iterating the mapping and starting from some point near x_*, we asymptotically approach the latter, as $|\mu|^r$, where r is the number of iterations. If $|\mu| > 1$, however, the fixed point is unstable.

Let us consider the loss of stability of periodic flow when the multiplier passes through -1. The equation $\mu = -1$ signifies that the initial perturbation changes sign after a time T_0, remaining the same in magnitude: after a further time T_0 it returns to its original value. Thus a passage of μ through -1 near a limit cycle with period T_0 creates a new limit cycle with period $2T_0$ (a *period-doubling bifurcation*).† Figure 20 gives a conventional representation of two successive such bifurcations; the continuous curves in diagrams *a* and *b* show the stable limit cycles $2T_0$ and $4T_0$, the broken curves the limit cycles that have become unstable.

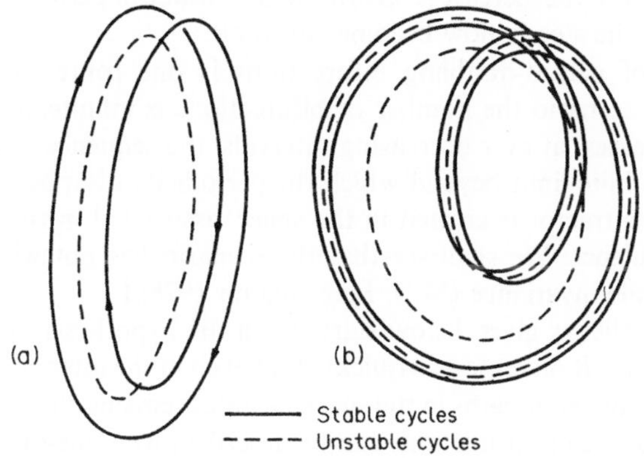

FIG. 20

If we arbitrarily take the fixed point of the Poincare mapping as $x = 0$, the mapping near it which describes the period-doubling bifurcation may be expressed as the expansion

$$x_{j+1} = -[1+(\mathrm{R}-\mathrm{R}_1)]x_j + x_j^2 + \beta x_j^3, \tag{32.2}$$

where $\beta > 0$.‡ For $\mathrm{R} < \mathrm{R}_1$ the fixed point $x_* = 0$ is stable; for $\mathrm{R} > \mathrm{R}_1$ it is unstable. In order to see how the period-doubling occurs, we have to iterate the mapping (32.2) twice, i.e. consider it after two steps (two time units) and determine the fixed points of the re-formed mapping; if these exist and are stable, they correspond to the period-doubling cycle.

The twofold iteration of the transformation (32.2) gives (with the necessary accuracy in respect of the small quantities x_j and $\mathrm{R}-\mathrm{R}_1$) the mapping

$$x_{j+2} = x_j + 2(\mathrm{R}-\mathrm{R}_1)x_j - 2(1+\beta)x_j^3. \tag{32.3}$$

This always has the fixed point $x_* = 0$. When $\mathrm{R} < \mathrm{R}_1$, that point is the only one and is stable, with the multiplier $|dx_{j+2}/dx_j| < 1$; for flow with period 1 (in units of T_0) the time

† In this section the basic period (that of the first periodic flow) is denoted by T_0, not by T_1. The critical Reynolds numbers corresponding to successive period-doubling bifurcations will be denoted here by $\mathrm{R}_1, \mathrm{R}_2, \ldots,$ without the suffix cr (R_1 replaces the previous $\mathrm{R}_{\mathrm{cr},2}$).

‡ The coefficient of $\mathrm{R}-\mathrm{R}_1$ can be made equal to unity by appropriately redefining R, and that of x_j^2 can be made $+1$ by redefining x_j; we assume in (32.2) that this has been done.

interval 2 is also a period. When $R = R_1$, the multiplier is $+1$, and when $R > R_1$ the point $x_* = 0$ becomes unstable. At that stage, a pair of stable fixed points are formed,

$$x_*^{(1),(2)} = \pm\sqrt{\left[\frac{R - R_1}{1+\beta}\right]}, \tag{32.4}$$

corresponding to a stable limit cycle of the double period†; the transformation (32.3) leaves each of these points in position, while (32.2) changes each into the other. It must be emphasized that the single-period cycle does not disappear at this bifurcation, but remains a solution (unstable) of the equations of motion.

Near the bifurcation, the motion is still "almost periodic" with period unity: the points $x_*^{(1)}$ and $x_*^{(2)}$ at which the paths return are close together. The interval $x_*^{(1)} - x_*^{(2)}$ between them is a measure of the amplitude of the oscillations with period 2; it increases as $\sqrt{(R - R_1)}$, similarly to the increase (26.10) in the amplitude of periodic flow after it begins at the point where the steady flow becomes unstable.

The repetition of period-doubling bifurcations is one route to the formation of turbulence. In this scenario the number of bifurcations is infinite, and they follow one another (as R increases) at ever decreasing intervals; the sequence of critical values R_1, R_2, . . . tends to a finite limit beyond which the periodicity disappears altogether and a complex aperiodic attractor is created in the space, associated in this scenario with the formation of turbulence. We shall see that the scenario has noteworthy properties of universality and scale invariance (M. J. Feigenbaum 1978).‡

The quantitative theory given below starts from the hypothesis that the bifurcations follow one another (as R increases) so quickly that even in the intervals between them the region occupied by the set of paths in the space of states remains almost two-dimensional, and the whole sequence of bifurcations can be described by a one-dimensional Poincaré mapping dependent on a single parameter.

The choice of mapping used below can be justified as follows. In a considerable part of the range of variation of x, the mapping must be a stretching one with $|df(x;\lambda)/dx| > 1$; this allows instabilities to occur. The mapping must also bring back to a given range the paths that have left it, since otherwise the velocity fluctuations would increase without limit, which is impossible. The two requirements can be simultaneously satisfied only by non-monotonic functions $f(x;\lambda)$, that is, mappings (32.1) that are not one-to-one: the x_{j+1} values are uniquely determined by the preceding x_j, but not conversely. The simplest form of such a function has a single maximum, near which we put

$$x_{j+1} = f(x_j;\lambda) = 1 - \lambda x_j^2, \tag{32.5}$$

with λ a positive parameter which is to be regarded (in terms of fluid mechanics) as an increasing function of R.†† We shall arbitrarily take the segment $[-1, +1]$ as the range of

† To be called for brevity a *2-cycle*. The relevant fixed points will be called *cycle elements*.

‡ The sequence of period-doubling bifurcations (numbered below as 1, 2, . . .) need not begin with the first bifurcation of the periodic flow. It may in principle begin after the first few bifurcations with the appearance of incommensurate frequencies, when these have been locked by the mechanism discussed in §30.

†† The admissibility of mappings that are not one-to-one depends on the approximateness of the one-dimensional treatment. If all the paths were exactly on one surface Σ, so that the Poincaré mapping would be strictly one-dimensional, this non-uniqueness would be impossible, since it would imply that two paths with different x_j intersected at x_{j+1}. In the same sense, the approximateness is responsible for the possibility of a zero multiplier if the fixed point of the mapping is at an extremum of the mapping function; such a point may be described as "superstable", and is approached more rapidly than according to the above relationship.

variation of x; when λ is between 0 and 2, all iterations of the mapping (32.5) leave x in that range.

The transformation (32.5) has a fixed point at the root of $x_* = 1 - \lambda x_*^2$. This becomes unstable when $\lambda > \Lambda_1$, where Λ_1 is the value of λ for which the multiplier $\mu = -2\lambda x_* = -1$; from the two equations, we find $\Lambda_1 = 3/4$. This is the first critical value of λ, which determines the position of the first period-doubling bifurcation and the appearance of the 2-cycle. Let us now trace the appearance of subsequent bifurcations by means of an approximate technique of determining some qualitative features of the process, though this does not give exact values of the characteristic constants; exact statements will then be formulated.

Repetition of the transformation (32.5) gives

$$x_{j+2} = 1 - \lambda + 2\lambda^2 x_j^2 - \lambda^3 x_j^4. \tag{32.6}$$

Here we will neglect the term in x_j^4. The remaining equation is converted by the scale transformation†

$$x_j \to x_j/\alpha_0, \qquad \alpha_0 = 1/(1-\lambda)$$

to the form

$$x_{j+2} = 1 - \lambda_1 x_j^2,$$

which differs from (32.5) only in that λ is replaced by

$$\lambda_1 = \phi(\lambda) \equiv 2\lambda^2 (\lambda - 1). \tag{32.7}$$

Repeating this operation with the scale factors $\alpha_1 = 1/(1-\lambda_1)$, etc., gives a sequence of mappings having the same form:

$$x_{j+2^m} = 1 - \lambda_m x_j^2, \qquad \lambda_m = \phi(\lambda_{m-1}). \tag{32.8}$$

The fixed points of the mappings (32.8) correspond to 2^m-cycles.‡ Since they all have the same form as (32.5), we can deduce at once that the 2^m-cycles ($m = 1, 2, 3, \ldots$) become unstable when $\lambda_m = \Lambda_1 = 3/4$. The corresponding critical values Λ_m of the initial parameter λ are found by solving the coupled equations

$$\Lambda_1 = \phi(\Lambda_2), \quad \Lambda_2 = \phi(\Lambda_3), \quad \ldots, \quad \Lambda_{m-1} = \phi(\Lambda_m);$$

they are obtained graphically by the construction shown in Fig. 21. Evidently, as $m \to \infty$ the sequence of numbers converges to a finite limit Λ_∞, the root of $\Lambda_\infty = \phi(\Lambda_\infty)$; this is $\Lambda_\infty = (1 + \sqrt{3})/2 = 1{\cdot}37$. The scale factors also tend to a finite limit: $\alpha_m \to \alpha$, where $\alpha = 1/(1 - \Lambda_\infty) = -2{\cdot}8$.

It is easy to find how Λ_m approaches Λ_∞ when m is large. From the equation $\Lambda_m = \phi(\Lambda_{m+1})$ when $\Lambda_\infty - \Lambda_m$ is small, we find

$$\Lambda_\infty - \Lambda_{m+1} = (\Lambda_\infty - \Lambda_m)/\Delta, \tag{32.9}$$

† This is not possible when $\lambda = 1$ (and the fixed point of the mapping (32.6) coincides with the central extremum: $x_* = 0$). The value $\lambda = 1$ is, however, certainly not the next critical value λ_2 that is needed here.

‡ To avoid misunderstanding, it should be emphasized that after the scale transformations the mappings (32.8) must be defined over extended ranges $|x| \leqslant |\alpha_0 \alpha_1 \ldots \alpha_{m-1}|$, not $|x| \leqslant 1$ as in (32.5) and (32.6). However, in view of the terms neglected, the expressions (32.8) can in practice give a description only of the range near the central extrema of the mapping functions.

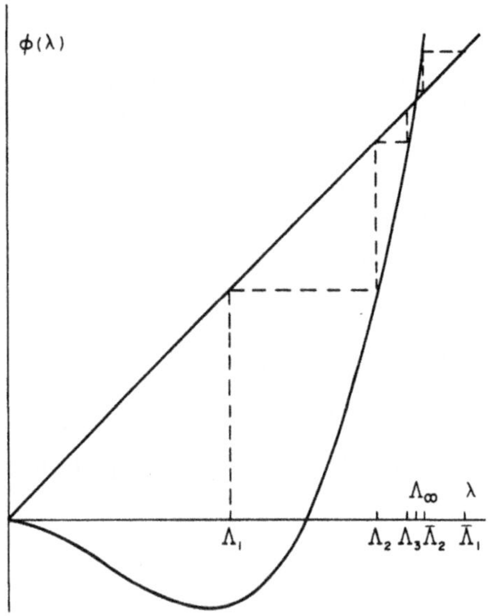

Fig. 21

where $\delta = \phi'(\Lambda_\infty) = 4 + \sqrt{3} = 5{\cdot}73$. Thus $\Lambda_\infty - \Lambda_m \propto \delta^m$, that is, Λ_m approaches the limit in geometrical progression. The same relation applies to the intervals between successive critical numbers: equation (32.9) can be written in the equivalent form

$$\Lambda_{m+2} - \Lambda_{m+1} = (\Lambda_{m+1} - \Lambda_m)/\delta. \tag{32.10}$$

As regards fluid dynamics, it has already been mentioned that λ is to be regarded as a function of the Reynolds number, and accordingly the latter has critical values which correspond to successive period-doubling bifurcations and tend to a finite limit R_∞. It is evident that for these values we have the same limiting relations (32.9), (32.10), with the same constant δ, as for Λ_m.

The above arguments illustrate the origin of the basic features of the process, namely the infinity of bifurcations, whose times of appearance converge to the limit Λ_∞ according to (32.9) and (32.10), and the existence of the scale factor α. The values thus found for the characteristic constants are not exact, however. The exact values (found by repeated computer iterations of the mapping (32.5)) of the convergence factor (*Feigenbaum number*) δ and the scale factor α are

$$\delta = 4{\cdot}6692\ldots, \quad \alpha = -2{\cdot}5029\ldots, \tag{32.11}$$

and the limiting value $\Lambda_\infty = 1{\cdot}401$.† The value of δ is comparatively large; the rapid convergence has the result that the limiting relations are very nearly satisfied after only a small number of period doublings.

A deficiency of the above derivation is that, when all powers of x_j^2 above the first are neglected, the mapping (32.8) yields only the fact that the next bifurcation occurs; it does

---

† The value of Λ_∞ is somewhat arbitrary, since it depends on how the parameter is used in the initial mapping, i.e. the function $f(x; \lambda)$ (the values of δ and α, however, do not depend on this).

not allow all the elements of the 2^m-cycle described by this mapping to be determined.† In reality, the iterated mappings (32.5) are polynomials in x_j^2 whose degree is doubled at each iteration. They are complicated functions of x_j with a rapidly increasing number of extrema lying symmetrically about $x_j = 0$ (which also always remains an extremum).

It is noteworthy that not only the values of δ and α but also the limiting form of the infinitely iterated mapping are in a certain sense independent of the form of the initial mapping $x_{j+1} = f(x_j; \lambda)$; it is sufficient that the function $f(x; \lambda)$ of one parameter be smooth with a single quadratic maximum (let this be at $x = 0$) – it need not even be symmetrical about the maximum at great distances from it. This *universality* increases considerably the degree of generality of the theory described. The exact formulation of the property is as follows.

Let us consider the mapping specified by $f(x)$, i.e. $f(x; \lambda)$ with a particular choice of λ (see below), normalized by the condition $f(0) = 1$. By applying this twice, we get the function $f[f(x)]$. We change the scale of this function and of x by a factor $\alpha_0 = 1/f(1)$, obtaining a new function

$$f_1(x) = \alpha_0 f[f(x/\alpha_0)],$$

for which again $f_1(0) = 1$. Repeating this operation, we find a sequence of functions connected by the recurrence formula‡

$$f_{m+1}(x) = \alpha_m f_m[f_m(x/\alpha_m)] \equiv \hat{T} f_m, \quad \alpha_m = 1/f_m(1). \tag{32.12}$$

If this sequence tends, as $m \to \infty$, to a definite limiting function $f_\infty(x) \equiv g(x)$, then the latter must be a "fixed function" of the operator $\hat{T}$ defined in (32.12), i.e. must satisfy the functional relation

$$g(x) = \hat{T} g \equiv \alpha g[g(x/\alpha)], \quad \alpha = 1/g(1), \quad g(0) = 1. \tag{32.13}$$

According to the assumed properties of the admissible functions $f(x)$, $g(x)$ must be smooth and have a quadratic extremum at $x = 0$; the specific form of $f(x)$ has no other influence on equation (32.13) or on the conditions imposed on its solution. We should emphasize that, after the scale transformations used in the derivation (with $|\alpha_m| > 1$), the solution of the equation is determined for all values of the variable x in it, from $-\infty$ to $+\infty$, and not only in the range $-1 \leqslant x \leqslant 1$. The function $g(x)$ is necessarily even, since the admissible functions $f(x)$ include even ones, and an even mapping certainly remains even, after any number of iterations.

Such a solution of equation (32.13) does in fact exist and is unique, although it cannot be derived in an analytical form; it is a function having an infinity of extrema and unlimited in magnitude, the constant α being determined along with $g(x)$. In practice, it is sufficient to derive the function in the range $[-1, 1]$, after which it can be continued outside the range by iterating the operation $\hat{T}$. Note that at each stage of iteration of $\hat{T}$ in (32.12) the values of $f_{m+1}(x)$ in the range $[-1, 1]$ are determined by those of $f_m(x)$ in a part of this range

† That is, all the 2^m points $x_*^{(1)}, x_*^{(2)}, \ldots$ which change successively into one another (and are periodic) when the mapping (31.5) is iterated, and are fixed (and stable) with respect to the 2^m-fold iterated mapping. To avoid any doubts, it may be noted that the derivatives dx_{j+2^m}/dx_j are necessarily the same at all points $x_*^{(1)}, x_*^{(2)}, \ldots$ (and therefore pass simultaneously through -1 at the next bifurcation); we shall not give here the proof of this property (which is evidently necessary).

‡ There is an obvious analogy between this procedure and the one used previously in deriving (32.8).

shortened by a factor $|\alpha_m| \cong |\alpha|$. This means that in the limit of many iterations, the determination of $g(x)$ in the range $[-1, 1]$ (and therefore on the whole of the x-axis) is governed by smaller and smaller parts of the initial function near its maximum, and herein lies the ultimate cause of the universality.†

The function $g(x)$ determines the structure of the aperiodic attractor formed by an infinite sequence of period doublings. This occurs at a parameter value $\lambda = \Lambda_\infty$ which is quite definite for a given function $f(x; \lambda)$. It is therefore clear that the functions formed from $f(x; \lambda)$ by repeated iteration of the transformation (32.12) do in fact converge to $g(x)$ only for this isolated value of λ. It follows in turn that the fixed function of the operator $\hat{T}$ is unstable with respect to small changes corresponding to small deviations of λ from the value Λ_∞. The study of this instability enables us to determine the universal constant δ, again independent of the specific form of $f(x)$.‡

The scale factor α determines the change (decrease) in the geometrical characteristics (in the space of states) of the attractor at each stage of period doubling; these characteristics are the distances between limit cycle elements on the x-axis. However, since each doubling is accompanied by a further increase in the number of cycle elements, this statement must be made more specific and precise. It is clear a priori that the scale cannot vary in the same way for the distances between every pair of points.†† For, if two adjacent points are transformed by an almost linear section of the mapping function, the distance between them is reduced by a factor $|\alpha|$; but if the transformation takes place by a section of the mapping function near its extremum, the distance is reduced by a factor α^2.

At the bifurcation ($\lambda = \Lambda_m$) each element (point) of the 2^m-cycle splits into two adjacent points, the distance between which gradually increases, but the points remain close over the whole range of variation of λ as far as the next bifurcation. If we follow the conversions of cycle elements into one another in the course of time, i.e. in successive mappings $x_{j+1} = f(x_j; \lambda)$, each component of the pair changes into the other after 2^m time units. This means that the distance between the points in the pair is a measure of the oscillation amplitude of the newly formed double period, and in this sense has especial physical interest.

Let us arrange all the elements of the 2^{m+1}-cycle in the order in which they are traversed in the course of time, and denote them by $x_{m+1}(t)$, where the time t, measured in units of the basic period T_0, takes integral values: $t/T_0 = 1, 2, \ldots, 2^{m+1}$. These elements are formed from those of the 2^m-cycle by splitting into pairs. The intervals between the points in each pair are

$$\xi_{m+1}(t) = x_{m+1}(t) - x_{m+1}(t + T_m), \tag{32.14}$$

where $T_m = 2^m T_0 = \frac{1}{2} T_{m+1}$ is the period of the 2^m-cycle, or half that of the 2^{m+1}-cycle. We

† The statement that there exists a unique solution of equation (32.13) is founded on computer simulation. The solution is sought, in the range $[-1, 1]$, as a polynomial of high degree in x^2; the accuracy of the simulation must increase with the width of the x value range (outside that mentioned) to which we wish to continue the function by iteration of $\hat{T}$. In the range $[-1, 1]$, $g(x)$ has one extremum, near which $g(x) = 1 - 1{\cdot}528x^2$ if it is taken to be a maximum, a choice which is arbitrary in view of the invariance of equation (32.13) under a change in the sign of g.

‡ See the original papers by M. J. Feigenbaum, *Journal of Statistical Physics* **19**, 25, 1978; **21**, 669, 1979.

†† These are distances in the unstretched range $[-1, 1]$ which is arbitrarily taken, from the start, as the range of x containing all cycle elements. Since α is negative, the bifurcations are accompanied by inversion of the positions of the elements relative to $x = 0$.

use the function $\sigma_m(t)$, a scale factor which determines the change in the intervals (32.14) from one cycle to the next†:

$$\xi_{m+1}(t)/\xi_m(t) = \sigma_m(t). \tag{32.15}$$

Evidently

$$\xi_{m+1}(t+T_m) = -\xi_{m+1}(t), \tag{32.16}$$

and therefore

$$\sigma_m(t+T_m) = -\sigma_m(t). \tag{32.17}$$

The function $\sigma_m(t)$ has complicated properties, but it can be shown that its limiting form for large m is very closely approximated by the simple expressions

$$\left.\begin{aligned} \sigma_m(t) &= 1/\alpha \text{ for } 0 < t < \tfrac{1}{2}T_m, \\ &= 1/\alpha^2 \text{ for } \tfrac{1}{2}T_m < t < T_m, \end{aligned}\right\} \tag{32.18}$$

with the appropriate choice of zero for t.‡

These formulae yield some conclusions as to the change in the flow frequency spectrum when period doubling occurs. In fluid dynamics terms, $x_m(t)$ is to be regarded as a characteristic of the fluid velocity. For a flow with period T, the spectrum of the function $x_m(t)$ of the continuous time t contains frequencies $k\omega_m$ ($k = 1, 2, 3, \ldots$), i.e. the fundamental frequency $\omega_m = 2\pi/T_m$ and its harmonics. After the period doubling, the flow is described by the function $x_{m+1}(t)$ with period $T_{m+1} = 2T_m$. Its spectrum contains not only the same frequencies $k\omega_m$ but also the subharmonics of ω_m, the frequencies $\frac{1}{2}l\omega_m$, $l = 1, 3, 5, \ldots$.

Let us write

$$x_{m+1}(t) = \tfrac{1}{2}\{\xi_{m+1}(t) + \eta_{m+1}(t)\},$$

where ξ_{m+1} is the difference (32.14) and

$$\eta_{m+1}(t) = x_{m+1}(t) + x_{m+1}(t+T_m).$$

The spectrum of $\eta_{m+1}(t)$ contains only the frequencies $k\omega_m$; the Fourier components for the subharmonics,

$$\frac{1}{T_{m+1}}\int_0^{T_{m+1}} \eta_{m+1}(t)e^{i\pi lt/T_m}\,dt = \frac{1}{2T_m}\int_0^{T_m} \{\eta_{m+1}(t) - \eta_{m+1}(t+T_m)\}\, e^{i\pi lt/T_m}\,dt$$

are zero, since $\eta_{m+1}(t+T_m) = \eta_{m+1}(t)$. On the other hand, in the first approximation the quantities $\eta_m(t)$ are unchanged in the bifurcation: $\eta_{m+1}(t) \cong \eta_m(t)$; this means that the strength of the oscillations with frequencies $k\omega_m$ also remains unchanged.

---

† Since the two cycles exist in different ranges of values of λ, $(\Lambda_{m-1}, \Lambda_m)$ and $(\Lambda_m, \Lambda_{m+1})$, and the quantities (32.14) vary considerably in these ranges, their significance in the definition (32.15) needs to be made more precise. We shall take them for the values of λ where the cycles are superstable (see the footnote following (32.5)); one such value occurs in the range where each cycle exists.

‡ We shall not give here the study of the properties of $\sigma_m(t)$, which is simple in principle but laborious; see M. J. Feigenbaum, *Los Alamos Science* 1, 4, 1980.

The spectrum of $\xi_{m+1}(t)$, on the other hand, contains only the subharmonics $\frac{1}{2}l\omega_m$, the new frequencies which appear at doubling stage $m+1$. The total strength of these components is given by the integral

$$I_{m+1} = \frac{1}{T_{m+1}} \int_0^{T_{m+1}} \xi_{m+1}{}^2(t)\,dt. \tag{32.19}$$

Expressing $\xi_{m+1}(t)$ in terms of $\xi_m(t)$, we can write

$$I_{m+1} = \frac{1}{2T_m} \cdot 2 \int_0^{T_m} \sigma_m{}^2(t)\, \xi_m{}^2(t)\,dt.$$

With (32.16)–(32.18),

$$I_{m+1} = \tfrac{1}{2}\left(\frac{1}{\alpha^2} + \frac{1}{\alpha^4}\right)\frac{1}{T_m} \int_0^{T_m} \xi_m{}^2(t)\,dt$$

$$= \tfrac{1}{2}\left(\frac{1}{\alpha^2} + \frac{1}{\alpha^4}\right) I_m,$$

and finally

$$I_m/I_{m+1} = 10{\cdot}8. \tag{32.20}$$

Thus the strength of the new components which appear after a period-doubling bifurcation exceeds the one for the next bifurcation by a definite factor independent of the bifurcation number (M. J. Feigenbaum 1979).†

Let us now consider the development of the flow properties when λ increases further beyond Λ_∞ (the Reynolds number $R > R_\infty$), in the turbulent range. Since, at the moment of its formation (at $\lambda = \Lambda_\infty$), the aperiodic attractor is described by a one-dimensional Poincaré mapping, we can suppose that even for values of λ slightly above Λ_∞ it is permissible to treat the properties of the attractor in terms of such a mapping.

The attractor formed by an infinite sequence of period doublings is at its appearance not a strange attractor as defined in §31: the 2^∞-cycle occurring as the limit of stable 2^m-cycles when $m \to \infty$ is also stable. The points of this attractor form on the interval $[-1, 1]$ an uncountable Cantorian set. Its measure on this interval, i.e. the total "length" of its elements, is zero; its dimension is between 0 and 1, and is found to be 0·54.‡

When $\lambda > \Lambda_\infty$, the attractor becomes a strange attractor, i.e. an attracting set of unstable paths. On the interval $[-1, 1]$, the points belonging to it occupy ranges whose total length is not zero. These ranges are the traces on the sectional plane σ of a continuous two-dimensional band which makes a large number of turns and is closed. In this connection, it should be remembered that the one-dimensional treatment is approximate. In reality, the

† This applies not only to the total strength of the subharmonics but also to each of them. For each subharmonic that appears after bifurcation m there are two (one to the right and one to the left) after bifurcation $m+1$. The ratio of strengths of the individual peaks that appear after two successive bifurcations is therefore twice (32.20). A more exact value of this quantity is 10·48. This is found by analysing the state at the point $\lambda = \Lambda_\infty$ itself by means of the universal function $g(x)$; at this point, all frequencies are already present, and the problem corresponding to that raised in the last footnote but one does not arise. See M. Nauenberg and J. Rudnick, *Physical Review* **B 24**, 493, 1981.

‡ See P. Grassberger, *Journal of Statistical Physics* **26**, 173, 1981.

band has a small but non-zero thickness. The segments forming its cross-section are therefore really strips with non-zero width. Across this width, the strange attractor has the layered Cantorian structure described in §31.† This structure will not be of interest, and we shall return to a discussion in terms of the one-dimensional Poincaré mapping.

The general development of the strange attractor as λ increases beyond Λ_∞ is as follows. For a given $\lambda > \Lambda_\infty$ the attractor occupies a number of ranges in the interval $[-1, 1]$; the spaces between these ranges are the attraction domains, and contain the elements of unstable cycles with periods not exceeding some 2^m. When λ increases, the rate of divergence of the paths on the strange attractor increases, and it "expands", successively absorbing the cycles with periods $2^m, 2^{m+1}, \ldots$; the number of ranges occupied by the attractor decreases, and their lengths increase. Thus the number of turns of the band mentioned above is successively halved, while their widths increase. There is then a sort of reverse cascade of successive simplifications of the attractor. The absorption of an unstable 2^m-cycle by the attractor is called a *reverse doubling bifurcation.* Figure 22 illustrates this process for two successive reverse bifurcations. In Fig. 22a, the band makes four turns and the reverse bifurcation converts it into one with two (Fig. 22b); the final bifurcation gives a band with only one turn and closed after a twist (Fig. 22c).

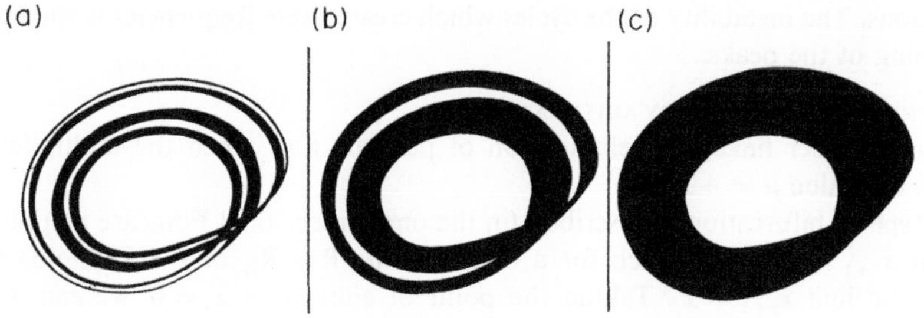

FIG. 22

Let the values of λ corresponding to successive reverse doubling bifurcations be denoted by $\bar{\Lambda}_{m+1}$, arranged in the order $\bar{\Lambda}_m > \bar{\Lambda}_{m+1}$. We shall show that they are in geometrical progression with the same universal factor δ as for forward bifurcations.

Before the final (as λ increases) reverse bifurcation, the attractor occupies two ranges separated by a gap containing the fixed point x_* of the mapping (32.5), which corresponds to an unstable cycle with period 1:

$$x_* = \frac{\sqrt{(1+4\lambda)}-1}{2\lambda}.$$

The bifurcation takes place at the value $\lambda = \bar{\Lambda}_1$, when this point is reached by the limits of the expanding attractor. Figure 22b shows that the outer limit of the attractor band becomes the inner limit after one loop and the boundary of the gap between turns after another. It follows that $\lambda = \bar{\Lambda}_1$ is given by the condition $x_{j+2} = x_*$, where

$$x_{j+2} = 1 - \lambda(1-\lambda)^2$$

† The dimension of the attractor in this direction is much less than unity, but it is not a universal property, and depends on the specific mapping.

is the result of twice iterating the mapping over the point $x_j = 1$, the limit of the attractor; $\bar{\Lambda}_1 = 1{\cdot}543$. The previous reverse bifurcations $\bar{\Lambda}_2, \bar{\Lambda}_3, \ldots$ can be approximately determined in succession by means of the recurrence relation between $\bar{\Lambda}_{m+1}$ and $\bar{\Lambda}_m$. This approximate relation is derived by the same method as was used above to deal with the sequence of forward doubling bifurcations, and has the form $\bar{\Lambda}_m = \phi(\bar{\Lambda}_{m+1})$ with the same function $\phi(\Lambda)$ from (32.7). The corresponding graphical construction is shown in the upper part of Fig. 21. Since $\phi(\Lambda)$ is the same for the forward and reverse bifurcation sequences, so is the expression governing the convergence of the sequences of numbers Λ_m and $\bar{\Lambda}_m$ (from below and above respectively) to their common limit $\Lambda_\infty \equiv \bar{\Lambda}_\infty$:

$$\bar{\Lambda}_{m+1} - \Lambda_\infty = (1/\delta)(\bar{\Lambda}_m - \Lambda_\infty). \qquad (32.21)$$

The development of the strange attractor properties for $\lambda > \Lambda_\infty$ is accompanied by corresponding changes in the frequency spectrum. The randomness of the flow is represented in the spectrum by the presence of a "noise" component whose strength increases with the width of the attractor. Against this background there are discrete peaks corresponding to the fundamental frequency of the unstable cycles and their harmonics and subharmonics; at successive reverse bifurcations, the relevant subharmonics disappear, in the opposite order to that of their appearance in the sequence of forward bifurcations. The instability of the cycles which create these frequencies is shown by the broadening of the peaks.

Transition to turbulence by alternation

Let us consider finally the elimination of periodic flow when the multiplier passes through the value $\mu = +1$.

This type of bifurcation is described (in the one-dimensional Poincaré mapping) by a function $x_{j+1} = f(x_j; \mathrm{R})$, which for a certain value $\mathrm{R} = \mathrm{R}_{cr}$ of the Reynolds number touches the line $x_{j+1} = x_j$. Taking the point of contact as $x_j = 0$, we can write the expansion of the mapping function near it as†

$$x_{j+1} = (\mathrm{R} - \mathrm{R}_{cr}) + x_j + x_j^2. \qquad (32.22)$$

When $\mathrm{R} < \mathrm{R}_{cr}$ (see Fig. 23), there are two fixed points

$$x_*^{(1),(2)} = \mp\sqrt{(\mathrm{R}_{cr} - \mathrm{R})},$$

of which $x_*^{(1)}$ corresponds to stable and $x_*^{(2)}$ to unstable periodic flow. When $\mathrm{R} = \mathrm{R}_{cr}$, the multiplier at both points is $+1$, the two periodic flows coalesce, and when $\mathrm{R} > \mathrm{R}_{cr}$ they disappear, the fixed points passing into the complex domain.

When $\mathrm{R} - \mathrm{R}_{cr}$ is small, the curve (32.22) and the straight line $x_{j+1} = x_j$ are close together (near $x_j = 0$). In this range of x, therefore, each iteration of the mapping (32.22) moves the trace of the path only slightly, and many steps are needed for it to cover the whole range. In other words, over a comparatively long interval of time the path is regular and almost periodic in the space of states. Such a path corresponds to regular laminar flow in physical space. This yields another theoretically possible scenario for the onset of turbulence (P. Manneville and Y. Pomeau 1980).

It can be imagined that the particular region of the mapping function is adjacent to regions which randomize the paths, corresponding in the space of states to a set of locally

† The coefficient of $\mathrm{R} - \mathrm{R}_{cr}$ and the positive coefficient of x_j^2 can be made equal to unity by appropriate definitions of R and x_j, and this is assumed in (32.22).

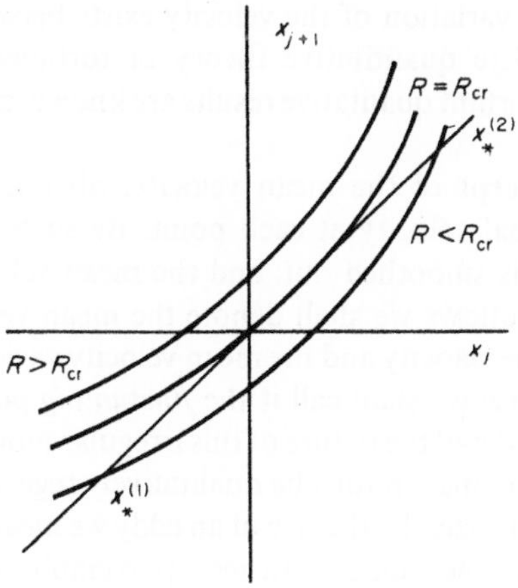

FIG. 23

unstable paths. This set is not itself an attractor, however, and in the course of time the point representing the system moves out of the set. When $R < R_{cr}$, the path reaches a stable cycle, and periodic laminar flow is established in physical space. When $R > R_{cr}$, there is no stable cycle, and a motion arises in which the turbulent periods alternate with laminar ones, the scenario therefore being called the transition to turbulence by alternation.

No general conclusions can be drawn as to the duration of the turbulent periods. The dependence of the laminar period duration on $R - R_{cr}$ is easily ascertained, however. To do so, we write the difference equation (32.22) as a differential equation. Since x_j changes only slightly in one mapping step, we replace $x_{j+1} - x_j$ by the derivative dx/dt with respect to the continuous variable t:

$$dx/dt = R - R_{cr} + x^2. \tag{32.23}$$

Let us find the time τ needed to traverse the segment between the points x_1 and x_2 lying on either side of $x = 0$ at distances much greater than $R - R_{cr}$ but still within the range where the expansion (32.22) is valid. We have

$$\tau = \frac{1}{\sqrt{(R - R_{cr})}}[\tan^{-1}\{x/\sqrt{(R - R_{cr})}\}]_{x_1}^{x_2},$$

whence

$$\tau \propto 1/\sqrt{(R - R_{cr})}; \tag{32.24}$$

this gives the required dependence. Thus the duration of the laminar periods decreases as $R - R_{cr}$ increases.

This scenario leaves unresolved both the way in which its starting-point is approached and the nature of the turbulence that occurs.

§33. Fully developed turbulence

Turbulent flow at fairly large Reynolds numbers is characterized by the presence of an extremely irregular disordered variation of the velocity with time at each point. This is called *fully developed turbulence.* The velocity continually fluctuates about some mean value. A similar irregular variation of the velocity exists between points in the flow at a given instant. No complete quantitative theory of turbulence has yet been evolved. Nevertheless, several important qualitative results are known, and the present section gives an account of these.

We introduce the concept of the mean velocity, obtained by averaging over long intervals of time the actual velocity at each point. By such an averaging the irregular variation of the velocity is smoothed out, and the mean velocity varies smoothly from point to point. In what follows we shall denote the mean velocity by $\mathbf{u}$. The difference $\mathbf{v}' = \mathbf{v} - \mathbf{u}$ between the true velocity and the mean velocity varies irregularly in the manner characteristic of turbulence; we shall call it the *fluctuating part* of the velocity.

Let us consider in more detail the nature of this irregular motion which is superposed on the mean flow. This motion may in turn be qualitatively regarded as the superposition of *turbulent eddies* of different sizes; by the size of an eddy we mean the order of magnitude of the distances over which the velocity varies appreciably. As the Reynolds number increases, large eddies appear first; the smaller the eddies, the later they appear. For very large Reynolds numbers, eddies of every size from the largest to the smallest are present. An important part in any turbulent flow is played by the largest eddies, whose size (the *fundamental* or *external* scale of turbulence) is of the order of the dimensions of the region in which the flow takes place; in what follows we shall denote by l this order of magnitude for any given turbulent flow. These large eddies have the largest amplitudes. The velocity in them is comparable with the variation of the mean velocity over the distance l; we shall denote by Δu the order of magnitude of this variation. We are speaking here of the order of magnitude, not of the mean velocity itself, but of its variation, since it is this variation Δu which characterizes the velocity of the tubulent flow. The mean velocity itself can have any magnitude, depending on the frame of reference used.† The frequencies corresponding to these eddies are of the order of u/l, the ratio of the mean velocity u (and not its variation Δu) to the dimension l. For the frequency determines the period with which the flow pattern is repeated when observed in some fixed frame of reference. Relative to such a frame, however, the whole pattern moves with the fluid at a velocity of the order of u.

The small eddies, on the other hand, which correspond to large frequencies, participate in the turbulent flow with much smaller amplitudes. They may be regarded as a fine detailed structure superposed on the fundamental large turbulent eddies. Only a comparatively small part of the total kinetic energy of the fluid resides in the small eddies.

From the picture of turbulent flow given above, we can draw a conclusion regarding the manner of variation of the fluctuating velocity from point to point at any given instant. Over large distances (comparable with l), the variation of the fluctuating velocity is given by the variation in the velocity of the large eddies, and is therefore comparable with Δu. Over small distances (compared with l), it is determined by the small eddies, and is therefore small (compared with Δu) (but large compared with the variation of the mean velocity over these small distances). The same kind of picture is obtained if we observe the variation of the velocity with time at any given point. Over short time intervals (compared with $T \sim l/u$), the velocity does not vary appreciably; over long intervals, it varies by a quantity of the order of Δu.

The length l appears as a characteristic dimension in the Reynolds number R, which determines the properties of a given flow. Besides this Reynolds number, we can introduce

† It seems that in fact the size of the largest eddies is actually somewhat less than l, and their velocity is somewhat less than Δu.

the qualitative concept of the Reynolds numbers for turbulent eddies of various sizes. If λ is the order of magnitude of the size of a given eddy, and v_λ the order of magnitude of its velocity, then the corresponding Reynolds number is defined as $R_\lambda \sim v_\lambda \lambda/\nu$. This number decreases with the size of the eddy.

For large Reynolds numbers R, the Reynolds numbers R_λ of the large eddies are also large. Large Reynolds numbers, however, are equivalent to small viscosities. We therefore conclude that, for the large eddies which are the basis of any turbulent flow, the viscosity is unimportant. It follows from this that there is no appreciable dissipation of energy in the large eddies.

The viscosity of the fluid becomes important only for the smallest eddies, whose Reynolds number is comparable with unity. We denote the size of these eddies by λ_0, which we shall determine later in this section. It is in these small eddies, which are unimportant as regards the general pattern of a turbulent flow, that the dissipation of energy occurs.

We thus arrive at the following conception of energy dissipation in turbulent flow (L. Richardson 1922). The energy passes from the large eddies to smaller ones, practically no dissipation occurring in this process. We may say that there is a continuous flow of energy from large to small eddies, i.e. from small to large frequencies. This flow of energy is dissipated in the smallest eddies, where the kinetic energy is transformed into heat. For a steady state to be maintained, it is of course necessary that external energy sources should be present which continually supply energy to the large eddies.

Since the viscosity of the fluid is important only for the smallest eddies, we may say that none of the quantities pertaining to eddies of sizes $\lambda \gg \lambda_0$ can depend on ν (more exactly, these quantities cannot be changed if ν varies but the other conditions of the motion are unchanged). This circumstance reduces the number of quantities which determine the properties of turbulent flow, and the result is that similarity arguments, involving the dimensions of the available quantities, become very important in the investigation of turbulence.

Let us apply these arguments to determine the order of magnitude of the energy dissipation in turbulent flow. Let ε be the mean dissipation of energy per unit time per unit mass of fluid.† We have seen that this energy is derived from the large eddies, whence it is gradually transferred to smaller eddies until it is dissipated in eddies of size $\sim \lambda_0$. Hence, although the dissipation is ultimately due to the viscosity, the order of magnitude of ε can be determined only by those quantities which characterize the large eddies. These are the fluid density ρ, the dimension l and the velocity Δu. From these three quantities we can form only one having the dimensions of ε, namely erg/g sec = cm^2/sec^3. Thus we find

$$\varepsilon \sim (\Delta u)^3/l, \tag{33.1}$$

and this determines the order of magnitude of the energy dissipation in turbulent flow.

In some respects a fluid in turbulent motion may be qualitatively described as having a "turbulent viscosity" ν_{turb} which differs from the true kinematic viscosity ν. Since ν_{turb} characterizes the properties of the turbulent flow, its order of magnitude must be determined by ρ, Δu and l. The only quantity that can be formed from these and has the dimensions of kinematic viscosity is $l\Delta u$, and therefore

$$\nu_{turb} \sim l\Delta u. \tag{33.2}$$

† In this chapter ε denotes the mean dissipation of energy, and not the internal energy of the fluid.

The ratio of the turbulent viscosity to the ordinary viscosity is consequently

$$\nu_{\text{turb}}/\nu \sim R \tag{33.3}$$

i.e. it increases with the Reynolds number.†

The energy dissipation ε is expressed in terms of ν_{turb} by

$$\varepsilon \sim \nu_{\text{turb}}(\Delta u/l)^2 \tag{33.4}$$

in accordance with the usual definition of viscosity. Whereas ν determines the energy dissipation in terms of the space derivatives of the true velocity, ν_{turb} relates it to the gradient ($\sim \Delta u/l$) of the mean velocity.

We may also apply similarity arguments to determine the order of magnitude Δp of the variation of pressure over the region of turbulent flow. The only quantity having the dimensions of pressure which can be formed from ρ, l and Δu is $\rho(\Delta u)^2$. Hence we must have

$$\Delta p \sim \rho(\Delta u)^2. \tag{33.5}$$

Let us now consider the properties of the turbulence as regards eddy sizes λ which are small compared with the fundamental eddy size l. We shall refer to these properties as *local* properties of the turbulence. We shall consider fluid that is far from all solid surfaces (more precisely, that is at distances from them large compared with λ).

It is natural to assume that such small-scale turbulence, far from solid bodies, is homogeneous and isotropic. The latter property means that, over regions whose dimensions are small compared with l, the properties of the turbulent flow are independent of direction; in particular, they do not depend on the direction of the mean velocity. It must be emphasized that here, and everywhere in the present section, when we speak of the properties of the turbulent flow in a small region of the fluid, we mean the relative motion of the fluid particles in that region, and not the absolute motion of the region as a whole, which is due to the larger eddies.

It is found that several important results concerning the local properties of turbulence can be obtained immediately from similarity arguments. These results are due to A. N. Kolmogorov and to A. M. Obukhov (1941). To obtain them, we shall first determine which parameters can be involved in the properties of turbulent flow over regions small compared with l but large compared with the distances λ_0 at which the viscosity of the fluid begins to be important. It is these intermediate distances which we shall discuss below. The parameters in question are the fluid density ρ and another quantity characterizing any turbulent flow, the energy ε dissipated per unit time per unit mass of fluid. We have seen that ε is the energy flux which continually passes from larger to smaller eddies. Hence, although the energy dissipation is ultimately due to the viscosity of the fluid and occurs in the smallest eddies, the quantity ε determine the properties of larger eddies. It is natural to suppose that (for given ρ and ε) the local properties of the turbulence are independent of the dimension l and velocity Δu of the flow as a whole. The fluid viscosity ν also cannot appear in any of the quantities in which we are at present interested (we recall that we are concerned with distances $\lambda \gg \lambda_0$).

† In reality, however, a fairly large numerical coefficient should be included. This is because, as mentioned above, l and Δu may differ quite considerably from the actual scale and velocity of the turbulent flow. The ratio ν_{turb}/ν may be more accurately written $\nu_{\text{turb}}/\nu \sim \mathrm{R}/\mathrm{R}_{\text{cr}}$, which formula takes into account the fact that ν_{turb} and ν must in reality be comparable in magnitude not for $\mathrm{R} \sim 1$, but for $\mathrm{R} \sim \mathrm{R}_{\text{cr}}$.

Let us determine the order of magnitude v_λ of the turbulent velocity variation over distances of the order of λ. It must be determined only by ε and, of course, the distance λ itself.† From these two quantities we can form only one having the dimensions of velocity, namely $(\varepsilon\lambda)^{1/3}$. Hence we can say that the relation

$$v_\lambda \propto (\varepsilon\lambda)^{1/3} \tag{33.6}$$

must hold. We thus find that the velocity variation over a small distance is proportional to the cube root of the distance (*Kolmogorov and Obukhov's law*). The quantity v_λ may also be regarded as the velocity of turbulent eddies whose size is of the order of λ: the variation of the mean velocity over small distances is small compared with the variation of the fluctuating velocity over those distances, and may be neglected.

The relation (33.6) may be obtained in another way by expressing a constant quantity, the dissipation ε, in terms of quantities characterizing the eddies of size λ; ε must be proportional to the squared gradient of the velocity v_λ and to the appropriate turbulent viscosity coefficient $\nu_{\mathrm{turb},\lambda} \propto v_\lambda\lambda$:

$$\varepsilon \propto \nu_{\mathrm{turb},\lambda}(v_\lambda/\lambda)^2 \propto v_\lambda{}^3/\lambda,$$

whence we obtain (33.6).

Let us now put the problem somewhat differently, and determine the order of magnitude v_τ of the velocity variation at a given point over a time interval τ which is short compared with the time $T \sim l/u$ characterizing the flow as a whole. To do this, we notice that, since there is a net mean flow, any given portion of the fluid is displaced, during the interval τ, over a distance of the order of τu, u being the mean velocity. Hence the portion of fluid which is at a given point at time τ will have been at a distance τu from that point at the initial instant. We can therefore obtain the required quantity v_τ by direct substitution of τu for λ in (33.6):

$$v_\tau \propto (\varepsilon\tau u)^{1/3}. \tag{33.7}$$

The quantity v_τ must be distinguished from v_τ', the variation in velocity of a portion of fluid as it moves about. This variation can evidently depend only on ε, which determines the local properties of the turbulence, and of course on τ itself. Forming the only combination of ε and τ that has the dimensions of velocity, we obtain

$$v_\tau' \propto (\varepsilon\tau)^{1/2}. \tag{33.8}$$

Unlike the velocity variation at a given point, it is proportional to the square root of τ, not to the cube root. It is easy to see that, for τ small compared with T, v_τ' is always less than v_τ.‡

Using the expression (33.1) for ε, we can rewrite (33.6) and (33.7) as

$$\left.\begin{aligned} v_\lambda &\propto \Delta u(\lambda/l)^{1/3}, \\ v_\tau &\propto \Delta u(\tau/T)^{1/3}. \end{aligned}\right\} \tag{33.9}$$

This form shows clearly the similarity property of local turbulence: the small-scale characteristics of different turbulent flows are the same apart from the scale of measurement of lengths and velocities (or, equivalently, lengths and times).††

† The dimensions of ε are erg/g sec = cm^2/sec^3, and do not include mass; the only quantity involving the mass dimension is the density ρ. The latter is therefore not involved in quantities whose dimensions do not include mass.

‡ The inequality $v_\tau' \ll v_\tau$ has in essence been assumed in the derivation of (33.7).

†† In this connection, the term *self-similarity* is often used in recent literature.

Let us now find at what distances the fluid viscosity begins to be important. These distances λ_0 also determine the order of magnitude of the size of the smallest eddies in the turbulent flow (called the "internal scale" of the turbulence, in contradistinction to the "external scale" l). To determine λ_0, we form the local Reynolds number $R_\lambda \sim v_\lambda \lambda/\nu \sim \Delta u \cdot \lambda^{4/3}/\nu l^{1/3} \sim R(\lambda/l)^{4/3}$, with the Reynolds number $R \sim l\Delta u/\nu$ for the flow as a whole. The order of magnitude of λ_0 is that for which $R_{\lambda_0} \sim 1$. Hence we find

$$\lambda_0 \sim l/R^{\frac{3}{4}}. \tag{33.10}$$

The same expression can be obtained by forming from ε and ν the only combination having the dimensions of length, namely

$$\lambda_0 \sim (\nu^3/\varepsilon)^{\frac{1}{4}}. \tag{33.11}$$

Thus the internal scale of the turbulence decreases rapidly with increasing R. For the corresponding velocity we have

$$v_{\lambda_0} \sim \Delta u/R^{\frac{1}{4}}; \tag{33.12}$$

this also decreases when R increases.†

The range of scales $\lambda \sim l$ is called the *energy range*; the majority of the kinetic energy of the fluid is concentrated there. Values $\lambda \lesssim \lambda_0$ form the *dissipation range*, where the kinetic energy is dissipated. For very large values of R, these two ranges are quite far apart, and between them lies the *inertial range*, in which $\lambda_0 \ll \lambda \ll l$; the results derived in this section are valid there.

Kolmogorov and Obukhov's law can be expressed in an equivalent spatial spectrum form. We replace the scales λ by corresponding wave numbers $k \sim 1/\lambda$ of the eddies; let $E(k)dk$ be the kinetic energy per unit mass of fluid in eddies with k values in the range dk. The function $E(k)$ has the dimensions $\mathrm{cm}^3/\mathrm{sec}^2$; the combination of ε and k having these dimensions gives

$$E(k) \propto \varepsilon^{2/3} k^{-5/3}. \tag{33.13}$$

The equivalence of this expression and (33.6) is easily seen by noting that $v_\lambda{}^2$ gives the order of magnitude of the total energy in eddies with all scales of the order of λ or less. The same result is reached by integration of (33.13):

$$\int_k^\infty E(k)dk \propto \varepsilon^{2/3}/k^{2/3} \sim (\varepsilon\lambda)^{2/3} \sim v_\lambda{}^2.$$

Together with the spatial scales of the turbulent eddies, we can also consider their time characteristics (frequencies). The lower end of the frequency spectrum of turbulent motion is at frequencies $\sim u/l$. The upper end is

$$\omega_0 \sim u/\lambda_0 \sim uR^{3/4}/l, \tag{33.14}$$

corresponding to the internal scale of turbulence. The inertial range corresponds to frequencies

$$u/l \ll \omega \ll (u/l)R^{3/4}.$$

† Formulae (33.10)–(33.12) give the manner of variation of the relevant quantities with R. Quantitatively, it would be more correct to replace R in them by R/R_{cr}.

The inequality $\omega \gg u/l$ signifies that as regards the local properties of turbulence the unperturbed flow may be treated as steady. The energy distribution in the frequency spectrum in the inertial range is found from (33.13) by substituting $k \sim \omega/u$:

$$E(\omega) \propto (u\varepsilon)^{2/3}\omega^{-5/3}, \qquad (33.15)$$

where $E(\omega)\mathrm{d}\omega$ is the energy in the frequency range $\mathrm{d}\omega$.

The frequency ω gives the time repetition period in the region of space concerned, as observed from a fixed frame of reference. It is to be distinguished from the frequency ω' which gives the flow repetition period in a given portion of fluid moving in space. The energy distribution in this frequency spectrum cannot depend on u, and must be determined only by ε and the frequency ω' itself. Again using dimensional arguments, we find

$$E(\omega') \propto \varepsilon/\omega'^2. \qquad (33.16)$$

This is in the same relationship to (33.15) as (33.8) is to (33.7).

Turbulent mixing causes a gradual separation of fluid particles that were originally close together. Let us consider two particles at a distance λ that is small (in the inertial range). Again, by dimensional arguments, the rate of change of this distance with time is

$$\mathrm{d}\lambda/\mathrm{d}t \propto (\varepsilon\lambda)^{1/3}. \qquad (33.17)$$

Integration of this shows that the time τ over which two particles initially at a distance λ_1 move apart to a distance $\lambda_2 \gg \lambda_1$ is in order of magnitude

$$\tau \sim \lambda_2^{4/3}/\varepsilon^{1/3}. \qquad (33.18)$$

Note that the process is self-accelerating: the rate of separation increases with λ. This occurs because only eddies with scales $\lesssim \lambda$ contribute to the separation of particles at a distance λ; the larger eddies carry both particles and do not cause them to separate.†

Finally, let us consider the properties of the flow in regions whose dimension λ is small compared with λ_0. In such regions the flow is regular and its velocity varies smoothly. Hence we can expand v_λ in a series of powers of λ and, retaining only the first term, obtain $v_\lambda = \text{constant} \times \lambda$. The order of magnitude of the constant is v_{λ_0}/λ_0, since for $\lambda \sim \lambda_0$ we must have $v_\lambda \sim v_{\lambda_0}$. Thus

$$v_\lambda \sim v_{\lambda_0}\lambda/\lambda_0 \sim \Delta u \cdot \mathrm{R}^{\frac{1}{4}}\lambda/l. \qquad (33.19)$$

This formula may also be obtained by equating two expressions for the energy dissipation ε: the expression $(\Delta u)^3/l$ (33.1), which determines ε in terms of quantities characterizing the large eddies, and the expression $\nu(v_\lambda/\lambda)^2$, which determines ε in terms of the velocity gradient for the eddies in which the energy dissipation actually occurs.

§34. The velocity correlation functions

Formula (33.6) determines qualitatively the *correlation of velocities* in local turbulence, i.e. the relation between the velocities at two neighbouring points. Let us now introduce

† These results can be applied to particles suspended in the fluid, which are passively conveyed by its motion.

functions which will serve to characterize this correlation quantitatively.† One is the rank-two correlation tensor

$$B_{ik} = \langle (v_{2i} - v_{1i})(v_{2k} - v_{1k}) \rangle, \tag{34.1}$$

where $\mathbf{v}_2$ and $\mathbf{v}_1$ are the fluid velocities at two neighbouring points, and the angle brackets denote an average with respect to time. The radius vector from point 1 to point 2 will be denoted by $\mathbf{r} = \mathbf{r}_2 - \mathbf{r}_1$. In discussing local turbulence, we shall suppose this distance much less than the fundamental scale l, but not necessarily much greater than the internal scale λ_0 of the turbulence.

The velocity variation over short distances is due to the small eddies. The properties of local turbulence, however, do not depend on the averaged flow. We can therefore simplify the study of the correlation functions of local turbulence by considering instead an idealized case of turbulent flow in which there is isotropy and homogeneity not only on small scales (as in local turbulence) but also on every scale; the averaged velocity is then zero. This completely isotropic and homogeneous turbulence‡ can be regarded as occurring in a fluid subjected to vigorous shaking and then left to itself. Such a flow will, of course, necessarily decay in the course of time, and so the components of the correlation tensor are also time-dependent.†† The relations derived below between the various correlation functions apply to homogeneous isotropic turbulence on every scale, and to local turbulence at distances $r \ll l$.

Since local turbulence is isotropic, the tensor B_{ik} cannot depend on any direction in space. The only vector that can appear in the expression for B_{ik} is the radius vector $\mathbf{r}$. The general form of such a symmetrical tensor of rank two is

$$B_{ik} = A(r)\delta_{ik} + B(r)n_i n_k, \tag{34.2}$$

where $\mathbf{n}$ is a unit vector in the direction of $\mathbf{r}$.

To see the significance of the functions A and B, we take the coordinate axes so that one of them is in the direction of $\mathbf{n}$, denoting the velocity component along this axis by v_r and the component perpendicular to $\mathbf{n}$ by v_t. The correlation tensor component B_{rr} is then the mean square relative velocity of two fluid particles along the line joining them. Similarly, B_{tt} is the mean square transverse velocity of one particle relative to the other. Since $n_r = 1$, $n_t = 0$, we have from (34.2)

$$B_{rr} = A + B, \qquad B_{tt} = A, \qquad B_{rt} = 0$$

The expression (34.2) may now be written as

$$B_{ik} = B_{tt}(r)(\delta_{ik} - n_i n_k) + B_{rr}(r) n_i n_k. \tag{34.3}$$

Expanding the parentheses in the definition (34.1) gives

$$B_{ik} = \langle v_{1i} v_{1k} \rangle + \langle v_{2i} v_{2k} \rangle - \langle v_{1i} v_{2k} \rangle - \langle v_{1k} v_{2i} \rangle.$$

Because of the homogeneity, the mean values of $v_i v_k$ at points 1 and 2 are the same, and because of the isotropy, $\langle v_{1i} v_{2k} \rangle$ is unaltered when points 1 and 2 are interchanged (i.e. when $\mathbf{r} = \mathbf{r}_2 - \mathbf{r}_1$ changes sign); thus

$$\langle v_{1i} v_{1k} \rangle = \langle v_{2i} v_{2k} \rangle = \tfrac{1}{3} \langle v^2 \rangle \delta_{ik}, \quad \langle v_{1i} v_{2k} \rangle = \langle v_{2i} v_{1k} \rangle.$$

† Correlation functions were first used in the dynamics of turbulence by L. V. Keller and A. A. Fridman (1924).

‡ The concept is due to G. I. Taylor (1935).

†† The averaging in the definition (34.1) must here, strictly speaking, be taken not as time averaging but as averaging over all possible positions of the points 1 and 2 (for a given distance between them) at a given instant.

Hence

$$B_{ik} = \tfrac{2}{3}\langle v^2 \rangle \delta_{ik} - 2b_{ik}, \quad b_{ik} = \langle v_{1i} v_{2k} \rangle. \tag{34.4}$$

The symmetrical auxiliary tensor b_{ik} tends to zero as $r \to \infty$; for the turbulent flow velocities at infinitely distant points may be regarded as statistically independent, so that the mean value of their product reduces to the product of the means of each factor separately, which are zero by hypothesis.

We differentiate (34.4) with respect to the coordinates of point 2:

$$\frac{\partial B_{ik}}{\partial x_{2k}} = -2\frac{\partial b_{ik}}{\partial x_{2k}} = -2\left\langle v_{1i}\frac{\partial v_{2k}}{\partial x_{2k}}\right\rangle$$

By the equation of continuity, $\partial v_{2k}/\partial x_{2k} = 0$, and so

$$\partial B_{ik}/\partial x_{2k} = 0.$$

Since B_{ik} is a function only of $\mathbf{r} = \mathbf{r}_2 - \mathbf{r}_1$, differentiation with respect to x_{2k} is equivalent to that with respect to x_k. Substituting (34.3) for B_{ik}, we easily find

$$B_{rr}' + (2/r)(B_{rr} - B_{tt}) = 0,$$

where the prime denotes differentiation with respect to r. Thus the longitudinal and transverse correlation functions are related by

$$B_{tt} = \frac{1}{2r}\frac{\mathrm{d}}{\mathrm{d}r}(r^2 B_{rr}). \tag{34.5}$$

According to (33.6), the velocity difference over a distance r in the inertial range is proportional to $r^{1/3}$. Hence the correlation functions B_{rr} and B_{tt} are proportional to $r^{2/3}$ in that range. We then get from (34.5) the simple relation

$$B_{tt} = \tfrac{4}{3} B_{rr} \qquad (\lambda_0 \ll r \ll l). \tag{34.6}$$

For distances $r \ll \lambda_0$, the velocity difference is proportional to r, and therefore B_{rr} and B_{tt} are proportional to r^2. Formula (34.5) then gives

$$B_{tt} = 2B_{rr} \qquad (r \ll \lambda_0). \tag{34.7}$$

For these distances, B_{tt} and B_{rr} can also be expressed in terms of the mean energy dissipation ε. We write $B_{rr} = ar^2$, where a is a constant, and combine (34.3), (34.4) and (34.7) to find

$$b_{ik} = \tfrac{1}{3}\langle v^2 \rangle \delta_{ik} - ar^2 \delta_{ik} + \tfrac{1}{2} a x_i x_k.$$

Differentiating this relation, we have

$$\left\langle \frac{\partial v_{1i}}{\partial x_{1l}}\frac{\partial v_{2i}}{\partial x_{2l}} \right\rangle = 15a, \quad \left\langle \frac{\partial v_{1i}}{\partial x_{1l}}\frac{\partial v_{2l}}{\partial x_{2i}} \right\rangle = 0.$$

Since these hold for arbitrarily small r, we can put $\mathbf{r}_1 = \mathbf{r}_2$, obtaining

$$\langle (\partial v_i/\partial x_l)^2 \rangle = 15a, \langle (\partial v_i/\partial x_l)(\partial v_l/\partial x_i) \rangle = 0.$$

According to (16.3), we have for the mean energy dissipation

$$\varepsilon = \tfrac{1}{2}\nu \left\langle \left(\frac{\partial v_i}{\partial x_k} + \frac{\partial v_k}{\partial x_i}\right)^2 \right\rangle = \nu \left\{ \left\langle \left(\frac{\partial v_i}{\partial x_k}\right)^2 \right\rangle + \left\langle \frac{\partial v_i}{\partial x_k}\frac{\partial v_k}{\partial x_i} \right\rangle \right\} = 15a\nu,$$

whence $a = \varepsilon/15\nu$.† We thus obtain the following final expressions for the correlation functions in terms of the energy dissipation:

$$B_{tt} = 2\varepsilon r^2/15\nu, \qquad B_{rr} = \varepsilon r^2/15\nu \tag{34.8}$$

(A. N. Kolmogorov 1941).

We next define the rank-three tensor

$$B_{ikl} = \langle (v_{2i} - v_{1i})(v_{2k} - v_{1k})(v_{2l} - v_{1l}) \rangle \tag{34.9}$$

and the auxiliary tensor

$$b_{ik,l} = \langle v_{1i} v_{1k} v_{2l} \rangle = -\langle v_{2i} v_{2k} v_{1l} \rangle. \tag{34.10}$$

The latter is symmetrical in the first pair of suffixes; the second equation (34.10) results from the fact that interchanging points 1 and 2 is equivalent to changing the sign of **r**, i.e. inverting the coordinates, and therefore changes the sign of the rank-three tensor. When $r = 0$ and the points 1 and 2 coincide, $b_{ik,l}(0) = 0$: the mean value of the product of an odd number of fluctuating velocity components is zero. Expanding the parentheses in the definition (34.9) gives B_{ikl} in terms of $b_{ik,l}$:

$$B_{ikl} = 2(b_{ik,l} + b_{il,k} + b_{lk,i}). \tag{34.11}$$

As $r \to \infty$, the tensor $b_{ik,l}$ and therefore B_{ikl} tend to zero.

Isotropy shows that $b_{ik,l}$ must be expressible in terms of the unit tensor δ_{ik} and the components of the unit tensor **n**. The general form of such a tensor symmetrical in the first pair of suffixes is

$$b_{ik,l} = C(r)\delta_{ik} n_l + D(r)(\delta_{il} n_k + \delta_{kl} n_i) + F(r) n_i n_k n_l. \tag{34.12}$$

Differentiating this with respect to the coordinates of point 2 and using the equation of continuity, we find

$$\partial b_{ik,l}/\partial x_{2l} = \langle v_{1i} v_{1k} \partial v_{2l}/\partial x_{2l} \rangle = 0.$$

Substitution of (34.12) leads, after a simple calculation, to the two equations

$$[r^2(3C + 2D + F')]' = 0, \qquad C' + 2(C + D)/r = 0.$$

Integration of the former gives

$$3C + 2D + F = \text{constant}/r^2.$$

When $r = 0$, C, D and F must be zero; the constant is therefore zero, and $3C + 2D + F = 0$. The two equations found then give

$$D = -C - \tfrac{1}{2} rC', \qquad F = rC' - C. \tag{34.13}$$

Substitution of these in (34.12) and thence in (34.11) gives $B_{ikl} = -2(rC' + C)(\delta_{ik} n_l + \delta_{il} n_k + \delta_{kl} n_i) + 6(rC' - C) n_i n_k n_l$. Again taking one of the coordinate axes to be parallel to **n**, we find as the components of B_{ikl}

$$B_{rrr} = -12C, \quad B_{rtt} = -2(C + rC'), \quad B_{rrt} = B_{ttt} = 0. \tag{34.14}$$

† For isotropic turbulence, the mean dissipation is related to the mean square vorticity by the simple formula

$$\langle (\mathbf{curl}\, \mathbf{v})^2 \rangle = \tfrac{1}{2} \left\langle \left(\frac{\partial v_i}{\partial x_k} - \frac{\partial v_k}{\partial x_i} \right)^2 \right\rangle = \varepsilon/\nu.$$

From this, we see that the non-zero correlation functions B_{rtt} and B_{rrr} are related by

$$B_{rtt} = \frac{1}{6}\frac{\mathrm{d}}{\mathrm{d}r}(rB_{rrr}). \tag{34.15}$$

We shall also need an expression for $b_{ik,l}$ in terms of the components of B_{ikl}. From (34.12)–(34.14),

$$b_{ik,l} = -\frac{1}{12}B_{rrr}\delta_{ik}n_l + \frac{1}{24}(rB_{rrr}' + 2B_{rrr})(\delta_{il}n_k + \delta_{kl}n_i) - \frac{1}{12}(rB_{rrr}' - B_{rrr})n_i n_k n_l. \tag{34.16}$$

The relations (34.5) and (34.15) follow from the continuity equation alone. With the Navier–Stokes equation, we can derive a relation between the correlation tensors B_{ik} and B_{ikl} (T. von Kármán and L. Howarth 1938; A. N. Kolmogorov 1941).

To do so, we calculate the derivative $\partial b_{ik}/\partial t$ (a completely homogeneous and isotropic turbulent flow, it will be remembered, decays in the course of time). Expressing the derivatives $\partial v_{1i}/\partial t$ and $\partial v_{2k}/\partial t$ by means of the Navier–Stokes equation, we find

$$\frac{\partial}{\partial t}\langle v_{1i}v_{2k}\rangle = -\frac{\partial}{\partial x_{1l}}\langle v_{1i}v_{1l}v_{2k}\rangle - \frac{\partial}{\partial x_{2l}}\langle v_{1i}v_{2k}v_{2l}\rangle - \frac{1}{\rho}\frac{\partial}{\partial x_{1i}}\langle p_1 v_{2k}\rangle -$$
$$-\frac{1}{\rho}\frac{\partial}{\partial x_{2k}}\langle p_2 v_{1i}\rangle + \nu\triangle_1\langle v_{1i}v_{2k}\rangle + \nu\triangle_2\langle v_{1i}v_{2k}\rangle. \tag{34.17}$$

The correlation function for the pressure and the velocity is

$$\langle p_1 \mathbf{v}_2\rangle = 0. \tag{34.18}$$

For isotropy implies that this function must have the form $f(r)\mathbf{n}$. And, from the equation of continuity,

$$\mathrm{div}_2\langle p_1\mathbf{v}_2\rangle = \langle p_1 \mathrm{div}_2\mathbf{v}_2\rangle = 0.$$

The only vector having the form $f(r)\mathbf{n}$ and zero divergence is constant $\times\ \mathbf{n}/r^2$, and this would not be finite at $r = 0$; the constant must therefore be zero.

Now replacing the derivatives with respect to x_{1i} and x_{2i} in (34.17) by those with respect to $-x_i$ and x_i, we get

$$\frac{\partial}{\partial t}b_{ik} = \frac{\partial}{\partial x_l}(b_{il,k} + b_{kl,i}) + 2\nu\triangle b_{ik}. \tag{34.19}$$

Here we have to substitute b_{ik} and $b_{ik,l}$ from (34.4) and (34.16). The time derivative of the kinetic energy per unit mass, $\frac{1}{2}\langle v^2\rangle$, is just the energy dissipation $-\varepsilon$. Hence

$$\frac{\partial}{\partial t}(\tfrac{1}{3}\langle v^2\rangle) = -\tfrac{2}{3}\varepsilon.$$

A straightforward but lengthy calculation gives†

$$-\tfrac{2}{3}\varepsilon - \tfrac{1}{2}\frac{\partial B_{rr}}{\partial t} = \frac{1}{6r^4}\frac{\partial}{\partial r}(r^4 B_{rrr}) - \frac{\nu}{r^4}\frac{\partial}{\partial r}\left(r^4\frac{\partial B_{rr}}{\partial r}\right). \tag{34.20}$$

† The result of the calculation corresponds to (34.20) with the operator $1 + \frac{1}{5}r\partial/\partial r$ applied to each side, but since the only solution of $f + \frac{1}{5}r\partial f/\partial r = 0$ finite when $r = 0$ is $f = 0$, the operator may be omitted.

The value of B_{rr} varies considerably with time only over an interval corresponding to the fundamental scale of turbulence ($\sim l/u$). In relation to local turbulence the unperturbed flow may be regarded as steady, as already mentioned in §33. This means that for local turbulence it is sufficiently accurate to neglect the derivative $\partial B_{rr}/\partial t$ on the left-hand side of (34.20) in comparison with ε. Multiplying the resulting equation by r^4 and integrating with respect to r we find, since the correlation functions are zero when $r = 0$, the following relation between B_{rr} and B_{rrr}:

$$B_{rrr} = -\tfrac{4}{5}\varepsilon r + 6\nu \frac{\mathrm{d}B_{rr}}{\mathrm{d}r} \tag{34.21}$$

(A. N. Kolmogorov 1941). This is valid when r is either greater or less than λ_0. When $r \gg \lambda_0$, the viscosity term is small, and we have simply

$$B_{rrr} = -\tfrac{4}{5}\varepsilon r. \tag{34.22}$$

If we substitute in (34.21) for $r \ll \lambda_0$ the expression (34.8) for B_{rr}, the result is zero, because in this case we must have $B_{rrr} \propto r^3$, and so the first-order terms must cancel.

The one equation (34.20) relates two independent functions B_{rr} and B_{rrr}, and therefore does not by itself enable us to find these. The presence of the correlation functions of two orders is due to the non-linearity of the Navier–Stokes equation. For the same reason, calculating the time derivative of the third-order correlation function would give an equation containing also a fourth-order one, and so on. This leads to an infinite sequence of equations. It is not possible to arrive in this way at a closed system of equations without making some additional assumptions.

One further general remark† should be made. It might be thought that the possibility exists in principle of obtaining a universal formula, applicable to any turbulent flow, which should give B_{rr} and B_{tt} for all distances r that are small compared with l. In fact, however, there can be no such formula, as we see from the following argument. The instantaneous value of $(v_{2i} - v_{1i})(v_{2k} - v_{1k})$ might in principle be expressed as a universal function of the energy dissipation ε at the instant considered. When we average these expressions, however, an important part will be played by the manner of variation of ε over times of the order of the periods of the large eddies (with size $\sim l$), and this variation is different for different flows. The result of the averaging therefore cannot be universal.‡

Loĭtsyanskiĭ's integral

We can rewrite equation (34.20) with b_{rr} and $b_{rr,r}$ in place of B_{rr} and B_{rrr}:

$$\frac{\partial b_{rr}}{\partial t} = \frac{1}{r^4}\frac{\partial}{\partial r}\left[2\nu r^4 \frac{\partial b_{rr}}{\partial r} + r^4 b_{rr,r}\right]. \tag{34.23}$$

We multiply this by r^4 and integrate over r from 0 to ∞. The expression in square brackets is zero when $r = 0$. Assuming that it tends also to zero as $r \to \infty$, we find

$$\Lambda \equiv \int_0^\infty r^4 b_{rr} \mathrm{d}r = \text{constant} \tag{34.24}$$

† Due to L. D. Landau (1944).

‡ The question whether fluctuations of ε should be reflected in the form of the correlation functions in the inertial range can scarcely be resolved with certainty until we have a consistent theory of turbulence; it has been posed by A. N. Kolmogorov (*Journal of Fluid Mechanics* **13**, 82, 1962) and A. M. Obukhov (*ibid.* 77). Existing attempts to apply relevant corrections to Kolmogorov and Obukhov's law are based on hypotheses about the statistical properties of the dissipation, whose correctness it is difficult to assess.

(L. G. Loĭtsyanskiĭ 1939). The integral converges if b_{rr} decreases at infinity faster than r^{-5}, and is in fact constant if $b_{rr,r}$ decreases faster than r^{-4}.

The functions b_{rr} and b_{tt} are related by a formula similar to (34.5) for B_{rr} and B_{tt}. We therefore have (under the same conditions)

$$\int_0^\infty b_{tt} r^4 \, dr = -\frac{3}{2} \int_0^\infty b_{rr} r^4 \, dr.$$

Since $b_{rr} + 2b_{tt} = \langle \mathbf{v}_1 \cdot \mathbf{v}_2 \rangle$, the integral (34.24) can be put in the form

$$\Lambda = -\frac{1}{4\pi} \int r^2 \langle \mathbf{v}_1 \cdot \mathbf{v}_2 \rangle \, dV, \tag{34.25}$$

where $dV = d^3(x_1 - x_2)$. This integral is closely related to the angular momentum of a fluid in a state of homogeneous and isotropic turbulence. It can be shown (though we shall not pause to do so) that the square of the total angular momentum $\mathbf{M}$ of the fluid in some large volume V within an infinite fluid is $M^2 = 4\pi\rho^2 \Lambda V$; the increase of $\mathbf{M}$ as $\sqrt{V}$ and not as V occurs because $\mathbf{M}$ is the sum of a large number of statistically independent terms (the angular momenta of separate small portions of fluid) with zero mean values.

The value of M^2 in a given volume V may vary because of the interaction with surrounding regions of the fluid. If this interaction decreased sufficiently rapidly with increasing distance, it would be a surface effect for the part of the fluid considered. The times during which M^2 could change considerably would then increase with the dimensions of V; these times and dimensions are to be regarded as very large, and in this sense M^2 would be conserved.

The condition stated is closely related to the conditions, formulated in deriving (34.24) from (34.23), for a sufficiently rapid decrease in the correlation functions. In incompressible fluid theory, however, it is doubtful whether they are satisfied. The physical point lies in the infinite speed of propagation of perturbations in an incompressible fluid. Mathematically, this is shown by the integral form of the fluid pressure dependence on the velocity distribution: if the right-hand side of (15.11) is regarded as given, the solution is

$$p(\mathbf{r}) = \frac{\rho}{4\pi} \int \frac{\partial^2 v_i(\mathbf{r}') v_k(\mathbf{r}')}{\partial x'_i \partial x'_k} \frac{dV'}{|\mathbf{r} - \mathbf{r}'|}.$$

As a result, any local perturbation of the velocity instantaneously affects the pressure in all space, and the pressure affects the acceleration of the fluid, and therefore the subsequent change in the velocity.

A natural way of formulating the problem is as follows. At the initial instant ($t = 0$), let an isotropic turbulent flow be set up, in which the functions $b_{ik}(r, t)$ and $b_{ik,l}(r, t)$ decrease exponentially with increasing distance. Expressing the pressure in terms of the velocities by means of the above formula, we can then use the equations of motion of the fluid in an attempt to determine the dependence of the time derivatives of the correlation functions at $t = 0$ on the distance as $r \to \infty$. This determines also the dependence of the correlation functions themselves on r for $t > 0$. The investigation yields the following results.†

† See I. Proudman and W. H. Reid, *Philosophical Transactions of the Royal Society* A **247**, 163, 1954; G. K. Batchelor and I. Proudman, *ibid.* **248**, 369, 1956. These researches have also been described by A. S. Monin and A. M. Yaglom, *Statistical Fluid Mechanics: Mechanics of Turbulence*, Vol. 2, §15.5, 15.6, Cambridge (Mass.) 1975.

For $t > 0$, $b_{rr}(r,t)$ decreases at infinity at least as r^{-6}, and perhaps exponentially. Loĭtsyanskiĭ's integral is therefore convergent. The decrease of $b_{rr,r}$ is only as r^{-4}, and Λ is therefore not conserved. Its time derivative is some non-zero negative (since $b_{rr,r}$ is found empirically to be negative) function of time. This function is entirely governed by inertial forces. It is reasonable to suppose that, as the turbulence decays, these forces become less important, and in the final stage they may be neglected in comparison with the viscous forces. Thus Λ decreases (the angular momentum "spreads" uniformly through infinite space), tending to a constant limit which it reaches in the final stage of turbulence.

It is therefore possible to determine for this stage the law of time variation of the fundamental scale l and characteristic velocity v of the turbulence. An estimate of the integral (34.25) gives $\Lambda \sim v^2 l^5 = \text{constant}$. Another relation is obtained by estimating the rate of energy decrease by viscous dissipation. The dissipation ε is proportional to the square of the velocity gradients; estimating these as v/l, we find $\varepsilon \sim \nu(v/l)^2$. Equating it to the derivative $\partial(v^2)/\partial t \sim v^2/t$, where t is reckoned from the start of the final stage of turbulence, we have $l \sim (\nu t)^{1/2}$ and so

$$v = \text{constant} \times t^{-5/4} \tag{34.26}$$

(M. D. Millionshchikov 1939).

Correlation function spectrum

As well as the coordinate representation of the correlation functions discussed above, there is a spectral (wave vector) representation of these functions that has methodological and physical interest. It is obtained by expansion as a Fourier space integral:

$$B_{ik}(\mathbf{r}) = \int B_{ik}(\mathbf{k}) e^{i\mathbf{k}\cdot\mathbf{r}} d^3k/(2\pi)^3,$$

$$B_{ik}(\mathbf{k}) = \int B_{ik}(\mathbf{r}) e^{-i\mathbf{k}\cdot\mathbf{r}} d^3x;$$

the spectral correlation function is denoted by the same symbol B_{ik} with a different independent variable, the wave vector $\mathbf{k}$. Since in isotropic turbulence $B_{ik}(-\mathbf{r}) = B_{ik}(\mathbf{r})$, we have $B_{ik}(\mathbf{k}) = B_{ik}(-\mathbf{k}) = B_{ik}{}^*(\mathbf{k})$, and the spectral functions $B_{ik}(\mathbf{k})$ are therefore real.

As $r \to \infty$, the functions $B_{ik}(\mathbf{r})$ tend to a finite limit given by the first term in (34.4). Accordingly, their Fourier components contain a delta function:

$$B_{ik}(\mathbf{k}) = \tfrac{2}{3}(2\pi)^3\, \delta(\mathbf{k}) \langle v^2 \rangle - 2b_{ik}(\mathbf{k}). \tag{34.27}$$

The components with $\mathbf{k} \neq 0$ are the same for the functions B_{ik} and $-2b_{ik}$.

Differentiation with respect to the coordinates x_l in the coordinate representation is equivalent to multiplication by ik_l in the spectral representation. The continuity equation $\partial b_{ik}(\mathbf{r})/\partial x_i = 0$ therefore reduces in the spectral representation to the condition that the tensor $b_{ik}(\mathbf{k})$ be transverse to the wave vector:

$$k_i b_{ik}(\mathbf{k}) = 0. \tag{34.28}$$

Because of the isotropy, the tensor $b_{ik}(\mathbf{k})$ must be expressible in terms of $\mathbf{k}$ and the unit tensor δ_{ik} only. The general form of such a symmetrical tensor satisfying the condition (34.28) is

$$b_{ik}(\mathbf{k}) = F^{(2)}(k)\,(\delta_{ik} - k_i k_k/k^2), \tag{34.29}$$

where $F^{(2)}(k)$ is a real function of the wave number.

The spectral representation of the rank-three correlation tensor is found similarly, $B_{ikl}(\mathbf{k})$ being expressed in terms of $b_{ik,l}(\mathbf{k})$ by (34.11); these tensors do not contain a delta function. The continuity equation $\partial b_{ik,l}(\mathbf{r})/\partial x_l = 0$ gives the condition that $b_{ik,l}(\mathbf{k})$ be transverse as regards the third suffix:

$$k_l b_{ik,l}(\mathbf{k}) = 0. \tag{34.30}$$

The general form of such a tensor is

$$b_{ik,l}(\mathbf{k}) = iF^{(3)}(k)\{\delta_{il}k_k/k + \delta_{kl}k_i/k - 2k_i k_k k_l/k^3\}. \tag{34.31}$$

Since $b_{ik,l}(-\mathbf{r}) = -b_{ik,l}(\mathbf{r})$, the spectral functions $b_{ik,l}(\mathbf{k})$ are imaginary; a factor i has been included in (34.31), so as to make $F^{(3)}(k)$ real.

Equation (34.19) in the spectral representation is

$$\frac{\partial}{\partial t} b_{ik}(\mathbf{k}) = ik_l[b_{il,k}(\mathbf{k}) + b_{kl,i}(\mathbf{k})] - 2\nu k^2 b_{ik}(\mathbf{k}).$$

Substitution of (34.29) and (34.31) gives

$$\partial F^{(2)}(k,t)/\partial t = -2kF^{(3)}(k,t) - 2\nu k^2 F^{(2)}(k,t). \tag{34.32}$$

The function $F^{(2)}(\mathbf{k})$ has an important physical significance. To understand this, let us approach the definition of the spectral correlation function at a somewhat earlier stage.†

We use the customary Fourier expansion of the fluctuating velocity $\mathbf{v}(\mathbf{r})$ itself:

$$\mathbf{v}(\mathbf{r}) = \int \mathbf{v}_{\mathbf{k}} e^{i\mathbf{k}\cdot\mathbf{r}} d^3k/(2\pi)^3, \qquad \mathbf{v}_{\mathbf{k}} = \int \mathbf{v}(\mathbf{r}) e^{-i\mathbf{k}\cdot\mathbf{r}} d^3x.$$

The latter integral is in fact divergent, since $\mathbf{v}(\mathbf{r})$ does not tend to zero at infinity. This is unimportant, however, in the formal derivations below, whose purpose is to calculate the mean squares, which are certainly finite.

The correlation tensor $b_{ik}(\mathbf{r})$ is expressed in terms of the velocity Fourier components by the integral

$$b_{il}(\mathbf{r}) = \iint \langle v_{i\mathbf{k}} v_{l\mathbf{k}'} \rangle e^{i(\mathbf{k}\cdot\mathbf{r}_2 + \mathbf{k}'\cdot\mathbf{r}_1)} d^3k\, d^3k'/(2\pi)^6. \tag{34.33}$$

For this to be a function only of $\mathbf{r} = \mathbf{r}_2 - \mathbf{r}_1$, the integrand must contain a delta function of $\mathbf{k} + \mathbf{k}'$, i.e. must be

$$\langle v_{i\mathbf{k}} v_{l\mathbf{k}'} \rangle = (2\pi)^3 (v_i v_l)_{\mathbf{k}} \delta(\mathbf{k} + \mathbf{k}'). \tag{34.34}$$

This relation is to be regarded as a definition of a quantity here symbolically denoted by $(v_i v_l)_{\mathbf{k}}$. Substituting (34.34) in (34.33) and eliminating the delta function by integration over d^3k', we find

$$b_{il}(\mathbf{r}) = \int (v_i v_k)_{\mathbf{k}} e^{i\mathbf{k}\cdot\mathbf{r}} d^3k(2\pi)^3;$$

that is, the $(v_i v_l)_{\mathbf{k}}$ are the Fourier components of $b_{il}(\mathbf{r})$, and are therefore symmetrical in i and l, and real. In particular, $b_{ii}(\mathbf{k}) = (\mathbf{v}^2)_{\mathbf{k}}$, and we can now say that this quantity is

† The following arguments are a paraphrase of the proof given in *SP* 1, §122.

positive, as is evident from its relation (34.34) to the positive quantity $\langle \mathbf{v}_{\mathbf{k}}\mathbf{v}_{\mathbf{k}'}\rangle = \langle |\mathbf{v}_{\mathbf{k}}|^2\rangle$, the mean square modulus of the fluctuating velocity Fourier component.

The value of the correlation function $b_{ii}(\mathbf{r})$ for $\mathbf{r} = 0$ determines the mean square velocity of the fluid at any point in space. It is expressed in terms of the spectral function by

$$\langle \mathbf{v}^2\rangle = b_{ii}(\mathbf{r}=0) = \int b_{ii}(\mathbf{k})\,\mathrm{d}^3k/(2\pi)^3$$

or, substituting $b_{ii}(\mathbf{k})$ from (34.29),

$$\tfrac{1}{2}\langle \mathbf{v}^2\rangle = \int F^{(2)}(k)\,\mathrm{d}^3k/(2\pi)^3$$

$$= \int_0^\infty F^{(2)}(k)\cdot 4\pi k^2\,\mathrm{d}k/(2\pi)^3. \tag{34.35}$$

The meaning of this expression is clear from the foregoing: the positive quantity $F^{(2)}(k)/(2\pi)^3$ is the spectral density of the kinetic energy per unit mass of the fluid in **k**-space. The energy in the fluctuations whose wave number is in the range $\mathrm{d}k$ is $E(k)\mathrm{d}k$, where

$$E(k) = k^2F^{(2)}(k)/2\pi^2. \tag{34.36}$$

The first term on the right of (34.32) arises as the Fourier component of the first term on the right of (34.19). When $r \to 0$, the latter reduces to the derivative

$$\langle v_{1k}\frac{\partial}{\partial x_{1l}}v_{1i}v_{1l}\rangle + \langle v_{1i}\frac{\partial}{\partial x_{1l}}v_{1k}v_{1l}\rangle = \frac{\partial}{\partial x_{1l}}\langle v_{1i}v_{1k}v_{1l}\rangle$$

and is zero on account of the homogeneity. In the spectral representation, this means that

$$\int \mathbf{k}\,F^{(3)}(k)\,\mathrm{d}^3k = 0, \tag{34.37}$$

so that $F^{(3)}(k)$ has variable sign.

Equation (34.32) has a simple meaning: it represents the energy balance of the various spectral components in the turbulent flow. The second term on the right is negative; it gives the energy loss due to dissipation. The first term (due to the non-linear term in the Navier–Stokes equation) describes the energy redistribution in the spectrum, i.e. the energy transfer from the components with smaller k to those with larger k. The energy density $E(k)$ has a maximum at $k \sim 1/l$; the majority of the total energy of the turbulent flow is concentrated near the maximum (in the energy range, §33). The energy dissipation density $2\nu k^2E(k)$ is greatest for $k \sim 1/\lambda_0$; the majority of the total dissipation is concentrated in the dissipation range. At very high Reynolds numbers, these two regions are far apart and the inertial range lies between them.

Integrating (34.32) over $\mathrm{d}^3k/(2\pi)^3$ gives on the left the time derivative of the total kinetic energy of the fluid; this is equal to the total energy dissipation $-\varepsilon$. We thus find the following "normalization condition" for $E(k)$:

$$2\nu \int_0^\infty k^2 E(k,t)\, dk = \varepsilon. \tag{34.38}$$

In the inertial range of wave numbers ($1/l \ll k \ll 1/\lambda_0$), the spectral functions, like the correlation functions in the coordinate representation, may be regarded as time-independent. According to (33.13), we have in this range

$$E(k) = C_1 \varepsilon^{2/3} k^{-5/3}, \tag{34.39}$$

where C_1 is a constant coefficient, related to the coefficient C in the correlation function

$$B_{rr}(r) = C(\varepsilon r)^{2/3} \tag{34.40}$$

by $C_1 = 0{\cdot}76C$; see the Problem. The empirical values are $C \cong 2$, $C_1 \cong 1{\cdot}5$.† Then

$$|B_{rrr}|/B_{rr}{}^{3/2} = 4/5C^{3/2} \cong 0{\cdot}3.$$

PROBLEM

Relate the coefficients C_1 and C in formulae (34.39) and (34.40) for the correlation function and the spectral density of energy in the inertial range.

SOLUTION. The functions

$$B_{ii}(r) = 2B_{tt}(r) + B_{rr}(r) = (11/3)B_{rr}(r)$$

(from (34.6)) and

$$B_{ii}(k) = -2b_{ii}(k) = -4F^{(2)}(k) = -8\pi^2 E(k)/k^2$$

($k \neq 0$) are related through the Fourier integral

$$B_{ii}(k) = \int B_{ii}(r) e^{-i\mathbf{k}\cdot\mathbf{r}} d^3x.$$

If the wave number is in the inertial range ($1/l \ll k \ll 1/\lambda_0$), the oscillatory factor cuts off the integral at an upper limit $r \sim 1/k \ll l$. At small distances, the integral converges, since $B_{ii}(r) \to 0$ as $r \to 0$. In practice, therefore, the integral is governed by distances that lie in the inertial range ($\lambda_0 \ll r \ll l$), and we can substitute in it $B_{rr}(r)$ from (34.40), at the same time extending the integration to all space. In the integral

$$I = \int r^{3/2} e^{-i\mathbf{k}\cdot\mathbf{r}} d^3x,$$

we first integrate over the directions of $\mathbf{r}$, obtaining

$$I = \frac{4\pi}{k} \operatorname{im} \int_0^\infty r^{5/3} e^{ikr} dr = \frac{4\pi}{k^{11/3}} \int_0^\infty \xi^{5/3} e^{i\xi} d\xi.$$

The remaining integral is found by rotating the contour of integration in the complex ξ-plane from the right-hand half of the real axis to the upper half of the imaginary axis. The result is

$$I = -\frac{4\pi}{k^{11/3}} \frac{10\pi}{9\Gamma(1/3)}.$$

Combining these expressions, we have finally

$$C_1 = \frac{55}{27\Gamma(1/3)} C = 0{\cdot}76C.$$

---

† The majority of the experiments relate to turbulence in the atmosphere or the ocean. The Reynolds numbers in these measurements were as high as 3×10^8.

§35. The turbulent region and the phenomenon of separation

Turbulent flow is in general rotational. However, the distribution of the vorticity in the fluid has certain peculiarities in turbulent flow (for very large R): in "steady" turbulent flow past bodies, the whole volume of the fluid can usually be divided into two separate regions. In one of these the flow is rotational, while in the other the vorticity is zero, and we have potential flow. Thus the vorticity is non-zero only in a part of the fluid (though not in general only in a finite part).

That such a limited region of rotational flow can exist is a consequence of the fact that turbulent flow may be regarded as the motion of an ideal fluid, satisfying Euler's equations.† We have seen (§8) that, for the motion of an ideal fluid, the law of conservation of circulation holds. In particular, if at any point on a streamline the curl of the velocity is zero, then the same is true at every point on that streamline. Conversely, if at any point on a streamline **curl v** $\neq 0$, then it does not vanish anywhere on the streamline. Hence it is clear that the existence of limited regions of rotational and irrotational flow is compatible with the equations of motion if the region of rotational flow is such that the streamlines within it do not penetrate into the region outside it. Such a distribution of the vorticity will be stable, and it will remain zero beyond the surface of separation.

One of the properties of the region of rotational turbulent flow is that the exchange of fluid between this region and the surrounding space can occur only in one direction. The fluid can enter this region from the region of potential flow, but can never leave it.

We should emphasize that the arguments given here cannot, of course, be regarded as affording a rigorous proof of the statements made. However, the existence of limited regions of rotational turbulent flow seems to be confirmed by experiment.

The flow is turbulent both in the rotational and in the irrotational region. The nature of the turbulence, however, is totally different in the two regions. To elucidate the reason for this difference, we may point out the following general property of potential flow, which obeys Laplace's equation $\triangle \phi = 0$. Let us suppose that the flow is periodic in the xy-plane, so that ϕ involves x and y through a factor having the form $e^{ik_1x+ik_2y}$. Then

$$\partial^2\phi/\partial x^2 + \partial^2\phi/\partial y^2 = -(k_1{}^2+k_2{}^2)\phi = -k^2\phi,$$

and, since the sum of the second derivatives must be zero, the second derivative of ϕ with respect to z must equal ϕ multiplied by a positive coefficient: $\partial^2\phi/\partial z^2 = k^2\phi$. The dependence of ϕ on z is then given by a damping factor of the form e^{-kz} for $z > 0$ (the unlimited increase given by e^{kz} is clearly impossible). Thus, if the potential flow is periodic in some plane, it must be damped in the direction perpendicular to that plane. Moreover, the greater k_1 and k_2 (i.e. the smaller the period of the flow in the xy-plane), the more rapidly the flow is damped along the z-axis. All these arguments remain qualitatively valid in cases where the motion is not strictly periodic, but has only some periodic quality.

From this the following result is obtained. Outside the region of rotational flow, the turbulent eddies must be damped, and must be so more rapidly for the smaller eddies. In other words, the small eddies do not penetrate very far into the region of potential flow. Consequently, only the largest eddies are important in this region; they are damped at distances of the order of the (transverse) dimension of the rotational region, which is just the fundamental scale of turbulence in this case. At distances greater than this dimension there is practically no turbulence, and the flow may be regarded as laminar.

† The applicability of these equations to turbulent flow ends at distances of the order of λ_0. The sharp boundary between rotational and irrotational flow is therefore defined only to within such distances.

We have seen that the energy dissipation in turbulent flow occurs in the smallest eddies; the large eddies do not involve appreciable dissipation, which is why Euler's equation is applicable to them. From what has been said above, we reach the important result that the energy dissipation occurs mainly in the region of rotational turbulent flow, and hardly at all outside that region.

Bearing in mind all these properties of the rotational and irrotational turbulent flow, we shall henceforward, for brevity, call the region of rotational turbulent flow simply the *region of turbulent flow* or the *turbulent region*. In the following sections we shall discuss the form of this region in various cases.

The turbulent region must be bounded in some direction by part of the surface of the body past which the flow takes place. The line bounding this part of the surface is called the *line of separation*. From it begins the surface of separation between the turbulent fluid and the remainder. The formation of a turbulent region in flow past a body is called the *phenomenon of separation*.

The form of the turbulent region is determined by the properties of the flow in the main body of the fluid (i.e. not in the immediate neighbourhood of the surface). A complete theory of turbulence (which does not yet exist) would have to make it possible, in principle, to determine the form of this region by using the equations of motion for an ideal fluid, given the position of the line of separation on the surface of the body. The actual position of the line of separation, however, is determined by the properties of the flow in the immediate neighbourhood of the surface (known as the *boundary layer*), where the viscosity plays a vital part (see §40).

In referring (in subsequent sections) to a free boundary of the turbulent region, we shall of course mean its time-averaged position. The instantaneous position of the boundary is a highly irregular surface; these irregular distortions and their time variation are due mainly to the large eddies and accordingly extend to depths comparable with the fundamental scale of the turbulence. The irregular movement of the boundary surface has the result that a point in the flow fixed in space and not too far from the average position of the surface is alternately on opposite sides of the boundary. When the flow pattern is observed at such a point, there will be alternate periods where small-scale turbulence is present and absent.†

§36. The turbulent jet

The form of the turbulent region, and some other basic properties of it, can be established in certain cases by simple similarity arguments. These cases include, among others, various kinds of free turbulent jet in a space filled with fluid (L. Prandtl 1925).

As a first example, let us consider the turbulent region formed when a flow is separated at an angle formed by two infinite intersecting planes (shown in cross-section in Fig. 24). For laminar flow (Fig. 3, §10), the flow along one side of the angle (*AO*, say) would turn smoothly and flow along the other side away from the angle (*OB*). In turbulent flow, the pattern is totally different.

The flow along one side of the angle now does not turn on reaching the vertex, but continues in its former direction. A flow appears along the other side in the direction *BO*.

† This is called the alternation (or intermittency) of turbulence. It is to be distinguished from the similar property of the flow structure within a turbulent region, called by the same name. The available models of such phenomena will not be discussed here.

The two flows mix in the turbulent region;† the boundaries of this region are shown, dashed, in cross-section in Fig. 24. The origin of this region can be seen as follows. Let us imagine a flow in which a uniform stream along *AO* continues in the same direction, occupying the whole space above the plane *AO* and its continuation into the fluid to the right, while the fluid below this plane is at rest. In other words, we have a surface of separation (the plane *AO* produced) between fluid moving with constant velocity and stationary fluid. Such a surface of discontinuity, however, is unstable, and cannot exist in practice (see §29). This instability leads to mixing and the formation of a turbulent region. The flow along *BO* arises because fluid must enter the turbulent region from outside.

Let us determine the form of the turbulent region. We take the *x*-axis in the direction shown in Fig. 24, the origin being at *O*. We denote by Y_1 and Y_2 the distances from the *xz*-plane to the upper and lower boundaries of the turbulent region, and require to determine Y_1 and Y_2 as functions of *x*. This can easily be done from similarity considerations. Since the planes are infinite in all directions, there are no constant parameters at our disposal having the dimensions of length. Hence it follows that Y_1, Y_2 can only be directly proportional to the distance *x*:

$$Y_1 = x \tan \alpha_1, \qquad Y_2 = x \tan \alpha_2. \tag{36.1}$$

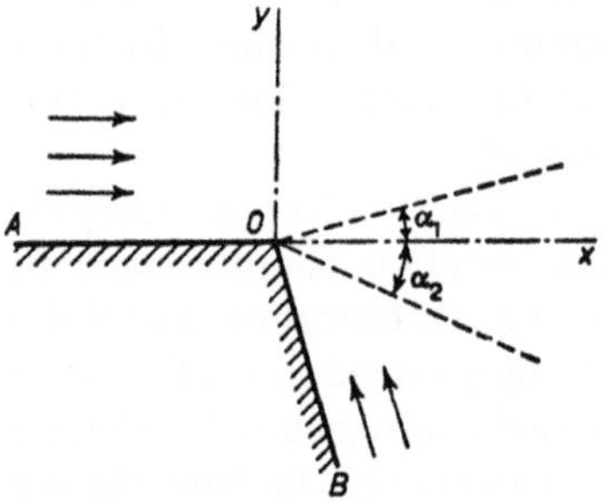

FIG. 24

The proportionality coefficients are simply numerical constants; we write them as $\tan \alpha_1$, $\tan \alpha_2$, so that α_1 and α_2 are the angles between the two boundaries of the turbulent region and the *x*-axis. Thus the turbulent region is bounded by two planes intersecting along the vertex of the angle.

The values of α_1, α_2 depend only on the size of the angle, and not, for example, on the velocity of the main stream. They cannot be calculated theoretically; the experimental results for flow round a right angle are $\alpha_1 = 5°$, $\alpha_2 = 10°$.‡

The velocities of the flows along the two sides of the angle are not the same; their ratio is a definite number, again depending only on the size of the angle. When the angle is not close to zero, one of the velocities is considerably the greater, namely that of the main stream, which is in the same direction (*AO*) as the turbulent region. For example, in flow round a right angle, the velocity along the plane *AO* is thirty times that along *BO*.

---

† We recall that, outside the turbulent region, there is irrotational turbulent flow which gradually becomes laminar as we move away from the boundaries of this region.

‡ Here, and elsewhere, we refer to experimental results on the velocity distribution in a transverse cross-section of the turbulent jet, reduced by means of calculations based on a semi-empirical theory (see the final note to the present section).

We may also mention that the difference between the fluid pressures on the two sides of the turbulent region is very small. For example, in flow round a right angle it is found that $p_1 - p_2 = 0{\cdot}003\rho U_1^2$, where U_1 is the velocity of the main stream (along AO), p_1 the pressure in that stream, and p_2 the pressure in the stream along BO.

In the limiting case of flow round an angle of zero, we have simply the edge of a plate with fluid moving along both sides. The angle $\alpha_1 + \alpha_2$ of the turbulent region is zero, i.e. there is no turbulent region; the velocities of the flows along the two sides of the plate become equal. As the angle AOB increases, a point is reached when the plane BO forms the lower boundary of the turbulent region; the angle AOB is by then obtuse. As the angle increases further, the turbulent region continues to be bounded by the plane BO on one side. Here we have simply a separation, with the line of separation along the vertex of the angle. The angle of the turbulent region remains finite.

As a second example, let us consider the problem of a turbulent jet of fluid issuing from the end of a narrow tube into an infinite space filled with the same fluid. The problem of laminar flow in such a "submerged jet" has been solved in §23. At distances (the only ones we shall consider) large compared with the dimensions of the mouth of the tube, the jet is axially symmetrical, whatever the actual shape of the opening.

Let us determine the form of the turbulent region in the jet. We take the axis of the jet as the x-axis, and denote by R the radius of the turbulent region; we require to determine R as a function of x (which is measured from the end of the tube). As in the previous example, this function is easily determined directly from dimensional considerations. At distances large compared with the dimensions of the mouth of the tube, the actual shape and size of the opening cannot affect the form of the jet. Hence we have at our disposal no characteristic parameters with the dimensions of length. It therefore follows as before that R must be proportional to x:

$$R = x \tan \alpha, \tag{36.2}$$

where the numerical constant $\tan \alpha$ is the same for all jets. Thus the turbulent region is a cone; the experimental value of the angle 2α is about 25 degrees (Fig. 25).†

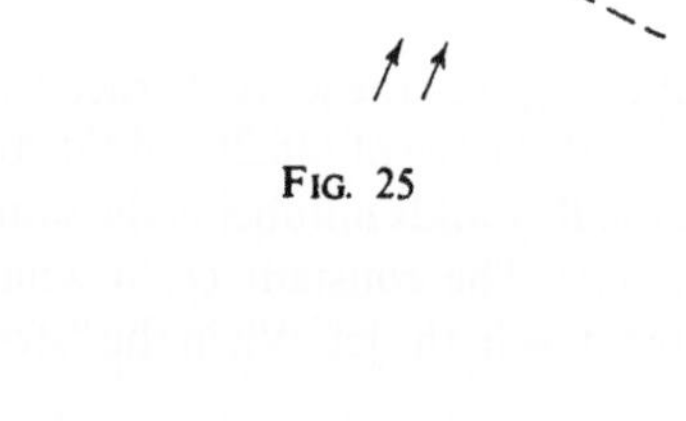

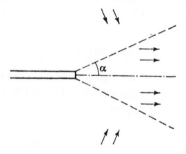

FIG. 25

† Formula (36.2) gives $R = 0$ for $x = 0$; that is, the coordinate x is measured from the point where the jet would start from a point source. This need not coincide with the actual position of the outlet aperture, but may be behind it by a distance of the same order of magnitude as is needed to establish the dependence (36.2). Since we are interested in the asymptotic form for large x, this difference may be neglected.

The flow in the jet is mainly axial. Because there are no parameters having the dimensions of length or velocity which could describe the flow in the jet,† the longitudinal velocity u_x (time-averaged) in it must have a distribution

$$u_x(r, x) = u_0(x) f[r/R(x)], \tag{36.3}$$

where r is the distance from the jet axis and u_0 is the velocity on the axis. Thus the velocity profiles in different cross-sections of the jet differ only as regards the scales of measurement of distance and speed; the jet structure is said to be *self-similar*. The function $f(\xi)$, equal to 1 when $\xi = 0$, decreases rapidly as the argument increases. It is equal to $\frac{1}{2}$ for $\xi = 0{\cdot}4$, and reaches $\sim 0{\cdot}01$ at the boundary of the turbulent region. The transverse velocity has about the same order of magnitude over the cross-section of the turbulent region, and at the boundary of the region is about $-0{\cdot}025\, u_0$ (being directed into the jet there). This transverse velocity is responsible for the inflow into the turbulent region. The flow outside the turbulent region can be found theoretically; see Problem 1.

The dependence of the velocity in the jet on the distance x can be determined from the following simple arguments. The total momentum flux in the jet through a spherical surface centred at its point of emergence must remain constant when the radius of the surface is varied. The momentum flux density in the jet is of the order of ρu^2, where u is of the order of some mean velocity in the jet. The area of the part of the jet cross-section where the velocity is appreciably different from zero is of the order of R^2. Hence the total momentum flux is $P \sim \rho u^2 R^2$. Substituting (36.2), we get

$$u \sim \sqrt{(P/\rho)}\, (1/x), \tag{36.4}$$

that is, the velocity diminishes inversely as the distance from the point of emergence.

The mass Q of fluid which passes per unit time through a cross-section of the turbulent region of the jet is of the order of $\rho u R^2$. Substituting (36.2) and (36.4), we find that $Q = \text{constant} \times x$: we write an equals sign because, if two quantities which vary within wide limits are always of the same order of magnitude, they must be proportional. The proportionality factor is conveniently expressed not in terms of the momentum flux P but in terms of the mass Q_0 of fluid which issues from the tube per unit time. At distances of the order of the linear dimensions a of the tube aperture, we must have $Q \sim Q_0$. Thus the constant is $\sim Q_0/a$, and

$$Q = \beta Q_0 x/a, \tag{36.5}$$

where β is a numerical coefficient which depends only on the form of the aperture. If the latter is circular with radius a, the empirical value is $\beta \cong 1{\cdot}5$. Thus the discharge through the cross-section of the turbulent region increases with x, and fluid is drawn into the turbulent region.‡

The flow in any section of the length of the jet is characterized by the Reynolds number for that section, defined as uR/ν. By virtue of (36.2) and (36.4), however, the product uR is constant along the jet, so that the Reynolds number is the same for all such sections. It can be taken, for instance, as $Q_0/\rho a\nu$. The constant Q_0/a which appears here is the only parameter which determines the flow in the jet. When the "strength" Q_0 of the jet increases

† Note once more that we are considering fully developed turbulence in the jet, and the viscosity therefore should not appear in the formulae concerned.

‡ The total flux through any infinite plane across the jet is infinite, i.e. a jet issuing into an infinite space carries with it an infinite amount of fluid.

(the value of a remaining constant), the Reynolds number eventually reaches a critical value, after which the flow simultaneously becomes turbulent along the whole length of the jet.†

PROBLEMS

PROBLEM 1. Determine the mean flow in the jet outside the turbulent region.

SOLUTION. We take spherical polar coordinates r, θ, ϕ, with the polar axis along the axis of the jet, and the origin at its point of emergence. Because the jet is axially symmetrical, the component u_ϕ of the mean velocity is zero, while u_θ and u_r are functions only of r and θ. The same arguments as were used in the problem of the laminar jet (§23) show that u_θ and u_r must have the forms $u_\theta = f(\theta)/r$, $u_r = F(\theta)/r$. Outside the turbulent region we have potential flow, i.e. $\mathbf{curl}\, \mathbf{u} = 0$, so that $\partial u_r/\partial\theta - \partial(ru_\theta)/\partial r = 0$. But ru_θ is independent of r, so that $\partial u_r/\partial\theta = (1/r)\mathrm{d}F/\mathrm{d}\theta = 0$, whence $F = \text{constant} = -b$, say, or

$$u_r = -b/r. \tag{1}$$

From the equation of continuity,

$$\frac{1}{r^2}\frac{\partial}{\partial r}(r^2 u_r) + \frac{1}{r\sin\theta}\frac{\partial}{\partial\theta}(u_\theta \sin\theta) = 0,$$

we then obtain

$$f = \frac{\text{constant} - b\cos\theta}{\sin\theta}.$$

The constant of integration must be $-b$ if the velocity is not infinite for $\theta = \pi$ (it does not matter that f is infinite for $\theta = 0$, since the solution in question refers only to the space outside the turbulent region, whereas $\theta = 0$ lies inside that region). Thus

$$u_\theta = -\frac{b(1+\cos\theta)}{r\sin\theta} = -\frac{b}{r}\cot\tfrac{1}{2}\theta. \tag{2}$$

The component of the velocity in the direction of the jet (u_x) and its absolute magnitude are

$$u_x = \frac{b}{r} = \frac{b\cos\theta}{x}, \qquad u = \frac{b}{r\sin\frac{1}{2}\theta}. \tag{3}$$

The constant b can be related to the constant $B = \beta Q_0/a$ in (36.5). Let us consider a segment of the cone formed by the turbulent region, bounded by two infinitely close cross-sections of the cone. The mass of fluid entering this segment per unit time is $\mathrm{d}Q = -2\pi r\rho\sin\alpha\cdot u_\theta \mathrm{d}r = 2\pi b\rho(1+\cos\alpha)\mathrm{d}r$, while from formula (36.5) we have $\mathrm{d}Q = B\,\mathrm{d}x = B\cos\alpha\,\mathrm{d}r$. Comparing the two expressions, we obtain

$$b = \frac{B\cos\alpha}{2\pi\rho(1+\cos\alpha)}. \tag{4}$$

At the boundary of the turbulent region, the velocity $\mathbf{u}$ is directed into this region, making an angle $\frac{1}{2}(\pi-\alpha)$ with the positive direction of the x-axis.

† In order to make more detailed calculations for various kinds of turbulent flow, it is customary to employ certain "semi-empirical" theories, based on assumptions concerning the dependence of the turbulent viscosity coefficient on the gradient of the mean velocity. For example, in Prandtl's theory it is assumed that (for plane flow)

$$\nu_{\text{turb}} = l^2 |\partial u_x/\partial y|,$$

where the dependence of l (called the *mixing length*) on the coordinates is chosen in accordance with the results of similarity arguments; for instance, in free turbulent jets we put $l = cx$, c being an empirical constant. Such theories usually give good agreement with experiment, and are therefore useful for interpolatory calculations. However, it is not possible to give universal values to the empirical constants which characterize each theory; for example, the value of the ratio of the mixing length l to the transverse dimension of the turbulent region has to be chosen diffierently in various particular cases. It should also be mentioned that good agreement with experimental results can be obtained with various expressions for the turbulent viscosity.

Let us compare the mean velocity $\bar{u}_x$ inside the turbulent region (defined as $\bar{u}_x = Q/\pi\rho R^2 = B/\pi\rho x \tan^2 \alpha$) with the velocity $(u_x)_{\text{pot}}$ at the boundary of the region. Taking the first equation (3) with $\theta = \alpha$, we find

$$(u_x)_{\text{pot}}/\bar{u}_x = \tfrac{1}{2}(1 - \cos \alpha).$$

For $\alpha = 12°$, this ratio is 0·011, i.e. the velocity at the boundary of the turbulent region is small compared with the mean velocity inside the region.

PROBLEM 2. Determine the law of variation of size and velocity in a submerged turbulent jet issuing from an infinitely long thin slit.

SOLUTION. By the same reasoning as for the axial jet, we conclude that the turbulent region is bounded by two planes intersecting along the slit, i.e. the half-width of the jet is $Y = x \tan \alpha$. The momentum flux in the jet (per unit length of the slit) is of the order of $\rho u^2 Y$. The dependence of the mean velocity u on x is therefore given by $u \sim \text{constant}/\sqrt{x}$. The discharge through a cross-section of the turbulent region is $Q \sim \rho u Y$, whence $Q = \text{constant} \times \sqrt{x}$. The local Reynolds number $R = uY/\nu$ increases in the same way with x. The experimental data give a value $2\alpha \cong 25°$ for the angle of a plane jet, about the same as for a circular jet.

§37. The turbulent wake

For Reynolds numbers considerably above the critical value, in flow past a solid body, a long region of turbulent flow is formed behind the body. This is called the *turbulent wake*. At distances large compared with the dimension of the body, simple arguments enable us to determine the form of this wake and the way in which the fluid velocity decreases there (L. Prandtl 1926).

As in the investigation of the laminar wake in §21, we denote by $\mathbf{U}$ the velocity of the incident stream, and take the direction of $\mathbf{U}$ as the x-axis. The fluid velocity at any point, averaged over the turbulent fluctuations, is written as $\mathbf{U} + \mathbf{u}$. Denoting by a some mean width of the wake, we shall find a as a function of x. If there is no lift, then at large distances from the body the wake is axially symmetrical and circular in cross-section; in this case, a may be the radius of the wake. If a lift force is present, a direction is selected in the yz-plane, and the wake is not axially symmetrical at any distance from the body.

The longitudinal fluid velocity component in the wake is of the order of U, while the transverse component is of the order of some mean value u of the turbulent velocity. The angle between the streamlines and the x-axis is therefore of the order of u/U. The boundary of the wake is, as we know, the boundary beyond which the streamlines of the rotational turbulent flow cannot pass. Hence it follows that the angle between the boundary of the longitudinal cross-section of the wake and the x-axis is also of the order of u/U. This means that we can write

$$da/dx \sim u/U. \tag{37.1}$$

Next we use formulae (21.1), (21.2), which determine the forces on the body in terms of integrals of the fluid velocity in the wake (the velocity now being interpreted as its mean value). The region of integration in these integrals is of the order of a^2. Hence an estimate of the integral gives $F \sim \rho U u a^2$, where F is of the order of the drag or the lift. Thus

$$u \sim F/\rho U a^2. \tag{37.2}$$

Substituting in (37.1), we find $da/dx \sim F/\rho U^2 a^2$, from which we have by integration

$$a \sim (Fx/\rho U^2)^{\frac{1}{3}}. \tag{37.3}$$

Thus the width of the wake increases as the cube root of the distance from the body. For the velocity u, we have from (37.2) and (37.3)

$$u \sim (FU/\rho x^2)^{\frac{1}{3}}, \tag{37.4}$$

i.e. the mean fluid velocity in the wake is inversely proportional to $x^{\frac{2}{3}}$.

The flow in any section of the wake is characterized by the Reynolds number $R \sim au/\nu$. Substituting (37.3) and (37.4), we obtain

$$R \sim F/\nu\rho U a \sim (F^2/\rho^2 U x \nu^3)^{\frac{1}{3}}.$$

We see that this number is not constant along the wake, unlike what we found for the turbulent jet. At sufficiently large distances from the body, R becomes so small that the flow in the wake is no longer turbulent. Beyond this point we have the laminar wake, whose properties have been investigated in §21.

In §21 formulae have been obtained which describe the flow outside the wake and far from the body. These formulae hold for flow outside the turbulent wake as well as outside the laminar wake.

We may mention here some general properties of the velocity distribution round the body. Both inside and outside the turbulent wake, the velocity (by which we always mean **u**) decreases away from the body. However, the longitudinal velocity u_x falls off more rapidly ($\sim 1/x^2$) outside the wake than inside it. Far from the body, therefore, we may suppose u_x to be zero outside the wake. We may say that u_x falls from some maximum value on the axis of the wake to zero at the boundary of the wake. The transverse components u_y, u_z at the boundary are of the same order of magnitude as they are inside the wake, diminishing rapidly as we move away from the wake at a given distance from the body.

§38. Zhukovskiĭ's theorem

The velocity distribution round a body, described at the end of the last section, does not hold for exceptional cases where the thickness of the wake formed behind the body is very small compared with its width. A wake of this kind is formed in flow past bodies whose thickness (in the y-direction) is small compared with their width (in the z-direction); the length (in the direction of flow, the x-direction) may be of any magnitude. That is, we are considering flow past bodies whose cross-section transverse to the flow is very elongated. These bodies include, in particular, *wings*, i.e. bodies whose width, or *span*, is large in comparison with their other dimensions.

It is clear that, in such a case, there is no reason why the velocity component u_y perpendicular to the plane of the turbulent wake should fall off appreciably at distances of the order of the thickness of the wake. On the contrary, this component will now be of the same order of magnitude inside the wake and at considerable distances from it, of the order of the span. Here, of course, we assume that the lift is not zero, since otherwise the transverse velocity practically vanishes.

Let us consider the vertical lift force F_y resulting from such a flow. According to formula (21.2), it is given by the integral

$$F_y = -\rho U \iint u_y \, \mathrm{d}y \, \mathrm{d}z, \tag{38.1}$$

where, on account of the nature of the distribution of u_y, the integration must now be taken over the whole transverse plane. Furthermore, since the thickness of the wake (in the y-direction) is small, while the velocity u_y inside the wake is not large compared with its value outside, we can with sufficient accuracy take the integration over y to be over the region outside the wake, writing

$$\int_{-\infty}^{\infty} u_y \, dy \cong \int_{y_1}^{\infty} u_y \, dy + \int_{-\infty}^{y_2} u_y \, dy,$$

where y_1 and y_2 are the coordinates of the boundaries of the wake (Fig. 26).

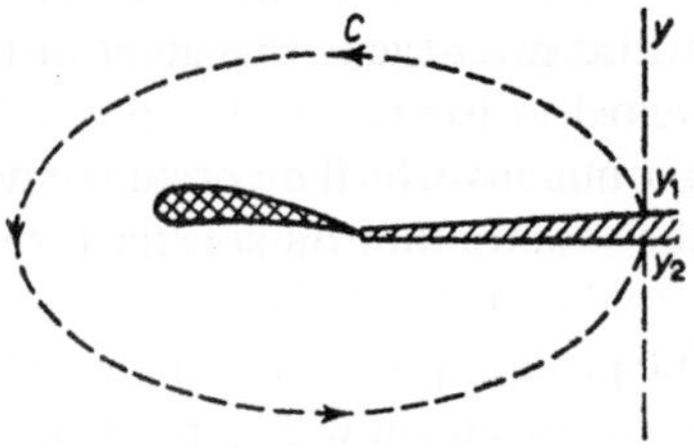

FIG. 26

Outside the wake, however, we have potential flow, and $u_y = \partial\phi/\partial y$; bearing in mind that $\phi = 0$ at infinity, we therefore obtain

$$\int u_y \, dy = \phi_2 - \phi_1,$$

where ϕ_2 and ϕ_1 are the values of the potential on the two sides of the wake. We may say that $\phi_2 - \phi_1$ is the discontinuity of the potential at the surface of discontinuity which may be substituted for a thin wake. The derivative $u_y = \partial\phi/\partial y$ must remain continuous. A discontinuity in the velocity component normal to the surface of the wake would mean that some quantity of fluid flows into the wake; in the approximation in which the thickness of the wake is neglected, however, this inflow must be zero. Thus we replace the wake by a surface of tangential discontinuity. Next, in the same approximation, the pressure also must be continuous at the wake. Since the variation of the pressure is given in the first approximation, according to Bernoulli's equation, by $\rho U u_x = \rho U \partial\phi/\partial x$, it follows that the derivative $\partial\phi/\partial x$ must also be continuous. The derivative $\partial\phi/\partial z$ (the velocity along the wing) is in general discontinuous, however.

Since the derivative $\partial\phi/\partial x$ is continuous, the discontinuity $\phi_2 - \phi_1$ depends only on z, and not on the coordinate x along the wake. Thus we have the following formula for the lift:

$$F_y = -\rho U \int (\phi_2 - \phi_1) \, dz. \tag{38.2}$$

The integration over z may be taken over the width of the wake (of course, $\phi_2 - \phi_1 \equiv 0$ outside the wake).

This formula can be put in a somewhat different form. To do so, we notice that, using well-known properties of an integral of the gradient of a scalar, we can write the difference $\phi_2 - \phi_1$ as a contour integral

$$\oint \mathbf{grad}\, \phi \cdot \mathrm{d}\mathbf{l} = \oint (u_y \mathrm{d}y + u_x \mathrm{d}x),$$

taken along a contour which starts from the point y_1, encircles the body, and ends at the point y_2, thus passing at every point through the region of potential flow. Since the wake is thin we can, without changing the integral except by quantities of higher order, close this contour by means of the short segment from y_2 to y_1. Denoting by Γ the velocity circulation round the closed contour C enclosing the body (Fig. 26), we have

$$\Gamma = \oint \mathbf{u} \cdot \mathrm{d}\mathbf{l} = \phi_2 - \phi_1, \tag{38.3}$$

and for the lift force the formula

$$F_y = -\rho U \int \Gamma\, \mathrm{d}z. \tag{38.4}$$

The sign of the velocity circulation is always chosen to be that obtained for a counter-clockwise path. The sign in formula (38.3) also depends on the chosen direction of flow. We always suppose that the flow is in the positive direction of the x-axis (from left to right).

The relation between the lift and the circulation given by formula (38.4) constitutes *Zhukovskiĭ's theorem*, first derived by N. E. Zhukovskiĭ in 1906. Cf. §46 for the application of this theorem to streamlined wings.

PROBLEMS

PROBLEM 1. Determine the manner of widening of the turbulent wake formed in transverse flow past a cylinder with infinite length.

SOLUTION. The drag f_x per unit length of the cylinder is of the order of $\rho U u Y$. Combining this with the relation (37.1), we find the width Y of the wake to be

$$Y = A\sqrt{(x f_x/\rho U^2)}, \tag{1}$$

where A is a constant. The mean velocity u in the wake falls off in accordance with $u \sim \sqrt{(f_x/\rho x)}$. The Reynolds number $\mathrm{R} \sim Yu/\nu \sim f_x/\rho U \nu$ is independent of x, and there is therefore no laminar wake.

We may mention that, according to experimental results, the constant coefficient in (1) is $A = 0{\cdot}9$ (Y being the half-width of the wake; if Y is taken as the distance at which the velocity u_x falls to half its maximum value (at the centre of the wake), then $A = 0{\cdot}4$).

PROBLEM 2. Determine the flow outside the wake formed in transverse flow past a body of infinite length.

SOLUTION. Outside the wake we have potential flow; we shall denote the potential by Φ to distinguish it from the angle ϕ in the system of cylindrical polar coordinates which we take, with the z-axis along the length of the body. As in (21.16), we conclude that we must have

$$\oint \mathbf{u} \cdot \mathrm{d}\mathbf{f} = \oint \mathbf{grad}\, \Phi \cdot \mathrm{d}\mathbf{f} = f_x/\rho U,$$

where now the integration is over the surface of a cylinder with large radius and unit length with its axis in the x-direction, and f_x is the drag per unit length of the body. The solution of the two-dimensional Laplace's equation $\Delta\Phi = 0$ that satisfies this condition is $\Phi = (f_x/2\pi\rho U) \log r$. Next, we have for the lift, by formula (38.2), $f_y = \rho U (\Phi_1 - \Phi_2)$. The solution of Laplace's equation that diminishes least rapidly with increasing distance and has a discontinuity of the plane $\phi = 0$ is $\Phi = \text{constant} \times \phi = -\phi f_y/2\pi\rho U$, the constant being determined by $\phi_2 - \phi_1 = 2\pi$. The flow is given by the sum of these two solutions, i.e.

$$\Phi = \frac{1}{2\pi\rho U}(f_x \log r - \phi f_y). \tag{2}$$

The cylindrical components of the velocity **u** are

$$u_r = \partial\Phi/\partial r = f_x/2\pi\rho U r, \qquad u_\phi = (1/r)\partial\Phi/\partial\phi = -f_y/2\pi\rho U r. \tag{3}$$

The velocity **u** is at a constant angle $\tan^{-1}(f_y/f_x)$ to the r-direction.

PROBLEM 3. Determine the manner of bending of the wake behind a body with infinite length when there is a lift force.

SOLUTION. If there is a lift force, the wake (regarded as a surface of discontinuity) is curved in the xy-plane. The function $y = y(x)$ which determines this is given by the equation $dx/(u_x + U) = dy/u_y$. Substituting, by (3), $u_y \cong -f_y/2\pi\rho U x$ and neglecting u_x in comparison with U, we obtain

$$dy/dx = -f_y/2\pi\rho U^2 x,$$

whence

$$y = \text{constant} - (f_y/2\pi\rho U^2)\log x.$$

CHAPTER IV

BOUNDARY LAYERS

§39. The laminar boundary layer

We have several times mentioned the fact that very large Reynolds numbers are equivalent to very small viscosities, and consequently a fluid may be regarded as ideal if R is large. However, this approximation can never be used when the flow in question occurs near solid walls. The boundary conditions for an ideal fluid require only the normal velocity component to vanish; the component tangential to the surface in general remains finite. For a viscous fluid, however, the velocity at a solid wall must vanish entirely.

From this we can conclude that, for large Reynolds numbers, the decrease of the velocity to zero occurs almost exclusively in a thin layer adjoining the wall. This is called the *boundary layer*, and is thus characterized by the presence in it of considerable velocity gradients. The flow in the boundary layer may be either laminar or turbulent. In this section we shall consider the properties of the laminar boundary layer. The boundary of the layer is not, of course, sharp; the transition from the laminar flow in it to the main stream of fluid is continuous.

The rapid decrease of the velocity in the boundary layer is due ultimately to the viscosity, which cannot be neglected even if R is large. Mathematically, this appears in the fact that the velocity gradients in the boundary layer are large, and therefore the viscosity terms in the equations of motion, which contain space derivatives of the velocity, are large even if ν is small.†

Let us derive the equations of motion of the fluid in a laminar boundary layer. For simplicity, we consider two-dimensional flow along a plane portion of the surface. This plane is taken as the xz-plane, with the x-axis in the direction of flow. The velocity distribution is independent of z, and the velocity has no z-component.

The exact Navier–Stokes equations and the equation of continuity are then

$$v_x \frac{\partial v_x}{\partial x} + v_y \frac{\partial v_x}{\partial y} = -\frac{1}{\rho}\frac{\partial p}{\partial x} + \nu\left(\frac{\partial^2 v_x}{\partial x^2} + \frac{\partial^2 v_x}{\partial y^2}\right), \tag{39.1}$$

$$v_x \frac{\partial v_y}{\partial x} + v_y \frac{\partial v_y}{\partial y} = -\frac{1}{\rho}\frac{\partial p}{\partial y} + \nu\left(\frac{\partial^2 v_y}{\partial x^2} + \frac{\partial^2 v_y}{\partial y^2}\right), \tag{39.2}$$

$$\frac{\partial v_x}{\partial x} + \frac{\partial v_y}{\partial y} = 0. \tag{39.3}$$

The flow is supposed steady, and the time derivatives are therefore omitted.

† The concept and basic equations of the laminar boundary layer theory were formulated by L. Prandtl (1904).

Since the boundary layer is thin, it is clear that the flow in it takes place mainly parallel to the surface, i.e. the velocity v_y is small compared with v_x (as is seen immediately from the equation of continuity).

The velocity varies rapidly along the y-axis, an appreciable change in it occurring at distances of the order of the thickness δ of the boundary layer. Along the x-axis, on the other hand, the velocity varies slowly, an appreciable change in it occurring only over distances of the order of a length l characteristic of the problem (the dimension of the body, say). Hence the y-derivatives of the velocity are large in comparison with the x-derivatives. It follows that, in equation (39.1), the derivative $\partial^2 v_x/\partial x^2$ may be neglected in comparison with $\partial^2 v_x/\partial y^2$; comparing (39.1) with (39.2), we see that the derivative $\partial p/\partial y$ is small in comparison with $\partial p/\partial x$ (the ratio being of the same order as v_y/v_x). In the approximation considered we can put simply

$$\partial p/\partial y = 0, \tag{39.4}$$

i.e. suppose that there is no transverse pressure gradient in the boundary layer. In other words the pressure in the boundary layer is equal to the pressure $p(x)$ in the main stream, and is a given function of x for the purpose of solving the boundary-layer problem. In equation (39.1) we can now write, instead of $\partial p/\partial x$, the total derivative $\mathrm{d}p(x)/\mathrm{d}x$; this derivative can be expressed in terms of the velocity $U(x)$ of the main stream. Since we have potential flow outside the boundary layer, Bernoulli's equation, $p+\frac{1}{2}\rho U^2 = \text{constant}$, holds, whence $(1/\rho)\mathrm{d}p/\mathrm{d}x = -U\,\mathrm{d}U/\mathrm{d}x$.

Thus we obtain the equations of motion in the laminar boundary layer in the form of *Prandtl's equations*:

$$v_x\frac{\partial v_x}{\partial x} + v_y\frac{\partial v_x}{\partial y} - \nu\frac{\partial^2 v_x}{\partial y^2} = -\frac{1}{\rho}\frac{\mathrm{d}p}{\mathrm{d}x} = U\frac{\mathrm{d}U}{\mathrm{d}x}, \tag{39.5}$$

$$\frac{\partial v_x}{\partial x} + \frac{\partial v_y}{\partial y} = 0. \tag{39.6}$$

The boundary conditions on these equations are that the velocity be zero at the wall:

$$v_x = v_y = 0 \quad \text{for } y = 0. \tag{39.7}$$

Away from the wall, the longitudinal velocity must tend asymptotically to that of the main stream:

$$v_x = U(x) \quad \text{for } y \to \infty. \tag{39.8}$$

It is not necessary to specify a separate condition for v_y at infinity.

It can easily be shown that equations (39.5) and (39.6), though derived for flow along a plane wall, remain valid in the more general case of any two-dimensional flow (transverse flow past a cylinder with infinite length and arbitrary cross-section). Here x is the distance measured along the circumference of the cross-section from some point on it, and y is the distance from the surface along the normal.

Let U_0 be a velocity characteristic of the problem (for example, the velocity of the main stream at infinity). Instead of the coordinates x, y and the velocities v_x, v_y, we introduce the dimensionless variables x', y', v'_x, v'_y:

$$x = lx', \qquad y = ly'/\sqrt{\mathrm{R}}, \qquad v_x = U_0 v'_x, \qquad v_y = U_0 v'_y/\sqrt{\mathrm{R}} \tag{39.9}$$

(and correspondingly $U = U_0 U'$), where $\mathrm{R} = U_0 l/\nu$. Then the equations (39.5) and (39.6) take the form

$$\left.\begin{aligned} v'_x \frac{\partial v'_x}{\partial x'} + v'_y \frac{\partial v'_x}{\partial y'} - \frac{\partial^2 v'_x}{\partial y'^2} &= U' \frac{\mathrm{d}U'}{\mathrm{d}x'}, \\ \frac{\partial v'_x}{\partial x'} + \frac{\partial v'_y}{\partial y'} &= 0. \end{aligned}\right\} \tag{39.10}$$

These equations (and the boundary conditions on them) do not involve the viscosity. This means that their solutions are independent of the Reynolds number. Thus we reach the important result that, when the Reynolds number is changed, the whole flow pattern in the boundary layer simply undergoes a similarity transformation, longitudinal distances and velocities remaining unchanged, while transverse distances and velocities vary as $1/\sqrt{\mathrm{R}}$.

Next, we can say that the dimensionless velocities v'_x, v'_y obtained by solving equations (39.10) must be of the order of unity, since they do not depend on R. From formulae (39.9) we can therefore conclude that

$$v_y \sim U_0/\sqrt{\mathrm{R}}, \tag{39.11}$$

i.e. the ratio of the transverse and longitudinal velocities is inversely proportional to $\sqrt{\mathrm{R}}$. The same is true of the boundary layer thickness δ: in the dimensionless coordinates x', y' we have $\delta' \cong 1$, and hence in the coordinates x and y

$$\delta \sim l/\sqrt{\mathrm{R}}. \tag{39.12}$$

Let us apply the equations for the boundary layer to the case of plane-parallel flow along a semi-infinite flat plate (H. Blasius 1908). Let the plane of the plate be the xz half-plane with $x > 0$ (the leading edge of the plate thus being the line $x = 0$). The velocity of the main stream in this case is constant (U = constant). The equations (39.5) and (39.6) become

$$v_x \frac{\partial v_x}{\partial x} + v_y \frac{\partial v_x}{\partial y} = \nu \frac{\partial^2 v_x}{\partial y^2}, \qquad \frac{\partial v_x}{\partial x} + \frac{\partial v_y}{\partial y} = 0. \tag{39.13}$$

In the solution of Prandtl's equations, we have seen that v_x/U and $v_y\sqrt{(l/U\nu)}$ can only be functions of $x' = x/l$ and $y' = y\sqrt{(U/l\nu)}$. The problem of a semi-infinite plate has no characteristic length l, however. Hence v_x/U can depend only on a combination of x' and y' which does not involve l, namely $y'/\sqrt{x'} = y\sqrt{(U/\nu x)}$. Similarly, the product $v'_y\sqrt{x'}$ must be a function of $y'/\sqrt{x'}$.

In order to take into account immediately the relation between v_x and v_y expressed by the equation of continuity, we use the stream function ψ as defined by (10.9):

$$v_x = \partial\psi/\partial y, \quad v_y = -\partial\psi/\partial x. \tag{39.14}$$

The above-mentioned properties of $v_x(x, y)$ and $v_y(x, y)$ correspond to a stream function

$$\psi = \sqrt{(\nu x U)}\, f(\xi), \qquad \xi = y\sqrt{(U/\nu x)}. \tag{39.15}$$

Then

$$v_x = Uf'(\xi), \qquad v_y = \tfrac{1}{2}\sqrt{(\nu U/x)}\,(\xi f' - f). \tag{39.16}$$

An important conclusion can be drawn without determining quantitatively the function $f(\xi)$. The chief characteristic of the flow in the boundary layer is the distribution of the longitudinal velocity v_x in it (since v_y is small). The velocity v_x increases from zero at the surface of the plate to a definite fraction of U for a given value of ξ. Hence we can conclude that the thickness of the boundary layer in flow along a plate (defined as the value of y for which v_x/U reaches a certain value ~ 1) is given in order of magnitude by

$$\delta \sim \sqrt{(\nu x/U)}. \qquad (39.17)$$

Thus, as we move away from the edge of the plate, δ increases as the square root of the distance from the edge.

Substituting (39.16) in the first equation (39.13), we get an equation for $f(\xi)$:

$$ff'' + 2f''' = 0. \qquad (39.18)$$

The boundary conditions (39.7) and (39.8) become

$$f(0) = f'(0) = 0, \qquad f'(\infty) = 1; \qquad (39.19)$$

the velocity distribution is evidently symmetrical about the plane $y = 0$, and it is therefore sufficient to consider the side $y > 0$. Equation (39.18) has to be solved numerically; a graph of the function $f'(\xi)$ thus obtained is shown in Fig. 27. We see that $f'(\xi)$ tends very rapidly to its limiting value of unity. The limiting form of $f(\xi)$ itself for small ξ is

$$f(\xi) = \tfrac{1}{2}\alpha\xi^2 + O(\xi^5), \qquad \alpha = 0{\cdot}332; \qquad (39.20)$$

there cannot be terms in ξ^3 or ξ^4, as is easily seen from (39.18). The limiting form for large ξ is

$$f(\xi) = \xi - \beta, \qquad \beta = 1{\cdot}72; \qquad (39.21)$$

it can be shown that the error in this expression in exponentially small.

The frictional force on unit area of the surface of the plate is

$$\sigma_{xy} = \eta(\partial v_x/\partial y)_{y=0} = \eta\sqrt{(U^3/\nu x)}f''(0)$$

or

$$\sigma_{xy} = 0{\cdot}332\sqrt{(\eta\rho U^3/x)}. \qquad (39.22)$$

If the plate has a length l (in the x-direction), then the total frictional force on it per unit length along the edge is

$$F = 2\int_0^l \sigma_{xy}\,dx = 1{\cdot}328\sqrt{(\eta\rho l U^3)}. \qquad (39.23)$$

The factor 2 is due to the fact that the plate has two sides exposed to the fluid.† The frictional force is proportional to the $\frac{3}{2}$ power of the velocity of the main stream. Formula (39.23) can be applied, of course, only to long plates, for which $R = Ul/\nu$ is sufficiently large. Instead of the force, it is customary to define the *drag coefficient* as the dimensionless ratio

$$C = F/\tfrac{1}{2}\rho U^2 \cdot 2l. \qquad (39.24)$$

† The boundary layer approximation is not valid near the leading edge of the plate, where $\delta \gtrsim x$. This, however, is not important in calculating the total force F, because the integral converges rapidly at the lower limit.

By (39.23), this quantity for laminar flow past a plate is inversely proportional to the square root of the Reynolds number:

$$C = 1{\cdot}328/\sqrt{\mathrm{R}}. \tag{39.25}$$

As an exactly definable characteristic of the boundary layer thickness, we can use the *displacement thickness* δ^* defined by

$$U\delta^* = \int_0^\infty (U - v_x)\,\mathrm{d}y. \tag{39.26}$$

Substitution of v_x from (39.16) gives

$$\delta^* = \sqrt{(\nu x/U)}\int_0^\infty (1-f')\,\mathrm{d}\xi$$

$$= \sqrt{(\nu x/U)}\,[\xi - f(\xi)]_{\xi\to\infty},$$

and, with the limiting expression (39.21),

$$\delta^* = \beta\sqrt{(\nu x/U)} = 1{\cdot}72\sqrt{(\nu x/U)}. \tag{39.27}$$

The expression on the right of the definition (39.26) is the amount by which the discharge in the boundary layer is less than in a homogeneous flow with velocity U. We can therefore say that δ^* is the distance by which the flow is displaced outwards from the plate because of the retardation of the fluid in the boundary layer. This displacement has the result that the transverse velocity v_y in the boundary layer tends, as $y \to \infty$, not to zero but to the non-zero value

$$v_y = \tfrac{1}{2}\sqrt{(\nu U/x)}\,[\xi f' - f]_{\xi\to\infty} = \tfrac{1}{2}\beta\sqrt{(\nu U/x)} = 0{\cdot}86\sqrt{(\nu U/x)}. \tag{39.28}$$

The quantitative formulae obtained above relate, of course, only to flow along a flat plate. The qualitative results, however, such as (39.11) and (39.12), hold for flow past bodies of any shape; in such cases l is the dimension of the body in the direction of flow.

We may make special mention of two cases of the boundary layer. If we have a plane disk, with large radius, rotating in the fluid about an axis perpendicular to its plane, then to

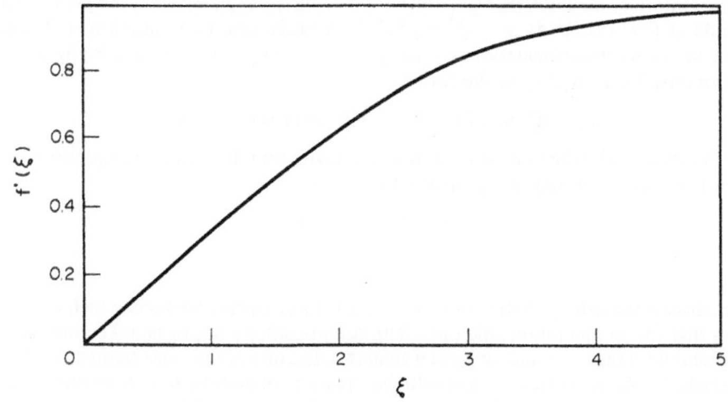

FIG. 27

estimate the thickness of the boundary layer we must replace U in (39.17) by Ωx, where Ω is the angular velocity of rotation. We then find

$$\delta \sim \sqrt{(\nu/\Omega)}. \tag{39.29}$$

We see that the thickness of the boundary layer may be regarded as a constant over the surface of the disk, in accordance with the exact solution of this problem obtained in §23. The moment of the frictional forces on the disk, as obtained from the equations for the boundary layer, is of course (23.4), since this formula is exact and therefore holds for laminar flow with any value of R.

Finally, let us consider the laminar boundary layer formed at the walls of a pipe near the point of entry of fluid. The fluid usually enters the pipe with a velocity distribution which is almost constant over the cross-section, and the velocity falls to zero entirely within the boundary layer. As we move away from the entrance to the pipe, the fluid layers nearer the axis are retarded. Since the mass of fluid that passes each cross-section is the same, the inner part of the stream, where the velocity is almost uniform, must be accelerated as its diameter is reduced. This continues until a Poiseuille velocity distribution is asymptotically reached; this distribution is thus found only at some distance from the entrance to the pipe. It is easy to determine the order of magnitude of the length l of the "inlet section". It is given by the fact that, a distance l from the entrance, the thickness of the boundary layer is of the same order of magnitude as the radius a of the pipe, so that the boundary layer fills almost the whole cross-section. Putting in (39.17) $x \sim l$ and $\delta \sim a$, we obtain

$$l \sim a^2U/\nu \sim a\mathrm{R}. \tag{39.30}$$

Thus the length of the inlet section is proportional to the Reynolds number.†

PROBLEMS

PROBLEM 1. Determine the thickness of the boundary layer near a stagnation point (see §10).

SOLUTION. Near the stagnation point the fluid velocity (outside the boundary layer) is proportional to the distance x from that point, so that we can put $U = cx$. By estimating the magnitudes of the terms in the equations (39.5) and (39.6) we find $\delta \sim \sqrt{(\nu/c)}$. Thus the thickness of the boundary layer near the stagnation point is finite.

PROBLEM 2. Determine the flow in the boundary layer in a converging channel (§23) between two non-parallel planes (K. Pohlhausen 1921).

SOLUTION. Considering the boundary layer along one of the planes, we measure the coordinate x along that plane from the point O (Fig. 8, §23). For an ideal fluid we should have the velocity $U = Q/\alpha x\rho$; this simply expresses the conservation of the discharge Q in the flow, α being the angle between the planes. Thus we have on the right-hand side of (39.5) $U\mathrm{d}U/\mathrm{d}x = -Q^2/\alpha^2\rho^2x^3$. It is easily seen that equations (39.5) and (39.6) then become invariant under the transformation $x \to ax$, $y \to ay$, $v_x \to v_x/a$, $v_y \to v_y/a$, a being any constant. This means that we can look for v_x and v_y in the forms

$$v_x = (Q/\alpha\rho x)f(\xi), \quad v_y = (Q/\alpha\rho x)f_1(\xi), \quad \xi = y/x,$$

which is likewise invariant under the transformation mentioned. From the continuity equation (39.6) we find that $f_1 = \xi f$, and then (39.5) gives the following equation for $f(\xi)$:

$$(\rho\nu\alpha/Q)f'' = 1 - f^2. \tag{1}$$

† We shall not discuss the theory of the boundary layer for a compressible fluid, which is considerably more complicated than that for an incompressible fluid. The compressibility has to be taken into account when the velocity is comparable with that of sound, or greater than this. Because of the then considerable heating of the gas and the body past which it flows, we have to deal with the equations of motion in the boundary layer together with the equation of heat transfer in it. It may also be necessary to take account of the temperature dependences of the viscosity and thermal conductivity of the gas.

The boundary conditions (39.8) show that we must have $f(0) = 0$, $f(\infty) = 1$. The first integral of (1) is

$$(\rho v\alpha/2Q)f'^2 = f - \tfrac{1}{3}f^3 + \text{constant}.$$

Since, as $\xi \to \infty$, f tends to unity, we see that f' also tends to a definite limit, and it is clear that this can only be zero. The constant is thus determined, and

$$(\rho v\alpha/2Q)f'^2 = -\tfrac{1}{3}(f-1)^2(f+2). \tag{2}$$

Since the right-hand side is negative in the range $0 \leqslant f \leqslant 1$, we must have $Q < 0$: a boundary layer of the type under consideration is formed only in converging flow (with large Reynolds numbers $R = |Q|/\rho\alpha v$), not in diverging flow, in agreement with the results of §23. A further integration gives finally

$$f = 3\tanh^2[\tanh^{-1}\sqrt{(2/3)} + \sqrt{(\tfrac{1}{2}R)}\xi] - 2. \tag{3}$$

The thickness of the boundary layer $\delta \sim x/\sqrt{R}$. The derivative $f'(0) = 2\sqrt{(\tfrac{1}{3}R)}$, as is seen from (2). The frictional force per unit wall area is therefore

$$\sigma_{xy} = (\eta U/x)f'(0) = \sqrt{(4U^3\eta\rho/3x)} = (2/x^2)\sqrt{(\eta|Q|^3/3\alpha^3\rho^2)}.$$

§40. Flow near the line of separation

In describing the line of separation (§35) we have already mentioned that the actual position of this line on the surface of the body is determined by the properties of the flow in the boundary layer. We shall see below that, from a mathematical point of view, the line of separation is a line whose points are singular points of the solutions of (Prandtl's) equations of motion in the boundary layers. The problem is to determine the properties of these solutions near such a line of singularities.†

We know already that, from the line of separation, there begins a surface which extends into the fluid and marks off the region of turbulent flow. The flow is rotational throughout the turbulent region, whereas in the absence of separation it would be rotational only in the boundary layer, where the viscosity is important; the curl of the velocity would be zero in the main stream. Hence we can say that separation causes this quantity to penetrate from the boundary layer into the fluid. By the conservation of circulation, however, this penetration can occur only by the direct mixing of fluid moving near the surface (in the boundary layer) with the main stream. In other words, the flow in the boundary layer must be separated from the surface of the body, the streamlines consequently leaving the surface layer and entering the interior of the fluid. This phenomenon is therefore called *separation* or *separation of the boundary layer*.

The equations of motion in the boundary layer lead, as we have seen, to the result that the tangential velocity component (v_x) in the boundary layer is large compared with the component (v_y) normal to the surface of the body. This relation between v_x and v_y derives from our basic assumptions regarding the nature of the flow in the boundary layer, and must necessarily be found wherever Prandtl's equations have physically meaningful solutions. Mathematically, it is found at all points not lying in the immediate neighbourhood of singular points. But, if $v_y \ll v_x$ it follows that the fluid moves along the surface of the body, and moves away from the surface only very slightly, so that there can be no separation. We therefore reach the conclusion that separation can occur only on a line whose points are singularities of the solution of Prandtl's equations.

The nature of these singularities also follows immediately. For, as we approach the line of separation, the flow deviates from the boundary layer towards the interior of the fluid.

† The treatment of the problem given here, due to L. D. Landau (1944), is somewhat different from that usually given.

In other words, the normal velocity component ceases to be small compared with the tangential component, and is now of at least the same order of magnitude. We have seen (cf. (39.11)) that the ratio v_y/v_x is of the order of $1/\sqrt{R}$, so that an increase of v_y to the point where $v_y \sim v_x$ means an increase by a factor of $\sqrt{R}$. Hence, for sufficiently large Reynolds numbers (which, of course, we are considering) we may suppose that v_y increases by an infinite factor. If we use Prandtl's equations in dimensionless form (see (39.10)), the situation just described is formally equivalent to an infinite value of the dimensionless velocity v'_y on the line of separation.

In order to simplify the subsequent discussion a little, we shall consider the two-dimensional problem of transverse flow past a body with infinite length. As usual, x is the coordinate along the surface in the direction of flow, while y is the distance from the surface of the body. Instead of a line of separation, we now have a point of separation, namely the intersection of the line of separation with the xy-plane; in the coordinates used, this is the point $x = \text{constant} \equiv x_0$, $y = 0$. Let $x < x_0$ be the region in front of the point of separation.

According to the above results, we have for all† y

$$v_y(x_0, y) = \infty. \tag{40.1}$$

In Prandtl's equations, however, v_y is a kind of parameter, which is usually of no interest (on account of its smallness) in investigating the flow in the boundary layer. Hence it is necessary to ascertain the properties of the function v_x near the line of separation.

It is clear from (40.1) that, for $x = x_0$, the derivative $\partial v_y/\partial y$ also becomes infinite. From the equation of continuity,

$$\partial v_x/\partial x + \partial v_y/\partial y = 0, \tag{40.2}$$

it then follows that $(\partial v_x/\partial x)_{x=x_0}$ is infinite, or

$$\partial x/\partial v_x = 0 \tag{40.3}$$

for $v_x = v_0$, where x is regarded as a function of v_x and y, and $v_0(y) = v_x(x_0, y)$. Near the point of separation, the differences $v_x - v_0$ and $x_0 - x$ are small, and we can expand $x_0 - x$ in powers of $v_x - v_0$ (for a given y). From (40.3), the first-order term in this expansion must vanish identically, and we have as far as terms of the second order $x_0 - x = f(y)(v_x - v_0)^2$, or

$$v_x = v_0(y) + \alpha(y)\sqrt{(x_0 - x)}, \tag{40.4}$$

where $\alpha = 1/\sqrt{f}$ is some function of y alone. Putting now

$$\frac{\partial v_y}{\partial y} = -\frac{\partial v_x}{\partial x} = \frac{\alpha(y)}{2\sqrt{(x_0 - x)}}$$

and integrating, we have for v_y

$$v_y = \beta(y)/\sqrt{(x_0 - x)}, \tag{40.5}$$

where $\beta(y)$ is another function of y.

† Except $y = 0$, where we must always have $v_y = 0$ in accordance with the boundary conditions at the surface of the body.

Next, we use the first equation (39.5):

$$v_x\frac{\partial v_x}{\partial x}+v_y\frac{\partial v_x}{\partial y}=\nu\frac{\partial^2 v_x}{\partial y^2}-\frac{1}{\rho}\frac{dp}{dx}. \tag{40.6}$$

The derivative $\partial^2 v_x/\partial y^2$ does not become infinite for $x = x_0$, as we see from (40.2). The same is true of dp/dx, which is determined by the flow outside the boundary layer. Both terms on the left-hand side of equation (40.6) become infinite, however. In the first approximation we can therefore write for the region near the point of separation $v_x\partial v_x/\partial x + v_y\partial v_x/\partial y = 0$. With the equation of continuity (40.2), we can rewrite this as

$$v_x\frac{\partial v_y}{\partial y}-v_y\frac{\partial v_x}{\partial y}=v_x{}^2\frac{\partial}{\partial y}\left(\frac{v_y}{v_x}\right)=0.$$

Since the velocity v_x does not in general vanish for $x = x_0$, it follows that the ratio v_y/v_x is independent of y. From (40.4) and (40.5), we have to within terms of higher order

$$\frac{v_y}{v_x}=\frac{\beta(y)}{v_0(y)\sqrt{(x_0-x)}}.$$

If this is a function of x alone, we must have $\beta(y) = \frac{1}{2}Av_0(y)$, where A is a numerical constant. Thus

$$v_y=\frac{Av_0(y)}{2\sqrt{(x_0-x)}}. \tag{40.7}$$

Finally, noticing that α and β in (40.4) and (40.5) obey the relation $\alpha = 2\beta'$, we obtain $\alpha = A\,dv_0/dy$, so that

$$v_x = v_0(y) + A(dv_0/dy)\sqrt{(x_0-x)}. \tag{40.8}$$

Formulae (40.7) and (40.8) determine v_x and v_y as functions of x near the point of separation. We see that each can be expanded in this region in powers of $\sqrt{(x_0-x)}$, the expansion of v_y beginning with the -1 power, so that v_y becomes infinite as $(x_0-x)^{-\frac{1}{2}}$ for $x \to x_0$. For $x > x_0$, i.e. beyond the point of separation, the expansions (40.7) and (40.8) are physically meaningless, since the square roots become imaginary; this means that the solutions of Prandtl's equations which give the flow up to the point of separation cannot meaningfully be continued beyond that point.

From the boundary conditions at the surface of the body, we must always have $v_x = v_y = 0$ for $y = 0$. We therefore conclude from (40.7) and (40.8) that

$$v_0(0) = 0, \qquad (dv_0/dy)_{y=0} = 0. \tag{40.9}$$

Thus we have the important result (due to Prandtl) that, at the point of separation itself ($x = x_0$, $y = 0$), not only the velocity v_x but also its first derivative with respect to y is zero.

It must be emphasized that the equation $\partial v_x/\partial y = 0$ on the line of separation holds only when v_y becomes infinite for that value of x. If the constant A in (40.7) happens to be zero, so that $v_y(x_0, y) \neq \infty$, then the point $x = x_0$, $y = 0$ at which the derivative $\partial v_x/\partial y$ vanishes would have no other particular properties, and would not be a point of separation. A can vanish, however, only by chance, and such an event is therefore unlikely. In practice a point on the surface of the body at which $\partial v_x/\partial y = 0$ is always a point of separation.

If there is no separation at the point $x = x_0$ (i.e. if $A = 0$), then for $x > x_0$ we have $(\partial v_x/\partial y)_{y=0} < 0$, i.e. v_x becomes negative (with increasing absolute magnitude) as we move away from the surface, y being still small. That is, the fluid beyond the point $x = x_0$ moves, in the lower parts of the boundary layer, in the direction opposite to that of the main stream; there is a "back-flow" of fluid at this point. It must be emphasized that from such arguments we cannot conclude that there is necessarily a point of separation where $\partial v_x/\partial y = 0$; the whole flow pattern with the "back-flow" might lie (as it does for $A = 0$) entirely within the boundary layer and not enter the main stream, whereas it is characteristic of separation that the flow enters the main body of the fluid.

It has been shown in the previous section that the flow pattern in the boundary layer is similar for different Reynolds numbers, and, in particular the scale in the x-direction remains unchanged. It follows from this that the value x_0 of the coordinate x for which the derivative $(\partial v_x/\partial y)_{y=0}$ is zero is the same for all R. Thus we have the important result that the position of the point of separation on the surface of the body is independent of the Reynolds number (so long as the boundary layer remains laminar, of course; see §45).

Let us also ascertain the properties of the pressure distribution $p(x)$ near the point of separation. For $y = 0$ the left-hand side of equation (40.6) is zero together with v_x and v_y, and there remains

$$\nu(\partial^2 v_x/\partial y^2)_{y=0} = (1/\rho)\mathrm{d}p/\mathrm{d}x. \tag{40.10}$$

It is clear from this that the sign of $\mathrm{d}p/\mathrm{d}x$ is the same as that of $(\partial^2 v_x/\partial y^2)_{y=0}$. When $(\partial v_x/\partial y)_{y=0} > 0$ we can say nothing regarding the sign of the second derivative. However, since v_x is positive and increases away from the surface (in front of the point of separation), we must always have $(\partial^2 v_x/\partial y^2)_{y=0} > 0$ at $x = x_0$ itself, where $\partial v_x/\partial y = 0$. Hence we conclude that

$$(\mathrm{d}p/\mathrm{d}x)_{x=x_0} > 0, \tag{40.11}$$

i.e. the fluid near the point of separation moves from the lower pressure to the higher pressure. The pressure gradient is related to the gradient of the velocity $U(x)$ outside the boundary layer by $(1/\rho)\mathrm{d}p/\mathrm{d}x = -U\,\mathrm{d}U/\mathrm{d}x$. Since the positive direction of the x-axis is the same as the direction of the main stream, $U > 0$, and therefore

$$(\mathrm{d}U/\mathrm{d}x)_{x=x_0} < 0, \tag{40.12}$$

i.e. the velocity U decreases in the direction of flow near the point of separation.

From the results obtained above we can deduce that there must be separation somewhere on the surface of the body. For there is on both the front and the back of the body a point (the stagnation point) at which the fluid velocity is zero for potential flow of an ideal fluid. Consequently, for some value of x, the velocity $U(x)$ must begin to decrease, and finally it becomes zero. It is clear, however, that the fluid moving over the surface of the body is retarded more strongly closer to the surface (i.e. for smaller y). Hence, before the velocity $U(x)$ is zero at the outer limit of the boundary layer, the velocity in the immediate neighbourhood of the surface must be zero. Mathematically, this evidently means that the derivative $\partial v_x/\partial y$ must always vanish (and therefore there must be separation) for some x less than the value for which $U(x) = 0$.

In flow past bodies of any form the calculations can be carried out in an entirely similar manner, and they lead to the result that the derivatives $\partial v_x/\partial y$, $\partial v_z/\partial y$ of the two velocity components v_x and v_z tangential to the surface of the body vanish on the line of separation (the y-axis, as before, is along the normal to the portion of the surface considered).

We may give a simple argument which demonstrates the necessity of separation in cases where the fluid would otherwise have a rapid increase of pressure (and therefore a rapid decrease in the velocity U) in the direction of its flow past the body. Over a small distance $\Delta x = x_2 - x_1$, let the pressure p increase rapidly from p_1 to p_2 ($p_2 \gg p_1$). Over the same distance Δx, the fluid velocity U outside the boundary layer falls from its initial value U_1 to a considerably smaller value U_2 determined by Bernoulli's equation:

$$\tfrac{1}{2}(U_1{}^2 - U_2{}^2) = (p_2 - p_1)/\rho.$$

Since p is independent of y, the pressure increase $p_2 - p_1$ is the same at all distances from the surface. If the pressure gradient $dp/dx \sim (p_2 - p_1)/\Delta x$ is sufficiently high, the term $\nu\partial^2 v_x/\partial y^2$ involving the viscosity may be omitted from the equation of motion (40.6) (if, of course, y is not small). Then, to estimate the change in the velocity v in the boundary layer, we can use Bernoulli's equation, putting $\tfrac{1}{2}(v_2{}^2 - v_1{}^2) = -(p_2 - p_1)/\rho$, or, from the equation previously obtained, $v_2{}^2 = v_1{}^2 - (U_1{}^2 - U_2{}^2)$. The velocity v_1 in the boundary layer is less than that of the main stream, and we can select a value of y for which $v_1{}^2 < U_1{}^2 - U_2{}^2$. The velocity v_2 is then imaginary, showing that Prandtl's equations have no physically significant solutions. In fact, there must be separation in the distance Δx, as a result of which the pressure gradient is reduced.

An interesting case of the appearance of separation is given by flow at an angle formed by two intersecting solid surfaces. For laminar potential flow outside an angle (Fig. 3), the fluid velocity at the vertex of the angle would become infinite (see §10, Problem 6), increasing in the stream approaching the vertex and diminishing in the stream leaving the vertex. In reality, the rapid decrease in velocity (and corresponding increase in pressure) beyond the vertex would lead to separation, the line of separation being the line of intersection of the surfaces. The resulting flow pattern is that discussed in §36.

In laminar flow inside an angle (Fig. 4), the fluid velocity is zero at the vertex. In this case the velocity diminishes (and the pressure increases) in the flow approaching the vertex. The result is in general the appearance of separation, the line of separation being upstream from the vertex of the angle.

PROBLEM

Determine the order of magnitude of the least possible increase Δp in the pressure which can occur (in the main stream) over a distance Δx and cause separation.

SOLUTION. Let y be a distance from the surface of the body at which, firstly, Bernoulli's equation can be applied and, secondly, the squared velocity $v^2(y)$ in the boundary layer is less than the change $|\Delta U^2|$ in the squared velocity outside that layer. For $v(y)$ we can write, in order of magnitude, $v(y) \cong y\, dv/dy \sim Uy/\delta$, where $\delta \sim \sqrt{(\nu l/U)}$ is the thickness of the boundary layer and l the dimension of the body. Equating, in order of magnitude, the two terms on the right-hand side of equation (40.6), we find

$$(1/\rho)\Delta p/\Delta x \sim \nu v(y)/y^2 \sim \nu U/\delta y.$$

From the condition $v^2 = |\Delta U^2| = (2/\rho)\Delta p$ we have $U^2 y^2/\delta^2 \sim \Delta p/\rho$. Eliminating y, we finally obtain

$$\Delta p \sim \rho U^2(\Delta x/l)^{\frac{1}{3}}.$$

§41. Stability of flow in the laminar boundary layer

Laminar flow in the boundary layer, like any other laminar flow, becomes to some extent unstable at sufficiently large Reynolds numbers. The manner of the loss of stability in the boundary layer is similar to that which occurs for flow in a pipe (§28).

The Reynolds number for flow in the boundary layer varies over the surface of the body. For example, in flow along a plate we could define the Reynolds number as $R_x = Ux/\nu$, where x is the distance from the leading edge of the plate, and U the fluid velocity outside the boundary layer. A more suitable definition for the boundary layer, however, is one in which the length parameter directly characterizes the thickness of the layer; such, for instance, is the displacement thickness δ^* defined as in (39.26):

$$\mathrm{R}_\delta = U\delta^*/\nu = 1{\cdot}72\sqrt{\mathrm{R}_x}; \qquad (41.1)$$

the numerical factor is for a boundary layer on a flat surface.

Because the change in the layer thickness with distance is comparatively slow, and the transverse velocity in the layer is small, in investigating the stability of flow in a small portion of the layer, we may consider a plane-parallel flow with a velocity profile that does not vary along the x-axis.† Then, from a mathematical point of view, the problem is analogous to that of the stability of flow between two parallel planes discussed in §29. The only difference is in the form of the velocity profile: instead of a symmetrical profile with $v = 0$ on both sides, we now have an unsymmetrical profile in which the velocity varies from zero at the surface of the body to some given value U, the velocity of the flow outside the boundary layer. The investigation leads to the following results (W. Tollmien 1929; H. Schlichting 1933; C. C. Lin 1945).

The form of the neutral curve in the $\omega\mathrm{R}$-plane (see §28) depends on the form of the velocity profile in the boundary layer. If the velocity profile has no point of inflexion, and the velocity v_x increases monotonically with the curve $v_x = v_x(y)$ everywhere convex upwards (Fig. 28a), then the boundary of the stable region is completely similar in form to that which is obtained for flow in a pipe: there is a minimum value $\mathrm{R} = \mathrm{R}_{\mathrm{cr}}$ at which amplified perturbations first appear, and for $\mathrm{R} \to \infty$ both branches of the curve are asymptotic to the axis of abscissae (Fig. 29a). For the velocity profile which occurs in the boundary layer on a flat plate, the critical Reynolds number is found by calculation to be $\mathrm{R}_{\delta,\mathrm{cr}} \cong 420$.‡

A velocity profile of the kind shown in Fig. 28a cannot occur if the fluid velocity outside the boundary layer decreases downstream. In this case the velocity profile must have a point of inflexion. For, let us consider a small portion of the surface, which we may regard as plane, and let x be again the coordinate in the direction of flow, and y the distance from the wall. From (40.10) we have

$$\nu(\partial^2 v_x/\partial y^2)_{y=0} = (1/\rho)\,\mathrm{d}p/\mathrm{d}x = -U\,\partial U/\partial x,$$

whence we see that, if U decreases downstream ($\partial U/\partial x < 0$), we must have $\partial^2 v_x/\partial y^2 > 0$ near the surface, i.e. the curve $v_x = v_x(y)$ is concave upwards. As y increases, the velocity v_x must tend asymptotically to the finite limit U. It is then clear from geometrical considerations that the curve must become convex upwards, and therefore must have a point of inflexion (Fig. 28b). In this case the form of the curve defining the stable region is slightly changed: the two branches have different asymptotes for $\mathrm{R} \to \infty$, one tending as before to the axis of abscissae and the other to a non-zero value of ω (Fig. 29b). The presence of a point of inflexion also reduces considerably the value of R_{cr}.

† In so doing, of course, we pass over the question of the effect which the curvature of the surface may have on the stability of the boundary layer. There is also some inconsistency in the approximations made, because the only plane-parallel flows (with the velocity profile depending on only one coordinate) that satisfy the Navier–Stokes equation are those with a linear profile (17.1) or a parabolic profile (17.4), whereas Euler's equation is satisfied by a plane-parallel flow with any profile. Thus the main stream flow considered in the theory of boundary layer stability is not, strictly speaking, a solution of the equations of motion.

‡ For $\mathrm{R}_\delta \to \infty$, ω tends to zero, on the two branches I and II of the neutral curve, as $\mathrm{R}_\delta^{-1/2}$ and $\mathrm{R}_\delta^{-1/5}$ respectively. The point $\mathrm{R} = \mathrm{R}_{\mathrm{cr}}$ corresponds to a frequency $\omega_{\mathrm{cr}} = 0{\cdot}15\,U/\delta^*$ and a wave number $k_{\mathrm{cr}} = 0{\cdot}36/\delta^*$

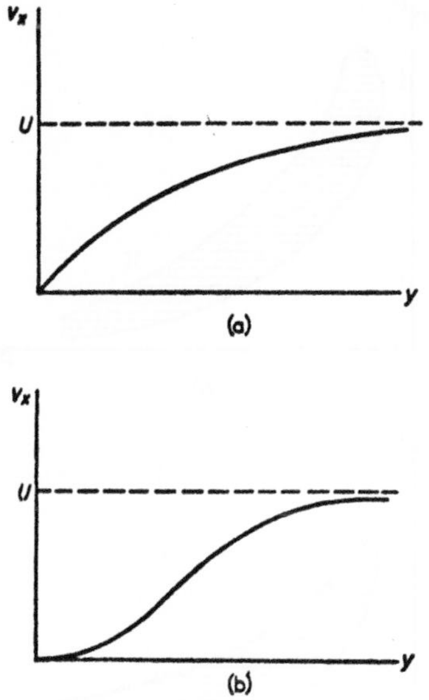

FIG. 28

The fact that the Reynolds number increases along the boundary layer makes the behaviour of the perturbations as they are carried downstream somewhat unusual. Let us consider flow along a flat plate, and suppose that a perturbation with a given frequency ω occurs at some point in the boundary layer. Its propagation downstream corresponds to a movement in Fig. 29a to the right along a horizontal line ω = constant. The perturbation is at first damped: then, on reaching branch I of the stability curve, it begins to be amplified. This continues until branch II is reached, whereupon the perturbation is again damped. The total amplification coefficient for the perturbation during its passage through the region of instability increases very rapidly as this region moves towards large R (i.e. as the corresponding horizontal segment between branches I and II moves downwards).

There is as yet no complete answer regarding the (absolute or convected) instability of the boundary layer under infinitesimal perturbations. For a velocity profile with no point of inflexion, the instability is convected for R values where both branches of the neutral curve (Fig. 29a) are close to the abscissa axis; the proof is the same as for plane Poiseuille flow in §28. For lower values of R and for velocity profiles having a point of inflexion, the problem is still unsolved.

Since the Reynolds number varies along the boundary layer, the whole layer does not become turbulent immediately, but only that part of it for which R_δ exceeds a certain value. For a given velocity of the incident flow, this means that turbulence begins at a particular distance from the leading edge, which becomes smaller as the velocity increases. The experimental results show that the point where turbulence begins in the boundary layer also depends considerably on the strength of the perturbation in the incident flow. As this decreases, the onset of turbulence moves to higher values of R_δ.

There is a fundamental difference between the neutral curves in Figs. 29a and 29b. The fact that, as $R_\delta \to \infty$, the frequency on the upper branch tends to a non-zero limit signifies that the flow becomes unstable for any viscosity, however small, whereas for a curve as in Fig. 29a perturbations with any non-zero frequency decay as $\nu \to 0$. This difference is

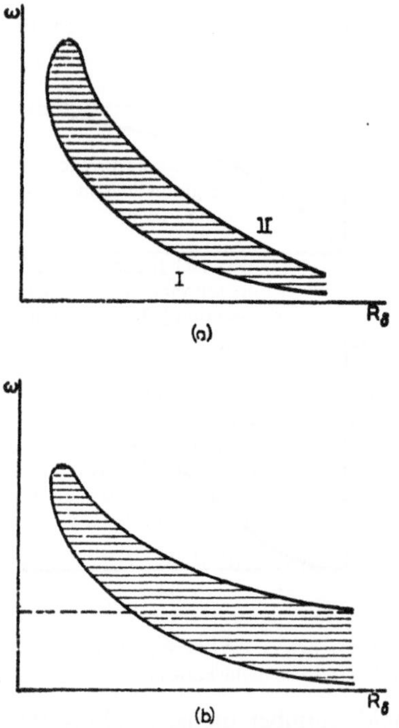

FIG. 29

caused by the presence or absence of a point of inflexion on the velocity profile $v_x = v(y)$. Its origin may be traced mathematically by considering the stability problem for an ideal fluid (Rayleigh 1880).

We substitute, in the equation (10.10) for two-dimensional flow of an ideal fluid, the stream function in the form

$$\psi = \psi_0(y) + \psi_1(x, y, t),$$

where ψ_0 is the stream function for the unperturbed flow, so that $\psi_0' = v(y)$; ψ_1 is a small perturbation, which we seek in the form

$$\psi_1 = \phi(y)e^{i(kx-\omega t)}$$

Substitution in (10.10) gives the following linearized equation† for ψ_1:

$$(v-\omega/k)(\phi''-k^2\phi)-v''\phi = 0. \tag{41.2}$$

If the flow is bounded (in the y-direction) by a solid wall, $\phi = 0$ there (since $v_y = 0$); if the flow is unlimited in width on one or both sides, a similar condition must be applied at infinity, where the flow is uniform. We shall regard k as a given real quantity; the frequency ω is then determined from the eigenvalues of the boundary-value problem for equation (41.2).

† Any function $\psi_0(y)$ satisfies (10.10) identically; cf. the first footnote to this section.

We divide (41.2) by $v-\omega/k$, multiply by ϕ^*, and integrate with respect to y between the limits of the flow y_1 and y_2. Integration by parts of the product $\phi^*\phi''$ gives

$$\int_{y_1}^{y_2} (|\phi'|^2 + k^2|\phi|^2)\,dy + \int_{y_1}^{y_2} \frac{v''|\phi|^2}{v-\omega/k}\,dy = 0. \tag{41.3}$$

The first term is always real. Assuming the frequency to be complex and separating the imaginary part of the equation, we get

$$\operatorname{im}\omega \int_{y_1}^{y_2} \frac{v''|\phi|^2}{|v-\omega/k|^2}\,dy = 0. \tag{41.4}$$

In order to have $\operatorname{im}\omega \neq 0$, the integral must be zero, and this certainly implies that v'' is zero somewhere in the range of integration. Thus instability can occur (when $\nu = 0$) only for velocity profiles having a point of inflexion).†

Physically, this instability is due to the resonance-type interaction between the oscillations of the medium and the movement of its particles in the main stream; in this sense, it is analogous to the Landau damping (or amplification in the case of instability) of oscillations in a collisionless plasma (*PK*, §30).‡

According to (41.2), the natural oscillations (if any) of the flow are associated with the part of it where $v''(y) \neq 0$.†† It is convenient to examine the mechanism of oscillation amplification for the case where the velocity profile has an oscillation source localized in one layer of the flow. Let us take a profile $v(y)$ whose curvature is small everywhere except near a point $y = y_0$. Replacing this simply by a kink in the profile, we get a term $A\delta(y-y_0)$ in $v''(y)$. This makes the main contribution to the integral in (41.3). We will describe the flow by means of coordinates in which the source is at rest, i.e. $v(y_0) = 0$, as shown in Fig. 30. Separating the real part of equation (41.3), we have

$$\int_{y_1}^{y_2} (|\phi'|^2 + k^2|\phi|^2)\,dy - \frac{A|\phi(y_0)|^2 \operatorname{re}(\omega/k)}{|\omega/k|^2} = 0.$$

Let $A > 0$, as in Fig. 30. Since the first term in this equation is certainly positive, we must then have re $(\omega/k) > 0$, the phase velocity of the wave being towards the right. The resonance point y_r at which the phase velocity is the same as the local flow velocity, $v(y_r) = \operatorname{re}(\omega/k)$, is to the right of y_0. Fluid particles moving near the resonance point and overtaking the wave transfer energy to it; those leaving the wave absorb energy from it; the wave is amplified (there is instability) if there are more of the first particles than of the

---

† It should be noted that the formulation of the stability problem with the exact equation $\nu = 0$ is physically not quite correct. It ignores the fact that a real fluid necessarily has a small but non-zero viscosity. This causes various mathematical difficulties: some solutions disappear, because of the lower order of the differential equation for ϕ, and new ones appear which do not occur when $\nu \neq 0$. The latter effect is related to the singularity of equation (41.2) (which is absent when $\nu \neq 0$): at the point where $v(y) = \omega/k$, the coefficient of the highest-order derivative in the equation is zero.

‡ This analogy was noted by A. V. Timofeev (1979) and by A. A. Andronov and A. L. Fabrikant (1979). The discussion below is as given by Timofeev.

†† When $v''(y) \equiv 0$, equation (41.2) has no solutions satisfying the necessary boundary conditions.

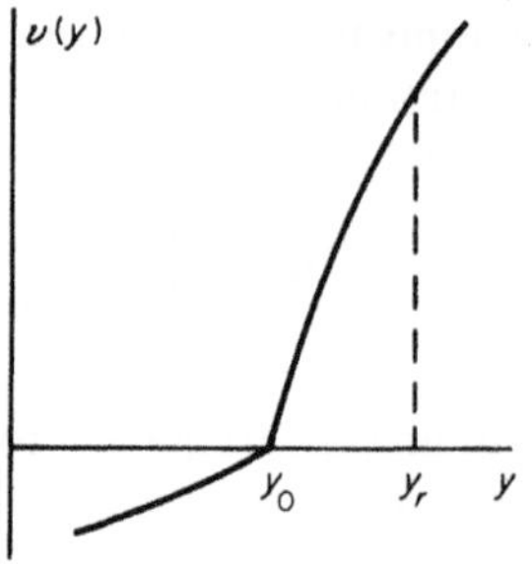

FIG. 30

second.† Because the fluid is assumed incompressible, the number of particles reaching an element dy of the width of the flow is proportional to dy; thus the number of particles with velocities in the range dv is proportional to $dy = (dy/dv)dv = dv/v'(y)$, so that $1/v'(y)$ acts as a velocity distribution function. Consequently, for instability to occur, it is necessary that the function $1/v'(y)$ should increase, and $v'(y)$ decrease, as we pass the point y_r from left to right. That is, we must have $v''(y_r) < 0$; since v'' is positive at y_0, the velocity profile must have a point of inflexion somewhere between y_0 and y_r.

The case where $A < 0$ can be treated similarly, and gives the same result; here, the phase velocity of the wave and the velocity of the resonance particles are to the left.

§42. The logarithmic velocity profile

Let us consider plane-parallel turbulent flow along an unbounded plane surface; the term "plane-parallel" applies, of course, to the time average of the flow.‡ We take the direction of the flow as the x-axis, and the plane of the surface as the xz-plane, so that y is the distance from the surface. The y and z components of the mean velocity are zero: $u_x = u$, $u_y = u_z = 0$. There is no pressure gradient, and all quantities depend on y only.

We denote by σ the frictional force on unit area of the surface; this force is clearly in the x-direction. The quantity σ is just the momentum transmitted by the fluid to the surface per unit time; it is the constant flux of the x-component of momentum, which is in the negative y-direction, and gives the amount of momentum continuously transmitted from the layers of fluid remote from the surface to those nearer it.

The existence of this momentum flux is due, of course, to the presence of a gradient, in the y-direction, of the mean velocity u. If the fluid moved with the same velocity at every point, there would be no momentum flux. The converse problem can also be stated: given some definite value of σ, what must be the motion of a fluid with a given density ρ to give rise to a momentum flux σ? With a view to deriving the asymptotic behaviour for very large Reynolds numbers, we again start from the supposition that this behaviour will not explicitly involve the fluid viscosity ν, although the latter becomes important for very small distances y; see below.

---

† The flow in the wave is steady with respect to the resonance particles; the energy exchange between them and the wave is therefore not zero when averaged over time (as it is for other particles relative to which the flow in the wave oscillates). It may also be noted that the above-mentioned direction of energy exchange corresponds to a tendency for the velocity gradient in the flow to decrease, and in this sense is equivalent to allowing for a very small viscosity.

‡ The results given in §§42–44 are due to T. von Kármán (1930) and L. Prandtl (1932).

Thus the value of the velocity gradient $\mathrm{d}u/\mathrm{d}y$ at any distance from the wall must be determined by the constant parameters ρ, σ, and of course the distance y itself. The only combination of ρ, σ and y that has the right dimensions is $\sqrt{(\sigma/\rho y^2)}$. Hence we must have

$$\mathrm{d}u/\mathrm{d}y = v_*/\kappa y, \tag{42.1}$$

with the quantity v_* (having the dimensions of velocity), which is convenient later, defined by

$$\sigma = \rho v_*^{\,2}. \tag{42.2}$$

and κ a numerical constant, the *von Kármán constant*, whose value cannot be calculated theoretically and must be determined experimentally. It is found to be†

$$\kappa = 0{\cdot}4. \tag{42.3}$$

Integration of (42.1) gives

$$u = (v_*/\kappa)(\log y + c), \tag{42.4}$$

where c is a constant of integration. To determine this constant we cannot use the ordinary boundary conditions at the surface, since for $y = 0$ the first term in (42.4) becomes infinite. The reason for this is that the above expression is really inapplicable at very small distances y from the surface, since the effect of the viscosity then becomes important, and cannot be neglected. There are also no conditions at infinity, since for $y = \infty$ the expression (42.4) again becomes infinite. This is because, in the idealized conditions which we have imposed, the surface is unbounded, and its influence therefore extends to infinitely great distances.

Before determining the constant c, we may first point out the following important property of the flow considered: contrary to what usually happens, it has no characteristic constant parameters of length which might give the fundamental scale of the turbulence. This scale is therefore determined by the distance y itself: the scale of turbulent flow at a distance y from the surface is of the order of y. The fluctuating velocity of the turbulence is of the order of v_*. This also follows at once from dimensional arguments, since v_* is the only quantity having the dimensions of velocity which can be formed from the quantities σ, ρ, y at our disposal. It should be emphasized that, whereas the mean velocity decreases with decreasing y, the fluctuating velocity remains of the same order of magnitude at all distances from the surface. This result is in accordance with the general rule that the order of magnitude of the fluctuating velocity is determined by the variation Δu of the mean velocity (§33). In the present case, there is no characteristic length l over which the variation of the mean velocity could be taken; Δu must now be defined, reasonably, as the change in u when the distance y changes appreciably. According to (42.4), such a change in y causes a change in the velocity u that is just of the order of v_*.

At sufficiently small distances from the surface, the viscosity of the fluid begins to be important; we denote the order of magnitude of these distances by y_0, which can be determined as follows. The scale of the turbulence at these distances is of the order of y_0, and the velocity is of the order of v_*. Hence the Reynolds number which characterizes the

† This value, and that of another constant in (42.8), are obtained from measurements of the velocity profile near the walls of pipes and rectangular channels, and in the boundary layer on flat surfaces.

flow at distances of the order of y_0 is $\mathrm{R} \sim v_* y_0/\nu$. The viscosity begins to be important when R becomes of the order of unity. Hence we find that

$$y_0 \sim \nu/v_*, \tag{42.5}$$

and this determines y_0.

At distances from the surface small compared with y_0, the flow is determined by ordinary viscous friction. The velocity distribution here can be obtained directly from the usual formula for viscous friction: $\sigma = \rho\nu\, du/dy$, whence

$$u = \sigma y/\rho\nu = v_*^2 y/\nu. \tag{42.6}$$

Thus, immediately adjoining the surface, there is a thin layer of fluid in which the mean velocity varies linearly with y; the velocity is small throughout this layer, varying from zero at the surface itself to values of the order of v_* for $y \sim y_0$. We shall call this layer the *viscous sublayer*. There is no sharp boundary between it and the rest of the flow, of course, and in this sense the concept is a purely qualitative one. It must be emphasized that the flow in the viscous sublayer is turbulent.†

We shall not be further interested in the flow in the viscous sublayer. Its presence has to be taken into account only in making the appropriate choice of the constant of integration in (42.4). This constant must be chosen so that the velocity becomes of the order of v_* at distances of the order of y_0. For this to be so, we must take $c = -\log y_0$, so that

$$u = (v_*/\kappa)\log(yv_*/\nu). \tag{42.7}$$

This formula determines (for a certain range of y) the velocity distribution in the turbulent stream which flows along the surface. This distribution is called the *logarithmic velocity profile*.‡

The argument of the logarithm in formula (42.7) should include a numerical coefficient. As written, it has only "logarithmic" accuracy. This means that the argument of the logarithm is supposed so large that the logarithm itself is large. The introduction of a small numerical coefficient in the argument of the logarithm in (42.7) is equivalent to adding a term of the form constant $\times\, v_*$, where the constant is of the order of unity; in the logarithmic approximation, such a term is negligible in comparison with that containing the large logarithm. In practice, however, the argument of the logarithm in the expressions derived here and below is still not very large, and therefore the accuracy of the logarithmic approximation is not high. The accuracy of the formulae can be improved by including an empirical numerical factor in the argument of the logarithm, or, equivalently, adding an empirical constant to the logarithm. For example, a more accurate expression for the velocity profile is

$$\begin{aligned} u &= v_*[2{\cdot}5\log(yv_*/\nu) + 5{\cdot}1] \\ &= 2{\cdot}5 v_* \log(yv_*/0{\cdot}13\,\nu). \end{aligned} \tag{42.8}$$

The two formulae (42.6) and (42.8) have the form

$$u = v_* f(\xi), \qquad \xi = yv_*/\nu, \tag{42.9}$$

---

† In this respect the name "laminar sublayer" still sometimes used is unsuitable. The resemblance to laminar flow lies only in the fact that the mean velocity is distributed according to the same law as the true velocity would be for a laminar flow under the same conditions.

The fluctuating flow in the viscous sublayer has some peculiar features that have not yet been given an adequate theoretical explanation.

‡ This simple derivation of the logarithmic profile is due to L. D. Landau (1944).

where $f(\xi)$ is a universal function. This is a direct consequence of the fact that ξ is the only dimensionless combination that can be formed from the available parameters ρ, σ, ν and the variable y. For this reason, such a dependence must hold at all distances from the surface, including the region intermediate between the ranges of applicability of (42.6) and (42.8). Figure 31 shows a graph of $f(\xi)$ on a decimal log scale. The continuous curves 1 and 2 correspond to (42.6) and (42.8) respectively; the dashed curve is the empirical dependence in the intermediate region, which occurs for values of ξ between about 5 and about 30.

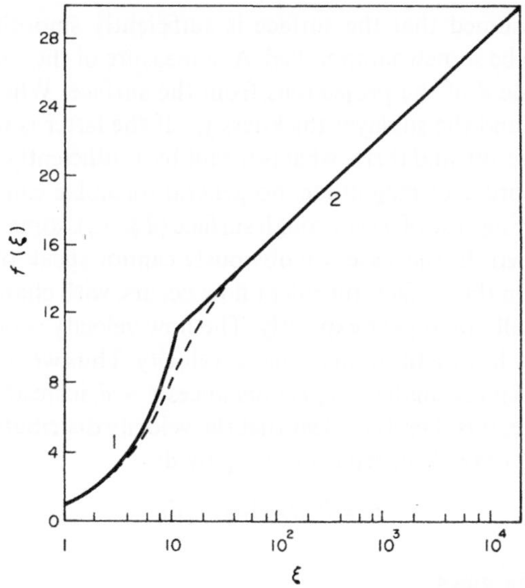

FIG. 31

It is not difficult to determine the energy dissipation in this turbulent flow. The value of σ is the mean value of the component Π_{xy} of the momentum flux density tensor. Outside the viscous sublayer, the viscosity term may be omitted, leaving $\Pi_{xy} = \rho v_x v_y$. With the fluctuating velocity $\mathbf{v}'$, and noting that the mean velocity is in the x-direction, we have $v_x = u + v'_x$, $v_y = v'_y$. Then†

$$\sigma = \rho \langle v_x v_y \rangle = \rho \langle v'_x v'_y \rangle + \rho u \langle v'_y \rangle$$
$$= \rho \langle v'_x v'_y \rangle. \qquad (42.10)$$

The energy flux density in the y-direction is $(p + \frac{1}{2}\rho v^2) v_y$, the viscosity term being again omitted. Putting $v^2 = (u + v'_x)^2 + v'^2_y + v'^2_z$ and averaging the whole expression, we get

$$\langle p' v'_y \rangle + \tfrac{1}{2}\rho \langle v'^2_x v'_y + v'^3_y + v'^2_z v'_y \rangle + \rho u \langle v'_x v'_y \rangle.$$

Here only the last term need be retained. The reason is that the fluctuating velocity is of the order of v_*, and hence, to logarithmic accuracy, it is small compared with u. The turbulent

† The momentum flux tensor for the transfer by turbulent eddies is called the *Reynolds stress tensor*; the concept is due to O. Reynolds (1895).

fluctuations in the pressure are $p' \sim \rho v_*^2$, and so we can, to the same accuracy, neglect the first term in the above expression. Thus we have for the mean energy flux density

$$\langle q \rangle = \rho u \langle v'_x v'_y \rangle = u\sigma. \tag{42.11}$$

As the surface is approached, this flux decreases, because energy is dissipated. The derivative $\mathrm{d}\langle q \rangle/\mathrm{d}y$ gives the dissipation per unit volume of the fluid; dividing it by ρ, we get the dissipation per unit mass:

$$\varepsilon = v_*^3/\kappa y = (1/\kappa y)(\sigma/\rho)^{3/2}. \tag{42.12}$$

So far, we have assumed that the surface is sufficiently smooth. If it is rough, the formulae derived may be somewhat modified. As a measure of the roughness, we can take the order of magnitude d of the projections from the surface. What is important is the comparative size of d and the sublayer thickness y_0. If the latter is much the greater, the roughness is not significant, and this is what is meant by a sufficiently smooth surface. If y_0 and d have the same order of magnitude, no general formulae can be obtained.

In the opposite limiting case of a very rough surface ($d \gg y_0$), some general relationships can again be established. In this case, we obviously cannot speak of a viscous sublayer. Near the projections on the surface, turbulent flow occurs, with characteristics ρ, σ, d; the viscosity ν, as usual, will not appear explicitly. The flow velocity is of the order of v_*, the only available quantity having the dimensions of velocity. Thus we see that in flow along a rough surface the velocity is small ($\sim v_*$) at distances $y \sim d$, instead of $y \sim y_0$ as for flow along a smooth surface. It is therefore clear that the velocity distribution will be given by a formula obtained from (42.7) on replacing ν/v_* by d:

$$u = (v_*/\kappa)\log(y/d). \tag{42.13}$$

§43. **Turbulent flow in pipes**

Let us now apply the above results to turbulent flow in a pipe. Near the walls of the pipe (at distances small compared with its radius a), the surface may be approximately regarded as plane, and the velocity distribution must be given by formula (42.7) or (42.8). Since the function $\log y$ varies only slowly, we can use formula (42.7) to logarithmic accuracy to give the mean velocity U of the flow in the pipe if we replace y in that formula by a:

$$U = (v_*/\kappa)\log(av_*/\nu). \tag{43.1}$$

By U we mean the volume of fluid that passes through a cross-section of the pipe per unit time, divided by the cross-sectional area: $U = Q/\rho\pi a^2$.

In order to relate the velocity U to the pressure gradient $\Delta p/l$ which maintains the flow (Δp being the pressure difference between the ends of the pipe, and l its length), we notice that the force on a cross-section of the flow is $\pi a^2 \Delta p$. This force overcomes the friction at the walls. Since the frictional force per unit area of the wall is $\sigma = \rho v_*^2$, the total frictional force is $2\pi a l \rho v_*^2$. Equating the two forces, we have

$$\Delta p/l = 2\rho v_*^2/a. \tag{43.2}$$

Equations (43.1) and (43.2) determine, through the parameter v_*, the relation between the velocity of flow in the pipe and the pressure gradient. This relation is called the *resistance law* of the pipe. Expressing v_* in terms of $\Delta p/l$ by (43.2), and substituting in (43.1), we obtain the resistance law in the form

$$U = \sqrt{(a\Delta p/2\kappa^2\rho l)}\log[(a/\nu)\sqrt{(a\Delta p/2\rho l)}]. \tag{43.3}$$

In this formula it is customary to introduce what is called the *resistance coefficient* of the pipe, a dimensionless quantity defined as

$$\lambda = \frac{2a\Delta p/l}{\frac{1}{2}\rho U^2}. \tag{43.4}$$

The dependence of λ on the dimensionless Reynolds number $\mathrm{R} = 2aU/\nu$ is given in implicit form by the equation

$$1/\sqrt{\lambda} = 0{\cdot}88 \log(\mathrm{R}\sqrt{\lambda}) - 0{\cdot}85. \tag{43.5}$$

We have here substituted for κ the value (42.3) and added to the logarithm an empirically determined constant.† The resistance coefficient determined by this formula is a slowly decreasing function of the Reynolds number. For comparison, we give the resistance law for laminar flow in a pipe. Introducing the resistance coefficient in formula (17.10), we obtain

$$\lambda = 64/\mathrm{R}. \tag{43.6}$$

In laminar flow the resistance coefficient diminishes with increasing Reynolds number more rapidly than in turbulent flow.

Figure 32 shows a logarithmic graph of λ as a function of R. The steep straight line corresponds to laminar flow (formula (43.6)), and the less steep curve (which is almost a straight line also) to turbulent flow. The transition from the first line to the second occurs, as the Reynolds number increases, at the point where the flow becomes turbulent; this may occur for various Reynolds numbers, depending on the actual conditions (the intensity of the perturbations). The resistance coefficient increases abruptly at the transition point.

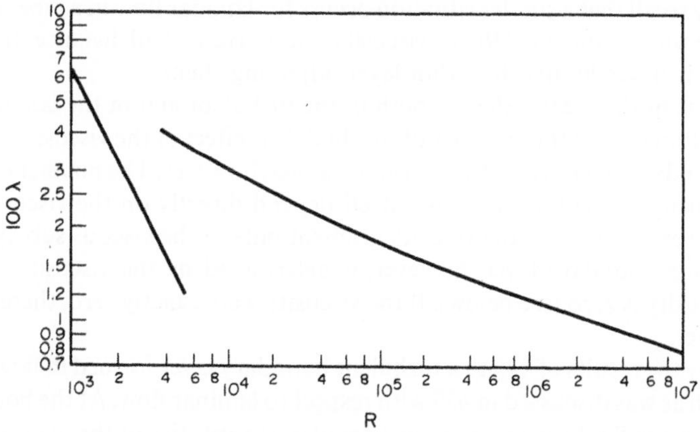

FIG. 32

† The coefficient of the logarithm in this formula is given to correspond with that in formula (42.8) for the logarithmic velocity profile. Only in this case does formula (43.5) have the theoretical significance of being a limiting formula for turbulent flow at sufficiently large values of the Reynolds number. If the values of the two constants appearing in formula (43.5) are chosen arbitrarily, it can only be a purely empirical formula for the dependence of λ on R. In that case, however, there would be no reason to prefer it to any other simpler empirical formula which adequately represents the experimental results.

The formulae above are for smooth-walled pipes. Similar ones for pipes with very rough walls are obtained on simply replacing ν/v_* by d; cf. (42.13). The resistance is then, instead of (43.3),

$$U = \sqrt{(a\Delta p/2\kappa^2\rho l)}\log(a/d). \tag{43.7}$$

The argument of the logarithm is now a constant, and does not involve the pressure gradient as (43.3) did. We see that the mean velocity is now simply proportional to the square root of the pressure gradient in the pipe. If we introduce the resistance coefficient, (43.7) becomes

$$\lambda = 8\kappa^2/\log^2(a/d) = 1\cdot 3/\log^2(a/d), \tag{43.8}$$

i.e. λ is a constant and does not depend on the Reynolds number.

§44. The turbulent boundary layer

The fact that we have obtained a logarithmic velocity distribution which formally holds in all space for plane-parallel turbulent flow is due to our having considered flow along a surface with infinite area. In flow along the surface of a finite body, only the motion at short distances from the surface—in the boundary layer—has a logarithmic profile. The thickness of the boundary layer increases along the surface of the body in the direction of flow, according to a law which we shall determine below. This explains why, for flow in a pipe, the logarithmic profile holds for the whole cross-section of the pipe. The thickness of the boundary layer at the wall of the pipe increases away from the point of entry of the fluid. At some finite distance from this point, the boundary layer fills almost the whole cross-section of the pipe. Hence, if we suppose the pipe sufficiently long and ignore its inlet section, the flow in the whole pipe will be of the same kind as in the turbulent boundary layer. We may recall that a similar situation occurs for laminar flow in a pipe. This is always in accordance with formula (17.9); the viscosity is important at all distances from the walls, and its effect is never limited to a thin layer adjoining them.

The decrease in the mean velocity, both in the turbulent and in the laminar boundary layer, is due ultimately to the viscosity of the fluid. The effect of the viscosity appears in the turbulent boundary layer in a rather unusual manner, however. The manner of variation of the mean velocity in the layer does not itself depend directly on the viscosity; the latter appears in the expression for the velocity gradient only in the viscous sublayer. The total thickness of the boundary layer, however, is determined by the viscosity, and vanishes when the viscosity is zero (see below). If the viscosity were exactly zero, there would be no boundary layer.

Let us apply the results of §43 to a turbulent boundary layer formed in flow along a thin flat plate, such as was discussed in §39 with respect to laminar flow. At the boundary of the turbulent layer, the fluid velocity is almost equal to the velocity of the main stream, which we denote by U. To determine this velocity at the boundary we can, however, use formula (42.7) with logarithmic accuracy, putting the thickness δ of the boundary layer instead of y.† Equating the two expressions, we obtain

$$U = (v_*/\kappa)\log(v_*\delta/\nu). \tag{44.1}$$

† In practice, the logarithmic profile is not observed over the whole thickness of the boundary layer. The last 20–25% of the velocity increase at the outside of the layer occurs faster than logarithmically. These deviations seem to be due to irregular oscillations of the layer boundary; cf. the discussion of turbulent region boundaries at the end of §35.

Here U is a constant parameter for a given flow; the thickness δ, however, varies along the plate, and v_* is therefore also a slowly varying function of x. Formula (44.1) is inadequate to determine these functions; we need some other equation, relating v_* and δ to x.

To obtain this, we use the same arguments as in deriving formula (37.3) for the width of the turbulent wake. As there, the derivative $d\delta/dx$ must be of the order of the ratio of the velocity along the y-axis to that along the x-axis at the boundary of the layer. The latter velocity is of the order of U, while the former is due to the fluctuating velocity, and is therefore of the order of v_*. Thus $d\delta/dx \sim v_*/U$, whence

$$\delta \sim v_* x/U. \tag{44.2}$$

Formulae (44.1) and (44.2) together determine v_* and δ as functions of the distance x.† These functions, however, cannot be written explicitly. We shall express δ in terms of an auxiliary quantity. Since v_* is a slowly varying function of x, it is seen from (44.2) that the thickness of the layer varies essentially as x. We may recall that the thickness of the laminar boundary layer increases as $\sqrt{x}$, i.e. more slowly than that of the turbulent boundary layer.

Let us determine the dependence on x of the frictional force σ acting on unit area of the plate. This dependence is given by two formulae:

$$\sigma = \rho {v_*}^2, \qquad U = (v_*/\kappa)\log({v_*}^2 x/U\nu).$$

The latter is obtained by substituting (44.2) in (44.1), and is valid to logarithmic accuracy. We introduce a drag coefficient c (referred to unit area of the plate), defined as the dimensionless ratio

$$c = 2\sigma/\rho U^2 = 2(v_*/U)^2. \tag{44.3}$$

Then, eliminating v_* from the two equations given, we obtain the following equation, which gives (to logarithmic accuracy) c as an implicit function of x:

$$\sqrt{(2\kappa^2/c)} = \log(c\mathrm{R}_x), \qquad \mathrm{R}_x = Ux/\nu. \tag{44.4}$$

The drag coefficient c given by this formula is a slowly decreasing function of the distance x.

Let us express the thickness of the boundary layer in terms of the function $c(x)$. We have $v_* = \sqrt{(\sigma/\rho)} = U\sqrt{(\frac{1}{2}c)}$. Substituting in (44.2), we find

$$\delta = \text{constant} \times x\sqrt{c}. \tag{44.5}$$

The empirical value of the constant is about 0·3.

Similarly, we can derive expressions for the turbulent boundary layer on a rough surface. According to (42.13), (44.1) is then replaced by

$$U = (v_*/\kappa)\log(\delta/d),$$

where d is the size of the projections on the surface. Substituting δ from (44.2), we get

$$U = (v_*/\kappa)\log(xv_*/Ud),$$

or, with the drag coefficient (44.3),

$$\sqrt{(2\kappa^2/c)} = \log(x\sqrt{c}/d). \tag{44.6}$$

† Here x must, strictly speaking, be reckoned as approximately the distance from the point where the laminar layer becomes turbulent.

§45. The drag crisis

From the results obtained in the previous sections we can draw important conclusions concerning the law of drag for large Reynolds numbers, i.e. the relation between the drag force acting on the body and the value of R when the latter is large.

The flow pattern for large R (the only case we shall discuss) has already been described, and is as follows. Throughout the main body of the fluid (i.e. everywhere except in the boundary layer, which does not here concern us) the fluid may be regarded as ideal, with potential flow everywhere except in the turbulent wake. The width of the wake depends on the position of the line of separation on the surface of the body. It is important to note that, although this position is determined by the properties of the boundary layer, it is found to be independent of the Reynolds number, as we have seen in §40. Thus we can say that the whole flow pattern for large Reynolds numbers is almost independent of the viscosity, i.e. of R (so long as the boundary layer remains laminar; see below).

Hence it follows that the drag also must be independent of the viscosity. There remain at our disposal only three quantities: the velocity U of the main stream, the fluid density ρ and the dimension l of the body. From these we can construct only one quantity having the dimensions of force, namely $\rho U^2 l^2$. Instead of the squared linear dimension of the body l^2, we introduce, as is customarily done, the proportional quantity S, the area of a cross-section transverse to the direction of flow, putting

$$F = \text{constant} \times \rho U^2 S, \tag{45.1}$$

where the constant is a number depending only on the shape of the body. Thus the drag must be (for large R) proportional to the cross-sectional area of the body and to the square of the main-stream velocity. We may recall for comparison that, for very small R ($\ll 1$), the drag is proportional to the linear dimension of the body and to the velocity itself ($F \sim \nu\rho l U$; see §20).†

It is customary, as we have said, to introduce, in place of the drag force F, the drag coefficient C defined by $C = F/\frac{1}{2}\rho U^2 S$. This is a dimensionless quantity, and can depend only on R. Formula (45.1) becomes

$$C = \text{constant}, \tag{45.2}$$

i.e. the drag coefficient depends only on the shape of the body.

The above behaviour of the drag force cannot continue to arbitrarily large Reynolds numbers. The reason is that, for sufficiently large R, the laminar boundary layer (on the surface of the body as far as the line of separation) becomes unstable and hence turbulent. However, the whole boundary layer does not become turbulent, but only some part of it. The surface of the body may therefore be divided into three parts: at the front there is a laminar boundary layer, then a turbulent layer, and finally the region beyond the line of separation.

The onset of turbulence in the boundary layer has an important effect on the whole pattern of flow in the main stream. It leads to a considerable displacement of the line of separation towards the rear of the body (i.e. downstream), so that the turbulent wake beyond the body is contracted, as shown in Fig. 33, where the wake region is shaded.‡ The

† The flow past a bubble of gas is a special case, where the drag remains proportional to U even for large R; see Problem.

‡ For example, in transverse flow past a long cylinder, the onset of turbulence in the boundary layer moves the point of separation from 95° to 60° (where the azimuthal angle on the cylinder is measured from the direction of flow).

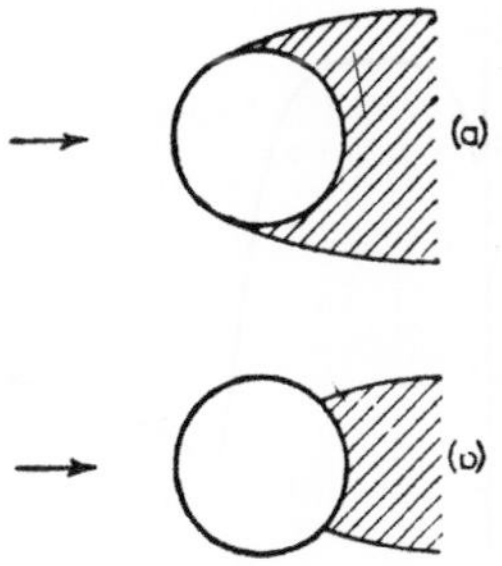

FIG. 33

contraction of the turbulent wake leads to a reduction of the drag force. Thus the onset of turbulence in the boundary layer at large Reynolds numbers is accompanied by a decrease in the drag coefficient, which falls off by a considerable factor over a relatively narrow range of Reynolds numbers near 10^5. We shall call this phenomenon the *drag crisis*. The decrease in the drag coefficient is so great that the drag itself, which for constant C is proportional to the square of the velocity, actually diminishes with increasing velocity in this range of Reynolds numbers.†

It may be mentioned that the degree of turbulence in the main stream affects the drag crisis; the greater the incident turbulence, the sooner the boundary layer becomes turbulent (i.e. the smaller is R when this happens). The decrease in the drag coefficient therefore begins at a smaller Reynolds number, and extends over a wider range of R.

Figures 34 and 35 give experimentally obtained graphs showing the drag coefficient as a function of the Reynolds number $R = Ud/\nu$ for a sphere with diameter d; Fig. 34 is plotted

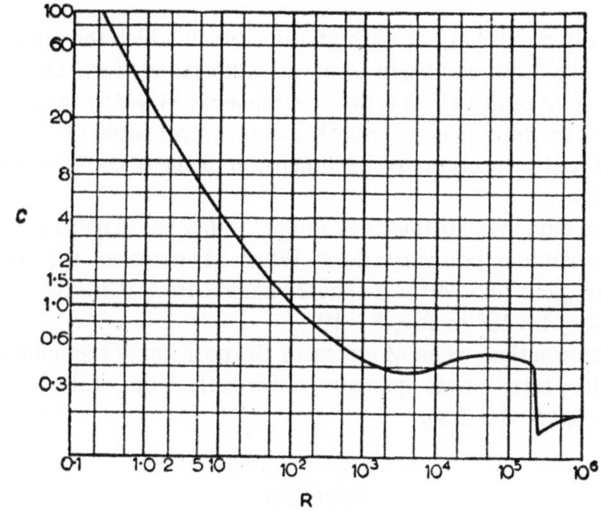

FIG. 34

† The first departure from steady flow past a sphere (R in the neighbourhood of 50) is not accompanied by any discontinuity of the drag. This is because the transition is continuous in the case of soft self-excitation. A change in the nature of the flow could occur only if there were a kink on the C(R) curve.

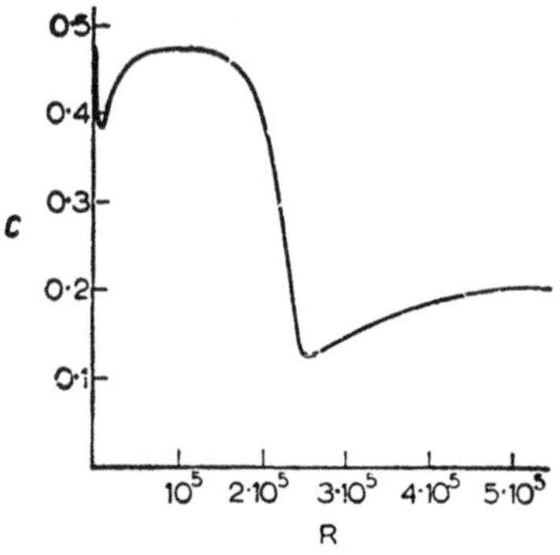

FIG. 35

logarithmically. For very small R ($\ll 1$), the drag coefficient decreases according to $C = 24/\mathrm{R}$ (Stokes' formula). The decrease in C continues more slowly as far as $\mathrm{R} \cong 5 \times 10^3$, where C reaches a minimum, beyond which it increases somewhat. In the range of Reynolds numbers 2×10^4 to 2×10^5, the law (45.2) holds, i.e. C is almost constant. The drag crisis occurs for R between 2×10^5 and 3×10^5, and the drag coefficient diminishes by a factor of 4 or 5.

For comparison, we may give an example of flow in which there is no critical Reynolds number. Let us consider flow past a flat disk in the direction perpendicular to its plane. In this case the location of the separation is obvious from purely geometrical considerations: it is clear that separation occurs at the edge of the disk and does not move from there. Hence, as R increases, the drag coefficient of the disk remains constant, and there is no drag crisis.

It must be borne in mind that, for the high velocities at which the drag crisis occurs, the compressibility of the fluid may begin to have a noticeable effect. The parameter which characterizes the extent of this effect is the *Mach number* $\mathrm{M} = U/c$, where c is the velocity of sound; if $\mathrm{M} \ll 1$, the fluid may be regarded as incompressible (§10). Since, of the two numbers M and R, only one contains the dimension of the body, these two numbers can vary independently.

The experimental data indicate that the compressibility has in general a stabilizing effect on the flow in the laminar boundary layer. When M increases, the critical value of R increases. For example, when M for a sphere changes from 0·3 to 0·7, the drag crisis is postponed from $\mathrm{R} \cong 4 \times 10^5$ to $\mathrm{R} \cong 8 \times 10^5$.

We may also mention that, when M increases, the position of the point of separation in the laminar boundary layer moves upstream, towards the front of the body, and this must lead to some increase in the drag.

PROBLEM

Determine the drag force on a gas bubble moving in a liquid at large Reynolds numbers.

SOLUTION. At the boundary between the liquid and the gas the tangential fluid velocity component does not vanish, but its normal derivative does (we neglect the viscosity of the gas). Hence the velocity gradient near the boundary will not be particularly high, and there will be no boundary layer in the sense of §39; there will therefore be no separation over almost the whole surface of the bubble. In calculating the energy dissipation from the volume integral (16.3) we can therefore use in all space the velocity distribution corresponding to potential flow past a sphere (§10, Problem 2), neglecting the surface layer of liquid and the very narrow turbulent wake. Using the formula obtained in §16, Problem, we find

$$\dot{E}_{\text{kin}} = -\eta \int \left(\frac{\partial v^2}{\partial r}\right)_{r=R} 2\pi R^2 \sin\theta d\theta = -12\pi\eta RU^2.$$

Hence we see that the required dissipative drag is $F = 12\pi\eta RU$.

The range of applicability of this formula is actually not large, since, when the velocity increases sufficiently, the bubble ceases to be spherical.

§46. Flow past streamlined bodies

The question may be asked what should be the shape of a body (with a given cross-sectional area, say) for the drag on it resulting from motion in a fluid to be as small as possible. It is clear from the above that, for this to be so, the separation must be as far back as possible: the separation must occur near the rear end of the body, so that the turbulent wake is as narrow as possible. We know already that the appearance of separation is facilitated by the presence of a rapid downstream increase in the pressure along the body. Hence the body must have a shape such that the variation in pressure along it, where the pressure is increasing, takes place as slowly and smoothly as possible. This can be achieved by giving the body a shape elongated in the direction of flow, tapering smoothly to a point downstream, so that the flows along the two sides of the body meet smoothly without having to go round any corners or turn through a considerable angle from the direction of the main stream. At the front end the body must be rounded; if there were an angle here, the fluid velocity at its vertex would become infinite (see §10, Problem 6), and consequently the pressure would increase rapidly downstream, with separation inevitably resulting.

All these requirements are closely satisfied by shapes of the kind shown in Fig. 36. The profile shown in Fig. 36b may be, for example, the cross-section of an elongated solid of revolution, or the cross-section of a body with a large span (we conventionally call such a body a *wing*). The cross-sectional profile of a wing may be unsymmetrical, as in Fig. 36a. In flow past a body with this shape, separation occurs only in the immediate neighbourhood of the pointed end, and consequently the drag coefficient is relatively small. Such bodies are said to be *streamlined*.

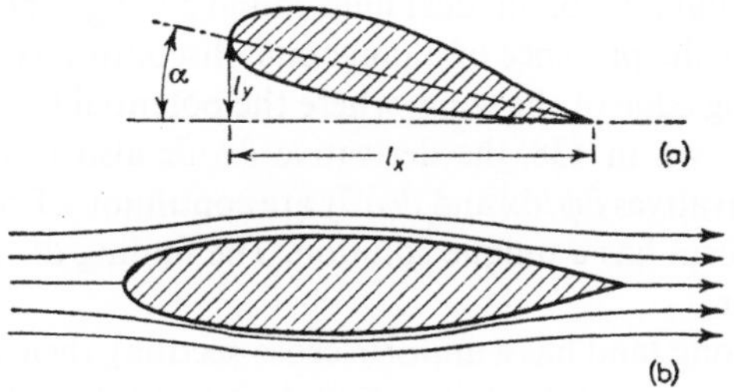

FIG. 36

The direct friction of the fluid on the surface in the boundary layer is important in the drag on streamlined bodies. This effect for non-streamlined bodies (which were considered in the previous section) is relatively small and therefore, in practice, of no significance. In the opposite limiting case of flow parallel to a flat disk, the effect becomes the only source of drag (§39).

In flow past a streamlined wing inclined to the main stream at a small angle α, called the *angle of attack* (Fig. 36), a large lift force F_y is developed, while the drag F_x remains small, and the ratio F_y/F_x may therefore reach large values ($\sim 10-100$). This continues, however, only while the angle of attack is small (usually $\lesssim 10°$). For larger angles the drag rises very rapidly, and the lift decreases. This is explained by the fact that, at large angles of attack, the body ceases to be streamlined: the point of separation moves a considerable way towards the front of the body, and the wake consequently becomes wider. It must be borne in mind that the limiting case of a very thin body, i.e. a flat plate, is streamlined only for a very small angle of attack; separation occurs at the leading edge of the plate when it is inclined at even a small angle to the main stream.

The angle of attack α is, by definition, measured from the position of the wing for which the lift force is zero. For small angles of attack, we can expand the lift as a series of powers of α. Taking only the first term, we can suppose that the force F_y is proportional to α. Next, by the same dimensional arguments as for the drag force, the lift must be proportional to ρU^2. Introducing also the span l_z of the wing, we can write

$$F_y = \text{constant} \times \rho U^2 \alpha l_x l_z, \tag{46.1}$$

where the numerical constant depends only on the shape of the wing and not, in particular, on the angle of attack. For very long wings, the lift may be supposed proportional to the span, in which case the constant depends only on the shape of the cross-section of the wing.

Instead of the lift on the wing, the *lift coefficient* is often used; it is defined as

$$C_y = F_y/\tfrac{1}{2}\rho U^2 l_x l_z. \tag{46.2}$$

For very long wings, according to what was said above, the lift coefficient is proportional to the angle of attack, and depends on neither the velocity nor the span:

$$C_y = \text{constant} \times \alpha. \tag{46.3}$$

To calculate the lift on a streamlined wing by means of Zhukovskiĭ's formula, it is necessary to determine the velocity circulation Γ. This is done as follows. We have potential flow everywhere outside the wake. In the present case, the wake is very thin, and occupies on the surface of the wing only a very small area near its pointed trailing edge. Hence, to determine the velocity distribution (and therefore the circulation Γ), we can solve the problem of potential flow of an ideal fluid round a wing. The existence of the wake is taken into account by the presence of a tangential discontinuity, extending into the fluid from the sharp trailing edge of the wing, where the potential has a discontinuity $\phi_2 - \phi_1 = \Gamma$. As has been shown in §38, the derivative $\partial\phi/\partial z$ also has a discontinuity on this surface, while the derivatives $\partial\phi/\partial x$ and $\partial\phi/\partial y$ are continuous. For a wing with finite span, the problem in this form has a unique solution. The finding of the exact solution is very complicated, however.

If the wing is very long (and has a uniform cross-section), then, regarding it as infinite in the z-direction, we may regard the flow as two-dimensional (in the xy-plane). It is evident from symmetry that the velocity $v_z = \partial\phi/\partial z$ along the wing must be zero. In this case, therefore, we must seek a solution in which only the potential has a discontinuity, its derivatives being continuous; in other words, there is no surface of tangential discontinuity, and we have simply a many-valued function $\phi(x, y)$, which receives a finite increment Γ when we go round a closed contour enclosing the profile of the wing. In this form, however, the problem of two-dimensional flow has no unique solution, since it admits solutions for any given discontinuity of the potential. To obtain a unique result, we must require the fulfilment of another condition (S. A. Chaplygin 1909).

This condition consists in requiring that the fluid velocity shall not become infinite at the sharp trailing edge of the wing; in this connection we may recall that, when an ideal fluid flows round an angle, the fluid velocity in general becomes infinite, according to a power law, at the vertex of the angle (§10, Problem 6). We can say that the condition stated implies that the jets coming from the two sides of the wing must meet smoothly without turning through an angle. When this condition is fulfilled, of course, the solution of the problem of potential flow gives a pattern very like the true one, where the velocity is everywhere finite and separation occurs only at the trailing edge. The solution now becomes unique and, in particular, the circulation Γ needed to calculate the lift force has a definite value.

§47. **Induced drag**

An important part of the drag on a streamlined wing (with finite span) is formed by the drag due to the dissipation of energy in the thin turbulent wake. This is called the *induced drag.*

It has been shown in §21 how we may calculate the drag force due to the wake by considering the flow far from the body. Formula (21.1), however, is not applicable in the present case. According to that formula, the drag is given by the integral of v_x over the cross-section of the wake, i.e. the discharge through the wake. On account of the thinness of the wake beyond a streamlined wing, however, the discharge is small in the present case, and may be neglected in the approximation used below.

As in §21, we write the force F_x as the difference between the total fluxes of the x-component of momentum through the planes $x = x_1$ and $x = x_2$ passing respectively far behind and far in front of the body. Writing the three velocity components as $U + v_x$, v_y, v_z, we have for the component Π_{xx} of the momentum flux density the expression $\Pi_{xx} = p + \rho(U + v_x)^2$, so that the drag force is

$$F_x = \left(\iint\limits_{x=x_2} - \iint\limits_{x=x_1} \right) [p + \rho(U + v_x)^2] \mathrm{d}y\, \mathrm{d}z. \tag{47.1}$$

On account of the thinness of the wake, we can neglect, in the integral over the plane $x = x_1$, the integral over the cross-section of the wake, and so integrate only over the region outside the wake. In that region, however, we have potential flow, and Bernoulli's equation $p + \frac{1}{2}\rho(\mathbf{U} + \mathbf{v})^2 = p_0 + \frac{1}{2}\rho U^2$ holds, whence

$$p = p_0 - \rho U v_x - \tfrac{1}{2}\rho(v_x^2 + v_y^2 + v_z^2). \tag{47.2}$$

Here we cannot neglect the quadratic terms as we did in §21, since it is these terms which determine the required drag force in the case under consideration. Substituting (47.2) in (47.1), we obtain

$$F_x = \left(\iint\limits_{x=x_2} - \iint\limits_{x=x_1} \right) [p_0 + \rho U^2 + \rho U v_x + \tfrac{1}{2}\rho(v_x^2 - v_y^2 - v_z^2)] \mathrm{d}y\, \mathrm{d}z.$$

The difference of the integrals of the constant $p_0 + \rho U^2$ is zero; the difference of the integrals of $\rho U v_x$ is likewise zero, since the mass fluxes

$$\iint \rho v_x \mathrm{d}y\mathrm{d}z$$

through the front and back planes must be the same (we neglect the discharge through the wake in the approximation here considered). Next, if we take the plane $x = x_2$ sufficiently far in front of the body, the velocity v on this plane is very small, so that the integral of $\frac{1}{2}\rho(v_x^2 - v_y^2 - v_z^2)$ over this plane may be neglected. Finally, in flow past a stremlined wing, the velocity v_x outside the wake is small compared with v_y and v_z. Hence we can neglect v_x^2 compared with $v_y^2 + v_z^2$ in the integral over the plane $x = x_1$. Thus we obtain

$$F_x = \tfrac{1}{2}\rho \iint (v_y^2 + v_z^2)\mathrm{d}y\mathrm{d}z, \tag{47.3}$$

where the integration is over a plane x = constant lying at a great distance behind the body, the cross-section of the wake being excluded from the region of integration.†

The drag on a streamlined wing calculated in this way can be expressed in terms of the velocity circulation Γ which determines the lift also. To do this, we first of all notice that, at sufficiently great distances from the body, the velocity depends only slightly on the coordinate x, and so we can regard $v_y(y, z)$ and $v_z(y, z)$ as the velocity of a two-dimensional flow, supposed independent of x. It is convenient to use as an auxiliary quantity the stream function (§10), so that $v_z = \partial\psi/\partial y$, $v_y = -\partial\psi/\partial z$. Then

$$F_x = \tfrac{1}{2}\rho \iint \left[\left(\frac{\partial\psi}{\partial y}\right)^2 + \left(\frac{\partial\psi}{\partial z}\right)^2\right]\mathrm{d}y\mathrm{d}z,$$

where the integration over the vertical coordinate y is from $+\infty$ to y_1 and from y_2 to $-\infty$, where y_1 and y_2 are the coordinates of the upper and lower boundaries of the wake (see Fig. 26, §38). Since we have potential flow (**curl v** = 0) outside the wake, $\partial^2\psi/\partial y^2 + \partial^2\psi/\partial z^2 = 0$. Using the two-dimensional Green's formula, we thus find

$$F_x = -\tfrac{1}{2}\rho \oint \psi(\partial\psi/\partial n)\mathrm{d}l,$$

where the integral is taken along a contour bounding the region of integration in the original integral, and $\partial/\partial n$ denotes differentiation in the direction of the outward normal to the contour. At infinity $\psi = 0$, and so the integral is taken round the cross-section of the wake by the yz-plane, giving

$$F_x = \tfrac{1}{2}\rho \int \psi\left[\left(\frac{\partial\psi}{\partial y}\right)_2 - \left(\frac{\partial\psi}{\partial y}\right)_1\right]\mathrm{d}z.$$

Here the integration is over the width of the wake, and the difference in the brackets is the

† Formula (47.3) may give the impression that the velocities v_y, v_z do not decrease in order of magnitude as x increases. This is true so long as the thickness of the wake is small compared with its width, as we have assumed in deriving formula (47.3). At very large distances behind the wing, the wake finally becomes so thick that it becomes approximately circular in cross-section. At this point, formula (47.3) is invalid, and v_y, v_z diminish rapidly with increasing x.

discontinuity of the derivative $\partial\psi/\partial y$ across the wake. Since $\partial\psi/\partial y = v_z = \partial\phi/\partial z$, we have

$$\left(\frac{\partial\psi}{\partial y}\right)_2 - \left(\frac{\partial\psi}{\partial y}\right)_1 = \left(\frac{\partial\phi}{\partial z}\right)_2 - \left(\frac{\partial\phi}{\partial z}\right)_1 = \frac{\mathrm{d}\Gamma}{\mathrm{d}z},$$

so that

$$F_x = \tfrac{1}{2}\rho \int \psi(\mathrm{d}\Gamma/\mathrm{d}z)\mathrm{d}z.$$

Finally, we use a formula from potential theory,

$$\psi = -\frac{1}{2\pi}\int\left[\left(\frac{\partial\psi}{\partial n}\right)_2 - \left(\frac{\partial\psi}{\partial n}\right)_1\right]\log r\,\mathrm{d}l,$$

where the integration is along a plane contour, r is the distance from $\mathrm{d}l$ to the point where ψ is to be found, and the expression in brackets is the given discontinuity of the derivative of ψ in the direction normal to the contour.† In our case the contour of integration is a segment of the z-axis, so that we can write the value of the function $\psi(y, z)$ on the z-axis as

$$\psi(0, z) = \frac{1}{2\pi}\left[\left(\frac{\partial\psi}{\partial y}\right)_1 - \left(\frac{\partial\psi}{\partial y}\right)_2\right]\log|z - z'|\mathrm{d}z'$$

$$= -\frac{1}{2\pi}\int\frac{\mathrm{d}\Gamma(z')}{\mathrm{d}z'}\log|z - z'|\mathrm{d}z'.$$

Finally, substituting this in F_x, we obtain the following formula for the induced drag:

$$F_x = -\frac{\rho}{4\pi}\int_0^l\int_0^l\frac{\mathrm{d}\Gamma(z)}{\mathrm{d}z}\frac{\mathrm{d}\Gamma(z')}{\mathrm{d}z'}\log|z - z'|\mathrm{d}z\mathrm{d}z' \quad (47.4)$$

(L. Prandtl 1918). The span of the wing is here denoted by $l_z = l$, and the origin of z is at one end of the wing.

If all the dimensions in the z-direction are increased by some factor (Γ remaining constant), the integral (47.4) remains constant.‡ This shows that the total induced drag on the wing remains of the same order of magnitude when its span is increased. In other words, the induced drag per unit length of the wing decreases with increasing length.†† Unlike the drag, the total lift force

$$F_y = -\rho U\int\Gamma\mathrm{d}z \quad (47.5)$$

---

† This formula gives, in two-dimensional potential theory, the potential due to a charged plane contour with a charge density

$$[(\partial\psi/\partial n)_2 - (\partial\psi/\partial n)_1]/2\pi.$$

‡ To avoid misunderstanding, we should mention that it does not matter that the logarithm in the integrand is increased by a constant when the unit of length is changed. For the integral which differs from that in (47.4) by having a constant instead of $\log|z - z'|$ is zero, since

$$\int(\mathrm{d}\Gamma/\mathrm{d}z)\mathrm{d}z = \Gamma,$$

and the definite integral is zero because Γ vanishes at the edges of the wake.

†† In the limit of infinite span, the induced drag per unit length is zero. In reality, a small amount of drag remains, determined by the discharge through the wake (i.e. the integral $\int\int v_x\,dy\,dz$), which we have neglected in deriving formula (47.3). This drag includes both the frictional drag and the remaining part due to dissipation in the wake.

increases almost linearly with the span of the wing, and the lift per unit length is constant.

The following method is convenient for the practical calculation of the integrals (47.4) and (47.5). Instead of the coordinate z, we introduce a new variable θ, defined by

$$z = \tfrac{1}{2}l(1-\cos\theta) \qquad (0 \leqslant \theta \leqslant \pi). \tag{47.6}$$

The distribution of the velocity circulation is written as a Fourier series:

$$\Gamma = -2Ul \sum_{n=1}^{\infty} A_n \sin n\theta. \tag{47.7}$$

The condition that $\Gamma = 0$ at the ends of the wing ($z = 0$ and l, or $\theta = 0$ and π) is then fulfilled.

Substituting the expression (47.7) in (47.5) and effecting the integration (using the orthogonality of the functions $\sin\theta$ and $\sin n\theta$ for $n \neq 1$), we obtain $F_y = \tfrac{1}{2}\rho U^2 \pi l^2 A_1$. Thus the lift force depends only on the first coefficient in the expansion (47.7). For the lift coefficient (46.2) we have

$$C_y = \pi\lambda A_1, \tag{47.8}$$

where $\lambda = l/l_x$ is the ratio of span to width of the wing.

To calculate the drag, we rewrite formula (47.4), integrating once by parts:

$$F_x = \frac{\rho}{4\pi}\int_0^l\int_0^l \Gamma(z)\frac{d\Gamma(z')}{dz'}\frac{dz'dz}{z-z'}. \tag{47.9}$$

It is easily seen that the integral over z' must be taken as a principal value. An elementary calculation, with the substitution (47.7),† leads to the following formula for the induced drag coefficient:

$$C_x = \pi\lambda \sum_{n=1}^{\infty} nA_n^2. \tag{47.10}$$

The drag coefficient for a wing is defined as

$$C_x = F_x/\tfrac{1}{2}\rho U^2 l_x l_z, \tag{47.11}$$

being referred, like the lift coefficient, to unit area in the xz-plane.

† In integrating over z' we need the integral

$$P\int_0^\pi \frac{\cos n\theta'}{\cos\theta' - \cos\theta}d\theta' = \frac{\pi\sin n\theta}{\sin\theta}.$$

In integrating over z we use the fact that

$$\int_0^\pi \sin n\theta \sin m\theta \, d\theta = \tfrac{1}{2}\pi \quad (m = n),$$
$$= 0 \quad (m \neq n).$$

PROBLEM

Determine the least value of the induced drag for a given lift and a given span $l_z = l$.

SOLUTION. It is clear from formulae (47.8) and (47.10) that the least value of C_x for given C_y (i.e. for given A_1) is obtained if all A_n for $n \neq 1$ are zero. Then

$$C_{x,\min} = C_y^2/\pi\lambda. \tag{1}$$

The distribution of velocity circulation over the span is given by the formula

$$\Gamma = -\frac{4}{\pi l} U l_x C_y \sqrt{[z(l-z)]}. \tag{2}$$

If the span is sufficiently large, then the flow round any cross-section of the wing is approximately two-dimensional flow round a wing with infinite length and the same cross-section. In this case we can say that the circulation distribution (2) is obtained for a wing whose shape in the xz-plane is an ellipse with semi-axes $\frac{1}{2}l_x$ and $\frac{1}{2}l$.

§48. The lift of a thin wing

The problem of calculating the lift force on a wing amounts, by Zhukovskiĭ's theorem, to that of finding the velocity circulation Γ. A general solution of the latter problem can be given for a thin streamlined wing with infinite span, the cross-section being the same at every point. The method of solution given below is due to M. V. Keldysh and L. I. Sedov (1939).

Let $y = \zeta_1(x)$ and $y = \zeta_2(x)$ be the equations of the lower and upper parts of the cross-sectional profile (Fig. 37). We suppose this profile to be thin, only slightly curved, and inclined at a small angle of attack to the main stream (the x-axis); that is, both ζ_1, ζ_2 themselves and their derivatives ζ_1', ζ_2' are small, i.e. the normal to the profile contour is everywhere almost parallel to the y-axis. Under these conditions, we may suppose the perturbation $\mathbf{v}$ in the fluid velocity, caused by the presence of the wing, to be everywhere (except in a small region near the rounded leading edge of the wing) small compared with the main-stream velocity U. The boundary condition at the surface of the wing is $v_y/U = \zeta'$ for $y = \zeta$. By virtue of the assumptions made, we can suppose this condition to hold for $y = 0$, and not for $y = \zeta$. Then we must have on the axis of abscissae between $x = 0$ and $x = l_x \equiv a$

$$v_y = U\zeta_2'(x) \quad \text{for } y \to 0+, \qquad v_y = U\zeta_1'(x) \quad \text{for } y \to 0-. \tag{48.1}$$

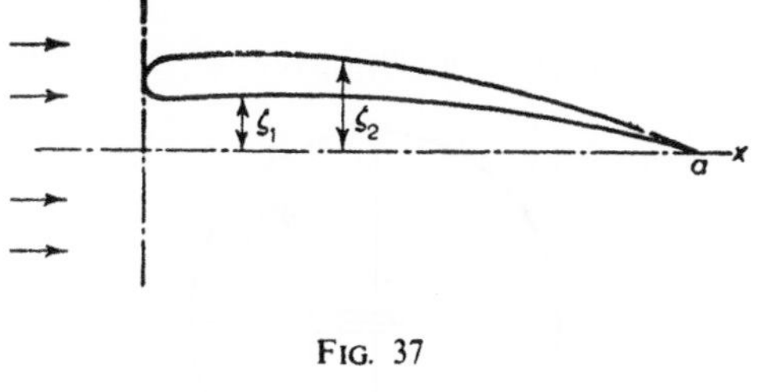

FIG. 37

In order to apply the methods of the theory of functions of a complex variable, we introduce the complex velocity $dw/dz = v_x - iv_y$ (cf. §10), which is an analytic function of the variable $z = x + iy$. In the present case this function must satisfy the conditions

$$\left.\begin{aligned} \operatorname{im}(dw/dz) &= -U\zeta_2'(x) \quad \text{for} \quad y \to 0+, \\ \operatorname{im}(dw/dz) &= -U\zeta_1'(x) \quad \text{for} \quad y \to 0-, \end{aligned}\right\} \tag{48.2}$$

on the segment $(0, a)$ of the axis of abscissae.

To solve the above problem, we first represent the required velocity distribution $\mathbf{v}(x, y)$ as a sum $\mathbf{v} = \mathbf{v}^+ + \mathbf{v}^-$ of two distributions having the following symmetry properties:

$$\left.\begin{aligned} v^-_x(x, -y) &= v^-_x(x, y), \qquad & v^-_y(x, -y) &= -v^-_y(x, y), \\ v^+_x(x, -y) &= -v^+_x(x, y), \qquad & v^+_y(x, -y) &= v^+_y(x, y). \end{aligned}\right\} \tag{48.3}$$

These properties of the separate distributions $\mathbf{v}^-$ and $\mathbf{v}^+$ do not violate the equation of continuity or that of potential flow, and, since the problem is linear, the two distributions may be sought separately.

The complex velocity is correspondingly represented as a sum

$$w' = w'_+ + w'_-,$$

and the boundary conditions on the segment $(0, a)$ for the two terms of the sum are

$$\left.\begin{aligned} [\operatorname{im} w'_+]_{y\to 0+} &= [\operatorname{im} w'_+]_{y \to 0-} = -\tfrac{1}{2}U(\zeta_1' + \zeta_2'), \\ [\operatorname{im} w'_-]_{y\to 0+} &= -[\operatorname{im} w'_-]_{y \to 0-} = \tfrac{1}{2}U(\zeta_1' - \zeta_2'). \end{aligned}\right\} \tag{48.4}$$

The function w'_- can be determined at once by Cauchy's formula:

$$w'_-(z) = \frac{1}{2\pi i}\oint_L \frac{w'_-(\xi)}{\xi - z}\,d\xi,$$

where the integration in the plane of the complex variable ξ is along a circle L with small radius centred at the point $\xi = z$ (Fig. 38). The contour L can be replaced by a circle C' with infinite radius and a contour C traversed clockwise; the latter can be deformed into the segment $(0, a)$ twice over. The integral along C' is zero, since $w'(z)$ vanishes at infinity. The integral along C gives

$$w'_- = -\frac{U}{2\pi}\int_0^a \frac{\zeta_2'(\xi) - \zeta_1'(\xi)}{\xi - z}\,d\xi. \tag{48.5}$$

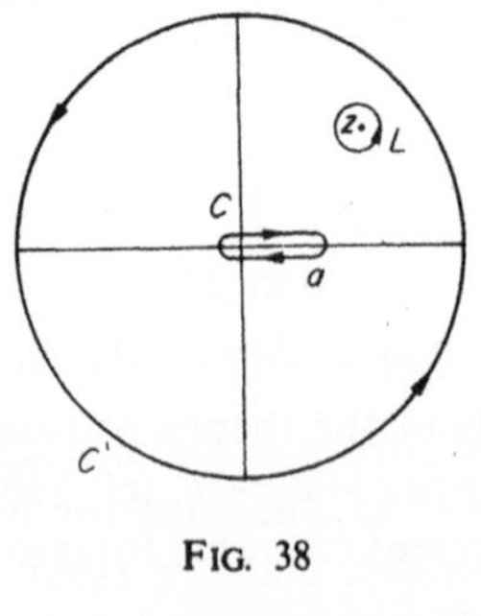

FIG. 38

Here we have used the boundary values (48.4) of the imaginary part of w'_- on the segment (0, a), and the fact that, by the symmetry conditions (48.3), the real part of w'_- is continuous across this segment.

To find the function w'_+, we have to apply Cauchy's formula, not to this function itself, but to the product $w'_+(z)g(z)$, where $g(z) = \sqrt{[z/(z-a)]}$, and the square root is taken with the plus sign for $z = x > a$. On the segment (0, a) of the real axis, the function $g(z)$ is purely imaginary and discontinuous: $g(x+i0) = -g(x-i0) = -i\sqrt{[x/(a-x)]}$. It is clear from these properties of the function $g(z)$ that the imaginary part of the product gw'_+ is discontinuous across the segment (0, a), while the real part is continuous, as with the function w'_-. Hence we have, exactly as in the derivation of formula (48.5),

$$w'_+(z)g(z) = -\frac{U}{2\pi}\int_0^a \frac{\zeta_1'(\xi)+\zeta_2'(\xi)}{\xi - z} g(\xi + i0)\mathrm{d}\xi.$$

Collecting the above expressions, we have the following formula for the velocity distribution in flow past a thin wing:

$$\frac{\mathrm{d}w}{\mathrm{d}z} = -\frac{U}{2\pi i}\sqrt{\frac{z-a}{z}}\int_0^a \frac{\zeta_1'(\xi)+\zeta_2'(\xi)}{\xi - z}\sqrt{\frac{\xi}{a-\xi}}\mathrm{d}\xi - \frac{U}{2\pi}\int_0^a \frac{\zeta_2'(\xi)-\zeta_1'(\xi)}{\xi - z}\mathrm{d}\xi. \tag{48.6}$$

Near the rounded leading edge (i.e. for $z \to 0$), this expression in general becomes infinite, the approximation used above being invalid in this region. Near the pointed trailing edge (i.e. for $z \to a$), the first term in (48.6) is finite, but the second term becomes infinite, though only logarithmically.† This logarithmic singularity is due to the approximation used, and is removed by a more exact treatment; there is no power-law divergence at the trailing edge, in accordance with the Chaplygin condition. The fulfilment of this condition is achieved by an appropriate choice of the function $g(z)$ used above.

Formula (48.6) immediately enables us to determine the velocity circulation Γ round the wing profile. According to the general rule (see §10), Γ is given by the residue of the function $w'(z)$ at its simple pole $z = 0$. The required residue is easily found as the coefficient of $1/z$ in an expansion of $w'(z)$ in powers of $1/z$ about the point at infinity: $\mathrm{d}w/\mathrm{d}z = \Gamma/2\pi iz + \ldots$, and Γ is given by the simple forumula

$$\Gamma = U\int_0^a (\zeta_1' + \zeta_2')\sqrt{\frac{\xi}{a-\xi}}\mathrm{d}\xi. \tag{48.7}$$

We may point out that only the sum of the functions ζ_1 and ζ_2 appears here. The lift force is unchanged if the thin wing is replaced by a bent plate whose shape is given by the function $\frac{1}{2}(\zeta_1 + \zeta_2)$.

For example, for a wing in the form of a thin plate with infinite length, inclined at a small angle of attack α, we have $\zeta_1 = \zeta_2 = \alpha(a-x)$, and formula (48.7) gives $\Gamma = -\pi\alpha a U$. The lift coefficient for such a wing is $C_y = -\rho U\Gamma/\frac{1}{2}\rho U^2 a = 2\pi\alpha$.

† This divergence disappears if ζ_1 and ζ_2 vanish as $(a-x)^k$, $k > 1$, near the trailing edge, i.e. if the point at the trailing edge is a cusp.

CHAPTER V

THERMAL CONDUCTION IN FLUIDS

§49. The general equation of heat transfer

It has been mentioned at the end of §2 that a complete system of equations of fluid dynamics must contain five equations. For a fluid in which processes of thermal conduction and internal friction occur, one of these equations is, as before, the equation of continuity, and Euler's equations are replaced by the Navier–Stokes equations. The fifth equation for an ideal fluid is the equation of conservation of entropy (2.6). In a viscous fluid this equation does not hold, of course, since irreversible processes of energy dissipation occur in it.

In an ideal fluid the law of conservation of energy is expressed by equation (6.1):

$$\frac{\partial}{\partial t}(\tfrac{1}{2}\rho v^2 + \rho\varepsilon) = -\operatorname{div}[\rho\mathbf{v}(\tfrac{1}{2}v^2 + w)].$$

The expression on the left is the rate of change of energy in unit volume of the fluid, while that on the right is the divergence of the energy flux density. In a viscous fluid the law of conservation of energy still holds, of course: the change per unit time in the total energy of the fluid in any volume must still be equal to the total flux of energy through the surface bounding that volume. The energy flux density, however, now has a different form. Besides the flux $\rho\mathbf{v}(\frac{1}{2}v^2 + w)$ due to the simple transfer of mass by the motion of the fluid, there is also a flux due to processes of internal friction. This latter flux is given by the vector $\mathbf{v}\cdot\boldsymbol{\sigma}'$, with components $v_i\sigma'_{ik}$ (see §16). There is, moreover, another term that must be included in the energy flux. If the temperature of the fluid is not constant throughout its volume, there will be, besides the two means of energy transfer indicated above, a transfer of heat by what is called *thermal conduction*. This signifies the direct molecular transfer of energy from points where the temperature is high to those where it is low. It does not involve macroscopic motion, and occurs even in a fluid at rest.

We denote by $\mathbf{q}$ the heat flux density due to thermal conduction. The flux $\mathbf{q}$ is related to the variation of temperature through the fluid. This relation can be written down at once in cases where the temperature gradient in the fluid is not large; in phenomena of thermal conduction we are almost always concerned with such cases. We can then expand $\mathbf{q}$ as a series of powers of the temperature gradient, taking only the first terms of the expansion. The constant term is evidently zero, since $\mathbf{q}$ must vanish when $\mathbf{grad}\, T$ does so. Thus we have

$$\mathbf{q} = -\kappa\, \mathbf{grad}\, T. \tag{49.1}$$

The constant κ is called the *thermal conductivity*. It is always positive, as we see at once from the fact that the energy flux must be from points at a high temperature to those at a low temperature, i.e. $\mathbf{q}$ and $\mathbf{grad}\, T$ must be in opposite directions. The coefficient κ is in general a function of temperature and pressure.

Thus the total energy flux in a fluid when there is viscosity and thermal conduction is $\rho\mathbf{v}(\frac{1}{2}v^2+w)-\mathbf{v}\cdot\boldsymbol{\sigma}'-\kappa\,\mathbf{grad}\,T$. Accordingly, the general law of conservation of energy is given by the equation

$$\frac{\partial}{\partial t}(\tfrac{1}{2}\rho v^2+\rho\varepsilon) = -\operatorname{div}[\rho\mathbf{v}(\tfrac{1}{2}v^2+w)-\mathbf{v}\cdot\boldsymbol{\sigma}'-\kappa\,\mathbf{grad}\,T]. \quad (49.2)$$

This equation could be taken to complete the system of fluid-mechanical equations of a viscous fluid. It is convenient, however, to put it in another form by transforming it with the aid of the equations of motion. To do so, we calculate the time derivative of the energy in unit volume of fluid, starting from the equations of motion. We have

$$\frac{\partial}{\partial t}(\tfrac{1}{2}\rho v^2+\rho\varepsilon) = \tfrac{1}{2}v^2\frac{\partial\rho}{\partial t}+\rho\mathbf{v}\cdot\frac{\partial\mathbf{v}}{\partial t}+\rho\frac{\partial\varepsilon}{\partial t}+\varepsilon\frac{\partial\rho}{\partial t}.$$

Substituting for $\partial\rho/\partial t$ from the equation of continuity and for $\partial\mathbf{v}/\partial t$ from the Navier–Stokes equation, we have

$$\frac{\partial}{\partial t}(\tfrac{1}{2}\rho v^2+\rho\varepsilon) = -\tfrac{1}{2}v^2\operatorname{div}(\rho\mathbf{v})-\rho\mathbf{v}\cdot\mathbf{grad}\,\tfrac{1}{2}v^2-\mathbf{v}\cdot\mathbf{grad}\,p+$$

$$+v_i\frac{\partial\sigma'_{ik}}{\partial x_k}+\rho\frac{\partial\varepsilon}{\partial t}-\varepsilon\operatorname{div}(\rho\mathbf{v}).$$

Using now the thermodynamic relation $\mathrm{d}\varepsilon = T\,\mathrm{d}s-p\,\mathrm{d}V = T\,\mathrm{d}s+(p/\rho^2)\,\mathrm{d}\rho$, we find

$$\frac{\partial\varepsilon}{\partial t} = T\frac{\partial s}{\partial t}+\frac{p}{\rho^2}\frac{\partial\rho}{\partial t} = T\frac{\partial s}{\partial t}-\frac{p}{\rho^2}\operatorname{div}(\rho\mathbf{v}).$$

Substituting this and introducing the heat function $w = \varepsilon+p/\rho$, we obtain

$$\frac{\partial}{\partial t}(\tfrac{1}{2}\rho v^2+\rho\varepsilon) = -(\tfrac{1}{2}v^2+w)\operatorname{div}(\rho\mathbf{v})-\rho\mathbf{v}\cdot\mathbf{grad}\,\tfrac{1}{2}v^2-\mathbf{v}\cdot\mathbf{grad}\,p+$$

$$+\rho T\frac{\partial s}{\partial t}+v_i\frac{\partial\sigma'_{ik}}{\partial x_k}.$$

Next, from the thermodynamic relation $\mathrm{d}w = T\,\mathrm{d}s+\mathrm{d}p/\rho$ we have $\mathbf{grad}\,p = \rho\,\mathbf{grad}\,w-\rho T\,\mathbf{grad}\,s$. The last term on the right of the above equation can be written

$$v_i\frac{\partial\sigma'_{ik}}{\partial x_k} = \frac{\partial}{\partial x_k}(v_i\sigma'_{ik})-\sigma'_{ik}\frac{\partial v_i}{\partial x_k} \equiv \operatorname{div}(\mathbf{v}\cdot\boldsymbol{\sigma}')-\sigma'_{ik}\frac{\partial v_i}{\partial x_k}.$$

Substituting these expressions, and adding and subtracting $\operatorname{div}(\kappa\,\mathbf{grad}\,T)$, we obtain

$$\frac{\partial}{\partial t}(\tfrac{1}{2}\rho v^2+\rho\varepsilon) = -\operatorname{div}[\rho\mathbf{v}(\tfrac{1}{2}v^2+w)-\mathbf{v}\cdot\boldsymbol{\sigma}'-\kappa\,\mathbf{grad}\,T]+$$

$$+\rho T\left(\frac{\partial s}{\partial t}+\mathbf{v}\cdot\mathbf{grad}\,s\right)-\sigma'_{ik}\frac{\partial v_i}{\partial x_k}-\operatorname{div}(\kappa\,\mathbf{grad}\,T). \quad (49.3)$$

Comparing this expression for the time derivative of the energy in unit volume with (49.2), we have

$$\rho T\left(\frac{\partial s}{\partial t}+\mathbf{v}\cdot\mathbf{grad}\,s\right) = \sigma'_{ik}\frac{\partial v_i}{\partial x_k}+\operatorname{div}(\kappa\,\mathbf{grad}\,T). \quad (49.4)$$

This equation is called the *general equation of heat transfer*. If there is no viscosity or thermal conduction, the right-hand side is zero, and the equation of conservation of entropy (2.6) for an ideal fluid is obtained.

The following interpretation of equation (49.4) should be noticed. The expression on the left is just the total time derivative ds/dt of the entropy, multiplied by ρT. The quantity ds/dt gives the rate of change of the entropy of a unit mass of fluid as it moves about in space, and $T\,ds/dt$ is therefore the quantity of heat gained by this unit mass in unit time, so that $\rho T\,ds/dt$ is the quantity of heat gained per unit volume. We see from (49.4) that the amount of heat gained by unit volume of the fluid is therefore

$$\sigma'_{ik}\,\partial v_i/\partial x_k + \operatorname{div}(\kappa\,\mathbf{grad}\,T).$$

The first term here is the energy dissipated into heat by viscosity, and the second is the heat conducted into the volume concerned.

We expand the term $\sigma'_{ik}\,\partial v_i/\partial x_k$ in (49.4) by substituting the expression (15.3) for σ'_{ik}. We have

$$\sigma'_{ik}\frac{\partial v_i}{\partial x_k} = \eta\frac{\partial v_i}{\partial x_k}\left(\frac{\partial v_i}{\partial x_k}+\frac{\partial v_k}{\partial x_i}-\tfrac{2}{3}\delta_{ik}\frac{\partial v_l}{\partial x_l}\right)+\zeta\frac{\partial v_i}{\partial x_k}\delta_{ik}\frac{\partial v_l}{\partial x_l}.$$

It is easy to verify that the first term may be written as

$$\tfrac{1}{2}\eta\left(\frac{\partial v_i}{\partial x_k}+\frac{\partial v_k}{\partial x_i}-\tfrac{2}{3}\delta_{ik}\frac{\partial v_l}{\partial x_l}\right)^2,$$

and the second is

$$\zeta\frac{\partial v_i}{\partial x_k}\delta_{ik}\frac{\partial v_l}{\partial x_l} = \zeta\frac{\partial v_i}{\partial x_i}\frac{\partial v_l}{\partial x_l} \equiv \zeta(\operatorname{div}\mathbf{v})^2.$$

Thus equation (49.4) becomes

$$\rho T\left(\frac{\partial s}{\partial t}+\mathbf{v}\cdot\mathbf{grad}\,s\right) = \operatorname{div}(\kappa\,\mathbf{grad}\,T)+\tfrac{1}{2}\eta\left(\frac{\partial v_i}{\partial x_k}+\frac{\partial v_k}{\partial x_i}-\tfrac{2}{3}\delta_{ik}\frac{\partial v_l}{\partial x_l}\right)^2 + \zeta(\operatorname{div}\mathbf{v})^2. \tag{49.5}$$

The entropy of the fluid increases as a result of the irreversible processes of thermal conduction and internal friction. Here, of course, we mean not the entropy of each volume element of fluid separately, but the total entropy of the whole fluid, equal to the integral

$$\int \rho s\,dV.$$

The change in entropy per unit time is given by the derivative

$$d\left[\int \rho s\,dV\right]/dt = \int[\partial(\rho s)/\partial t]\,dV.$$

Using the equation of continuity and equation (49.5) we have

$$\frac{\partial(\rho s)}{\partial t} = \rho\frac{\partial s}{\partial t}+s\frac{\partial \rho}{\partial t} = -s\operatorname{div}(\rho\mathbf{v})-\rho\mathbf{v}\cdot\mathbf{grad}\,s+\frac{1}{T}\operatorname{div}(\kappa\,\mathbf{grad}\,T)+$$

$$+\frac{\eta}{2T}\left(\frac{\partial v_i}{\partial x_k}+\frac{\partial v_k}{\partial x_i}-\tfrac{2}{3}\delta_{ik}\frac{\partial v_l}{\partial x_l}\right)^2+\frac{\zeta}{T}(\operatorname{div}\mathbf{v})^2.$$

The first two terms on the right together give $-\operatorname{div}(\rho s\mathbf{v})$. The volume integral of this is transformed into the integral of the entropy flux $\rho s\mathbf{v}$ over the surface. If we consider an unbounded volume of fluid at rest at infinity, the bounding surface can be removed to infinity; the integrand in the surface integral is then zero, and so is the integral itself. The integral of the third term on the right is transformed as follows:

$$\int \frac{1}{T}\operatorname{div}(\kappa\,\mathbf{grad}\,T)\,\mathrm{d}V = \int \operatorname{div}\left(\frac{\kappa\,\mathbf{grad}\,T}{T}\right)\mathrm{d}V + \int \frac{\kappa(\mathbf{grad}\,T)^2}{T^2}\,\mathrm{d}V.$$

Assuming that that the fluid temperature tends sufficiently rapidly to a constant value at infinity, we can transform the first integral into one over an infinitely remote surface, on which $\mathbf{grad}\,T = 0$ and the integral therefore vanishes.

The result is

$$\frac{\mathrm{d}}{\mathrm{d}t}\int \rho s\,\mathrm{d}V = \int \frac{\kappa(\mathbf{grad}\,T)^2}{T^2}\,\mathrm{d}V + \int \frac{\eta}{2T}\left(\frac{\partial v_i}{\partial x_k} + \frac{\partial v_k}{\partial x_i} - \tfrac{2}{3}\delta_{ik}\frac{\partial v_l}{\partial x_l}\right)^2 \mathrm{d}V + \int \frac{\zeta}{T}(\operatorname{div}\mathbf{v})^2\,\mathrm{d}V. \tag{49.6}$$

The first term on the right is the rate of increase of entropy owing to thermal conduction, and the other two terms give the rate of increase due to internal friction. The entropy can only increase, i.e. the sum on the right of (49.6) must be positive. In each term, the integrand may be non-zero even if the other two integrals vanish. Thus each integral separately must always be positive. Hence it follows that the second viscosity coefficient ζ is positive, as well as κ and η, which we already know are positive.

It has been tacitly assumed in the derivation of formula (49.1) that the heat flux depends only on the temperature gradient, and not on the pressure gradient. This assumption, which is not evident à priori, can now be justified as follows. If $\mathbf{q}$ contained a term proportional to $\mathbf{grad}\,p$, the expression (49.6) for the rate of change of entropy would include another term having the product $\mathbf{grad}\,p \cdot \mathbf{grad}\,T$ in the integrand. Since the latter might be either positive or negative, the time derivative of the entropy would not necessarily be positive, which is impossible.

Finally, the above arguments must also be refined in the following respect. Strictly speaking, in a system which is not in thermodynamic equilibrium, such as a fluid with velocity and temperature gradients, the usual definitions of thermodynamic quantities are no longer meaningful, and must be modified. The necessary definitions are, firstly, that ρ, ε and $\mathbf{v}$ are defined as before: ρ and $\rho\varepsilon$ are the mass and internal energy per unit volume, and $\mathbf{v}$ is the momentum of unit mass of fluid. The remaining thermodynamic quantities are then defined as being the same functions of ρ and ε as they are in thermal equilibrium. The entropy $s = s(\rho, \varepsilon)$, however, is no longer the true thermodynamic entropy: the integral

$$\int \rho s\,\mathrm{d}V$$

will not, strictly speaking, be a quantity that must increase with time. Nevertheless, it is easy to see that, for small velocity and temperature gradients, s is the same as the true entropy in the approximation here used. For, if there are gradients present, they in general lead to additional terms (besides $s(\rho, \varepsilon)$) in the entropy. The results given above, however, can be altered only by terms linear in the gradients (for instance, a term proportional to the scalar $\operatorname{div}\mathbf{v}$). Such terms would necessarily take both positive and negative values. But they ought to be negative definite, since the equilibrium value $s = s(\rho, \varepsilon)$ is the maximum possible value. Hence the expansion of the entropy in powers of the small gradients can contain (apart from the zero-order term) only terms of the second and higher orders.

Similar remarks should have been made in §15 (cf. the first footnote to that section), since the presence of even a velocity gradient implies the absence of thermodynamic equilibrium. The pressure p which appears in the expression for the momentum flux density tensor in a viscous fluid must be taken to be the same function $p = p(\rho, \varepsilon)$ as in thermal equilibrium. In this case p will not, strictly speaking, be the pressure in the usual sense, viz. the normal force on a surface element. Unlike what happens for the entropy (see above), there is here a resulting difference of the first order with respect to the small gradient; we have seen that the normal component of the force includes, besides p, a term proportional to div **v** (in an incompressible fluid, this term is zero, and the difference is then of higher order).

Thus the three coefficients η, ζ, κ which appear in the equations of motion of a viscous conducting fluid completely determine the mechanical properties of the fluid in the approximation considered (i.e. when the higher-order space derivatives of velocity, temperature, etc. are neglected). The introduction of any further terms (for example, the inclusion in the mass flux density of terms proportional to the gradient of density or temperature) has no physical meaning, and would mean at least a change in the definition of the basic quantities; in particular, the velocity would no longer be the momentum of unit mass of fluid.†

§50. Thermal conduction in an incompressible fluid

The general equation of thermal conduction (49.4) or (49.5) can be considerably simplified in certain cases. If the fluid velocity is small compared with the velocity of sound, the pressure variations occurring as a result of the motion are so small that the variation in the density (and in the other thermodynamic quantities) caused by them may be neglected. However, a non-uniformly heated fluid is still not completely incompressible in the sense used previously. The reason is that the density varies with the temperature; this variation cannot in general be neglected, and therefore, even at small velocities, the density of a non-uniformly heated fluid cannot be supposed constant. In determining the derivatives of thermodynamic quantities in this case, it is therefore necessary to suppose the pressure constant, and not the density. Thus we have

$$\frac{\partial s}{\partial t} = \left(\frac{\partial s}{\partial T}\right)_p \frac{\partial T}{\partial t}, \qquad \mathbf{grad}\, s = \left(\frac{\partial s}{\partial T}\right)_p \mathbf{grad}\, T,$$

and, since $T(\partial s/\partial T)_p$ is the specific heat at constant pressure, c_p, we obtain $T\partial s/\partial t = c_p \partial T/\partial t$, $T\,\mathbf{grad}\, s = c_p\, \mathbf{grad}\, T$. Equation (49.4) becomes

† Worse still, the inclusion of such terms may violate the necessary conservation laws. It must be borne in mind that, whatever the definitions used, the mass flux density **j** must always be the momentum of unit volume of fluid. For **j** is defined by the equation of continuity,

$$\partial\rho/\partial t + \operatorname{div}\mathbf{j} = 0;$$

multiplying this by **r** and integrating over the fluid volume, we have

$$\mathrm{d}\left(\int \rho\mathbf{r}\,\mathrm{d}V\right)/\mathrm{d}t = \int \mathbf{j}\,\mathrm{d}V,$$

and since the integral $\int\rho\mathbf{r}\,\mathrm{d}V$ determines the position of the centre of mass, it is clear that the integral $\int\mathbf{j}\,\mathrm{d}V$ is the momentum.

$$\rho c_p\left(\frac{\partial T}{\partial t}+\mathbf{v}\cdot\mathbf{grad}\,T\right)=\operatorname{div}(\kappa\,\mathbf{grad}\,T)+\sigma'_{ik}\frac{\partial v_i}{\partial x_k}. \tag{50.1}$$

If the density is to be supposed constant in the equations of motion for a non-uniformly heated fluid, it is necessary that the fluid velocity should be small compared with that of sound, and also that the temperature differences in the fluid should be small. We emphasize that we mean the actual values of the temperature differences, not the temperature gradient. The fluid may then be supposed incompressible in the usual sense; in particular, the equation of continuity is simply $\operatorname{div}\mathbf{v}=0$. Supposing the temperature differences small, we neglect also the temperature variation of η, κ and c_p, supposing them constant. Writing the term $\sigma'_{ik}\,\partial v_i/\partial x_k$ as in (49.5), we obtain the equation of heat transfer in an incompressible fluid in the following comparatively simple form:

$$\frac{\partial T}{\partial t}+\mathbf{v}\cdot\mathbf{grad}\,T=\chi\triangle T+\frac{\nu}{2c_p}\left(\frac{\partial v_i}{\partial x_k}+\frac{\partial v_k}{\partial x_i}\right)^2, \tag{50.2}$$

where $\nu=\eta/\rho$ is the kinematic viscosity, and we have written κ in terms of the *thermometric conductivity*, defined as

$$\chi=\kappa/\rho c_p. \tag{50.3}$$

The equation of heat transfer is particularly simple for an incompressible fluid at rest, in which the transfer of energy takes place entirely by thermal conduction. Omitting the terms in (50.2) which involve the velocity, we have simply

$$\partial T/\partial t=\chi\triangle T. \tag{50.4}$$

This equation is called in mathematical physics the *equation of thermal conduction* or *Fourier's equation.* It can, of course, be obtained much more simply without using the general equation of heat transfer in a moving fluid. According to the law of conservation of energy, the amount of heat absorbed in some volume in unit time must equal the total heat flux into this volume through the surface surrounding it. As we know, such a law of conservation can be expressed as an "equation of continuity" for the amount of heat. This equation is obtained by equating the amount of heat absorbed in unit volume in unit time to minus the divergence of the heat flux density. The former is $\rho c_p\,\partial T/\partial t$; we must take the specific heat c_p, since the pressure is of course constant throughout a fluid at rest. Equating this to $-\operatorname{div}\mathbf{q}=\kappa\triangle T$, we have equation (50.4).

It must be mentioned that the applicability of the thermal conduction equation (50.4) to fluids is actually very limited. The reason is that, in fluids in a gravitational field, even a small temperature gradient usually results in considerable motion (*convection*; see §56). Hence we can actually have a fluid at rest with a non-uniform temperature distribution only if the direction of the temperature gradient is opposite to that of the gravitational force, or if the fluid is very viscous. Nevertheless, a study of the equation of thermal conduction in the form (50.4) is very important, since processes of thermal conduction in solids are described by an equation of the same form. We shall therefore consider it in more detail in §§51 and 52.

If the temperature distribution in a non-uniformly heated medium at rest is maintained constant in time (by means of some external source of heat), the equation of thermal conduction becomes

$$\triangle T=0. \tag{50.5}$$

Thus a steady temperature distribution in a medium at rest satisfies Laplace's equation. In the more general case where κ cannot be regarded a constant, we have in place of (50.5) the equation

$$\operatorname{div}(\kappa\,\mathbf{grad}\,T) = 0. \tag{50.6}$$

If the fluid contains external sources of heat (for example, heating by an electric current), the equation of thermal conduction must correspondingly contain another term. Let Q be the quantity of heat generated by these sources in unit volume of the fluid per unit time; Q is, in general, a function of the coordinates and of the time. Then the heat balance equation, i.e. the equation of thermal conduction, is

$$\rho c_p \,\partial T/\partial t = \kappa \triangle T + Q. \tag{50.7}$$

Let us write down the boundary conditions on the equation of thermal conduction which hold at the boundary between two media. First of all, the temperatures of the two media must be equal at the boundary:

$$T_1 = T_2. \tag{50.8}$$

Furthermore, the heat flux out of one medium must equal the heat flux into the other medium. Taking a coordinate system in which the part of the boundary considered is at rest, we can write this condition as $\kappa_1\,\mathbf{grad}\,T_1\cdot\mathbf{df} = \kappa_2\,\mathbf{grad}\,T_2\cdot\mathbf{df}$ for each surface element df. Putting $\mathbf{grad}\,T\cdot\mathbf{df} = (\partial T/\partial n)\,df$, where $\partial T/\partial n$ is the derivative of T along the normal to the surface, we obtain the boundary condition in the form

$$\kappa_1\,\partial T_1/\partial n = \kappa_2\,\partial T_2/\partial n. \tag{50.9}$$

If there are on the surface of separation external sources of heat which generate an amount of heat $Q^{(s)}$ on unit area in unit time, then (50.9) must be replaced by

$$\kappa_1\,\partial T_1/\partial n - \kappa_2\,\partial T_2/\partial n = Q^{(s)}. \tag{50.10}$$

In physical problems concerning the distribution of temperature in the presence of heat sources, the strength of the latter is usually given as a function of temperature. If the function $Q(T)$ increases sufficiently rapidly with T, it may be impossible to establish a steady temperature distribution in a body whose boundaries are maintained in fixed conditions (e.g. at a given temperature). The loss of heat through the outer surface of the body is proportional to some mean value of the temperature difference $T - T_0$ between the body and the external medium, regardless of the law of heat generation within the body; it is clear that, if the generation of heat increases sufficiently rapidly with temperature, the loss of heat may be inadequate to achieve an equilibrium state.

There may then be a *thermal explosion*: if the rate of an exothermic combustion reaction increases sufficiently rapidly with temperature, the impossibility of a steady distribution leads to a rapid non-steady ignition of the substance and an acceleration of the reaction (N. N. Semenov 1923). The rate of explosive combustion reactions, and therefore the rate of heat generation, depend on the temperature roughly as $e^{-U/T}$, with a large activation energy U. To investigate the conditions for a thermal explosion to occur, we must consider the course of the reaction when the ignition of the substance is comparatively slow, and therefore use the expansion

$$\frac{1}{T} \cong \frac{1}{T_0} - \frac{T-T_0}{T_0{}^2},$$

where T_0 is the external temperature. The problem thus leads to a study of the equation of thermal conduction with a volume density of heat sources

$$Q = Q_0 \, e^{\alpha(T-T_0)} \tag{50.11}$$

(D.A. Frank-Kamenetskiĭ 1939); see Problem 1.

PROBLEMS

PROBLEM 1. Heat sources with the strength (50.11) are distributed in a layer of material bounded by two parallel planes, which are kept at a constant temperature. Find the condition for a steady temperature distribution to be possible (D. A. Frank-Kamenetskiĭ 1939).†

SOLUTION. The equation for steady heat conduction is here

$$\kappa \, d^2T/dx^2 = -Q_0 e^{\alpha(T-T_0)},$$

with the boundary conditions $T = T_0$ for $x = 0$ and $x = 2l$ ($2l$ being the thickness of the layer). We introduce the dimensionless variables $\tau = \alpha(T-T_0)$ and $\xi = x/l$. Then

$$\tau'' + \lambda e^{\tau} = 0, \qquad \lambda = Q_0 \alpha l^2/\kappa.$$

Integrating this equation once (after multiplying by $2\tau'$), we find

$$\tau'^2 = 2\lambda(e^{\tau_0} - e^{\tau}),$$

where τ_0 is a constant, which is evidently the maximum value of τ; by symmetry, this value must be attained halfway through the layer, i.e. for $\xi = 1$. Hence a second integration, with the condition $\tau = 0$ for $\xi = 0$, gives

$$\frac{1}{\sqrt{(2\lambda)}} \int_0^{\tau_0} \frac{d\tau}{\sqrt{(e^{\tau_0} - e^{\tau})}} = \int_0^1 d\xi = 1.$$

Effecting the integration, we have

$$e^{-\frac{1}{2}\tau_0} \cosh^{-1} e^{\frac{1}{2}\tau_0} = \sqrt{(\tfrac{1}{2}\lambda)}. \tag{1}$$

The function $\lambda(\tau_0)$ determined by this equation has a maximum $\lambda = \lambda_{cr}$ for a definite value $\tau_0 = \tau_{0,cr}$; if $\lambda > \lambda_{cr}$, there is no solution satisfying the boundary conditions.‡ The numerical values are $\lambda_{cr} = 0{\cdot}88$, $\tau_{0,cr} = 1{\cdot}2$.††

PROBLEM 2. A sphere is immersed in a fluid at rest, in which a constant temperature gradient is maintained. Determine the resulting steady temperature distribution in the fluid and the sphere.

SOLUTION. The temperature distribution satisfies the equation $\triangle T = 0$ in all space, with the boundary conditions

$$T_1 = T_2, \qquad \kappa_1 \partial T_1/\partial r = \kappa_2 \partial T_2/\partial r$$

for $r = R$ (where R is the radius of the sphere; quantities with the suffixes 1 and 2 refer to the sphere and the fluid respectively), and $\mathbf{grad}\, T = \mathbf{A}$ at infinity, where $\mathbf{A}$ is the given temperature gradient. By the symmetry of the problem, $\mathbf{A}$ is the only vector which can determine the required solution. Such solutions of Laplace's equation are constant $\times\, \mathbf{A}\cdot\mathbf{r}$ and constant $\times\, \mathbf{A}\cdot\mathbf{grad}\,(1/r)$. Noticing also that the solution must remain finite at the centre of the sphere, we seek the temperatures T_1 and T_2 in the forms

$$T_1 = c_1 \mathbf{A}\cdot\mathbf{r}, \qquad T_2 = c_2 \mathbf{A}\cdot\mathbf{r}/r^3 + \mathbf{A}\cdot\mathbf{r}.$$

The constants c_1 and c_2 are determined from the conditions for $r = R$, the result being

$$T_1 = \frac{3\kappa_2}{\kappa_1 + 2\kappa_2} \mathbf{A}\cdot\mathbf{r}, \qquad T_2 = \left[1 + \frac{\kappa_2 - \kappa_1}{\kappa_1 + 2\kappa_2}\left(\frac{R}{r}\right)^3\right] \mathbf{A}\cdot\mathbf{r}.$$

† A more detailed discussion of related topics is given by Frank-Kamenetskiĭ in *Diffusion and Heat Transfer in Chemical Kinetics*, New York 1969.

‡ Only the smaller of the two roots of equation (1) for $\lambda < \lambda_{cr}$ corresponds to a stable temperature distribution.

†† The corresponding values for a spherical region (with radius l) are $\lambda_{cr} = 3{\cdot}32$, $\tau_{0,cr} = 1{\cdot}47$, and for an infinite cylinder $\lambda_{cr} = 2{\cdot}00$, $\tau_{0,cr} = 1{\cdot}36$.

§51. Thermal conduction in an infinite medium

Let us consider thermal conduction in an infinite medium at rest. The most general problem of this kind is as follows. The temperature distribution is given in all space at the initial instant $t = 0$:

$$T = T_0(\mathbf{r}) \text{ for } t = 0,$$

where $T_0(\mathbf{r})$ is a given function of the coordinates. It is required to determine the temperature distribution at all subsequent instants.

We expand the required function $T(\mathbf{r}, t)$ as a Fourier integral with respect to the coordinates:

$$T(\mathbf{r}, t) = \int T_{\mathbf{k}}(t) \exp(i\mathbf{k}\cdot\mathbf{r})\, d^3k/(2\pi)^3, \quad T_{\mathbf{k}}(t) = \int T(\mathbf{r}, t) \exp(-i\mathbf{k}\cdot\mathbf{r})\, d^3x. \tag{51.1}$$

For each Fourier component $T_{\mathbf{k}} \exp(i\mathbf{k}\cdot\mathbf{r})$, equation (50.4) gives

$$dT_{\mathbf{k}}/dt + k^2\chi T_{\mathbf{k}} = 0.$$

This equation gives $T_{\mathbf{k}}$ as a function of time:

$$T_{\mathbf{k}} = \exp(-k^2\chi t) T_{0\mathbf{k}}.$$

Since we must have $T = T_0(\mathbf{r})$ for $t = 0$, it is clear that the $T_{0\mathbf{k}}$ are the expansion coefficients of the function T_0 as a Fourier integral:

$$T_{0\mathbf{k}} = \int T_0(\mathbf{r}') \exp(-i\mathbf{k}\cdot\mathbf{r}')\, d^3x'.$$

Thus

$$T = \iint T_0(\mathbf{r}') \exp(-k^2\chi t) \exp[i\mathbf{k}\cdot(\mathbf{r}-\mathbf{r}')]\, d^3x'\, d^3k/(2\pi)^3.$$

The integral over k is the product of three simple integrals, each having the form

$$\int_{-\infty}^{\infty} \exp(-\alpha\xi^2) \cos\beta\xi\, d\xi = \sqrt{(\pi/\alpha)} \exp(-\beta^2/4\alpha),$$

where ξ is one component of $\mathbf{k}$; the similar integral with sin in place of cos is zero, since the sine function is odd. Thus we have finally

$$T(\mathbf{r}, t) = \frac{1}{8(\pi\chi t)^{3/2}} \int T_0(\mathbf{r}') \exp\{-[(\mathbf{r}-\mathbf{r}')^2]/4\chi t\}\, d^3x'. \tag{51.2}$$

This formula gives the complete solution of the problem; it determines the temperature distribution at any instant in terms of the given initial distribution.

If the initial temperature distribution is a function of only one coordinate, x, then we can integrate over y' and z' in (51.2) and obtain

$$T(x, t) = \frac{1}{2\sqrt{(\pi\chi t)}} \int_{-\infty}^{\infty} T_0(x') \exp[-(x-x')^2/4\chi t]\, dx'. \tag{51.3}$$

At time $t = 0$, let the temperature be zero in all space except for one point (the origin), where it is infinite in such a way that the total quantity of heat (proportional to $\int T_0(\mathbf{r})\,d^3x$) is finite. Such a distribution can be represented by a delta function:

$$T_0(\mathbf{r}) = \text{constant} \times \delta(\mathbf{r}). \tag{51.4}$$

The integration in formula (51.2) then amounts to replacing $\mathbf{r}'$ by zero, the result of which is

$$T(\mathbf{r}, t) = \text{constant} \times \frac{1}{8(\pi\chi t)^{\frac{3}{2}}} \exp(-r^2/4\chi t). \tag{51.5}$$

In the course of time, the temperature at the point $r = 0$ decreases as $t^{-\frac{3}{2}}$. The temperature in the surrounding space rises correspondingly, and the region where the temperature is appreciably different from zero expands (Fig. 39). The manner of this expansion is determined principally by the exponential factor in (51.5). The order of magnitude l of the dimension of this region is given by

$$l \sim \sqrt{(\chi t)}, \tag{51.6}$$

i.e. l increases as the square root of the time.

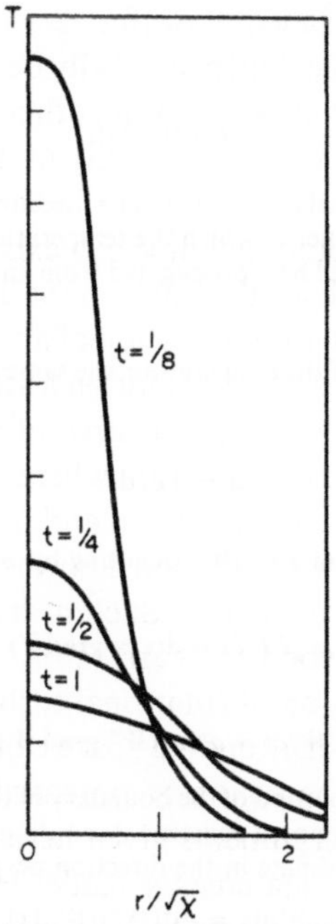

FIG. 39

Similarly, if at the initial instant a finite amount of heat is concentrated on the plane $x = 0$, the subsequent temperature distribution is

$$T(x, t) = \text{constant} \times \frac{1}{2\sqrt{(\pi\chi t)}} \exp(-x^2/4\chi t). \tag{51.7}$$

Formula (51.6) can also be interpreted in a somewhat different way. Let l be the order of magnitude of the dimension of a body. Then we can say that, if the body is heated non-uniformly, the order of magnitude τ of the time required for the temperature to become more or less the same throughout the body is

$$\tau \sim l^2/\chi. \tag{51.8}$$

The time τ, which may be called the *relaxation time* for thermal conduction, is proportional to the square of the dimension of the body, and inversely proportional to the thermometric conductivity.

The thermal conduction process described by the formulae obtained above has the property that the effect of any perturbation is propagated instantaneously through all space. It is seen from formula (51.5) that the heat from a point source is propagated in such a manner that, even at the next instant, the temperature of the medium is zero only at infinity. This property holds also for a medium in which the thermometric conductivity χ depends on the temperature, provided that χ does not vanish anywhere. If, however, χ is a function of temperature which vanishes when $T = 0$, the propagation of heat is retarded, and at each instant the effect of a given perturbation extends only to a finite region of space (we suppose that the temperature outside this region can be taken as zero). This result, as well as the solution of the following Problems, is due to Ya. B. Zel'dovich and A. S. Kompaneets (1950).

PROBLEMS

PROBLEM 1. The specific heat and thermal conductivity of a medium vary as powers of the temperature, while its density is constant. Determine the manner in which the temperature tends to zero near the boundary of the region which, at a given instant, has received heat propagated from an arbitrary source (the temperature outside that region being zero).

SOLUTION. If κ and c_p vary as powers of the temperature, the same is true of the thermometric conductivity χ and of the heat function

$$w = \int c_p \, dT$$

(we omit a constant in w). Hence we can put $\chi = aW^n$, denoting by $W = \rho w$ the heat function per unit volume. Then the thermal conduction equation

$$\rho c_p \partial T/\partial t = \operatorname{div}(\kappa \,\mathbf{grad}\, T)$$

becomes

$$\partial W/\partial t = a \operatorname{div}(W^n \,\mathbf{grad}\, W). \tag{1}$$

During a short interval of time, a small portion of the boundary of the region may be regarded as plane, and its rate of displacement in space, v, may be supposed constant. Accordingly, we seek a solution of equation (1) in the form $W = W(x - vt)$, where x is the coordinate in the direction perpendicular to the boundary. We have

$$-v \, dW/dx = a \, d(W^n \, dW/dx)/dx, \tag{2}$$

whence we find, after two integrations, that W vanishes as

$$W \propto |x|^{1/n}, \tag{3}$$

where $|x|$ is the distance from the boundary of the heated region. This also confirms our conclusion that, if $n > 0$, the heated region has a boundary outside which W and T are zero. If $n \leqslant 0$, then equation (2) has no solution vanishing at a finite distance, i.e. the heat is distributed through all space at every instant.

PROBLEM 2. A medium like that described in Problem 1 has, at the initial instant, an amount of heat Q per unit area concentrated on the plane $x = 0$, while $T = 0$ everywhere else. Determine the temperature distribution at subsequent instants.

SOLUTION. In the one-dimensional case, equation (1) is

$$\frac{\partial W}{\partial t} = a\frac{\partial}{\partial x}\left(W^n\frac{\partial W}{\partial x}\right). \tag{4}$$

From the parameters Q and a and variables x and t at our disposal, we can form only one dimensionless combination,

$$\xi = x/(Q^n at)^{1/(2+n)}; \tag{5}$$

Q and a have the dimensions erg/cm^2 and (cm^2/sec) (cm^3/erg)^n. Hence the required function $W(x, t)$ must have the form

$$W = (Q^2/at)^{1/(2+n)} f(\xi), \tag{6}$$

where the dimensionless function $f(\xi)$ is multiplied by a quantity having the dimensions erg/cm^3. With this substitution, equation (4) gives

$$(2+n)\frac{\mathrm{d}}{\mathrm{d}\xi}\left(f^n\frac{\mathrm{d}f}{\mathrm{d}\xi}\right) + \xi\frac{\mathrm{d}f}{\mathrm{d}\xi} + f = 0.$$

This ordinary differential equation has a simple solution which satisfies the conditions of the problem, namely

$$f(\xi) = [\tfrac{1}{2}n(\xi_0{}^2 - \xi^2)/(2+n)]^{1/n}, \tag{7}$$

where ξ_0 is a constant of integration.

For $n > 0$, this formula gives the temperature distribution in the region between the planes $x = \pm x_0$ corresponding to the equation $\xi = \pm\xi_0$; outside this region, $W = 0$. Hence it follows that the heated region expands with time in a manner given by $x_0 = \text{constant} \times t^{1/(2+n)}$. The constant ξ_0 is determined by the condition that the total amount of heat be constant:

$$Q = \int_{-x_0}^{x_0} W\,\mathrm{d}x = Q\int_{-\xi_0}^{\xi_0} f(\xi)\,\mathrm{d}\xi, \tag{8}$$

whence we have

$$\xi_0{}^{2+n} = \frac{(2+n)^{1+n}2^{1-n}}{n\pi^{n/2}}\,\frac{\Gamma^n(\tfrac{1}{2}+1/n)}{\Gamma^n(1/n)}. \tag{9}$$

For $n = -\nu < 0$, we write the solution in the form

$$f(\xi) = \left[\frac{\nu}{2(2-\nu)}(\xi_0{}^2 + \xi^2)\right]^{-1/\nu} \tag{10}$$

Here the heat is distributed through all space, and at large distances W decreases as $x^{-2/\nu}$. This solution is valid only for $\nu < 2$; for $\nu \geqslant 2$, the normalization integral (8) (which now extends to $\pm\infty$) diverges, which means physically that the heat is conducted instantaneously to infinity. For $\nu < 2$, the constant ξ_0 in (10) is given by

$$\xi_0{}^{2-\nu} = \frac{2(2-\nu)\pi^{\nu/2}}{\nu}\,\frac{\Gamma^\nu(1/\nu - \tfrac{1}{2})}{\Gamma^\nu(1/\nu)}. \tag{11}$$

Finally, for $n \to 0$ we have $\xi_0 \to 2/\sqrt{n}$, and the solution given by formulae (5)–(7) is

$$W = \lim_{n\to 0}\left\{\frac{Q}{2\sqrt{(\pi at)}}\left(1 - n\frac{x^2}{4at}\right)^{1/n}\right\} = \frac{Q}{2\sqrt{(\pi at)}}\exp(-x^2/4at),$$

in agreement with formula (51.7).

§52. Thermal conduction in a finite medium

In problems of thermal conduction in a finite medium, the initial temperature distribution does not suffice to determine a unique solution, and the boundary conditions at the surface of the medium must also be given.

Let us consider thermal conduction in a half-space ($x > 0$), beginning with the case where a given constant temperature is maintained on the bounding plane $x = 0$. We may arbitrarily take this temperature as zero, i.e. measure the temperature at other points relative to it. At the initial instant, the temperature distribution throughout the medium is given, as before. The boundary and initial conditions are therefore

$$T = 0 \quad \text{for} \quad x = 0; \quad T = T_0(x, y, z) \quad \text{for} \quad t = 0 \quad \text{and} \quad x > 0. \tag{52.1}$$

The solution of the thermal conduction equation with these conditions can, by means of the following device, be reduced to the solution for a medium infinite in all directions. We imagine the medium to extend on both sides of the plane $x = 0$, the temperature distribution for $t = 0$ and $x < 0$ being given by $-T_0$. That is, the temperature distribution at the initial instant is given in all space by an odd function of x:

$$T_0(-x, y, z) = -T_0(x, y, z). \tag{52.2}$$

It follows from equation (52.2) that $T_0(0, y, z) = -T_0(0, y, z) = 0$, i.e. the necessary boundary condition (52.1) is automatically satisfied for $t = 0$, and it is evident from symmetry that it will continue to be satisfied for all t.

Thus the problem is reduced to the solution of equation (50.4) in an infinite medium with an initial function $T_0(x, y, z)$ which satisfies (52.2), and without boundary conditions. Hence we can use the general formula (51.2). We divide the range of integration over x' in (51.2) into two parts, from $-\infty$ to 0 and from 0 to ∞. Using the relation (52.2), we then have

$$T(\mathbf{r}, t) = \frac{1}{8(\pi\chi t)^{\frac{3}{2}}} \int\limits_{-\infty}^{\infty} \int\limits_{-\infty}^{\infty} \int\limits_{0}^{\infty} T_0(\mathbf{r}') \times$$
$$\times \{\exp[-(x-x')^2/4\chi t] - \exp[-(x+x')^2/4\chi t]\} \times$$
$$\times \exp\{-[(y-y')^2 + (z-z')^2]/4\chi t\}\, dx'\, dy'\, dz'. \tag{52.3}$$

This formula gives the solution of the problem, since it determines the temperature throughout the medium.

If the initial temperature distribution is a function of x only, formula (52.3) becomes

$$T(x, t) = \frac{1}{2\sqrt{(\pi\chi t)}} \int\limits_{0}^{\infty} T_0(x')\{\exp[-(x-x')^2/4\chi t] - \exp[-(x+x')^2/4\chi t]\}\, dx'. \tag{52.4}$$

As an example, let us consider the case where the initial temperature is a given constant everywhere except at $x = 0$. Without loss of generality, this constant may be taken as -1. The temperature on the plane $x = 0$ is always zero. The appropriate solution is obtained at once by substituting $T_0(x) = -1$ in (52.4). The integral in (52.4) is the sum of two integrals, in each of which we change the variables as in $\xi = (x'-x)/2\sqrt{(\chi t)}$. We then obtain for $T(x, t)$ the expression

$$T(x, t) = \tfrac{1}{2}\{\operatorname{erf}[-x/2\sqrt{(\chi t)}] - \operatorname{erf}[x/2\sqrt{(\chi t)}]\},$$

where the function erf x is defined as

$$\operatorname{erf} x = \frac{2}{\sqrt{\pi}} \int\limits_{0}^{x} e^{-\xi^2}\, d\xi, \tag{52.5}$$

and is called the *error function* (we notice that $\operatorname{erf}\infty = 1$). Since $\operatorname{erf}(-x) = -\operatorname{erf} x$, we have finally

$$T(x, t) = -\operatorname{erf}[x/2\sqrt{(\chi t)}]. \tag{52.6}$$

Figure 40 shows a graph of the function erf x. The temperature distribution becomes more uniform in space in the course of time. This occurs in such a way that any given value of the temperature "moves" proportionally to $\sqrt{t}$. This last result is obviously true. For the problem under consideration is characterized by only one parameter, the initial temperature difference T_0 between the boundary plane and the remaining space; in the above discussion, this difference was arbitrarily taken as unity. From the parameters T_0 and χ and variables x and t at our disposal we can form only one dimensionless combination, $x/\sqrt{(\chi t)}$; hence it is clear that the required temperature distribution must be given by a function having the form $T = T_0 f(x/\sqrt{(\chi t)})$.

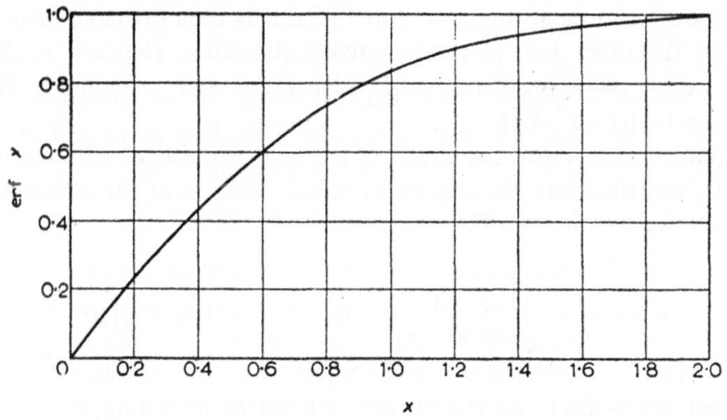

FIG. 40

Let us now consider a case where the surface bounding the medium is a thermal insulator. That is, there is no heat flux at the plane $x = 0$, so that we must have $\partial T/\partial x = 0$. We thus have the following boundary and initial conditions:

$$\partial T/\partial x = 0 \text{ for } x = 0; \quad T = T_0(x, y, z) \text{ for } t = 0,\ x > 0. \tag{52.7}$$

To find the solution we proceed as in the previous problem. That is, we again imagine the medium to extend on both sides of the plane $x = 0$, the initial temperature distribution being this time symmetrical about the plane. In other words, we now suppose that $T_0(x, y, z)$ is an even function of x:

$$T_0(-x, y, z) = T_0(x, y, z). \tag{52.8}$$

Then $\partial T_0(x, y, z)/\partial x = -\partial T_0(-x, y, z)/\partial x$, and $\partial T_0/\partial x = 0$ for $x = 0$. It is evident from symmetry that this condition will continue to be satisfied for all t.

Repeating the calculations given above, but using (52.8) in place of (52.2), we have the general solution of the problem in a form which differs from (52.3) or (52.4) only in having the sum instead of the difference of the two exponentials.

Let us now consider problems with boundary conditions of a different type, which also enable the equation of thermal conduction to be solved in a general form. Let a heat flux (a given function of time) enter a medium through its bounding plane $x = 0$. The boundary and initial conditions are then

$$-\kappa\,\partial T/\partial x = q(t) \text{ for } x = 0;\ T = 0 \text{ for } t = -\infty,\ x > 0, \tag{52.9}$$

where $q(t)$ is a given function.

We first solve an auxiliary problem, in which $q(t) = \delta(t)$. It is easy to see that this problem is physically equivalent to that of the propagation of heat in an infinite medium from a point source which generates a given amount of heat. For the boundary condition $-\kappa\partial T/\partial x = \delta(t)$ for $x = 0$ signifies physically that a unit of heat enters through each unit area of the plane $x = 0$ at the instant $t = 0$. In the problem where the condition is $T = 2\delta(x)/\rho c_p$ for $t = 0$, an amount of heat

$$\int \rho c_p T \,\mathrm{d}x = 2$$

is concentrated on this area at time $t = 0$; half of this is then propagated in the positive x-direction, and the other half in the negative x-direction. Hence it is clear that the solutions of the two problems are identical, and we find from (51.7) $\kappa T(x, t) = \sqrt{(\chi/\pi t)}\exp(-x^2/4\chi t)$.

Since the equations are linear, the effects of the heat entering at different moments are simply additive, and therefore the required general solution of the equation of thermal conduction with the conditions (52.9) is

$$\kappa T(x, t) = \int_{-\infty}^{t} \sqrt{\frac{\chi}{\pi(t-\tau)}}\, q(\tau) \exp[-x^2/4\chi(t-\tau)]\,\mathrm{d}\tau. \tag{52.10}$$

In particular, the temperature on the plane $x = 0$ varies according to

$$\kappa T(0, t) = \int_{-\infty}^{t} \sqrt{\frac{\chi}{\pi(t-\tau)}}\, q(\tau)\,\mathrm{d}\tau. \tag{52.11}$$

Using these results, we can obtain at once the solution of another problem, in which the temperature T on the plane $x = 0$ is a given function of time:

$$T = T_0(t) \text{ for } x = 0; \quad T = 0 \text{ for } t = -\infty, \quad x > 0. \tag{52.12}$$

To do so, we notice that, if some function $T(x, t)$ satisfies the equation of thermal conduction, then so does its derivative $\partial T/\partial x$. Differentiating (52.10) with respect to x, we obtain

$$-\kappa\frac{\partial T(x, t)}{\partial x} = \int_{-\infty}^{t} \frac{x q(\tau)}{2\sqrt{[\pi\chi(t-\tau)^3]}} \exp[-x^2/4\chi(t-\tau)]\,\mathrm{d}\tau.$$

This function satisfies the equation of thermal conduction and (by (52.9)) its value for $x = 0$ is $q(t)$; it therefore gives the required solution of the problem whose conditions are (52.12). Writing $T(x, t)$ instead of $-\kappa\partial T/\partial x$, and $T_0(t)$ instead of $q(t)$, we thus have

$$T(x, t) = \frac{x}{2\sqrt{(\pi\chi)}} \int_{-\infty}^{t} \frac{T_0(\tau)}{(t-\tau)^{\frac{3}{2}}} \exp[-x^2/4\chi(t-\tau)]\,\mathrm{d}\tau. \tag{52.13}$$

The heat flux $q = -\kappa\partial T/\partial x$ through the bounding plane $x = 0$ is found by a simple calculation to be

$$q(t) = \frac{\kappa}{\sqrt{(\pi\chi)}} \int_{-\infty}^{t} \frac{\mathrm{d}T_0(\tau)}{\mathrm{d}\tau} \frac{\mathrm{d}\tau}{\sqrt{(t-\tau)}}. \tag{52.14}$$

This formula is the inverse of (52.11).

The solution is easily obtained for the important problem where the temperature on the bounding plane $x = 0$ is a given periodic function of time: $T = T_0 e^{-i\omega t}$ for $x = 0$. It is clear that the temperature distribution in all space will also depend on the time through a factor $e^{-i\omega t}$. Since the one-dimensional equation of thermal conduction is formally identical with the equation (24.3) which determines the motion of a viscous fluid above an oscillating plane, we can immediately write down the required temperature distribution by analogy with (24.5):

$$T = T_0 \exp[-x\sqrt{(\omega/2\chi)}]\exp\{i[x\sqrt{(\omega/2\chi)} - \omega t]\}. \tag{52.15}$$

We see that the oscillations of the temperature on the bounding surface are propagated from it as *thermal waves* which are rapidly damped in the interior of the medium.

Another kind of thermal-conduction problem comprises those concerning the rate at which the temperature is equalized in a non-uniformly heated finite body whose surface is maintained in given conditions. To solve these problems by general methods, we seek a solution of the equation of thermal conduction in the form $T = T_n(\mathbf{r})e^{-\lambda_n t}$, with λ_n a constant. For the function T_n we have the equation

$$\chi \triangle T_n = -\lambda_n T_n. \tag{52.16}$$

This equation, with given boundary conditions, has non-zero solutions only for certain λ_n, its *eigenvalues*. All the eigenvalues are real and positive, and the corresponding functions $T_n(x,y,z)$ form a complete set of orthogonal functions. Let the temperature distribution at the initial instant be given by the function $T_0(x,y,z)$. Expanding this as a series of functions T_n,

$$T_0(\mathbf{r}) = \sum c_n T_n(\mathbf{r}),$$

we obtain the required solution in the form

$$T(\mathbf{r},t) = \sum c_n T_n(\mathbf{r}) \exp(-\lambda_n t). \tag{52.17}$$

The rate of equalization of the temperature is evidently determined mainly by the term corresponding to the smallest λ_n, which we call λ_1. The "equalization time" may be defined as $\tau = 1/\lambda_1$.

PROBLEMS

PROBLEM 1. Determine the temperature distribution around a spherical surface (with radius R) whose temperature is a given function $T_0(t)$ of time.

SOLUTION. The thermal-conduction equation for a centrally symmetrical temperature distribution is, in spherical polar coordinates, $\partial T/\partial t = (\chi/r)\partial^2(rT)/\partial r^2$. The substitution $rT(r,t) = F(r,t)$ reduces this to $\partial F/\partial t = \chi\partial^2 F/\partial r^2$, which is the ordinary one-dimensional thermal-conduction equation. Hence the required solution can be found at once from (52.13), and is

$$T(r,t) = \frac{R(r-R)}{2r\sqrt{(\pi\chi)}} \oint_{-\infty}^{t} \frac{T_0(\tau)}{(t-\tau)^{3/2}} \exp[-(r-R)^2/4\chi(t-\tau)]\mathrm{d}\tau.$$

PROBLEM 2. The same as Problem 1, but for the case where the temperature of the spherical surface is $T_0 e^{-i\omega t}$.

SOLUTION. Similarly to (52.15), we obtain

$$T = T_0 \exp(-i\omega t)(R/r)\exp[-(1-i)(r-R)\sqrt{(\omega/2\chi)}].$$

PROBLEM 3. Determine the temperature equalization time for a cube with side a whose surface is (a) maintained at a temperature $T = 0$, (b) an insulator.

SOLUTION. In case (a) the smallest value of λ is given by the following solution of equation (52.16):

$$T_1 = \sin(\pi x/a)\sin(\pi y/a)\sin(\pi z/a)$$

(the origin being at one corner of the cube), when $\tau = 1/\lambda_1 = a^2/3\pi^2\chi$. In case (b) we have $T_1 = \cos(\pi x/a)$ (or the same function of y or z), when $\tau = a^2/\pi^2\chi$.

PROBLEM 4. The same as Problem 3, but for a sphere with radius R.

SOLUTION. The smallest value of λ is given by the centrally symmetrical solution of (52.16) $T_1 = (1/r)\sin kr$; in case (a), $k = \pi/R$, and $\tau = 1/\chi k^2 = R^2/\chi\pi^2$. In case (b) k is the smallest non-zero root of the equation $kR = \tan kR$, whence we find $kR = 4{\cdot}493$ and $\tau = 0{\cdot}050\ R^2/\chi$.

§53. The similarity law for heat transfer

The processes of heat transfer in a fluid are more complex than those in solids, because the fluid may be in motion. A heated body immersed in a moving fluid cools considerably more rapidly than one in a fluid at rest, where the heat transfer is accomplished only by conduction. The motion of a non-uniformly heated fluid is called *convection*.

We shall suppose that the temperature differences in the fluid are so small that its physical properties may be supposed independent of temperature, but are at the same time so large that we can neglect in comparison with them the temperature changes caused by the heat from the energy dissipation by internal friction (see §55). Then the viscosity term in equation (50.2) may be omitted, leaving

$$\partial T/\partial t + \mathbf{v}\cdot\mathbf{grad}\, T = \chi \triangle T, \tag{53.1}$$

where $\chi = \kappa/\rho c_p$ is the thermometric conductivity. This equation, together with the Navier–Stokes equation and the equation of continuity, completely determines the convection in the conditions considered.

In what follows we shall be interested only in steady convective flow.† Then all the time derivatives are zero, and we have the following fundamental equations:

$$\mathbf{v}\cdot\mathbf{grad}\, T = \chi \triangle T, \tag{53.2}$$

$$(\mathbf{v}\cdot\mathbf{grad})\mathbf{v} = -\mathbf{grad}(p/\rho) + \nu \triangle \mathbf{v}, \quad \operatorname{div}\mathbf{v} = 0. \tag{53.3}$$

This system of equations, in which the unknown functions are $\mathbf{v}$, T and p/ρ, contains only two constant parameters, ν and χ. Furthermore, the solution of these equations depends also, through the boundary conditions, on some characteristic length l, velocity U, and temperature difference $T_1 - T_0$. The first two of these are given as usual by the dimension of the solid bodies which appear in the problem and the velocity of the main stream, while the third is given by the temperature difference between the fluid and these bodies.

In forming dimensionless quantities from the parameters at our disposal, the question arises of the dimensions to be ascribed to the temperature. To resolve this, we notice that

† In order that the convection can be steady, it is, strictly speaking, necessary that the solid bodies adjoining the fluid should contain sources of heat which maintain these bodies at a constant temperature.

the temperature is determined by equation (53.2), which is linear and homogeneous in T. Hence the temperature can be multiplied by any constant and still satisfy the equations. In other words, the unit of measurement of temperature can be chosen arbitrarily. The possibility of this transformation of the temperature can be formally allowed for by giving it a dimension of its own, unrelated to those of the other quantities. This can be measured in degrees, the usual unit of temperature.

Thus convection in the above-mentioned conditions is characterized by five parameters, whose dimensions are $\nu = \chi = \text{cm}^2/\text{sec}$, $U = \text{cm/sec}$, $l = \text{cm}$, $T_1 - T_0 = \text{deg}$. From these we can form two independent dimensionless combinations. These may be the Reynolds number $\mathrm{R} = Ul/\nu$ and the *Prandtl number*, defined as

$$\mathrm{P} = \nu/\chi. \tag{53.4}$$

Any other dimensionless combination can be expressed in terms of R and P.†

The Prandtl number is just a constant of the material, and does not depend on the properties of the flow. For gases it is always of the order of unity. The value of P for liquids varies more widely. For very viscous liquids, it may be very large. The following are values of P at 20°C for various substances:

Air	0·733
Water	6·75
Alcohol	16·6
Glycerine	7250
Mercury	0·044

As in §19, we can now conclude that, in steady convection (of the type described), the temperature and velocity distributions have the form

$$\frac{T - T_0}{T_1 - T_0} = f\left(\frac{\mathbf{r}}{l}, \mathrm{R}, \mathrm{P}\right), \quad \frac{\mathbf{v}}{U} = \mathbf{f}\left(\frac{\mathbf{r}}{l}, \mathrm{R}\right). \tag{53.5}$$

The dimensionless function which gives the temperature distribution depends on both R and P as parameters, but the velocity distribution depends only on R, since it is determined by equations (53.3), which do not involve the conductivity. Two convective flows are similar if their Reynolds and Prandtl numbers are the same.

The heat transfer between solid bodies and the fluid is usually characterized by the *heat transfer coefficient* α, defined by

$$\alpha = q/(T_1 - T_0), \tag{53.6}$$

where q is the heat flux density through the surface and $T_1 - T_0$ is a characteristic temperature difference between the solid body and the fluid. If the temperature distribution in the fluid is known, the heat transfer coefficient is easily found by calculating the heat flux density $q = -\kappa \partial T/\partial n$ at the boundary of the fluid (the derivative being taken along the normal to the surface).

The heat transfer coefficient is not a dimensionless quantity. A dimensionless quantity which characterizes the heat transfer is the *Nusselt number*:

$$\mathrm{N} = \alpha l/\kappa. \tag{53.7}$$

† The *Péclet number* is sometimes used; it is defined as $Ul/\chi = \mathrm{RP}$.

It follows from similarity arguments that, for any given type of convective flow, the Nusselt number is a definite function of the Reynolds and Prandtl numbers only:

$$\mathrm{N} = f(\mathrm{R},\mathrm{P}). \tag{53.8}$$

This function is very simple for convection at sufficiently small Reynolds numbers. These correspond to small velocities. Hence, in the first approximation, we can neglect the velocity term in equation (53.2), so that the temperature distribution is determined by the equation $\triangle T = 0$, i.e. the ordinary equation of steady thermal conduction in a medium at rest. The heat transfer coefficient can then depend on neither the velocity nor the viscosity, and so we must have simply

$$\mathrm{N} = \text{constant}, \tag{53.9}$$

and in calculating the constant the fluid may be supposed at rest.

PROBLEM

Determine the temperature distribution in a fluid moving in Poiseuille flow along a pipe with circular cross-section, when the temperature of the walls varies linearly along the pipe.

SOLUTION. The conditions of the flow are the same at every cross-section of the pipe, and we can look for the temperature distribution in the form $T = Az + f(r)$, where Az is the wall temperature; we use cylindrical polar coordinates, with the z-axis along the axis of the pipe. For the velocity we have, by (17.9), $v_z = v = 2\bar{v}(1 - r^2/R^2)$, where $\bar{v}$ is the mean velocity. Substituting in (53.2), we find

$$\frac{1}{r}\frac{\mathrm{d}}{\mathrm{d}r}\left(r\frac{\mathrm{d}f}{\mathrm{d}r}\right) = \frac{2\bar{v}A}{\chi}\left[1 - \left(\frac{r}{R}\right)^2\right].$$

The solution of this equation which is finite for $r = 0$ and zero for $r = R$ is

$$f(r) = -\frac{\bar{v}AR^2}{2\chi}\left[\frac{3}{4} - \left(\frac{r}{R}\right)^2 + \frac{1}{4}\left(\frac{r}{R}\right)^4\right].$$

The heat flux density is

$$q = \kappa[\partial T/\partial r]_R = \tfrac{1}{2}\rho c_p \bar{v} R A.$$

It is independent of the thermal conductivity.

§54. Heat transfer in a boundary layer

The temperature distribution in a fluid at very high Reynolds numbers exhibits properties similar to those of the velocity distribution. Very large values of R are equivalent to a very small viscosity. But since the number $\mathrm{P} = \nu/\chi$ is not small, the thermometric conductivity χ must be supposed small, as well as ν. This corresponds to the fact that, for sufficiently high velocities, the fluid may be approximately regarded as an ideal fluid, and in an ideal fluid both internal friction and thermal conduction are absent.

This viewpoint, however, must again be abandoned in a boundary layer, since neither the boundary condition of no slip nor that of equal temperatures would be satisfied. In the boundary layer, therefore, there occurs both a rapid decrease of the velocity and a rapid change of the fluid temperature to a value equal to the temperature of the solid surface. The boundary layer is characterized by the presence of large gradients of both velocity and temperature.

It is easy to see that, in flow past a heated body (with R large), the heating of the fluid occurs almost exclusively in the wake, while outside the wake the fluid temperature does not change. For, when R is large, the processes of thermal conduction in the main stream are unimportant. Hence the temperature varies only in the region reached by fluid that has been heated in the boundary layer. We know (see §35) that the streamlines from the boundary layer enter the main stream only beyond the line of separation, where they go into the region of the turbulent wake. From the wake, however, the streamlines do not emerge at all. Thus the fluid which flows past the surface of the heated body in the boundary layer goes entirely into the wake and remains there. We see that the heat becomes distributed through the regions where the vorticity is non-zero.

In the turbulent region itself, a very considerable exchange of heat occurs, which is due to the intensive mixing of the fluid characteristic of any turbulent flow. This mechanism of heat transfer may be called *turbulent conduction* and characterized by a coefficient χ_{turb}, in the same way as we introduced the turbulent viscosity η_{turb} in §33. The turbulent thermometric conductivity is defined, in order of magnitude, by the same formula as ν_{turb} (33.2): $\chi_{\text{turb}} \sim l \triangle u$.

Thus the processes of heat transfer in laminar and in turbulent flow are fundamentally different. In the limiting case of very small viscosity and thermal conductivity, in laminar flow, the processes of heat transfer are absent, and the fluid temperature is constant at every point in space. In turbulent flow, however, even in the same limiting case, heat transfer occurs and rapidly equalizes the temperatures in various parts of the stream.

Let us begin by considering heat transfer in a laminar boundary layer. The equations of motion (39.13) are unaltered. A similar simplification must now be performed for equation (53.2). Written explicitly, this equation is (since all quantities are independent of the coordinate z)

$$v_x \frac{\partial T}{\partial x} + v_y \frac{\partial T}{\partial y} = \chi \left(\frac{\partial^2 T}{\partial x^2} + \frac{\partial^2 T}{\partial y^2} \right).$$

On the right-hand side we may neglect the derivative $\partial^2 T/\partial x^2$ in comparison with $\partial^2 T/\partial y^2$, leaving

$$v_x \frac{\partial T}{\partial x} + v_y \frac{\partial T}{\partial y} = \chi \frac{\partial^2 T}{\partial y^2}. \tag{54.1}$$

By comparing this equation with the first of (39.13) we see that, if the Prandtl number is of the order of unity, then the order of magnitude δ of the thickness of the layer in which the velocity v_x decreases and the temperature T varies will again be given by the formulae obtained in §39, i.e. it will be inversely proportional to $\sqrt{\mathrm{R}}$. The heat flux $q = -\kappa \partial T/\partial n$ is equal, in order of magnitude, to $\kappa(T_1 - T_0)/\delta$. Hence we conclude that q, and therefore the Nusselt number, are proportional to $\sqrt{\mathrm{R}}$. The dependence of N on P is not determined. Thus we have

$$\mathrm{N} = \sqrt{\mathrm{R}}\, f(\mathrm{P}). \tag{54.2}$$

From this it follows, in particular, that the heat transfer coefficient α is inversely proportional to the square root of the dimension l of the body.

Let us now consider heat transfer in a turbulent boundary layer. Here it is convenient, as in §42, to take an infinite plane-parallel turbulent stream flowing along an infinite plane surface. The transverse temperature gradient $\mathrm{d}T/\mathrm{d}y$ in such a flow can be determined from the same kind of dimensional argument as we used to find the velocity gradient

du/dy. We denote by q the heat flux density along the y-axis caused by the temperature gradient. This flux is a constant (independent of y), like the momentum flux σ, and can likewise be regarded as a given parameter which determines the properties of the flow. Furthermore, we have as parameters also the density ρ and the specific heat c_p per unit mass. Instead of σ we use as parameter v_*; q and c_p have the dimensions erg/cm^{2}sec = g/sec^3 and erg/g deg = cm^2/sec^2 deg. The viscosity and thermal conductivity cannot appear explicitly in dT/dy when R is sufficiently large.

Because of the homogeneity of the equations as regards the temperature, already mentioned in §53, the temperature can be changed by any factor without violating the equations. When the temperature is changed in this way, however, the heat flux must change by the same factor. Hence q and T must be proportional. From q, v_*, ρ, c_p and y we can form only one quantity proportional to q and having the dimension deg cm, namely $q/\rho c_p v_* y$. Thus we must have $dT/dy = \beta q/\kappa\rho c_p v_* y$, where β is a numerical constant which must be determined by experiment.† Hence

$$T = (\beta q/\kappa\rho c_p v_*)(\log y + c). \tag{54.3}$$

Thus the temperature, like the velocity, varies logarithmically. The constant of integration c which appears here must be determined from the conditions in the viscous sublayer, as in the derivation of (42.7). The temperature difference between the fluid at a given point and the wall (which we arbitrarily take to be at zero temperature) is composed of the temperature change across the turbulent layer and that across the viscous sublayer. The logarithmic law (54.3) determines only the first of these. Hence, if we write (54.3) in the form $T = (\beta q/\kappa\rho c_p v_*)[\log(yv_*/\nu) + \text{constant}]$, including in the argument of the logarithm a factor equal to the thickness y_0, then the constant (multiplied by the coefficient of the bracket) must be the change in temperature across the viscous sublayer. This change, of course, depends on the coefficients ν and χ also. Since the constant is dimensionless, it must be some function of P, which is the only dimensionless combination of the quantities ν, χ, ρ, v_* and c_p (q cannot appear, since T must be proportional to q, which already occurs in the coefficient). Thus we find the temperature distribution to be

$$T = (\beta q/\kappa\rho c_p v_*)[\log(yv_*/\nu) + f(\text{P})] \tag{54.4}$$

(L. D. Landau 1944). The empirical value of β here is about 0·9. The value of f for air is $f(0{\cdot}7) \cong 1{\cdot}5$.

Using formula (54.4), we can calculate the heat transfer for turbulent flow in a pipe, along a flat plate, etc. We shall not pause to do this here.

TURBULENT TEMPERATURE FLUCTUATIONS

In referring to the temperature of a turbulent fluid, we have of course meant the time average. The actual temperature at any point in space undergoes very irregular variations with time, similar to those of the velocity.

We shall suppose that a significant change in the mean temperature occurs over the same distances l (the fundamental scale of the turbulence) as for the mean flow velocity. The same concepts and arguments involving similarity as were used in §33 in discussing the local properties of turbulence can be applied to the small-scale ($\lambda \ll l$) temperature fluctuations. Here it will be assumed that P $\sim$ 1; otherwise, it may be necessary to use two internal scales, determined by ν and by χ. The inertial range of scales is then also the

† Here κ is the von Kármán constant appearing in the logarithmic velocity profile (42.4). With this definition, β is the ratio ν_{turb}/χ_{turb}, where ν_{turb} and χ_{turb} are the coefficients in $q = \rho c_p \chi_{turb} dT/dy$, $\sigma = \rho\nu_{turb} du/dy$.

convective range: the equalization of temperatures within it takes place by mechanical mixing of "fluid particles" at different temperatures without involving true thermal conduction, and the properties of the temperature fluctuations in this range are independent of the large-scale flow also. Let us determine how the temperature differences T_λ depend on the distance λ in the inertial range (A. M. Obukhov 1949).

The energy dissipation by thermal conduction, per unit volume, is $\kappa(\mathbf{grad}\, T)^2/T$; compare (49.6), or (79.1) below. Dividing this by ρc_p, we have $\chi(\mathbf{grad}\, T)^2/T \equiv \phi/T$, which determines the rate at which the temperature is lowered by dissipation; assuming the turbulent temperature fluctuations to be relatively small, we can replace T in the denominator by a constant mean temperature. The quantity ϕ thus defined is another parameter (besides ε) which determines the local properties of turbulence in a non-uniformly heated fluid.

With the method described in §33 (see after (33.1)), we express ϕ in terms of quantities which relate to fluctuations with scale λ:

$$\phi \sim \chi_{\text{turb},\lambda}(T_\lambda/\lambda)^2.$$

Substituting from (33.2) and (33.6)

$$\chi_{\text{turb},\lambda} \sim \nu_{\text{turb},\lambda} \sim \lambda v_\lambda, \quad v_\lambda \propto (\varepsilon\lambda)^{1/3},$$

we get the required result

$$T_\lambda^2 \propto \phi\varepsilon^{-1/3}\lambda^{2/3}. \tag{54.5}$$

Thus, when $\lambda \gg \lambda_0$, the temperature fluctuations, like the velocity fluctuations, are proportional to $\lambda^{1/3}$.

At distances $\lambda \lesssim \lambda_0$, however, the temperature is equalized by true thermal conduction. At scales $\lambda \ll \lambda_0$, the temperature varies smoothly. By the same reasoning as for the velocity (cf. (33.19)), the differences T_λ are then proportional to λ.

PROBLEMS

PROBLEM 1. Determine the limiting form of the dependence of the Nusselt number on the Prandtl number in a laminar boundary layer when P and R are large.

SOLUTION. For large P, the distance δ' over which the temperature changes is small compared with the thickness δ of the layer in which the velocity v_x diminishes. δ' may be called the thickness of the temperature boundary layer. The order of magnitude of δ' may be obtained from an estimate of the terms in equation (54.1). Over the distance from $y = 0$ to $y \sim \delta'$, the temperature varies by an amount of the order of the total temperature difference $T_1 - T_0$ between the fluid and the solid body, while the velocity v_x varies over this distance by an amount of the order of $U\delta'/\delta$ (since the total change, of the order of U, occurs over a distance δ). Hence, for $y \sim \delta'$, the terms in equation (54.1) are, in order of magnitude,

$$\chi\partial^2 T/\partial y^2 \sim \chi(T_1 - T_0)/\delta'^2 \text{ and } v_x \partial T/\partial x \sim U\delta'(T_1 - T_0)/l\delta.$$

If the two expressions are comparable, we have $\delta'^3 \sim \chi l\delta/U$. Substituting $\delta \sim l/\sqrt{R}$, we obtain $\delta' \sim l/\mathrm{R}^{\frac{1}{2}}\mathrm{P}^{\frac{1}{3}} \sim \delta/\mathrm{P}^{\frac{1}{3}}$. Thus, for large P, the thickness of the temperature boundary layer decreases, relative to that of the velocity boundary layer, inversely as the cube root of P.

The heat flux $q = -\kappa\partial T/\partial y \sim \kappa(T_1 - T_0)/\delta'$, and the required limiting law of heat transfer is found to be†

$$\mathrm{N} = \text{constant} \times \mathrm{R}^{\frac{1}{2}}\mathrm{P}^{\frac{1}{3}}.$$

† For the values of the thermal conductivity actually found, the Prandtl number does not reach the values for which this limiting law holds. Such laws can, however, be applied to convective diffusion; this obeys the same equations as convective heat transfer, but with the temperature replaced by the concentration of the solute, and the heat flux by the flux of solute, the "diffusion Prandtl number" being defined as $\mathrm{P}_D = \nu/D$, where D is the diffusion coefficient. For example, for solutions in water and similar liquids, P_D reaches values of the order of 10^3, while for very viscous solvents it is 10^6 or more.

PROBLEM 2. Determine the limiting form of the function $f(\mathrm{P})$, in the logarithmic temperature distribution (54.4), for large values of P.

SOLUTION. According to what was said in §42, the transverse velocity in the viscous sublayer is of the order of $v_*(y/y_0)^2$, while the scale of the turbulence is of the order of y^2/y_0. The turbulent thermometric conductivity is therefore

$$\chi_{\text{turb}} \sim v_* y_0 (y/y_0)^4 \sim \nu (y/y_0)^4$$

(where we have used the relation (42.5)); χ_{turb} is comparable in magnitude with the ordinary coefficient χ at distances $y_1 \sim y_0 \mathrm{P}^{-\frac{1}{4}}$. Since χ_{turb} increases very rapidly with y, it is clear that most of the temperature change in the viscous sublayer occurs over distances from the wall of the order of y_1, and may be supposed proportional to y_1, being in order of magnitude $qy_1/\kappa \sim qy_0/\kappa\mathrm{P}^{\frac{1}{4}} \sim q\mathrm{P}^{\frac{3}{4}}/\rho c_p v_*$. Comparing with formula (54.4), we see that function $f(\mathrm{P})$ is a numerical constant times $\mathrm{P}^{\frac{3}{4}}$.

PROBLEM 3. Derive a relation between the local correlation functions

$$B_{TT} = \langle (T_2 - T_1)^2 \rangle, \qquad B_{iTT} = \langle (v_{2i} - v_{1i})(T_2 - T_1)^2 \rangle$$

in a non-uniformly heated turbulent flow (A. M. Yaglom 1949).

SOLUTION. The calculations are similar to those in §34. Together with B_{TT} and B_{iTT}, we use the auxiliary functions

$$b_{TT} = \langle T_1 T_2 \rangle, \qquad b_{iTT} = \langle v_{1i} T_1 T_2 \rangle,$$

and to facilitate the calculations we regard the turbulence as completely homogeneous and isotropic. Then

$$B_{TT} = 2\langle T^2 \rangle - 2b_{TT}, \; B_{iTT} = 4b_{iTT}; \tag{1}$$

the mean values $\langle v_{1i} T_1 T_2 \rangle = -\langle v_{2i} T_1 T_2 \rangle$, and mean values of the type $\langle v_{1i} T_2{}^2 \rangle$ are zero, because the fluid is incompressible—compare the derivation of (34.18). With the equations

$$\partial T/\partial t + (\mathbf{v}\cdot\mathbf{grad})T = \chi \triangle T, \quad \operatorname{div}\mathbf{v} = 0,$$

we calculate the derivative

$$\partial b_{TT}/\partial t = -2\partial b_{iTT}/\partial x_{1i} + 2\chi \triangle_1 b_{TT}. \tag{2}$$

The isotropy and homogeneity also mean that

$$b_{iTT} = n_i b_{rTT}, \tag{3}$$

where $\mathbf{n}$ is a unit vector parallel to $\mathbf{r} = \mathbf{r}_2 - \mathbf{r}_1$; b_{rTT} and b_{TT} depend only on r. Using (1) and (3), we can put (2) in the form

$$\begin{aligned} -2\phi - \partial B_{TT}/\partial t &= \tfrac{1}{2}\operatorname{div}(\mathbf{n}B_{rTT}) - \chi\triangle B_{TT} \\ &= \frac{1}{2r^2}\frac{\partial}{\partial r}(r^2 B_{rTT}) - \frac{\chi}{r^2}\frac{\partial}{\partial r}\left(r^2 \frac{\partial B_{TT}}{\partial r}\right), \end{aligned}$$

where $\phi = -\frac{1}{2}\partial\langle T^2 \rangle/\partial t$ as in the text. Since the local turbulence may be regarded as steady, we neglect $\partial B_{TT}/\partial t$. Integration of the resulting equation with respect to r gives the required relation, analogous to (34.21),

$$B_{rTT} - 2\chi \mathrm{d}B_{TT}/\mathrm{d}r = -\tfrac{4}{3}r\phi. \tag{4}$$

When $r \gg \lambda_0$, the term in χ is small, since from (54.5) $B_{TT} \propto r^{\frac{2}{3}}$. Then, from (4),

$$B_{rTT} \cong -\tfrac{4}{3}r\phi.$$

At distances $r \ll \lambda_0$, we have $B_{TT} \propto r^2$, and B_{rTT} may be neglected; in this case,

$$B_{TT} \cong \tfrac{1}{3}r^2\phi/\chi.$$

§55. Heating of a body in a moving fluid

A thermometer immersed in a fluid at rest indicates a temperature equal to that of the fluid. If the fluid is in motion, however, the thermometer indicates a somewhat higher temperature. This is because the fluid brought to rest at the surface of the thermometer is heated by internal friction.

The general problem may be formulated as follows. A body of arbitrary shape is immersed in a moving fluid; thermal equilibrium is established after a sufficient length of time, and it is required to determine the temperature difference $T_1 - T_0$ then existing between the body and the fluid.

The solution of this problem is given by equation (50.2), in which, however, we cannot now neglect the term containing the viscosity as we did in (53.1); it is this term which is responsible for the effect under consideration. Thus we have for a steady state

$$\mathbf{v}\cdot\mathbf{grad}\,T = \chi\triangle T + \frac{\nu}{2c_p}\left(\frac{\partial v_i}{\partial x_k} + \frac{\partial v_k}{\partial x_i}\right)^2. \tag{55.1}$$

This must be supplemented by the equations of motion (53.3) of the fluid itself and also, strictly speaking, by the equation of thermal conduction in the body. In the limiting case where the body has a sufficiently small thermal conductivity, we can neglect the latter and suppose the temperature at any point on the surface of the body to be simply equal to the fluid temperature at that point, obtained by solving equation (55.1) with the boundary condition $\partial T/\partial n = 0$, i.e. the condition that there be no heat flux through the surface of the body. In the opposite limiting case where the body has a sufficiently large thermal conductivity, we can use the approximate condition that the temperature should be the same at every point of its surface; the derivative $\partial T/\partial n$ will not then in general vanish over the whole surface, and we must require only that the total heat flux through the surface of the body (i.e. the integral of $\partial T/\partial n$ over the surface) should be zero. In both these limiting cases the thermal conductivity of the body does not appear explicitly in the solution of the problem, and we shall suppose in what follows that one of these cases holds.

Equations (55.1) and (53.3) contain the constant parameters χ, ν and c_p, and their solutions involve also the dimension l of the body and the velocity U of the main stream. (The temperature difference $T_1 - T_0$ is not now an arbitrary parameter, but must itself be determined by solving the equations.) From these parameters we can construct two independent dimensionless quantities, which we take to be R and P. Then we can say that the required temperature difference $T_1 - T_0$ is equal to some quantity having the dimensions of temperature (which we take to be U^2/c_p), multiplied by a function of R and P:

$$T_1 - T_0 = (U^2/c_p) f(\mathrm{R},\mathrm{P}). \tag{55.2}$$

It is easy to determine the form of this function for very small Reynolds numbers, i.e. for sufficiently small velocities U. In this case the term $\mathbf{v}\cdot\mathbf{grad}\,T$ in (55.1) is small compared with $\chi\triangle T$, so that this equation becomes

$$\chi\triangle T = -\frac{\nu}{2c_p}\left(\frac{\partial v_i}{\partial x_k} + \frac{\partial v_k}{\partial x_i}\right)^2. \tag{55.3}$$

The temperature and velocity vary considerably over distances of the order of l. Hence an estimate of the two sides of equation (55.3) gives $\chi(T_1 - T_0)/l^2 \sim \nu U^2/c_p l^2$. Thus we conclude that, for small R,

$$T_1 - T_0 = \text{constant} \times \mathrm{P}U^2/c_p, \tag{55.4}$$

where the numerical constant depends on the shape of the body. It should be noticed that the temperature difference is proportional to the square of the velocity U.

Some general conclusions concerning the form of the function $f(\mathrm{P},\mathrm{R})$ in (55.2) can be drawn in the opposite limiting case of large R, when the velocity and the temperature vary only in a narrow boundary layer. Let δ and δ' be the distances over which the velocity and temperature respectively vary; δ and δ' differ by a factor depending on P. The amount of heat evolved in the boundary layer in unit time owing to the viscosity of the fluid is given by (16.3). This integral per unit area of the surface is of the order of $\nu\rho(U^2/\delta^2)\delta = \nu\rho U^2/\delta$. The same amount of heat must be lost from the body, and it is therefore equal to the heat flux $q = -\kappa\partial T/\partial n \sim \chi c_p\rho(T_1 - T_0)/\delta'$. Comparing the two expressions, we find

$$T_1 - T_0 = (U^2/c_p)f(\mathrm{P}). \tag{55.5}$$

Thus, in this case, the function f is independent of R, but its dependence on P remains undetermined.

PROBLEMS

PROBLEM 1. Determine the temperature distribution in a fluid moving in Poiseuille flow in a pipe with circular cross-section whose walls are maintained at constant temperature T_0.

SOLUTION. In cylindrical polar coordinates, with the z-axis along the axis of the pipe, we have $v_z = v = 2\bar{v}[1-(r/R)^2]$, where $\bar{v}$ is the mean velocity of the flow. Substitution in (55.3) gives the equation

$$\frac{1}{r}\frac{\mathrm{d}}{\mathrm{d}r}\left(r\frac{\mathrm{d}T}{\mathrm{d}r}\right) = -\frac{16\bar{v}^2}{R^4}\frac{\nu}{\chi c_p}r^2.$$

The solution finite at $r = 0$ and equal to T_0 for $r = R$ is

$$T - T_0 = \bar{v}^2\frac{\mathrm{P}}{c_p}\left[1-\left(\frac{r}{R}\right)\right]^4.$$

PROBLEM 2. Determine the temperature difference between a solid sphere and a fluid moving past it at small Reynolds numbers. The thermal conductivity of the sphere is supposed large.

SOLUTION. We take spherical polar coordinates r, θ, ϕ, with the origin at the centre of the sphere and the polar axis in the direction of the velocity of the main stream. Calculating the components of the tensor $\partial v_i/\partial x_k + \partial v_k/\partial x_i$ by means of formulae (15.20) and (20.9), for the velocity of flow past a sphere, we obtain equation (55.3) in the form

$$\frac{1}{r^2}\frac{\partial}{\partial r}\left(r^2\frac{\partial T}{\partial r}\right) + \frac{1}{r^2\sin\theta}\frac{\partial}{\partial\theta}\left(\sin\theta\frac{\partial T}{\partial\theta}\right)$$
$$= -A(R/r)^4[\cos^2\theta\{3-6(R/r)^2+2(R/r)^4\}+(R/r)^4],$$

where $A = 9u^2\mathrm{P}/4c_p$. We look for $T(r,\theta)$ in the form $T = f(r)\cos^2\theta + g(r)$, and, separating the part which depends on θ, find two equations for f and g:

$$r^2f'' + 2rf' - 6f = -A[3(R/r)^2 - 6(R/r)^4 + 2(R/r)^6],$$
$$r^2g'' + 2rg' + 2f = -A(R/r)^6.$$

From the first we obtain

$$f = A[\tfrac{3}{4}(R/r)^2 + (R/r)^4 - \tfrac{1}{12}(R/r)^6] + c_1 (R/r)^3;$$

the term having the form constant $\times r^2$ is omitted, since it does not vanish at infinity. The second equation then gives

$$g = -\tfrac{1}{2}A[\tfrac{3}{2}(R/r)^2 + \tfrac{1}{3}(R/r)^4 + \tfrac{1}{18}(R/r)^6] - \tfrac{1}{3}c_1 (R/r)^3 + c_2 R/r + c_3.$$

The constants c_1, c_2, c_3 are determined from the conditions

$$T = \text{constant and} \int (\partial T/\partial r) r^2 \sin\theta \, d\theta = 0$$

for $r = R$, which are equivalent to $f(R) = 0$ and $g'(R) + \frac{1}{3}f'(R) = 0$; also $T = T_0$ at infinity. Thus $c_1 = -5A/3$, $c_2 = 2A/3$, $c_3 = T_0$. The temperature difference between $T_1 = T(R)$ and T_0 is found to be $T_1 - T_0 = 5u^2 \mathrm{P}/8c_p$. It may be noted that the temperature distribution obtained actually satisfies the condition $\partial T/\partial r = 0$ for $r = R$, i.e. $f'(R) = g'(R) = 0$. Hence it is also the solution of the same problem for a sphere with small thermal conductivity.

§56. Free convection

We have seen in §3 that, if there is mechanical equilibrium in a fluid in a gravitational field, the temperature distribution can depend only on the altitude z: $T = T(z)$. If the temperature distribution does not satisfy this condition, but is a function of the other coordinates also, then mechanical equilibrium in the fluid is not possible. Furthermore, even if $T = T(z)$, mechanical equilibrium may still be impossible if the vertical temperature gradient is directed downwards and its magnitude exceeds a certain value (§4).

The absence of mechanical equilibrium results in the appearance of internal currents in the fluid, which tend to mix the fluid and bring it to a constant temperature. Such motion in a gravitational field is called *free convection.*

Let us derive the equations describing this convection. We shall suppose the fluid incompressible. This means that the pressure is supposed to vary only slightly through the fluid, so that the density change due to changes in pressure may be neglected. For example, in the atmosphere, where the pressure varies with height, this assumption means that we shall not consider columns of air of great height, in which the density varies considerably over the height of the column. The density change due to non-uniform heating of the fluid, of course, cannot be neglected; it results in the forces which bring about the convection.

We write the variable temperature in the form $T = T_0 + T'$, where T_0 is some constant mean temperature from which the variation T' is reckoned. We shall suppose that T' is small compared with T_0.

We write the fluid density also in the form $\rho = \rho_0 + \rho'$, with ρ_0 a constant. Since the temperature variation T' is small, the resulting density change ρ' is also small, and we can write

$$\rho' = (\partial\rho_0/\partial T)_p T' = -\rho_0 \beta T'. \tag{56.1}$$

Here $\beta = -(1/\rho)\partial\rho/\partial T$ is the thermal-expansion coefficient of the fluid.†

In the pressure $p = p_0 + p'$, p_0 is not constant. It is the pressure corresponding to mechanical equilibrium, when the temperature and density are constant and equal to T_0

† We shall assume that $\beta > 0$.

and ρ_0 respectively. It varies with height according to the hydrostatic equation

$$p_0 = \rho_0 \mathbf{g}\cdot\mathbf{r} + \text{constant} = -\rho_0 g z + \text{constant}, \tag{56.2}$$

the coordinate z being measured vertically upwards.

In a fluid column with height h, the hydrostatic pressure drop is $\rho_0 g h$. This causes a density change $\sim \rho g h/c^2$, where c is the velocity of sound; see (64.4). According to the condition stated, this change must be negligible, not only in comparison with the density itself, but also in comparison with the thermal change (56.1). That is, we must have

$$gh/c^2 \ll \beta\Theta, \tag{56.3}$$

where Θ is a characteristic temperature difference.

We start by transforming the Navier–Stokes equation, which has, in the presence of a gravitational field, the form

$$\partial\mathbf{v}/\partial t + (\mathbf{v}\cdot\mathbf{grad})\mathbf{v} = -(1/\rho)\mathbf{grad}\,p + \nu\triangle\mathbf{v} + \mathbf{g};$$

this is obtained by adding the force $\mathbf{g}$ per unit mass to the right-hand side of equation (15.7). We now substitute $p = p_0 + p'$, $\rho = \rho_0 + \rho'$; to the first order of small quantities, we have

$$\frac{\mathbf{grad}\,p}{\rho} = \frac{\mathbf{grad}\,p_0}{\rho_0} + \frac{\mathbf{grad}\,p'}{\rho_0} - \frac{\mathbf{grad}\,p_0}{\rho_0^{\,2}}\rho',$$

or, substituting (56.1) and (56.2),

$$\frac{\mathbf{grad}\,p}{\rho} = \mathbf{g} + \frac{\mathbf{grad}\,p'}{\rho_0} + \mathbf{g}T'\beta.$$

With this expression, the Navier–Stokes equation gives

$$\partial\mathbf{v}/\partial t + (\mathbf{v}\cdot\mathbf{grad})\mathbf{v} = -\mathbf{grad}(p'/\rho) + \nu\triangle\mathbf{v} - \beta T'\mathbf{g}, \tag{56.4}$$

where the suffix has been dropped from ρ_0. In the thermal conduction equation (50.2), the viscosity term can be shown to be small in free convection compared with the other terms, and may therefore be omitted. We thus obtain

$$\partial T'/\partial t + \mathbf{v}\cdot\mathbf{grad}\,T' = \chi\triangle T'. \tag{56.5}$$

Equations (56.4) and (56.5), together with the equation of continuity div $\mathbf{v} = 0$, form a complete system of equations governing free convection (A. Oberbeck 1879, J. Boussinesq 1903).

For steady flow, the equations of convection become

$$(\mathbf{v}\cdot\mathbf{grad})\mathbf{v} = -\mathbf{grad}\,(p'/\rho) - \beta T'\mathbf{g} + \nu\triangle\mathbf{v}, \tag{56.6}$$

$$\mathbf{v}\cdot\mathbf{grad}\,T' = \chi\triangle T', \tag{56.7}$$

$$\operatorname{div}\mathbf{v} = 0. \tag{56.8}$$

This system of five equations for the unknown functions $\mathbf{v}$, p'/ρ and T' contains three parameters, ν, χ and βg. Moreover, the solution will involve the characteristic length h and the temperature difference Θ. There is here no characteristic velocity, since there is no flow due to external forces, and the whole motion of the fluid is due to its non-uniform heating. From these quantities we can form two independent dimensionless combinations (the temperature is to be regarded as having a dimension of its own; see §53), which are usually taken to be the Prandtl number $\mathrm{P} = \nu/\chi$ and the *Rayleigh number*†

$$\mathcal{R} = \beta g \Theta h^3/\nu\chi. \tag{56.9}$$

The Prandtl number depends only on the properties of the fluid; the Rayleigh number is the chief characteristic of the convection as such.

The similarity law for free convection is

$$\mathbf{v} = (\nu/h)\mathbf{f}(\mathbf{r}/h, \mathcal{R}, \mathrm{P}), \quad T = \Theta f(\mathbf{r}/h, \mathcal{R}, \mathrm{P}). \tag{56.10}$$

Two flows are similar if their Rayleigh and Prandtl numbers are the same. Convective heat transfer under gravity is again described by the Nusselt number (53.7), which is now a function of $\mathcal{R}$ and P only.

Convective flow may be either laminar or turbulent. The onset of turbulence is governed by the Rayleigh number: the convection becomes turbulent when $\mathcal{R}$ is very large.

PROBLEMS

PROBLEM 1. Reduce to the solution of ordinary differential equations the determination of the Nusselt number for free convection on a flat vertical wall. It is assumed that the velocity and the temperature differences are appreciably different from zero only in a thin boundary layer adjoining the surface of the wall (E. Pohlhausen 1921).

SOLUTION. We take the origin on the lower edge of the wall, the x-axis vertical, and the y-axis perpendicular to the wall. The pressure in the boundary layer does not vary along the y-axis (cf. §39), and therefore is everywhere equal to the hydrostatic pressure $p_0(x)$, i.e. $p' = 0$. With the usual accuracy of boundary-layer theory, equations (56.6)–(56.8) become

$$\left.\begin{aligned} v_x\frac{\partial v_x}{\partial x} + v_y\frac{\partial v_x}{\partial y} &= \nu\frac{\partial^2 v_x}{\partial y^2} + \beta g(T - T_0), \\ v_x\frac{\partial T}{\partial x} + v_y\frac{\partial T}{\partial y} &= \chi\frac{\partial^2 T}{\partial y^2}, \\ \frac{\partial v_x}{\partial x} + \frac{\partial v_y}{\partial y} &= 0, \end{aligned}\right\} \tag{1}$$

with the boundary conditions $v_x = v_y = 0$ and $T = T_1$ for $y = 0$ (T_1 being the temperature of the wall), $v_x = 0$ and $T = T_0$ for $y = \infty$ (T_0 being the fluid temperature at a great distance). These equations can be converted into ordinary differential equations by introducing as the independent variable

$$\xi = \mathrm{G}^{\frac{1}{4}} y/(4xh^3)^{\frac{1}{4}}, \quad \mathrm{G} = \beta g(T_1 - T_0)h^3/\nu^2, \tag{2}$$

where h is the height of the wall. We put

$$\left.\begin{aligned} v_x &= (2\nu/h^{3/2})\sqrt{(\mathrm{G}x)}\phi'(\xi), \\ T - T_0 &= (T_1 - T_0)\theta(\xi). \end{aligned}\right\} \tag{3}$$

The last equation (1) then gives

$$v_y = \nu \mathrm{G}^{\frac{1}{4}}(\xi\phi' - 3\phi)/(4xh^3)^{\frac{1}{4}},$$

and the first two give equations for $\phi(\xi)$ and $\theta(\xi)$:

$$\phi''' + 3\phi\phi'' - 2\phi'^2 + \theta = 0, \quad \theta'' + 3\mathrm{P}\phi\theta' = 0. \tag{4}$$

† The *Grashof number* $\mathrm{G} = \beta g\Theta h^3/\nu^2 = \mathcal{R}/\mathrm{P}$ is also sometimes used.

It follows from (3) and (4) that the boundary layer thickness $\delta \sim (xh^3/\mathrm{G})^{\frac{1}{4}}$. The condition $\delta \ll h$ for the solution to be valid is satisfied when G is sufficiently large.

The total heat flux per unit area of the wall is

$$q = -\frac{1}{h}\int_0^h \kappa\left(\frac{\partial T}{\partial y}\right)_{y=0} \mathrm{d}x$$
$$= -\tfrac{4}{3}\kappa\theta'(0, \mathrm{P})(T_1 - T_0)(\mathrm{G}/4h)^{\frac{1}{4}}.$$

The Nusselt number is

$$\mathrm{N} = f(\mathrm{P})\mathrm{G}^{\frac{1}{4}},$$

where $f(\mathrm{P})$ is determined by solving the equations (4).

PROBLEM 2. A hot turbulent submerged jet of gas is bent round by a gravitational field; find its shape (G. N. Abramovich 1938).

SOLUTION. Let T' be some mean value (over the cross-section of the jet) of the temperature difference between the jet and the surrounding gas, u some mean velocity of the gas in the jet, and l the distance along the jet from its point of entry; l is supposed large compared with the dimensions of the aperture by which the jet enters. The condition of constant heat flux Q along the jet is $Q \sim \rho c_p T' u R^2 =$ constant and, since the radius of a turbulent jet is proportional to l (cf. §36), we have

$$T'ul^2 = \text{constant} \sim Q/\rho c_p; \tag{1}$$

we notice that, in the absence of the gravitational field, $u \propto 1/l$ (see (36.3)) and it then follows from (1) that $T' \propto 1/l$.

The momentum flux vector through the cross-section of the jet is proportional to $\rho u^2 R^2 \mathbf{n} \sim \rho u^2 l^2 \mathbf{n}$, where $\mathbf{n}$ is a unit vector along the jet. Its horizontal component is constant along the jet:

$$u^2 l^2 \cos\theta = \text{constant}, \tag{2}$$

where θ is the angle between $\mathbf{n}$ and the horizontal, while the change in the vertical component is due to the "lift force" on the jet. This force is proportional to

$$\rho\beta g T' R^2 \sim \rho\beta g T' l^2 \sim \beta g Q/c_p u.$$

Hence we have

$$\mathrm{d}(l^2 u^2 \sin\theta)/\mathrm{d}l \sim \beta g Q/\rho c_p u. \tag{3}$$

It then follows from (2) that $\mathrm{d}(\tan\theta)/\mathrm{d}l = \text{constant} \times l \cos^{\frac{1}{2}}\theta$, whence we obtain finally

$$\int_{\theta_0}^{\theta} \frac{\mathrm{d}\theta}{\cos^{5/2}\theta} = \text{constant} \times l^2, \tag{4}$$

where θ_0 gives the direction of the emergent jet.

In particular, if θ does not vary appreciably along the jet, (4) gives $\theta - \theta_0 = \text{constant} \times l^2$. This means that the jet is a cubical parabola, in which the deviation d from a straight line is $d = \text{constant} \times l^3$.

PROBLEM 3. A turbulent jet of heated gas (i.e. one with a large Rayleigh number) rises from a fixed hot body. Determine the variation of the velocity and temperature in the jet with height (Ya. B. Zel'dovich 1937).

SOLUTION. As in the preceding case, the radius of the jet is proportional to the distance from its source, and we have, analogously to (1) of Problem 2, $T'uz^2 =$ constant, and instead of (3) $\mathrm{d}(z^2u^2)/\mathrm{d}z = \text{constant}/u$, where z is the height above the body, supposed large compared with the dimension of the body. Integrating, we find $u \propto z^{-\frac{1}{3}}$, and for the temperature $T' \propto z^{-5/3}$.

PROBLEM 4. The same as Problem 3, but for a laminar convective jet rising freely (Ya. B. Zel'dovich 1937).

SOLUTION. Together with the relation $T'uR^2 =$ constant, which expresses the constancy of the heat flux, we have $u^2/z \sim \nu u/R^2 \sim \beta g T'$, which follows from equation (56.6). From these relations we find the following variation of the radius, velocity and temperature with height: $R \propto \sqrt{z}$, $u =$ constant, $T' \propto 1/z$. It may be noticed that the number $\mathcal{R} \propto T'R^3 \propto \sqrt{z}$, i.e. increases with height, and the jet must therefore become turbulent at a certain altitude.

§57. Convective instability of a fluid at rest

If the Rayleigh number is gradually increased in a given configuration of a fluid and solid walls, a point is reached when the fluid at rest becomes unstable with respect to infinitesimal perturbations.† This gives rise to convection; the transition from pure thermal conduction in the fluid at rest to the convective regime takes place continuously. The dependence of the Nusselt number on $\mathcal{R}$ therefore has only a kink at the transition, not a discontinuity.

The theoretical determination of the critical value $\mathcal{R}_{cr}$ is to be carried out as described in §26. The treatment will be repeated here for the case now in question.

We write

$$T' = T'_0 + \tau, \quad p' = p'_0 + \rho w, \tag{57.1}$$

where T'_0 and p'_0 refer to the fluid at rest; τ and w are perturbations. T'_0 and p'_0 satisfy the equations

$$\triangle T'_0 = d^2T'_0/dz^2 = 0, \quad dp'_0/dz = \rho\beta g T'_0.$$

The first of these gives $T'_0 = -Az$, where A is a constant; in the case considered here, where the fluid is heated from below, $A > 0$.

In equations (56.4) and (56.5), the small quantities are $\mathbf{v}$ (the unperturbed velocity is zero), τ and w. Omitting quadratic terms and considering perturbations which vary with time as $e^{-i\omega t}$, we get the equations

$$-i\omega\mathbf{v} = -\mathbf{grad}\, w + \nu\triangle\mathbf{v} - \beta\tau\mathbf{g},$$

$$-i\omega\tau - Av_z = \chi\triangle\tau, \quad \operatorname{div}\mathbf{v} = 0.$$

It is useful to write these in dimensionless form, measuring lengths in terms of h, frequencies of ν/h^2, velocities of ν/h, pressures of $\rho\nu^2/h^2$, and temperatures of $Ah\nu/\chi$. In the rest of this section and in the Problems, all symbols denote the appropriate dimensionless quantities. The equations become

$$-i\omega\mathbf{v} = -\mathbf{grad}\, w + \triangle\mathbf{v} + \mathcal{R}\tau\mathbf{n}, \tag{57.2}$$

$$-i\omega\tau \mathrm{P} = \triangle\tau + v_z, \tag{57.3}$$

$$\operatorname{div}\mathbf{v} = 0, \tag{57.4}$$

$\mathbf{n}$ being a unit vector in the z-direction, vertically upwards. The dimensionless parameters $\mathcal{R}$ and P now appear explicitly. If the solid surfaces bounding the fluid are kept at constant temperatures, the following conditions must be satisfied there‡:

$$\mathbf{v} = 0, \quad \tau = 0. \tag{57.5}$$

Equations (57.2) – (57.4) with the boundary conditions (57.5) determine the spectrum of eigenfrequencies ω. When $\mathcal{R} < \mathcal{R}_{cr}$, their imaginary parts $\gamma \equiv \operatorname{im}\omega < 0$, and the perturbations are damped. The value of $\mathcal{R}_{cr}$ is given by the point at which, with increasing $\mathcal{R}$, an eigenfrequency with $\gamma > 0$ first appears; at $\mathcal{R} = \mathcal{R}_{cr}$, γ passes through zero.

---

† This is not to be confused with the convected instability discussed in §28.

‡ We are considering the simplest boundary conditions, appropriate to perfectly conducting walls. When the conductivity of the walls is finite, the equation of heat transfer in the wall has to be included. Cases where the fluid has a free surface will also not be discussed. Here it would, strictly speaking, be necessary to take into account the deformation of the surface by the perturbation and the resulting surface tension forces.

The problem of convective instability of a fluid at rest has the particular feature that all the eigenvalues $i\omega$ are real, so that the perturbations decay or are amplified monotonically, without oscillations. Accordingly, the stable flow resulting from the instability of the fluid at rest is steady. We shall show this for a fluid occupying a closed cavity with the boundary conditions (57.5) at its walls.†

We multiply equations (57.2) and (57.3) by $\mathbf{v}^*$ and τ^* respectively, and integrate them over the volume of the cavity. On integrating by parts‡ for the terms $\mathbf{v}^* \cdot \triangle \mathbf{v}$ and $\tau^* \triangle \tau$, and noting that the integrals over the cavity surface are zero on account of the boundary conditions, we find

$$\left.\begin{aligned} -i\omega \int |\mathbf{v}|^2 \, dV &= \int (-|\mathbf{curl}\,\mathbf{v}|^2 + \mathscr{R}\tau v^*_z)\, dV, \\ -i\omega \mathrm{P} \int |\tau|^2 \, dV &= \int (-|\mathbf{grad}\,\tau|^2 + \tau^* v_z)\, dV. \end{aligned}\right\} \tag{57.6}$$

Subtracting from these the complex conjugate equations gives

$$-i(\omega + \omega^*) \int |\mathbf{v}|^2 \, dV = \mathscr{R} \int (\tau v^*_z - \tau^* v_z)\, dV,$$

$$-i(\omega + \omega^*)\, \mathrm{P} \int |\tau|^2 \, dV = -\int (\tau v^*_z - \tau^* v_z)\, dV.$$

Lastly, we multiply the second equation by $\mathscr{R}$ and add, obtaining

$$\mathrm{re}\, \omega \int (|\mathbf{v}|^2 + \mathscr{R}\mathrm{P}\, |\tau|^2)\, dV = 0.$$

Since the integral is positive definite, it follows that $\mathrm{re}\,\omega = 0$, as was to be proved.†† When $A < 0$ (the fluid is heated from above), formally corresponding to $\mathscr{R} < 0$, the integral can be zero, and $i\omega$ may be complex.

Let us now return to the equations (57.6). Multiplying the second equation by $\mathscr{R}$ and adding, we find for the growth rate $\gamma = -i\omega$ the expression

$$-\gamma = J/N, \tag{57.7}$$

† We follow V. S. Sorokin (1953) in this derivation and in the subsequent formulation of the variational principle.

‡ Using the equations

$$\mathbf{v}^* \cdot \triangle \mathbf{v} = -\mathbf{v}^* \cdot \mathbf{curl}\,\mathbf{curl}\,\mathbf{v} = \mathrm{div}\,(\mathbf{v}^* \times \mathbf{curl}\,\mathbf{v}) - |\mathbf{curl}\,\mathbf{v}|^2,$$

$$\tau^* \triangle \tau = \mathrm{div}\,(\tau^*\, \mathbf{grad}\,\tau) - |\mathbf{grad}\,\tau|^2,$$

$$\mathbf{v} \cdot \mathbf{grad}\, w = \mathrm{div}\,(w\mathbf{v}).$$

†† Mathematically, this proof amounts to showing that equations (57.2)–(57.4) are self-adjoint. Physically, the result can be interpreted as follows. Let a perturbation cause a fluid element to move upwards, say. It is then surrounded by cooler fluid, and its temperature is reduced by conduction, but remains above that of the environment. The buoyancy force on it is therefore upwards, and it continues to move in the same direction, more slowly or more quickly according to the relation between the temperature gradient and the dissipative coefficients. In either case, there is no "restoring force" and therefore no oscillations. When a free surface is present, a restoring force is provided by surface tension, which seeks to smooth the deformed surface; if this force is taken into account, the statements made are no longer correct.

where

$$\left.\begin{aligned} J &= \int [(\mathbf{curl\ v})^2 + \mathscr{R}(\mathbf{grad}\ \tau)^2 - 2\mathscr{R}\tau v_z]\,\mathrm{d}V, \\ N &= \int (\mathbf{v}^2 + \mathscr{R}\mathrm{P}\tau^2)\,\mathrm{d}V; \end{aligned}\right\} \tag{57.8}$$

v and τ are assumed to be real. It is well known that the eigenvalue problem for self-adjoint linear differential operators allows a variational formulation based on expressions having the form (57.7), (57.8). Regarding J and N as functionals of **v** and τ, we make J an extremum under the constraints div $\mathbf{v} = 0$ and $N = 1$, the latter acting as a "normalization condition". Following the general rules of the variational calculus, we form the variational equation

$$\delta J + \gamma\delta N - \int 2w\delta(\mathrm{div}\ \mathbf{v})\,\mathrm{d}V = 0, \tag{57.9}$$

where the constant γ and the function $w(\mathbf{r})$ act as undetermined Lagrange multipliers. Calculating the variations (with integration by parts, using the boundary conditions (57.5)) and equating to zero the expressions with the independent variations $\delta\mathbf{v}$ and $\delta\tau$, we in fact get equations (57.2) and (57.3). The value of J calculated from this variational problem determines, by (57.7), the lowest value of $-\gamma = -\gamma_1$, that is, the growth rate of the most rapidly amplified perturbations (or the decay rate of the least rapidly damped ones, depending on the sign of γ).

According to its derivation, the critical value $\mathscr{R}_{\mathrm{cr}}$ defines the limit of stability with respect to infinitesimal perturbations. For the case of convective instability of a fluid at rest, however, this value proves to be also the limit of stability with respect to any finite perturbations.† That is, when $\mathscr{R} < \mathscr{R}_{\mathrm{cr}}$ there are no solutions of the equations of motion, other than the state of rest, which do not decay in the course of time. We shall now prove this result (V. S. Sorokin 1954).

For finite perturbations, the equations of motion have to be written in the form

$$\left.\begin{aligned} \partial\mathbf{v}/\partial t &= -\mathbf{grad}\,w + \triangle\mathbf{v} + \mathscr{R}\tau\mathbf{n} - (\mathbf{v}\cdot\mathbf{grad})\,\mathbf{v}, \\ \mathrm{P}\partial\tau/\partial t &= \triangle\tau + v_z - \mathrm{P}\mathbf{v}\cdot\mathbf{grad}\,\tau, \end{aligned}\right\} \tag{57.10}$$

which differ from (57.2) and (57.3) by containing non-linear terms. We carry out with these equations just the same operations as we did with (57.2) and (57.3) when deriving (57.6) and (57.7). Since div $\mathbf{v} = 0$, the non-linear terms reduce to divergences:

$$\mathbf{v}\cdot(\mathbf{v}\cdot\mathbf{grad})\mathbf{v} = \mathrm{div}(\tfrac{1}{2}v^2\mathbf{v}), \quad \tau(\mathbf{v}\cdot\mathbf{grad})\tau = \mathrm{div}(\tfrac{1}{2}\tau^2\mathbf{v}),$$

and give zero when integrated. We therefore arrive at the relation

$$\tfrac{1}{2}\mathrm{d}N/\mathrm{d}t = -J,$$

which differs from $\gamma N = -J$ (57.7) only by having the time derivative instead of the product γN. According to the variational principle formulated above, $-J \leqslant \gamma_1 N$ for any functions **v** and τ. Hence

$$\mathrm{d}N(t)/\mathrm{d}t \leqslant 2\gamma_1 N(t),$$

† When referring to perturbations with finite amplitude we mean here those for which the non-linear terms in (56.4) and (56.5) cannot be neglected, while at the same time the conditions imposed when deriving these equations are still satisfied.

whence

$$N(t) \leqslant N(0)e^{2\gamma_1 t}. \tag{57.11}$$

Below the critical value ($\mathscr{R} < \mathscr{R}_{cr}$), all growth rates are negative in the linear theory, including the largest one γ_1. It therefore follows from (57.11) that $N(t) \to 0$ as $t \to \infty$, and, since the integrand in N is positive definite, the functions $\mathbf{v}$ and τ tend to zero also.

Let us now return to the calculation of $\mathscr{R}_{cr}$. Since all the eigenvalues $i\omega$ are real, the equation $\gamma = 0$ for $\mathscr{R} = \mathscr{R}_{cr}$ implies that $\omega = 0$. The value of $\mathscr{R}_{cr}$ is then found as the smallest eigenvalue of $\mathscr{R}$ in the equations

$$\left.\begin{aligned} &\triangle \mathbf{v} - \mathbf{grad}\, w + \mathscr{R}\tau\mathbf{n} = 0, \\ &\triangle\tau = -v_z, \quad \operatorname{div}\mathbf{v} = 0; \end{aligned}\right\} \tag{57.12}$$

this problem too can have a variational formulation (see Problem 2). Note that P does not appear in the equations (57.12) or in the boundary conditions. Thus the critical Rayleigh number which they yield for a given configuration of the fluid and the solid walls is independent of the fluid substance.

The simplest problem, which is also of theoretical importance†, is that of the stability of a layer of fluid between two infinite horizontal planes, of which the upper one is maintained at a lower temperature than the other.‡

Here it is convenient to reduce (57.12) to a single equation. Taking the **curl curl** = **grad** div $-\triangle$ of the first equation, then the z-component, and using the other two equations, we get

$$\triangle^3\tau = \mathscr{R}\triangle_2\tau, \tag{57.13}$$

where $\triangle_2 = \partial^2/\partial x^2 + \partial^2/\partial y^2$ is the two-dimensional Laplacian. The boundary conditions on the two planes are

$$\tau = 0, \quad v_z = 0, \quad \partial v_z/\partial z = 0 \text{ and } 1;$$

the last is equivalent to $v_x = v_y = 0$ for all x and y, by the equation of continuity. From the second equation (57.12), the conditions on v_z can be replaced by conditions on higher derivatives of τ:

$$\frac{\partial^2\tau}{\partial z^2} = 0, \quad \frac{\partial^3\tau}{\partial z^3} - k^2\frac{\partial\tau}{\partial z} = 0.$$

We seek τ in the form

$$\tau = f(z)\phi(x, y), \quad \phi = e^{i\mathbf{k}\cdot\mathbf{r}}, \tag{57.14}$$

where $\mathbf{k}$ is a vector in the xy-plane, obtaining for $f(z)$ the equation

$$\left(\frac{d^2}{dz^2} - k^2\right)^3 f + \mathscr{R}k^2 f = 0.$$

The general solution is a linear combination of $\cosh\mu z$ and $\sinh\mu z$, where $\mu^2 = k^2 - \mathscr{R}^{\frac{1}{3}}k^{\frac{2}{3}}\sqrt[3]{1}$, with the three different values of the cube root. The coefficients in this

† First proposed experimentally by H. Bénard (1900) and discussed theoretically by Rayleigh (1916).

‡ It has been shown by A. Pellew and R. V. Southwell (1940) that $i\omega$ is real in this case.

combination are determined by the boundary conditions, which lead to a system of algebraic equations, and the condition for these to be compatible yields a transcendental equation whose roots determine the functions $k = k_n(\mathscr{R})$, $n = 1, 2, \ldots$. The inverse functions $\mathscr{R} = \mathscr{R}_n(k)$ have minima for certain values of k, and the smallest of these gives $\mathscr{R}_{cr}$.† The value is found to be 1708, and the corresponding value of the wave number k_{cr} is 3·12 in units of $1/h$ (H. Jeffreys 1928).

Thus a horizontal layer of fluid with thickness h and a downward temperature gradient A becomes unstable when‡

$$\beta g A h^3/\nu\chi > 1708 \tag{57.15}$$

When $\mathscr{R} > \mathscr{R}_{cr}$, there is steady convective flow, periodic in the xy-plane. The whole space between the planes is divided into adjacent identical cells, in each of which the fluid moves in closed paths without passing from one cell to another. The outlines of these cells on the bounding planes form some kind of lattice. The value of k_{cr} determines the periodicity of this lattice but not its symmetry; the linearized equations of motion allow in (57.14) any function $\phi(x, y)$ that satisfies the equation $(\triangle_2 - k^2)\phi = 0$. The uncertainty cannot be eliminated in the linear theory. There must evidently be a "two-dimensional" structure of the flow, having in the xy-plane only a one-dimensional periodicity, as a system of parallel bands.††

PROBLEMS

PROBLEM 1. Find the critical Rayleigh number for the occurrence of convection in a fluid in a vertical cylindrical pipe along which a constant temperature gradient is maintained; the pipe walls are (a) perfectly conducting or (b) perfectly insulating (G. A. Ostroumov 1946).

SOLUTION. We seek a solution of (57.2) – (57.4) in which the convective velocity **v** is everywhere parallel to the axis of the pipe (the z-axis) and the flow pattern does not vary along this axis, i.e. $v_z = v$, τ and $\partial w/\partial z$ depend only on the coordinates in the plane of the pipe cross-section.§ The equations become

$$\partial w/\partial x = \partial w/\partial y = 0, \quad \triangle_2 v = -\mathscr{R}\tau + \partial w/\partial z, \quad \triangle_2 \tau = v,$$

where $\mathscr{R} = \beta g A R^4/\chi\nu$ and R is the pipe radius. From the first two, it follows that $\partial w/\partial z$ = constant; eliminating τ from the others, we find $\triangle_2{}^2 v = \mathscr{R}v$. On the walls of the pipe ($r = 1$) we must have $v = 0$ and $\tau = 0$ for case a, $\partial\tau/\partial r = 0$ for case b. In addition, the total mass flux through a cross-section of the pipe must be zero.

The equation has solutions of the form $J_n(kr)\cos n\phi$ and $I_n(kr)\cos n\phi$, where J_n and I_n are Bessel functions with real and imaginary argument respectively, $k^4 = \mathscr{R}$, r and ϕ are polar coordinates in the pipe cross-section. The onset of convection corresponds to the solution for which $\mathscr{R}$ is least. This is found to be the solution with $n = 1$:

$$v = v_0 \cos\phi\,[J_1(kr)I_1(k) - I_1(kr)J_1(k)],$$

$$\tau = (v_0/\mathscr{R}^{\frac{1}{2}})\cos\phi\,[J_1(kr)I_1(k) + I_1(kr)J_1(k)],$$

† The details of the calculations are given by G. Z. Gershuni and E. M. Zhukhovitskii in *Convective Stability of Incompressible Fluids*, Jerusalem 1976, and also in the books by Chandrasekhar and by Drazin and Reid cited in §27.

‡ For a given value of A, this condition is always satisfied when h is sufficiently large. To avoid misunderstanding, it should be mentioned that we are concerned here only with values of h for which the fluid density does not vary significantly in the gravitational field. The condition is therefore not applicable to tall columns of fluid. For these, the condition derived in §4 should be used, and shows that convection may be absent for any column height if the temperature gradient is not too great.

†† The theoretical indications are that just above $\mathscr{R}_{cr}$ only this structure is stable with respect to small perturbations; "three-dimensional" prismatic structures are unstable. The experimental results depend considerably on the conditions used, including the shape and size of the side walls, and are not definite. The three-dimensional hexagonal structure seems to be due to the influence of surface tension at the upper free surface and the temperature dependence of the viscosity; in the theory given here, ν has of course been treated as a constant.

§ The equations also have solutions periodic in the z-direction, which contain a factor e^{ikz}. These, however, all give higher values of $\mathscr{R}_{cr}$. It should be noted that the solution under consideration with $k = 0$ also satisfies the exact (not linearized) equations (57.10), since the non-linear terms $(\mathbf{v}\cdot\mathbf{grad})\mathbf{v}$ and $\mathbf{v}\cdot\mathbf{grad}\,\tau$ vanish identically.

with the gradient $\partial w/\partial z = 0$. The flow described by these formulae is antisymmetrical about a vertical plane through the pipe axis bisecting the cavity; the fluid descends in one half and rises in the other. The solution shown satisfies the condition $v = 0$ for $r = 1$. In case a, the condition $\tau = 0$ leads to $J_1(k) = 0$; its smallest root gives $\mathscr{R}_{cr} = k^4 = 216$. In case b, the condition $\partial\tau/\partial r = 0$ leads to

$$\frac{J_0(k)}{J_1(k)} + \frac{I_0(k)}{I_1(k)} = \frac{2}{k}.$$

The smallest root gives $\mathscr{R}_{cr} = 68$.

PROBLEM 2. Formulate a variational principle for the problem of $\mathscr{R}$ eigenvalues with equations (57.12).

SOLUTION. We put the equations in a more symmetrical form by replacing τ by a new function $\tilde{\tau} = \tau\sqrt{\mathscr{R}}$, again changing the unit of temperature measurement. Then

$$\sqrt{\mathscr{R}}\tilde{\tau}\mathbf{n} = \mathbf{grad}\, w - \triangle\mathbf{v}, \quad \sqrt{\mathscr{R}}\, v_z = -\triangle\tilde{\tau}, \quad \operatorname{div}\mathbf{v} = 0.$$

Proceeding as in the derivation of (57.7), we find $\sqrt{\mathscr{R}} = J/N$, where

$$J = \tfrac{1}{2}\int[(\mathbf{curl}\,\mathbf{v})^2 + (\mathbf{grad}\,\tilde{\tau})^2]\mathrm{d}V, \quad N = \int v_z\tilde{\tau}\,\mathrm{d}V;$$

N is positive, as is easily seen by converting it to $\mathscr{R}^{-\frac{1}{2}}\int(\mathbf{grad}\,\tilde{\tau})^2\,\mathrm{d}V$. The variational principle is formulated as requiring an extremum of J under the constraints $\operatorname{div}\mathbf{v} = 0$ and $N = 1$. The minimum of J determines the smallest eigenvalue of $\sqrt{\mathscr{R}}$.

CHAPTER VI

DIFFUSION

§58. The equations of fluid dynamics for a mixture of fluids

Throughout the above discussion it has been assumed that the fluid is completely homogeneous. If we are concerned with a mixture of fluids whose composition is different at different points, then the equations of fluid dynamics are considerably modified.

We shall discuss here only mixtures with two components. The composition of the mixture is described by the concentration c, defined as the ratio of the mass of one component to the total mass of the fluid in a given volume element.

In the course of time, the distribution of the concentration through the fluid will in general change. This change occurs in two ways. Firstly, when there is macroscopic motion of the fluid, any given small portion of it moves as a whole, its composition remaining unchanged. This results in a purely mechanical mixing of the fluid; although the composition of each moving portion of it is unchanged, the concentration of the fluid at any point in space varies with time. If we ignore any processes of thermal conduction and internal friction which may also be taking place, this change in concentration is a thermodynamically reversible process, and does not result in the dissipation of energy.

Secondly, a change in composition can occur by the molecular transfer of the components from one part of the fluid to another. Th equalization of the concentration by this direct change of composition of every small portion of fluid is called *diffusion*. Diffusion is an irreversible process, and is, like thermal conduction and viscosity, one of the sources of energy dissipation in a mixture of fluids.

We denote by ρ the total density of the fluid. The equation of continuity for the total mass of the fluid is, as before,

$$\partial\rho/\partial t + \operatorname{div}(\rho\mathbf{v}) = 0. \tag{58.1}$$

It signifies that the total mass of fluid in any volume can vary only by the movement of fluid into or out of that volume. It must be emphasized that, strictly speaking, the concept of velocity itself must be redefined for a mixture of fluids. By writing the equation of continuity in the form (58.1), we have defined the velocity, as before, as the total momentum of unit mass of fluid.

The Navier–Stokes equation (15.5) is also unchanged. We shall now derive the remaining equations of fluid dynamics for a mixture of fluids.

In the absence of diffusion, the composition of any given fluid element would remain unchanged as it moved about. This means that the total derivative $\mathrm{d}c/\mathrm{d}t$ would be zero, i.e. the equation $\mathrm{d}c/\mathrm{d}t = \partial c/\partial t + \mathbf{v}\cdot\mathbf{grad}\, c = 0$ would hold. This equation can be written, using (58.1), as

$$\partial(\rho c)/\partial t + \operatorname{div}(\rho c\mathbf{v}) = 0,$$

i.e. as an equation of continuity for one of the components of the mixture (ρc being the mass of that component in unit volume). In the integral form

$$\frac{\partial}{\partial t}\int \rho c \, \mathrm{d}V = -\oint \rho c \mathbf{v} \cdot \mathbf{df}$$

it shows that the rate of change of the amount of this component in any volume is equal to the amount of the component transported through the surface of that volume by the motion of the fluid.

When diffusion occurs, besides the flux $\rho c \mathbf{v}$ of the component in question as it moves with the fluid, there is another flux which results in the transfer of the components even when the fluid as a whole is at rest. Let $\mathbf{i}$ be the density of this diffusion flux, i.e. the amount of the component transported by diffusion through unit area in unit time.† Then we have for the rate of change of the amount of the component in any volume

$$\frac{\partial}{\partial t}\int \rho c \, \mathrm{d}V = -\oint \rho c \mathbf{v} \cdot \mathbf{df} - \oint \mathbf{i} \cdot \mathbf{df},$$

or, in differential form,

$$\partial(\rho c)/\partial t = -\operatorname{div}(\rho c \mathbf{v}) - \operatorname{div} \mathbf{i}. \tag{58.2}$$

Using (58.1), we can rewrite this equation of continuity for one component in the form

$$\rho(\partial c/\partial t + \mathbf{v} \cdot \mathbf{grad}\, c) = -\operatorname{div} \mathbf{i}. \tag{58.3}$$

To derive another equation, we repeat the arguments given in §49, bearing in mind that the thermodynamic quantities for the fluid are now functions of the concentration also. In calculating the derivative $\partial(\frac{1}{2}\rho v^2 + \rho\varepsilon)/\partial t$ (in §49) by means of the equations of motion, we had to transform the terms $\rho \partial\varepsilon/\partial t$ and $-\mathbf{v} \cdot \mathbf{grad}\, p$. This transformation must now be modified, because the thermodynamic relations for the energy and the heat function now contain an additional term involving the differential of the concentration:

$$\mathrm{d}\varepsilon = T\,\mathrm{d}s + (p/\rho^2)\mathrm{d}\rho + \mu\,\mathrm{d}c,$$

$$\mathrm{d}w = T\,\mathrm{d}s + (1/\rho)\mathrm{d}p + \mu\,\mathrm{d}c,$$

where μ is an appropriately defined chemical potential of the mixture.‡ Accordingly, an

† The sum of the flux densities for the two components must be $\rho \mathbf{v}$. If the flux density for one component is $\rho c \mathbf{v} + \mathbf{i}$, that for the other component is therefore $\rho(1-c)\mathbf{v} - \mathbf{i}$.

‡ It is known from thermodynamics (see *SP* 1, §85) that, for a mixture of two substances,

$$\mathrm{d}\varepsilon = T\,\mathrm{d}s - p\,\mathrm{d}V + \mu_1\,\mathrm{d}n_1 + \mu_2\,\mathrm{d}n_2,$$

where n_1, n_2 are the numbers of particles of the two substances in unit mass of the mixture, and μ_1, μ_2 are the chemical potentials of the substances. The numbers n_1, n_2 satisfy the relation $n_1 m_1 + n_2 m_2 = 1$, where m_1 and m_2 are the masses of the two kinds of particle. If we introduce as a variable the concentration $c = n_1 m_1$, we have

$$\mathrm{d}\varepsilon = T\,\mathrm{d}s - p\,\mathrm{d}V + \left(\frac{\mu_1}{m_1} - \frac{\mu_2}{m_2}\right)\mathrm{d}c.$$

Comparing this with the relation given in the text, we see that the chemical potential μ is related to μ_1 and μ_2 by

$$\mu = \frac{\mu_1}{m_1} - \frac{\mu_2}{m_2}.$$

additional term $\rho\mu\partial c/\partial t$ appears in the derivative $\rho\partial\varepsilon/\partial t$. Writing the second thermodynamic relation in the form

$$dp = \rho\, dw - \rho T\, ds - \rho\mu\, dc,$$

we see that the term $-\mathbf{v}\cdot\mathbf{grad}\, p$ will contain an additional term $\rho\mu\mathbf{v}\cdot\mathbf{grad}\, c$.

Thus we must add $\rho\mu(\partial c/\partial t + \mathbf{v}\cdot\mathbf{grad}\, c) = -\mu\, \mathrm{div}\, \mathbf{i}$ to the expression (49.3). The result is

$$\frac{\partial}{\partial t}(\tfrac{1}{2}\rho v^2 + \rho\varepsilon) = -\mathrm{div}[\rho\mathbf{v}(\tfrac{1}{2}v^2 + w) - \mathbf{v}\cdot\boldsymbol{\sigma}' + \mathbf{q}] + $$

$$+\rho T\left(\frac{\partial s}{\partial t} + \mathbf{v}\cdot\mathbf{grad}\, s\right) - \sigma'_{ik}\frac{\partial v_i}{\partial x_k} + \mathrm{div}\,\mathbf{q} - \mu\,\mathrm{div}\,\mathbf{i}. \tag{58.4}$$

We have replaced $-\kappa\,\mathbf{grad}\, T$ by a heat flux $\mathbf{q}$, which may depend not only on the temperature gradient but also on the concentration gradient (see the next section). The sum of the last two terms on the right can be written

$$\mathrm{div}\,\mathbf{q} - \mu\,\mathrm{div}\,\mathbf{i} = \mathrm{div}(\mathbf{q} - \mu\mathbf{i}) + \mathbf{i}\cdot\mathbf{grad}\,\mu.$$

The expression $\rho\mathbf{v}(\frac{1}{2}v^2 + w) - \mathbf{v}\cdot\boldsymbol{\sigma}' + \mathbf{q}$ which is the operand of the divergence operator in (58.4) is, by the definition of $\mathbf{q}$, the total energy flux in the fluid. The first term is the reversible energy flux, due simply to the movement of the fluid as a whole, while the sum $-\mathbf{v}\cdot\boldsymbol{\sigma}' + \mathbf{q}$ is the irreversible flux. When there is no macroscopic motion, the viscosity flux $\mathbf{v}\cdot\boldsymbol{\sigma}'$ is zero, and the heat flux is simply $\mathbf{q}$.

The equation of conservation of energy is

$$\frac{\partial}{\partial t}(\tfrac{1}{2}\rho v^2 + \rho\varepsilon) = -\mathrm{div}[\rho\mathbf{v}(\tfrac{1}{2}v^2 + w) - \mathbf{v}\cdot\boldsymbol{\sigma}' + \mathbf{q}]. \tag{58.5}$$

Subtracting from (58.4), we obtain the required equation

$$\rho T\left(\frac{\partial s}{\partial t} + \mathbf{v}\cdot\mathbf{grad}\, s\right) = \sigma'_{ik}\frac{\partial v_i}{\partial x_k} - \mathrm{div}(\mathbf{q} - \mu\mathbf{i}) - \mathbf{i}\cdot\mathbf{grad}\,\mu, \tag{58.6}$$

which is a generalization of (49.4).

We have thus obtained a complete system of equations of fluid dynamics for a mixture of fluids. The number of equations in this system is one more than for a single fluid, since there is one more unknown function, namely the concentration. The equations are the equation of continuity (58.1), the Navier–Stokes equations, the equation of continuity (58.2) for one component, and equation (58.6), which determines the change in entropy. It must be noticed that equations (58.2) and (58.6) as they stand determine only the form of the corresponding equations of fluid dynamics, since they involve the undetermined fluxes $\mathbf{i}$ and $\mathbf{q}$. These equations become determinate only when $\mathbf{i}$ and $\mathbf{q}$ are replaced by expressions in terms of the gradients of concentration and temperature. The corresponding expressions will be obtained in §59.

For the rate of change of the total entropy of the fluid, a calculation entirely similar to that of §49, but using (58.6) in place of (49.4), gives the result

$$\frac{\partial}{\partial t}\int \rho s\, dV = -\int \frac{(\mathbf{q} - \mu\mathbf{i})\cdot\mathbf{grad}\, T}{T^2}\, dV - \int \frac{\mathbf{i}\cdot\mathbf{grad}\,\mu}{T}\, dV + \ldots, \tag{58.7}$$

where we have omitted, for brevity, the viscosity terms.

§59. Coefficients of mass transfer and thermal diffusion

The diffusion flux **i** and the heat flux **q** are due to the presence of concentration and temperature gradients in the fluid. It should not be thought, however, that **i** depends only on the concentration gradient and **q** only on the temperature gradient. On the contrary, each of these fluxes depends, in general, on both gradients.

If the concentration and temperature gradients are small, we can suppose that **i** and **q** are linear functions of **grad** μ and **grad** T. The fluxes **q** and **i** are independent of the pressure gradient (for given **grad** μ and **grad** T), for the same reason as that given with regard to **q** in §49. Accordingly, we write **i** and **q** as

$$\mathbf{i} = -\alpha\,\mathbf{grad}\,\mu - \beta\,\mathbf{grad}\,T, \quad \mathbf{q} = -\delta\,\mathbf{grad}\,\mu - \gamma\,\mathbf{grad}\,T + \mu\mathbf{i}.$$

There is a simple relation between the coefficients β and δ, which is a consequence of a *symmetry principle for the kinetic coefficients*. This symmetry principle is as follows (see *SP*1, §120).

Let us consider some closed system, and let $x_1, x_2, \ldots$ be some quantities characterizing the state of the system. Their equilibrium values are determined by the fact that, in statistical equilibrium, the entropy S of the whole system must be a maximum, i.e. we must have $X_a = 0$ for all a, where X_a denotes the derivative

$$X_a = -\partial S/\partial x_a. \tag{59.1}$$

We assume that the system is in a state near to equilibrium. This means that all the x_a are very little different from their equilibrium values, and the X_a are small. Processes will occur in the system which tend to bring it into equilibrium. The quantities x_a are functions of time, and their rate of change is given by the time derivatives $\dot{x}_a$; we express the latter as functions of X_a, and expand these functions in series. As far as terms of the first order we have

$$\dot{x}_a = -\sum_b \gamma_{ab} X_b. \tag{59.2}$$

Onsager's symmetry principle for the kinetic coefficients states that the γ_{ab} (called the *kinetic coefficients*) are symmetrical with respect to the suffixes a and b:

$$\gamma_{ab} = \gamma_{ba}. \tag{59.3}$$

The rate of change of the entropy S is

$$\dot{S} = -\sum_a X_a \dot{x}_a.$$

Now let the x_a themselves be different at different points of the system, i.e. each volume element have its own values of the x_a. That is, we suppose the x_a to be functions of the coordinates. Then, in the expression for $\dot{S}$, besides summing over a we must integrate over the volume of the system:

$$\dot{S} = -\int \sum_a X_a \dot{x}_a \, dV. \tag{59.4}$$

It is usually true that the values of the $\dot{x}_a$ at any given point depend only on the values of the X_a at that point. In this case we can write down the relation between $\dot{x}_a$ and X_a for each point in the system, and obtain the same formulae as previously.

In the problem under consideration we take as the $\dot{x}_a$ the components of the vectors **i** and $\mathbf{q}-\mu\mathbf{i}$. Then we see from a comparison of (58.7) and (59.4) that the X_a are respectively the components of the vectors $(1/T)\,\mathbf{grad}\,\mu$ and $(1/T^2)\,\mathbf{grad}\,T$. The kinetic coefficients γ_{ab} are the coefficients of these vectors in the equations

$$\mathbf{i} = -\alpha T\left(\frac{\mathbf{grad}\,\mu}{T}\right) - \beta T^2\left(\frac{\mathbf{grad}\,T}{T^2}\right),$$

$$\mathbf{q}-\mu\mathbf{i} = -\delta T\left(\frac{\mathbf{grad}\,\mu}{T}\right) - \gamma T^2\left(\frac{\mathbf{grad}\,T}{T^2}\right).$$

By the symmetry of the kinetic coefficients, we must have $\beta T^2 = \delta T$, or $\delta = \beta T$. This is the required relation.

We can therefore write the fluxes **i** and **q** as

$$\left.\begin{aligned} \mathbf{i} &= -\alpha\,\mathbf{grad}\,\mu - \beta\,\mathbf{grad}\,T,\\ \mathbf{q} &= -\beta T\,\mathbf{grad}\,\mu - \gamma\,\mathbf{grad}\,T + \mu\mathbf{i}, \end{aligned}\right\} \tag{59.5}$$

with only three independent coefficients α, β, γ. It is convenient to eliminate $\mathbf{grad}\,\mu$ from the expression for the heat flux, replacing it by **i** and $\mathbf{grad}\,T$. Then we have

$$\begin{aligned} \mathbf{i} &= -\alpha\,\mathbf{grad}\,\mu - \beta\,\mathbf{grad}\,T,\\ \mathbf{q} &= (\mu + \beta T/\alpha)\mathbf{i} - \kappa\,\mathbf{grad}\,T, \end{aligned} \tag{59.6}$$

where

$$\kappa = \gamma - \beta^2 T/\alpha. \tag{59.7}$$

If the diffusion flux **i** is zero, we have *pure thermal conduction*. For this to be so, T and μ must satisfy the equation $\alpha\,\mathbf{grad}\,\mu + \beta\,\mathbf{grad}\,T = 0$, or $\alpha\,\mathrm{d}\mu + \beta\,\mathrm{d}T = 0$. The integration of this equation gives a relation of the form $f(c,\ T) = 0$ which does not contain the coordinates explicitly. (The chemical potential is a function of the pressure, as well as of c and T, but in equilibrium the pressure is constant.) This relation determines the dependence of the concentration on the temperature which must hold if there is no diffusion flux. Moreover, for $\mathbf{i} = 0$ we have from (59.7)

$$\mathbf{q} = -\kappa\,\mathbf{grad}\,T,$$

so that κ is just the thermal conductivity.

Let us now change to the usual variables p, T and c. We have

$$\mathbf{grad}\,\mu = (\partial\mu/\partial c)_{p,T}\,\mathbf{grad}\,c + (\partial\mu/\partial T)_{c,p}\,\mathbf{grad}\,T + (\partial\mu/\partial p)_{c,T}\,\mathbf{grad}\,p.$$

In the last term we use the thermodynamic relation

$$\mathrm{d}\phi = -s\,\mathrm{d}T + V\,\mathrm{d}p + \mu\,\mathrm{d}c, \tag{59.8}$$

where ϕ is the Gibbs free energy per unit mass, and V is the specific volume, obtaining

$$(\partial\mu/\partial p)_{c,T} = \partial^2\phi/\partial p\,\partial c = (\partial V/\partial c)_{p,T}.$$

Substituting **grad** μ in (59.6) and putting

$$\left.\begin{aligned} D &= \frac{\alpha}{\rho}\left(\frac{\partial\mu}{\partial c}\right)_{p,T}, \\ \rho k_T D/T &= \alpha(\partial\mu/\partial T)_{c,p} + \beta, \end{aligned}\right\} \tag{59.9}$$

$$k_p = p(\partial V/\partial c)_{p,T}/(\partial\mu/\partial c)_{p,T}, \tag{59.10}$$

we obtain

$$\mathbf{i} = -\rho D[\mathbf{grad}\, c + (k_T/T)\,\mathbf{grad}\, T + (k_p/p)\,\mathbf{grad}\, p], \tag{59.11}$$

$$\mathbf{q} = [k_T(\partial\mu/\partial c)_{p,T} - T(\partial\mu/\partial T)_{p,c} + \mu]\mathbf{i} - \kappa\,\mathbf{grad}\, T. \tag{59.12}$$

The coefficient D is called the *diffusion coefficient* or *mass transfer coefficient*; it gives the diffusion flux when only a concentration gradient is present. The diffusion flux due to the temperature gradient is given by the *thermal diffusion coefficient* $k_T D$; the dimensionless quantity k_T is called the *thermal diffusion ratio*.

The last term in (59.11) need be taken into account only when there is a considerable pressure gradient in the fluid (caused by an external field, say). The coefficient $k_p D$ may be called the *barodiffusion coefficient*; we shall discuss it further at the end of this section.

In a single fluid there is, of course, no diffusion flux. Hence it is clear that k_T and k_p must vanish in each of the two limiting cases $c = 0$ and $c = 1$.

The condition that the entropy must increase places certain restrictions on the coefficients in formulae (59.6). Substituting these formulae in the expression (58.7) for the rate of change of the entropy, we find

$$\frac{\partial}{\partial t}\int \rho s\,\mathrm{d}V = \int \frac{\kappa(\mathbf{grad}\, T)^2}{T^2}\mathrm{d}V + \int \frac{\mathbf{i}^2}{\alpha T}\mathrm{d}V + \dots. \tag{59.13}$$

Hence it is clear that, besides the condition $\kappa > 0$ which we already know, we must have also $\alpha > 0$. Bearing in mind that the derivative $(\partial\mu/\partial c)_{p,T}$ is always positive according to one of the thermodynamic inequalities (see *SP*1, §96), we therefore find that the diffusion coefficient must be positive: $D > 0$. The quantities k_T and k_p, however, may be either positive or negative.

We shall not pause to write out the lengthy general equations obtained by substituting the above expressions for **i** and **q** in (58.3) and (58.6). We shall take only the case where there is no significant pressure gradient, while the concentration and temperature of the fluid vary so little that the coefficients in the expressions (59.11) and (59.12) may be supposed constant, although they are in general functions of c and T. Furthermore, we shall suppose that there is no macroscopic motion in the fluid except that which may be caused by the temperature and concentration gradients. The velocity of this motion is proportional to the gradients, and the terms in equations (58.3) and (58.6) which involve the velocity are therefore quantities of the second order, and may be neglected. The term $-\mathbf{i}\cdot\mathbf{grad}\,\mu$ in (58.6) is also of the second order. Thus we have $\rho\partial c/\partial t + \mathrm{div}\,\mathbf{i} = 0$, $\rho T\partial s/\partial t + \mathrm{div}(\mathbf{q} - \mu\mathbf{i}) = 0$.

Substituting for **i** and **q** the expressions (59.11) and (59.12) (without the term in **grad** p), and transforming the derivative $\partial s/\partial t$ as follows:

$$\frac{\partial s}{\partial t} = \left(\frac{\partial s}{\partial T}\right)_{c,p}\frac{\partial T}{\partial t} + \left(\frac{\partial s}{\partial c}\right)_{T,p}\frac{\partial c}{\partial t} = \frac{c_p}{T}\frac{\partial T}{\partial t} - \left(\frac{\partial\mu}{\partial T}\right)_{p,c}\frac{\partial c}{\partial t}$$

(since by (59.8) $(\partial s/\partial c)_{p,T} = -\partial^2\phi/\partial c\partial T = -(\partial\mu/\partial T)_{p,c}$), we obtain after a simple calculation

$$\partial c/\partial t = D[\triangle c + (k_T/T)\triangle T], \tag{59.14}$$

$$\partial T/\partial t - (k_T/c_p)(\partial\mu/\partial c)_{p,T}\partial c/\partial t = \chi\Delta T. \tag{59.15}$$

This system of linear equations determines the temperature and concentration distributions in the fluid.

There is a particularly important case where the concentration is small. When the concentration tends to zero, the diffusion coefficient tends to a finite constant, but the thermal diffusion coefficient tends to zero. Hence k_T is small for small concentrations, and we can neglect the term $k_T\triangle T$ in (59.14), which then becomes the diffusion equation

$$\partial c/\partial t = D\triangle c. \tag{59.16}$$

The boundary conditions on the solution of (59.16) are different in different cases. At the surface of a body insoluble in the fluid the normal component of the diffusion flux $\mathbf{i} = -\rho D\,\mathbf{grad}\,c$ must vanish, i.e. we must have $\partial c/\partial n = 0$. If, however, there is diffusion from a body which dissolves in the fluid, equilibrium is rapidly established near its surface, and the concentration in the fluid adjoining the body is the saturation concentration c_0; the diffusion out of this layer takes place more slowly than the process of solution. The boundary condition at such a surface is therefore $c = c_0$. Finally, if a solid surface absorbs the diffusing substance incident on it, the boundary condition is $c = 0$; an example of such a case is found in the study of chemical reactions at the surface of a solid.

Since the equations of pure diffusion (59.16) and of thermal conduction are of exactly the same form, we can immediately apply all the formulae derived in §§51 and 52 to the case of diffusion, simply replacing T by c and χ by D. The boundary condition for a thermally insulating surface corresponds to that for an insoluble surface, while a surface maintained at a constant temperature corresponds to a soluble surface from which diffusion takes place.

In particular, we can write down, by analogy with (51.5), the following solution of the diffusion equation:

$$c(\gamma, t) = \frac{M}{8\rho(\pi D t)^{\frac{3}{2}}}\exp(-r^2/4Dt). \tag{59.17}$$

This gives the distribution of the solute at any time, if at time $t = 0$ it is all concentrated at the origin (M being the total amount of the solute).

There is an important comment to be made regarding the above discussion. The expressions (59.5), or (59.11) and (59.12), are the first non-vanishing terms in an expansion of the fluxes in terms of the derivatives of the thermodynamic quantities. It is known from kinetic theory (see *PK*, §§5, 6, 14) that such an expansion is microscopically (for gases) one in powers of l/L, the ratio of the molecular mean free path l to the characteristic distance L for the problem. Including terms in higher-order derivatives would imply including quantities of higher order in this ratio. The terms next after those shown in (59.5), formed from derivatives of the scalars μ and T, would involve third-order derivatives, $\mathbf{grad}\,\triangle\mu$ and $\mathbf{grad}\,\triangle T$; there are certainly much less than those already included, in the ratio $(l/L)^2$.

The expressions for the fluxes may, however, also contain terms involving velocity derivatives. With the first-order derivatives $\partial v_i/\partial x_k$ we can construct only tensor quantities; these form the viscous stress tensor which appears in the momentum flux density tensor.

Vectors can be formed from the second-order derivatives. For example, the diffusion flux density vector contains terms

$$\mathbf{i}' = \rho\lambda_1 \triangle \mathbf{v} + \rho\lambda_2 \,\mathbf{grad}\,\mathrm{div}\,\mathbf{v}. \tag{59.18}$$

The condition that these terms be small in comparison with those which already appear in (59.11) and (59.12) leads to further conditions on the validity of the latter. For example, if it is meaningful to retain the **grad** p term in (59.11) while omitting the terms (59.18), we must have

$$D(p_2 - p_1)/pL \gg \lambda U/L^2,$$

where $p_2 - p_1$ is a characteristic pressure drop over the distance L, and U is a characteristic velocity drop; in this estimate, we have put $k_p \sim 1$ (see Problem). According to kinetic theory, D and λ can be expressed in terms of quantities describing the thermal motion of the gas molecules. It is evident from dimensional considerations that $\lambda/D \sim l/v_T$, where v_T is the mean thermal velocity of the molecules. Using also the fact that the gas pressure $p \sim \rho v_T^2$, we obtain the condition

$$p_2 - p_1 \gg \rho v_T U l/L. \tag{59.19}$$

This is by no means necessarily satisfied. On the contrary, in the important case of steady flow at low Reynolds numbers, the **grad** p and $\triangle \mathbf{v}$ terms in the diffusion flux have the same order of magnitude (Yu. M. Kagan 1962). For this flow, the pressure gradient is related to the velocity derivatives by (20.1):

$$(1/\rho)\,\mathbf{grad}\,p = \nu \triangle \mathbf{v}; \tag{59.20}$$

we assume that the gas may be regarded as incompressible. The kinematic viscosity is $\nu \sim v_T l$, and this equation therefore gives

$$p_2 - p_1 \sim \rho \nu U/L \sim \rho v_T U l/L,$$

instead of the inequality (59.19). Since $\triangle \mathbf{v}$ is expressible directly in terms of **grad** p by (20.1), the need to include the **grad** p and $\triangle \mathbf{v}$ terms at the same time signifies that the barodiffusion coefficient k_p is replaced by an effective coefficient

$$(k_p)_{\text{eff}} = k_p - p\lambda_1/\rho\nu D. \tag{59.21}$$

Note that it is therefore a kinetic quantity and not a purely thermodynamic one like k_p in (59.10).

PROBLEM

Determine the barodiffusion coefficient for a mixture of two perfect gases.

SOLUTION. We have for the specific volume $V = kT(n_1 + n_2)/p$ (the notation is that used in the second footnote to §58), and the chemical potentials are (see *SP* 1, §93)

$$\mu_1 = f_1(p, T) + T \log[n_1/(n_1 + n_2)],$$

$$\mu_2 = f_2(p, T) + T \log[n_2/(n_1 + n_2)].$$

The numbers n_1 and n_2 are expressed in terms of the concentration of the first component by $n_1 m_1 = c$, $n_2 m_2 = 1 - c$. A calculation using formula (59.10) gives

$$k_p = (m_2 - m_1)c(1-c)\left[\frac{1-c}{m_2} + \frac{c}{m_1}\right].$$

§60. Diffusion of particles suspended in a fluid

Under the influence of the molecular motion in a fluid, particles suspended in the fluid move in an irregular manner (called the *Brownian motion*). Let one such particle be at the origin at the initial instant. Its subsequent motion may be regarded as a diffusion, in which the concentration is represented by the probability of finding the particle in any particular volume element. To determine this probability, therefore, we can use the solution (59.17) of the diffusion equation. The possibility of this procedure is due to the fact that, for diffusion in weak solutions (i.e. when $c \ll 1$, which is when the diffusion equation can be used in the form (59.16)), the particles of the solute hardly affect one another, and so the motion of each particle can be considered independently.

Let $w(r, t)\,dr$ be the probability of finding the particle at a distance between r and $r + dr$ from the origin at time t. Putting in (59.17) $M/\rho = 1$ and multiplying by the volume $4\pi r^2\,dr$ of the spherical shell, we find

$$w(r, t)\,dr = \frac{1}{2\sqrt{(\pi D^3 t^3)}}\exp(-r^2/4Dt)r^2\,dr. \tag{60.1}$$

Let us determine the mean square distance from the origin at time t. We have

$$\overline{r^2} = \int_0^\infty r^2 w(r, t)\,dr. \tag{60.2}$$

The result, using (60.1), is

$$\overline{r^2} = 6Dt. \tag{60.3}$$

Thus the mean distance travelled by the particle during any time is proportional to the square root of the time.

The diffusion coefficient for particles suspended in a fluid can be calculated from what is called their *mobility*. Let us suppose that some constant external force $\mathbf{f}$ (the force of gravity, for example) acts on the particles. In a steady state, the force acting on each particle must be balanced by the drag force exerted by the fluid on a moving particle. When the velocity is small, the drag force is proportional to it and is $\mathbf{v}/b$, say, where b is a constant. Equating this to the external force $\mathbf{f}$, we have

$$\mathbf{v} = b\mathbf{f}, \tag{60.4}$$

i.e. the velocity acquired by the particle under the action of the external force is proportional to that force. The constant b is called the *mobility*, and can in principle be calculated from the equations of fluid dynamics. For example, for spherical particles with radius R, the drag force is $6\pi\eta Rv$ (see (20.14)), and therefore the mobility is

$$b = 1/6\pi\eta R. \tag{60.5}$$

For non-spherical particles, the drag depends on the direction of motion; it can be written in the form $a_{ik}v_k$, where a_{ik} is a symmetrical tensor (see (20.15)). To calculate the mobility we have to average over all orientations of the particle; if a_1, a_2, a_3 are the principal values of the symmetrical tensor a_{ik}, then

$$b = \frac{1}{3}\left(\frac{1}{a_1} + \frac{1}{a_2} + \frac{1}{a_3}\right) \tag{60.6}$$

The mobility b is simply related to the diffusion coefficient D. To derive this relation, we write down the diffusion flux $\mathbf{i}$, which contains the usual term $-\rho D\,\mathbf{grad}\,c$ due to the concentration gradient (we suppose the temperature constant), and also a term involving the velocity acquired by the particle owing to the external forces. This latter term is evidently $\rho c\mathbf{v} = \rho cb\mathbf{f}$. Thus†

$$\mathbf{i} = -\rho D\,\mathbf{grad}\,c + \rho cb\mathbf{f}. \tag{60.7}$$

This can be rewritten as

$$\mathbf{i} = -\frac{\rho D}{(\partial\mu/\partial c)_{T,p}}\,\mathbf{grad}\,\mu + \rho cb\mathbf{f},$$

where μ is now the chemical potential of the suspended particles (which act as the solute). The dependence of this potential on the concentration (in a weak solution) is

$$\mu = T\log c + \psi(p, T)$$

(see *SP* 1, §87), so that

$$\mathbf{i} = -(\rho Dc/T)\,\mathbf{grad}\,\mu + \rho cb\mathbf{f}.$$

In thermodynamic equilibrium, there is no diffusion, and $\mathbf{i}$ must be zero. On the other hand, when an external field is present, the condition of equilibrium requires $\mu + U$ to be constant throughout the solution, where U is the potential energy of a suspended particle in that field. Then $\mathbf{grad}\,\mu = -\mathbf{grad}\,U = -\mathbf{f}$, and the equation $\mathbf{i} = 0$ gives

$$D = Tb. \tag{60.8}$$

This is *Einstein's relation* between the diffusion coefficient and the mobility.

Substituting (60.5) in (60.8), we find the following expression for the diffusion coefficient for spherical particles:

$$D = T/6\pi\eta R. \tag{60.9}$$

Besides the translatory Brownian motion and diffusion of suspended particles, we may consider also their rotary Brownian motion and diffusion. Just as the translatory diffusion coefficient is calculated in terms of the drag force, so the rotary diffusion coefficient can be expressed in terms of the moment of the forces on a particle executing a rotary movement in the fluid.

PROBLEMS

PROBLEM 1. Particles execute Brownian motion in a fluid bounded on one side by a plane wall; particles incident on the wall "adhere" to it. Determine the probability that a particle which is at a distance x_0 from the wall at time $t = 0$ will have adhered to it after a time t.

SOLUTION. The probability distribution $w(x, t)$ (where x is the distance from the wall) is determined by the diffusion equation, with the boundary condition $w = 0$ for $x = 0$ and the initial condition $w = \delta(x - x_0)$ for $t = 0$. Such a solution is given by formula (52.4) when T is replaced by w, χ by D, and $w_0(x')$ in the integrand by $\delta(x' - x_0)$. We then obtain

$$w(x, t) = \frac{1}{2\sqrt{(\pi Dt)}}\{\exp[-(x - x_0)^2/4Dt] - \exp[-(x + x_0)^2/4Dt]\}.$$

† Here c may be defined as the number of suspended particles per unit mass of the fluid, and $\mathbf{i}$ as their number flux density.

The probability of adhering to the wall per unit time is given by the diffusion flux $D\partial w/\partial x$ for $x = 0$, and the required probability $W(t)$ over the time t is

$$W(t) = D\int_0^t [\partial w/\partial x]_{x=0}\,dt.$$

Substituting for w, we find

$$W(t) = 1 - \text{erf}\,[x_0/2\sqrt{(Dt)}].$$

PROBLEM 2. Determine the order of magnitude of the time τ during which a particle suspended in a fluid turns through a large angle about its axis.

SOLUTION. The required time τ is that during which a particle in Brownian motion moves over a distance of the order of its linear dimension a. According to (60.3) we have $\tau \sim a^2/D$, and by (60.9) $D \sim T/\eta a$. Thus $\tau \sim \eta a^3/T$.

CHAPTER VII

SURFACE PHENOMENA

§61. Laplace's formula

In this chapter we shall study the phenomena which occur near the surface separating two continuous media (in reality, of course, the media are separated by a narrow transitional layer, but this is so thin that it may be regarded as a surface). If the surface of separation is curved, the pressures near it in the two media are different. To determine the pressure difference (called the *surface pressure*), we write down the condition that the two media be in thermodynamic equilibrium together, taking into account the properties of the surface of separation.

Let the surface of separation undergo an infinitesimal displacement. At each point of the undisplaced surface we draw the normal. The length of the segment of the normal lying between the points where it intersects the displaced and undisplaced surfaces is denoted by $\delta\zeta$. Then a volume element between the two surfaces is $\delta\zeta\, \mathrm{d}f$, where $\mathrm{d}f$ is a surface element. Let p_1 and p_2 be the pressures in the two media, and let $\delta\zeta$ be reckoned positive if the displacement of the surface is towards medium 2 (say). Then the work necessary to bring about the above change in volume is

$$\int (-p_1 + p_2)\, \delta\zeta\, \mathrm{d}f.$$

The total work δR done in displacing the surface is obtained by adding to this the work connected with the change in area of the surface. This part of the work is proportional to the change δf in the area of the surface, and is $\alpha\delta f$, where α is called the *surface-tension coefficient*. Thus the total work is

$$\delta R = -\int (p_1 - p_2)\, \delta\zeta\, \mathrm{d}f + \alpha\delta f. \tag{61.1}$$

The condition of thermodynamic equilibrium is, of course, that δR be zero.

Next, let R_1 and R_2 be the principal radii of curvature at a given point of the surface; we reckon R_1 and R_2 as positive if they are drawn into medium 1. Then the elements of length $\mathrm{d}l_1$ and $\mathrm{d}l_2$ on the surface in its principal sections receive increments $(\delta\zeta/R_1)\mathrm{d}l_1$ and $(\delta\zeta/R_2)\mathrm{d}l_2$ respectively when the surface undergoes an infinitesimal displacement; here $\mathrm{d}l_1$ and $\mathrm{d}l_2$ are regarded as elements of the circumference of circles with radii R_1 and R_2. Hence the surface element $\mathrm{d}f = \mathrm{d}l_1\, \mathrm{d}l_2$ becomes, after the displacement,

$$\mathrm{d}l_1 (1 + \delta\zeta/R_1)\mathrm{d}l_2 (1 + \delta\zeta/R_2) \cong \mathrm{d}l_1\, \mathrm{d}l_2 (1 + \delta\zeta/R_1 + \delta\zeta/R_2),$$

i.e. it changes by $\delta\zeta \mathrm{d}f (1/R_1 + 1/R_2)$. Hence we see that the total change in area of the surface of separation is

$$\delta f = \int \delta\zeta \left(\frac{1}{R_1} + \frac{1}{R_2}\right) \mathrm{d}f. \tag{61.2}$$

Substituting these expressions in (61.1) and equating to zero, we obtain the equilibrium condition in the form

$$\int \delta\zeta\left\{(p_1 - p_2) - \alpha\left(\frac{1}{R_1} + \frac{1}{R_2}\right)\right\} \mathrm{d}f = 0.$$

This condition must hold for every infinitesimal displacement of the surface, i.e. for all $\delta\zeta$. Hence the expression in braces must be identically equal to zero:

$$p_1 - p_2 = \alpha\left(\frac{1}{R_1} + \frac{1}{R_2}\right). \tag{61.3}$$

This is *Laplace's formula*, which gives the surface pressure.† We see that, if R_1 and R_2 are positive, $p_1 - p_2 > 0$. This means that the pressure is greater in the medium whose surface is convex. If $R_1 = R_2 = \infty$, i.e. the surface of separation is plane, the pressure is the same in either medium, as we should expect.

Let us apply formula (61.3) to investigate the mechanical equilibrium of two adjoining media. We assume that no external forces act, either on the surface of separation or on the media themselves. Then the pressure is constant in each body. Using formula (61.3), we can therefore write the equation of equilibrium as

$$\frac{1}{R_1} + \frac{1}{R_2} = \text{constant}. \tag{61.4}$$

Thus the sum of the curvatures must be a constant over any free surface of separation. If the whole surface is free, the condition (61.4) means that it must be spherical (for instance, the surface of a small drop, for which the effect of gravity may be neglected). If, however, the surface is supported along some curve (for instance, a film of liquid on a solid frame), its shape is less simple.

When the condition (61.4) is applied to the equilibrium of thin films supported on a solid frame, the constant on the right must be zero. For the sum $1/R_1 + 1/R_2$ must be the same everywhere on the free surface of the film, while on opposite sides of the film it must have opposite signs, since, if one side is convex, the other side is concave, and the radii of curvature are the same with opposite signs. Hence it follows that the equilibrium condition for a thin film is

$$\frac{1}{R_1} + \frac{1}{R_2} = 0. \tag{61.5}$$

Let us now consider the equilibrium condition on the surface of a medium in a gravitational field. We assume for simplicity that medium 2 is simply the atmosphere, whose pressure may be regarded as constant over the surface, and that medium 1 is an incompressible fluid. Then we have $p_2 = \text{constant}$, while p_1, the fluid pressure, is by (3.2) $p_1 = \text{constant} - \rho g z$, the coordinate z being measured vertically upwards. Thus the equilibrium condition becomes

$$\frac{1}{R_1} + \frac{1}{R_2} + \frac{g\rho z}{\alpha} = \text{constant}. \tag{61.6}$$

† The proof given here differs from that in *SP* 1, §156, essentially only in that here we are considering a surface of separation having any shape, not necessarily spherical.

It should be mentioned that, to determine the equilibrium form of the surface of the fluid in particular cases, it is usually convenient to use the condition of equilibrium, not in the form (61.6), but by directly solving the variational problem of minimizing the total free energy. The internal free energy of an incompressible fluid depends only on the volume of the fluid, and not on the shape of its surface. The latter affects, firstly, the surface free energy $\int \alpha \, df$ and, secondly, the energy in the external field (gravity), which is $g\rho \int z \, dV$. Thus the equilibrium condition can be written

$$\alpha \int df + g\rho \int z \, dV = \text{minimum}. \tag{61.7}$$

The minimum is to be determined subject to the condition

$$\int dV = \text{constant}, \tag{61.8}$$

which expresses the fact that the volume of the fluid is constant.

The constants α, ρ and g appear in the equilibrium conditions (61.6) and (61.7) only in the form $\alpha/g\rho$. This ratio has the dimensions cm^2. The length

$$a = \sqrt{(2\alpha/g\rho)} \tag{61.9}$$

is called the *capillary constant* for the substance concerned.† The shape of the fluid surface is determined by this quantity alone. If the capillary constant is large compared with the dimension of the medium, we may neglect gravity in determining the shape of the surface.

In order to find the shape of the surface from the condition (61.4) or (61.6), we need formulae which determine the radii of curvature, given the shape of the surface. These formulae are obtained in differential geometry, but in the general case they are somewhat complicated. They are considerably simplified when the surface deviates only slightly from a plane. We shall derive the appropriate formula directly, without using the general results of differential geometry.

Let $z = \zeta(x, y)$ be the equation of the surface; we suppose that ζ is everywhere small, i.e. that the surface deviates only slightly from the plane $z = 0$. As is well known, the area f of the surface is given by the integral

$$f = \int \sqrt{\left[1 + \left(\frac{\partial \zeta}{\partial x}\right)^2 + \left(\frac{\partial \zeta}{\partial y}\right)^2\right]} \, dx \, dy,$$

or, for small ζ, approximately by

$$f = \int \left[1 + \frac{1}{2}\left(\frac{\partial \zeta}{\partial x}\right)^2 + \frac{1}{2}\left(\frac{\partial \zeta}{\partial y}\right)^2\right] dx \, dy. \tag{61.10}$$

The variation δf is

$$\delta f = \int \left\{\frac{\partial \zeta}{\partial x}\frac{\partial \delta \zeta}{\partial x} + \frac{\partial \zeta}{\partial y}\frac{\partial \delta \zeta}{\partial y}\right\} dx \, dy.$$

† For water (e.g.), $a = 0{\cdot}39$ cm at 20°C.

Integrating by parts, we find

$$\delta f = -\int \left(\frac{\partial^2 \zeta}{\partial x^2} + \frac{\partial^2 \zeta}{\partial y^2} \right) \delta \zeta \, dx \, dy.$$

Comparing this with (61.2), we obtain

$$\frac{1}{R_1} + \frac{1}{R_2} = -\left(\frac{\partial^2 \zeta}{\partial x^2} + \frac{\partial^2 \zeta}{\partial y^2} \right) \tag{61.11}$$

This is the required formula; it determines the sum of the curvatures of a slightly curved surface.

When three adjoining media are in equilibrium, the surfaces of separation are such that the resultant of the surface-tension forces is zero on the common line of intersection. This condition implies that the surfaces of separation must intersect at angles (called *angles of contact*) determined by the values of the surface-tension coefficients.

Finally, let us consider the question of the boundary conditions that must be satisfied at the boundary between two fluids in motion, when the surface-tension forces are taken into account. If the latter forces are neglected, we have at the boundary between the fluids $n_k(\sigma_{2,ik} - \sigma_{1,ik}) = 0$, which expresses the equality of the forces of viscous friction on the surface of each fluid. When the surface tension is included, we have to add on the right-hand side a force determined in magnitude by Laplace's formula and directed along the normal:

$$n_k \sigma_{2,ik} - n_k \sigma_{1,ik} = \alpha \left(\frac{1}{R_1} + \frac{1}{R_2} \right) n_i. \tag{61.12}$$

This equation can also be written

$$(p_1 - p_2) n_i = (\sigma'_{1,ik} - \sigma'_{2,ik}) n_k + \alpha \left(\frac{1}{R_1} + \frac{1}{R_2} \right) n_i. \tag{61.13}$$

If the two fluids are both ideal, the viscous stresses σ'_{ik} are zero, and we return to the simple equation (61.3).

The condition (61.13), however, is still not completely general. The reason is that the surface-tension coefficient α may not be constant over the surface (for example, on account of a variation in temperature). Then, besides the normal force (which is zero for a plane surface), there is another force tangential to the surface. Just as there is a body force $-\mathbf{grad}\, p$ per unit volume in cases where the pressure is not uniform, so we have here a tangential force $\mathbf{f}_t = \mathbf{grad}\, \alpha$ per unit area of the surface of separation. In this case we take the positive gradient, because the surface-tension forces tend to reduce the area of the surface, whereas the pressure forces tend to increase the volume. Adding this force to the right-hand side of equation (61.13), we obtain the boundary condition

$$\left[p_1 - p_2 - \alpha \left(\frac{1}{R_1} + \frac{1}{R_2} \right) \right] n_i = (\sigma'_{1,ik} - \sigma'_{2,ik}) n_k + \frac{\partial \alpha}{\partial x_i}; \tag{61.14}$$

the unit normal vector $\mathbf{n}$ is directed into medium 1. We notice that this condition can be satisfied only for a viscous fluid: in an ideal fluid, $\sigma'_{ik} = 0$ and the left-hand side of equation (61.14) is a vector along the normal, while the right-hand side is in this case a tangential vector. This equality cannot hold, except of course in the trivial case where both sides are zero.

PROBLEMS

Problem 1. Determine the shape of a film of liquid supported on two circular frames with their centres on a line perpendicular to their planes, which are parallel; Fig. 41 shows a cross-section of the film.

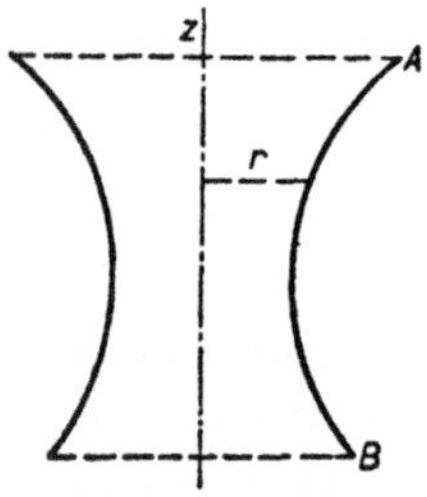

Fig. 41

Solution. The problem amounts to that of finding the surface having the smallest area that can be formed by the revolution about the line $r = 0$ of a curve $z = z(r)$ which passes between two given points A and B. The area of a surface of rotation is

$$f = 2\pi \int_{z_1}^{z_2} F(r, r')\,\mathrm{d}z, \quad F = r(1 + r'^2)^{\frac{1}{2}},$$

where $r' \equiv \mathrm{d}r/\mathrm{d}z$. The first integral of Euler's equation for the problem of minimizing such an integral (with F independent of z) is

$$F - r'\,\partial F/\partial r' = \text{constant}.$$

In the present case, this gives

$$r = c_1(1 + r'^2)^{\frac{1}{2}},$$

whence we have by integration $r = c_1 \cosh[(z - c_2)/c_1]$. Thus the required surface (called a *catenoid*) is that formed by the revolution of a catenary. The constants c_1 and c_2 must be chosen so that the curve $r(z)$ passes through the given points A and B. The value of c_2 depends only on the choice of the origin of z. For the constant c_1, however, two values are obtained, of which the larger must be chosen (the smaller does not give a minimum of the integral).

When the distance h between the frames increases, it reaches a value for which the equation for the constant c_1 no longer has a real root. For greater values of h, only the shape consisting of one film on each frame is stable. For example, for two frames with equal radius R the catenoid form is impossible for a distance h between the frames greater than $1{\cdot}33R$.

Problem 2. Determine the shape of the surface of a fluid in a gravitational field and bounded on one side by a vertical plane wall. The angle of contact between the fluid and the wall is θ (Fig. 42).

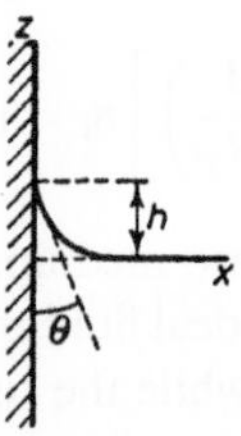

Fig. 42

SOLUTION. We take the coordinate axes as shown in Fig. 42. The plane $x = 0$ is the plane of the wall, and $z = 0$ is the plane of the fluid surface far from the wall. The radii of curvature of the surface $z = z(x)$ are $R_1 = \infty$, $R_2 = -(1+z'^2)^{\frac{3}{2}}/z''$, so that equation (61.6) becomes

$$\frac{2z}{a^2} - \frac{z''}{(1+z'^2)^{\frac{3}{2}}} = \text{constant}, \tag{1}$$

where a is the capillary constant. For $x = \infty$ we must have $z = 0$, $1/R_2 = 0$, and the constant is therefore zero. A first integral of the resulting equation is

$$\frac{1}{\sqrt{(1+z'^2)}} = A - \frac{z^2}{a^2}. \tag{2}$$

From the condition at infinity ($z = z' = 0$ for $x = \infty$) we have $A = 1$. A second integration gives

$$x = -\frac{a}{\sqrt{2}}\cosh^{-1}\frac{\sqrt{2}a}{z} + a\sqrt{\left(2 - \frac{z^2}{a^2}\right)} + x_0.$$

The constant x_0 must be chosen so that, at the surface of the wall ($x = 0$), we have $z' = -\cot\theta$ or, by (2), $z = h$, where $h = a\sqrt{(1-\sin\theta)}$ is the height to which the fluid rises at the wall itself.

PROBLEM 3. Determine the shape of the surface of a fluid rising between two parallel vertical flat plates (Fig. 43).

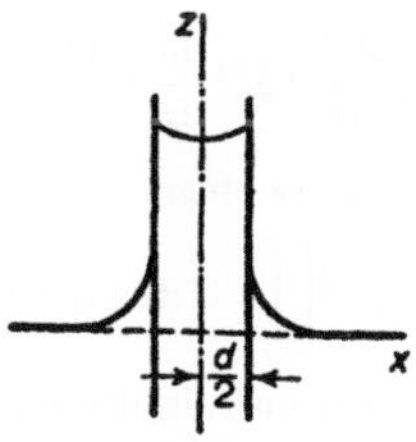

FIG. 43

SOLUTION. We take the yz-plane half-way between the two plates, and the xy-plane to coincide with the fluid surface far from the plates. In equation (1) of Problem 2, which gives the condition of equilibrium and is therefore valid everywhere on the surface of the fluid (both between the plates and elsewhere), the conditions at $x = \infty$ again give the constant as zero. In the integral (2), the constant A is now different according as $|x| > \frac{1}{2}d$ or $|x| < \frac{1}{2}d$ (the function $z(x)$ having a discontinuity for $|x| = \frac{1}{2}d$). For the space between the plates, the conditions are $z' = 0$ for $x = 0$ and $z' = \cot\theta$ for $x = \frac{1}{2}d$, where θ is the angle of contact. According to (2) we have for the heights $z_0 = z(0)$ and $z_1 = z(\frac{1}{2}d)$: $z_0 = a\sqrt{(A-1)}$, $z_1 = a\sqrt{(A-\sin\theta)}$. Integrating (2), we obtain

$$x = \int\limits_{z_0}^{z} \frac{(A - z^2/a^2)\,\mathrm{d}z}{\sqrt{[1-(A-z^2/a^2)^2]}} = \tfrac{1}{2}a \int\limits_{0}^{a\sqrt{(A-\cos\xi)}} \frac{\cos\xi\,\mathrm{d}\xi}{\sqrt{(A-\cos\xi)}},$$

where ξ is a new variable related to z by $z = a\sqrt{(A-\cos\xi)}$. This is an elliptic integral, and cannot be expressed in terms of elementary functions. The constant A is found from the condition that $z = z_1$ for $x = \frac{1}{2}d$, or

$$d = a\int\limits_{0}^{\frac{1}{2}\pi-\theta} \frac{\cos\xi\,\mathrm{d}\xi}{\sqrt{(A-\cos\xi)}}.$$

The formulae obtained above give the shape of the fluid surface in the space between the plates. As $d \to 0$, A tends to infinity. Hence we have for $d \ll a$

$$d \cong \frac{a}{\sqrt{A}}\int\limits_{0}^{\frac{1}{2}\pi-\theta} \cos\xi\,\mathrm{d}\xi = \frac{a}{\sqrt{A}}\cos\theta,$$

or $A = (a^2/d^2)\cos^2\theta$. The height to which the fluid rises is $z_0 \cong z_1 \cong (a^2/d)\cos\theta$; this formula can also be obtained directly, of course.

PROBLEM 4. A thin non-uniformly heated layer of fluid rests under gravity on a horizontal plane solid surface; its temperature is a given function of the coordinate x in the plane, and (because the layer is thin) may be supposed independent of the coordinate z across the layer. The non-uniform heating results in the occurrence of a steady flow, and its thickness ζ consequently varies in the x-direction. Determine the function $\zeta(x)$.

SOLUTION. The fluid density ρ and the surface tension α are, together with the temperature, known functions of x. The fluid pressure $p = p_0 + \rho g(\zeta - z)$, where p_0 is the atmospheric pressure (the pressure on the free surface); the variation of pressure due to the curvature of the surface may be neglected. The fluid velocity in the layer may be supposed everywhere parallel to the x-axis. The equation of motion is

$$\eta \partial^2 v/\partial z^2 = \partial p/\partial x = g[\mathrm{d}(\rho\zeta)/\mathrm{d}x - z\,\mathrm{d}\rho/\mathrm{d}x]. \tag{1}$$

On the solid surface ($z = 0$) we have $v = 0$, while on the free surface ($z = \zeta$) the boundary condition (61.14) must be fulfilled; in this case it is $\eta[\mathrm{d}v/\mathrm{d}z]_{z=\zeta} = \mathrm{d}\alpha/\mathrm{d}x$. Integrating equation (1) with these conditions, we obtain

$$\eta v = gz(\zeta - \tfrac{1}{2}z)\,\mathrm{d}(\rho\zeta)/\mathrm{d}x - \tfrac{1}{6}gz(3\zeta^2 - z^2)\,\mathrm{d}\rho/\mathrm{d}x - z\,\mathrm{d}\alpha/\mathrm{d}x. \tag{2}$$

Since the flow is steady, the total mass flux through a cross-section of the layer must be zero:

$$\int_0^\zeta v\,\mathrm{d}z = 0.$$

Substituting (2), we find

$$\tfrac{1}{3}\rho\frac{\mathrm{d}\zeta^2}{\mathrm{d}x} + \tfrac{1}{4}\zeta^2\frac{\mathrm{d}\rho}{\mathrm{d}x} = \frac{1}{g}\frac{\mathrm{d}\alpha}{\mathrm{d}x},$$

which determines the function $\zeta(x)$. Integrating, we obtain

$$g\zeta^2 = 3\rho^{-\frac{3}{4}}\left[\int \rho^{-\frac{1}{4}}\,\mathrm{d}\alpha + \text{constant}\right]. \tag{3}$$

If the temperature (and therefore ρ and α) varies only slightly, then (3) can be written

$$\zeta^2 = \zeta_0^2(\rho_0/\rho)^{\frac{3}{4}} + 3(\alpha - \alpha_0)/\rho g,$$

where ζ_0 is the value of ζ at a point where $\rho = \rho_0$ and $\alpha = \alpha_0$.

§62. Capillary waves

Fluid surfaces tend to assume an equilibrium shape, both under the action of the force of gravity and under that of surface-tension forces. In studying waves on the surface of a fluid in §12, we did not take the latter forces into account. We shall see below that capillarity has an important effect on gravity waves with short wavelength.

As in §12, we suppose the amplitude of the oscillations small compared with the wavelength. For the velocity potential we have as before the equation $\triangle\phi = 0$. The condition at the surface of the fluid is now different, however: the pressure difference between the two sides of the surface is not zero, as we supposed in §12, but is given by Laplace's formula (61.3).

We denote by ζ the z coordinate of a point on the surface. Since ζ is small, we can use the expression (61.11), and write Laplace's formula as

$$p - p_0 = -\alpha\left(\frac{\partial^2\zeta}{\partial x^2} + \frac{\partial^2\zeta}{\partial y^2}\right).$$

Here p is the pressure in the fluid near the surface, and p_0 is the constant external pressure. For p we substitute, according to (12.2),

$$p = -\rho g\zeta - \rho\partial\phi/\partial t,$$

obtaining

$$\rho g \zeta + \rho \frac{\partial \phi}{\partial t} - \alpha \left(\frac{\partial^2 \zeta}{\partial x^2} + \frac{\partial^2 \zeta}{\partial y^2} \right) = 0;$$

for the same reasons as in §12, we can omit the constant p_0 if we redefine ϕ. Differentiating this relation with respect to t, and replacing $\partial \zeta / \partial t$ by $\partial \phi / \partial z$, we obtain the boundary condition on the potential ϕ:

$$\rho g \frac{\partial \phi}{\partial z} + \rho \frac{\partial^2 \phi}{\partial t^2} - \alpha \frac{\partial}{\partial z} \left(\frac{\partial^2 \phi}{\partial x^2} + \frac{\partial^2 \phi}{\partial y^2} \right) = 0 \quad \text{for} \quad z = 0. \tag{62.1}$$

Let us consider a plane wave propagated in the direction of the x-axis. As in §12, we obtain a solution in the form $\phi = Ae^{kz} \cos(kx - \omega t)$. The relation between k and ω is now obtained from the boundary condition (62.1), and is

$$\omega^2 = gk + \alpha k^3 / \rho \tag{62.2}$$

(W. Thomson 1871).

We see that, for long wavelengths such that $k \ll \sqrt{(g\rho/\alpha)}$, or $k \ll 1/a$ (where a is the capillary constant), the effect of capillarity may be neglected, and we have a pure gravity wave. In the opposite case of short wavelengths, the effect of gravity may be neglected. Then

$$\omega^2 = \alpha k^3 / \rho. \tag{62.3}$$

Such waves are called *capillary waves* or *ripples*. Intermediate cases are referred to as *capillary gravity waves*.

Let us also determine the characteristic oscillations of a spherical drop of incompressible fluid under the action of capillary forces. The oscillations cause the surface of the drop to deviate from the spherical form. As usual, we shall suppose the amplitude of the oscillations to be small.

We begin by determining the value of the sum $1/R_1 + 1/R_2$ for a surface slightly different from that of a sphere. Here we proceed as in the derivation of formula (61.11). The area of a surface given in spherical polar coordinates† r, θ, ϕ by a function $r = r(\theta, \phi)$ is

$$f = \int_0^{2\pi} \int_0^{\pi} \sqrt{\left[r^2 + \left(\frac{\partial r}{\partial \theta} \right)^2 + \frac{1}{\sin^2 \theta} \left(\frac{\partial r}{\partial \phi} \right)^2 \right]} r \sin \theta \, d\theta \, d\phi. \tag{62.4}$$

A spherical surface is given by $r = \text{constant} \equiv R$ (where R is the radius of the sphere), and a neighbouring surface by $r = R + \zeta$, where ζ is small. Substituting in (62.4), we obtain approximately

$$f = \int_0^{2\pi} \int_0^{\pi} \left\{ (R + \zeta)^2 + \frac{1}{2} \left[\left(\frac{\partial \zeta}{\partial \theta} \right)^2 + \frac{1}{\sin^2 \theta} \left(\frac{\partial \zeta}{\partial \phi} \right)^2 \right] \right\} \sin \theta \, d\theta \, d\phi.$$

† In the remainder of this section ϕ denotes the azimuthal angle, and we denote the velocity potential by ψ.

Let us find the variation δf in the area when ζ changes. We have

$$\delta f = \int_0^{2\pi}\int_0^{\pi}\left\{2(R+\zeta)\delta\zeta + \frac{\partial\zeta}{\partial\theta}\frac{\partial\delta\zeta}{\partial\theta} + \frac{1}{\sin^2\theta}\frac{\partial\zeta}{\partial\phi}\frac{\partial\delta\zeta}{\partial\phi}\right\}\sin\theta\, d\theta\, d\phi.$$

Integrating the second term by parts with respect to θ, and the third by parts with respect to ϕ, we obtain

$$\delta f = \int_0^{2\pi}\int_0^{\pi}\left\{2(R+\zeta) - \frac{1}{\sin\theta}\frac{\partial}{\partial\theta}\left(\sin\theta\frac{\partial\zeta}{\partial\theta}\right) - \frac{1}{\sin^2\theta}\frac{\partial^2\zeta}{\partial\phi^2}\right\}\delta\zeta\sin\theta\, d\theta\, d\phi.$$

If we divide the expression in braces by $R(R+2\zeta)$, the resulting coefficient of $\delta\zeta df \cong \delta\zeta R(R+2\zeta)\sin\theta\, d\theta\, d\phi$ in the integrand is, by formula (61.2), just the required sum of the curvatures, correct to terms of the first order in ζ. Thus we find

$$\frac{1}{R_1}+\frac{1}{R_2} = \frac{2}{R} - \frac{2\zeta}{R^2} - \frac{1}{R^2}\left\{\frac{1}{\sin^2\theta}\frac{\partial^2\zeta}{\partial\phi^2} + \frac{1}{\sin\theta}\frac{\partial}{\partial\theta}\left(\sin\theta\frac{\partial\zeta}{\partial\theta}\right)\right\}. \tag{62.5}$$

The first term corresponds to a spherical surface, for which $R_1 = R_2 = R$.

The velocity potential ψ satisfies Laplace's equation $\triangle\psi = 0$, with a boundary condition at $r = R$ like that for a plane surface:

$$\rho\frac{\partial\psi}{\partial t} + \alpha\left\{\frac{2}{R} - \frac{2\zeta}{R^2} - \frac{1}{R^2}\left[\frac{1}{\sin\theta}\frac{\partial}{\partial\theta}\left(\sin\theta\frac{\partial\zeta}{\partial\theta}\right) + \frac{1}{\sin^2\theta}\frac{\partial^2\zeta}{\partial\phi^2}\right]\right\} + p_0 = 0.$$

The constant $p_0 + 2\alpha/R$ can again be omitted; differentiating with respect to time and putting $\partial\zeta/\partial t = v_r = \partial\psi/\partial r$, we have finally the boundary condition on ψ:

$$\rho\frac{\partial^2\psi}{\partial t^2} - \frac{\alpha}{R^2}\left\{2\frac{\partial\psi}{\partial r} + \frac{\partial}{\partial r}\left[\frac{1}{\sin\theta}\frac{\partial}{\partial\theta}\left(\sin\theta\frac{\partial\psi}{\partial\theta}\right) + \frac{1}{\sin^2\theta}\frac{\partial^2\psi}{\partial\phi^2}\right]\right\} = 0$$

$$\text{for} \quad r = R. \tag{62.6}$$

We shall seek a solution in the form of a stationary wave: $\psi = e^{-i\omega t}f(r, \theta, \phi)$, where the function f satisfies Laplace's equation, $\triangle f = 0$. As is well known, any solution of Laplace's equation can be represented as a linear combination of what are called *volume spherical harmonic functions* $r^l Y_{lm}(\theta, \phi)$, where $Y_{lm}(\theta, \phi)$ are Laplace's spherical harmonics: $Y_{lm}(\theta, \phi) = P_l^m(\cos\theta)e^{im\phi}$. Here $P_l^m(\cos\theta) = \sin^m\theta\, d^m P_l(\cos\theta)/d(\cos\theta)^m$ is what is called an *associated Legendre function*, $P_l(\cos\theta)$ being the Legendre polynomial of order l. As is well known, l takes all integral values from zero upwards, while m takes the values 0, ± 1, $\pm 2, \ldots, \pm l$.

Accordingly, we seek a particular solution of the problem in the form

$$\psi = Ae^{-i\omega t}r^l P_l^m(\cos\theta)e^{im\phi}. \tag{62.7}$$

The frequency ω must be such as to satisfy the boundary condition (62.6). Substituting the expression (62.7) and using the fact that the spherical harmonics Y_{lm} satisfy

$$\frac{1}{\sin\theta}\frac{\partial}{\partial\theta}\left(\sin\theta\frac{\partial Y_{lm}}{\partial\theta}\right) + \frac{1}{\sin^2\theta}\frac{\partial^2 Y_{lm}}{\partial\phi^2} + l(l+1)Y_{lm} = 0,$$

we find (cancelling ψ)

$$\rho\omega^2 + l\alpha[2 - l(l+1)]/R^3 = 0,$$

or

$$\omega^2 = \alpha l(l-1)(l+2)/\rho R^3 \tag{62.8}$$

(Rayleigh 1879).

This formula gives the eigenfrequencies of capillary oscillations of a spherical drop. We see that it depends only on l, and not on m. To a given l, however, there correspond $2l+1$ different functions (62.7). Thus each of the frequencies (62.8) corresponds to $2l+1$ different oscillations. Independent oscillations having the same frequency are said to be *degenerate*; in this case we have $(2l+1)$-fold degeneracy.

The expression (62.8) vanishes for $l = 0$ and $l = 1$. The value $l = 0$ would correspond to radial oscillations, i.e. to spherically symmetrical pulsations of the drop; in an incompressible fluid such oscillations are clearly impossible. For $l = 1$ the motion is simply a translatory motion of the drop as a whole. The smallest possible frequency of oscillations of the drop corresponds to $l = 2$, and is

$$\omega_{\min} = \sqrt{(8\alpha/\rho R^3)}. \tag{62.9}$$

PROBLEMS

PROBLEM 1. Determine the frequency as a function of the wave number for capillary gravity waves on the surface of a liquid with depth h.

SOLUTION. Substituting in the condition (62.1) $\phi = A\cos(kx - \omega t)\cosh k(z+h)$ (cf. §12, Problem 1), we obtain $\omega^2 = (gk + \alpha k^3/\rho)\tanh kh$. For $kh \gg 1$ we return to formula (62.2), while for long waves ($kh \ll 1$) we have $\omega^2 = ghk^2 + \alpha hk^4/\rho$.

PROBLEM 2. Determine the damping coefficient for capillary waves.

SOLUTION. Substituting (62.3) in (25.5), we find $\gamma = 2\eta k^2/\rho = 2\eta\omega^{4/3}/\rho^{1/3}\alpha^{2/3}$.

PROBLEM 3. Find the condition for the stability of a horizontal tangential discontinuity in a gravitational field, taking account of surface tension (the fluids on the two sides of the surface of discontinuity being supposed different (W. Thomson 1871).

SOLUTION. Let U be the velocity of the upper fluid relative to the lower. On the original flow we superpose a perturbation periodic in the horizontal direction, and seek the velocity potential in the form

$$\phi = Ae^{kz}\cos(kx - \omega t) \text{ in the lower fluid,}$$

$$\phi' = A'e^{-kz}\cos(kx - \omega t) + Ux \text{ in the upper fluid.}$$

For the lower fluid we have on the surface of discontinuity $v_z = \partial\phi/\partial z = \partial\zeta/\partial t$, where ζ is a vertical coordinate in the surface of discontinuity, and for the upper fluid

$$v'_z = \partial\phi'/\partial z = U\partial\zeta/\partial x + \partial\zeta/\partial t.$$

The condition of equal pressures in the two fluids at the surface of discontinuity is

$$\rho\partial\phi/\partial t + \rho g\zeta - \alpha\partial^2\zeta/\partial x^2 = \rho'\partial\phi'/\partial t + \rho' g\zeta + \tfrac{1}{2}\rho'(v'^2 - U^2);$$

only terms of the first order in A' need be retained in expanding the expression $v'^2 - U^2$. We seek the displacement ζ in the form $\zeta = a\sin(kx - \omega t)$. Substituting ϕ, ϕ' and ζ in the above three conditions for $z = 0$, we obtain three equations from which a, A and A' can be eliminated, leaving

$$\omega = \frac{k\rho' U}{\rho+\rho'} \pm \sqrt{\left[\frac{kg(\rho-\rho')}{\rho+\rho'} - \frac{k^2\rho\rho' U^2}{(\rho+\rho')^2} + \frac{\alpha k^3}{\rho+\rho'}\right]}.$$

In order that this expression should be real for all k, it is necessary that

$$U^4 \leqslant 4\alpha g(\rho-\rho')(\rho+\rho')^2/\rho^2\rho'^2.$$

If this condition does not hold, there are complex ω with a positive imaginary part, and the motion is unstable.

§63. The effect of adsorbed films on the motion of a liquid

The presence on the surface of a liquid of a film of adsorbed material may have a considerable effect on the hydrodynamical properties of the surface. The reason is that, when the shape of the surface changes with the motion of the liquid, the film is stretched or compressed, i.e. the surface concentration of the adsorbed substance is changed. These changes result in the appearance of additional forces which have to be taken into account in the boundary conditions at the free surface.

Here we shall consider only adsorbed films of substances which may be regarded as insoluble in the liquid. This means that the substance is entirely on the surface, and does not penetrate into the liquid. If the adsorbed substance is appreciably soluble, it is necessary to take into account the diffusion of it between the surface film and the volume of the liquid when the concentration of the film varies.

When the adsorbed material is present, the surface-tension coefficient α is a function of the surface concentration of the material (the amount of it per unit surface area), which we denote by γ. If γ varies over the surface, then the coefficient α is also a function of the coordinates in the surface. The boundary condition at the surface of the liquid therefore includes a tangential force, which we have already discussed at the end of §61 (equation (61.14)). In the present case, the gradient of α can be expressed in terms of the surface concentration gradient, so that the tangential force on the surface is

$$\mathbf{f}_t = (\partial\alpha/\partial\gamma)\,\mathbf{grad}\,\gamma. \tag{63.1}$$

It has been mentioned in §61 that the boundary condition (61.14), in which this force is taken into account, can be satisfied only for a viscous fluid. Hence it follows that, in cases where the viscosity of the liquid is small, and unimportant as regards the phenomenon under consideration, the presence of the film can be ignored.

To determine the motion of a liquid covered by a film we must add to the equations of motion, with the boundary condition (61.14), a further equation, since we now have another unknown quantity, the surface concentration γ. This further equation is an equation of continuity, expressing the fact that the total amount of adsorbed material in the film is unchanged. The actual form of the equation depends on the shape of the surface. If the latter is plane, then the equation is evidently

$$\partial\gamma/\partial t + \partial(\gamma v_x)/\partial x + \partial(\gamma v_y)/\partial y = 0, \tag{63.2}$$

where all quantities have their values at the surface (taken as the xy-plane).

The solution of problems of the motion of a liquid covered by an adsorbed film is considerably simplified in cases where the film may be supposed incompressible, i.e. we may assume that the area of any surface element of the film remains constant during the motion.

An example of the important hydrodynamic effects of an adsorbed film is given by the motion of a gas bubble in a viscous liquid. If there is no film on the surface of the bubble, the gas inside it moves also, and the drag force exerted on the bubble by the liquid is not the same as the drag on a solid sphere with the same radius (see §20, Problem 2). If, however, the bubble is covered by a film of adsorbed material, it is clear from symmetry that the film remains at rest when the bubble moves. For a motion in the film could occur only along meridian lines on the bubble surface, and the result would be that material would continually accumulate at one of the poles (since the adsorbed material does not penetrate into the liquid or the gas); this is impossible. Besides the velocity of the film, the gas velocity at the surface of the bubble must also be zero, and with this boundary condition the gas in the bubble must be entirely at rest. Thus a bubble covered by a film moves like a solid sphere and, in particular, the drag on it (for small Reynolds numbers) is given by Stokes' formula. This result is due to V. G. Levich.

PROBLEMS

Problem 1. Two vessels are joined by a long deep channel with width a and length l, with plane parallel walls. The surface of the liquid in the system is covered by an adsorbed film, and the surface concentrations γ_1 and γ_2 of the film in the two vessels are different. There results a motion near the surface of the liquid in the channel. Determine the amount of film material transported by this motion.

Solution. We take the plane of one wall of the channel as the xz-plane, and the surface of the liquid as the xy-plane, so that the x-axis is along the channel; the liquid is in the region $z < 0$. There is no pressure gradient, so that the equation of steady flow is (cf. §17)

$$\frac{\partial^2 v}{\partial y^2}+\frac{\partial^2 v}{\partial z^2}=0, \tag{1}$$

where v is the liquid velocity, which is evidently in the x-direction. There is a concentration gradient $d\gamma/dx$ along the channel. At the surface of the liquid in the channel we have the boundary condition

$$\eta \partial v/\partial z = d\alpha/dx \quad \text{for} \quad z=0. \tag{2}$$

At the channel walls the liquid must be at rest, i.e.

$$v=0 \quad \text{for} \quad y=0 \quad \text{and} \quad y=a. \tag{3}$$

The channel depth is supposed infinite, and so

$$v=0 \quad \text{for} \quad z\to-\infty. \tag{4}$$

Particular solutions of equation (1) which satisfy the conditions (3) and (4) are

$$v_n = \text{constant}\times \exp[(2n+1)\pi z/a]\sin(2n+1)\pi y/a,$$

with n integral. The condition (2) is satisfied by the sum

$$v=\frac{4a}{\eta\pi^2}\frac{d\alpha}{dx}\sum_{n=0}^{\infty}\frac{\exp[(2n+1)\pi z/a]\sin(2n+1)\pi y/a}{(2n+1)^2}.$$

The amount of film material transferred per unit time is

$$Q=\int_0^a \gamma[v]_{z=0}\,dy=\frac{8a^2}{\eta\pi^3}\left(\sum_{n=0}^{\infty}\frac{1}{(2n+1)^3}\right)\gamma\frac{d\alpha}{dx},$$

the motion being in the direction of α increasing. The value of Q must obviously be constant along the channel. Hence we can write

$$\gamma\frac{d\alpha}{dx}=\text{constant}\equiv\frac{1}{l}\int_0^l \gamma\frac{d\alpha}{dx}dx=\frac{1}{l}\int_{\alpha_2}^{\alpha_1}\gamma\,d\alpha,$$

where $\alpha_1=\alpha(\gamma_1)$, $\alpha_2=\alpha(\gamma_2)$, and we assume that $\alpha_1>\alpha_2$. Thus we have finally

$$Q=\frac{8a^2}{\eta l\pi^3}\left(\sum_{n=0}^{\infty}\frac{1}{(2n+1)^3}\right)\int_{\alpha_2}^{\alpha_1}\gamma\,d\alpha=0{\cdot}27\frac{a^2}{\eta l}\int_{\alpha_2}^{\alpha_1}\gamma\,d\alpha.$$

Problem 2. Determine the damping coefficient for capillary waves on the surface of a liquid covered by an adsorbed film.

Solution. If the viscosity of the liquid is not too great, the stretching (tangential) forces exerted on the film by the liquid are small, and the film may therefore be regarded as incompressible. Accordingly, we can calculate the energy dissipation as if it took place at a solid wall, i.e. from formula (24.14). Writing the velocity potential in the form

$$\phi = \phi_0 e^{ikx - i\omega t} e^{-kz},$$

we obtain for the dissipation per unit area of the surface

$$\overline{\dot{E}}_{\text{kin}} = -\sqrt{(\tfrac{1}{8}\rho\eta\omega)}\,|k\phi_0|^2.$$

The total energy (also per unit area) is

$$\overline{E} = \rho \int \overline{v^2}\, dz = \tfrac{1}{2}\rho|k\phi_0|^2/k.$$

The damping coefficient is (using (62.3))

$$\gamma = \frac{\omega^{7/6}\eta^{1/2}}{2\sqrt{2}\alpha^{1/3}\rho^{1/6}} = \frac{k^{7/4}\eta^{1/2}\alpha^{1/4}}{2\sqrt{2}\rho^{3/4}}.$$

The ratio of this quantity to the damping coefficient for capillary waves on a clean surface (§62, Problem 2) is $(\alpha\rho/k\eta^2)^{1/4}/4\sqrt{2}$, and is large compared with unity unless the wavelength is extremely short. Thus the presence of an adsorbed film on the surface of a liquid leads to a marked increase in the damping coefficient.

CHAPTER VIII

SOUND

§64. Sound waves

We proceed now to the study of the flow of compressible fluids, and begin by investigating small oscillations; an oscillatory motion with small amplitude in a compressible fluid is called a *sound wave.* At each point in the fluid, a sound wave causes alternate compression and rarefaction.

Since the oscillations are small, the velocity $\mathbf{v}$ is small also, so that the term $(\mathbf{v}\cdot\mathbf{grad})\mathbf{v}$ in Euler's equation may be neglected. For the same reason, the relative changes in the fluid density and pressure are small. We can write the variables p and ρ in the form

$$p = p_0 + p', \quad \rho = \rho_0 + \rho', \tag{64.1}$$

where ρ_0 and p_0 are the constant equilibrium density and pressure, and ρ' and p' are their variations in the sound wave ($\rho' \ll \rho_0, p' \ll p_0$). The equation of continuity $\partial\rho/\partial t + \operatorname{div}(\rho\mathbf{v}) = 0$, on substituting (64.1) and neglecting small quantities of the second order (ρ', p' and $\mathbf{v}$ being of the first order), becomes

$$\partial\rho'/\partial t + \rho_0 \operatorname{div}\mathbf{v} = 0. \tag{64.2}$$

Euler's equation

$$\partial\mathbf{v}/\partial t + (\mathbf{v}\cdot\mathbf{grad})\mathbf{v} = -(1/\rho)\mathbf{grad}\,p$$

reduces, in the same approximation, to

$$\partial\mathbf{v}/\partial t + (1/\rho_0)\mathbf{grad}\,p' = 0. \tag{64.3}$$

The condition that the linearized equations of motion (64.2) and (64.3) should be applicable to the propagation of sound waves is that the velocity of the fluid particles in the wave should be small compared with the velocity of sound: $v \ll c$. This condition can be obtained, for example, from the requirement that $\rho' \ll \rho_0$ (see formula (64.12) below).

Equations (64.2) and (64.3) contain the unknown functions $\mathbf{v}$, p' and ρ'. To eliminate one of these, we notice that a sound wave in an ideal fluid is, like any other motion in an ideal fluid, adiabatic. Hence the small change p' in the pressure is related to the small change ρ' in the density by

$$p' = (\partial p/\partial\rho_0)_s\rho'. \tag{64.4}$$

Substituting for ρ' according to this equation in (64.2), we find

$$\partial p'/\partial t + \rho_0(\partial p/\partial\rho_0)_s \operatorname{div}\mathbf{v} = 0. \tag{64.5}$$

The two equations (64.3) and (64.5), with the unknowns $\mathbf{v}$ and p', give a complete description of the sound wave.

In order to express all the unknowns in terms of one of them, it is convenient to introduce the velocity potential by putting $\mathbf{v} = \mathbf{grad}\,\phi$. We have from equation (64.3)

$$p' = -\rho\,\partial\phi/\partial t, \tag{64.6}$$

which relates p' and ϕ (here, and henceforward, we omit for brevity the suffix in p_0 and ρ_0). We then obtain from (64.5) the equation

$$\partial^2\phi/\partial t^2 - c^2\triangle\phi = 0, \tag{64.7}$$

which the potential ϕ must satisfy; here we have introduced the notation

$$c = \sqrt{(\partial p/\partial\rho)_s}. \tag{64.8}$$

An equation having the form (64.7) is called a *wave equation*. Applying the gradient operator to (64.7), we find that each of the three components of the velocity $\mathbf{v}$ satisfies an equation having the same form, and on differentiating (64.7) with respect to time we see that the pressure p' (and therefore ρ') also satisfies the wave equation.

Let us consider a sound wave in which all quantities depend on only one coordinate (x, say). That is, the flow is completely homogeneous in the yz-plane. Such a wave is called a *plane wave*. The wave equation (64.7) becomes

$$\partial^2\phi/\partial x^2 - (1/c^2)\partial^2\phi/\partial t^2 = 0. \tag{64.9}$$

To solve this equation, we replace x and t by the new variables $\xi = x - ct$, $\eta = x + ct$. It is easy to see that in these variables (64.9) becomes $\partial^2\phi/\partial\eta\partial\xi = 0$. Integrating this equation with respect to ξ, we find $\partial\phi/\partial\eta = F(\eta)$, where $F(\eta)$ is an arbitrary function of η. Integrating again, we obtain $\phi = f_1(\xi) + f_2(\eta)$, where f_1 and f_2 are arbitrary functions of their arguments. Thus

$$\phi = f_1(x - ct) + f_2(x + ct). \tag{64.10}$$

The distribution of the other quantities (p', ρ', $\mathbf{v}$) in a plane wave is given by functions having the same form.

To be definite, we shall discuss the density, $\rho' = f_1(x - ct) + f_2(x + ct)$. For example, let $f_2 = 0$, so that $\rho' = f_1(x - ct)$. The meaning of this solution is evident: in any plane x = constant the density varies with time, and at any given time it is different for different x, but it is the same for coordinates x and times t such that $x - ct$ = constant, or x = constant + ct. This means that, if at some instant $t = 0$ and at some point the fluid density has a certain value, then after a time t the same value of the density is found at a distance ct along the x-axis from the original point. The same is true of all the other quantities in the wave. Thus the pattern of motion is propagated through the medium in the x-direction with a velocity c; c is called the *velocity of sound*.

Thus $f_1(x - ct)$ represents what is called a *travelling plane wave* propagated in the positive direction of the x-axis. It is evident that $f_2(x + ct)$ represents a wave propagated in the opposite direction.

Of the three components of the velocity $\mathbf{v} = \mathbf{grad}\,\phi$ in a plane wave, only $v_x = \partial\phi/\partial x$ is not zero. Thus the fluid velocity in a sound wave is in the direction of propagation. For this reason sound waves in a fluid are said to be *longitudinal*.

In a travelling plane wave, the velocity $v_x = v$ is related to the pressure p' and the density ρ' in a simple manner. Putting $\phi = f(x - ct)$, we find $v = \partial\phi/\partial x = f'(x - ct)$ and $p' = -\rho\partial\phi/\partial t = \rho c f'(x - ct)$. Comparing the two expressions, we find

$$v = p'/\rho c. \tag{64.11}$$

Substituting here from (64.4) $p' = c^2\rho'$, we find the relation between the velocity and the density variation:

$$v = c\rho'/\rho. \tag{64.12}$$

We may mention also the relation between the velocity and the temperature oscillations in a sound wave. We have $T' = (\partial T/\partial p)_s p'$ and, using the well-known thermodynamic formula $(\partial T/\partial p)_s = (T/c_p)(\partial V/\partial T)_p$ and formula (64.11), we obtain

$$T' = c\beta T v/c_p, \tag{64.13}$$

where $\beta = (1/V)(\partial V/\partial T)_p$ is the coefficient of thermal expansion.

Formula (64.8) gives the velocity of sound in terms of the adiabatic compressibility of the fluid. This is related to the isothermal compressibility by the thermodynamic formula

$$(\partial p/\partial \rho)_s = (c_p/c_v)(\partial p/\partial \rho)_T. \tag{64.14}$$

Let us calculate the velocity of sound in a perfect gas. The equation of state is $pV = p/\rho = RT/\mu$, where R is the gas constant and μ the molecular weight. We obtain for the velocity of sound the expression

$$c = \sqrt{(\gamma RT/\mu)}, \tag{64.15}$$

where γ denotes the ratio c_p/c_v. Since γ usually depends only slightly on the temperature, the velocity of sound in the gas may be supposed proportional to the square root of the temperature.† For a given temperature it does not depend on the pressure.‡

What are called *monochromatic waves* are a very important case. Here all quantities are just periodic (harmonic) functions of the time. It is usually convenient to write such functions as the real part of a complex quantity (see the beginning of §24). For example, we put for the velocity potential

$$\phi = \mathrm{re}[\phi_0(x, y, z)e^{-i\omega t}], \tag{64.16}$$

where ω is the frequency of the wave. The function ϕ_0 satisfies the equation

$$\triangle\phi_0 + (\omega^2/c^2)\phi_0 = 0, \tag{64.17}$$

which is obtained by substituting (64.16) in (64.7).

Let us consider a monochromatic travelling plane wave, propagated in the positive direction of the x-axis. In such a wave, all quantities are functions of $x - ct$ only, and so the potential is of the form

$$\phi = \mathrm{re}\{A \exp[-i\omega(t - x/c)]\}, \tag{64.18}$$

where A is a constant called the *complex amplitude*. Writing this as $A = ae^{i\alpha}$ with real constants a and α, we have

$$\phi = a\cos(\omega x/c - \omega t + \alpha). \tag{64.19}$$

The constant a is called the *amplitude* of the wave, and the argument of the cosine is called the *phase*. We denote by $\mathbf{n}$ a unit vector in the direction of propagation. The vector

$$\mathbf{k} = (\omega/c)\mathbf{n} = (2\pi/\lambda)\mathbf{n} \tag{64.20}$$

† It is useful to note that the velocity of sound in a gas has the same order of magnitude as the mean thermal velocity of the molecules.

‡ The expression $c^2 = p/\rho$ for the velocity of sound in a gas was first derived by Newton (1687). The need for the factor γ was shown by Laplace.

is called the *wave vector*, and its magnitude k the *wave number*. In terms of this vector (64.18) can be written

$$\phi = \mathrm{re}\{A \exp[i(\mathbf{k}\cdot\mathbf{r} - \omega t)]\}. \tag{64.21}$$

Monochromatic waves are very important, because any wave whatsoever can be represented as a sum of superposed monochromatic plane waves with various wave vectors and frequencies. This decomposition of a wave into monochromatic waves is simply an expansion as a Fourier series or integral (called also *spectral resolution*). The terms of this expansion are called the *monochromatic components* or *Fourier components* of the wave.

PROBLEMS

PROBLEM 1. Determine the velocity of sound in a nearly homogeneous two-phase system consisting of a vapour with small liquid droplets suspended in it (a "wet vapour"), or a liquid with small vapour bubbles in it. The wavelength of the sound is supposed large compared with the size of the inhomogeneities in the system.

SOLUTION. In a two-phase system, p and T are not independent variables, but are related by the equation of equilibrium of the phases. A compression or rarefaction of the system is accompanied by a change from one phase to the other. Let x be the fraction (by mass) of phase 2 in the system. We have

$$\left.\begin{aligned} s &= (1-x)s_1 + xs_2, \\ V &= (1-x)V_1 + xV_2, \end{aligned}\right\} \tag{1}$$

where the suffixes 1 and 2 distinguish quantities pertaining to the pure phases 1 and 2. To calculate the derivative $(\partial V/\partial p)_s$, we transform it from the variables p, s to p, x, obtaining $(\partial V/\partial p)_s = (\partial V/\partial p)_x - (\partial V/\partial x)_p(\partial s/\partial p)_x/(\partial s/\partial x)_p$. The substitution (1) then gives

$$\left(\frac{\partial V}{\partial p}\right)_s = x\left[\frac{\mathrm{d}V_2}{\mathrm{d}p} - \frac{V_2 - V_1}{s_2 - s_1}\frac{\mathrm{d}s_2}{\mathrm{d}p}\right] + (1-x)\left[\frac{\mathrm{d}V_1}{\mathrm{d}p} - \frac{V_2 - V_1}{s_2 - s_1}\frac{\mathrm{d}s_1}{\mathrm{d}p}\right]. \tag{2}$$

The velocity of sound is obtained from (1) and (2), using formula (64.8).

Expanding the total derivatives with respect to the pressure, introducing the latent heat of the transition from phase 1 to phase 2 ($q = T(s_2 - s_1)$), and using the Clapeyron–Clausius equation for the derivative $\mathrm{d}p/\mathrm{d}T$ along the curve of equilibrium ($\mathrm{d}p/\mathrm{d}T = q/T(V_2 - V_1)$; see *SP*1, §82), we obtain the expression in the first brackets in (2) in the form

$$\left(\frac{\partial V_2}{\partial p}\right)_T + \frac{2T}{q}\left(\frac{\partial V_2}{\partial T}\right)_p (V_2 - V_1) - \frac{Tc_{p2}}{q^2}(V_2 - V_1)^2.$$

The second bracket is transformed similarly.

Let phase 1 be the liquid and phase 2 the vapour; we suppose the latter to be a perfect gas, and neglect the specific volume V_1 in comparison with V_2. If $x \ll 1$ (a liquid containing some bubbles of vapour), the velocity of sound is found to be

$$c = q\mu p V_1/RT\sqrt{(c_{p1}T)}, \tag{3}$$

where R is the gas constant and μ the molecular weight. This velocity is in general very small; thus, when vapour bubbles form in a liquid (*cavitation*), the velocity of sound undergoes a sudden sharp decrease.

If $1 - x \ll 1$ (a vapour containing some droplets of liquid), we obtain

$$\frac{1}{c^2} = \frac{\mu}{RT} - \frac{2}{q} + \frac{c_{p2}T}{q^2}. \tag{4}$$

Comparing this with the velocity of sound in the pure gas (64.15), we find that here also the addition of a second phase reduces the value of c, though by no means so markedly.

As x increases from 0 to 1, the velocity of sound increases monotonically from the value (3) to the value (4). For $x = 0$ and $x = 1$ it changes discontinuously as we go from a one-phase system to a two-phase system. This has the result that, for values of x very close to zero or unity, the usual linear theory of sound is no longer applicable, even when the amplitude of the sound wave is small; the compressions and rarefactions produced by the wave are in this case accompanied by a change between a one-phase and a two-phase system, and the essential assumption of a constant velocity of sound no longer holds good.

PROBLEM 2. Determine the velocity of sound in a gas heated to such a high temperature that the pressure of equilibrium black-body radiation becomes comparable with the gas pressure.

SOLUTION. The pressure is $p = nT + \frac{1}{3}aT^4$, and the entropy is

$$s = (1/m)\log(T^{\frac{3}{2}}/n) + aT^3/n.$$

In these expressions the first terms relate to the particles, and the second terms to the radiation; n is the number density of particles, m their mass, and $a = 4\pi^2/45\hbar^3c^3$ (see *SP*1, §63).† The density of matter is not affected by the black-body radiation, so that $\rho = mn$. The velocity of sound, denoted here by u to distinguish it from that of light, is given by

$$u^2 = \frac{\partial(p, s)}{\partial(\rho, s)} = \frac{\partial(p, s)}{\partial(n, T)} \bigg/ \frac{\partial(\rho, s)}{\partial(n, T)},$$

where the derivatives have been written in Jacobian form. Evaluating the Jacobians, we have

$$u^2 = \frac{5T}{3m}\left[1 + \frac{2a^2T^6}{5n(n+2aT^3)}\right].$$

§65. The energy and momentum of sound waves

Let us derive an expression for the energy of a sound wave. According to the general formula, the energy in unit volume of the fluid is $\rho\varepsilon + \frac{1}{2}\rho v^2$. We now substitute $\rho = \rho_0 + \rho'$, $\varepsilon = \varepsilon_0 + \varepsilon'$, where the primed letters denote the deviations of the respective quantities from their values when the fluid is at rest. The term $\frac{1}{2}\rho' v^2$ is a quantity of the third order. Hence, if we take only terms up to the second order, we have

$$\rho_0\varepsilon_0 + \rho'\frac{\partial(\rho\varepsilon)}{\partial\rho_0} + \tfrac{1}{2}\rho'^2\frac{\partial^2(\rho\varepsilon)}{\partial{\rho_0}^2} + \tfrac{1}{2}\rho_0 v^2.$$

The derivatives are taken at constant entropy, since the sound wave is adiabatic. From the thermodynamic relation $d\varepsilon = Tds - pdV = Tds + (p/\rho^2)d\rho$ we have $[\partial(\rho\varepsilon)/\partial\rho]_s = \varepsilon + p/\rho = w$, and the second derivative is

$$[\partial^2(\rho\varepsilon)/\partial\rho]_s = (\partial w/\partial\rho)_s = (\partial w/\partial p)_s(\partial p/\partial\rho)_s = c^2/\rho.$$

Thus the energy in unit volume of the fluid is

$$\rho_0\varepsilon_0 + w_0\rho' + \tfrac{1}{2}c^2\rho'^2/\rho_0 + \tfrac{1}{2}\rho_0 v^2.$$

The first term ($\rho_0\varepsilon_0$) in this expression is the energy in unit volume when the fluid is at rest, and does not relate to the sound wave. The second term ($w_0\rho'$) is the change in energy due to the change in the mass of fluid in unit volume. This term disappears in the total energy, which is obtained by integrating the energy over the whole volume of the fluid: since the total mass of fluid is unchanged, we have

$$\int \rho' dV = 0.$$

Thus the total change in the energy of the fluid caused by the sound wave is given by the integral

$$\int (\tfrac{1}{2}\rho_0 v^2 + \tfrac{1}{2}c^2\rho'^2/\rho_0)dV.$$

The integrand may be regarded as the density E of sound energy:

$$E = \tfrac{1}{2}\rho_0 v^2 + \tfrac{1}{2}c^2\rho'^2/\rho_0. \tag{65.1}$$

---

† The temperature is in energy units, as elsewhere in this book.

This expression takes a simpler form for a travelling plane wave. In such a wave $\rho' = \rho_0 v/c$, and the two terms in (65.1) are equal, so that

$$E = \rho_0 v^2. \tag{65.2}$$

In general this relation does not hold. A similar formula can be obtained only for the (time) average of the total sound energy. It follows immediately from a well-known general theorem of mechanics, that the mean total potential energy of a system executing small oscillations is equal to the mean total kinetic energy. Since the latter is, in the case considered,

$$\tfrac{1}{2}\int \rho_0 \overline{v^2}\,\mathrm{d}V,$$

we find that the mean total sound energy is

$$\int \bar{E}\,\mathrm{d}V = \int \rho_0 \overline{v^2}\,\mathrm{d}V. \tag{65.3}$$

Next, let us consider some volume of a fluid in which sound is propagated, and determine the flux of energy through the closed surface bounding this volume. The energy flux density in the fluid is, by (6.3), $\rho\mathbf{v}(\tfrac{1}{2}v^2 + w)$. In the present case we can neglect the term in v^2, which is of the third order. Hence the energy flux density in the sound wave is $\rho w\mathbf{v}$. Substituting $w = w_0 + w'$, we have $\rho w\mathbf{v} = w_0\rho\mathbf{v} + \rho w'\mathbf{v}$. For a small change w' in the heat function we have $w' = (\partial w/\partial p)_s p' = p'/\rho$, and $\rho w\mathbf{v} = w_0\rho\mathbf{v} + p'\mathbf{v}$. The total energy flux through the surface in question is

$$\oint (w_0\rho\mathbf{v} + p'\mathbf{v})\cdot \mathrm{d}\mathbf{f}.$$

The first term here is the energy flux due to the change in the mass of fluid in the volume considered. We have already omitted the corresponding term $w_0\rho'$ (which gives zero on integration over an infinite volume) in the energy density. Hence, to find the energy flux, whose density is given by (65.1), we should omit this term, and the energy flux is simply

$$\oint p'\mathbf{v}\cdot \mathrm{d}\mathbf{f}.$$

We see that the sound energy flux density is represented by the vector

$$\mathbf{q} = p'\mathbf{v}. \tag{65.4}$$

It is easy to verify that the expected relation

$$\partial E/\partial t + \operatorname{div}(p'\mathbf{v}) = 0 \tag{65.5}$$

holds. In this form the equation gives the law of conservation of the sound energy, with the vector (65.4) taking the part of the energy flux density.

In a travelling plane wave (left to right) the pressure variation is related to the velocity by $p' = c\rho_0 v$, where $v \equiv v_x$ is taken with the appropriate sign. Introducing the unit vector $\mathbf{n}$ in the direction of propagation of the wave, we obtain

$$\mathbf{q} = c\rho_0 v^2\mathbf{n} = cE\mathbf{n}. \tag{65.6}$$

Thus the energy flux density in a plane sound wave equals the energy density multiplied by the velocity of sound, a result which was to be expected.

Let us now consider a sound wave which, at any given instant, occupies a finite region of space nowhere bounded by solid walls (a *wave packet*), and determine the total momentum of the fluid in the wave. The momentum of unit volume of fluid is equal to the mass flux density $\mathbf{j} = \rho\mathbf{v}$. Substituting $\rho = \rho_0 + \rho'$, we have $\mathbf{j} = \rho_0\mathbf{v} + \rho'\mathbf{v}$. The density change is related to the pressure change by $\rho' = p'/c^2$. Using (65.4), we therefore obtain

$$\mathbf{j} = \rho_0\mathbf{v} + \mathbf{q}/c^2. \tag{65.7}$$

If the viscosity of the fluid is not significant in the phenomena under consideration, we can assume potential flow in a sound wave, and write $\mathbf{v} = \mathbf{grad}\,\phi$; it should be emphasized that this result is not a consequence of the approximations made in deriving the linear equations of motion in §64, since a solution such that $\mathbf{curl}\,\mathbf{v} = 0$ is an exact solution of Euler's equations. We therefore have $\mathbf{j} = \rho_0\,\mathbf{grad}\,\phi + \mathbf{q}/c^2$. The total momentum in the wave equals the integral $\int \mathbf{j}\,\mathrm{d}V$ over the volume occupied by the wave. The integral of $\mathbf{grad}\,\phi$ can be transformed into a surface integral,

$$\int \mathbf{grad}\,\phi\,\mathrm{d}V = \oint \phi\,\mathrm{d}\mathbf{f},$$

and is zero, since ϕ is zero outside the volume occupied by the wave packet. Thus the total momentum of the wave packet is

$$\int \mathbf{j}\,\mathrm{d}V = (1/c^2)\int \mathbf{q}\,\mathrm{d}V. \tag{65.8}$$

This quantity is not, in general, zero. The existence of a non-zero total momentum means that there is a transfer of matter. We therefore conclude that the propagation of a sound-wave packet is accompanied by the transfer of fluid. This is a second-order effect (since $\mathbf{q}$ is a second-order quantity).

Finally, let us consider a region of space unlimited in length but finite in cross-section (a *wave train* with finite aperture), and calculate the mean value of the pressure change p' in a sound wave. In the first approximation, corresponding to the usual linearized equations of motion, p' is a function which periodically changes sign, and the mean value of p' is zero. This result, however, may cease to hold if we go to higher approximations. If we take only second-order quantities, p' can be expressed in terms of quantities calculated from the linear sound equations, so that it is not necessary to solve directly the non-linear equations of motion obtained when terms of higher order are taken into account.

A characteristic property of the sound in question is that the difference between the velocity potentials ϕ at different points remains finite when the distance between them increases without limit (and the same is true of the difference in the values of ϕ at a given point in space at different times): this difference is

$$\phi_2 - \phi_1 = \oint_1^2 \mathbf{v}\cdot\mathrm{d}\mathbf{l},$$

which can be taken along any path between the points 1 and 2, and the property stated is obvious if we note that a path can be chosen which lies along the wave train but outside it.†

† Essentially similar arguments have been used in deriving (65.8) from the proposition that $\phi = 0$ everywhere far from a wave packet.

We therefore start from Bernoulli's equation: $w+\frac{1}{2}v^2+\partial\phi/\partial t=$ constant, and average it with respect to time. The mean value of the time derivative $\overline{\partial\phi/\partial t}$ is zero.† Putting also $w=w_0+w'$ and including w_0 in the constant, we obtain $\overline{w'}+\frac{1}{2}\overline{v^2}=$ constant. Since the constant is the same in all space, and w' and v are zero far from the wave train, the constant must evidently be zero, so that

$$\overline{w'}+\tfrac{1}{2}\overline{v^2}=0. \tag{65.9}$$

We next expand w' in powers of p', and take only the terms up to the second order:

$$w'=(\partial w/\partial p)_s p'+\tfrac{1}{2}(\partial^2 w/\partial p^2)_s p'^2;$$

since $(\partial w/\partial p)_s=1/\rho$, we have

$$w'=\frac{p'}{\rho_0}-\frac{p'^2}{2\rho_0{}^2}\left(\frac{\partial\rho}{\partial p}\right)_s=\frac{p'}{\rho_0}-\frac{p'^2}{2c^2\rho_0{}^2}.$$

Substituting this in (65.9) gives

$$\overline{p'}=-\tfrac{1}{2}\rho_0\overline{v^2}+\overline{p'^2}/2\rho_0c^2=-\tfrac{1}{2}\rho_0\overline{v^2}+\overline{\rho'^2}\,c^2/2\rho_0, \tag{65.10}$$

which determines the required mean value. The expression on the right is a second-order quantity, and is calculated by using the ρ' and v obtained from the solution of the linearized equations of motion. The mean density is

$$\overline{\rho'}=(\partial\rho/\partial p)_s\overline{p'}+\tfrac{1}{2}(\partial^2\rho/\partial p^2)_s\overline{p'^2}. \tag{65.11}$$

Since the cross-section of the wave train is finite, it cannot be regarded as exactly a plane wave, but if the linear size of the cross-section is sufficiently great relative to the sound wavelength, there may be a very close approximation to a plane wave. In a travelling plane wave, $v=c\rho'/\rho_0$, so that $\overline{v^2}=c^2\overline{\rho'^2}/\rho_0{}^2$, and the expression (65.10) is zero, i.e. the mean pressure variation in a plane wave is an effect of higher order than the second. The density variation $\overline{\rho'}=\frac{1}{2}(\partial^2\rho/\partial p^2)_s\overline{p'^2}$ is not zero, however.‡ In the same approximation, we have for the mean value of the momentum flux density tensor in a travelling plane wave $\overline{p'}+\overline{\rho v_i v_k}=\overline{p'}+\rho_0\overline{v_i v_k}$. The first term is zero. In the second term, we introduce the unit vector $\mathbf{n}$ in the direction of propagation of the wave (the same as the direction of $\mathbf{v}$, apart from sign), and, using (65.2), obtain for the momentum flux density

$$\overline{\Pi}_{ik}=\overline{E}n_i n_k. \tag{65.12}$$

If the wave is propagated in the x-direction, only the component $\overline{\Pi}_{xx}=\overline{E}$ is not zero. Thus, in this approximation, there is only an x-component of the mean momentum flux, and this is transmitted in the x-direction.

It should be emphasized once more in connection with the discussion in the preceding paragraph that the wave train has a limited cross-section. For a strictly plane wave, the results would not be valid; in particular, $\overline{p'}$ might not be zero even in the quadratic approximation (see §101, Problem 4). This arises, formally, because for a strictly plane

† By the general definition of the mean value, we have for the mean derivative of any function $f(t)$

$$\overline{\mathrm{d}f/\mathrm{d}t}=\frac{1}{2T}\int_{-T}^{T}\frac{\mathrm{d}f}{\mathrm{d}t}\mathrm{d}t=\frac{f(T)-f(-T)}{2T}.$$

‡ We may mention that the derivative $(\partial^2\rho/\partial p^2)_s$ is in fact always negative, and therefore $\overline{\rho'}<0$ in a travelling wave.

wave (which cannot be "by-passed") it is not in general correct to say that the potential ϕ is finite in all space or at all times. The physical difference is the result of the possible occurrence (in a wave train with a limited cross-section) of a transverse flow which equalizes the mean pressure.

§66. Reflection and refraction of sound waves

When a sound wave is incident on the boundary between two different fluid media, it undergoes reflection and refraction. The motion in the first medium is a combination of two waves (the incident wave and the *reflected wave*), whereas in the second medium there is only one, the *refracted wave.*

The relation between these three waves is determined by the boundary conditions at the surface of separation.

Let us consider the reflection and refraction of a monochromatic longitudinal wave at a plane surface separating two media, which we take as the yz-plane. It is easy to see that all three waves have the same frequency ω and the same components k_y, k_z of the wave vector, but not the same component k_x perpendicular to the plane of separation. For, in an infinite homogeneous medium, a monochromatic wave with constant $\mathbf{k}$ and ω satisfies the equations of motion. The presence of a boundary introduces only some boundary conditions, which in the case considered apply at $x = 0$, i.e. do not depend on the time or on the coordinates y, z. Hence the dependence of the solution on t, y and z remains the same in all space and time, i.e. ω, k_y, and k_z are the same as in the incident wave.

From this result we can immediately derive the relations which give the directions of propagation of the reflected and refracted waves. Let the plane of the incident wave be the xy-plane. Then $k_z = 0$ in the incident wave, and the same must be true of the reflected and refracted waves. Thus the directions of propagation of the three waves are coplanar.

Let θ be the angle between the direction of propagation of the wave and the x-axis. Then, from the equality of $k_y = (\omega/c)\sin\theta$ for the incident and reflected waves, it follows that

$$\theta_1 = \theta_1', \tag{66.1}$$

i.e. the angle of incidence θ_1 is equal to the angle of reflection θ_1'. From a similar equation for the incident and refracted waves it follows that

$$\sin\theta_1/\sin\theta_2 = c_1/c_2, \tag{66.2}$$

which relates the angle of incidence θ_1 to the angle of refraction θ_2 (c_1 and c_2 being the velocities of sound in the two media).

In order to obtain a quantitative relation between the intensities of the three waves, we write the respective velocity potentials as

$$\phi_1 = A_1 \exp[i\omega\{(x/c_1)\cos\theta_1 + (y/c_1)\sin\theta_1 - t\}],$$

$$\phi_1' = A_1' \exp[i\omega\{(-x/c_1)\cos\theta_1 + (y/c_1)\sin\theta_1 - t\}],$$

$$\phi_2 = A_2 \exp[i\omega\{(x/c_2)\cos\theta_2 + (y/c_2)\sin\theta_2 - t\}].$$

On the surface of separation ($x = 0$) the pressure ($p = -\rho\partial\phi/\partial t$) and the normal velocities ($v_x = \partial\phi/\partial x$) in the two media must be equal; these conditions lead to the equations

$$\rho_1(A_1 + A_1') = \rho_2 A_2, \quad \frac{\cos\theta_1}{c_1}(A_1 - A_1') = \frac{\cos\theta_2}{c_2}A_2.$$

The *reflection coefficient* R is defined as the ratio of the (time) average energy flux densities in the reflected and incident waves. Since the energy flux density in a plane wave is $c\rho v^2$, we have $R = c_1\rho_1\overline{v_1'^2}/c_1\rho_1\overline{v_1^2} = |A_1'|^2/|A_1|^2$. A simple calculation gives

$$R = \left(\frac{\rho_2 \tan\theta_2 - \rho_1 \tan\theta_1}{\rho_2 \tan\theta_2 + \rho_1 \tan\theta_1}\right)^2. \tag{66.3}$$

The angles θ_1 and θ_2 are related by (66.2); expressing θ_2 in terms of θ_1, we can put the reflection coefficient in the form

$$R = \left[\frac{\rho_2 c_2 \cos\theta_1 - \rho_1\sqrt{(c_1^2 - c_2^2 \sin^2\theta_1)}}{\rho_2 c_2 \cos\theta_1 + \rho_1\sqrt{(c_1^2 - c_2^2 \sin^2\theta_1)}}\right]^2 \tag{66.4}$$

For normal incidence ($\theta_1 = 0$), this formula gives simply

$$R = \left(\frac{\rho_2 c_2 - \rho_1 c_1}{\rho_2 c_2 + \rho_1 c_1}\right)^2. \tag{66.5}$$

For an angle of incidence such that

$$\tan^2\theta_1 = \frac{\rho_2^2 c_2^2 - \rho_1^2 c_1^2}{\rho_1^2(c_1^2 - c_2^2)} \tag{66.6}$$

the reflection coefficient is zero, i.e. the wave is totally refracted. This can happen if $c_1 > c_2$ but $\rho_2 c_2 > \rho_1 c_1$, or if both inequalities are reversed.

PROBLEM

Determine the pressure exerted by a sound wave on the boundary separating two fluids.

SOLUTION. The sum of the total energy fluxes in the reflected and refracted waves must equal the incident energy flux. Taking the energy flux per unit area of the surface of separation, we can write this condition in the form $c_1\overline{E}_1\cos\theta_1 = c_1\overline{E}_1'\cos\theta_1 + c_2\overline{E}_2\cos\theta_2$, where E_1, E_1' and E_2 are the energy densities in the three waves. Introducing the reflection coefficient $R = \overline{E}_1'/\overline{E}_1$, we therefore have

$$\overline{E}_2 = \frac{c_1 \cos\theta_1}{c_2 \cos\theta_2}(1-R)\overline{E}_1.$$

The required pressure p is determined as the x-component of the momentum lost per unit time by the sound wave (per unit area of the boundary). Using the expression (65.12) for the momentum flux density tensor in a sound wave, we find

$$p = \overline{E}_1\cos^2\theta_1 + \overline{E}_1'\cos^2\theta_1 - \overline{E}_2\cos^2\theta_2.$$

Substituting for $\overline{E}_2$, introducing R and using (66.2), we obtain

$$p = \overline{E}_1 \sin\theta_1 \cos\theta_1[(1+R)\cot\theta_1 - (1-R)\cot\theta_2].$$

For normal incidence ($\theta_1 = 0$), we find, using (66.5),

$$p = 2\overline{E}_1\left[\frac{\rho_1^2 c_1^2 + \rho_2^2 c_2^2 - 2\rho_1\rho_2 c_1^2}{(\rho_1 c_1 + \rho_2 c_2)^2}\right].$$

§67. Geometrical acoustics

A plane wave has the distinctive property that its direction of propagation and its amplitude are the same in all space. An arbitrary sound wave, of course, does not possess this property. However, cases can occur where a sound wave that is not plane may still be regarded as plane in any small region of space. For this to be so it is necessary that the amplitude and the direction of propagation should vary only slightly over distances of the order of the wavelength.

If this condition holds, we can introduce the idea of *rays*, these being lines such that the tangent to them at any point is in the same direction as the direction of propagation; and we can say that the sound is propagated along the rays, and ignore its wave nature. The study of the laws of propagation of sound in such cases is the task of *geometrical acoustics*. We may say that geometrical acoustics corresponds to the limit of short wavelengths, $\lambda \to 0$.

Let us derive the basic equation of geometrical acoustics, which determines the direction of the rays. We write the wave velocity potential as

$$\phi = ae^{i\psi} \tag{67.1}$$

In the case where the wave is not plane but geometrical acoustics can be applied, the amplitude a is a slowly varying function of the coordinates and the time, while the wave phase ψ is "almost linear" (we recall that in a plane wave $\psi = \mathbf{k}\cdot\mathbf{r} - \omega t + \alpha$, with constant $\mathbf{k}$ and ω). Over small regions of space and short intervals of time, the phase ψ may be expanded in series; up to terms of the first order we have

$$\psi = \psi_0 + \mathbf{r}\cdot\mathbf{grad}\,\psi + t\,\partial\psi/\partial t.$$

In accordance with the fact that, in any small region of space (and during short intervals of time), the wave may be regarded as plane, we define the wave vector and the frequency at each point as

$$\mathbf{k} = \partial\psi/\partial\mathbf{r} \equiv \mathbf{grad}\,\psi, \qquad \omega = -\partial\psi/\partial t. \tag{67.2}$$

The quantity ψ is called the *eikonal*.

In a sound wave we have $\omega^2/c^2 = k^2 = k_x{}^2 + k_y{}^2 + k_z{}^2$. Substituting (67.2), we obtain the basic equation of geometrical acoustics:

$$\left(\frac{\partial\psi}{\partial x}\right)^2 + \left(\frac{\partial\psi}{\partial y}\right)^2 + \left(\frac{\partial\psi}{\partial z}\right)^2 - \frac{1}{c^2}\left(\frac{\partial\psi}{\partial t}\right)^2 = 0. \tag{67.3}$$

If the fluid is not homogeneous, the coefficient $1/c^2$ is a function of the coordinates.

As we know from mechanics, the motion of material particles can be determined by means of the Hamilton–Jacobi equation, which, like (67.3), is a first-order partial differential equation. The quantity analogous to ψ is the action S of the particle, and the derivatives of the action determine the momentum $\mathbf{p} = \partial S/\partial\mathbf{r}$ and the Hamilton's function (the energy) $H = -\partial S/\partial t$ of the particle; these formulae are similar to (67.2). We know, also, that the Hamilton–Jacobi equation is equivalent to Hamilton's equations

$$\dot{\mathbf{p}} = -\partial H/\partial\mathbf{r}, \qquad \mathbf{v} \equiv \dot{\mathbf{r}} = \partial H/\partial\mathbf{p}.$$

From the above analogy between the mechanics of a material particle and geometrical acoustics, we can write down similar equations for rays:

$$\dot{\mathbf{k}} = -\partial\omega/\partial\mathbf{r}, \qquad \dot{\mathbf{r}} = \partial\omega/\partial\mathbf{k}. \tag{67.4}$$

In a homogeneous isotropic medium $\omega = ck$ with c constant, so that $\dot{\mathbf{k}} = 0$, $\dot{\mathbf{r}} = c\mathbf{n}$ ($\mathbf{n}$ being a unit vector in the direction of $\mathbf{k}$), i.e. the rays are propagated in straight lines with a constant frequency ω, as we should expect.

The frequency, of course, remains constant along a ray in all cases where the propagation of sound occurs under steady conditions, i.e. the properties of the medium at each point in space do not vary with time. For the total time derivative of the frequency, which gives its rate of variation along a ray, is $d\omega/dt = \partial\omega/\partial t + \dot{\mathbf{r}}\cdot\partial\omega/\partial\mathbf{r} + \dot{\mathbf{k}}\cdot\partial\omega/\partial\mathbf{k}$. On substituting (67.4), the last two terms cancel, and in a steady state $\partial\omega/\partial t = 0$, so that $d\omega/dt = 0$.

In steady propagation of sound in an inhomogeneous medium at rest $\omega = ck$, where c is a given function of the coordinates. The equations (67.4) give

$$\dot{\mathbf{r}} = c\mathbf{n}, \qquad \dot{\mathbf{k}} = -k\,\mathbf{grad}\,c. \tag{67.5}$$

The magnitude of the vector $\mathbf{k}$ varies along a ray simply according to $k = \omega/c$ (with ω constant). To determine the change in direction of $\mathbf{n}$ we put $\mathbf{k} = \omega\mathbf{n}/c$ in the second of (67.5): $\omega\dot{\mathbf{n}}/c - (\omega\mathbf{n}/c^2)(\dot{\mathbf{r}}\cdot\mathbf{grad}\,c) = -k\,\mathbf{grad}\,c$, whence $d\mathbf{n}/dt = -\mathbf{grad}\,c + \mathbf{n}(\mathbf{n}\cdot\mathbf{grad}\,c)$. Introducing the element of length along the ray $dl = c\,dt$, we can rewrite this equation

$$d\mathbf{n}/dl = -(1/c)\,\mathbf{grad}\,c + \mathbf{n}(\mathbf{n}\cdot\mathbf{grad}\,c)/c. \tag{67.6}$$

This equation determines the form of the rays; $\mathbf{n}$ is a unit vector tangential to a ray.†

If equation (67.3) is solved, and the eikonal ψ is a known function of coordinates and time, we can then find also the distribution of sound intensity in space. In steady conditions, it is given by the equation $\operatorname{div}\mathbf{q} = 0$ ($\mathbf{q}$ being the sound energy flux density), which must hold in all space except at sources of sound. Putting $\mathbf{q} = cE\mathbf{n}$, where E is the sound energy density (see (65.6)), and remembering that $\mathbf{n}$ is a unit vector in the direction of $\mathbf{k} = \mathbf{grad}\,\psi$, we obtain the equation

$$\operatorname{div}(cE\,\mathbf{grad}\,\psi/|\mathbf{grad}\,\psi|) = 0, \tag{67.7}$$

which determines the distribution of E in space.

The second formula (67.4) gives the velocity of propagation of the waves from the known dependence of the frequency on the components of the wave vector. This is an important formula, which holds not only for sound waves, but for all waves (for example, we have already applied it to gravity waves in §12). We shall give here another derivation of this formula, which puts in evidence the meaning of the velocity which it defines. Let us consider a *wave packet*, which occupies some finite region of space. We assume that its spectral composition includes monochromatic components whose frequencies lie in only a small range; the same is true of the components of their wave vectors. Let ω be some mean frequency of the wave, and $\mathbf{k}$ a mean wave vector. Then, at some initial instant, the wave is described by a function having the form

$$\phi = \exp(i\mathbf{k}\cdot\mathbf{r})f(\mathbf{r}). \tag{67.8}$$

The function $f(\mathbf{r})$ is appreciably different from zero only in a region which is small (though it is large compared with the wavelength $1/k$). Its expansion as a Fourier integral contains, by the above assumptions, components having the form $\exp(i\mathbf{r}\cdot\Delta\mathbf{k})$, where $\Delta\mathbf{k}$ is small.

Thus each monochromatic component is, at the initial instant, proportional to

$$\phi_{\mathbf{k}} = \text{constant} \times \exp[i(\mathbf{k}+\Delta\mathbf{k})\cdot\mathbf{r}]. \tag{67.9}$$

† As we know from differential geometry, the derivative $d\mathbf{n}/dl$ along the ray is equal to $\mathbf{N}/R$, where $\mathbf{N}$ is a unit vector along the principal normal and R is the radius of curvature of the ray. The expression on the right-hand side of (67.6) is, apart from a factor $1/c$, the derivative of the velocity of sound along the principal normal; hence we can write the equation as $1/R = -(1/c)\mathbf{N}\cdot\mathbf{grad}\,c$. The rays bend towards the region where c is smaller.

The corresponding frequency is $\omega(\mathbf{k}+\Delta\mathbf{k})$ (we recall that the frequency is a function of the wave vector). Hence the same component at time t has the form

$$\phi_{\mathbf{k}} = \text{constant} \times \exp\left[i(\mathbf{k}+\Delta\mathbf{k})\cdot\mathbf{r} - i\omega(\mathbf{k}+\Delta\mathbf{k})t\right].$$

We use the fact that $\Delta\mathbf{k}$ is small, and put $\omega(\mathbf{k}+\Delta\mathbf{k}) \cong \omega + (\partial\omega/\partial\mathbf{k})\cdot\Delta\mathbf{k}$. Then $\phi_{\mathbf{k}}$ becomes

$$\phi_{\mathbf{k}} = \text{constant} \times \exp\left[i(\mathbf{k}\cdot\mathbf{r} - \omega t)\right] \exp\left[i\Delta\mathbf{k}\cdot(\mathbf{r} - t\partial\omega/\partial\mathbf{k})\right]. \tag{67.10}$$

If we now sum all the monochromatic components, with all the $\Delta\mathbf{k}$ that occur in the wave packet, we see from (67.9) and (67.10) that the result is

$$\phi = \exp\left[i(\mathbf{k}\cdot\mathbf{r} - \omega t)\right] f(\mathbf{r} - t\partial\omega/\partial\mathbf{k}), \tag{67.11}$$

where f is the same function as in (67.8). A comparison with (67.8) shows that, after a time t, the amplitude distribution has moved as a whole through a distance $t\partial\omega/\partial\mathbf{k}$; the exponential coefficient of f in (67.11) affects only the phase. Consequently, the velocity of the wave is

$$\mathbf{U} = \partial\omega/\partial\mathbf{k}. \tag{67.12}$$

This formula gives the velocity of propagation for any dependence of ω on $\mathbf{k}$. When $\omega = ck$, with c constant, it of course gives the usual result $U = \omega/k = c$. In general, when $\omega(\mathbf{k})$ is an arbitrary function, the velocity of propagation is a function of the frequency, and the direction of propagation may not be the same as that of the wave vector.

The velocity defined by (67.12) is called the *group velocity* of the wave, and the ratio ω/k the *phase velocity*. However, it must be borne in mind that the phase velocity does not correspond to any actual physical propagation.

Regarding the derivation given here it should be noted that the motion of the wave packet without change of form, expressed by (67.11), is approximate, and results from the assumption that the range $\Delta\mathbf{k}$ is small. In general, when U depends on ω, a wave packet is "smoothed out" during its propagation, and the region of space which it occupies increases in size. It can be shown that the amount of this smoothing out is proportional to the squared magnitude of the range $\Delta\mathbf{k}$ of the wave vectors which occur in the composition of the wave packet.

PROBLEM

Determine the altitude variation in the amplitude of sound propagated in an isothermal atmosphere under gravity.

SOLUTION. In an isothermal atmosphere (regarded as a perfect gas) the velocity of sound is constant. The energy flux density evidently decreases along a ray in inverse proportion to the square of the distance r from the source: $c\rho v^2 \propto 1/r^2$. Hence it follows that the amplitude of the velocity fluctuations in the sound wave varies along a ray inversely as $r\sqrt{\rho}$; according to the barometric formula, $\rho \propto \exp(-\mu g z/RT)$, where z is the altitude, μ the molecular weight of the gas and R the gas constant.

§68. Propagation of sound in a moving medium

The relation $\omega = ck$ between the frequency and the wave number is valid only for a monochromatic sound wave propagated in a medium at rest. It is not difficult to obtain a similar relation for a wave propagated in a moving medium (and observed in a fixed system of coordinates).

Let us consider a homogeneous flow with velocity $\mathbf{u}$. We take a fixed system K of coordinates x, y, z, and also a system K' of coordinates x', y', z' moving with velocity $\mathbf{u}$ relative to K. In the system K' the fluid is at rest, and a monochromatic wave has the usual form $\phi = \text{constant} \times \exp[i(\mathbf{k}\cdot\mathbf{r}' - kct)]$. The position vector $\mathbf{r}'$ in the system K' is related to the position vector $\mathbf{r}$ in the system K by $\mathbf{r}' = \mathbf{r} - \mathbf{u}t$. Hence, in the fixed system of coordinates, the wave has the form $\phi = \text{constant} \times \exp\{i[\mathbf{k}\cdot\mathbf{r} - (kc + \mathbf{k}\cdot\mathbf{u})t]\}$. The coefficient of t in the exponent is the frequency ω of the wave. Thus the frequency in a moving medium is related to the wave vector $\mathbf{k}$ by

$$\omega = ck + \mathbf{u}\cdot\mathbf{k}. \tag{68.1}$$

The velocity of propagation is

$$\partial\omega/\partial\mathbf{k} = c\mathbf{k}/k + \mathbf{u}; \tag{68.2}$$

this is the vector sum of the velocity c in the direction of $\mathbf{k}$ and the velocity $\mathbf{u}$ with which the sound is "carried along" by the moving fluid.

Let us next determine the sound wave energy density in the moving medium. The total instantaneous energy density is

$$\tfrac{1}{2}(\rho + \rho')(\mathbf{u} + \mathbf{v})^2 + \tfrac{1}{2}c^2\rho'^2/\rho = \tfrac{1}{2}\rho u^2 + \tfrac{1}{2}\rho' u^2 + \rho\mathbf{v}\cdot\mathbf{u} + (\tfrac{1}{2}\rho v^2 + \rho'\mathbf{u}\cdot\mathbf{v} + c^2\rho'^2/2\rho)$$

(cf. (65.1); the suffix 0 to the unperturbed quantities is omitted). The first term here is the energy of the unperturbed flow. The next two are first-order small quantities, but on averaging over time they give second-order quantities related to the energy of the mean flow due to the wave. All these are to be omitted, and the required energy density of the sound wave as such is given by the last three terms, in the brackets. The velocity and the pressure change in a plane wave in the moving medium are related by

$$(\omega - \mathbf{k}\cdot\mathbf{u})\mathbf{v} = \mathbf{k}c^2\rho'/\rho,$$

which follows from the linearized Euler's equation

$$\partial\mathbf{v}/\partial t + (\mathbf{u}\cdot\mathbf{grad})\,\mathbf{v} = -(1/\rho)\,\mathbf{grad}\,p.$$

With (68.1), we have finally as the sound energy density in the moving medium

$$E = E_0\omega/(\omega - \mathbf{k}\cdot\mathbf{u}), \tag{68.3}$$

where $E_0 = c^2\rho'^2/\rho = p'^2/\rho c^2$ is the energy density in the frame of reference moving with the medium.†

Using formula (68.1), we can investigate what is called the *Doppler effect*: the frequency of sound, as received by an observer moving relative to the source, is not the same as the frequency of oscillation of the source.

Let sound emitted by a source at rest (relative to the medium) be received by an observer moving with velocity $\mathbf{u}$. In a system K' at rest relative to the medium we have $k = \omega_0/c$, where ω_0 is the frequency of oscillation of the source. In a system K moving with the observer, the medium moves with velocity $-\mathbf{u}$, and the frequency of the sound is, by (68.1), $\omega = ck - \mathbf{u}\cdot\mathbf{k}$. Introducing the angle θ between the direction of the velocity $\mathbf{u}$ and that of the wave vector $\mathbf{k}$, and putting $k = \omega_0/c$, we find that the frequency of the sound received by the moving observer is

$$\omega = \omega_0[1 - (u/c)\cos\theta]. \tag{68.4}$$

† Equation (68.3) can be interpreted from a quantum standpoint as meaning simply that the number of sound quanta (phonons) $N = E/\hbar\omega = E_0/\hbar(\omega - \mathbf{k}\cdot\mathbf{u})$ is independent of the choice of reference frame.

The opposite case, to a certain extent, is the propagation in a medium at rest of a sound wave emitted from a moving source. Let **u** be now the velocity of the source. We change from the fixed system of coordinates to a system K' moving with the source; in the system K', the fluid moves with velocity $-\mathbf{u}$. In K', where the source is at rest, the frequency of the emitted sound wave must equal the frequency ω_0 of the oscillations of the source. Changing the sign of **u** in (68.1) and introducing the angle θ between the directions of **u** and **k**, we have $\omega_0 = ck[1-(u/c)\cos\theta]$. In the original fixed system K, however, the frequency and the wave number are related by $\omega = ck$. Thus we find

$$\omega = \omega_0/[1-(u/c)\cos\theta]. \tag{68.5}$$

This formula gives the relation between the frequency ω_0 of the oscillations of a moving source and the frequency ω of the sound heard by an observer at rest.

If the source is moving away from the observer, the angle θ between its velocity and the direction to the observer lies in the range $\frac{1}{2}\pi < \theta \leqslant \pi$, so that $\cos\theta < 0$. It then follows from (68.5) that, if the source is moving away from the observer, the frequency of the sound heard is less than ω_0.

If, on the other hand, the source is approaching the observer, then $0 \leqslant \theta < \frac{1}{2}\pi$, so that $\cos\theta > 0$, and the frequency $\omega > \omega_0$ increases with u. For $u\cos\theta > c$, according to formula (68.5) ω becomes negative, which means that the sound heard by the observer actually reaches him in the reverse order, i.e. sound emitted by the source at any given instant arrives earlier than sound emitted at previous instants.

As has been mentioned at the beginning of §67, the approximation of geometrical acoustics corresponds to the case of short wavelengths, i.e. large wave numbers. For this to be so the frequency of the sound must in general be large. In the acoustics of moving media, however, the latter condition need not be fulfilled if the velocity of the medium exceeds that of sound. For in this case k can be large even when the frequency is zero; from (68.1) we have for $\omega = 0$ the equation

$$ck = -\mathbf{u}\cdot\mathbf{k}, \tag{68.6}$$

and this has solutions if $u > c$. Thus, in a medium moving with supersonic velocities, there can be steady small perturbations described (if k is sufficiently large) by geometrical acoustics. This means that such perturbations are propagated along rays.

Let us consider, for example, a homogeneous supersonic stream moving with constant velocity **u**, whose direction we take as the x-axis. The vector **k** is taken to lie in the xy-plane, and its components are related by

$$(u^2-c^2)k_x^{\ 2} = c^2k_y^{\ 2}, \tag{68.7}$$

which is obtained by squaring both sides of equation (68.6). To determine the form of the rays, we use the equations of geometrical acoustics (67.4), according to which $\dot{x} = \partial\omega/\partial k_x$, $\dot{y} = \partial\omega/\partial k_y$. Dividing one of these equations by the other, we have $\mathrm{d}y/\mathrm{d}x = (\partial\omega/\partial k_y)/(\partial\omega/\partial k_x)$. This relation, however, is, by the rule of differentiation for implicit functions, just the derivative $-\partial k_x/\partial k_y$ taken at a constant frequency (in this case zero). Thus the equation which gives the form of the rays from the known relation between k_x and k_y is

$$\mathrm{d}y/\mathrm{d}x = -\partial k_x/\partial k_y. \tag{68.8}$$

Substituting (68.7), we obtain

$$\mathrm{d}y/\mathrm{d}x = \pm c/\sqrt{(u^2-c^2)}.$$

For constant u this equation represents two straight lines intersecting the x-axis at angles $\pm\alpha$, where $\sin\alpha = c/u$.

We shall return to a detailed study of these rays in gas dynamics, where they are very important.

PROBLEMS

PROBLEM 1. Derive an equation giving the form of sound rays propagated in a steadily moving medium with a velocity distribution $\mathbf{u}(x, y, z)$, when $u \ll c$ everywhere. It is assumed that the velocity $\mathbf{u}$ varies appreciably only over distances large compared with the wavelength of the sound.

SOLUTION. Substituting (68.1) in (67.4), we obtain the equations of propagation of the rays in the form

$$\dot{\mathbf{k}} = -(\mathbf{k}\cdot\mathbf{grad})\mathbf{u} - \mathbf{k}\times\mathbf{curl}\,\mathbf{u}, \quad \dot{\mathbf{r}} \equiv \mathbf{v} = c\mathbf{k}/k + \mathbf{u}.$$

Using these equations, and also

$$d\mathbf{u}/dt \equiv \partial\mathbf{u}/\partial t + (\mathbf{v}\cdot\mathbf{grad})\mathbf{u} = (\mathbf{v}\cdot\mathbf{grad})\mathbf{u} \cong (c/k)(\mathbf{k}\cdot\mathbf{grad})\mathbf{u},$$

we calculate the derivative $d(k\mathbf{v})/dt$, retaining only terms as far as the first order in $\mathbf{u}$. The result is $d(k\mathbf{v})/dt = -kv\,\mathbf{n}\times\mathbf{curl}\,\mathbf{u}$, when $\mathbf{n}$ is a unit vector in the direction of $\mathbf{v}$. But $d(k\mathbf{v})/dt = \mathbf{n}d(kv)/dt + kv\,d\mathbf{n}/dt$. Since $\mathbf{n}$ and $d\mathbf{n}/dt$ are perpendicular (because $\mathbf{n}^2 = 1$, and therefore $\mathbf{n}\cdot\dot{\mathbf{n}} = 0$), it follows from the above equations that $\dot{\mathbf{n}} = -\mathbf{n}\times\mathbf{curl}\,\mathbf{u}$. Introducing the element of length along the ray $dl = c\,dt$, we can write finally

$$d\mathbf{n}/dl = -\mathbf{n}\times\mathbf{curl}\,\mathbf{u}/c. \qquad (1)$$

This equation determines the form of the rays; $\mathbf{n}$ is a unit tangential vector (and is no longer in the same direction as $\mathbf{k}$).

PROBLEM 2. Determine the form of sound rays in a moving medium with a velocity distribution $u_x = u(z)$, $u_y = u_z = 0$.

SOLUTION. Expanding equation (1), Problem 1, we find $dn_x/dl = (n_z/c)du/dz$, $dn_y/dl = 0$; the equation for n_z need not be written down, since $\mathbf{n}^2 = 1$. The second equation gives $n_y = \text{constant} \equiv n_{y,0}$. In the first equation we write $n_z = dz/dl$, and then we have by integration $n_x = n_{x,0} + u(z)/c$. These formulae give the required solution.

Let us assume that the velocity u is zero for $z = 0$ and increases upwards ($du/dz > 0$). If the sound is propagated "against the wind" ($n_x < 0$), its path is curved upwards; if it is propagated "with the wind" ($n_x > 0$), its path is curved downwards. In the latter case a ray leaving the point $z = 0$ at a small angle to the x-axis (i.e. with $n_{x,0}$ close to unity) rises only to a finite altitude $z = z_{max}$, which can be calculated as follows. At the altitude z_{max} the ray is horizontal, i.e. $n_z = 0$. Hence we have

$$n_x^2 + n_y^2 \cong n_{x,0}^2 + n_{y,0}^2 + 2n_{x,0}u/c = 1,$$

so that $2n_{x,0}u(z_{max})/c = n_{z,0}^2$, whence we can determine z_{max} from the given function $u(z)$ and the initial direction $\mathbf{n}_0$ of the ray.

PROBLEM 3. Obtain the expression of Fermat's principle for sound rays in a steadily moving medium.

SOLUTION. Fermat's principle is that the integral

$$\oint \mathbf{k}\cdot d\mathbf{l},$$

taken along a ray between two given points, is a minimum; $\mathbf{k}$ is supposed expressed as a function of the frequency ω and the direction $\mathbf{n}$ of the ray (see *Fields*, §53). This function can be found by eliminating v and k from the relations $\omega = ck + \mathbf{u}\cdot\mathbf{k}$ and $v\mathbf{n} = c\mathbf{k}/k + \mathbf{u}$. Fermat's principle then takes the form

$$\delta\oint \{\sqrt{[(c^2 - u^2)dl^2 + (\mathbf{u}\cdot d\mathbf{l})^2]} - \mathbf{u}\cdot d\mathbf{l}\}/(c^2 - u^2) = 0.$$

In a medium at rest, this integral reduces to the usual one, $\oint dl/c$.

§69. Characteristic vibrations

Hitherto we have discussed only oscillatory motion in infinite media, and we have seen, in particular, that in such media waves with any frequency can be propagated.

The situation is very different when we consider a fluid in a vessel with finite dimensions. The equations of motion themselves (the wave equations) are of course unchanged, but they must now be supplemented by boundary conditions to be satisfied at the solid walls or at the free surface of the fluid. We shall consider here only what are called *free vibrations*, i.e. those which occur in the absence of variable external forces. Vibrations occurring as a result of external forces are called *forced vibrations*.

The equations of motion for a finite fluid do not have solutions satisfying the appropriate boundary conditions for every frequency. Such solutions exist only for a series of definite frequencies ω. In other words, in a medium with finite volume, free vibrations can occur only with certain frequencies. These are called the *characteristic frequencies* of the fluid in the vessel concerned.

The actual values of the characteristic frequencies depend on the size and shape of the vessel. In any given case there is an infinite number of characteristic frequencies. To find them, it is necessary to examine the equations of motion with the appropriate boundary conditions.

The order of magnitude of the first (i.e. smallest) characteristic frequency can be seen at once from dimensional considerations. The only parameter having the dimensions of length which appears in the problem is the linear dimension l of the body. Hence it is clear that the wavelength λ_1 corresponding to the first characteristic frequency must be of the order of l, and the order of magnitude of the frequency ω_1 itself is obtained by dividing the velocity of sound by the wavelength. Thus

$$\lambda_1 \sim l, \quad \omega_1 \sim c/l. \tag{69.1}$$

Let us ascertain the nature of the motion in characteristic vibrations. If we seek a solution of the wave equation for the velocity potential (say) which is periodic in time, having the form $\phi = \phi_0(x, y, z)e^{-i\omega t}$, then we find for ϕ_0 the equation

$$\triangle\phi_0 + (\omega^2/c^2)\phi_0 = 0. \tag{69.2}$$

In an infinite medium, where no boundary conditions need be applied, this equation has both real and complex solutions. In particular, it has a solution proportional to $e^{i\mathbf{k}\cdot\mathbf{r}}$, which gives the velocity potential in the form

$$\phi = \text{constant} \times \exp[i(\mathbf{k}\cdot\mathbf{r} - \omega t)].$$

Such a solution represents a wave propagated with a definite velocity—a *travelling wave*.

For a medium with finite volume, however, complex solutions cannot in general exist. This can be seen as follows. The equation satisfied by ϕ_0 is real, and the boundary conditions are real also. Hence, if $\phi_0(x, y, z)$ is a solution of the equations of motion, the complex conjugate function ϕ_0^* is also a solution. Since, however, the solution of the equations for given boundary conditions is in general unique† apart from a constant factor, we must have $\phi_0^* = \text{constant} \times \phi_0$, where the constant is complex and its modulus is unity. Thus ϕ_0 must have the form $\phi_0 = f(x, y, z)e^{-i\alpha}$, the function f and the constant α being real. The potential ϕ thus has the form (taking the real part of $\phi_0 e^{-i\omega t}$)

$$\phi = f(x, y, z)\cos(\omega t + \alpha), \tag{69.3}$$

i.e. it is the product of some function of the coordinates and a simple periodic function of the time.

† This may not be true when the vessel is highly symmetrical in form (e.g. a sphere).

This solution has properties entirely different from those of a travelling wave. In the latter, the phase $\mathbf{k}\cdot\mathbf{r}-\omega t+\alpha$ of the oscillations at different points in space is different at any given instant, except only at points separated by a distance equal to the wavelength. In the wave represented by (69.3), all points are oscillating in the same phase $\omega t+\alpha$ at any given instant. Such a wave is obviously not "propagated"; it is called a *stationary wave.* Thus the characteristic vibrations are stationary waves.

Let us consider a stationary plane sound wave, in which all quantities are functions of one coordinate only (x, say) and of time. Writing the general solution of $\partial^2\phi_0/\partial x^2+\omega^2\phi_0/c^2=0$ in the form $\phi_0=a\cos(\omega x/c+\beta)$, we have $\phi=a\cos(\omega t+\alpha)\cos(\omega x/c+\beta)$. By an appropriate choice of the origins of x and t, we can make α and β zero, so that

$$\phi = a\cos\omega t\cos\omega x/c. \tag{69.4}$$

For the velocity and pressure in the wave we have

$$v=\partial\phi/\partial x=-(a\omega/c)\cos\omega t\sin\omega x/c;$$

$$p'=-\rho\partial\phi/\partial t=\rho\omega\sin\omega t\cos\omega x/c.$$

At the points $x=0,\ \pi c/\omega,\ 2\pi c/\omega,\ \ldots$, which are at a distance $\pi c/\omega=\frac{1}{2}\lambda$ apart, the velocity v is always zero; these points are called *nodes* of the velocity. The points midway between them ($x=\pi c/2\omega,\ 3\pi c/2\omega,\ \ldots$) are those at which the amplitude of the time variations of the velocity is greatest. These are called *antinodes.* The pressure p' evidently has nodes and antinodes in the reverse positions. Thus, in a stationary plane wave, the nodes of the pressure are the antinodes of the velocity, and vice versa.

An interesting case of characteristic vibrations is that of the vibrations of a gas in a vessel having a small aperture (a *resonator*). In a closed vessel the smallest characteristic frequency is, as we know, of the order of c/l, where l is the linear dimension of the vessel. When there is a small aperture, however, new characteristic vibrations with considerably smaller frequency appear. These are due to the fact that, if there is a pressure difference between the gas in the vessel and that outside, this difference can be equalized by the motion of gas into or out of the vessel. Thus oscillations appear which involve an exchange of gas between the resonator and the outside medium. Since the aperture is small, this exchange takes place only slowly, and hence the period of the oscillations is large, and the frequency correspondingly small (see Problem 2). The frequencies of the ordinary vibrations occurring in a closed vessel are practically unchanged by the presence of a small aperture.

PROBLEMS

PROBLEM 1. Determine the characteristic frequencies of sound waves in a fluid contained in a cuboidal vessel.

SOLUTION. We seek a solution of the equation (69.2) in the form

$$\phi_0=\text{constant}\times\cos qx\cos ry\cos sz,$$

where $q^2+r^2+s^2=\omega^2/c^2$. At the walls of the vessel we have the conditions $v_x=\partial\phi/\partial x=0$ for $x=0$ and a, $\partial\phi/\partial y=0$ for $y=0$ and b, $\partial\phi/\partial z=0$ for $z=0$ and c, where a, b, c are the sides of the cuboid. Hence we find $q=m\pi/a$, $r=n\pi/b$, $s=p\pi/c$, where m, n, p are any integers. Thus the characteristic frequencies are

$$\omega^2=c^2\pi^2(m^2/a^2+n^2/b^2+p^2/c^2).$$

PROBLEM 2. A narrow tube with cross-sectional area S and length l is fixed to the aperture of a resonator. Determine the characteristic frequency.

SOLUTION. Since the tube is narrow, in considering oscillations accompanied by the movement of gas into and out of the resonator we can suppose that only the gas in the tube has an appreciable velocity, while the gas in the vessel is almost at rest. The mass of gas in the tube is $S\rho l$, and the force on it is $S(p_0 - p)$, where p and p_0 are the gas pressures inside and outside the resonator respectively. Hence we must have $S\rho l\dot{v} = S(p - p_0)$, where v is the gas velocity in the tube. The time derivative of the pressure is given by $\dot{p} = c^2\dot{\rho}$, and the decrease per unit time in the gas density in the resonator $(-\dot{\rho})$ can be supposed equal to the mass of gas leaving the resonator per unit time $(S\rho v)$ divided by the volume V of the resonator. Thus we have $\dot{p} = -c^2 S\rho v/V$, whence

$$\ddot{p} = -c^2 S\rho\dot{v}/V = -c^2 S(p - p_0)/lV.$$

This equation gives $p - p_0 = \text{constant} \times \cos\omega_0 t$, where the characteristic frequency $\omega_0 = c\sqrt{(S/lV)}$. This is small compared with c/L (where L is the linear dimension of the vessel), and the wavelength is therefore large compared with L.

In solving this problem we have supposed that the linear amplitude of the oscillations of gas in the tube is small compared with its length l. If this were not so, the oscillations would be accompanied by the outflow of a considerable fraction of the gas in the tube, and the linear equation of motion used above would be inapplicable.

§70. Spherical waves

Let us consider a sound wave in which the distribution of density, velocity, etc., depends only on the distance from some point, i.e. is spherically symmetrical. Such a wave is called a *spherical wave*.

Let us determine the general solution of the wave equation which represents a spherical wave. We take the wave equation for the velocity potential: $\triangle\phi - (1/c^2)\partial^2\phi/\partial t^2 = 0$. Since ϕ is a function only of the distance r from the centre and of the time t, we have, using the expression for the Laplacian in spherical polar coordinates,

$$\frac{\partial^2\phi}{\partial t^2} = c^2\frac{1}{r^2}\frac{\partial}{\partial r}\left(r^2\frac{\partial\phi}{\partial r}\right). \tag{70.1}$$

We write $\phi = f(r, t)/r$. Substituting, we have the following equation for f: $\partial^2 f/\partial t^2 = c^2\partial^2 f/\partial r^2$. This is just the ordinary one-dimensional wave equation, with the radius r as the coordinate. The solution of this equation has, as we know, the form $f = f_1(ct - r) + f_2(ct + r)$, where f_1 and f_2 are arbitrary functions. Thus the general solution of equation (70.1) has the form

$$\phi = \frac{f_1(ct - r)}{r} + \frac{f_2(ct + r)}{r}. \tag{70.2}$$

The first term is an outgoing wave, propagated in all directions from the origin. The second term is a wave coming in to the centre. Unlike a plane wave, whose amplitude remains constant, a spherical wave has an amplitude which decreases inversely as the distance from the centre. The intensity in the wave is given by the square of the amplitude, and falls off inversely as the square of the distance, as it should, since the total energy flux in the wave is distributed over a surface whose area increases as r^2

The variable parts of the pressure and density are related to the potential by $p' = -\rho\partial\phi/\partial t$, $\rho' = -(\rho/c^2)\partial\phi/\partial t$, and their distribution is determined by formulae having the same form as (70.2). The (radial) velocity distribution, however, being given by the gradient of the potential, has the form

$$v = \frac{\partial}{\partial r}\left\{\frac{f_1(ct - r) + f_2(ct + r)}{r}\right\}. \tag{70.3}$$

If there is no source of sound at the origin, the potential (70.2) must remain finite for $r = 0$. For this to be so we must have $f_1(ct) = -f_2(ct)$, i.e.

$$\phi = \frac{f(ct-r) - f(ct+r)}{r} \tag{70.4}$$

(a stationary spherical wave). If there is a source at the origin, on the other hand, the potential of the outgoing wave from it is $\phi = f(ct-r)/r$; it need not remain finite at $r = 0$, since the solution holds only for the region outside sources.

A monochromatic stationary spherical wave has the form

$$\phi = Ae^{-i\omega t}\frac{\sin kr}{r}, \tag{70.5}$$

where $k = \omega/c$. An outgoing monochromatic spherical wave is given by

$$\phi = Ae^{i(kr-\omega t)}/r. \tag{70.6}$$

It is useful to note that this expression satisfies the differential equation

$$\triangle\phi + k^2\phi = -4\pi Ae^{-i\omega t}\delta(\mathbf{r}), \tag{70.7}$$

where on the right-hand side we have the delta function $\delta(\mathbf{r}) = \delta(x)\delta(y)\delta(z)$. For $\delta(\mathbf{r}) = 0$ everywhere except at the origin, and we return to the homogeneous equation (70.1); and, integrating (70.7) over the volume of a small sphere including the origin (where the expression (70.6) reduces to $Ae^{-i\omega t}/r$) we obtain $-4\pi Ae^{-i\omega t}$ on each side.

Let us consider an outgoing spherical wave, occupying a spherical shell outside which the medium is either at rest or very nearly so; such a wave can originate from a source which emits during a finite interval of time only, or from some region where there is a sound disturbance (cf. the end of §72, and §74, Problem 4). Before the wave arrives at any given point, the potential is $\phi \equiv 0$. After the wave has passed, the motion must die away; this means that ϕ must become constant. In an outgoing spherical wave, however, the potential is a function having the form $\phi = f(ct-r)/r$; such a function can tend to a constant only if the function f is zero identically. Thus the potential must be zero both before and after the passage of the wave.† From this we can draw an important conclusion concerning the distribution of condensations and rarefactions in a spherical wave.

The variation of pressure in the wave is related to the potential by $p' = -\rho\partial\phi/\partial t$. From what has been said above, it is clear that, if we integrate p' over all time for a given r, the result is zero:

$$\int_{-\infty}^{\infty} p'\,dt = 0. \tag{70.8}$$

This means that, as the spherical wave passes through a given point, both condensations ($p' > 0$) and rarefactions ($p' < 0$) will be observed at that point. In this respect a spherical wave is markedly different from a plane wave, which may consist of condensations or rarefactions only.

† Unlike what happens for a plane wave, where we can have $\phi = \text{constant} \neq 0$ after the wave has passed.

A similar pattern will be observed if we consider the manner of variation of p' with distance at a given instant; instead of the integral (70.8) we now consider another which also vanishes, namely

$$\int_0^\infty rp' \, \mathrm{d}r = 0. \tag{70.9}$$

PROBLEMS

PROBLEM 1. At the initial instant, the gas inside a sphere with radius a is compressed so that $\rho' = \text{constant} \equiv \Delta$; outside this sphere, $\rho' = 0$. The initial velocity is zero in all space. Determine the subsequent motion.

SOLUTION. The initial conditions on the potential $\phi(r, t)$ are $\phi(r, 0) = 0$, $\dot{\phi}(r, 0) = F(r)$, where $F(r) = 0$ for $r > a$ and $F(r) = -c^2\Delta/\rho$ for $r < a$. We seek ϕ in the form (70.4). From the initial conditions we obtain $f(-r) - f(r) = 0, f'(-r) - f'(r) = rF(r)/c$. Hence $f'(r) = -f'(-r) = -rF(r)/2c$. Finally, substituting the value of $F(r)$, we find the following expressions for the derivative $f'(\xi)$ and the function $f(\xi)$ itself:

$$\text{for } |\xi| > a, \quad f'(\xi) = 0, \quad f(\xi) = 0;$$

$$\text{for } |\xi| < a, \quad f'(\xi) = c\xi\Delta/2\rho, \quad f(\xi) = c(\xi^2 - a^2)\Delta/4\rho,$$

which give the solution of the problem. If we consider a point with $r > a$, i.e. outside the region of the initial compression, we have for the density

for $t < (r-a)/c$, $\quad \rho' = 0$;
for $(r-a)/c < t < (r+a)/c$, $\quad \rho' = \frac{1}{2}(r - ct)\Delta/r$;
for $t > (r+a)/c$, $\quad \rho' = 0$.

The wave passes the point considered during a time interval $2a/c$; in other words, the wave has the form of a spherical shell with thickness $2a$, which at time t lies between the spheres with radii $ct - a$ and $ct + a$. Within this shell the density varies linearly; in the outer part ($r > ct$), the gas is compressed ($\rho' > 0$), while in the inner part ($r < ct$) it is rarefied ($\rho' < 0$).

PROBLEM 2. Determine the characteristic frequencies of centrally symmetrical sound oscillations in a spherical vessel with radius a.

SOLUTION. From the boundary condition $\partial\phi/\partial r = 0$ for $r = a$ (where ϕ is given by (70.5)) we find $\tan ka = ka$, which determines the characteristic frequencies. The first (lowest) frequency is $\omega_1 = 4{\cdot}49\, c/a$.

§71. Cylindrical waves

Let us now consider a wave in which the distribution of all quantities is homogeneous in some direction (which we take as the z-axis) and has complete axial symmetry about that direction. This is called a *cylindrical wave*, and in it we have $\phi = \phi(R, t)$, where R denotes the distance from the z-axis. Let us determine the general form of such an axially symmetrical solution of the wave equation. This can be done by starting from the general spherically symmetrical solution (70.2). R is related to r by $r^2 = R^2 + z^2$, so that ϕ as given by formula (70.2) depends on z when R and t are given. A function which depends on R and t only and still satisfies the wave equation can be obtained by integrating (70.2) over all z from $-\infty$ to ∞, or equally well from 0 to ∞. We can convert the integration over z to one over r. Since $z = \sqrt{(r^2 - R^2)}$, $\mathrm{d}z = r\,\mathrm{d}r/\sqrt{(r^2 - R^2)}$. When z varies from 0 to ∞, r varies from R to ∞. Hence we find the general axially symmetrical solution to be

$$\phi = \int_R^\infty \frac{f_1(ct - r)}{\sqrt{(r^2 - R^2)}}\mathrm{d}r + \int_R^\infty \frac{f_2(ct + r)}{\sqrt{(r^2 - R^2)}}\mathrm{d}r, \tag{71.1}$$

where f_1 and f_2 are arbitrary functions. The first term is an outgoing cylindrical wave, and the second an ingoing one.

Substituting in these integrals $ct \pm r = \xi$, we can rewrite formula (71.1) as

$$\phi = \int\limits^{ct-R} \frac{f_1(\xi)\,d\xi}{\sqrt{[(ct-\xi)^2 - R^2]}} + \int\limits_{ct+R}^{\infty} \frac{f_2(\xi)\,d\xi}{\sqrt{[(\xi-ct)^2 - R^2]}}. \tag{71.2}$$

We see that the value of the potential at time t at the point R in the outgoing cylindrical wave is determined by the values of f_1 at times from $-\infty$ to $t - R/c$; similarly, the values of f_2 which affect the ingoing wave are those at times from $t + R/c$ to infinity.

As in the spherical case, stationary waves are obtained when $f_1(\xi) = -f_2(\xi)$. It can be shown that a stationary cylindrical wave can also be represented in the form

$$\phi = \int\limits_{ct-R}^{ct+R} \frac{F(\xi)\,d\xi}{\sqrt{[R^2 - (\xi - ct)^2]}}, \tag{71.3}$$

where $F(\xi)$ is another arbitrary function.

Let us derive an expression for the potential in a monochromatic cylindrical wave. The wave equation for the potential $\phi(R, t)$ in cylindrical polar coordinates is

$$\frac{1}{R}\frac{\partial}{\partial R}\left(R\frac{\partial \phi}{\partial R}\right) - \frac{1}{c^2}\frac{\partial^2 \phi}{\partial t^2} = 0.$$

In a monochromatic wave $\phi = e^{-i\omega t} f(R)$, and we have for the function $f(R)$ the equation $f'' + f'/R + k^2 f = 0$. This is Bessel's equation of order zero. In a stationary cylindrical wave, ϕ must remain finite for $R = 0$; the appropriate solution is $J_0(kR)$, where J_0 is a Bessel function of the first kind. Thus, in a stationary cylindrical wave,

$$\phi = Ae^{-i\omega t} J_0(kR). \tag{71.4}$$

For $R = 0$ the function J_0 tends to unity, so that the amplitude tends to the finite limit A. At large distances R, J_0 may be replaced by its asymptotic expression, and ϕ then takes the form

$$\phi = A\sqrt{\frac{2}{\pi}} \frac{\cos(kR - \frac{1}{4}\pi)}{\sqrt{(kR)}} e^{-i\omega t} \tag{71.5}$$

The solution corresponding to a monochromatic outgoing travelling wave is

$$\phi = Ae^{-i\omega t} H_0^{(1)}(kR), \tag{71.6}$$

where $H_0^{(1)}$ is the Hankel function of the first kind, of order zero. For $R \to 0$ this function has a logarithmic singularity:

$$\phi \cong (2iA/\pi)\log(kR)e^{-i\omega t} \tag{71.7}$$

At large distances we have the asymptotic formula

$$\phi = A\sqrt{\frac{2}{\pi}} \frac{\exp[i(kR - \omega t - \frac{1}{4}\pi)]}{\sqrt{(kR)}} \tag{71.8}$$

We see that the amplitude of a cylindrical wave diminishes (at large distances) inversely as the square root of the distance from the axis, and the intensity therefore decreases as $1/R$.

This result is obvious, since the total energy flux is distributed over a cylindrical surface, whose area increases proportionally to R as the wave is propagated.

An outgoing cylindrical wave differs from a spherical or plane wave in the important respect that it has a forward front but no backward front: once the sound disturbance has reached a given point, it does not cease, but diminishes comparatively slowly as $t \to \infty$. Suppose that the function $f_1(\xi)$ in the first term of (71.2) is different from zero only in some finite range $\xi_1 \leqslant \xi \leqslant \xi_2$. Then, at times such that $ct > R + \xi_2$, we have

$$\phi = \int_{\xi_1}^{\xi_2} \frac{f_1(\xi)\mathrm{d}\xi}{\sqrt{[(ct-\xi)^2 - R^2]}}.$$

As $t \to \infty$, this expression tends to zero as

$$\phi = \frac{1}{ct} \int_{\xi_1}^{\xi_2} f_1(\xi)\mathrm{d}\xi,$$

i.e. inversely as the time.

Thus the potential in an outgoing cylindrical wave, due to a source which operates only for a finite time, vanishes, though slowly, as $t \to \infty$. This means that, as in the spherical case, the integral of p' over all time is zero:

$$\int_{-\infty}^{\infty} p'\mathrm{d}t = 0. \tag{71.9}$$

Hence a cylindrical wave, like a spherical wave, must necessarily include both condensations and rarefactions.

§72. The general solution of the wave equation

We shall now derive a general formula giving the solution of the wave equation in an infinite fluid for any initial conditions, i.e. giving the velocity and pressure distribution in the fluid at any instant in terms of their initial distribution.

We first obtain some auxiliary formulae. Let $\phi(x, y, z, t)$ and $\psi(x, y, z, t)$ be any two solutions of the wave equation which vanish at infinity. We consider the integral

$$I = \int (\phi\dot{\psi} - \psi\dot{\phi})\mathrm{d}V,$$

taken over all space, and calculate its time derivative. Since ϕ and ψ satisfy the equations $\triangle\phi - \ddot{\phi}/c^2 = 0$ and $\triangle\psi - \ddot{\psi}/c^2 = 0$, we have

$$\mathrm{d}I/\mathrm{d}t = \int (\phi\ddot{\psi} - \psi\ddot{\phi})\mathrm{d}V = c^2 \int (\phi\triangle\psi - \psi\triangle\phi)\mathrm{d}V$$

$$= c^2 \int \operatorname{div}(\phi\,\mathbf{grad}\,\psi - \psi\,\mathbf{grad}\,\phi)\mathrm{d}V.$$

The last integral can be transformed into an integral over an infinitely distant surface, and is therefore zero. Thus we conclude that $\mathrm{d}I/\mathrm{d}t = 0$, i.e. I is independent of time:

$$I \equiv \int(\phi\dot{\psi} - \psi\dot{\phi})\,\mathrm{d}V = \text{constant}. \tag{72.1}$$

Next, let us consider the following particular solution of the wave equation:

$$\psi = \delta[r - c(t_0 - t)]/r \tag{72.2}$$

(where r is the distance from some given point O, t_0 is some definite instant, and δ denotes the delta function), and calculate the integral of ψ over all space. We have

$$\int\psi\,\mathrm{d}V = \int_0^\infty \psi\cdot 4\pi r^2\,\mathrm{d}r = 4\pi\int_0^\infty r\delta[r - c(t_0 - t)]\,\mathrm{d}r.$$

The argument of the delta function is zero for $r = c(t_0 - t)$ (we assume that $t_0 > t$). Hence, from the properties of the delta function, we find

$$\int\psi\,\mathrm{d}V = 4\pi c(t_0 - t). \tag{72.3}$$

Differentiating this equation with respect to time, we obtain

$$\int\dot{\psi}\,\mathrm{d}V = -4\pi c. \tag{72.4}$$

We now substitute for ψ, in the integral (72.1), the function (72.2), and take ϕ to be the required general solution of the wave equation. According to (72.1), I is a constant; using this, we write down the expressions for I at the instants $t = 0$ and $t = t_0$, and equate the two. For $t = t_0$ the two functions ψ and $\dot{\psi}$ are each different from zero only for $r = 0$. Hence, on integrating, we can put $r = 0$ in ϕ and $\dot{\phi}$ (i.e. take their values at the point O), and take ϕ and $\dot{\phi}$ outside the integral:

$$I = \phi(x, y, z, t_0)\int\dot{\psi}\,\mathrm{d}V - \dot{\phi}(x, y, z, t_0)\int\psi\,\mathrm{d}V,$$

where x, y, z are the coordinates of O. According to (72.3) and (72.4), the second term is zero for $t = t_0$, and the first term gives

$$I = -4\pi c\phi(x, y, z, t_0).$$

Let us now calculate I for $t = 0$. Putting $\dot{\psi} = \partial\psi/\partial t = -\partial\psi/\partial t_0$, and denoting by ϕ_0 the value of the function ϕ for $t = 0$, we have

$$I = -\int\left(\phi_0\frac{\partial\psi}{\partial t_0} + \dot{\phi}_0\psi\right)\mathrm{d}V = -\frac{\partial}{\partial t_0}\int\phi_0\psi_{t=0}\,\mathrm{d}V - \int\dot{\phi}_0\psi_{t=0}\,\mathrm{d}V.$$

We write the element of volume as $\mathrm{d}V = r^2\mathrm{d}r\mathrm{d}o$, where $\mathrm{d}o$ is an element of solid angle, and then we obtain, by the properties of the delta function,

$$\int\phi_0\psi_{t=0}\,\mathrm{d}V = \int\phi_0 r\delta(r - ct_0)\,\mathrm{d}r\,\mathrm{d}o = ct_0\int\phi_{0,r=ct_0}\,\mathrm{d}o;$$

the integral of $\dot{\phi}_0\psi$ is similar. Thus

$$I = -\frac{\partial}{\partial t_0}\left(ct_0\int \phi_{0,r=ct_0}\,\mathrm{d}o\right) - ct_0\int \dot{\phi}_{0,r=ct_0}\,\mathrm{d}o.$$

Finally, equating the two expressions for I and omitting the suffix zero in t_0, we obtain

$$\phi(x,y,z,t) = \frac{1}{4\pi}\left\{\frac{\partial}{\partial t}\left(t\int \phi_{0,r=ct}\,\mathrm{d}o\right) + t\int \dot{\phi}_{0,r=ct}\,\mathrm{d}o\right\}. \tag{72.5}$$

This formula, called *Poisson's formula*, gives the spatial distribution of the potential at any instant in terms of the distribution of the potential and its time derivative (or, equivalently, in terms of the velocity and pressure distribution) at some initial instant. We see that the value of the potential at time t is determined by the values of ϕ and $\dot{\phi}$ at time $t = 0$ on the surface of a sphere centred at O, with radius ct.

Let us suppose that, at the initial instant, ϕ_0 and $\dot{\phi}_0$ are different from zero only in some finite region of space, bounded by a closed surface C (Fig. 44). We consider the values of ϕ at subsequent instants at some point O. These values are determined by the values of ϕ_0 and $\dot{\phi}_0$ at a distance ct from O. The spheres with radius ct pass through the region within the surface C only for $d/c \leqslant t \leqslant D/c$, where d and D are the least and greatest distances from the point O to the surface C. At other instants, the integrands in (72.5) are zero. Thus the motion at O begins at time $t = d/c$ and ceases at time $t = D/c$. The wave propagated from the region inside C has a forward front and a backward front. The motion begins when the forward front arrives at the point in question, while on the backward front particles previously oscillating come to rest.

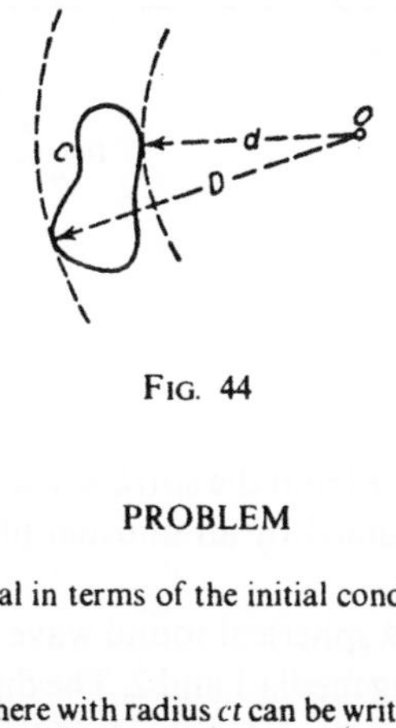

FIG. 44

PROBLEM

Derive the formula giving the potential in terms of the initial conditions for a wave depending on only two coordinates, x and y.

SOLUTION. An element of area of a sphere with radius ct can be written $\mathrm{d}f = c^2t^2\,\mathrm{d}o$, where $\mathrm{d}o$ is an element of solid angle. The projection of $\mathrm{d}f$ on the xy-plane is $\mathrm{d}x\,\mathrm{d}y = \mathrm{d}f\sqrt{[(ct)^2-\rho^2]}/ct$, where ρ is the distance of the point x, y from the centre of the sphere. Comparing the two expressions, we can write $\mathrm{d}o = \mathrm{d}x\,\mathrm{d}y/ct\sqrt{[(ct)^2-\rho^2]}$. Denoting by x, y the coordinates of the point where we seek the value of ϕ, and by ξ, η the coordinates of a variable point in the region of integration, we can therefore replace $\mathrm{d}o$ in the general formula (72.5) by $\mathrm{d}\xi\,\mathrm{d}\eta/ct\sqrt{[(ct)^2-(x-\xi)^2-(y-\eta)^2]}$, doubling the resulting expression because $\mathrm{d}x\,\mathrm{d}y$ is the projection of two elements of area on opposite sides of the xy-plane. Thus

$$\phi(x,y,z,t) = \frac{1}{2\pi c}\frac{\partial}{\partial t}\iint \frac{\phi_0(\xi,\eta)\,\mathrm{d}\xi\,\mathrm{d}\eta}{\sqrt{[(ct)^2-(x-\xi)^2-(y-\eta)^2]}} + \\ + \frac{1}{2\pi c}\iint \frac{\dot{\phi}_0(\xi,\eta)\,\mathrm{d}\xi\,\mathrm{d}\eta}{\sqrt{[(ct)^2-(x-\xi)^2-(y-\eta)^2]}},$$

where the integration is over a circle centred at O, with radius ct. If ϕ_0 and $\dot{\phi}_0$ are zero except in a finite region C of the xy-plane (or, more exactly, except in a cylindrical region with its generators parallel to the z-axis), the oscillations at the point O (Fig. 44) begin at time $t = d/c$, where d is the least distance from O to a point in the region. After this time, however, circles with radius $ct > d$ centred at O will always enclose part or all of the region C, and ϕ will tend only asymptotically to zero. Thus, unlike three-dimensional waves, the two-dimensional waves here considered have a forward front but no backward front (cf. §71).

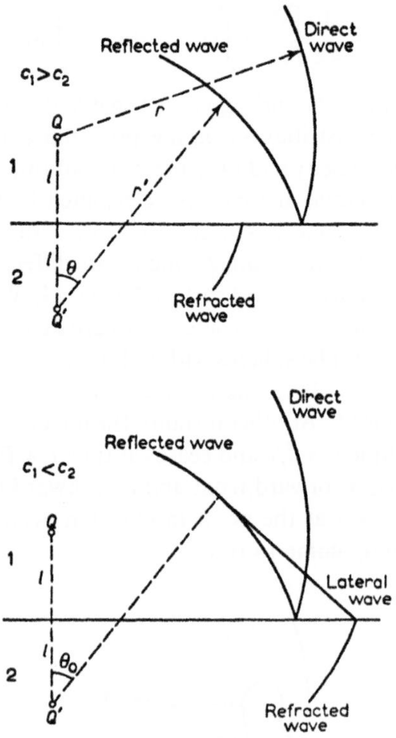

FIG. 45

§73. The lateral wave

The reflection of a spherical wave from the surface separating two media is of particular interest in that it may be accompanied by an unusual phenomenon, the appearance of a *lateral wave*.

Let Q (Fig. 45) be the source of a spherical sound wave in medium 1, at a distance l from the infinite plane surface separating media 1 and 2. The distance l is arbitrary, and need not be large compared with the wavelength λ. Let the densities of the two media be ρ_1, ρ_2, and the velocities of sound in them c_1, c_2. We suppose first that $c_1 > c_2$; then, at distances from the source large compared with λ, the motion in medium 1 will be a superposition of two outgoing waves. One of these is the spherical wave emitted by the source (the *direct wave*); its potential is

$$\phi_1{}^0 = e^{ikr}/r, \tag{73.1}$$

where r is the distance from the source, and the amplitude is arbitrarily taken to be unity. We shall, for brevity, omit the factor $e^{-i\omega t}$ from all expressions in the present section.

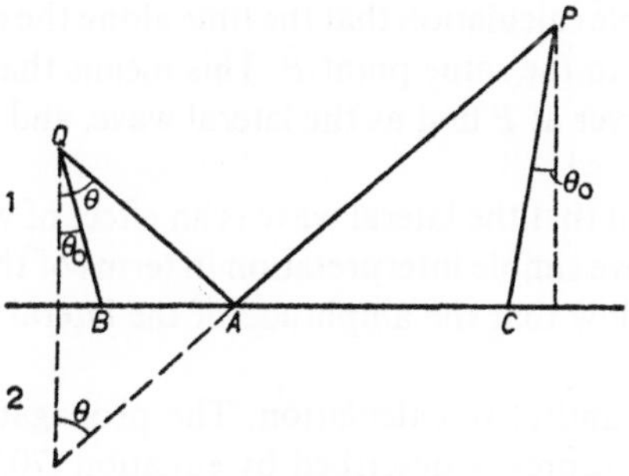

FIG. 46

The wave surfaces of the second (*reflected*) wave are spheres centred at Q', the image of the source Q in the plane of separation; this is the locus of points P reached at a given time by rays which leave Q simultaneously and are reflected from the plane in accordance with the laws of geometrical acoustics (in Fig. 46, the ray QAP with angles of incidence and reflection θ is shown). The amplitude of the reflected wave decreases inversely as the distance r' from the point Q' (which is sometimes called an *imaginary source*), but depends also on the angle θ, as if each ray were reflected with the coefficient corresponding to the reflection of a plane wave at the given angle of incidence θ. In other words, at large distances the reflected wave is given by the formula

$$\phi_1' = \frac{e^{ikr'}}{r'}\left[\frac{\rho_2 c_2 \cos\theta - \rho_1\sqrt{(c_1^2 - c_2^2\sin^2\theta)}}{\rho_2 c_2 \cos\theta + \rho_1\sqrt{(c_1^2 - c_2^2\sin^2\theta)}}\right]; \tag{73.2}$$

cf. formula (66.4) for the reflection coefficient for a plane wave. This formula, which is clearly valid for large r', can be rigorously derived by the method shown below.

A more interesting case is that where $c_1 < c_2$. Here, besides the ordinary reflected wave (73.2), another wave appears in the first medium. The chief properties of this wave can be seen from the following simple considerations.

The ordinary reflected ray QAP (Fig. 46) obeys Fermat's principle in the sense that it is the quickest path from Q to P, among paths lying entirely in medium 1 and involving a single reflection. When $c_1 < c_2$, however, Fermat's principle is also satisfied by another path, where the ray is incident on the boundary at the critical angle of total internal reflection θ_0 ($\sin\theta_0 = c_1/c_2$), then is propagated in medium 2 along the boundary, and finally returns to medium 1 at the angle θ_0. The path is $QBCP$ in Fig. 46, and it is evident that $\theta > \theta_0$. It is easy to see that this path also has the extremal property: the time taken to traverse it is less than for any other path from Q to P lying partly in medium 2.

The geometrical locus of points P reached at the same time by rays which simultaneously leave Q along the path QB, and then return to medium 1 at various points C, is evidently a conical surface whose generators are perpendicular to lines drawn from the imaginary source Q' at an angle θ_0.

Thus, if $c_1 < c_2$, together with the ordinary reflected wave, which has a spherical front, there is propagated in medium 1 another wave, which has a conical front extending from the plane of separation (where it meets the refracted wave front in medium 2) to the point where it touches the spherical front of the reflected wave; this occurs along the line of intersection with a cone having semi-angle θ_0 and axis QQ' (Fig. 45). This conical wave is called the *lateral wave*.

It is easy to see by a simple calculation that the time along the path $QBCP$ (Fig. 46) is less than along the path QAP to the same point P. This means that a sound signal from the source Q reaches an observer at P first as the lateral wave, and only later as the ordinary reflected wave.

It must be borne in mind that the lateral wave is an effect of wave acoustics, despite the fact that it follows the above simple interpretation in terms of the concepts of geometrical acoustics. We shall see below that the amplitude of the lateral wave tends to zero in the limit $\lambda \to 0$.

Let us now make a quantitative calculation. The propagation of a monochromatic sound wave from a point source is described by equation (70.7):

$$\triangle\phi + k^2\phi = -4\pi\delta(\mathbf{r}-\mathbf{l}), \tag{73.3}$$

where $k = \omega/c$ and $\mathbf{l}$ is the position vector of the source. The coefficient of the delta function is chosen so that the direct wave has the form (73.1). In what follows we take a system of coordinates with the xy-plane as the plane of separation and the z-axis along QQ', with the first medium in $z > 0$. At the surface of separation the pressure and the z-component of the velocity, or (equivalently) $\rho\phi$ and $\partial\phi/\partial z$, must be continuous.

Using the general Fourier method, we obtain the solution in the form

$$\phi = \frac{1}{4\pi^2}\int_{-\infty}^{\infty}\int_{-\infty}^{\infty} \phi_\kappa(z)\exp[i(\kappa_x x + \kappa_y y)]\,d\kappa_x d\kappa_y, \tag{73.4}$$

where

$$\phi_\kappa(z) = \int_{-\infty}^{\infty}\int_{-\infty}^{\infty} \phi\exp[-i(\kappa_x x + \kappa_y y)]\,dx\,dy. \tag{73.5}$$

From the symmetry relative to the xy-plane it is evident that ϕ_κ can depend only on the quantity $|\kappa| = \sqrt{(\kappa_x{}^2 + \kappa_y{}^2)}$. Using the formula

$$J_0(u) = \frac{1}{2\pi}\int_0^{2\pi} \cos(u\sin\phi)\,d\phi,$$

we can therefore write (73.4) as

$$\phi = \frac{1}{2\pi}\int_0^{\infty} \phi_\kappa(z) J_0(\kappa R)\kappa\,d\kappa, \tag{73.6}$$

where $R = \sqrt{(x^2+y^2)}$ is the cylindrical coordinate (the distance from the z-axis). It is convenient for the subsequent calculations to transform this formula into one in which the integral is taken from $-\infty$ to ∞, expressing the integrand in terms of the Hankel function $H_0{}^{(1)}(u)$. The latter has a logarithmic singularity at $u = 0$; if we agree to go from positive to negative real u by passing above the point $u = 0$ in the complex u-plane, then $H_0{}^{(1)}(-u) = H_0{}^{(1)}(ue^{i\pi}) = H_0{}^{(1)}(u) - 2J_0(u)$. Using this relation, we can rewrite (73.6) as

$$\phi = \frac{1}{4\pi}\int_{-\infty}^{\infty} \phi_\kappa(z) H_0{}^{(1)}(\kappa R)\kappa\,d\kappa. \tag{73.7}$$

From equation (73.3) we find for the function ϕ_κ the equation

$$\frac{d^2\phi_\kappa}{dz^2}-\left(\kappa^2-\frac{\omega^2}{c^2}\right)\phi_\kappa = -4\pi\delta(z-l). \tag{73.8}$$

The delta function on the right-hand side of the equation can be eliminated by imposing on the function $\phi_\kappa(z)$ (satisfying the homogeneous equation) the boundary conditions at $z = l$:

$$\left.\begin{aligned}[\phi_\kappa(z)]_{l+}-[\phi_\kappa(z)]_{l-} &= 0,\\ [d\phi_\kappa/dz]_{l+}-[d\phi_\kappa/dz]_{l-} &= -4\pi.\end{aligned}\right\} \tag{73.9}$$

The boundary conditions at $z = 0$ are

$$\left.\begin{aligned}[\rho\phi_\kappa]_{0+}-[\rho\phi_\kappa]_{0-} &= 0,\\ [d\phi_\kappa/dz]_{0+}-[d\phi_\kappa/dz]_{0-} &= 0.\end{aligned}\right\} \tag{73.10}$$

We seek a solution in the form

$$\left.\begin{aligned}\phi_\kappa &= Ae^{-\mu_1 z} && \text{for } z > l,\\ \phi_\kappa &= Be^{-\mu_1 z}+Ce^{\mu_1 z} && \text{for } l > z > 0,\\ \phi_\kappa &= De^{\mu_2 z} && \text{for } 0 > z.\end{aligned}\right\} \tag{73.11}$$

Here

$$\mu_1{}^2 = \kappa^2-k_1{}^2,\quad \mu_2{}^2 = \kappa^2-k_2{}^2\ (k_1 = \omega/c_1, k_2 = \omega/c_2),$$

and we must put

$$\left.\begin{aligned}\mu &= +\sqrt{(\kappa^2-k^2)} \text{ for } \kappa > k,\\ \mu &= -i\sqrt{(k^2-\kappa^2)} \text{ for } \kappa < k.\end{aligned}\right\} \tag{73.12}$$

The first of these is necessary so that ϕ should not increase without limit as $z \to \infty$, and the second so that ϕ should represent an outgoing wave. The conditions (73.9) and (73.10) give four equations which determine the coefficients A, B, C and D. A simple calculation gives

$$\left.\begin{aligned}B &= C\frac{\mu_1\rho_2-\mu_2\rho_1}{\mu_1\rho_2+\mu_2\rho_1}, & C &= \frac{2\pi e^{-l\mu_1}}{\mu_1},\\ D &= C\frac{2\rho_1\mu_1}{\mu_1\rho_2+\mu_2\rho_1}, & A &= B+Ce^{2l\mu_1}.\end{aligned}\right\} \tag{73.13}$$

For $\rho_2 = \rho_1$, $c_2 = c_1$ (i.e. when all space is occupied by one medium), B is zero and $A = Ce^{2l\mu_1}$; the corresponding term in ϕ is evidently the direct wave (73.1), and the reflected wave in which we are interested is therefore

$$\phi_1' = \frac{1}{4\pi}\int_{-\infty}^{\infty} B(\kappa)e^{-z\mu_1}H_0{}^{(1)}(\kappa R)\kappa d\kappa. \tag{73.14}$$

In this expression the path of integration has to be specified. It passes above the singular point $\kappa = 0$ (in the complex κ-plane), as already mentioned. The integrand also has singular points (branch points) at $\kappa = \pm k_1, \pm k_2$, where μ_1 or μ_2 vanishes. In accordance with the conditions (73.10), the contour must pass below the points $+k_1$, $+k_2$, and above the points $-k_1$, $-k_2$.

Let us investigate the resulting expression for large distances from the source. Replacing the Hankel function by its asymptotic expression, we obtain

$$\phi_1' = \int_C \frac{\mu_1\rho_2 - \mu_2\rho_1}{\mu_1(\mu_1\rho_2 + \mu_2\rho_1)} \sqrt{\frac{\kappa}{2i\pi R}} \exp[-(z+l)\mu_1 + i\kappa R]\,d\kappa. \tag{73.15}$$

Figure 47 shows the path of integration C for the case $c_1 > c_2$. The integral can be calculated by means of the saddle-point method. The exponent $i[(z+l)\sqrt{(k_1^2-\kappa^2)}+\kappa R]$ has an extremum at the point where

$$\kappa/\sqrt{(k_1^2-\kappa^2)} = R/(z+l) = r'\sin\theta/r'\cos\theta = \tan\theta,$$

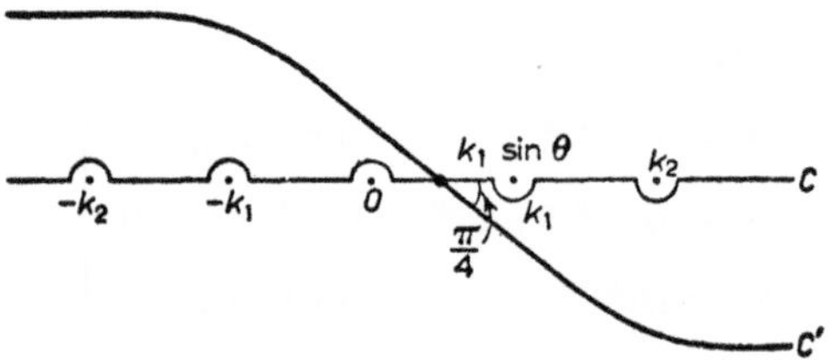

FIG. 47

i.e. $\kappa = k_1 \sin\theta$, where θ is the angle of incidence (see Fig. 45). On changing to the path of integration C' which passes through this point at an angle of $\pi/4$ to the axis of abscissae, we obtain formula (73.2).

In the case $c_1 < c_2$ (i.e. $k_1 > k_2$), the point $\kappa = k_1 \sin\theta$ lies between k_2 and k_1 if $\sin\theta > k_2/k_1 = c_1/c_2 = \sin\theta_0$, i.e. if $\theta > \theta_0$ (Fig. 45). In this case the contour C' must make a loop round the point k_2, and we have, besides the ordinary reflected wave (73.2), a wave ϕ_1'' given by the integral (73.15) taken around the loop, which we call C'' (Fig. 48). This is the lateral wave. The integral is easily calculated if the point $k_1 \sin\theta$ is not close to k_2, i.e. if the angle θ is not close to the internal-reflection angle θ_0.†

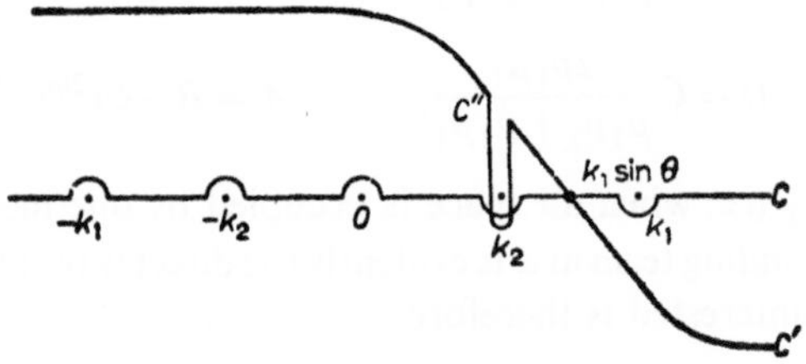

FIG. 48

† For an investigation of the lateral wave for all values of θ, see L. Brekhovskikh, *Zhurnal tekhnicheskoĭ fiziki* **18**, 455, 1948. This paper gives also the next term in the expansion of the ordinary reflected wave in powers of λ/R. We may mention here that, for angles θ close to θ_0 (in the case $c_1 < c_2$), the ratio of the correction term to the leading term falls off with distance as $(\lambda/R)^{\frac{1}{4}}$, and not as λ/R.

Near the point $\kappa = k_2$, μ_2 is small; we expand the coefficient of the exponential in the integrand of (73.15) in powers of μ_2. The zero-order term has no singularity at $\kappa = k_2$, and its integral round C'' is zero. Hence we have

$$\phi_1'' = -\int_{C''} \frac{2\mu_2\rho_1}{\mu_1^2\rho_2}\sqrt{\frac{\kappa}{2\pi i r}}\exp[-(z+l)\mu_1 + i\kappa R]\,d\kappa. \tag{73.16}$$

Expanding the exponent in powers of $\kappa - k_2$ and integrating round the loop C'', we have after a simple calculation the following expression for the potential of the lateral wave:

$$\phi_1'' = \frac{2i\rho_1 k_2 \exp[ik_1 r'\cos(\theta_0-\theta)]}{r'^2\rho_2 k_1^2\sqrt{[\cos\theta_0\sin\theta\sin^3(\theta_0-\theta)]}}. \tag{73.17}$$

In accordance with the previous results, the wave surfaces are the cones $r'\cos(\theta-\theta_0) = R\sin\theta_0 + (z+l)\cos\theta_0 =$ constant. In a given direction, the wave amplitude decreases inversely as the square of the distance r'. We see also that this wave disappears in the limit $\lambda \to 0$. For $\theta \to \theta_0$, the expression (73.17) ceases to be valid; in actual fact, the amplitude of the lateral wave in this range of θ decreases with distance as $r'^{-5/4}$.

§74. The emission of sound

A body oscillating in a fluid causes a periodic compression and rarefaction of the fluid near it, and thus produces sound waves. The energy carried away by these waves is supplied from the kinetic energy of the body. Thus we can speak of the emission of sound by oscillating bodies. In what follows we shall always suppose that the velocity u of the oscillating body is small compared with the velocity of sound. Since $u \sim a\omega$, where a is the linear amplitude of the oscillations, this means that $a \ll \lambda$.†

In the general case of a body of arbitrary shape oscillating in any manner, the problem of the emission of sound waves must be solved as follows. We take the velocity potential ϕ as the fundamental quantity; it satisfies the wave equation

$$\triangle\phi - (1/c^2)\partial^2\phi/\partial t^2 = 0. \tag{74.1}$$

At the surface of the body, the normal component of the fluid velocity must be equal to the corresponding component of the velocity $\mathbf{u}$ of the body:

$$\partial\phi/\partial n = u_n. \tag{74.2}$$

At large distances from the body, the wave must become an outgoing spherical wave. The solution of equation (74.1) which satisfies these boundary conditions and the condition at infinity determines the sound wave emitted by the body.

Let us consider the two limiting cases in more detail. We suppose first that the frequency of oscillation of the body is so large that the length of the emitted wave is very small compared with the dimension l of the body:

$$\lambda \ll l. \tag{74.3}$$

† The amplitude of the oscillations is in general supposed small in comparison with the dimensions of the body also, since otherwise we do not have potential flow near the body (cf. §9). This condition is unnecessary only for pure pulsations, when the solution (74.7) used below is really a direct deduction from the equation of continuity.

In this case we can divide the surface of the body into portions whose dimensions are so small that they may be approximately regarded as plane, but yet are large compared with the wavelength. Then we may suppose that each such portion emits a plane wave, in which the fluid velocity is simply the normal component u_n of the velocity of that portion of the surface. But the mean energy flux in a plane wave is (see §65) $c\rho \overline{v^2}$, where v is the fluid velocity in the wave. Putting $v = u_n$ and integrating over the whole surface of the body, we reach the result that the mean energy emitted per unit time by the body in the form of sound waves, i.e. the total intensity of the emitted sound, is

$$I = c\rho \oint \overline{u_n^2}\, \mathrm{d}f. \qquad (74.4)$$

It is independent of the frequency of the oscillations (for a given velocity amplitude).

Let us now consider the opposite limiting case, where the length of the emitted wave is large compared with the dimension of the body:

$$\lambda \gg l. \qquad (74.5)$$

Then we can neglect the term $(1/c^2)\partial^2\phi/\partial t^2$, in the general equation (74.1), near the body (at distances small compared with the wavelength). For this term is of the order of $\omega^2\phi/c^2 \sim \phi/\lambda^2$, whereas the second derivatives with respect to the coordinates are, in this region, of the order of ϕ/l^2.

Thus the flow near the body satisfies Laplace's equation, $\triangle\phi = 0$. This is the equation for potential flow of an incompressible fluid. Consequently the fluid near the body moves as if it were incompressible. Sound waves proper, i.e. compression and rarefaction waves, occur only at large distances from the body.

At distances of the order of the dimension of the body and smaller, the required solution of the equation $\triangle\phi = 0$ cannot be written in a general form, but depends on the actual shape of the oscillating body. At distances large compared with l, however (though still small compared with λ, so that the equation $\triangle\phi = 0$ remains valid), we can find a general form of the solution by using the fact that ϕ must decrease with increasing distance. We have already discussed such solutions of Laplace's equation in §11. As there, we write the general form of the solution as

$$\phi = -(a/r) + \mathbf{A}\cdot\mathbf{grad}(1/r), \qquad (74.6)$$

where r is the distance from an origin anywhere inside the body. Here, of course, the distances involved must be large compared with the dimension of the body, since we cannot otherwise restrict ourselves to the terms in ϕ which decrease least rapidly as r increases. We have included both terms in (74.6), although it must be borne in mind that the first term is sometimes absent (see below).

Let us ascertain in what cases this term $-a/r$ is non-zero. We found in §11 that a potential $-a/r$ results in a non-zero value $4\pi\rho a$ of the mass flux through a surface surrounding the body. In an incompressible fluid, however such a mass flux can occur only if the total volume of fluid enclosed within the surface changes. In other words, there must be a change in the volume of the body, as a result of which the fluid is either expelled from or "sucked into" the volume of space concerned. Thus the first term in (74.6) appears in cases where the emitting body undergoes pulsations during which its volume changes.

Let us suppose that this is so, and determine the total intensity of the emitted sound. The volume $4\pi a$ of the fluid which flows through the closed surface must, by the foregoing argument, be equal to the change per unit time in the volume V of the body, i.e. to the derivative $\mathrm{d}V/\mathrm{d}t$ (the volume V being a given function of the time): $4\pi a = \dot{V}$. Thus, at distances r such that $l \ll r \ll \lambda$, the motion of the fluid is given by the function $\phi = -\dot{V}(t)/4\pi r$. At distances $r \gg \lambda$, however (i.e. in the *wave region*), ϕ must represent an outgoing spherical wave, i.e. must have the form

$$\phi = -\frac{f(t-r/c)}{r}. \tag{74.7}$$

Hence we conclude at once that the emitted wave has, at all distances large compared with l, the form

$$\phi = -\frac{\dot{V}(t-r/c)}{4\pi r}, \tag{74.8}$$

which is obtained by replacing the argument t of $\dot{V}(t)$ by $t-r/c$.

The velocity $\mathbf{v} = \mathbf{grad}\,\phi$ is directed at every point along the position vector, and its magnitude is $v = \partial\phi/\partial r$. In differentiating (74.8) for distances $r \gg \lambda$, only the derivative of the numerator need be taken, since differentiation of the denominator would give a term of higher order in $1/r$, which we neglect. Since $\partial\dot{V}(t-r/c)/\partial r = -(1/c)\ddot{V}(t-r/c)$, we obtain

$$\mathbf{v} = \ddot{V}(t-r/c)\mathbf{n}/4\pi c r, \tag{74.9}$$

where $\mathbf{n}$ is a unit vector in the direction of $\mathbf{r}$.

The intensity of the sound is given by the square of the velocity, and is here independent of the direction of emission, i.e. the emission is isotropic. The mean value of the total energy emitted per unit time is

$$I = \rho c \oint \overline{v^2}\,\mathrm{d}f = (\rho/16c\pi^2)\oint(\overline{\ddot{V}^2}/r^2)\,\mathrm{d}f,$$

where the integration is taken over a closed surface surrounding the origin. Taking this surface to be a sphere with radius r, and noticing that the integrand depends only on the distance from the origin, we have finally

$$I = \rho\overline{\ddot{V}^2}/4\pi c. \tag{74.10}$$

This is the total intensity of the emitted sound. We see that it is given by the squared second time derivative of the volume of the body.

If the body executes harmonic pulsations with frequency ω, the second time derivative of the volume is proportional to the frequency and velocity amplitude of the oscillations, and its mean square is proportional to the square of the frequency. Thus the intensity of emission is proportional to the square of the frequency for a given velocity amplitude of points on the surface of the body. For a given amplitude of the oscillations, however, the velocity amplitude is itself proportional to the frequency, so that the intensity of emission is proportional to ω^4.

Let us now consider the emission of sound by a body oscillating without change of volume. Only the second term then remains in (74.6); we write it $\phi = \mathrm{div}[\mathbf{A}(t)/r]$. As in the preceding case, we conclude that the general form of the solution at all distances $r \gg l$ is $\phi = \mathrm{div}[\mathbf{A}(t-r/c)/r]$. That this expression is in fact a solution of the wave equation is seen immediately, since the function $\mathbf{A}(t-r/c)/r$ is a solution, and therefore so are its derivatives with respect to the coordinates. Again differentiating only the numerator, we obtain (for distances $r \gg \lambda$)

$$\phi = -\dot{\mathbf{A}}(t-r/c)\cdot\mathbf{n}/cr. \tag{74.11}$$

To calculate the velocity $\mathbf{v} = \mathbf{grad}\,\phi$, we need again differentiate only $\mathbf{A}$. Hence we have, by the familiar rules of vector analysis for differentiation with respect to a scalar argument,

$$\mathbf{v} = -\frac{\ddot{\mathbf{A}}(t-r/c)\cdot\mathbf{n}}{c^2 r}\mathbf{grad}\left(t-\frac{r}{c}\right),$$

and, substituting $\mathbf{grad}(t-r/c) = -(1/c)\,\mathbf{grad}\,r = -\mathbf{n}/c$, we have finally

$$\mathbf{v} = \mathbf{n}(\mathbf{n}\cdot\ddot{\mathbf{A}})/c^2 r. \tag{74.12}$$

The intensity is now proportional to the squared cosine of the angle between the direction of emission (i.e. the direction of $\mathbf{n}$) and the vector $\ddot{\mathbf{A}}$; this is called *dipole emission.* The total emission is given by the integral

$$I = \frac{\rho}{c^3}\oint\frac{\overline{(\mathbf{n}\cdot\ddot{\mathbf{A}})^2}}{r^2}\,\mathrm{d}f.$$

We again take the surface of integration to be a sphere with radius r, and use spherical polar coordinates with the polar axis in the direction of the vector $\ddot{\mathbf{A}}$. A simple integration gives finally for the total emission per unit time

$$I = \frac{4\pi\rho}{3c^3}\overline{\ddot{\mathbf{A}}^2}. \tag{74.13}$$

The components of the vector $\mathbf{A}$ are linear functions of the components of the velocity $\mathbf{u}$ of the body (see §11). Thus the intensity is here a quadratic function of the second time derivatives of the velocity components.

If the body executes harmonic oscillations with frequency ω, we conclude (reasoning as in the previous case) that the intensity is proportional to ω^4 for a given value of the velocity amplitude. For a given linear amplitude of the oscillations of the body, the velocity amplitude is proportional to the frequency, and therefore the intensity is proportional to ω^6.

In an entirely similar manner we can solve the problem of the emission of cylindrical sound waves by a cylinder with any cross-section pulsating or oscillating perpendicularly to its axis. We shall give here the corresponding formulae, with a view to later applications.

Let us first consider small pulsations of a cylinder, and let $S = S(t)$ be its (variable) cross-sectional area. At distances r from the axis of the cylinder such that $l \ll r \ll \lambda$, where l is the transverse dimension of the cylinder, we have similarly to (74.8)

$$\phi = [\dot{S}(t)/2\pi]\log fr, \tag{74.14}$$

where $f(t)$ is a function of time, and the coefficient of $\log fr$ is chosen so as to obtain the correct value for the mass flux through a coaxial cylindrical surface. In accordance with the formula for the potential of an outgoing cylindrical wave (the first term of formula (71.2)), we now conclude that at all distances $r \gg l$ the potential is given by

$$\phi = -\frac{c}{2\pi}\int_{-\infty}^{t-r/c}\frac{\dot{S}(t')\,\mathrm{d}t'}{\sqrt{[c^2(t-t')^2-r^2]}}. \tag{74.15}$$

As $r \to 0$ the leading term of this expression is the same as (74.14), and the function $f(t)$ in the latter equation is automatically determined (we suppose that the derivative $\dot{S}(t)$ tends sufficiently rapidly to zero as $t \to -\infty$). For very large values of r, on the other hand (the wave region), the values of $t-t' \sim r/c$ are the most important in the integral (74.15). We can therefore put, in the denominator of the integrand,

$$(t-t')^2 - r^2/c^2 \cong (2r/c)(t-t'-r/c),$$

obtaining

$$\phi = -\frac{c}{2\pi\sqrt{(2r)}} \int_{-\infty}^{t-r/c} \frac{\dot{S}(t')\,dt'}{\sqrt{[c(t-t')-r]}}. \tag{74.16}$$

Finally, the velocity $v = \partial\phi/\partial r$. To effect the differentiation, it is convenient to substitute $t-t'-r/c = \xi$:

$$\phi = -\frac{1}{2\pi}\sqrt{\frac{c}{2r}} \int_0^{\infty} \frac{\dot{S}(t-r/c-\xi)}{\sqrt{\xi}}\,d\xi;$$

the limits of integration are then independent of r. The factor $1/\sqrt{r}$ in front of the integral need not be differentiated, since this would give a term of higher order in $1/r$. Differentiating under the integral sign and then returning to the variable t', we obtain

$$v = \frac{1}{2\pi\sqrt{(2r)}} \int_{-\infty}^{t-r/c} \frac{\ddot{S}(t')\,dt'}{\sqrt{[c(t-t')-r]}}. \tag{74.17}$$

The intensity is given by the product $2\pi r\rho c v^2$. It should be noticed that here, unlike what happens for the spherical case, the intensity at any instant is determined by the behaviour of the function $S(t)$ at all times from $-\infty$ to $t-r/c$.

Finally, for translatory oscillations of an infinite cylinder in a direction perpendicular to its axis, the potential at distances r such that $l \ll r \ll \lambda$ has the form

$$\phi = \operatorname{div}(\mathbf{A}\log fr), \tag{74.18}$$

where $\mathbf{A}(t)$ is determined by solving Laplace's equation for the flow of an incompressible fluid past a cylinder. Hence we again conclude that, at all distances $r \gg l$,

$$\phi = -\operatorname{div} \int_{-\infty}^{t-r/c} \frac{\mathbf{A}(t')\,dt'}{\sqrt{[(t-t')^2 - r^2/c^2]}}. \tag{74.19}$$

In conclusion, we must make the following remark. We have here entirely neglected the effect of the viscosity of the fluid, and accordingly have supposed that there is potential flow in the emitted wave. In reality, however, we do not have potential flow in a fluid layer with thickness $\sim\sqrt{(\nu/\omega)}$ round the oscillating body (see §24). Hence, if the above formulae are to be applicable, it is necessary that the thickness of this layer should be small in comparison with the dimension l of the body:

$$\sqrt{(\nu/\omega)} \ll l. \tag{74.20}$$

This condition may not hold for small frequencies or small dimensions of the body.

PROBLEMS

PROBLEM 1. Determine the total intensity of sound emitted by a sphere executing small (harmonic) translatory oscillations with frequency ω, the wavelength being comparable in magnitude with the radius R of the sphere.

SOLUTION. We write the velocity of the sphere in the form $\mathbf{u} = \mathbf{u}_0 e^{-i\omega t}$; then ϕ depends on the time through a factor $e^{-i\omega t}$ also, and satisfies the equation $\triangle\phi + k^2\phi = 0$, where $k = \omega/c$. We seek a solution in the form $\phi = \mathbf{u}\cdot\mathbf{grad}\, f(r)$, the origin being taken at the instantaneous position of the centre of the sphere. For f we obtain the equation $\mathbf{u}\cdot\mathbf{grad}\ (\triangle f + k^2 f) = 0$, whence $\triangle f + k^2 f = \text{constant}$. Apart from an unimportant additive constant, we therefore have $f = Ae^{ikr}/r$. The constant A is determined from the condition $\partial\phi/\partial r = u_r$ for $r = R$, and the result is

$$\phi = \mathbf{u}\cdot\mathbf{r}e^{ik(r-R)}\left(\frac{R}{r}\right)^3 \frac{ikr-1}{2-2ikR-k^2R^2}.$$

Thus we have dipole emission. At sufficiently large distances from the sphere, we can neglect unity in comparison with ikr, and ϕ takes the form (74.11), the vector $\dot{\mathbf{A}}$ being

$$\dot{\mathbf{A}} = -\mathbf{u}e^{ik(r-R)}R^3\frac{i\omega}{2-2ikR-k^2R^2}.$$

Noticing that $\overline{(\text{re}\ \ddot{\mathbf{A}})^2} = \frac{1}{2}\overline{|\ddot{\mathbf{A}}|^2}$, we obtain for the total emission, by (74.13),

$$I = \frac{2\pi\rho}{3c^3}|\mathbf{u}_0|^2\frac{R^6\omega^4}{4+(\omega R/c)^4}.$$

For $\omega R/c \ll 1$, this expression becomes $I = \pi\rho R^6|\mathbf{u}_0|^2\omega^4/6c^3$, a result which could also be obtained by directly substituting in (74.13) the expression $\mathbf{A} = \frac{1}{2}R^3\mathbf{u}$ from §11, Problem 1. For $\omega R/c \gg 1$ we have $I = 2\pi\rho c R^2|\mathbf{u}_0|^2/3$, corresponding to formula (74.4).

The drag force acting on the sphere is obtained by integrating over the surface of the sphere the component of the pressure forces $(p' = -\rho(\dot{\phi}')_{r=R})$ in the direction of $\mathbf{u}$, and is

$$\mathbf{F} = \frac{4\pi}{3}\rho\omega R^3\mathbf{u}\frac{-k^3R^3+i(2+k^2R^2)}{4+k^4R^4};$$

see the end of §24 concerning the meaning of a complex drag force.

PROBLEM 2. The same as Problem 1, but for the case where the radius R of the sphere is comparable in magnitude with $\sqrt{(\nu/\omega)}$, whilst $\lambda \gg R$.

SOLUTION. If the dimension of the body is not large compared with $\sqrt{(\nu/\omega)}$, then the emitted wave must be investigated not from the equation $\triangle\phi = 0$, but from the equation of motion of an incompressible viscous fluid. The appropriate solution of this equation for a sphere is given by formulae (1) and (2) in §24, Problem 5. At great distances the first term in (1), which diminishes exponentially with r, may be omitted. The second term gives the velocity $\mathbf{v} = -b(\mathbf{u}\cdot\mathbf{grad})\mathbf{grad}(1/r)$. Comparison with (74.6) shows that

$$\mathbf{A} = -b\mathbf{u} = \tfrac{1}{2}R^3[1-3/(i-1)\kappa - 3/2i\kappa^2]\,\mathbf{u},$$

where $\kappa = R\sqrt{(\omega/2\nu)}$, i.e. $\mathbf{A}$ differs from the corresponding expression for an ideal fluid by the factor in brackets. The result is

$$I = \frac{\pi\rho R^6}{6c^3}\omega^4\left(1+\frac{3}{\kappa}+\frac{9}{2\kappa^2}+\frac{9}{2\kappa^3}+\frac{9}{4\kappa^4}\right)|\mathbf{u}_0|^2.$$

For $\kappa \gg 1$ this becomes the formula given in Problem 1, while for $\kappa \ll 1$ we obtain

$$I = 3\pi\rho R^2\nu^2\omega^2|\mathbf{u}_0|^2/2c^3,$$

i.e. the emission is proportional to the second, and not the fourth, power of the frequency.

PROBLEM 3. Determine the intensity of sound emitted by a sphere executing small (harmonic) pulsations with any frequency.

SOLUTION. We seek a solution of the form $\phi = (au/r)\,e^{ik(r-R)}$, R being the equilibrium radius of the sphere, and determine the constant a from the condition $[\partial\phi/\partial r]_{r=R} = u = u_0 e^{-i\omega t}$ (where u is the radial velocity of points on the surface of the sphere):

$$a = R^2/(ikR-1).$$

The intensity is $I = 2\pi\rho c|u_0|^2k^2R^4/(1+k^2R^2)$. For $kR \ll 1$, $I = 2\pi\rho\omega^2R^4|u_0|^2/c$, in accordance with (74.10), while for $kR \gg 1$, $I = 2\pi\rho cR^2|u_0|^2$, in accordance with (74.4).

Problem 4. Determine the nature of the wave emitted by a sphere (with radius R) executing small pulsations, when the radial velocity of points on the surface is any function $u(t)$ of the time.

Solution. We seek a solution in the form $\phi = f(t')/r$, where $t' = t-(r-R)/c$, and determine f from the boundary condition $\partial\phi/\partial r = u(t)$ for $r = R$. This gives the equation $\mathrm{d}f/\mathrm{d}t + cf(t)/R = -Rcu(t)$. Solving this linear equation and replacing t by t' in the solution for f, we obtain

$$\phi(r,t) = -\frac{cR}{r}e^{-ct'/R}\int_{-\infty}^{t'} u(\tau)e^{c\tau/R}\,\mathrm{d}\tau. \tag{1}$$

If the oscillations of the sphere cease at some instant, say $t = 0$ (i.e. $u(\tau) = 0$ for $\tau > 0$), then the potential at a distance r from the centre will have the form $\phi = \text{constant} \times e^{-ct/R}$ after the instant $t = (r-R)/c$, i.e. it will diminish exponentially.

Let T be the time during which the velocity $u(t)$ changes appreciably. If $T \gg R/c$, i.e. if the wavelength of the emitted waves $\lambda \sim cT \gg R$, then we can take the slowly varying factor $u(\tau)$ outside the integral in (1), replacing it by $u(t')$. For distances $r \gg R$, we then obtain $\phi = -(R^2/r)u(t-r/c)$, in accordance with formula (74.8). If, on the other hand, $T \ll R/c$, we obtain in a similar manner

$$\phi = -\frac{cR}{r}\int_{-\infty}^{t'} u(\tau)\,\mathrm{d}\tau, \quad v = \partial\phi/\partial r = (R/r)u(t'),$$

in accordance with formula (74.4).

Problem 5. Determine the motion of an ideal compressible fluid when a sphere with radius R executes in it an arbitrary translatory motion, with velocity small compared with that of sound.

Solution. We seek a solution in the form

$$\phi = \operatorname{div}[\mathbf{f}(t')/r], \tag{1}$$

where r is the distance from the origin, taken at the position of the centre of the sphere at the time $t' = t-(r-R)/c$; since the velocity $\mathbf{u}$ of the sphere is small compared with the velocity of sound, the movement of the origin may be neglected. The fluid velocity is

$$\mathbf{v} = \mathbf{grad}\,\phi = \frac{3(\mathbf{f}\cdot\mathbf{n})\mathbf{n}-\mathbf{f}}{r^3} + \frac{3(\mathbf{f}'\cdot\mathbf{n})\mathbf{n}-\mathbf{f}'}{cr^2} + \frac{(\mathbf{f}''\cdot\mathbf{n})\mathbf{n}}{c^2r}, \tag{2}$$

where $\mathbf{n}$ is a unit vector in the direction of $\mathbf{r}$, and the prime denotes differentiation with respect to the argument of $\mathbf{f}$. The boundary condition is $v_r = \mathbf{u}\cdot\mathbf{n}$ for $r = R$, whence $\mathbf{f}''(t) + (2c/R)\mathbf{f}'(t) + (2c^2/R^2)\mathbf{f}(t) = Rc^2\mathbf{u}(t)$. Solving this equation by variation of the parameters, we obtain for the function $\mathbf{f}(t)$ the general expression

$$\mathbf{f}(t) = cR^2e^{-ct/R}\int_{-\infty}^{t}\mathbf{u}(\tau)\sin\frac{c(t-\tau)}{R}e^{c\tau/R}\,\mathrm{d}\tau. \tag{3}$$

In substituting in (1), we must replace t by t'. The lower limit is taken as $-\infty$ so that $\mathbf{f}$ shall be zero for $t = -\infty$.

Problem 6. A sphere with radius R begins at time $t = 0$ to move with constant velocity $\mathbf{u}_0$. Determine the sound intensity emitted at the instant when the motion begins.

Solution. Putting in formula (3) of Problem 5 $\mathbf{u}(\tau) = 0$ for $\tau < 0$ and $\mathbf{u}(\tau) = \mathbf{u}_0$ for $\tau > 0$, and substituting in formula (2) (retaining only the last term, which decreases least rapidly with increasing r), we find the fluid velocity far from the sphere:

$$\mathbf{v} = -\mathbf{n}(\mathbf{n}\cdot\mathbf{u}_0)\frac{\sqrt{2}R}{r}e^{-ct'/R}\sin\left(\frac{ct'}{R}-\tfrac{1}{4}\pi\right),$$

where $t' > 0$. The total intensity diminishes with time according to

$$I = (8\pi/3)c\rho R^2{u_0}^2e^{-2ct'/R}\sin^2(ct'/R-\tfrac{1}{4}\pi).$$

The total amount of energy emitted is $\frac{1}{3}\pi\rho R^3{u_0}^2$.

Problem 7. Determine the intensity of sound emitted by an infinite cylinder, with radius R, executing harmonic pulsations with wavelength $\lambda \gg R$.

Solution. According to formula (74.14), we find first of all that, at distances $r \ll \lambda$ (in Problems 7 and 8 r is the distance from the axis of the cylinder), the potential is $\phi = Ru \log kr$, where $u = u_0 e^{-i\omega t}$ is the velocity of points on the surface of the cylinder. From a comparison with formulae (71.7) and (71.8), we now find that at large distances the potential has the form $\phi = -Ru\sqrt{(i\pi/2kr)}\,e^{ikr}$. The velocity is therefore

$$\mathbf{v} = Ru\sqrt{(\pi k/2ir)}\,\mathbf{n}e^{ikr},$$

where $\mathbf{n}$ is a unit vector perpendicular to the axis of the cylinder, and the intensity per unit length of the cylinder is $I = \frac{1}{2}\pi^2\rho\omega R^2|u_0|^2$.

Problem 8. Determine the intensity of sound emitted by a cylinder executing harmonic translatory oscillations in a direction perpendicular to its axis.

Solution. At distances $r \ll \lambda$ we have $\phi = -\mathrm{div}\,(R^2\mathbf{u}\log kr)$; cf. formula (74.18) and §10, Problem 3. Hence we conclude that at large distances

$$\phi = R^2\sqrt{(i\pi/2k)}\,\mathrm{div}\,(e^{ikr}\mathbf{u}/\sqrt{r}) = -R^2\mathbf{u}\cdot\mathbf{n}\sqrt{(\pi k/2ir)}\,e^{ikr},$$

whence the velocity is $\mathbf{v} = -kR^2\sqrt{(i\pi k/2r)}\,\mathbf{n}(\mathbf{u}\cdot\mathbf{n})\,e^{ikr}$. The intensity is proportional to the squared cosine of the angle between the directions of oscillation and emission. The total intensity is $I = (\pi^2/4c^2)\,\rho\omega^3R^4|\mathbf{u}_0|^2$.

Problem 9. Determine the intensity of sound emitted by a plane surface whose temperature varies periodically with frequency $\omega \ll c^2/\chi$, where χ is the thermometric conductivity of the fluid.

Solution. Let the variable part of the temperature of the surface be $T'_0 e^{-i\omega t}$. These temperature oscillations cause a damped thermal wave (52.15) in the fluid:

$$T' = T'_0 e^{-i\omega t} e^{-(1-i)\sqrt{(\omega/2\chi)}x},$$

and the fluid density therefore oscillates also: $\rho' = (\partial\rho/\partial T)_p T' = -\rho\beta T'$, where β is the coefficient of thermal expansion. This, in turn, results in the occurrence of a motion determined by the equation of continuity: $\rho\partial v/\partial x = -\partial\rho'/\partial t = -i\omega\rho\beta T'$. At the solid surface the velocity $v_x = v = 0$, and far from the surface it tends to the limit

$$v = -i\omega\beta\int_0^\infty T'\,\mathrm{d}x = \frac{1-i}{\sqrt{2}}\,\beta\sqrt{(\omega\chi)}\,T'_0 e^{-i\omega t}.$$

This value is reached at distances $\sim\sqrt{(\chi/\omega)}$, which are small compared with c/ω, and we thus have a boundary condition on the resulting sound wave. Hence we find the intensity per unit area of the surface to be $I = \frac{1}{2}c\rho\beta^2\omega\chi|T'_0|^2$.

Problem 10. A point source emitting a spherical wave is at a distance l from a solid wall which totally reflects sound and bounds a half-space occupied by fluid. Determine the ratio of the total intensity of sound emitted by the source to that which would be found in an infinite medium, and the dependence of the intensity on direction for large distances from the source.

Solution. The sum of the direct and reflected waves is given by a solution of the wave equation such that the normal velocity component $v_n = \partial\phi/\partial n$ is zero at the wall. Such a solution is

$$\phi = \left(\frac{e^{ikr}}{r} + \frac{e^{ikr'}}{r'}\right)e^{-i\omega t}$$

(we omit the constant factor, for brevity), where r is the distance from the source O (Fig. 49), and r' is the distance from a point O' which is the image of O in the wall. At large distances from the source we have $r' \cong r - 2l\cos\theta$, so that

$$\phi = \frac{e^{i(kr-\omega t)}}{r}(1 + e^{-2ikl\cos\theta}).$$

The dependence of the intensity on direction is given by a factor $\cos^2(kl\cos\theta)$.

To determine the total intensity, we integrate the energy flux $\bar{\mathbf{q}} = \overline{p'\mathbf{v}} = \overline{-\rho\dot{\phi}\,\mathbf{grad}\,\phi}$ (see (65.4)) over the surface of a sphere with arbitrarily small radius, centred at O. This gives $2\pi\rho k\omega(1 + [1/2kl]\sin 2kl)$. In an infinite medium, on the other hand, we should have simply a spherical wave $\phi = e^{i(kr-\omega t)}/r$, with a total energy flux $2\pi\rho k\omega$. Thus the required ratio of intensities is $1 + (1/2kl)\sin 2kl$.

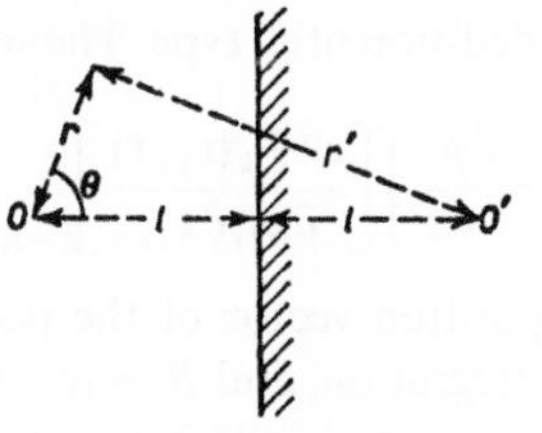

FIG. 49

PROBLEM 11. The same as Problem 10, but for a fluid bounded by a free surface.

SOLUTION. At the free surface the condition $p' = -\rho\dot{\phi} = 0$ must hold; in a monochromatic wave this is equivalent to $\phi = 0$. The corresponding solution of the wave equation is

$$\phi = \left(\frac{e^{ikr}}{r} - \frac{e^{ikr'}}{r'}\right)e^{-i\omega t}$$

At large distances from the source, the intensity is given by a factor $\sin^2(kl\cos\theta)$. The required ratio of intensities is $1-(1/2kl)\sin 2kl$.

§75. Sound excitation by turbulence

Turbulent velocity fluctuations also are a cause of sound excitation in the surrounding fluid. The present section will give the general theory of this effect (M. J. Lighthill 1952). We shall consider the case where the turbulence occupies a finite region V_0 surrounded by an infinite volume of fluid at rest. The turbulence itself is treated in terms of incompressible fluid theory, the density changes due to the fluctuations being neglected. This means that the velocity of the turbulent flow is assumed to be much less than that of sound (as was assumed throughout Chapter III).

We shall begin by deriving the general equation, taking into account not only the motion in the sound waves but also the flow in the turbulent region. The only difference from the derivation in §64 is that the non-linear term $(\mathbf{v}\cdot\mathbf{grad})\mathbf{v}$ must be retained: although v is much less than c, it is much greater than the fluid velocity in the sound wave. We therefore have instead of (64.3)

$$\partial\mathbf{v}/\partial t + (\mathbf{v}\cdot\mathbf{grad})\mathbf{v} + (1/\rho_0)\,\mathbf{grad}\,p' = 0.$$

Taking the divergence of this equation and using (64.5),

$$\partial p'/\partial t + \rho_0 c^2 \operatorname{div}\mathbf{v} = 0,$$

we obtain

$$\frac{1}{c^2}\frac{\partial^2 p'}{\partial t^2} - \triangle p' = \rho_0\frac{\partial}{\partial x_i}\left(v_k\frac{\partial v_i}{\partial x_k}\right).$$

The right-hand side of this equation can be transformed by means of the equation of continuity $\operatorname{div}\mathbf{v} = 0$ (the turbulence being regarded as incompressible), and the differentiation with respect to x_k taken outside the brackets. The final result is

$$\frac{1}{c^2}\frac{\partial^2 p'}{\partial t^2} - \triangle p' = \rho\frac{\partial^2 T_{ik}}{\partial x_i \partial x_k}, \quad T_{ik} = v_i v_k, \tag{75.1}$$

the suffix in ρ_0 being again omitted. Outside the turbulent region, the expression on the right is a second-order small quantity and may be omitted, so that we return to the wave equation of sound propagation. The non-zero right-hand side in the volume V_0 acts as a source of sound. In that volume, $\mathbf{v}$ is the velocity of the turbulent flow.

Equation (75.1) is of the retarded-potential type. The solution which describes emission from a source is

$$p'(\mathbf{r}, t) = \frac{\rho}{4\pi}\int\left[\frac{\partial^2 T_{ik}(\mathbf{r}_1, t)}{\partial x_{1i}\partial x_{1k}}\right]_{t-R/c}\frac{dV_1}{R}; \tag{75.2}$$

see *Fields*, §62. Here, $\mathbf{r}$ is the position vector of the point of observation, $\mathbf{r}_1$ that of a variable point in the region of integration, and $R = |\mathbf{r} - \mathbf{r}_1|$; the integrand is taken at the "retarded" time $t - R/c$. The integration in (75.2) is in practice to be taken only over the volume V_0 in which the integrand is non-zero.

The majority of the energy of turbulent flow is at frequencies $\sim u/l$ which correspond to the fundamental scale l of the turbulence; u is the characteristic velocity (see §33). These will evidently be also the main frequencies in the spectrum of sound waves excited. The corresponding wavelengths $\lambda \sim cl/u \gg l$.

To determine the emission intensity, it is sufficient to consider the sound at distances much greater than the wavelength λ (in the wave region); these are also much greater than the linear size of the source, i.e. of the turbulent region.† The factor $1/R$ in the integrand may be replaced in this region by $1/r$ and taken outside the integral (r being the distance from the point of observation to an origin taken somewhere inside the source); we thereby neglect terms that decrease faster than $1/r$, which in any case do not contribute to the intensity of waves going to infinity. Thus

$$p'(\mathbf{r}, t) = \frac{\rho}{4\pi r}\int\left[\frac{\partial^2 T_{ik}(\mathbf{r}_1, t)}{\partial x_{1i}\partial x_{1k}}\right]_{t-R/c} dV_1. \tag{75.3}$$

The derivatives in the integrand are taken before the evaluation at $t - R/c$, that is, only with respect to the first argument of the $T_{ik}(\mathbf{r}_1, t)$. They may be replaced by derivatives of the functions $T_{ik}(\mathbf{r}, t - R/c)$ taken with respect to both arguments, the derivatives with respect to the second argument being subtracted. The former are complete divergences, and their integrals give zero when transformed into integrals over distant closed surfaces, since $T_{ik} = 0$ outside the turbulent region. The derivatives with respect to the variable coordinates $\mathbf{r}_1$ which appear in the argument $t - R/c$ may be replaced by those with respect to the coordinates of the point of observation $\mathbf{r}$, since $\mathbf{r}$ and $\mathbf{r}_1$ occur only as the difference $R = |\mathbf{r} - \mathbf{r}_1|$. We thus obtain

$$p'(\mathbf{r}, t) = \frac{\rho}{4\pi r}\frac{\partial^2}{\partial x_i \partial x_k}\int T_{ik}(\mathbf{r}_1, t - R/c) dV_1. \tag{75.4}$$

The time $t - R/c$ differs from $t - r/c$ by $\sim l/c$. This, however, is small compared with the periods l/u of the fundamental turbulent eddies. This allows the argument $t - R/c$ in the integrand to be replaced by $\tau \equiv t - r/c$.‡ Then, differentiating under the integral sign and noting that $\partial r/\partial x_i = n_i$ (where $\mathbf{n}$ is a unit vector along $\mathbf{r}$), we obtain

$$p'(\mathbf{r}, t) = \frac{\rho}{4\pi c^2 r} n_i n_k \int \ddot{T}_{ik}(\mathbf{r}_1, \tau) dV_1. \tag{75.5}$$

where a dot denotes differentiation with respect to τ.

The tensor $\ddot{T}_{ik}$, like any symmetrical tensor with non-zero trace, can be put in the form

$$\ddot{T}_{ik} = (\ddot{T}_{ik} - \tfrac{1}{3}\ddot{T}_{ll}\delta_{ik}) + \tfrac{1}{3}\ddot{T}_{ll}\delta_{ik} \equiv Q_{ik} + Q\delta_{ik}, \tag{75.6}$$

† In referring to orders of magnitude we make no distinction between the fundamental scale l and the size of the turbulent region, although the latter may be noticeably larger.

‡ Here, we do not consider the emission spectrum, but take only the principal frequencies which determine the total intensity. Note also that the substitution in question could not have been made at an earlier stage, in (75.3), since the integral would then be zero.

where Q_{ik} is an "irreducible" tensor with zero trace, and Q is a scalar. Then the spherical wave (75.5) separates as a sum of two terms:

$$p'(\mathbf{r}, t) = \frac{\rho}{4\pi c^2 r}\left\{\int Q(\mathbf{r}_1, \tau)\,\mathrm{d}V_1 + n_i n_k \int Q_{ik}(\mathbf{r}_1, \tau)\,\mathrm{d}V_1\right\}, \tag{75.7}$$

which respectively represent the emission from monopole and quadrupole sources.

Let us next calculate the total emitted intensity. The sound energy flux density in the wave region is along $\mathbf{n}$ at every point, and its magnitude is $q = p'^2/c\rho$. The total intensity is found by multiplying q by $r^2\,do$ and integrating over all directions of $\mathbf{n}$.† In practice, however, we are interested not in the instantaneous fluctuating value of the intensity but in the time-averaged value (the turbulence being here assumed "steady"). The latter operation is carried out by writting the squares of the integrals as double integrals and averaging (denoted by angle brackets) under the integral signs. The result is

$$I = \frac{\rho_0}{60\pi c^5}\iint \langle Q(\mathbf{r}_1, \tau) Q(\mathbf{r}_2, \tau)\rangle\,\mathrm{d}V_1\,\mathrm{d}V_2 + {} + \frac{\rho_0}{30\pi c^5}\iint \langle Q_{ik}(\mathbf{r}_1, \tau) Q_{ik}(\mathbf{r}_2, \tau)\rangle\,\mathrm{d}V_1\,\mathrm{d}V_2. \tag{75.8}$$

The cross product of the two terms in (75.7) disappears on integration over directions, and so the total intensity is the sum of the monopole and quadrupole emissions. In the present case, these two parts have in general the same order of magnitude.

Let us estimate this order of magnitude (or rather, determine the dependence of I on the turbulent flow parameters). The tensor components $T_{ik} \sim u^2$, where u is the characteristic velocity of the turbulent flow. Each differentiation with respect to time multiplies this order of magnitude by the characteristic frequency u/l. Hence $Q \sim u^4/l^2$. The correlation between the turbulent fluctuation velocities at different points extends to distances $\sim l$. The quantity of energy emitted as sound by unit mass of the turbulent medium per unit time is therefore

$$\varepsilon_s \sim \frac{1}{c^5}\frac{u^8}{l^4}l^3 = \frac{u^8}{c^5 l}. \tag{75.9}$$

This emission intensity is thus proportional to the eighth power of the turbulent flow velocity.

The turbulent flow is maintained by power supplied from some external source. In the "steady" case, this is equal to the energy dissipated per unit time. The latter is, per unit mass, $\varepsilon_d \sim u^3/l$.‡ The acoustic efficiency may be defined as the ratio of the emitted power and the dissipated power:

$$\varepsilon_s/\varepsilon_d \sim (u/c)^5. \tag{75.10}$$

The high power of u/c has the result that when $u/c \ll 1$ the effectiveness of turbulence as a sound source is low.

---

† This integration is achieved by using the following expressions for the mean products of two and four components of $\mathbf{n}$:

$$\overline{n_i n_k} = \tfrac{1}{3}\delta_{ik}, \quad \overline{n_i n_k n_l n_m} = \tfrac{1}{15}(\delta_{ik}\delta_{lm} + \delta_{il}\delta_{km} + \delta_{im}\delta_{kl}).$$

‡ See (33.1). We here do not distinguish between u and Δu; the choice of the frame of reference in which the flow is considered is determined by the fact that the fluid is assumed to be at rest outside the turbulent region.

§76. The reciprocity principle

In deriving the equations of a sound wave in §64, it was assumed that the wave is propagated in a homogeneous medium. In particular, the density ρ_0 of the medium and the velocity of sound in it, c, were regarded as constants. In order to obtain some general relations applicable for an arbitrary inhomogeneous medium, we shall first derive the equation for the propagation of sound in such a medium.

We write the equation of continuity in the form $\mathrm{d}\rho/\mathrm{d}t + \rho\,\mathrm{div}\,\mathbf{v} = 0$. Since the propagation of sound is adiabatic, we have

$$\frac{\mathrm{d}\rho}{\mathrm{d}t} = \left(\frac{\partial \rho}{\partial p}\right)_s \frac{\mathrm{d}p}{\mathrm{d}t} = \frac{1}{c^2}\frac{\mathrm{d}p}{\mathrm{d}t} = \frac{1}{c^2}\left(\frac{\partial p}{\partial t} + \mathbf{v}\cdot\mathbf{grad}\,p\right),$$

and the equation of continuity becomes $\partial p/\partial t + \mathbf{v}\cdot\mathbf{grad}\,p + \rho c^2\,\mathrm{div}\,\mathbf{v} = 0$.

As usual, we put $\rho = \rho_0 + \rho'$, where ρ_0 is now a given function of the coordinates. In the equation $p = p_0 + p'$, however, we must put as before p_0 = constant, since the pressure must be constant throughout a medium in equilibrium (in the absence of an external field, of course). Thus we have to within second-order quantities $\partial p'/\partial t + \rho_0 c^2\,\mathrm{div}\,\mathbf{v} = 0$.

This equation is the same in form as equation (64.5), but the coefficient $\rho_0 c^2$ is a function of the coordinates. As in §64, we obtain Euler's equation in the form $\partial\mathbf{v}/\partial t = -(1/\rho_0)$ $\mathbf{grad}\,p'$. Eliminating $\mathbf{v}$, and omitting the suffix in ρ_0, we finally obtain the equation of propagation of sound in an inhomogeneous medium:

$$\mathrm{div}\,\frac{\mathbf{grad}\,p'}{\rho} - \frac{1}{\rho c^2}\frac{\partial^2 p'}{\partial t^2} = 0. \tag{76.1}$$

If the wave is monochromatic, with frequency ω, we have $\ddot{p}' = -\omega^2 p'$, so that

$$\mathrm{div}\,\frac{\mathbf{grad}\,p'}{\rho} + \frac{\omega^2}{\rho c^2}p' = 0. \tag{76.2}$$

Let us consider a sound wave emitted by a pulsating source of small dimension; we have seen in §74 that the emission is isotropic. We denote by A the point where the source is, and by $p_A(B)$ the pressure p' at a point B in the emitted wave.† If the same source is placed at B, it produces at A a pressure which we denote by $p_B(A)$. We shall derive the relation between $p_A(B)$ and $p_B(A)$.

To do so, we use equation (76.2), applying it first to the sound from a source at A and then to the sound from a source at B:

$$\mathrm{div}\,\frac{\mathbf{grad}\,p'_A}{\rho} + \frac{\omega^2}{\rho c^2}p'_A = 0, \quad \mathrm{div}\,\frac{\mathbf{grad}\,p'_B}{\rho} + \frac{\omega^2}{\rho c^2}p'_B = 0.$$

We multiply the first equation by p'_B and the second by p'_A and subtract. The result is

$$p'_B\,\mathrm{div}\,\frac{\mathbf{grad}\,p'_A}{\rho} - p'_A\,\mathrm{div}\,\frac{\mathbf{grad}\,p'_B}{\rho}$$

$$= \mathrm{div}\left(\frac{p'_B\,\mathbf{grad}\,p'_A}{\rho} - \frac{p'_A\,\mathbf{grad}\,p'_B}{\rho}\right) = 0.$$

† The dimension of the source must be small compared with the distance between A and B and with the wavelength.

We integrate this equation over the volume between an infinitely distant closed surface C and two small spheres C_A and C_B which enclose the points A and B respectively. The volume integral can be transformed into three surface integrals, and the integral over C is zero, since the sound field vanishes at infinity. Thus we obtain

$$\oint_{C_A+C_B} \left(p'_B \frac{\mathbf{grad}\, p'_A}{\rho} - p'_A \frac{\mathbf{grad}\, p'_B}{\rho} \right) \cdot \mathbf{df} = 0. \tag{76.3}$$

Inside the small sphere C_A, the pressure p'_A in the wave from a source at A falls off rapidly with the distance from A, and the gradient $\mathbf{grad}\, p'_A$ is therefore large. The pressure p'_B due to a source at B is a slowly varying function of the coordinates in the region near the point A, which is at a considerable distance from B, so that the gradient $\mathbf{grad}\, p'_B$ is relatively small. When the radius of the sphere C_A is sufficiently small, therefore, we can neglect the integral

$$\oint (p'_A/\rho)\, \mathbf{grad}\, p'_B \cdot \mathbf{df}$$

over C_A in comparison with

$$\oint (p'_B/\rho)\, \mathbf{grad}\, p'_A \cdot \mathbf{df},$$

and in the latter the almost constant quantity p'_B can be taken outside the integral and replaced by its value at the point A. Similar arguments hold for the integrals over the sphere C_B, and as a result we obtain from (76.3) the relation

$$p'_B(A) \oint_{C_A} \frac{\mathbf{grad}\, p'_A}{\rho} \cdot \mathbf{df} = p'_A(B) \oint_{C_B} \frac{\mathbf{grad}\, p'_B}{\rho} \cdot \mathbf{df}.$$

But $(1/\rho)$ $\mathbf{grad}$ $p' = -\partial \mathbf{v}/\partial t$, and this equation can therefore be rewritten

$$p'_B(A) \frac{\partial}{\partial t} \oint_{C_A} \mathbf{v}_A \cdot \mathbf{df} = p'_A(B) \frac{\partial}{\partial t} \oint_{C_B} \mathbf{v}_B \cdot \mathbf{df}.$$

The integral

$$\oint \mathbf{v}_A \cdot \mathbf{df}$$

over C_A is the volume of fluid flowing per unit time through the surface of the sphere C_A, i.e. it is the rate of change of the volume of the pulsating source of sound. Since the sources at A and B are identical, it is clear that

$$\oint_{C_A} \mathbf{v}_A \cdot \mathbf{df} = \oint_{C_B} \mathbf{v}_B \cdot \mathbf{df},$$

and consequently

$$p'_A(B) = p'_B(A). \tag{76.4}$$

This equation constitutes the *reciprocity principle*: the pressure at B due to a source at A is equal to the pressure at A due to a similar source at B. It should be emphasized that this result holds, in particular, for the case where the medium is composed of several different regions, each of which is homogeneous. When sound is propagated in such a medium, it is reflected and refracted at the surfaces separating the various regions. Thus the reciprocity principle is valid also in cases where the wave undergoes reflection and refraction on its path from A to B.

PROBLEM

Derive the reciprocity principle for dipole emission of sound by a source which oscillates without change of volume.

SOLUTION. In this case the integral

$$\oint_{C_A} \mathbf{v}_A \cdot d\mathbf{f} = 0, \tag{1}$$

and the next approximation must be taken in calculating the integrals in (76.3). To do so, we write, as far as the first-order terms,

$$p'_B = p'_B(A) + \mathbf{r} \cdot \mathbf{grad}\, p'_B, \tag{2}$$

where $\mathbf{r}$ is the radius vector from A. In the integral

$$\oint_{C_A} \left(p'_B \frac{\mathbf{grad}\, p'_A}{\rho} - p'_A \frac{\mathbf{grad}\, p'_B}{\rho} \right) \cdot d\mathbf{f}, \tag{3}$$

the two terms are now of the same order of magnitude. Substituting here for p'_B from (2) and using (1), we get

$$\oint_{C_A} \left\{ (\mathbf{r} \cdot \mathbf{grad}\, p'_B) \frac{\mathbf{grad}\, p'_A}{\rho} - p'_A \frac{\mathbf{grad}\, p'_B}{\rho} \right\} \cdot d\mathbf{f}.$$

Next, we take the almost constant quantity $\mathbf{grad}\, p'_B = -\rho \dot{\mathbf{v}}_B$ outside the integral, replacing it by its value at A:

$$\rho_A \dot{\mathbf{v}}_B(A) \cdot \oint_{C_A} \left\{ \frac{p'_A}{\rho} d\mathbf{f} - \mathbf{r} \left(\frac{\mathbf{grad}\, p'_A}{\rho} \cdot d\mathbf{f} \right) \right\},$$

where ρ_A is the density of the medium at the point A. To calculate this integral, we notice that near a source the fluid can be supposed incompressible (see §74), and hence we can write for the pressure inside the small sphere C_A, by (11.1), $p'_A = -\rho \dot{\phi} = \rho \dot{\mathbf{A}} \cdot \mathbf{r}/r^3$. In a monochromatic wave $\dot{\mathbf{v}} = -i\omega \mathbf{v}$, $\dot{\mathbf{A}} = -i\omega \mathbf{A}$; introducing also the unit vector $\mathbf{n}_A$ in the direction of the vector $\mathbf{A}$ for a source at A, we find that the integral (3) is proportional to $\rho_A \mathbf{v}_B(A) \cdot \mathbf{n}_A$. Similarly, the integral over the sphere C_B is proportional to $-\rho_B \mathbf{v}_A(B) \cdot \mathbf{n}_B$, with the same factor of proportionality. Equating the sum to zero, we find the required relation

$$\rho_A \mathbf{v}_B(A) \cdot \mathbf{n}_A = \rho_B \mathbf{v}_A(B) \cdot \mathbf{n}_B,$$

which expresses the reciprocity principle for dipole emission of sound.

§77. Propagation of sound in a tube

Let us now consider the propagation of a sound wave in a long narrow tube. By a "narrow" tube we mean one whose width is small compared with the wavelength. The cross-section of the tube may vary along its length in both shape and area. It is important, however, that this variation should occur fairly slowly: the cross-sectional area S must vary only slightly over distances of the order of the width of the tube.

Under these conditions we can suppose that all quantities (velocity, density, etc.) are constant over any transverse cross-section of the tube. The direction of propagation of the wave can be supposed to coincide with that of the axis of the tube at all points. The equation for the propagation of such a wave is most conveniently derived by a method similiar to that used in §12 in deriving the equation for the propagation of gravity waves in channels.

In unit time a mass $S\rho v$ of fluid passes through a cross-section of the tube. Hence the mass of fluid in the volume between two transverse cross-sections at a distance dx apart decreases in unit time by

$$(S\rho v)_{x+dx} - (S\rho v)_x = [\partial(S\rho v)/\partial x]dx,$$

the coordinate x being measured along the axis of the tube. Since the volume between the two cross-sections remains constant, the decrease must be due only to the change in density of the fluid. The change in density per unit time is $\partial\rho/\partial t$, and the corresponding decrease in the mass of fluid in the volume $S\,dx$ between the two cross-sections is $-S(\partial\rho/\partial t)dx$. Equating the two expressions, we obtain

$$S\partial\rho/\partial t = -\partial(S\rho v)/\partial x, \tag{77.1}$$

which is the equation of continuity for flow in a pipe.

Next, we write down Euler's equation, omitting the term quadratic in the velocity:

$$\partial v/\partial t = -(1/\rho)\partial p/\partial x. \tag{77.2}$$

We differentiate (77.1) with respect to time, regarding ρ on the right-hand side as independent of time, since the differentiation of ρ gives a term which involves $v\,\partial\rho/\partial t = v\,\partial\rho'/\partial t$ and is therefore of the second order of smallness. Thus $S\,\partial^2\rho/\partial t^2 = -\partial(S\rho\partial v/\partial t)/\partial x$. Here we substitute the expression (77.2) for $\partial v/\partial t$, and express the derivative of the density on the left-hand side in terms of the derivative of the pressure by $\ddot{\rho} = \ddot{p}/c^2$.

The result is the following equation for the propagation of sound in a tube:

$$\frac{1}{S}\frac{\partial}{\partial x}\left(S\frac{\partial p}{\partial x}\right) - \frac{1}{c^2}\frac{\partial^2 p}{\partial t^2} = 0. \tag{77.3}$$

In a monochromatic wave p depends on time through a factor $e^{-i\omega t}$, and (77.3) becomes

$$\frac{1}{S}\frac{\partial}{\partial x}\left(S\frac{\partial p}{\partial x}\right) + k^2 p = 0. \tag{77.4}$$

where $k = \omega/c$ is the wave number.†

Finally, let us consider the problem of the emission of sound from the open end of a tube. The pressure difference between the gas in the end of the tube and that in the space surrounding the tube is small compared with the pressure differences within the tube. Hence the boundary condition at the open end of the tube is, with sufficient accuracy, that the pressure p should vanish. The gas velocity v at the end of the tube is not zero; let its value be v_0. The product Sv_0 is the volume of gas leaving the tube per unit time.

We can now regard the open end of the tube as a source of gas with strength Sv_0. The problem of the emission from a tube thus becomes equivalent to that of the emission by a

† Here, and in the Problems, p denotes the variable part of the pressure, which we have previously denoted by p'.

pulsating body, which is solved by formula (74.10). In place of the time derivative $\dot{V}$ of the volume of the body we must now put Sv_0. Thus the total intensity of the sound emitted is

$$I = \rho S^2 \overline{\dot{v}_0^{\,2}}/4\pi c. \tag{77.5}$$

PROBLEMS

Problem 1. Determine the transmission coefficient for sound passing from a tube with cross-section S_1 into one with cross-section S_2.

Solution. In the first tube we have two waves, the incident wave $p_1 = a_1 e^{i(kx-\omega t)}$ and reflected wave $p_1' = a_1' e^{-i(kx+\omega t)}$. In the second tube we have the transmitted wave $p_2 = a_2 e^{i(kx-\omega t)}$. At the point where the tubes join ($x = 0$), the pressures must be equal, and so must the volumes Sv of gas passing from one tube to the other per unit time. These conditions give $a_1 + a_1' = a_2$, $S_1(a_1 - a_1') = S_2 a_2$, whence $a_2 = 2a_1 S_1/(S_1 + S_2)$. The ratio D of the energy flux in the transmitted wave to that in the incident wave is

$$\begin{aligned} D &= S_2\overline{|v_2|^2}/S_1\overline{|v_1|^2} \\ &= \frac{4S_1S_2}{(S_1+S_2)^2} = 1 - \left(\frac{S_2 - S_1}{S_2 + S_1}\right)^2 \end{aligned}$$

Problem 2. Determine the amount of energy emitted from the open end of a cylindrical tube.

Solution. In the boundary condition $p = 0$ at the open end of the tube, we can approximately neglect the emitted wave (we shall see that the intensity emitted from the end of the tube is small). Then we have the condition $p_1 = -p_1'$, where p_1 and p_1' are the pressures in the incident wave and in the wave reflected back into the tube; for the velocities we have correspondingly $v_1 = v_1'$, so that the total velocity at the end of the tube is $v_0 = v_1 + v_1' = 2v_1$. The energy flux in the incident wave is $cS\rho\overline{v_1^{\,2}} = \frac{1}{4}cS\rho\overline{v_0^{\,2}}$. Using (77.5), we obtain for the ratio of the emitted energy to the energy flux in the incident wave $D = S\omega^2/\pi c^2$. For a tube with circular cross-section (radius R) we have $D = R^2\omega^2/c^2$. Since, by hypothesis, $R \ll c/\omega$, it follows that $D \ll 1$.

Problem 3. One of the ends of a cylindrical pipe is covered by a membrane which executes a given oscillation and emits sound; the other end is open. Determine the way in which sound is emitted from the tube.

Solution. In the general solution

$$p = (ae^{ikx} + be^{-ikx})e^{-i\omega t}$$

we determine the constants a and b from the conditions $v = u = u_0 e^{-i\omega t}$, the given velocity of the membrane, at the closed end ($x = 0$), and $p = 0$ at the open end ($x = l$). These give $ae^{ikl} + be^{-ikl} = 0$, $a - b = c\rho u_0$. Determining a and b, we find the gas velocity at the open end of the tube to be $v_0 = u/\cos kl$. If the tube were absent, the intensity of the sound emitted by the oscillating membrane would be given by the mean square $S^2\overline{|\dot{u}|^2} = S^2\omega^2\overline{|u|^2}$, according to formula (74.10) with Su in place of $\dot{V}$; S is the cross-sectional area of the membrane. The emission from the end of the tube is proportional to $S^2\overline{|v_0|^2}\omega^2$. The amplification coefficient of the pipe is $A = S^2\overline{|v_0|^2}/S^2\overline{|u|^2} = 1/\cos^2 kl$. This becomes infinite for frequencies of oscillation of the membrane equal to the characteristic frequencies of the tube (*resonance*); in reality, of course, it remains finite because of effects which we have neglected (such as friction due to the emission of sound).

Problem 4. The same as Problem 3, but for a conical tube, with the membrane covering the smaller end.

Solution. The cross-section of the tube is $S = S_0 x^2$; let the values of the coordinate x which correspond to the smaller and larger ends be x_1, x_2, so that the length of the tube is $l = x_2 - x_1$. The general solution of equation (77.4) is $p = (1/x)(ae^{ikx} + be^{-ikx})e^{-i\omega t}$; a and b are determined from the conditions $v = u$ for $x = x_1$ and $p = 0$ for $x = x_2$. The amplification coefficient is found to be

$$A = \frac{S_0^{\,2}x_2^{\,4}\overline{|v_2|^2}}{S_0^{\,2}x_1^{\,4}\overline{|u|^2}} = \frac{k^2x_2^{\,2}}{(\sin kl + kx_1\cos kl)^2}.$$

Problem 5. The same as Problem 3, but for a tube whose cross-section varies exponentially along its length: $S = S_0 e^{\alpha x}$.

SOLUTION. Equation (77.4) becomes $\partial^2 p/\partial x^2 + \alpha\partial p/\partial x + k^2 p = 0$, whence

$$p = e^{-\frac{1}{2}\alpha x}(ae^{imx} + be^{-imx})e^{-i\omega t},$$

with $m = \sqrt{(k^2 - \frac{1}{4}\alpha^2)}$. Determining a and b from the conditions $v = u$ for $x = 0$ and $p = 0$ for $x = l$, we find the amplification coefficient.

$$A = \frac{S_0{}^2 e^{2\alpha l}\overline{|v_0|^2}}{S_0{}^2\overline{|u|^2}} = \frac{e^{\alpha l}}{[\frac{1}{2}(\alpha/m)\sin ml + \cos ml]^2}$$

for $k > \frac{1}{2}\alpha$ and

$$A = \frac{e^{\alpha l}}{[\frac{1}{2}(\alpha/m')\sinh m'l + \cosh m'l]^2}, \qquad m' = \sqrt{(\tfrac{1}{4}\alpha^2 - k^2)},$$

for $k < \frac{1}{2}\alpha$.

§78. Scattering of sound

If there is some body in the path of propagation of a sound wave, then the sound is *scattered*: besides the incident wave there appear other (scattered) waves, which are propagated in all directions from the scattering body. The scattering of a sound wave occurs simply on account of the presence of the body in its path. In addition, the incident wave causes the body itself to move, and this in turn brings about additional emission of sound by the body, i.e. further scattering. If, however, the density of the body is large compared with that of the medium in which the sound is propagated, and its compressibility is small, then the scattering due to the motion of the body forms only a small correction to the main scattering caused by the mere presence of the body. In what follows we shall neglect this correction, and therefore suppose the scattering body immovable.

We assume that the wavelength λ of the sound is large compared with the dimension l of the body; to calculate the properties of the scattered wave, we can then use formulae (74.8) and (74.11).† In doing so, we regard the scattered wave as being emitted by the body; the only difference is that, instead of a motion of the body in the fluid, we now have a motion of the fluid relative to the body. The two problems are clearly equivalent.

For the potential of the emitted wave we have obtained the expression $\phi = -\dot{V}/4\pi r - \dot{\mathbf{A}}\cdot\mathbf{r}/cr^2$. In this formula V was the volume of the body. In the present case, however, the volume of the body itself remains unchanged, and $\dot{V}$ must be taken not as the rate of change of the volume of the body, but as the volume of fluid which would enter, per unit time, the volume V_0 occupied by the body if the body were absent. For, in the presence of the body, this volume of fluid does not penetrate into V_0, which is equivalent to the emission of the same volume of fluid from V_0. The coefficient of $1/4\pi r$ in the first term of ϕ must, as we have seen in §74, be just the volume of fluid emitted from the origin per unit time. This volume is easily found. The change per unit time in the mass of fluid in a volume equal to that of the body is $V_0\dot{\rho}$, where $\dot{\rho}$ gives the rate of change of the fluid density in the incident sound wave (since the wavelength is large compared with the dimension of the body, the density ρ may be supposed constant over distances of the order of this dimension; hence we can write the rate of change of the mass of fluid in V_0 as $V_0\dot{\rho}$ simply, where $\dot{\rho}$ is the same throughout the volume V_0). The change in volume corresponding to a mass change $V_0\dot{\rho}$ is evidently $V_0\dot{\rho}/\rho$. Thus $\dot{V}$ in the expression for ϕ must be replaced by

† At the same time, the dimension of the body must be large in comparison with the displacement amplitude of fluid particles in the wave, since otherwise the fluid is not in general in potential flow.

$V_0\dot{\rho}/\rho$. In an incident plane wave, the variable part ρ' of the density is related to the velocity by $\rho' = \rho v/c$; hence $\dot{\rho} = \dot{\rho}' = \rho\dot{v}/c$, and we can replace $V_0\dot{\rho}/\rho$ by $V_0\dot{v}/c$.

When the body moves in the fluid, the vector **A** is determined by formulae (11.5), (11.6): $4\pi\rho A_i = m_{ik}u_k + \rho V_0 u_i$. We must now replace the velocity **u** of the body by the reversed velocity **v** of the fluid in the incident wave which it would have at the position of the body if the latter were absent. Thus

$$A_i = -m_{ik}v_k/4\pi\rho - V_0 v_i/4\pi. \tag{78.1}$$

We finally obtain for the potential of the scattered wave

$$\phi_{sc} = -V_0\dot{v}/4\pi cr - \dot{\mathbf{A}}\cdot\mathbf{r}/cr^2, \tag{78.2}$$

the vector **A** being given by formula (78.1). Hence we have for the velocity distribution in the scattered wave

$$\mathbf{v}_{sc} = V_0\ddot{v}\mathbf{n}/4\pi rc^2 + \mathbf{n}(\mathbf{n}\cdot\ddot{\mathbf{A}})/rc^2 \tag{78.3}$$

(see §74), **n** being a unit vector in the direction of scattering.

The mean amount of energy scattered per unit time into a given solid angle element d*o* is given by the energy flux, which is $c\rho\overline{\mathbf{v}_{sc}^2}\,do$. The total scattered intensity I_{sc} is obtained by integrating this expression over all directions. The integration of twice the product of the two terms in (78.3) gives zero, since this product is proportional to the cosine of the angle between the direction of scattering and the direction of propagation of the incident wave, and there remains (cf. (74.10) and (74.13))

$$I_{sc} = \frac{V_0{}^2\rho}{4\pi c^3}\overline{\ddot{v}^2} + \frac{4\pi\rho}{3c^3}\overline{\ddot{\mathbf{A}}^2}. \tag{78.4}$$

The scattering is generally characterized by what is called the *cross-section* dσ, which is the ratio of the (time) average energy scattered into a given solid-angle element to the mean energy flux density in the incident wave. The *total cross-section* σ is the integral of dσ over all directions of scattering, i.e. it is the ratio of the total scattered intensity to the incident energy flux density, and evidently has the dimensions of area.

The mean energy flux density in the incident wave is $c\rho\overline{\mathbf{v}^2}$. Hence the differential scattering cross-section is

$$d\sigma = (\overline{\mathbf{v}_{sc}{}^2}/\overline{\mathbf{v}^2})r^2 do. \tag{78.5}$$

The total cross-section is

$$\sigma = \frac{V_0{}^2}{4\pi c^4}\frac{\overline{\ddot{\mathbf{v}}^2}}{\overline{\mathbf{v}^2}} + \frac{4\pi}{3c^4}\frac{\overline{\ddot{\mathbf{A}}^2}}{\overline{\mathbf{v}^2}}. \tag{78.6}$$

For a monochromatic incident wave, the mean square second time derivative of the velocity is proportional to the fourth power of the frequency. Thus the cross-section for the scattering of sound by a body which is small compared with the wavelength is proportional to ω^4.

Finally, let us briefly discuss the opposite limiting case, where the wavelength of the scattered sound is small compared with the dimension of the body. In this case all the scattering, except for the scattering through very small angles, amounts to simple reflection from the surface of the body. The corresponding part of the total scattering cross-section is clearly equal to the area S of the cross-section of the body by a plane perpendicular to the direction of the incident wave. The scattering through small angles (of the order of λ/l), however, constitutes *diffraction* from the edges of the body. We shall not pause here to expound the theory of this phenomenon, which is entirely analogous to that of the diffraction of light (see *Fields*, §§60, 61). We shall only mention that, by Babinet's principle, the total intensity of diffracted sound is equal to the total intensity of reflected sound. Hence the diffraction part of the scattering cross-section is also equal to S, and the total cross-section is therefore $2S$.

PROBLEMS

PROBLEM 1. Determine the cross-section for the scattering of a plane sound wave by a solid sphere with radius R small compared with the wavelength.

SOLUTION. The velocity at a given point in a plane wave is $v = a\cos\omega t$. In the case of a sphere (see §11, Problem 1), the vector $\mathbf{A}$ is $-\frac{1}{2}R^3\mathbf{v}$. For the differential cross-section we obtain

$$d\sigma = \frac{\omega^4 R^6}{9c^4}(1-\tfrac{3}{2}\cos\theta)^2\, do,$$

where θ is the angle between the direction of the incident wave and the direction of scattering. The scattered intensity is greatest in the direction $\theta = \pi$, which is opposite to the direction of incidence. The total cross-section is

$$\sigma = (7\pi/9)(R^3\omega^2/c^2)^2. \tag{1}$$

Here (and also in Problems 3 and 4 below) it is assumed that the density ρ_0 of the sphere is large compared with the density ρ of the gas; if this were not so, it would be necessary to take account of the movement of the sphere by the pressure forces exerted on it by the oscillating gas.

PROBLEM 2. Determine the cross-section for the scattering of sound by a drop of fluid, taking into account the compressibility of the fluid and the motion of the drop caused by the incident wave.

SOLUTION. When the pressure of the gas surrounding the drop changes adiabatically by p', the volume of the drop is reduced by $(V_0/\rho_0)(\partial\rho_0/\partial p)_s p' = V_0 c\rho v/\rho_0 c_0^2$, where ρ_0 is the density of the drop, c_0 the velocity of sound in the fluid, and ρ the density of the gas. In the expressions (78.2) and (78.3), we must now replace $V_0\dot{v}/c$ by the difference $V_0(\dot{v}/c - \dot{v}c\rho/c_0^2\rho_0)$. Moreover, in the expression for $\mathbf{A}$ we must replace $-\mathbf{v}$ by the difference $\mathbf{u}-\mathbf{v}$, where $\mathbf{u}$ is the velocity acquired by the drop as a result of the action of the incident wave. For a sphere we have, using the results of §11, Problem 1, $\mathbf{A} = R^3\mathbf{v}(\rho-\rho_0)/(2\rho_0+\rho)$. Substituting these expressions, we have the cross-section

$$d\sigma = \frac{\omega^4 R^6}{9c^4}\left\{\left(1-\frac{c^2\rho}{c_0^2\rho_0}\right) - 3\cos\theta\frac{\rho_0-\rho}{2\rho_0+\rho}\right\}^2 do.$$

The total cross-section is

$$\sigma = \frac{4\pi\omega^4 R^6}{9c^4}\left\{\left(1-\frac{c^2\rho}{c_0^2\rho_0}\right)^2 + \frac{3(\rho_0-\rho)^2}{(2\rho_0+\rho)^2}\right\}.$$

PROBLEM 3. Determine the cross-section for the scattering of sound by a solid sphere with radius R much less than $\sqrt{(\nu/\omega)}$. The specific heat of the sphere is supposed so large that its temperature can be regarded as a constant.

SOLUTION. In this case we have to take into account the effect of the gas viscosity on the motion of the sphere, and the vector $\mathbf{A}$ must be modified as shown in §74, Problem 2. For $R\sqrt{(\omega/\nu)} \ll 1$ we have $\mathbf{A} = -3iR\nu\mathbf{v}/2\omega$.

The thermal conductivity of the gas also results in scattering of the same order. Let $T'_0 e^{-i\omega t}$ be the temperature variation at a given point in the sound wave. The temperature distribution near a sphere is (see §52, Problem 2)

$$T' = T'_0 e^{-i\omega t}[1-(R/r)e^{-(1-i)(r-R)\sqrt{(\omega/2\chi)}}]$$

(for $r = R$ we must have $T' = 0$). The amount of heat transferred from the gas to the sphere per unit time is (for $R\sqrt{(\omega/\chi)} \ll 1$) $q = 4\pi R^2\kappa[dT'/dr]_{r=R} = 4\pi R\kappa T'_0 e^{-i\omega t}$. This transfer of heat results in a change in the volume of the gas, which can be taken to affect the scattering like a corresponding effective change in the volume of the sphere, $\dot{V} = -4\pi R\chi\beta T'_0 e^{-i\omega t} = -4\pi R\chi(\gamma-1)v/c$, where β is the coefficient of thermal expansion of the gas and $\gamma = c_p/c_v$; we have used also formulae (64.13) and (79.2).

Taking account of both effects, we obtain the differential scattering cross-section

$$d\sigma = (\omega R/c^2)^2[\chi(\gamma-1)-\tfrac{3}{2}v\cos\theta]^2\,do.$$

The total cross-section is

$$\sigma = 4\pi(\omega R/c^2)^2[\chi^2(\gamma-1)^2+\tfrac{3}{4}v^2].$$

These formulae are valid only if the Stokes frictional force is small compared with the inertia force, i.e. $\eta R \ll M\omega$, where $M = 4\pi R^3\rho_0/3$ is the mass of the sphere; otherwise, the movement of the sphere by viscous forces becomes important.

PROBLEM 4. Determine the mean force on a solid sphere which scatters a plane sound wave ($\lambda \gg R$).

SOLUTION. The momentum transmitted per unit time from the incident wave to the sphere, i.e. the required force, is the difference between the momentum in the incident wave and the total momentum flux in the scattered wave. From the incident wave an energy flux $\sigma c\overline{E}_0$ is scattered, where E_0 is the energy density in the incident wave; the corresponding momentum flux is obtained by dividing by c, and is therefore $\sigma\overline{E}_0$. In the scattered wave, the momentum flux into the solid angle element do is $\overline{E}_{sc}r^2do = \overline{E}_0 d\sigma$; projecting this on the direction of propagation of the incident wave (which is obviously the direction of the required force), and integrating over all angles, we obtain

$$\overline{E}_0\int\cos\theta\,d\sigma.$$

Thus the force on the sphere is

$$F = \overline{E}_0\int(1-\cos\theta)d\sigma.$$

Substituting for $d\sigma$ from Problem 1, we obtain $F = 11\pi\omega^4R^6\overline{E}_0/9c^4$.

§79. Absorption of sound

The existence of viscosity and thermal conductivity results in the dissipation of energy in sound waves, and the sound is consequently *absorbed*, i.e. its intensity progressively diminishes. To calculate the rate of energy dissipation $\dot{E}_{\rm mech}$, we use the following general arguments. The mechanical energy is just the maximum amount of work that can be done in passing from a given non-equilibrium state to one of thermodynamic equilibrium. As we know from thermodynamics, the maximum work is obtained when the transition is reversible (i.e. without change of entropy), and is then $E_{\rm mech} = E_0 - E(S)$, where E_0 is the given initial value of the energy, and $E(S)$ is the energy in the equilibrium state with the same entropy S as the system had initially. Differentiating with respect to time, we obtain $\dot{E}_{\rm mech} = -\dot{E}(S) = -(\partial E/\partial S)\dot{S}$. The derivative of the energy with respect to the entropy is the temperature. Hence $\partial E/\partial S$ is the temperature which the system would have if it were in thermodynamic equilibrium (with the given value of the entropy). Denoting this temperature by T_0, we therefore have $\dot{E}_{\rm mech} = -T_0\dot{S}$.

We use for S the expression (49.6), which gives the rate of change of the entropy due to both thermal conduction and viscosity. Since the temperature T varies only slightly through the fluid, and differs little from T_0, it can be taken outside the integral, and T_0 can be written as T simply:

$$\dot{E}_{\rm mech} = -\frac{\kappa}{T}\int(\mathbf{grad}\ T)^2dV - \tfrac{1}{2}\eta\int\left(\frac{\partial v_i}{\partial x_k}+\frac{\partial v_k}{\partial x_i}-\tfrac{2}{3}\delta_{ik}\frac{\partial v_l}{\partial x_l}\right)^2dV - \zeta\int(\text{div}\,\mathbf{v})^2dV. \tag{79.1}$$

This formula generalizes formula (16.3) to the case of a compressible fluid which conducts heat.

Let the x-axis be in the direction of propagation of the sound wave. Then $v_x = v_0 \cos(kx - \omega t)$, $v_y = v_z = 0$. The last two terms in (79.1) give

$$-(\tfrac{4}{3}\eta + \zeta)\int\left(\frac{\partial v_x}{\partial x}\right)^2 \mathrm{d}V = -k^2(\tfrac{4}{3}\eta + \zeta)v_0^2\int \sin^2(kx - \omega t)\mathrm{d}V.$$

We are, of course, interested only in the time average; taking this average, we have $-k^2(\tfrac{4}{3}\eta + \zeta).\tfrac{1}{2}v_0^2 V_0$, where V_0 is the volume of the fluid.

Next we calculate the first term in (79.1). The deviation T' of the temperature in the sound wave from its equilibrium value is related to the velocity by formula (64.13), so that the temperature gradient is

$$\partial T/\partial x = (\beta c T/c_p)\partial v/\partial x = -(\beta c T/c_p)v_0 k \sin(kx - \omega t).$$

For the time average of the first term in (79.1) we obtain $-\kappa c^2 T\beta^2 v_0^2 k^2 V_0/2c_p^2$. Using the thermodynamic formulae

$$c_p - c_v = T\beta^2(\partial p/\partial\rho)_T = T\beta^2(c_v/c_p)(\partial p/\partial\rho)_s = T\beta^2 c^2 c_v/c_p, \tag{79.2}$$

we can rewrite this expression as $-\tfrac{1}{2}\kappa(1/c_v - 1/c_p)k^2 v_0^2 V_0$.

Collecting the above results, we find the mean value of the energy dissipation:

$$\overline{\dot{E}_{\text{mech}}} = -\tfrac{1}{2}k^2 v_0^2 V_0[(\tfrac{4}{3}\eta + \zeta) + \kappa(1/c_v - 1/c_p)]. \tag{79.3}$$

The total energy of the sound wave is

$$\bar{E} = \tfrac{1}{2}\rho v_0^2 V_0. \tag{79.4}$$

The damping coefficient derived in §25 for gravity waves gives the manner of decrease of the intensity with time. For sound, however, the problem is usually stated somewhat differently: a sound wave is propagated through a fluid, and its intensity decreases with the distance x traversed. It is evident that this decrease will occur according to a law $e^{-2\gamma x}$, and the amplitude will decrease as $e^{-\gamma x}$, where the *absorption coefficient* γ is defined by

$$\gamma = |\overline{\dot{E}_{\text{mech}}}|/2c\bar{E}. \tag{79.5}$$

Substituting here (79.3) and (79.4), we find the following expression for the sound absorption coefficient:

$$\gamma = \frac{\omega^2}{2\rho c^3}\left[(\tfrac{4}{3}\eta + \zeta) + \kappa\left(\frac{1}{c_v} - \frac{1}{c_p}\right)\right] \equiv a\omega^2. \tag{79.6}$$

We may point out that it is proportional to the square of the frequency of the sound.†

This formula is applicable so long as the absorption coefficient determined by it is small: the amplitude must decrease relatively little over distances of the order of a wavelength (i.e. we must have $\gamma c/\omega \ll 1$). The above derivation is essentially founded on this assumption, since we have calculated the energy dissipation by using the expression for an undamped

† M. A. Isakovich (1948) has shown that there must be a special absorption when sound is propagated in a two-phase system (an *emulsion*). Because of the different thermodynamic properties of the two components, their temperature changes during the passage of the sound wave will in general be different. The resulting heat exchange between the components leads to an additional absorption of sound. On account of the relative slowness of this heat exchange, a considerable dispersion of the sound takes place comparatively quickly.

sound wave. For gases this condition is in practice always satisfied. Let us consider, for example, the first term in (79.6). The condition $\gamma c/\omega \ll 1$ means that $\nu\omega/c^2 \ll 1$. It is known from the kinetic theory of gases, however, that the viscosity coefficient ν for a gas is of the order of product of the mean free path l and the mean thermal velocity of the molecules; the latter is of the same order as the velocity of sound in the gas, so that $\nu \sim lc$. Hence we have

$$\nu\omega/c^2 \sim l\omega/c \sim l/\lambda \ll 1, \tag{79.7}$$

since we know that $l \ll \lambda$. The thermal-conduction term in (79.6) gives the same result, since $\chi \sim \nu$.

In liquids, the condition of small absorption is always fulfilled when the problem of sound absorption, as stated here, is significant at all. The absorption over one wavelength can become large only if the viscous forces are comparable with the pressure forces which occur when the substance is compressed. In these conditions, however, the Navier–Stokes equation itself (with the viscosity coefficients independent of frequency) becomes invalid and a considerable dispersion of sound, due to processes of internal friction, occurs.†

For absorption of sound, the relation between the wave number and the frequency can evidently be written

$$k = \omega/c + ia\omega^2, \tag{79.8}$$

where a denotes the coefficient in (79.6). It is easy to see from this how the equation for a travelling sound wave must be modified in order to take absorption into account. To do so, we notice that, in the absence of absorption, the differential equation for (say) the pressure $p' = p'(x-ct)$ can be written $\partial p'/\partial x = -(1/c)\partial p'/\partial t$. The equation whose solution is $e^{i(kx-\omega t)}$, with k given by (79.8), must clearly be

$$\frac{\partial p'}{\partial x} = -\frac{1}{c}\frac{\partial p'}{\partial t} + a\frac{\partial^2 p'}{\partial t^2}. \tag{79.9}$$

If we replace t by $\tau + x/c$, this equation becomes

$$\partial p'/\partial x = a\partial^2 p'/\partial \tau^2,$$

i.e. a one-dimensional equation of thermal conduction.

The general solution of this equation can be written (see §51)

$$p'(x,\tau) = \frac{1}{2\sqrt{(\pi a x)}}\int p'_0(\tau')\exp[-(\tau'-\tau)^2/4ax]\mathrm{d}\tau', \tag{79.10}$$

where $p'_0(\tau) = p'(0,\tau)$. If the wave is emitted during a finite time interval, this expression becomes, at sufficiently large distances from the source,

$$p'(x,\tau) = \frac{1}{2\sqrt{(\pi a x)}}\exp(-\tau^2/4ax)\int p'_0(\tau')\mathrm{d}\tau'. \tag{79.11}$$

In other words, the wave profile at large distances is Gaussian. Its width is of the order of $\sqrt{(ax)}$, i.e. it increases as the square root of the distance travelled by the wave, while the

† A special case where strong absorption is possible but can be discussed by the usual methods is that of a gas with a thermal conductivity which is unusually large compared with its viscosity, on account of effects such as radiative transfer at very high temperatures (see Problem 3).

amplitude falls off inversely as $\sqrt{x}$. Hence we at once conclude that the total energy of the wave decreases as $1/\sqrt{x}$.

It is easy to derive analogous formulae for spherical waves; to do so, we must use the fact that for such a wave

$$\int p'\,dt = 0$$

(see (70.8)). Instead of (79.11) we now have

$$p'(r, \tau) = \text{constant} \times \frac{1}{r}\frac{\partial}{\partial \tau}\frac{\exp(-\tau^2/4ar)}{\sqrt{r}},$$

or

$$p'(r, \tau) = \text{constant} \times \frac{\tau}{r^{\frac{5}{2}}}\exp(-\tau^2/4ar). \tag{79.12}$$

Strong absorption must occur when a sound wave is reflected from a solid wall. The reason for this is the following (K. F. Herzfeld 1938; B. P. Konstantinov 1939). In a sound wave not only the density and the pressure, but also the temperature, undergo periodic oscillations about their mean values. Near a solid wall, therefore, there is a periodically fluctuating temperature difference between the fluid and the wall, even if the mean fluid temperature is equal to the wall temperature. At the wall itself, however, the temperatures of the wall and the adjoining fluid must be the same. As a result, a large temperature gradient is formed in a thin boundary layer of fluid, where the temperature changes rapidly from its value in the sound wave to the wall temperature. The presence of large temperature gradients, however, results in a large dissipation of energy by thermal conduction. For a similar reason, the fluid viscosity leads to strong absorption of sound when the wave is incident in an oblique direction. In this case the fluid velocity in the wave (in the direction of propagation) has a non-zero component tangential to the surface. At the surface itself, however, the fluid must completely "adhere". Hence a large tangential-velocity gradient† must occur in the boundary layer of fluid, resulting in a large viscous dissipation of energy (see Problem 1).

PROBLEMS

PROBLEM 1. Determine the fraction of energy that is absorbed when a sound wave is reflected from a solid wall. The density of the wall is supposed so large that the sound does not penetrate it, and the specific heat so large that the temperature of the wall may be supposed constant.

SOLUTION. We take the plane of the wall as the plane $x = 0$, and the plane of incidence as the xy-plane. Let the angle of incidence (which equals the angle of reflection) be θ. The change in density in the incident wave at any given point on the surface ($x = y = 0$, say) is $\rho'_1 = Ae^{-i\omega t}$. The reflected wave has the same amplitude, so that $\rho'_2 = \rho'_1$ at the wall. The actual change in the fluid density, since both waves (incident and reflected) are propagated simultaneously, is $\rho' = 2Ae^{-i\omega t}$. The fluid velocity in the wave is given by $\mathbf{v}_1 = c\rho'_1\mathbf{n}_1/\rho$, $\mathbf{v}_2 = c\rho'_2\mathbf{n}_2/\rho$. The total velocity on the wall, $\mathbf{v} = \mathbf{v}_1 + \mathbf{v}_2$, is therefore $v = v_y = 2A\sin\theta \times ce^{-i\omega t}/\rho$ (or, more precisely, this is what the velocity is found to be when the correct boundary conditions at the wall in the presence of viscosity are not applied). The actual variation of the velocity v_y near the wall is determined by formula (24.13), and the energy dissipation due to viscosity by formula (24.14), in which the above expression for v must be substituted for $v_0e^{-i\omega t}$

† The normal velocity component is zero at the boundary because of the boundary conditions, whether or not viscosity is present.

The deviation T' of the temperature from its mean value (which is the temperature of the wall), if calculated without using the correct boundary conditions at the wall, would be found to be (see (64.13)) $T' = 2Ac^2T\beta e^{-i\omega t}/c_p\rho$. In reality, however, the temperature distribution is determined by the equation of thermal conduction, with the boundary condition $T' = 0$ for $x = 0$, and is accordingly given by a formula entirely similar to (24.13).

On calculating the energy dissipation due to thermal conduction as the first term in formula (79.1), we obtain for the total energy dissipation per unit area of the wall

$$\bar{\dot{E}}_{\text{mech}} = -\frac{A^2c^2\sqrt{(2\omega)}}{\rho}\left[\sqrt{\chi}\left(\frac{c_p}{c_v}-1\right)+\sqrt{\nu}\sin^2\theta\right].$$

The mean energy flux density incident on unit area of the wall from the incident wave is $c\rho\overline{v_1^2}\cos\theta = (c^3A^2/2\rho)\cos\theta$. Hence the fraction of energy absorbed on reflection is

$$\frac{2\sqrt{(2\omega)}}{c\cos\theta}\left[\sqrt{\nu}\sin^2\theta+\sqrt{\chi}\left(\frac{c_p}{c_v}-1\right)\right].$$

This expression is valid only if its value is small (since in deriving it we have supposed the amplitudes of the incident and reflected waves to be the same). This condition means that the angle of incidence θ must not be too near $\frac{1}{2}\pi$.

PROBLEM 2. Determine the coefficient of absorption of sound propagated in a cylindrical pipe.

SOLUTION. The main contribution to the absorption is due to the presence of the walls. The absorption coefficient γ is equal to the energy dissipated at the walls per unit time and per unit length of the pipe, divided by twice the total energy flux through a cross-section of the pipe. A calculation similar to that given in Problem 1 leads to the result

$$\gamma = \frac{\sqrt{\omega}}{\sqrt{2}Rc}\left[\sqrt{\nu}+\sqrt{\chi}\left(\frac{c_p}{c_v}-1\right)\right],$$

where R is the radius of the pipe.

PROBLEM 3. Find the dispersion relation for sound propagated in a medium with very high thermal conductivity.

SOLUTION. In the presence of a large thermal conductivity the flow in a sound wave is not adiabatic. Hence, instead of the condition of constant entropy, we now have

$$\dot{s}' = \kappa\triangle T'/\rho T, \tag{1}$$

which is the linearized form of equation (49.4) without the viscosity terms. As a second equation we take

$$\ddot{\rho}' = \triangle p', \tag{2}$$

which is obtained by eliminating $\mathbf{v}$ from equations (64.2) and (64.3). Taking as the fundamental variables p' and T', we write ρ' and s' in the form

$$\rho' = (\partial\rho/\partial T)_p T' + (\partial\rho/\partial p)_T p', \qquad s' = (\partial s/\partial T)_p T' + (\partial s/\partial p)_T p'.$$

We substitute these expressions in (1) and (2), and then seek T' and p' in a form proportional to $e^{i(kx-\omega t)}$. The compatibility condition for the resulting two equations for p' and T' can (by using various relations between the derivatives of thermodynamic quantities) be brought to the form

$$k^4 - k^2\left(\frac{\omega^2}{c_T^2}+\frac{i\omega}{\chi}\right)+\frac{i\omega^3}{\chi c_s^2} = 0, \tag{3}$$

which gives the required relation between k and ω. We have here used the notation

$$c_s^2 = (\partial p/\partial\rho)_s, \quad c_T^2 = (\partial p/\partial\rho)_T = c_s^2/\gamma,$$

where $\gamma = c_p/c_v$ is the ratio of specific heats.

In the limiting case of low frequencies ($\omega \ll c^2/\chi$), equation (3) gives

$$k = \frac{\omega}{c_s} + i\frac{\omega^2\chi}{2c_s}\left(\frac{1}{c_T^2}-\frac{1}{c_s^2}\right),$$

which corresponds to the propagation of sound with the ordinary "adiabatic" velocity c_s and a small absorption coefficient which is the second term in (79.6). This is as it should be, since the condition $\omega \ll c^2/\chi$ means that, during one period, heat can be transmitted only over a distance $\sim \sqrt{(\chi/\omega)}$ (cf. (51.7)) which is small compared with the wavelength c/ω.

In the opposite limiting case of large frequencies, we find from (3)

$$k = \frac{\omega}{c_T} + i\frac{c_T}{2\chi c_s^2}(c_s^2 - c_T^2).$$

In this case the sound is propagated with the "isothermal" velocity c_T, which is always less than c_s. The absorption coefficient is again small compared with the reciprocal of the wavelength, and is independent of the frequency and inversely proportional to the thermal conductivity.†

PROBLEM 4. Determine the additional absorption, due to diffusion, of sound propagated in a mixture of two substances (I. G. Shaposhnikov and Z. A. Gol'dberg 1952).

SOLUTION. The mixture contains an additional source of absorption of sound because the temperature and pressure gradients occurring in the sound wave result in irreversible processes of thermal diffusion and barodiffusion (but there is evidently no mass-concentration gradient, and therefore no mass transfer). This absorption is given by the term

$$(1/T\rho D)(\partial\mu/\partial C)_{p,T}\int \mathbf{i}^2 \mathrm{d}V$$

in the rate of change of entropy (59.13); we here denote the concentration by C to distinguish it from c, the velocity of sound. The diffusion flux is

$$\mathbf{i} = -\rho D[(k_T/T)\,\mathbf{grad}\,T + (k_p/p)\,\mathbf{grad}\,p],$$

with k_p given by (59.10). A calculation similar to that given in §79, using various relations between the derivatives of thermodynamic quantities, leads to the result that there must be added to the expression (79.6) for the absorption coefficient a term

$$\gamma_D = \frac{D\omega^2}{2c\rho^2(\partial\mu/\partial C)_{p,T}}\left\{\left(\frac{\partial\rho}{\partial C}\right)_{p,T} + \frac{k_T}{c_p}\left(\frac{\partial\rho}{\partial T}\right)_{p,C}\left(\frac{\partial\mu}{\partial C}\right)_{p,T}\right\}^2$$

PROBLEM 5. Determine the cross-section for the absorption of sound by a sphere whose radius is small compared with $\sqrt{(\nu/\omega)}$.

SOLUTION. The total absorption is composed of the effects of the viscosity and thermal conductivity of the gas. The former is given by the work done by the Stokes frictional force when gas moving in a sound wave flows round a sphere; as in §78, Problem 3, it is assumed that the sphere is not moved by this force. The effect of conductivity is given by the amount of heat q transferred from the gas to the sphere per unit time (§78, Problem 3): the energy dissipation when an amount of heat q is transferred, the temperature difference between the gas (far from the sphere) and the sphere being T', is qT'/T. The total absorption cross-section is found to be

$$\sigma = \frac{2\pi R}{c}\left[3\nu + 2\chi\left(\frac{c_p}{c_v} - 1\right)\right].$$

§80. Acoustic streaming

One of the most interesting ways in which sound waves are affected by viscosity consists in the formation of steady vortex flow in a stationary sound wave when there are solid obstacles or solid boundary walls. This *acoustic streaming* occurs in the second approximation with respect to the wave amplitude; its characteristic feature is that the velocity in it (in the region outside a thin boundary layer) is independent of the viscosity, even though it originates from the viscosity (Rayleigh 1883).

---

† The second root of equation (3), which is quadratic in k^2, corresponds to thermal waves which are rapidly damped with increasing x. In the limit $\omega\chi \ll c^2$ this root gives

$$k = \sqrt{(i\omega/\chi)} = (1+i)\sqrt{(\omega/2\chi)},$$

in agreement with (52.15). In the case $\omega\chi \gg c^2$ we have

$$k = (1+i)\sqrt{(\omega c_v/2\chi c_p)}.$$

The properties of acoustic streaming are most typically seen when the characteristic length in the problem (the size of the obstacles or of the flow region) is much less than the sound wavelength λ, but much greater than the penetration depth $\delta = \sqrt{(2\nu/\omega)}$ for viscous waves (§24):

$$\lambda \gg l \gg \delta. \tag{80.1}$$

In view of the second condition, we can distinguish in the flow region a narrow *acoustic boundary layer* in which the velocity decreases from its value in the sound wave to zero at the solid surface. Since the velocity in this layer, as in the sound wave itself, is much less than that of sound, and the characteristic dimension δ is much less than λ according to (10.17), the flow there may be regarded as incompressible.

Let us consider the acoustic boundary layer at a plane solid wall (the xz-plane), assuming two-dimensional flow in the xy-plane (H. Schlichting 1932). The approximations resulting from the thinness of the boundary layer have been described in §39 and remain valid for the non-steady flow under consideration. The non-steadiness simply means that Prandtl's equation (39.5) includes time-derivative terms:

$$\frac{\partial v_x}{\partial t} + v_x \frac{\partial v_x}{\partial x} + v_y \frac{\partial v_x}{\partial y} - \nu \frac{\partial^2 v_x}{\partial y^2} = U \frac{\partial U}{\partial x} + \frac{\partial U}{\partial t}; \tag{80.2}$$

the derivative $\mathrm{d}p/\mathrm{d}x$ is expressed in terms of the flow velocity $U(x, t)$ outside the boundary layer by means of (9.3). In the present case,

$$U = v_0 \cos kx \cdot \cos \omega t = v_0 \cos kx \cdot \mathrm{re}\, e^{-i\omega t}, \tag{80.3}$$

where $k = \omega/c$; this corresponds to a plane stationary sound wave with frequency ω. The required velocity $\mathbf{v}$ in the boundary layer is expressed in terms of the stream function $\psi(x, y, t)$ by

$$v_x = \partial\psi/\partial y, \qquad v_y = -\partial\psi/\partial x,$$

and the continuity equation (39.6) is then satisfied automatically.

We shall solve equation (80.2) by successive approximations with respect to the small quantity v_0, the amplitude of the velocity fluctuations in the sound wave. In the first approximation, the quadratic terms are omitted altogether. The solution of the equation

$$\frac{\partial v^{(1)}{}_x}{\partial t} - \nu \frac{\partial^2 v^{(1)}{}_x}{\partial y^2} = -i\omega v_0 \cos kx \cdot e^{-i\omega t}$$

which satisfies the necessary conditions at $y = 0$ and $y = \infty$ is

$$v^{(1)}{}_x = \mathrm{re}\,\{v_0 \cos kx \cdot e^{-i\omega t}(1 - e^{-\kappa y})\},$$

where

$$\kappa = \sqrt{(-i\omega/\nu)} = (1 - i)/\delta. \tag{80.4}$$

The corresponding stream function (satisfying the condition $\psi^{(1)} = 0$ at $y = 0$, which is equivalent to $v^{(1)}{}_y = 0$) is

$$\psi^{(1)} = \mathrm{re}\,\{v_0 \cos kx \cdot \zeta^{(1)}(y) e^{-i\omega t}\}, \tag{80.5}$$

$$\zeta^{(1)}(y) = y + e^{-\kappa y}/\kappa.$$

In the next approximation, we write $\mathbf{v} = \mathbf{v}^{(1)} + \mathbf{v}^{(2)}$ and obtain for $\mathbf{v}^{(2)}$, from (80.2), the equation

$$\frac{\partial v^{(2)}{}_x}{\partial t} - \nu \frac{\partial v^{(2)}{}_x}{\partial y^2} = U\frac{\partial U}{\partial x} - v^{(1)}{}_x \frac{\partial v^{(1)}{}_x}{\partial x} - v^{(1)}{}_y \frac{\partial v^{(1)}{}_x}{\partial y}. \tag{80.6}$$

The right-hand side contains terms with frequencies $\omega + \omega = 2\omega$ and $\omega - \omega = 0$. The latter give rise to time-independent terms in $\mathbf{v}^{(2)}$, which are the ones representing the steady flow in question; we shall take $\mathbf{v}^{(2)}$ to mean only this part of the velocity. The corresponding part of the stream function is written as

$$\psi^{(2)} = (v_0{}^2/c) \sin 2kx \cdot \zeta^{(2)}(y), \tag{80.7}$$

and we obtain for $\zeta^{(2)}(y)$ the equation

$$\delta^2 \zeta^{(2)'''} = \tfrac{1}{2} - \tfrac{1}{2}|\zeta^{(1)'}|^2 + \tfrac{1}{2}\operatorname{re}(\zeta^{(1)}\zeta^{(1)''}), \tag{80.8}$$

the primes denoting differentiation with respect to y.

The solution of this equation must satisfy the conditions $\zeta^{(2)} = 0$, $\zeta^{(2)'}(0) = 0$, which are equivalent to $v^{(2)}{}_x = v^{(2)}{}_y = 0$ on the solid surface. The condition far from this surface can only be that $v^{(2)}{}_x$ tends to a finite value (not necessarily zero). Substitution of (80.5) in (80.8) and a twofold integration gives the following result for the derivative $\zeta^{(2)'}$:

$$\zeta^{(2)'}(y) = \tfrac{3}{8} - \tfrac{1}{8}e^{-2y/\delta} - e^{-y/\delta}\sin y/\delta -$$
$$-\tfrac{1}{4}e^{-y/\delta}\cos y/\delta + (y/4\delta)e^{-y/\delta}(\cos y/\delta - \sin y/\delta).$$

As $y \to \infty$, it tends to

$$\zeta^{(2)'}(\infty) = 3/8, \tag{80.9}$$

corresponding to a velocity

$$v^{(2)}{}_x(\infty) = (3v_0{}^2/8c)\sin 2kx. \tag{80.10}$$

This demonstrates the effect described at the beginning of the section. We see that outside the boundary layer there is (in the second approximation with respect to v_0) a steady flow whose velocity is independent of the viscosity. Its value (80.10) serves as a boundary condition for determining the main acoustic flow (see Problem).†

PROBLEM

Determine the acoustic streaming in the space between two plane-parallel walls (the planes $y = 0$ and $y = h$), where there is a stationary sound wave (80.3). The distance h between the planes, which acts as the characteristic length l, satisfies the conditions (80.1) (Rayleigh 1883).

SOLUTION. Since the velocity $v^{(2)}$ of the required steady flow is much less than that of sound, the flow may be regarded as incompressible. Moreover, since v_0 is assumed infinitesimal in the sound wave (and therefore so is $v^{(2)} \sim v_0{}^2/c$), the quadratic terms in the equation of motion may be neglected.‡ Then equation (15.12) for the stream function reduces to

$$\triangle^2\psi^{(2)} = \left(\frac{\partial^2}{\partial x^2} + \frac{\partial^2}{\partial y^2}\right)^2 \psi^{(2)} = 0$$

† The transverse velocity corresponding to the longitudinal velocity (80.10) is

$$v^{(2)}{}_y = -(3v_0{}^2k/4c)y\cos 2kx \ll v^{(2)}{}_x.$$

In solving the problem of flow outside the boundary layer, this arises automatically from the equation of continuity with the boundary condition $v^{(2)}{}_y = 0$ at $y = 0$.

‡ That is, the ratio v_0/c is assumed much smaller than any of the other small parameters of the problem; in particular, $v_0/c \ll \delta/h$.

(this arises from the viscosity term, but the viscosity itself does not appear in it). We shall seek $\psi^{(2)}$ in the form (80.7). According to the condition $h \ll \lambda$, the derivatives with respect to y are much larger than those with respect to x; neglecting the latter, we obtain for $\zeta^{(2)}(y)$ the equation

$$\zeta^{(2)''''} = 0 \tag{1}$$

From the obvious symmetry of the problem, the flow is symmetrical about the plane $y = \frac{1}{2}h$. Hence

$$v^{(2)}{}_x(x, y) = v^{(2)}{}_x(x, h-y), \quad v^{(2)}{}_y(x, y) = -v^{(2)}{}_y(x, h-y),$$

and therefore

$$\zeta^{(2)}(y) = -\zeta^{(2)}(h-y).$$

A solution of equation (1) having this property is

$$\zeta^{(2)}(y) = A(y-\tfrac{1}{2}h) + B(y-\tfrac{1}{2}h)^3.$$

The constants A and B are determined by the boundary conditions $\zeta^{(2)}(0) = 0$, $\zeta^{(2)'}(0) = 3/8$. This gives for the stream function

$$\psi^{(2)} = \frac{3v_0^2}{16c}\sin 2kx[-(y-\tfrac{1}{2}h) + (y-\tfrac{1}{2}h)^3/(\tfrac{1}{2}h)^2],$$

and thence the velocity distributions

$$v^{(2)}{}_x = -\frac{3v_0^2}{16c}\sin 2kx[1 - 3(y-\tfrac{1}{2}h)^2/(\tfrac{1}{2}h)^2],$$

$$v^{(2)}{}_y = \frac{3v_0^2k}{8c}\cos 2kx[y-\tfrac{1}{2}h - (y-\tfrac{1}{2}h)^3/(\tfrac{1}{2}h)^2].$$

The velocity $v^{(2)}{}_x$ changes sign at a distance $\frac{1}{2}h(1-1/\sqrt{3}) = 0{\cdot}423 \cdot \frac{1}{2}h$ from the wall.

The flow described by these expressions consists of two series of vortices lying symmetrically about the median plane $y = \frac{1}{2}h$ and periodic in the x-direction, with period $\frac{1}{2}\lambda$.

§81. Second viscosity

The second viscosity coefficient ζ (which we shall call simply the *second viscosity*) is usually of the same order of magnitude as the viscosity coefficient η. There are, however, cases where ζ can take values considerably exceeding η. As we know, the second viscosity appears in processes which are accompanied by a change in volume (i.e. in density) of the fluid. In compression or expansion, as in any rapid change of state, the fluid ceases to be in thermodynamic equilibrium, and internal processes are set up in it which tend to restore this equilibrium. These processes are usually so rapid (i.e. their relaxation time is so short) that the restoration of equilibrium follows the change in volume almost immediately unless, of course, the rate of change of volume is very large.

It may happen, nevertheless, that the relaxation times of the processes of restoration of equilibrium are long, i.e. they take place comparatively slowly. For instance, if we are concerned with a liquid or gas which is a mixture of substances between which a chemical reaction occurs, there is a state of chemical equilibrium, characterized by the concentrations of the substances in the mixture, for any given density and temperature. If, for example, we compress the fluid, the state of equilibrium is destroyed, and a reaction begins, as a result of which the concentrations of the substances tend to take the equilibrium values corresponding to the new density and temperature. If this reaction is not rapid, the restoration of equilibrium occurs relatively slowly and does not immediately follow the compression. The latter process is then accompanied by internal processes which tend towards the equilibrium state. But the processes which establish equilibrium are irreversible; they increase the entropy, and therefore involve energy dissipation. Hence, if the relaxation time of these processes is long, a considerable dissipation of energy occurs when the fluid is compressed or expanded, and, since this dissipation must be determined by the second viscosity, we reach the conclusion that ζ is large.†

The intensity of the dissipative processes, and therefore the value of ζ, depend of course on the relation between the rate of compression or expansion and the relaxation time. If, for example, we have compression or expansion due to a sound wave, the second viscosity will depend on the frequency of the wave. Thus the second viscosity is not just a constant characteristic of the material concerned, but depends on the frequency of the motion in which it appears. The dependence of ζ on the frequency is called its *dispersion.*

The following general method of discussing all these phenomena is due to L. I. Mandel'shtam and M. A. Leontovich (1937). Let ξ be some physical quantity characterizing the state of a body, and ξ_0 its value in the equilibrium state; ξ_0 is a function of density and temperature. For instance, in fluid mixtures ξ may be the concentration of one component, and then ξ_0 is the concentration in chemical equilibrium.

If the body is not in equilibrium, ξ will vary with time, tending to the value ξ_0. In states close to equilibrium the difference $\xi - \xi_0$ is small, and we can expand the rate of change $\dot{\xi}$ of ξ in a series of powers of this difference. The zero-order term is absent, since $\dot{\xi}$ must be zero in the equilibrium state, i.e. when $\xi = \xi_0$. Hence, as far as the first-order term, we have

$$\dot{\xi} = -(\xi - \xi_0)/\tau. \tag{81.1}$$

The proportionality coefficient must be negative, since otherwise ξ would not tend to a finite limit. The positive constant τ is of the dimensions of time, and may be regarded as the relaxation time for the process in question; the greater is τ, the more slowly the approach to equilibrium takes place.

In what follows we shall consider processes in which the fluid is subjected to a periodic adiabatic‡ compression and expansion, so that the variable part of the density (and of the other thermodynamic quantities) depends on the time through a factor $e^{-i\omega t}$; we are considering a sound wave in the fluid. Together with the density and other quantities, the position of equilibrium also varies, so that ξ_0 can be written as $\xi_0 = \xi_{00} + \xi_0'$, where ξ_{00} is the constant value of ξ_0 corresponding to the mean density, and ξ_0' is a periodic part, proportional to $e^{-i\omega t}$. Writing the true value ξ in the form $\xi = \xi_{00} + \xi'$, we conclude from equation (81.1) that ξ' also is a periodic function of time, related to ξ_0' by

$$\xi' = \xi_0'/(1 - i\omega\tau). \tag{81.2}$$

Let us calculate the derivative of the pressure with respect to the density for the process in question. The pressure must now be regarded as a function of the density and of the value of ξ in the state concerned, and also of the entropy, which we suppose constant and, for brevity, omit. Then

$$\partial p/\partial\rho = (\partial p/\partial\rho)_\xi + (\partial p/\partial\xi)_\rho \partial\xi/\partial\rho.$$

In accordance with (81.2), we substitute here

$$\frac{\partial\xi}{\partial\rho} = \frac{\partial\xi'}{\partial\rho} = \frac{1}{1-i\omega\tau}\frac{\partial\xi_0'}{\partial\rho} = \frac{1}{1-i\omega\tau}\frac{\partial\xi_0}{\partial\rho},$$

† A slow process which results in a large ζ is often the transfer of energy from translatory degrees of freedom of a molecule to vibrational (intramolecular) degrees of freedom.

‡ The change in the entropy (in states close to equilibrium) is of the second order of smallness. Hence, to the first order of accuracy, we can speak of an adiabatic process.

obtaining

$$\frac{\partial p}{\partial \rho} = \frac{1}{1-i\omega\tau}\left\{\left(\frac{\partial p}{\partial \rho}\right)_\xi + \left(\frac{\partial p}{\partial \xi}\right)_\rho \frac{\partial \xi_0}{\partial \rho} - i\omega\tau\left(\frac{\partial p}{\partial \rho}\right)_\xi\right\}.$$

The sum $(\partial p/\partial \rho)_\xi + (\partial p/\partial \xi)_\rho \partial \xi_0/\partial \rho$ is just the derivative of p with respect to ρ for a process which is so slow that the fluid remains in equilibrium; denoting it by $(\partial p/\partial \rho)_{\text{eq}}$, we have finally

$$\frac{\partial p}{\partial \rho} = \frac{1}{1-i\omega\tau}\left[\left(\frac{\partial p}{\partial \rho}\right)_{\text{eq}} - i\omega\tau\left(\frac{\partial p}{\partial \rho}\right)_\xi\right]. \tag{81.3}$$

Next, let p_0 be the pressure in a state of thermodynamic equilibrium; p_0 is related to the other thermodynamic quantities by the equation of state of the fluid, and is entirely determined when the density and entropy are given. The pressure p in a non-equilibrium state, however, differs from p_0, and is a function of ξ also. If the density is adiabatically increased by $\delta\rho$, the equilibrium pressure changes by $\delta p_0 = (\partial p/\partial \rho)_{\text{eq}}\delta\rho$, while the total increase in the pressure is $(\partial p/\partial \rho)\delta\rho$, with $\partial p/\partial \rho$ given by formula (81.3). Hence the difference $p - p_0$ between the true pressure and the equilibrium pressure, in a state where the density is $\rho + \delta\rho$, is

$$p - p_0 = \left[\frac{\partial p}{\partial \rho} - \left(\frac{\partial p}{\partial \rho}\right)_{\text{eq}}\right]\delta\rho = \frac{i\omega\tau}{1-i\omega\tau}\left[\left(\frac{\partial p}{\partial \rho}\right)_{\text{eq}} - \left(\frac{\partial p}{\partial \rho}\right)_\xi\right]\delta\rho.$$

We are here interested in the density changes due to the motion of the fluid. Then $\delta\rho$ is related to the velocity by the equation of continuity, which we write in the form $\mathrm{d}(\delta\rho)/\mathrm{d}t + \rho \operatorname{div} \mathbf{v} = 0$, where $\mathrm{d}/\mathrm{d}t$ denotes the total time derivative. In a periodic motion we have $\mathrm{d}(\delta\rho)/\mathrm{d}t = -i\omega\delta\rho$, and therefore $\delta\rho = (\rho/i\omega)\operatorname{div} \mathbf{v}$. Substituting this expression in $p - p_0$, we obtain

$$p - p_0 = \frac{\tau\rho}{1-i\omega\tau}({c_0}^2 - {c_\infty}^2)\operatorname{div} \mathbf{v}, \tag{81.4}$$

where we have used the notation

$${c_0}^2 = (\partial p/\partial \rho)_{\text{eq}}, \quad {c_\infty}^2 = (\partial p/\partial \rho)_\xi, \tag{81.5}$$

the significance of which will be explained below.

In order to relate these expressions to the viscosity of the fluid, we write down the stress tensor σ_{ik}. In this tensor the pressure appears in the term $-p\delta_{ik}$. Subtracting the pressure p_0 determined by the equation of state, we find that in a non-equilibrium state σ_{ik} contains an additional term

$$-(p - p_0)\delta_{ik} = \frac{\tau\rho}{1-i\omega\tau}({c_\infty}^2 - {c_0}^2)\delta_{ik}\operatorname{div} \mathbf{v}.$$

Comparing this with the general expression (15.2) and (15.3) for the stress tensor, in which $\operatorname{div} \mathbf{v}$ appears in the term $\zeta \operatorname{div} \mathbf{v}$, we conclude that the presence of slow processes tending to establish equilibrium is macroscopically equivalent to the presence of a second viscosity given by

$$\zeta = \tau\rho({c_\infty}^2 - {c_0}^2)/(1 - i\omega\tau). \tag{81.6}$$

These processes do not affect the ordinary viscosity η. For processes so slow that $\omega\tau \ll 1$, ζ is

$$\zeta_0 = \tau\rho({c_\infty}^2 - {c_0}^2); \tag{81.7}$$

it increases with the relaxation time τ, in accordance with what was said above. For large frequencies, ζ depends on the frequency, i.e. it exhibits dispersion.

Let us now consider the question of how the presence of processes with long relaxation times (we shall speak of chemical reactions) affects the propagation of sound in a fluid. To do so, we might start from the equation of motion of a viscous fluid, with ζ given by formula (81.6). It is simpler, however, to consider a motion in which viscosity is neglected but the pressure p is given by the above formulae instead of by the equation of state. The general relations which we obtained in §64 then remain formally applicable. In particular, the wave number and the frequency are still related by $k = \omega/c$, where $c = \sqrt{(\partial p/\partial\rho)}$, and the derivative $\partial p/\partial\rho$ is now given by (81.3); the quantity c, however, no longer denotes the velocity of sound, being complex. Thus we obtain

$$k = \omega\sqrt{[(1 - i\omega\tau)/({c_0}^2 - {c_\infty}^2 i\omega\tau)]}. \tag{81.8}$$

The "wave number" given by this formula is complex. The meaning of this fact is easily seen. In a plane wave, all quantities depend on the coordinate x (the x-axis being in the direction of propagation) through a factor e^{ikx}. Writing k in the form $k = k_1 + ik_2$ with k_1, k_2 real, we have $e^{ikx} = e^{ik_1x}\, e^{-k_2x}$, i.e. besides the periodic factor e^{ik_1x} we have a damping factor e^{-k_2x} (k_2 must, of course, be positive). Thus the complex nature of the wave number formally expresses the fact that the wave is damped, i.e. there is absorption of sound. The real part of the complex wave number gives the variation in phase of the wave with distance, and the imaginary part is the absorption coefficient.

It is not difficult to separate the real and imaginary parts of (81.8). In the general case of arbitrary ω the expressions for k_1 and k_2 are rather cumbersome, and we shall not write them out here. It is important that k_1 is a function of the frequency (as is k_2). Thus, if chemical reactions can occur in the fluid, the propagation of sound at sufficiently high frequencies is accompanied by dispersion.

In the limiting case of low frequencies ($\omega\tau \ll 1$), formula (81.8) gives to a first approximation $k = \omega/c_0$, corresponding to the propagation of sound with velocity c_0. This is as it should be, of course: the condition $\omega\tau \ll 1$ means that the period $1/\omega$ of the sound wave is large compared with the relaxation time, i.e. the establishment of chemical equilibrium follows the variations of density in the sound wave, and so the velocity of sound is determined by the equilibrium value of the derivative $\partial p/\partial\rho$. In the second approximation we have

$$k = \frac{\omega}{c_0} + \frac{i\omega^2\tau}{2{c_0}^3}({c_\infty}^2 - {c_0}^2), \tag{81.9}$$

i.e. damping occurs, with a coefficient proportional to the square of the frequency. Using (81.7), we can write the imaginary part of k in the form $k_2 = \omega^2\zeta_0/2\rho {c_0}^3$; this agrees with the ζ-dependent part of the absorption coefficient γ as given by (79.6), which was obtained without taking account of the dispersion.

In the opposite limiting case of high frequencies ($\omega\tau \gg 1$), we have in the first approximation $k = \omega/c_\infty$, i.e. the propagation of sound with velocity c_∞—again a natural result, since for $\omega\tau \gg 1$ we can suppose that no reaction occurs during a single period, and the velocity of sound must therefore be determined by the derivative $(\partial p/\partial\rho)_\xi$ taken at constant concentration. The second approximation gives

$$k = \frac{\omega}{c_\infty} + i\frac{c_\infty{}^2 - c_0{}^2}{2\tau c_\infty{}^3}. \tag{81.10}$$

The damping coefficient is independent of the frequency. As we go from $\omega \ll 1/\tau$ to $\omega \gg 1/\tau$, this coefficient increases monotonically to the constant value given by formula (81.10). It should be noted that the quantity k_2/k_1, which represents the amount of absorption over a distance of one wavelength, is small in both limiting cases ($k_2/k_1 \ll 1$); it has a maximum at some intermediate frequency, namely $\omega = \sqrt{(c_0/c_\infty)}/\tau$.

It is seen from (81.7) (e.g.) that

$$c_\infty > c_0, \tag{81.11}$$

since we must have $\zeta > 0$. The same result can be obtained by simple arguments based on Le Chatelier's principle. Let us suppose that the volume of the system is reduced, and the density increased, by some external agency. The system is thereby brought out of equilibrium, and according to Le Chatelier's principle processes must begin which tend to reduce the pressure. This means that $\partial p/\partial \rho$ will decrease, and, when the system returns to equilibrium, the value of $\partial p/\partial \rho = c^2$ will be less than in the non-equilibrium state.

In deriving all the above formulae we have assumed that there is only a single slow internal process of relaxation. Cases are also possible where several different such processes occur simultaneously. All the formulae can easily be generalized to cover such cases. Instead of a single quantity ξ, we now have several quantities $\xi_1, \xi_2, \ldots$ which characterize the state of the system, and a corresponding series of relaxation times $\tau_1, \tau_2, \ldots$. We choose the quantities ξ_n in such a way that each of the derivatives $\dot{\xi}_n$ depends only on the corresponding ξ_n, i.e. so that

$$\dot{\xi}_n = -(\xi_n - \xi_{n0})/\tau_n. \tag{81.12}$$

Calculations entirely similar to the above then give

$$c^2 = c_\infty{}^2 + \sum_n a_n/(1 - i\omega\tau_n), \tag{81.13}$$

where $c_\infty{}^2 = (\partial p/\partial \rho)_\xi$, and the constants a_n are

$$a_n = (\partial p/\partial \xi_n)(\partial \xi_n/\partial \rho)_{\text{eq}}. \tag{81.14}$$

If there is only one quantity ξ, formula (81.13) becomes (81.3), as it should.

CHAPTER IX

SHOCK WAVES

§82. Propagation of disturbances in a moving gas

When the velocity of a fluid in motion becomes comparable with or exceeds that of sound, effects due to the compressibility of the fluid become of prime importance. Such motions are in practice met with in gases. The dynamics of high-speed flow is therefore usually called *gas dynamics*.

It should be mentioned first of all that, in gas dynamics, the Reynolds numbers involved are almost always very large. For the kinematic viscosity of a gas is, as we know from the kinetic theory of gases, of the order of the mean free path l of the molecules multiplied by the mean velocity of their thermal motion; the latter is of the same order as the velocity of sound, so that $\nu \sim cl$. If the characteristic velocity in a problem of gas dynamics is also of the order of c, then the Reynolds number $\mathrm{R} \sim Lu/\nu \sim Lu/lc$, i.e. it is determined by the ratio of the dimension L to the mean free path l, which we know is very large.† As always occurs when R is very large, the viscosity has an important effect on the motion of the gas only in a very small region, and in what follows we shall (except where the contrary is specifically stated) regard the gas as an ideal fluid.

The flow of a gas is entirely different in nature according as it is *subsonic* or *supersonic*, i.e. the velocity is less than or greater than that of sound. One of the most important distinctive features of supersonic flow is the fact that there can occur in it what are called *shock waves*, whose properties we shall examine in detail in the following sections. Here we shall consider another characteristic property of supersonic flow, relating to the manner of propagation of small disturbances in the gas.

If a gas in steady motion receives a slight perturbation at any point, the effect of the perturbation is subsequently propagated through the gas with the velocity of sound (relative to the gas itself). The rate of propagation of the disturbance relative to a fixed system of coordinates is composed of two parts: firstly, the perturbation is "carried along" by the gas flow with velocity $\mathbf{v}$ and, secondly, it is propagated relative to the gas with velocity c in any direction $\mathbf{n}$. Let us consider, for simplicity, a uniform flow of gas with constant velocity $\mathbf{v}$, subjected to a small perturbation at some point O (fixed in space). The velocity $\mathbf{v}+c\mathbf{n}$ with which the perturbation is propagated from O (relative to the fixed system of coordinates) has different values for different directions of the unit vector $\mathbf{n}$. We obtain all its possible values by placing one end of the vector $\mathbf{v}$ at the point O and drawing a sphere with radius c centred at the other end. The vectors from O to points on this sphere give the possible magnitudes and directions of the velocity of propagation of the

† We shall not consider the problem of the motion of bodies in very rarefied gases, where the mean free path of the molecules is comparable with the dimension of the body. The problem is in essence not one of fluid dynamics, and must be examined by means of the kinetic theory of gases.

perturbation. Let us first suppose that $v < c$. Then the vector $\mathbf{v} + c\mathbf{n}$ can have any direction in space (Fig. 50a). That is, a disturbance which starts from any point in a subsonic flow will eventually reach every point in the gas. If, on the other hand, $v > c$, the direction of the vector $\mathbf{v} + c\mathbf{n}$ can lie, as we see from Fig. 50b, only in a cone with its vertex at O, which touches the sphere with its centre at the other end of the vector $\mathbf{v}$. If the aperture angle of the cone is 2α, then, as is seen from the figure,

$$\sin \alpha = c/v. \tag{82.1}$$

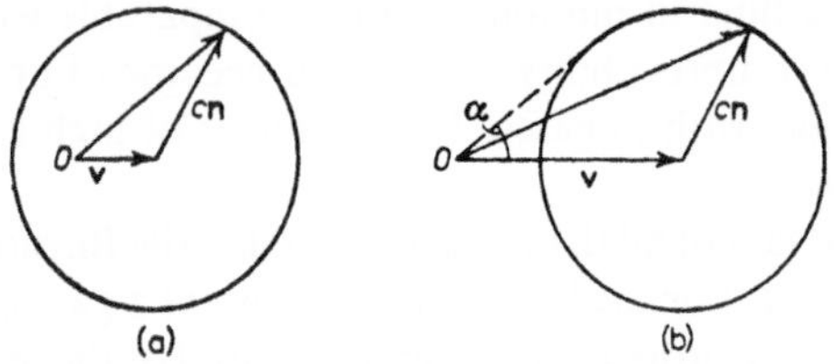

FIG. 50

Thus a disturbance starting from any point in a supersonic flow is propagated only downstream within a cone whose aperture angle decreases with the ratio c/v. A disturbance starting from O does not affect the flow outside this cone.

The angle α determined by equation (82.1) is called the *Mach angle*. The ratio v/c itself, which often occurs in gas dynamics, is the *Mach number* M:

$$\mathrm{M} = v/c. \tag{82.2}$$

The surface bounding the region reached by a disturbance starting from a given point is called the *Mach surface* or *characteristic surface*.

In the general case of an arbitrary steady flow, the Mach surface is not a cone throughout the volume. However, it can be asserted that, as before, this surface cuts the streamline through any point on it at the Mach angle. The value of the Mach angle varies from point to point with the velocities v and c. It should be emphasized here, incidentally, that, in flow with high velocities, the velocity of sound is different at different points: it varies with the thermodynamic quantities (pressure, density, etc.) of which it is a function.† The velocity of sound as a function of the coordinates is called the *local velocity of sound*.

The properties of supersonic flow described above give it a character quite different from that of subsonic flow. If a subsonic gas flow meets any obstacle (if, for instance, it flows past a body), the presence of this obstacle affects the flow in all space, both upstream and downstream; the effect of the obstacle is zero only asymptotically at an infinite distance from it. A supersonic flow, however, is incident "blindly" on an obstacle; the effect of the latter extends only downstream,‡ and in all the remaining part of space upstream the gas flows as if the obstacle were absent.

In the case of steady two-dimensional flow of a gas, the characteristic surfaces can be replaced by *characteristic lines* (or simply *characteristics*) in the plane of the flow. Through

† In the discussion of sound waves given in Chapter VIII, the velocity of sound could be regarded as constant.

‡ To avoid misunderstanding, we should mention that, if a shock wave is formed in front of the obstacle, this region is somewhat enlarged (see §122).

any point O in this plane there pass two characteristics (AA' and BB' in Fig. 51), which intersect the streamline through this point at the Mach angle. The downstream branches OA and OB of the characteristics may be said to *leave* the point O; they bound the region AOB of the flow where perturbations starting from O can take effect. The branches $B'O$ and $A'O$ may be said to *reach* the point O; the region $A'OB'$ between them is that which can affect the flow at O.

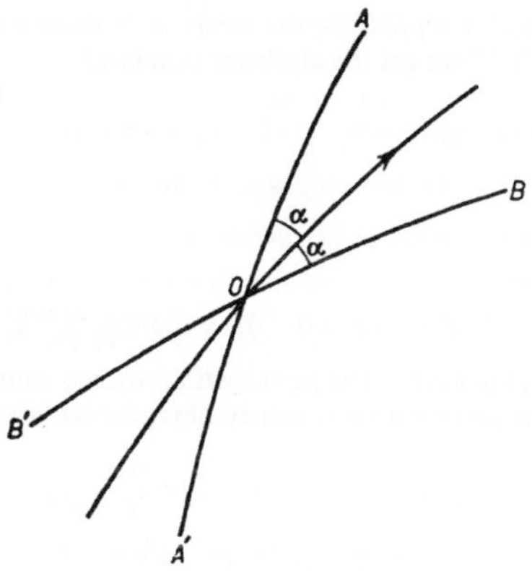

FIG. 51

The concept of characteristics (surfaces in the three-dimensional case) has also a somewhat different aspect. They are rays along which disturbances are propagated which satisfy the conditions of geometrical acoustics. If, for example, a steady supersonic gas flow meets a fairly small obstacle, then a steady perturbation of the gas flow will be found along the characteristics which leave this obstacle. The same result was reached in §68 from a study of the geometrical acoustics of moving media.

When we speak of a perturbation of the state of the gas, we mean a slight change in any of the quantities characterizing its state: the velocity, pressure, density, etc. The following remark should be made on this point. Perturbations in the values of the entropy of the gas (for constant pressure) and of its vorticity are not propagated with the velocity of sound. These perturbations, once having arisen, do not move relative to the gas; relative to a fixed system of coordinates they move with the gas at the velocity appropriate to each point. For the entropy, this is an immediate consequence of the law of conservation (in an ideal fluid), which states that the entropy of any given volume element in the gas remains constant as the element moves about. The same result for the vorticity follows from the conservation of circulation.

Thus we can say that, for perturbations of entropy and vorticity, the characteristics are the streamlines. This, of course, does not affect the general validity of the statements made above about regions of influence, since they were based only on the existence of a maximum velocity of propagation (that of sound) of disturbances relative to the gas itself.

PROBLEM

Find the relations between small changes in the velocity and in the thermodynamic quantities when there is an arbitrary small perturbation in a uniform gas flow.

SOLUTION. We denote the small changes in quantities during the perturbation by a prefixed δ (instead of a prime as in §64). In the approximation linear in these quantities, Euler's equation becomes

$$\partial\delta\mathbf{v}/\partial t + (\mathbf{v}\cdot\mathbf{grad})\,\delta\mathbf{v} + (1/\rho)\,\mathbf{grad}\,\delta p = 0 \tag{1}$$

where $\mathbf{v}$ is the constant unperturbed velocity; the equation of conservation of entropy is

$$\partial\delta s/\partial t + \mathbf{v}\cdot\mathbf{grad}\,\delta s = 0; \tag{2}$$

the equation of continuity is

$$\partial\delta p/\partial t + \mathbf{v}\cdot\mathbf{grad}\,\delta p + \rho c^2\,\mathrm{div}\,\delta\mathbf{v} = 0. \tag{3}$$

Here we have substituted $\delta\rho = \delta p/c^2 + (\partial\rho/\partial s)_p\,\delta s$; the terms in δs disappear in accordance with (2). For a perturbation having the form $e^{i\mathbf{k}\cdot\mathbf{r}-i\omega t}$ we get the algebraic equations

$$(\mathbf{v}\cdot\mathbf{k}-\omega)\delta s = 0, \quad (\mathbf{v}\cdot\mathbf{k}-\omega)\delta\mathbf{v} + \mathbf{k}\,\delta p/\rho = 0,$$
$$(\mathbf{v}\cdot\mathbf{k}-\omega)\delta p + \rho c^2\mathbf{k}\cdot\delta\mathbf{v} = 0.$$

These show that there are two possible types of perturbation.

In one type (entropy-vortex wave),

$$\omega = \mathbf{v}\cdot\mathbf{k}, \quad \delta s \neq 0, \quad \delta p = 0, \quad \delta\rho = (\partial\rho/\partial s)_p\delta s, \quad \mathbf{k}\cdot\delta\mathbf{v} = 0;$$

the vorticity $\mathbf{curl}\,\delta\mathbf{v} = i\mathbf{k}\times\delta\mathbf{v}$ is also not zero. The perturbations δs and $\delta\mathbf{v}$ in this wave are independent. The equation $\omega = \mathbf{v}\cdot\mathbf{k}$ signifies that the perturbation is carried along by the gas flow.

In the other type,

$$(\omega - \mathbf{v}\cdot\mathbf{k})^2 = c^2k^2, \quad \delta s = 0, \quad \delta p = c^2\delta\rho,$$
$$(\omega - \mathbf{v}\cdot\mathbf{k})\delta p = \rho c^2\mathbf{k}\cdot\delta\mathbf{v}, \quad \mathbf{k}\times\delta\mathbf{v} = 0.$$

This is a sound wave whose frequency is shifted by the Doppler effect. When the perturbation of one quantity in this wave is specified, those of all others are determined.

§83. Steady flow of a gas

We can obtain immediately from Bernoulli's equation a number of general results concerning adiabatic steady flow of a gas. The equation is, for steady flow, $w + \frac{1}{2}v^2 =$ constant along each streamline; if we have potential flow, then the constant is the same for every streamline, i.e. at every point in the fluid. If there is a point on some streamline at which the gas velocity is zero, then we can write Bernoulli's equation as

$$w + \tfrac{1}{2}v^2 = w_0, \tag{83.1}$$

where w_0 is the value of the heat function at the point where $v = 0$.

The equation of conservation of entropy for steady flow is $\mathbf{v}\cdot\mathbf{grad}\,s = v\partial s/\partial l = 0$, i.e. s is constant along each streamline. We can write this in a form analogous to (83.1):

$$s = s_0. \tag{83.2}$$

We see from equation (83.1) that the velocity v is greater at points where the heat function w is smaller. The maximum value of the velocity (on the streamline considered) is found at the point where w is least. For constant entropy, however, we have $\mathrm{d}w = \mathrm{d}p/\rho$; since $\rho > 0$, the differentials $\mathrm{d}w$ and $\mathrm{d}p$ have like signs, and so w and p vary in the same sense. We can therefore say that the velocity increases along a streamline when the pressure decreases, and vice versa.

The smallest possible values of the pressure and the heat function (in adiabatic flow) are obtained when the absolute temperature $T = 0$. The corresponding pressure is $p = 0$, and the value of w for $T = 0$ can be arbitrarily taken as the zero of energy; then $w = 0$ for $T = 0$. We can now deduce from (83.1) that the greatest possible value of the velocity (for given values of the thermodynamic quantities at the point where $v = 0$) is

$$v_{max} = \sqrt{(2w_0)}. \tag{83.3}$$

This velocity can be attained when a gas flows steadily out into a vacuum.†

Let us now consider how the mass flux density $j = \rho v$ varies along a streamline. From Euler's equation $(\mathbf{v}\cdot\mathbf{grad})\mathbf{v} = -(1/\rho)\mathbf{grad}\ p$, we find that the relation $v\,dv = -dp/\rho$ between the differentials dv and dp holds along a streamline. Putting $dp = c^2 d\rho$, we have

$$d\rho/dv = -\rho v/c^2 \tag{83.4}$$

and, substituting in $d(\rho v) = \rho\, dv + v\, d\rho$, we obtain

$$d(\rho v)/dv = \rho(1 - v^2/c^2). \tag{83.5}$$

From this we see that, as the velocity increases along a streamline, the mass flux density increases as long as the flow remains subsonic. In the supersonic range, however, the mass flux density diminishes with increasing velocity, and vanishes together with ρ when $v = v_{max}$ (Fig. 52). This important difference between subsonic and supersonic steady flows can be simply interpreted as follows. In a subsonic flow, the streamlines approach in the direction of increasing velocity. In a supersonic flow, however, they diverge in that direction.

The flux j has its maximum value j_* at the point where the gas velocity is equal to the local velocity of sound:

$$j_* = \rho_* c_*, \tag{83.6}$$

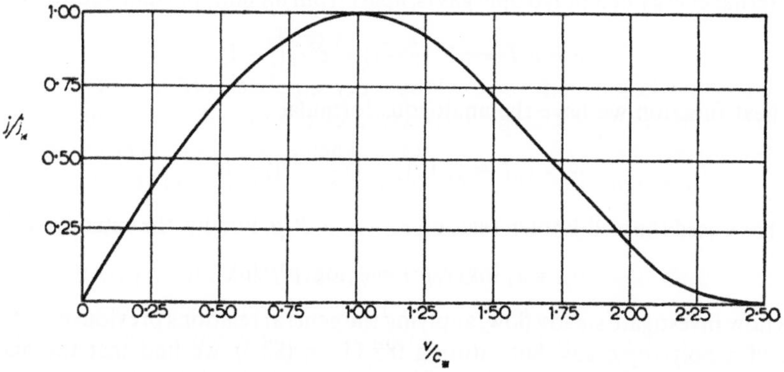

FIG. 52

where the asterisk suffix indicates values corresponding to this point. The velocity $v_* = c_*$ is called the *critical velocity*. In the general case of an arbitrary gas, the critical values of quantities can be expressed in terms of their values at the point $v = 0$, by solving the simultaneous equations

$$s_* = s_0, \qquad w_* + \tfrac{1}{2}c_*^2 = w_0. \tag{83.7}$$

† In reality, of course, when there is a sharp fall in temperature the gas must condense and form a two-phase "fog".

It is evident that, whenever $\mathrm{M} = v/c < 1$, we have also $v/c_* < 1$, and if $\mathrm{M} > 1$ then $v/c_* > 1$. Hence the ratio $\mathrm{M}_* = v/c_*$ serves in this case as a criterion analogous to M, and is more convenient, since c_* is a constant, unlike c, which varies along the stream.

In applications of the general equations of gas dynamics, the case of a perfect gas is of particular importance. For a perfect gas we shall always assume (except where otherwise specified) that the specific heat is a constant independent of temperature in the range considered. Such a gas is often called a *polytropic* gas, and we shall use this term in order to emphasize that the assumption made goes much further than that of a perfect gas. The relations between the thermodynamic quantities for a polytropic gas are given by very simple formulae, and this often allows a complete solution of the equations of gas dynamics. We shall give here, for reference, the formulae in question, since they will be needed several times in what follows.

The equation of state for a perfect gas is

$$pV = p/\rho = RT/\mu, \tag{83.8}$$

where $R = 8{\cdot}314 \times 10^7$ erg/deg mol is the gas constant, and μ the molecular weight of the gas. The velocity of sound in a perfect gas is, as shown in §64, given by

$$c^2 = \gamma RT/\mu = \gamma p/\rho, \tag{83.9}$$

where $\gamma = c_p/c_v$ is the ratio of specific heats, which always exceeds unity; for a polytropic gas it is constant. For monatomic gases $\gamma = 5/3$, and for diatomic gases $\gamma = 7/5$, at ordinary temperatures.†

The internal energy of a polytropic gas is, apart from an unimportant additive constant,

$$\varepsilon = c_v T = pV/(\gamma - 1) = c^2/\gamma(\gamma - 1). \tag{83.10}$$

For the heat function we have the analogous formulae

$$w = c_p T = \gamma pV/(\gamma - 1) = c^2/(\gamma - 1). \tag{83.11}$$

Here we have used the well-known relation $c_p - c_v = R/\mu$. Finally, the entropy of the gas is

$$s = c_v \log(p/\rho^\gamma) = c_p \log(p^{1/\gamma}/\rho). \tag{83.12}$$

Let us now investigate steady flow, applying the general relations previously obtained to the case of a polytropic gas. Substituting (83.11) in (83.3), we find that the maximum velocity of steady flow is

$$v_{\max} = c_0\sqrt{[2/(\gamma - 1)]}. \tag{83.13}$$

For the critical velocity we obtain from the second equation (83.7)

$$\frac{c_*^2}{\gamma - 1} + \tfrac{1}{2}c_*^2 = w_0 = \frac{c_0^2}{\gamma - 1},$$

† The name "polytropic" is derived from "polytropic process", i.e. one in which the pressure varies inversely as some power of the volume. For a gas with constant specific heats, such a process may be either isothermal, or adiabatic with pV^γ = constant (Poisson adiabatic). The specific-heat ratio γ is called the *adiabatic index*.

whence†

$$c_* = c_0\sqrt{[2/(\gamma+1)]}. \tag{83.14}$$

Bernoulli's equation (83.1), after substitution of the expression (83.11) for the heat function, gives the relation between the temperature and the velocity at any point on the streamline; similar relations for the pressure and density can then be obtained directly by means of the Poisson adiabatic equation:

$$\rho = \rho_0(T/T_0)^{1/(\gamma-1)}, \qquad p = p_0(\rho/\rho_0)^\gamma. \tag{83.15}$$

Thus we obtain the important results

$$\left.\begin{aligned} T &= T_0\left[1-\tfrac{1}{2}(\gamma-1)\frac{v^2}{c_0^2}\right] = T_0\left(1-\frac{\gamma-1}{\gamma+1}\frac{v^2}{c_*^2}\right),\\ \rho &= \rho_0\left[1-\tfrac{1}{2}(\gamma-1)\frac{v^2}{c_0^2}\right]^{1/(\gamma-1)} = \rho_0\left(1-\frac{\gamma-1}{\gamma+1}\frac{v}{c_*^2}\right)^{1/(\gamma-1)},\\ p &= p_0\left[1-\tfrac{1}{2}(\gamma-1)\frac{v^2}{c_0^2}\right]^{\gamma/(\gamma-1)} = p_0\left(1-\frac{\gamma-1}{\gamma+1}\frac{v^2}{c_*^2}\right)^{\gamma/(\gamma-1)} \end{aligned}\right\} \tag{83.16}$$

It is sometimes convenient to use these relations in a form which gives the velocity in terms of other quantities:

$$v^2 = \frac{2\gamma}{\gamma-1}\frac{p_0}{\rho_0}\left[1-\left(\frac{p}{p_0}\right)^{(\gamma-1)/\gamma}\right] = \frac{2\gamma}{\gamma-1}\frac{p_0}{\rho_0}\left[1-\left(\frac{\rho}{\rho_0}\right)^{\gamma-1}\right]. \tag{83.17}$$

We may also give the relation between the velocity of sound and the velocity v:

$$c^2 = c_0^2 - \tfrac{1}{2}(\gamma-1)v^2 = \tfrac{1}{2}(\gamma+1)c_*^2 - \tfrac{1}{2}(\gamma-1)v^2. \tag{83.18}$$

Hence we find that the numbers M and M_* are related by

$$M_*^2 = \frac{\gamma+1}{\gamma-1+2/M^2}; \tag{83.19}$$

when M varies from 0 to ∞, M_*^2 varies from 0 to $(\gamma+1)/(\gamma-1)$.

Finally, we may give expressions for the critical temperature, pressure and density: they are obtained by putting $v = c_*$ in formulae (83.16)‡:

$$\left.\begin{aligned} T_* &= 2T_0/(\gamma+1),\\ p_* &= p_0\left(\frac{2}{\gamma+1}\right)^{\gamma/(\gamma-1)},\\ \rho_* &= \rho_0\left(\frac{2}{\gamma+1}\right)^{1/(\gamma-1)} \end{aligned}\right\} \tag{83.20}$$

In conclusion, it should be emphasized that the results derived above are valid only for flow in which shock waves do not occur. When shock waves are present, equation (83.2)

† Figure 52 shows the ratio j/j_* as a function of v/c_* for air ($\gamma = 1{\cdot}4$, $v_{max} = 2{\cdot}45c_*$).

‡ For air, e.g., ($\gamma = 1{\cdot}4$)

$$c_* = 0{\cdot}913c_0,\quad p_* = 0{\cdot}528p_0,\quad \rho_* = 0{\cdot}634\rho_0,\quad T_* = 0{\cdot}833T_0.$$

does not hold; the entropy of the gas increases when a streamline passes through a shock wave. We shall see, however, that Bernoulli's equation (83.1) remains valid even when there are shock waves, since $w+\frac{1}{2}v^2$ is a quantity which is conserved across a surface of discontinuity (§85); formula (83.14), for example, therefore remains valid also.

PROBLEM

Express the temperature, pressure and density along a streamline in terms of the Mach number.

SOLUTION. Using the formulae obtained above, we find

$$T_0/T = 1+\tfrac{1}{2}(\gamma-1)\mathrm{M}^2, \qquad p_0/p = [1+\tfrac{1}{2}(\gamma-1)\mathrm{M}^2]^{\gamma/(\gamma-1)},$$
$$\rho_0/\rho = [1+\tfrac{1}{2}(\gamma-1)\mathrm{M}^2]^{1/(\gamma-1)}.$$

§84. Surfaces of discontinuity

In the preceding chapters we have considered only flows such that all quantities (velocity, pressure, density, etc.) vary continuously. Flows are also possible, however, for which discontinuities in the distribution of these quantities occur.

A discontinuity in a gas flow occurs over one or more surfaces; the quantities concerned change discontinuously as we cross such a surface, which is called a *surface of discontinuity*. In non-steady gas flow the surfaces of discontinuity do not in general remain fixed; here it should be emphasized, however, that the rate of motion of these surfaces bears no relation to the velocity of the gas flow itself. The gas particles in their motion may cross a surface of discontinuity.

Certain boundary conditions must be satisfied on surfaces of discontinuity. To formulate these conditions, we consider an element of the surface and use a coordinate system fixed to this element, with the x-axis along the normal.†

Firstly, the mass flux must be continuous: the mass of gas coming from one side must equal the mass leaving the other side. The mass flux through the surface element considered is ρv_x per unit area. Hence we must have $\rho_1 v_{1x} = \rho_2 v_{2x}$, where the suffixes 1 and 2 refer to the two sides of the surface of discontinuity.

The difference between the values of any quantity on the two sides of the surface will be denoted by enclosing it in square brackets; for example, $[\rho v_x] \equiv \rho_1 v_{1x} - \rho_2 v_{2x}$, and the condition just derived can be written

$$[\rho v_x] = 0. \tag{84.1}$$

Next, the energy flux must be continuous. It is given by (6.3). We therefore obtain the condition

$$[\rho v_x(\tfrac{1}{2}v^2 + w)] = 0. \tag{84.2}$$

Finally, the momentum flux must be continuous, i.e. the forces exerted on each other by the gases on the two sides of the surface of discontinuity must be equal. The momentum flux per unit area is (see §7) $pn_i + \rho v_i v_k n_k$. The normal vector $\mathbf{n}$ is along the x-axis. The continuity of the x-component of the momentum flux therefore gives the condition

$$[p + \rho v_x{}^2] = 0, \tag{84.3}$$

† If the flow is not steady, we consider an element of the surface during a short interval of time.

while that of the y and z components gives

$$[\rho v_x v_y] = 0, \qquad [\rho v_x v_z] = 0. \tag{84.4}$$

Equations (84.1)–(84.4) form a complete system of boundary conditions at a surface of discontinuity. From them we can immediately deduce the possibility of two types of surface of discontinuity.

In the first type, there is no mass flux through the surface. This means that $\rho_1 v_{1x} = \rho_2 v_{2x} = 0$. Since ρ_1 and ρ_2 are not zero, it follows that $v_{1x} = v_{2x} = 0$. The conditions (84.2) and (84.4) are then satisfied, and the condition (84.3) gives $p_1 = p_2$. Thus the normal velocity component and the gas pressure are continuous at the surface of discontinuity:

$$v_{1x} = v_{2x} = 0, \quad [p] = 0, \tag{84.5}$$

while the tangential velocities v_y, v_z and the density (as well as the other thermodynamic quantities except the pressure) may be discontinuous by any amount. We call this a *tangential discontinuity*.

In the second type, the mass flux is not zero, and v_{1x} and v_{2x} are therefore also not zero. We then have from (84.1) and (84.4)

$$[v_y] = 0, \quad [v_z] = 0, \tag{84.6}$$

i.e. the tangential velocity is continuous at the surface of discontinuity. The pressure, the density (and the other thermodynamic quantities) and the normal velocity, however, are discontinuous, their discontinuities being related by (84.1)–(84.3). In the condition (84.2) we can cancel ρv_x by (84.1), and replace v^2 by v_x^2 since v_y and v_z are continuous. Thus the following conditions must hold at the surface of discontinuity in this case:

$$\left.\begin{aligned} [\rho v_x] &= 0, \\ [\tfrac{1}{2}v_x^2 + w] &= 0, \\ [p + \rho v_x^2] &= 0. \end{aligned}\right\} \tag{84.7}$$

A discontinuity of this kind is called a *shock wave*, or simply a *shock*.

If we now return to the fixed coordinate system, we must everywhere replace v_x by the difference between the gas velocity component v_n normal to the surface of discontinuity and the velocity u of the surface itself, which is defined to be normal to the surface:

$$v_x = v_n - u. \tag{84.8}$$

The velocities v_n and u are taken in the fixed system. The velocity v_x is the velocity of the gas relative to the surface of discontinuity; we can also say that $-v_x = u - v_n$ is the rate of propagation of the surface relative to the gas. It should be noticed that, if v_x is discontinuous, this velocity has different values relative to the gas on the two sides of the surface.

We have already discussed (in §29) tangential discontinuities, at which the tangential velocity component is discontinuous, and we showed that, in an incompressible fluid, such discontinuities are unstable and must spread to form a turbulent region. A similar investigation for a compressible fluid shows that the same instability occurs, for any velocities (see Problem 1).

A particular case of tangential discontinuity is that of a *contact discontinuity*, where the velocity is continuous, but not the density (and therefore the other thermodynamic quantities, except the pressure). The above remarks on instability do not relate to discontinuities of this kind.

PROBLEMS

PROBLEM 1. Investigate the stability (with respect to infinitesimal perturbations) of tangential discontinuities in a homogeneous compressible medium (gas or liquid).

SOLUTION. The calculations are similar to those in §29 for an incompressible fluid. As there, the z-axis is taken to be normal to the surface.

In medium 2 (velocity $\mathbf{v}_2 = 0$, $z < 0$), the pressure satisfies the equation

$$\ddot{p}'_2 - c^2 \triangle p'_2 = 0$$

instead of Laplace's equation (29.2) for an incompressible fluid. We seek p'_2 in the form

$$p'_2 = \text{constant} \times \exp(-i\omega t + iqx + i\kappa_2 z),$$

where the wave number of the surface "ripples" is denoted by q, instead of k as in §29; if κ_2 is complex, it must be chosen so that im $\kappa_2 < 0$. The wave equation gives

$$\omega^2 = c^2(q^2 + \kappa_2{}^2). \tag{1}$$

Instead of (29.7), we now find by the same procedure

$$p'_2 = \zeta\rho\omega^2/i\kappa_2.$$

In medium 1, moving with velocity $\mathbf{v}_1 = \mathbf{v}$ ($z > 0$), we seek p'_1 in the form

$$p'_1 = \text{constant} \times \exp(-i\omega t + iqx - i\kappa_1 z).$$

To simplify the derivations, we first assume that $\mathbf{v}$ also is in the x-direction. The relation between ω, q and κ_1 is

$$(\omega - vq)^2 = c^2(q^2 + \kappa_1{}^2); \tag{2}$$

cf. (68.1). Instead of (29.6), we now have

$$p'_1 = -\zeta(\omega - qv)^2 \rho/i\kappa_1,$$

and the condition $p'_1 = p'_2$ gives

$$\frac{\kappa_1}{(\omega - qv)^2} + \frac{\kappa_2}{\omega^2} = 0. \tag{3}$$

The assumption made above concerning the direction of $\mathbf{v}$ can be avoided if we note that the unperturbed velocity appears in the original linearized continuity equation and Euler's equation only as $\mathbf{v}\cdot\mathbf{grad}$, in the terms $\mathbf{v}\cdot\mathbf{grad}\,p'$ and $(\mathbf{v}\cdot\mathbf{grad})\mathbf{v}'$ respectively. Hence, to change to an arbitrary direction of $\mathbf{v}$ (in the xy-plane) it is sufficient to replace v in (1)–(3) by $v\cos\phi$, where ϕ is the angle between $\mathbf{v}$ and $\mathbf{q}$; see the second footnote to §29.

Eliminating κ_1 and κ_2 from (1)–(3), we get the following dispersion relation for the perturbation frequency ω in terms of the wave number q:

$$\left[\frac{1}{\omega^2} - \frac{1}{(\omega - qv\cos\phi)^2}\right]\left[\frac{1}{c^2q^2} - \frac{1}{\omega^2} - \frac{1}{(\omega - qv\cos\phi)^2}\right] = 0. \tag{4}$$

The zero of the first factor,

$$\omega = \tfrac{1}{2}qv\cos\phi, \tag{5}$$

is always real. The zeros of the second factor are

$$\omega = \tfrac{1}{2}qv\cos\phi \pm q\sqrt{[\tfrac{1}{4}v^2\cos^2\phi + c^2 \pm c\sqrt{(c^2 + v^2\cos^2\phi)}]}; \tag{6}$$

they are real only if $v\cos\phi > v_k$, where

$$v_k = c\sqrt{8}. \tag{7}$$

Thus, when $v\cos\phi < v_k$, the dispersion relation has a pair of complex conjugate roots, one of which has im $\omega > 0$; the corresponding perturbations cause instability. When $v < v_k$, these are perturbations with any angle ϕ; when $v > v_k$, only those with $\cos\phi < v_k/k$ are unstable. The tangential discontinuity is therefore always unstable. The fact of instability (though not the perturbations which cause it) is evident from that which occurs in an incompressible fluid, coupled with the fact that v appears in the dispersion relation only in the combination

$v\cos\phi$: whatever the value of v, there must be angles ϕ for which $v\cos\phi \ll c$, and the fluid therefore behaves like an incompressible one with respect to such perturbations.†

PROBLEM 2. A plane sound wave is incident on a tangential discontinuity in a homogeneous compressible medium. Determine the intensity of the waves reflected and refracted by the discontinuity (J. W. Miles 1957; H. S. Ribner 1957).

SOLUTION. We take the coordinate axes as in Problem 1, the velocity **v** (in medium 1; $z > 0$) being in the x-direction. Let the sound wave be incident from the medium at rest (medium 2; $z < 0$), the direction of its wave vector **k** being specified by the angle θ between **k** and the z-axis and the angle ϕ between the projection **q** of **k** on the xy-plane and the velocity **v**:

$$k_x = q\cos\phi, \quad k_y = q\sin\phi, \quad k_z = (\omega/c)\cos\theta,$$
$$q = (\omega/c)\sin\theta = k\sin\theta,$$

with $0 < \theta < \frac{1}{2}\pi$ (the wave is incident in the positive z-direction). In medium 2, we seek the pressure in the form

$$p'_2 = \exp(ik_x x + ik_y y - i\omega t)[\exp ik_z z + A\exp(-ik_z z)],$$

where A is the reflected wave amplitude, the incident wave amplitude being arbitrarily taken as unity. In medium 1, there is just the refracted wave

$$p'_1 = B\exp(ik_x x + ik_y y + i\kappa z - i\omega t),$$

where κ satisfies the equation

$$(\omega - vk_x)^2 = c^2(k_x{}^2 + k_y{}^2 + \kappa^2);$$

cf. (2). The amplitudes A and B are found from the continuity conditions for the pressure and the vertical displacement of the fluid particles on either side of the discontinuity: $p'_1 = p'_2$ for $z = 0$, $\zeta_1 = \zeta_2 \equiv \zeta$. This gives two equations:

$$1 + A = B, \qquad \frac{\kappa}{(\omega - vk_x)^2}B = \frac{k_z}{\omega^2}(1 - A),$$

whence

$$A = \frac{(\omega - vk_x)^2/\kappa - \omega^2/k_z}{(\omega - vk_x)^2/\kappa + \omega^2/k_z}, \qquad B = \frac{2(\omega - vk_x)^2/\kappa}{(\omega - vk_x)^2/\kappa + \omega^2/k_z}, \tag{8}$$

and the problem is thus solved. The sign of κ,

$$\kappa^2 = (\omega/c)^2[(1 - \mathrm{M}\sin\theta\cos\phi)^2 - \sin^2\theta], \quad \mathrm{M} = v/c,$$

must be chosen in accordance with the limiting conditions for $z \to \infty$: the velocity of the refracted wave is away from the discontinuity, i.e.

$$U_z = \partial\omega/\partial\kappa = c^2\kappa/(\omega - vk_x) > 0. \tag{9}$$

These formulae show that three types of reflection are possible.

(1) When $\mathrm{M}\cos\phi < \operatorname{cosec}\theta - 1$, κ is real, and since $\omega - vk_x > 0$ it follows from (9) that $\kappa > 0$. Then, from (8), $|A| < 1$, and the reflected wave is weaker.

(2) When $\operatorname{cosec}\theta - 1 < \mathrm{M}\cos\phi < \operatorname{cosec}\theta + 1$, κ is imaginary and $|A| = 1$; there is total internal reflection of the sound wave.

(3) When $\mathrm{M}\cos\phi > \operatorname{cosec}\theta + 1$, which can occur only if $\mathrm{M} > 2$, κ is again real, but we must now take $\kappa < 0$. Then, from (8), $|A| > 1$, and the reflected wave is stronger. Moreover, the denominator of the expressions (8) may be zero for certain angles of incidence; the reflection coefficient then becomes infinite. Since this denominator is, apart from the notation, the same as the left-hand side of (3) in Problem 1, we can conclude immediately that the "resonance" angles of incidence are given by equations (5) and (6), the latter for $\mathrm{M} > \sqrt{8}$. In turn the infinite reflection (and transmission) coefficient, i.e. the non-zero amplitude of the reflected wave when the incident wave amplitude tends to zero, signifies that there can be spontaneous emission of sound by a surface of discontinuity: a perturbation (ripples) once formed on it will continue to emit sound waves indefinitely, being neither damped nor amplified; the energy carried away by the emitted sound is drawn from the whole of the medium in motion.

The energy flux density (averaged over time) in the refracted wave is

$$\bar{q}_2 = U_z\bar{E}_2 = \frac{c^2\kappa}{\omega - vk_x}\frac{\omega}{\omega - vk_x}\frac{|B|^2}{2\rho c^2},$$

---

† The value (7) was derived by L. D. Landau (1944). The need to allow for the non-collinearity of **v** and **q** in this problem was noted by S. I. Syrovat'skiĭ (1954).

with E_2 from (68.3). In case 3 we have $\kappa < 0$ and therefore $\bar{q}_2 < 0$: energy reaches the discontinuity from the moving medium, which acts as a source of amplification. When sound is spontaneously emitted, this incoming energy is equal to the energy carried away by the wave into the medium at rest.

In the solution given here, the instability of the surface of discontinuity is not taken into account. The fact that this statement of the problem is formally correct is the result of the linear independence of the sound waves and the unstable surface waves (which are damped as $z \to \pm\infty$). Physical correctness demands compliance with certain special (e.g. initial) conditions such that the surface waves are sufficiently weak.

§85. The shock adiabatic

Let us now investigate shock waves in detail.† We have seen that, in this type of discontinuity, the tangential component of the gas velocity is continuous. We can therefore take a coordinate system in which the surface element considered is at rest, and the tangential component of the gas velocity is zero on both sides.‡ Then we can write the normal component v_x as v simply, and the conditions (84.7) take the form

$$\rho_1 v_1 = \rho_2 v_2 \equiv j, \tag{85.1}$$

$$p_1 + \rho_1 v_1^2 = p_2 + \rho_2 v_2^2, \tag{85.2}$$

$$w_1 + \tfrac{1}{2}v_1^2 = w_2 + \tfrac{1}{2}v_2^2, \tag{85.3}$$

where j denotes the mass flux density at the surface of discontinuity. In what follows we shall always take j positive, with the gas going from side 1 to side 2. That is, we call gas 1 the one into which the shock wave moves, and gas 2 that which remains behind the shock. We call the side of the shock wave towards gas 1 the *front* of the shock, and that towards gas 2 the *back*.

We shall derive a series of relations which follow from the above conditions. Using the specific volumes $V_1 = 1/\rho_1$, $V_2 = 1/\rho_2$, we obtain from (85.1)

$$v_1 = jV_1, \qquad v_2 = jV_2 \tag{85.4}$$

and, substituting in (85.2),

$$p_1 + j^2 V_1 = p_2 + j^2 V_2, \tag{85.5}$$

or

$$j^2 = (p_2 - p_1)/(V_1 - V_2). \tag{85.6}$$

This formula, together with (85.4), relates the rate of propagation of a shock wave to the pressures and densities of the gas on the two sides of the surface.

Since j^2 is positive, we see that either $p_2 > p_1$, $V_1 > V_2$, or $p_2 < p_1$, $V_1 < V_2$; we shall see below that only the former case can actually occur.

We may note the following useful formula for the velocity difference $v_1 - v_2$. Substituting (85.6) in $v_1 - v_2 = j(V_1 - V_2)$, we obtain ††

$$v_1 - v_2 = \sqrt{[(p_2 - p_1)(V_1 - V_2)]}. \tag{85.7}$$

† One comment on terminology is needed. The shock wave is the discontinuity surface itself. Some authors, however, call this the shock front, while by the shock wave they mean the surface together with the gas flow behind it.

‡ This coordinate system is used everywhere in this chapter except §92.

A shock wave at rest is often called a *compression discontinuity*. If the shock is perpendicular to the direction of flow, we have a *normal shock*, otherwise an *oblique shock*.

†† Here we write the positive square root, since, as we shall see later (§87), we must have $v_1 - v_2 > 0$.

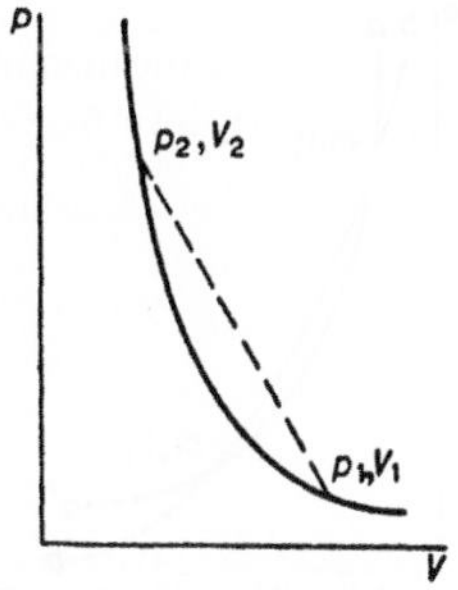

FIG. 53

Next, we write (85.3) in the form

$$w_1 + \tfrac{1}{2}j^2 V_1{}^2 = w_2 + \tfrac{1}{2}j^2 V_2{}^2 \tag{85.8}$$

and, substituting j^2 from (85.6), obtain

$$w_1 - w_2 + \tfrac{1}{2}(V_1 + V_2)(p_2 - p_1) = 0. \tag{85.9}$$

If we replace the heat function w by $\varepsilon + pV$, where ε is the internal energy, we can write this relation as

$$\varepsilon_1 - \varepsilon_2 + \tfrac{1}{2}(V_1 - V_2)(p_1 + p_2) = 0. \tag{85.10}$$

These relations hold between the thermodynamic quantities on the two sides of the surface of discontinuity.

For given p_1, V_1, equation (85.9) or (85.10) gives the relation between p_2 and V_2. This relation is called the *shock adiabatic* or the *Hugoniot adiabatic* (W. J. M. Rankine 1870; H. Hugoniot 1885). It is represented graphically in the pV-plane (Fig. 53) by a curve passing through the given point (p_1, V_1) corresponding to the state of gas 1 in front of the shock wave, which we shall call the *initial point*. It should be noted that the shock adiabatic cannot intersect the vertical line $V = V_1$ except at the initial point. For the existence of another intersection would mean that two different pressures satisfying (85.10) correspond to the same volume. For $V_1 = V_2$, however, we have from (85.10) also $\varepsilon_1 = \varepsilon_2$, and when the volumes and energies are the same the pressures must be the same. Thus the line $V = V_1$ divides the shock adiabatic into two parts, each of which lies entirely on one side of the line. Similarly, the shock adiabatic meets the horizontal line $p = p_1$ only at the point (p_1, V_1).

Let aa' (Fig. 54) be the shock adiabatic through the point (p_1, V_1) as initial point. We take any point (p_2, V_2) on it and draw through that point another adiabatic bb', for which (p_2, V_2) is an initial point. It is evident that the pair of values (p_1, V_1) satisfies the equation of this adiabatic also. The adiabatics aa' and bb' therefore intersect at the two points (p_1, V_1) and (p_2, V_2). It must be emphasized that the adiabatics are not identical, as would happen for Poisson adiabatics through a given point. This is a consequence of the fact that the equation of the shock adiabatic cannot be written in the form $f(p, V) = \text{constant}$, where f is some function, whereas the Poisson adiabatic, for example, can be written $s(p, V) = \text{constant}$. The Poisson adiabatics for a given gas form a one-parameter family of curves, but the shock adiabatic is determined by two parameters, the initial values p_1 and V_1. This has also the following important result: if two (or more) successive shock waves

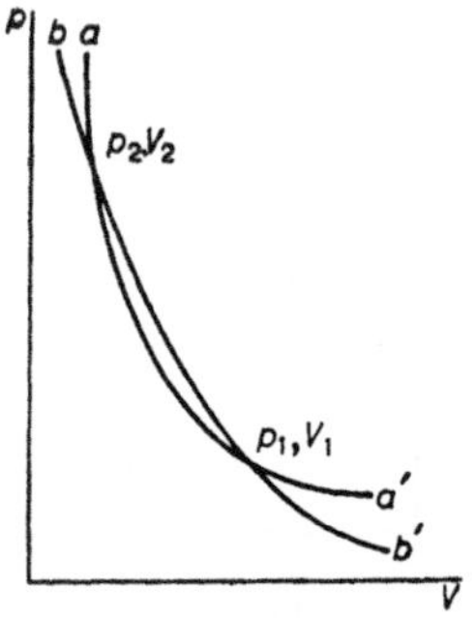

FIG. 54

take a gas from state 1 to state 2 and from there to state 3, the transition from state 1 to state 3 cannot in general be effected by the passage of any one shock wave.

For a given initial thermodynamic state of the gas (i.e. for given p_1 and V_1), the shock wave is defined by only one parameter; for instance, if the pressure p_2 behind the shock is given, then V_2 is determined by the Hugoniot adiabatic, and the flux density j and the velocities v_1 and v_2 are then given by formulae (85.4) and (85.6). It should be mentioned, however, that we are here considering the shock wave in a coordinate system in which the gas is moving normal to the surface. If the shock wave may be situated obliquely to the direction of flow, another parameter is needed; for example, the value of the velocity component tangential to the surface.

The following convenient graphical interpretation of formula (85.6) may be mentioned. If the point (p_1, V_1) on the shock adiabatic (Fig. 53) is joined by a chord to any other point (p_2, V_2) on it, then $(p_2 - p_1)/(V_2 - V_1) = -j^2$ is just the slope of this chord relative to the axis of abscissae. Thus j, and therefore the velocity of the shock wave, are determined at each point of the shock adiabatic by the slope of the chord joining that point to the initial point.

Like the other thermodynamic quantities, the entropy is discontinuous at a shock wave. By the law of increase of entropy, the entropy of a gas can only increase during its motion. Hence the entropy s_2 of the gas which has passed through the shock wave must exceed its initial entropy s_1:

$$s_2 > s_1. \tag{85.11}$$

We shall see below that this condition places very important restrictions on the manner of variation of all quantities in a shock wave.

The following fact should be emphasized. The presence of shock waves results in an increase in entropy in those flows which can be regarded as motions of an ideal fluid in all space, the viscosity and thermal conductivity being zero. The increase in entropy signifies that the motion is irreversible, i.e. energy is dissipated. Thus the discontinuities are a means by which energy can be dissipated in the motion of an ideal fluid. It follows that d'Alembert's paradox (§11) does not arise when bodies move in an ideal fluid in such a way as to cause shock waves. In such cases there is a drag force.

The true mechanism by which the entropy increases in shock waves lies, of course, in dissipative processes occurring in the very thin layers which actual shock waves are (see §93). It should be noticed, however, that the amount of this dissipation is entirely determined by the laws of conservation of mass, energy and momentum, when they are applied to the two sides of such layers; the width of the layers is just such as to give the increase in entropy required by these conservation laws.

The increase in entropy in a shock wave has another important effect on the motion: even if we have potential flow in front of the shock wave, the flow behind it is in general rotational. We shall return to this matter in §114.

§86. Weak shock waves

Let us consider a shock wave in which the discontinuity in every quantity is small; we call this a *weak shock wave*. We transform the relation (85.9) by expanding in powers of the small differences $s_2 - s_1$ and $p_2 - p_1$. We shall see that the first- and second-order terms in $p_2 - p_1$ then cancel; we must therefore carry the expansion with respect to $p_2 - p_1$ as far as the third order. In the expansion with respect to $s_2 - s_1$, only the first-order terms need be retained. We have

$$w_2 - w_1 = (\partial w/\partial s_1)_p(s_2 - s_1) + (\partial w/\partial p_1)_s(p_2 - p_1) + \\ + \tfrac{1}{2}(\partial^2 w/\partial p_1{}^2)_s(p_2 - p_1)^2 + \tfrac{1}{6}(\partial^3 w/\partial p_1{}^3)_s(p_2 - p_1)^3.$$

By the thermodynamic relation $dw = T ds + V dp$ we have for the derivatives

$$(\partial w/\partial s)_p = T, \qquad (\partial w/\partial p)_s = V.$$

Hence

$$w_2 - w_1 = T_1(s_2 - s_1) + V_1(p_2 - p_1) + \\ + \tfrac{1}{2}(\partial V/\partial p_1)_s(p_2 - p_1)^2 + \tfrac{1}{6}(\partial^2 V/\partial p_1{}^2)_s(p_2 - p_1)^3.$$

The volume V_2 need be expanded only with respect to $p_2 - p_1$, since the second term of equation (85.9) already contains the small difference $p_2 - p_1$, and an expansion with respect to $s_2 - s_1$ would give a term of the form $(s_2 - s_1)(p_2 - p_1)$, which is of no interest. Thus

$$V_2 - V_1 = (\partial V/\partial p_1)_s(p_2 - p_1) + \tfrac{1}{2}(\partial^2 V/\partial p_1{}^2)_s(p_2 - p_1)^2.$$

Substituting this expansion in (85.9), we obtain

$$s_2 - s_1 = \frac{1}{12T_1}\left(\frac{\partial^2 V}{\partial p_1{}^2}\right)_s(p_2 - p_1)^3. \tag{86.1}$$

Thus the discontinuity of entropy in a weak shock wave is of the third order of smallness relative to the discontinuity of pressure.

The adiabatic compressibility $-(\partial V/\partial p)_s$ almost always decreases with increasing pressure, i.e. the second derivative†

$$(\partial^2 V/\partial p^2)_s > 0. \tag{86.2}$$

It should be emphasized, however, that this is not a thermodynamic relation, and it is therefore possible in principle that the derivative might be negative.‡ We shall find several

---

† For a polytropic gas $(\partial^2 V/\partial p^2)_s = (\gamma + 1)V/\gamma^2 p^2$. This expression can be most simply obtained by differentiating the Poisson adiabatic equation $pV^\gamma = \text{constant}$.

‡ This may be true, for instance, near a gas–liquid critical point. The case where (86.2) does not hold can also be simulated by the shock adiabatic for a medium which has a phase transition (and for which the adiabatic therefore has a kink); see Ya. B. Zel'dovich and Yu. P. Raizer, *Physics of Shock Waves and High-Temperature Hydrodynamic Phenomena*, New York 1966, 1967, Chapter I §19, Chapter XI §20.

times in what follows that the sign of the derivative $(\partial^2 V/\partial p^2)_s$ is very important in gas dynamics. In future we shall assume it to be positive.

Let us draw through the point 1 (p_1, V_1) in the pV-plane two curves, the shock adiabatic and the Poisson adiabatic. The equation of the latter is $s_2 - s_1 = 0$. By comparing this with the equation (86.1) of the shock adiabatic near the point 1, we see that the two curves have contact of the second order at this point, both the first and the second derivatives being equal. In order to decide the relative position of the two curves near the point 1, we use the fact that, according to (86.1) and (86.2), we must have $s_2 > s_1$ on the shock adiabatic for $p_2 > p_1$, while on the Poisson adiabatic $s_2 = s_1$. The abscissa of a point on the shock adiabatic must therefore exceed that of a point on the Poisson adiabatic having the same ordinate p_2. This follows at once from the fact that, by the well-known thermodynamic formula $(\partial V/\partial s)_p = (T/c_p)(\partial V/\partial T)_p$, the entropy increases with the volume at constant pressure for all substances which expand on heating, i.e. which have $(\partial V/\partial T)_p$ positive. We can similarly deduce that, for $p_2 < p_1$, the abscissa of a point on the Poisson adiabatic exceeds that of the corresponding point on the shock adiabatic. Thus, near the point of contact, the two curves lie as shown in Fig. 55 (HH' being the shock adiabatic and PP' the Poisson adiabatic)†, both being concave upwards, by (86.2).

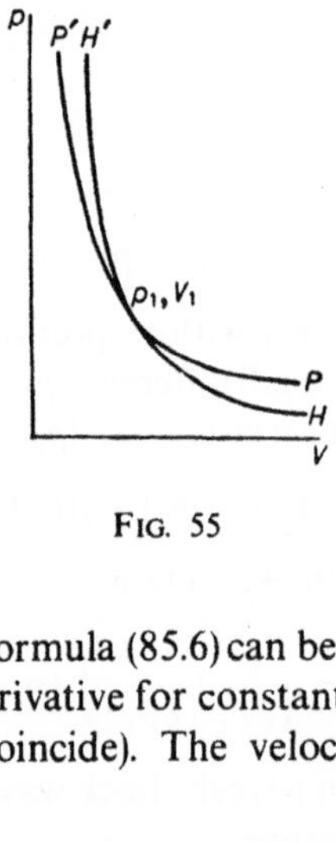

FIG. 55

For small $p_2 - p_1$ and $V_2 - V_1$, formula (85.6) can be written, in the first approximation, as $j^2 = -(\partial p/\partial V)_s$ (we take the derivative for constant entropy, since the tangents to the two adiabatics at the point 1 coincide). The velocities v_1 and v_2 are, in the same approximation, equal:

$$v_1 = v_2 = v = jV = \sqrt{[-V^2(\partial p/\partial V)_s]} = \sqrt{(\partial p/\partial \rho)_s}.$$

This is just the velocity of sound c. Thus the rate of propagation of weak shock waves is, in the first approximation, the velocity of sound:

$$v = c. \tag{86.3}$$

From the properties of the shock adiabatic near the point 1 derived above we can deduce a number of important consequences. Since we must have $s_2 > s_1$ in a shock wave, it follows that $p_2 > p_1$, i.e. the point 2 (p_2, V_2) must lie above the point 1. Moreover, since the chord 12 has a greater slope than the tangent to the adiabatic at the point 1 (Fig. 53), and

† If $(\partial V/\partial T)_p$ is negative, the relative position is reversed.

the slope of the tangent is equal to the derivative $(\partial p/\partial V_1)_{s_1}$, we have $j^2 > -(\partial p/\partial V_1)_{s_1}$. Multiplying both sides of this inequality by $V_1{}^2$, we find

$$j^2 V_1{}^2 = v_1{}^2 > -V_1{}^2(\partial p/\partial V_1)_{s_1} = (\partial p/\partial \rho_1)_{s_1} = c_1{}^2,$$

where c_1 is the velocity of sound corresponding to the point 1. Thus $v_1 > c_1$. Finally, from the fact that the chord 12 has a smaller slope than the tangent at the point 2, it follows in like manner that $v_2 < c_2$.†

When the derivative $(\partial^2 V/\partial p^2)_s$ is negative, the condition $s_2 > s_1$ for weak shock waves implies that $p_2 < p_1$, while the velocities again satisfy $v_1 > c_1$, $v_2 < c_2$.

§87. The direction of variation of quantities in a shock wave

The results of §86 show that, if the derivative $(\partial^2 V/\partial p^2)_s$ is assumed positive, it can be demonstrated very simply that for weak shocks the condition of increasing entropy ($s_2 > s_1$) necessarily means that

$$p_2 > p_1, \tag{87.1}$$

$$v_1 > c_1, \qquad v_2 < c_2. \tag{87.2}$$

From the remark made concerning (85.6) it follows that, if $p_2 > p_1$, then

$$V_1 > V_2, \tag{87.3}$$

and, since $v_1/V_1 = v_2/V_2 = j$, also‡

$$v_1 > v_2. \tag{87.4}$$

The inequalities (87.1) and (87.3) signify that, when the gas passes through the shock wave, it is compressed, the pressure and density increasing. The inequality $v_1 > c_1$ means that the shock wave moves supersonically relative to the gas ahead of it; clearly, therefore, no perturbations starting from the shock wave can penetrate into that gas. In other words, the presence of the shock has no effect on the state of gas in front of it.

We shall now show that all the inequalities (87.1)–(87.4) hold for shock waves with *any* intensity, if it is again assumed that $(\partial^2 V/\partial p^2)_s$ is positive.††

The quantity j^2 gives the slope of the chord from the initial point 1 on the shock adiabatic to any point 2 ($-j^2$ is the slope of this chord to the V-axis). We shall show first that the direction of variation of j^2 as the point 2 moves along the adiabatic is in a definite relation to that of the entropy s_2.

We differentiate the relations (85.5) and (85.8) with respect to the quantities pertaining to gas 2, assuming the state of gas 1 to be unchanged. This means that p_1, V_1 and w_1 are regarded as constants, while p_2, V_2, w_2 and j are differentiated. From (85.5) we obtain

$$\mathrm{d}p_2 + j^2 \mathrm{d}V_2 = (V_1 - V_2)\mathrm{d}(j^2), \tag{87.5}$$

† This argument is valid only near point 1, where the slope of the tangent to the shock adiabatic at point 2 differs from the derivative $(\partial p_2/\partial V_2)_{s_2}$ only by second-order small quantities.

‡ If we change to a frame of reference in which gas 1 (in front of the shock wave) is at rest, and the shock is moving, then the inequality $v_1 > v_2$ means that the gas behind the shock wave moves (with velocity $v_1 - v_2$) in the same direction as the shock itself.

†† These inequalities were derived for shock waves with any intensity in a polytropic gas by E. Jouguet (1904) and G. Zemplén (1905). The proof given below for any medium is due to L. D. Landau (1944).

and from (85.8)

$$dw_2 + j^2 V_2 dV_2 = \tfrac{1}{2}(V_1{}^2 - V_2{}^2)d(j^2),$$

or, expanding the differential dw_2,

$$T_2 ds_2 + V_2(dp_2 + j^2 dV_2) = \tfrac{1}{2}(V_1{}^2 - V_2{}^2)\,d(j^2).$$

Substituting in this equation from (87.5), we obtain

$$T_2 ds_2 = \tfrac{1}{2}(V_1 - V_2)^2 d(j^2). \tag{87.6}$$

Hence we see that

$$d(j^2)/ds_2 > 0, \tag{87.7}$$

i.e. j^2 increases with s_2.

We now show that there can be no point on the shock adiabatic at which it touches any line drawn from the point 1 (such as the point O is in Fig. 56). At such a point the slope of the chord from the point 1 is a minimum, and j^2 has a corresponding maximum, so that $d(j^2)/dp_2 = 0$. We see from (87.6) that in this case we also have $ds_2/dp_2 = 0$. Next, we calculate $d(j^2)/dp_2$ at any point on the shock adiabatic; substituting in (87.5) the differential dV_2 in the form $dV_2 = (\partial V_2/\partial p_2)_{s_2} dp_2 + (\partial V_2/\partial s_2)_{p_2} ds_2$ and ds_2 in the form given by (87.6), and dividing by dp_2, we obtain

$$\frac{d(j^2)}{dp_2} = \frac{1 + j^2(\partial V_2/\partial p_2)_{s_2}}{(V_1 - V_2)[1 - \tfrac{1}{2}(j^2/T_2)(V_1 - V_2)(\partial V_2/\partial s_2)_{p_2}]}. \tag{87.8}$$

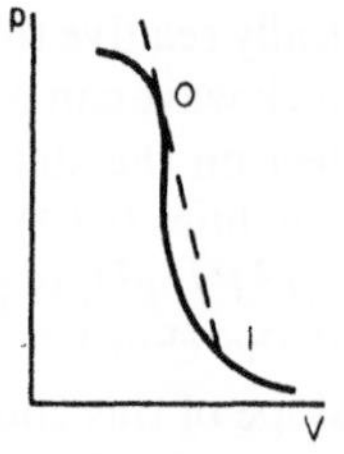

FIG. 56

Hence it follows that, for this to be zero, we must have

$$1 + j^2(\partial V_2/\partial p_2)_{s_2} = 1 - v_2{}^2/c_2{}^2 = 0,$$

i.e. $v_2 = c_2$; conversely, if $v_2 = c_2$, it follows that $d(j^2)/dp_2 = 0$; the only other possibility would be for both the numerator and the denominator in (87.8) to vanish, but these are two different functions of the point 2 on the shock adiabatic, and so their simultaneous vanishing would be an improbable accident.†

† It should be emphasized, to avoid misunderstanding, that $d(j^2)/dp_2$ itself is not a further independent function of the point 2, since it is determined by (87.8).

Thus, of the three equations

$$d(j^2)/dp_2 = 0, \qquad ds_2/dp_2 = 0, \qquad v_2 = c_2, \tag{87.9}$$

each implies the other two and all three would hold at the point O (Fig. 56). On account of the third equation, O will be called an *acoustic point*. Finally, we have for the derivative of v_2^2/c_2^2 at the point O

$$\frac{d}{dp_2}\left(\frac{v_2^2}{c_2^2}\right) = -\frac{d}{dp_2}\left[j^2\left(\frac{\partial V_2}{\partial p_2}\right)_{s_2}\right] = -j^2\left(\frac{\partial^2 V_2}{\partial p_2^2}\right)_{s_2}.$$

In view of the assumption that $(\partial^2 V/\partial p^2)_s$ is positive, we therefore have at the acoustic point

$$d(v_2/c_2)/dp_2 < 0. \tag{87.10}$$

It is now easy to show that such a point cannot exist on the shock adiabatic. At points just above the initial point 1, $v_2/c_2 < 1$ (see the end of §86). The equation $v_2/c_2 = 1$ can therefore be satisfied only by an increase in v_2/c_2; that is, at the acoustic point we should necessarily have $d(v_2/c_2)/dp_2 > 0$, whereas by (87.10) the converse is true. In a similar manner, we can show that the ratio v_2/c_2 also cannot become equal to unity on the part of the shock adiabatic below the point 1.

From the impossibility of the existence of acoustic points, which has just been demonstrated, we can at once deduce from the graph of the shock adiabatic that the slope of the chord from the point 1 (p_1, V_1) to the point 2 (p_2, V_2) decreases as point 2 moves up the curve, and j^2 correspondingly increases. From this property of the shock adiabatic and the inequality (87.7) it follows immediately that s_2 likewise increases, and the necessary condition $s_2 > s_1$ implies that $p_2 > p_1$ also.

It is also easy to see that, on the upper part of the shock adiabatic, the inequalities $v_2 < c_2, v_1 > c_1$ hold. The former follows at once from the fact that it holds near the point 1, and the ratio v_2/c_2 can never become equal to unity. The second inequality follows from the fact that every chord from the point 1 to a point 2 above it is steeper than the tangent to the adiabatic at the point 1, since the curve cannot behave as shown in Fig. 56.

The condition $s_2 > s_1$ and all three inequalities (87.1), (87.2) are therefore satisfied on the upper part of the shock adiabatic. On the lower part, however, none of these conditions holds. They are consequently equivalent, and if one is satisfied so are all the others.

In the preceding discussion we have everywhere assumed that the derivative $(\partial^2 V/\partial p^2)_s$ is positive. If this derivative could change sign, it would no longer be possible to draw from the necessity of $s_2 > s_1$ any general conclusions concerning inequalities for the other quantities.

§88. Evolutionary shock waves

The derivation of the inequalities (87.1)–(87.4) in §§86 and 87 involved a particular assumption concerning the thermodynamic properties of the medium, namely that $(\partial^2 V/\partial p^2)_s$ is positive. It is most important, however, to note that the inequalities

$$v_1 > c_1, \qquad v_2 < c_2 \tag{88.1}$$

for the velocities can be obtained by quite different arguments, which show that shock waves in which the inequalities (88.1) do not hold cannot exist, even if their existence would not be disproved by the purely thermodynamic arguments given above.†

The point is that we have still to discuss the subject of the stability of shock waves. The most general necessary condition for stability is that any infinitesimal perturbation of the initial state (at some instant $t = 0$) should cause only a quite definite infinitesimal change in the flow, at least during a sufficiently short interval t. The latter restriction means that the condition in question is not a sufficient one. For example, if an initial small perturbation grows, even exponentially as $e^{\gamma t}$ with a positive constant γ, it remains small for a time $t \lesssim 1/\gamma$, although ultimately it will disrupt the flow pattern concerned. A perturbation that does not satisfy the necessary condition stated is the splitting of a shock wave into two or more surfaces of discontinuity. It is evident that the change in the flow here is immediately not small, although for small t (when the two discontinuities have not moved far apart) it occupies only a short range of distances δx.

Any initial small perturbation is defined by some number of independent parameters. Its subsequent development is governed by a set of linearized boundary conditions which must be satisfied at the surface of discontinuity. The necessary condition for stability stated above will be satisfied if the number of these equations is the same as the number of unknown parameters in them; the boundary conditions then determine the subsequent development of the perturbation, which remains small for small $t > 0$. If the number of equations is greater or less than the number of independent parameters, then the problem of the small perturbation has either no solution or an infinity of solutions. Either case would show that the initial assumption (that the perturbation is small if t is small) is incorrect, and would therefore contradict the condition imposed. The condition thus formulated is called the condition for the flow to be *evolutionary*.

Let us suppose that a shock wave is subjected to an infinitesimal displacement in a direction perpendicular to its plane.‡ The displacement is accompanied by infinitesimal perturbations in the gas pressure, velocity, etc. on both sides of the surface of discontinuity. These perturbations near the shock are then propagated away from it with the velocity of sound (relative to the gas); this, however, does not apply to the perturbation in the entropy, which is transmitted only with the gas itself. Thus an arbitrary perturbation of the type in question can be regarded as consisting of sound disturbances propagated in gases 1 and 2 on both sides of the shock wave, and a perturbation of the entropy; the latter, which moves with the gas, will evidently occur only in gas 2 behind the shock. In each of the sound disturbances, the changes in the various quantities are related by certain formulae which follow from the equations of motion (as in any sound wave, §64), and therefore any such disturbance is specified by only one parameter.

Let us now compute the number of possible sound disturbances. It depends on the relative magnitudes of the gas velocities v_1, v_2 and the sound velocities c_1, c_2. We take the direction of motion of the gas (from 1 to 2) as the positive direction of the x-axis. The rate of propagation of the disturbance in gas 1 relative to the stationary shock wave is $u_1 = v_1 \pm c_1$, and in gas 2 it is $u_2 = v_2 \pm c_2$. Since these disturbances must be propagated away from the shock wave, it follows that $u_1 < 0$, $u_2 > 0$.

Let us suppose that $v_1 > c_1$, $v_2 < c_2$. Then it is clear that both values $u_1 = v_1 \pm c_1$ are positive, while only $v_2 + c_2$ of the two values of u_2 is positive. This means that the sound

† It should be mentioned at the same time that (at least for weak shock waves) these thermodynamic arguments lead to the conditions (88.1) also when $(\partial^2 V/\partial p^2)_s < 0$ and the shock wave is a rarefaction wave, not a compression wave, as has been noted at the end of §86.

‡ The following proof of the inequalities (88.1) is due to L.D. Landau (1944).

disturbances in which we are interested cannot exist in gas 1, while in gas 2 there can be only one, which is propagated relative to the gas with velocity c_2. The calculation in other cases is similar.

The result is shown in Fig. 57, where each arrow corresponds to one sound disturbance, propagated relative to the gas in the direction shown by the arrow. Each sound disturbance is defined, as stated above, by one parameter. Furthermore, in all four cases there are two other parameters, one determining the entropy perturbation propagated in gas 2 and one determining the displacement of the shock wave. For each of the four cases in Fig. 57, the number in a circle shows the total number of parameters, thus obtained, which define an arbitrary perturbation arising from the displacement of the shock wave.

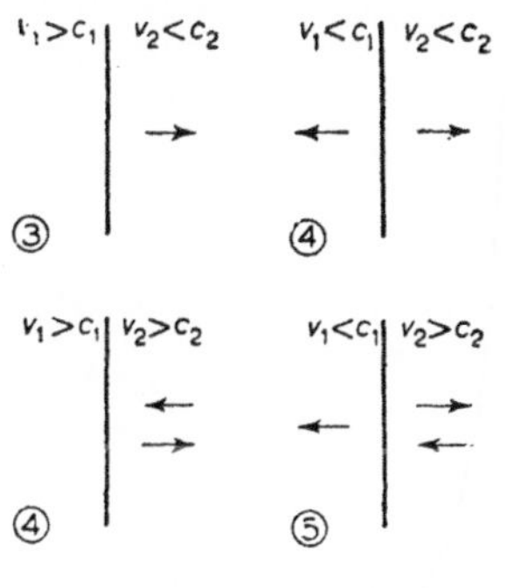

FIG. 57

The number of boundary conditions which must be satisfied by a perturbation on the surface of discontinuity is three (the continuity of the mass, energy and momentum fluxes). In all except the first of the cases shown in Fig. 57, the number of independent parameters available exceeds the number of equations. It is seen that only the shock waves which satisfy the conditions (88.1) are evolutionary. These conditions are therefore necessary ones for the existence of shock waves, whatever the thermodynamic properties of the gas. An artificially created discontinuity that did not satisfy these conditions would immediately disintegrate into other discontinuities.†

An evolutionary shock wave is stable with respect to the perturbation type considered and in the ordinary sense of the word. If the movement of the shock (and therefore the perturbation of all other quantities) is sought in a form proportional to $e^{-i\omega t}$, then it is evident a priori that the value of ω uniquely determined by the boundary conditions can only be zero, since the problem involves no parameters having the dimension of reciprocal time which might determine a non-zero value of ω.

We shall return in §90 to the question of shock wave stability.

§89. Shock waves in a polytropic gas

Let us apply the general relations obtained in the previous sections to shock waves in a polytropic gas. The heat function of such a gas is given by the simple formula (83.11).

† For each of the non-evolutionary cases shown in Fig. 57, the perturbation is under-determined: the number of arbitrary parameters exceeds the number of equations. In magnetohydrodynamics, shock waves may be non-evolutionary because the perturbation is either under- or over-determined (see *ECM*, §73).

Substituting this expression in (85.9), we have after a simple transformation

$$\frac{V_2}{V_1} = \frac{(\gamma+1)p_1+(\gamma-1)p_2}{(\gamma-1)p_1+(\gamma+1)p_2}. \tag{89.1}$$

Using this formula, we can determine any of the quantities p_1, V_1, p_2, V_2 from the other three. The ratio V_2/V_1 is a monotonically decreasing function of the ratio p_2/p_1, tending to the finite limit $(\gamma-1)(\gamma+1)$. The curve showing p_2 as a function of V_2 for given p_1, V_1 (the shock adiabatic) is represented in Fig. 58. It is a rectangular hyperbola with asymptotes $V_2/V_1 = (\gamma-1)/(\gamma+1)$, $p_2/p_1 = -(\gamma-1)/(\gamma+1)$. As we know, only the upper part of the curve, above the point $V_2/V_1 = p_2/p_1 = 1$, has any real significance; it is shown in Fig. 58 (for $\gamma = 1{\cdot}4$) by a continuous line.

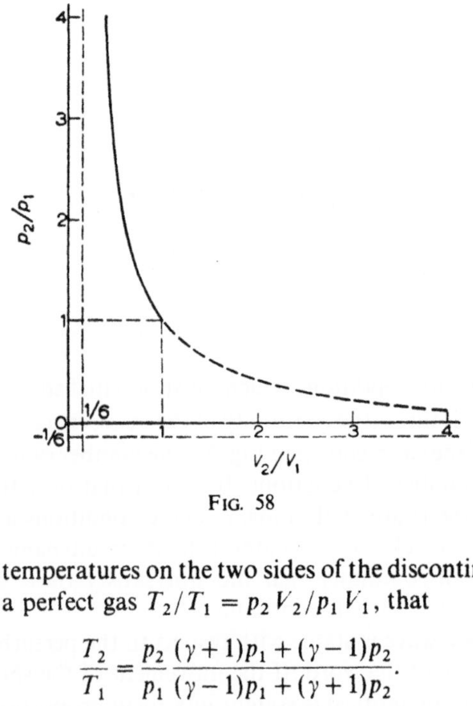

FIG. 58

For the ratio of the temperatures on the two sides of the discontinuity we find, from the equation of state for a perfect gas $T_2/T_1 = p_2V_2/p_1V_1$, that

$$\frac{T_2}{T_1} = \frac{p_2}{p_1}\frac{(\gamma+1)p_1+(\gamma-1)p_2}{(\gamma-1)p_1+(\gamma+1)p_2}. \tag{89.2}$$

For the flux density j we obtain from (85.6) and (89.1)

$$j^2 = \{(\gamma-1)p_1+(\gamma+1)p_2\}/2V_1, \tag{89.3}$$

and then for the velocities of propagation of the shock wave relative to the gas before and behind it

$$\left.\begin{aligned} v_1^2 &= \tfrac{1}{2}V_1\{(\gamma-1)p_1+(\gamma+1)p_2\} = \tfrac{1}{2}(c_1^2/\gamma)[\gamma-1+(\gamma+1)p_2/p_1], \\ v_2^2 &= \tfrac{1}{2}V_1\{(\gamma+1)p_1+(\gamma-1)p_2\}^2/\{(\gamma-1)p_1+(\gamma+1)p_2\} \\ &= \tfrac{1}{2}(c_2^2/\gamma)[\gamma-1+(\gamma+1)p_1/p_2], \end{aligned}\right\} \tag{89.4}$$

their difference being

$$v_1 - v_2 = \sqrt{(2V_1)}(p_2-p_1)/\sqrt{[(\gamma-1)p_1+(\gamma+1)p_2]}. \tag{89.5}$$

There are some formulae useful in applications, which express the ratios of densities, pressures and temperatures in a shock wave in terms of the Mach number $M_1 = v_1/c_1$. These formulae are easily derived from the foregoing results:

$$\rho_2/\rho_1 = v_1/v_2 = (\gamma+1)M_1^2/\{(\gamma-1)M_1^2+2\}, \tag{89.6}$$

$$p_2/p_1 = 2\gamma M_1^2/(\gamma+1)-(\gamma-1)/(\gamma+1), \tag{89.7}$$

$$T_2/T_1 = \{2\gamma M_1^2-(\gamma-1)\}\{(\gamma-1)M_1^2+2\}/(\gamma+1)^2 M_1. \tag{89.8}$$

The Mach number M_2 is given in terms of M_1 by

$$M_2^2 = \{2+(\gamma-1)M_1^2\}/\{2\gamma M_1^2-(\gamma-1)\}. \tag{89.9}$$

This is symmetrical in M_1 and M_2: it may be written in the form $2\gamma M_1^2 M_2^2 - (\gamma-1)(M_1^2+M_2^2) = 2$.

We can give limiting results for very strong shock waves, in which $(\gamma-1)p_2$ is very large compared with $(\gamma+1)p_1$. From (89.1) and (89.2) we have

$$V_2/V_1 = \rho_1/\rho_2 = (\gamma-1)/(\gamma+1), \qquad T_2/T_1 = (\gamma-1)p_2/(\gamma+1)p_1. \tag{89.10}$$

The ratio T_2/T_1 increases to infinity with p_2/p_1, i.e. the temperature discontinuity in a shock wave, like the pressure discontinuity, can be arbitrarily great. The density ratio, however, tends to a constant limit; e.g., for a monatomic gas the limit is $\rho_2 = 4\rho_1$, and for a diatomic gas $\rho_2 = 6\rho_1$. The velocities of propagation of a strong shock wave are

$$v_1 = \sqrt{\{\tfrac{1}{2}(\gamma+1)p_2 V_1\}}, \qquad v_2 = \sqrt{\{\tfrac{1}{2}(\gamma-1)^2 p_2 V_1/(\gamma+1)\}}. \tag{89.11}$$

They increase as the square root of the pressure p_2.

Lastly, there are relations for weak shock waves, which are the leading terms in expansions in powers of the small quantity $z \equiv (p_2-p_1)/p_1$:

$$\left.\begin{aligned} M_1 - 1 = 1 - M_2 = (\gamma+1)z/4\gamma, \qquad c_2/c_1 &= 1+(\gamma-1)z/2\gamma,\\ \rho_2/\rho_1 &= 1+z/\gamma-(\gamma-1)z^2/2\gamma. \end{aligned}\right\} \tag{89.12}$$

These are the terms giving the first correction to the acoustic approximation.

PROBLEMS

PROBLEM 1. Derive the formula $v_1 v_2 = c_*^2$, where c_* is the critical velocity (L. Prandtl).

SOLUTION. Since $w+\frac{1}{2}v^2$ is continuous at a shock wave, we can define a critical velocity which is the same for gases 1 and 2 by

$$\frac{\gamma p_1}{(\gamma-1)\rho_1}+\tfrac{1}{2}v_1^2 = \frac{\gamma p_2}{(\gamma-1)\rho_2}+\tfrac{1}{2}v_2^2 = \frac{\gamma+1}{2(\gamma-1)}c_*^2;$$

cf. (83.7). Determining p_2/ρ_2 and p_1/ρ_1 from these equations and substituting in

$$v_1 - v_2 = \frac{p_2}{\rho_2 v_2} - \frac{p_1}{\rho_1 v_1}$$

(obtained by combining (85.1) and (85.2)), we obtain

$$\frac{\gamma+1}{2\gamma}(v_1-v_2)\left(1-\frac{c_*^2}{v_1 v_2}\right) = 0.$$

Since $v_1 \neq v_2$, this gives the required relation.

PROBLEM 2. Determine the value of the ratio p_2/p_1, for given temperatures T_1, T_2 at a shock wave in a perfect gas with a variable specific heat.

SOLUTION. For such a gas, we can say only that w (like ε) is a function of temperature alone, and that p, V and T are related by the equation of state $pV = RT/\mu$. Solving equation (85.9) for p_2/p_1, we obtain

$$\frac{p_2}{p_1} = \frac{\mu}{RT_1}(w_2 - w_1) - \frac{T_2 - T_1}{2T_1} + \sqrt{\left\{\left[\frac{\mu(w_2 - w_1)}{RT_1} - \frac{T_2 - T_1}{2T_1}\right]^2 + \frac{T_2}{T_1}\right\}},$$

where $w_1 = w(T_1)$, $w_2 = w(T_2)$.

§90. Corrugation instability of shock waves

The conditions for a shock wave to be evolutionary are necessary, but not sufficient, to ensure that it is stable. It may be unstable with respect to perturbations having periodicity on the surface of discontinuity and thus forming "ripples" or "corrugations" on that surface; such perturbations have already been discussed in §29 for the case of tangential discontinuities.† We shall show how this topic may be investigated for shock waves in any medium (S. P. D'yakov 1954).

Let a shock wave be at rest on the plane $x = 0$, with the fluid passing through it from left to right, in the positive x-direction. Let the surface of discontinuity undergo a perturbation in which the points on the surface are displaced in the x-direction by a small amount

$$\zeta = \zeta_0 \exp(ik_y y - i\omega t), \tag{90.1}$$

where k_y is the wave number of the ripple. This surface ripple perturbs the flow behind the shock wave in the region $x > 0$; that in front of the discontinuity, in $x < 0$, is not perturbed, because of its supersonic velocity.

Any perturbation of the flow is composed of an entropy-vortex wave and a sound wave (see §82, Problem). In each, the dependence of quantities on time and coordinates is given by a factor having the form $\exp(i\mathbf{k}\cdot\mathbf{r} - i\omega t)$ with the same frequency ω as in (90.1). It is evident from symmetry that the wave vector $\mathbf{k}$ is in the xy-plane; its y-component is the same as k_y in (90.1), but the x-component is different for the two types of perturbation.

In the entropy-vortex wave, $\mathbf{k}\cdot\mathbf{v}_2 = \omega$, so that $k_x = \omega/v_2$, where v_2 is the unperturbed gas velocity beyond the discontinuity. In this wave there is no pressure perturbation; the specific volume perturbation arises from the entropy perturbation, $\delta V^{(\text{ent})} = (\partial V/\partial s)_p \delta s$, and the velocity perturbation satisfies the condition

$$\mathbf{k}\cdot\delta\mathbf{v}^{(\text{ent})} = (\omega/v_2)\delta v_x{}^{(\text{ent})} + k_y \delta v_y{}^{(\text{ent})} = 0. \tag{90.2}$$

In the sound wave in the moving gas, the relation between the frequency and the wave vector is $(\omega - \mathbf{k}\cdot\mathbf{v})^2 = c^2k^2$ (see (68.1)); k_x in this wave is therefore given by

$$(\omega - k_x v_2)^2 = c_2{}^2(k_x{}^2 + k_y{}^2). \tag{90.3}$$

The perturbations of the pressure, the specific volume, and the velocity are related by

$$\delta p^{(s)} = -(c_2/V_2)^2 \delta V^{(s)}, \tag{90.4}$$

$$(\omega - v_2 k_x)\delta\mathbf{v}^{(s)} = V_2 \mathbf{k}\delta p^{(s)}. \tag{90.5}$$

The total perturbation is a linear combination of perturbations of each type:

$$\delta v = \delta v^{(\text{ent})} + \delta v^{(s)}, \quad \delta V = \delta V^{(\text{ent})} + \delta V^{(s)}, \quad \delta p = \delta p^{(s)}. \tag{90.6}$$

† Instability with respect to such perturbations is called *corrugation instability*.

It has to satisfy certain boundary conditions on the perturbed surface of discontinuity.

Firstly, the tangential velocity component must be continuous on this surface, and the discontinuity of the normal component must be expressible in terms of the perturbed pressure and density by (85.7). These conditions are

$$\mathbf{v}_1 \cdot \mathbf{t} = (\mathbf{v}_2 + \delta\mathbf{v}) \cdot \mathbf{t},$$

$$\mathbf{v}_1 \cdot \mathbf{n} - (\mathbf{v}_2 + \delta\mathbf{v}) \cdot \mathbf{n} = \sqrt{[(p_2 - p_1 + \delta p)(V_1 - V_2 - \delta V)]},$$

where **t** and **n** are unit vectors along the tangent and normal to the surface of discontinuity (Fig. 59). As far as first-order small quantities, the components of these vectors in the xy-plane are $\mathbf{t}(ik\zeta, 1)$ and $\mathbf{n}(1, -ik\zeta)$; the expression $ik\zeta$ arises as the derivative $\partial\zeta/\partial y$. To the same accuracy, the boundary conditions for the velocity are

$$\delta v_y = ik\zeta(v_1 - v_2), \quad \delta v_x = \tfrac{1}{2}(v_2 - v_1)\left[\frac{\delta p}{p_2 - p_1} - \frac{\delta V}{V_1 - V_2}\right]. \tag{90.7}$$

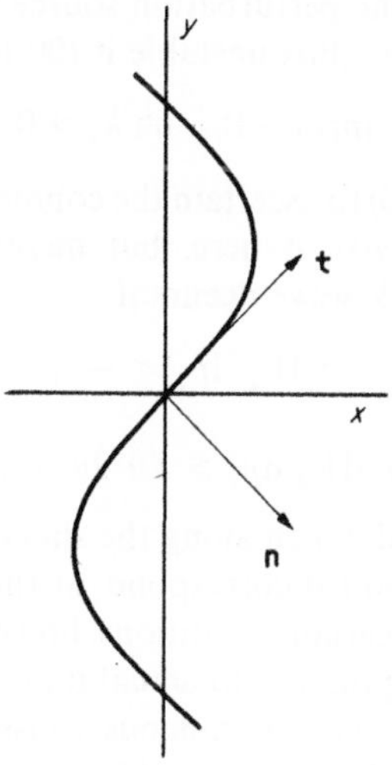

FIG. 59

Next, the perturbed values $p_2 + \delta p$ and $V_2 + \delta V$ must satisfy the same Hugoniot adiabatic equation as the unperturbed p_2 and V_2. This gives the relation between δp and δV:

$$\delta p = (\mathrm{d}p_2/\mathrm{d}V_2)\mathrm{d}V, \tag{90.8}$$

the derivative being taken along the adiabatic.

Lastly, one further relation arises from that between the mass flux through the surface of discontinuity and the pressure and density discontinuities there. For the unperturbed surface, this relation is given by (85.6); for the perturbed surface, the corresponding one is

$$\frac{1}{V_1{}^2}(\mathbf{v}_1 \cdot \mathbf{n} - \mathbf{u} \cdot \mathbf{n})^2 = \frac{p_2 - p_1 + \delta p}{V_1 - V_2 - \delta V},$$

where **u** is the velocity of points on the surface. In the first approximation with respect to small quantities we have $\mathbf{u}\cdot\mathbf{n} = -i\omega\zeta$; expanding the equation also in powers of δp and δV, we find

$$\frac{2i\omega}{v_1}\zeta = \frac{\delta p}{p_2 - p_1} + \frac{\delta V}{V_1 - V_2}. \tag{90.9}$$

The equations (90.2), (90.4), (90.5), and (90.7)–(90.9) form a set of eight linear algebraic equations for the eight quantities ζ, δp, $\delta V^{(\mathrm{ent})}$, $\delta V^{(s)}$, $\delta v_{x,y}{}^{(\mathrm{ent})}$ and $\delta v_{x,y}{}^{(s)}$.† The compatibility condition of these equations (expressed by the vanishing of the determinant of their coefficients) is

$$\frac{2\omega v_2}{v_1}\left(k_y{}^2 + \frac{\omega^2}{v_2{}^2}\right) - \left(\frac{\omega^2}{v_1 v_2} + k_x{}^2\right)(\omega - v_2 k_y)(1+h) = 0, \tag{90.10}$$

where for brevity $h = j^2(\mathrm{d}V_2/\mathrm{d}p_2)$ and j as usual denotes $v_1/V_1 = v_2/V_2$. In (90.10) k_x is to be taken as the function of k_y and ω that is determined by (90.3).

The instability condition is that perturbations exist which grow exponentially with time, and they must decay exponentially away from the surface of discontinuity (i.e. as $x \to \infty$); the latter condition means that the perturbation source is the shock wave itself, and not something outside it. The wave is thus unstable if (90.10) has solutions for which

$$\operatorname{im}\omega > 0, \quad \operatorname{im} k_x > 0. \tag{90.11}$$

The analysis of equation (90.10) to ascertain the conditions for such solutions to exist is fairly laborious. We shall not give it here, but mention only the final result.‡ The corrugation instability of a shock wave occurs if

$$j^2\,\mathrm{d}V_2/\mathrm{d}p_2 < -1, \tag{90.12}$$

or

$$j^2\,\mathrm{d}V_2/\mathrm{d}p_2 > 1 + 2v_2/c_2; \tag{90.13}$$

the derivative is, as already stated, taken along the shock adiabatic for given p_1 and V_1.††

The conditions (90.12) and (90.13) correspond to the existence of complex roots of (90.10), satisfying (90.11). Under certain conditions, however, this equation may also have roots with real ω and k_x, corresponding to actual undamped sound and entropy waves leaving the discontinuity, i.e. to the spontaneous emission of sound by the surface of discontinuity. This will be referred to as a special form of shock wave instability, although there is here no instability in the literal sense: the perturbation (ripples), once created on the surface, continues indefinitely to emit waves without being either damped or amplified; the energy carried away by the emitted waves is drawn from the whole of the medium in motion.§

To determine the conditions for this phenomenon to occur, we transform (90.10) by using the angle θ between $\mathbf{k}$ and the x-axis; then

$$\left.\begin{aligned} c_2 k_x = \omega_0\cos\theta, \quad c_2 k_y = \omega_0\sin\theta, \quad \omega = \omega_0[1 + (v_2/c_2)\cos\theta], \\ \omega_0{}^2 = c_2{}^2(k_x{}^2 + k_y{}^2), \end{aligned}\right\} \tag{90.14}$$

† All these equations are taken for $x = 0$, and the enumerated quantities in them may be regarded as the constant amplitudes, without the variable exponential factors.

‡ The analysis is reported by S. P. D'yakov, *Zhurnal éksperimental'noĭ i teoreticheskoĭ fiziki* **27**, 288, 1954. In §91, a less rigorous but more easily understandable derivation of the conditions (90.12) and (90.13) is given.

†† The derivation of (90.12) and (90.13) uses only the obligatory condition (88.1), not the inequality $p_2 > p_1$. These instability conditions therefore apply also to rarefaction shocks, which can exist if $(\partial^2 V/\partial p^2)_s < 0$.

§ Compare the analogous situation for tangential discontinuities (§84, Problem 2).

where ω_0 is the sound frequency in coordinates moving with the gas behind the shock wave, and we obtain an equation quadratic in $\cos\theta$:

$$\frac{v_2{}^2}{c_2{}^2}\left[\frac{4}{1+h}+\frac{v_1}{v_2}-1\right]\cos^2\theta+\frac{2v_2}{c_2}\left[\frac{3+(v_2/c_2)^2}{1+h}-1\right]\cos\theta+$$
$$+\frac{2[1+(v_2/c_2)^2]}{1+h}-\left(1+\frac{v_1 v_2}{c^2}\right)=0. \tag{90.15}$$

The velocity of the sound wave in the gas moving with velocity v_2, relative to the discontinuity surface at rest, is $v_2+c_2\cos\theta$. The sound wave leaves the surface if this sum is positive, i.e. if

$$-v_2/c_2<\cos\theta<1 \tag{90.16}$$

(values of $\cos\theta<0$ correspond to cases where $\mathbf{k}$ is towards the discontinuity but the transport of the sound wave by the moving gas makes it still a wave that leaves the discontinuity). The spontaneous emission of sound by the shock wave occurs if (90.15) has a root in this range. A simple analysis yields the following inequalities determining the range of the instability:†

$$\frac{1-v_2{}^2/c_2{}^2-v_1v_2/c_2{}^2}{1-v_2{}^2/c_2{}^2+v_1v_2/c_2{}^2}<j^2\frac{dV_2}{dp_2}<1+2v_2/c_2; \tag{90.17}$$

the lower and upper limits here correspond almost exactly to those in (90.16). The range (90.17) is adjacent to and extends the instability range (90.13).

The origin of shock wave instability in the range (90.17) can also be approached from a somewhat different standpoint by considering the reflection from the discontinuity surface of sound incident on it from the compressed-gas side. Since the shock is moving with supersonic velocity relative to the gas in front of it, the sound does not penetrate into that gas. In the gas behind the shock, we have not only the incident sound wave but also a reflected sound wave and an entropy-vortex wave (and ripples are formed on the discontinuity surface itself). The problem of determining the reflection coefficient is formulated similarly to the instability problem. The only difference is that the boundary conditions contain, as well as the amplitudes to be determined for the (reflected) waves leaving the discontinuity, the specified amplitude of the incident sound wave. Instead of a set of homogeneous algebraic equations, we now have a set of inhomogeneous equations in which the inhomogeneous terms involve the incident wave amplitude. The solution is given by expressions whose denominators contain the determinant of the homogeneous equations, whose vanishing gives the dispersion relation (90.10) for spontaneous perturbations. The fact that in the range (90.17) this equation has real roots for $\cos\theta$ means that there are certain values of the reflection angle (and therefore of the incidence angle) at which the reflection coefficient becomes infinite. This is another way of stating the possibility of spontaneous emission of sound, i.e. emission with no externally incident sound wave.

The same applies to the transmission coefficient for sound incident from the front on the surface of discontinuity. Here, there is no reflected wave; behind the surface, there are

† This instability also was noted by S. P. D'yakov (1954); the correct value of the lower limit in (90.17) was given by V. M. Kontorovich (1957).

transmitted sound and entropy-vortex waves. In the range (90.17), the transmission coefficient can become infinite.†

A few words may be said about some types of shock adiabatic, possible in principle, which have instability regions as discussed above.‡

The condition (90.12) requires the derivative dp_2/dV_2 to be negative, and the shock adiabatic at point 2 must be inclined to the abscissa axis less steeply than the chord 12 through that point; this is the opposite of the usual case (Fig. 53, §85). For this to be possible, the adiabatic must have the form shown in Fig. 60. The instability condition (90.12) is satisfied on the section *ab*.

The condition (90.13) requires dp_2/dV_2 to be positive, and the slope of the adiabatic must be sufficiently small. In Fig. 60, this is true on certain sections of the adiabatic immediately adjacent to the points *a* and *b*, thus extending the instability range. The condition (90.13) may also be satisfied on a section (*cd* in Fig. 61) of an adiabatic that does not have a section of the type *ab*.

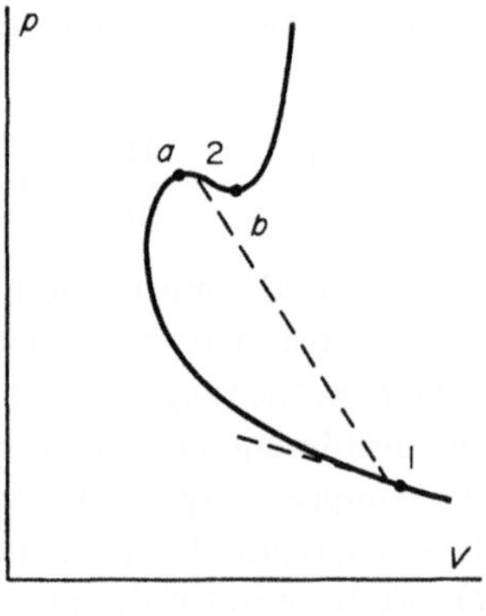

FIG. 60

The condition (90.17) is even less stringent than (90.13), and further extends the instability range on Hugoniot adiabatics having $dp_2/dV_2 > 0$. Moreover, the lower limit in (90.17) may be negative, so that instability of this type can in principle occur also on some sections of ordinary-type adiabatics having $dp_2/dV_2 < 0$ everywhere.

The ultimate behaviour of shock waves having corrugation instability is closely related to the following noteworthy fact. When the condition (90.12) or (90.13) is satisfied, the equations of fluid dynamics have more than one solution (C. S. Gardner 1963). For two states 1 and 2 of the medium, related by (85.1)–(85.3), the shock wave is usually the only solution of the (one-dimensional) problem of a flow which takes the medium from state 1 to state 2. It is found that, if one of the conditions (90.12) and (90.13) is satisfied, the solution of the problem is not unique: the transition from state 1 to state 2 can be brought

† The calculation of the sound reflection and transmission coefficients at a shock wave for any direction of incidence and in any media is given by S. P. D'yakov, *Soviet Physics JETP* **6**, 729, 739, 1958; V. M. Kontorovich, *ibid.* 1180; *Soviet Physics Acoustics* **5**, 320, 1959.

‡ In a polytropic gas $h = -(c_1/v_1)^2$, as is easily seen from the expressions derived in §89. None of the conditions (90.12), (90.13), and (90.17) is then satisfied, and so the shock wave is stable. Of course, weak shock waves in any medium are also stable.

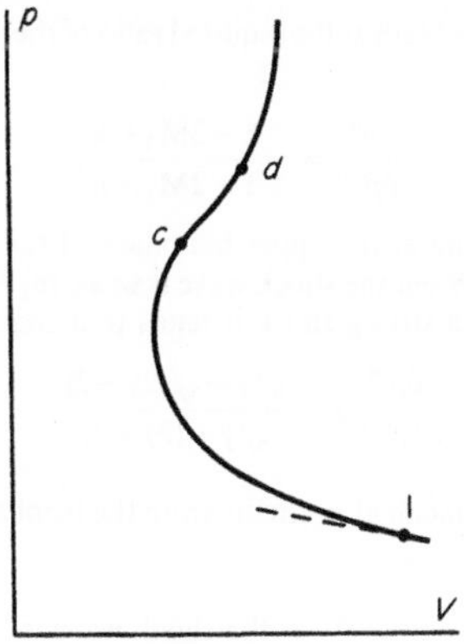

FIG. 61

about not only in a shock wave but also through a more complex system of waves. This second or decay solution consists of a weaker shock wave, a following contact discontinuity, and an isentropic non-steady rarefaction wave (see §99 below) propagated in the opposite direction relative to the gas behind the shock; in the shock wave, the entropy increases from s_1 to a value $s_3 < s_2$, and the further increase to s_2 takes place abruptly in the contact discontinuity. This is the picture for the type shown in Fig. 78b (§100); the inequality (86.2) is assumed satisfied.†

The question of what determines the choice of one of the two solutions in specific problems of fluid dynamics is not yet resolved. If the decay solution is chosen, this would mean that the instability of the shock wave with spontaneous amplification of surface ripples does not occur at all. It appears, however, that the choice may not be related to this instability, since the non-uniqueness of the solution is not limited by the conditions (90.12) and (90.13).‡

PROBLEMS

PROBLEM 1. A plane sound wave is incident normally from the rear (the compressed-gas side) on a shock wave. Determine the sound reflection coefficient.

SOLUTION. We consider the process in coordinates for which the shock wave is at rest and the gas moves through it in the positive x-direction; the incident sound wave is propagated in the negative x-direction. For normal incidence (and therefore normal reflection) the velocity in the reflected entropy wave is $\delta\mathbf{v}^{(\mathrm{ent})} = 0$. The pressure perturbation is $\delta p = \delta p^{(s)} + \delta p^{(0)}$, where the superscripts 0 and s refer to the incident and reflected sound waves respectively. The velocity $\delta v_x \equiv \delta v$ is

$$\delta v = (V_2/c_2)(\delta p^{(s)} - \delta p^{(0)});$$

the difference appears instead of the sum, because of the opposite directions of propagation of the two waves. The second boundary condition (90.7) has its previous form (but now with $\delta V = \delta V^{(0)} + \delta V^{(s)} + \delta V^{(\mathrm{ent})}$; with (90.8) and (85.6), it can be written as

$$\delta v = -\frac{1-h}{2j}(\delta p^{(s)} + \delta p^{(0)}).$$

† This is shown for the range (90.13) in Gardner's paper (*Physics of Fluids* **6**, 1366, 1963). A more general treatment including the range (90.12) has been given by N. M. Kuznetsov (*Soviet Physics JETP* **61**, 275, 1985), who also discusses shock adiabatics for which $(\partial^2 V/\partial p^2)_s$ is not positive and the decay solutions are made up of other sets of waves.

‡ It seems that the non-uniqueness range extends on the shock adiabatic somewhat beyond the limits of the instability range given by these conditions; see Kuznetsov's paper cited in the last footnote.

Equating the two expressions for δv, we obtain as the required ratio of the pressure amplitudes in the reflected and incident sound waves

$$\frac{\delta p^{(s)}}{\delta p^{(0)}} = -\frac{1-2\mathrm{M}_2-h}{1+2\mathrm{M}_2-h}, \tag{1}$$

where $\mathrm{M}_2 = v_2/c_2$. This becomes infinite at the upper boundary of the range (90.17).

For a polytropic gas, $h = -1/\mathrm{M}_1{}^2$. When the shock wave is weak ($p_2 - p_1 \ll p_1$), the ratio (1) tends to zero as $(p_2-p_1)^2$, and in the opposite case of a strong shock it tends to a constant,

$$\frac{\delta p^{(s)}}{\delta p^{(0)}} \cong -\frac{\sqrt{\gamma}-\sqrt{(2\gamma-2)}}{\sqrt{\gamma}+(2\gamma-2)}.$$

PROBLEM 2. A plane sound wave is incident normally from the front on a shock wave. Determine the sound transmission coefficient.†

SOLUTION. The perturbation in gas 1 in front of the shock wave is

$$\delta p_1 = \delta p^{(0)}, \quad \delta V_1 = \delta V^{(0)} = -(V_1{}^2/c_1{}^2)\delta p_1, \quad \delta v_1 = (V_1/c_1)\delta p_1,$$

and in gas 2 behind it

$$\delta p_2 = \delta p^{(s)}, \quad \delta V_2 = \delta V^{(s)} + \delta V^{(\mathrm{ent})}, \quad \delta v_2 = (V_2/c_2)\delta p_2;$$

the superscripts 0, s and ent refer to the incident sound wave, the transmitted sound wave, and the transmitted entropy wave. The perturbations δp_2 and δV_2 are related in a way which follows from the equation of the shock adiabatic: if this equation is written as $V_2 = V_2(p_2; p_1, V_1)$, then

$$\delta V_1 = \left(\frac{\partial V_2}{\partial p_2}\right)_H \delta p_2 + \left(\frac{\partial V_2}{\partial V_1}\right)_H \delta V_1 + \left(\frac{\partial V_2}{\partial p_1}\right)_H \delta p_1$$

$$= \left(\frac{\partial V_2}{\partial p_2}\right)_H \delta p_2 + \left[-\frac{V_1{}^2}{c_1{}^2}\left(\frac{\partial V_2}{\partial V_1}\right)_H + \left(\frac{\partial V_2}{\partial p_1}\right)_H\right]\delta p_1,$$

where the suffix H to the derivatives means that they are taken along the Hugoniot adiabatic.‡ The boundary condition (90.7) is now replaced by

$$\delta v_2 - \delta v_1 = -\tfrac{1}{2}(v_1 - v_2)\left[\frac{\delta p_2 - \delta p_1}{p_2 - p_1} - \frac{\delta V_2 - \delta V_1}{V_1 - V_2}\right]$$

$$= -\frac{1}{2j}[\delta p_2 - \delta p_1 - j^2(\delta V_2 - \delta V_1)].$$

Equating the two expressions for $\delta v_2 - \delta v_1$, we obtain as the required ratio of amplitudes in the transmitted and incident sound waves

$$\frac{\delta p^{(s)}}{\delta p^{(0)}} = \frac{(1+\mathrm{M}_1)^2 + q}{1+2\mathrm{M}_2-h}, \tag{2}$$

where h is as before and

$$q = j^2[-(V_1/c_1)^2(\partial V_2/\partial V_1)_H + (\partial V_2/\partial p_1)_H].$$

For a polytropic gas,

$$q = -\frac{\gamma-1}{\gamma+1}\frac{(\mathrm{M}_1{}^2-1)^2}{\mathrm{M}_1{}^2}$$

and the transmission coefficient is

$$\frac{\delta p^{(s)}}{\delta p^{(0)}} = \frac{(1+\mathrm{M}_1)^2}{1+2\mathrm{M}_2+1/\mathrm{M}_1{}^2}\left[1-\frac{\gamma-1}{\gamma+1}\left(1-\frac{1}{\mathrm{M}_1}\right)^2\right].$$

---

† This problem was discussed for a polytropic gas by D. I. Blokhintsev (1945) and J. M. Burgers (1946).

‡ The derivative $(\partial V_2/\partial p_2)_H$ is what was previously denoted by $\mathrm{d}V_2/\mathrm{d}p_2$ simply, it being implied that the derivative is taken at constant p_1 and V_1.

For a weak shock, this gives

$$\frac{\delta p^{(s)}}{\delta p^{(0)}} \cong 1 + \frac{\gamma+1}{2\gamma}\frac{p_2 - p_1}{p_1},$$

and in the opposite case of a strong shock

$$\frac{\delta p^{(s)}}{\delta p^{(0)}} \cong \frac{1}{\gamma + \sqrt{[2\gamma(\gamma-1)]}}\frac{p_2}{p_1}.$$

In either case, the pressure amplitude in the transmitted sound wave is greater than in the incident wave.

§91. Shock wave propagation in a pipe

Let us consider the propagation of a shock wave in a medium occupying a long pipe with variable cross-section. The aim is to ascertain the influence of the changing area of the shock wave on its velocity (G. B. Whitham 1958).

We shall assume that the area $S(x)$ of the pipe cross-section varies only slowly along its length (in the x-direction), i.e. only slightly over distances of the order of the pipe width. This enables us to apply the *hydraulic approximation* already used in §77: all quantities in the flow may be assumed constant over any cross-section of the tube, and the velocity axial; the flow is thus regarded as quasi-one-dimensional. Such a flow is described by the equations

$$\frac{\partial v}{\partial t} + v\frac{\partial v}{\partial x} + \frac{1}{\rho}\frac{\partial p}{\partial x} = 0, \tag{91.1}$$

$$\frac{\partial p}{\partial t} + v\frac{\partial p}{\partial x} - c^2\left(\frac{\partial \rho}{\partial t} + v\frac{\partial \rho}{\partial x}\right) = 0, \tag{91.2}$$

$$S\frac{\partial \rho}{\partial t} + \frac{\partial}{\partial x}(\rho v S) = 0. \tag{91.3}$$

The first is Euler's equation, the second is the adiabatic equation, and the third is the continuity equation in the form (77.1).

To elucidate the question raised, it is sufficient to consider a pipe in which the variation of $S(x)$ not only is slow but also remains relatively small in magnitude over the whole length. Then the flow perturbations due to the non-constancy of the cross-section will also be small, and equations (91.1)–(91.3) can be linearized. Lastly, initial conditions are to be imposed, which exclude the occurrence of any extraneous perturbations that might influence the motion of the shock wave, since we are interested only in the perturbations due to the change in $S(x)$. This can be achieved by assuming that the shock wave moves originally with constant velocity along a pipe with constant cross-section, which begins to vary only to the right of a certain point taken as $x = 0$.

The linearized equations (91.1)–(91.3) are

$$\frac{\partial \delta v}{\partial t} + v\frac{\partial \delta v}{\partial x} + \frac{1}{\rho}\frac{\partial \delta p}{\partial x} = 0,$$

$$\frac{\partial \delta p}{\partial t} + v\frac{\partial \delta p}{\partial x} - c^2\left(\frac{\partial \delta\rho}{\partial t} + v\frac{\partial \delta\rho}{\partial x}\right) = 0,$$

$$\frac{\partial \delta\rho}{\partial t} + v\frac{\partial \delta\rho}{\partial x} + \rho\frac{\partial \delta v}{\partial x} + \frac{\rho v}{S}\frac{\partial \delta S}{\partial x} = 0,$$

where the letters without prefix denote the constant values in the homogeneous flow in the uniform part of the pipe, and δ denotes the changes in these quantities in the pipe with variable cross-section. Multiplying the first and third equations by ρc and c^2 respectively, and then adding all three, we can write

$$\left[\frac{\partial}{\partial t}+(v+c)\frac{\partial}{\partial x}\right](\delta p+\rho c\delta v) = -\frac{\rho v c^2}{S}\frac{\partial \delta S}{\partial x}. \tag{91.4}$$

The general solution of this equation is given by the sum of the general solution of the homogeneous equation and a particular solution of the equation as it stands. The former is $F(x-vt-ct)$, where F is any function; this describes sound disturbances coming from the left. In the uniform part $x<0$, however, there are none, and we must therefore put $F\equiv 0$. The solution is then the integral of the inhomogeneous equation

$$\delta p+\rho c\,\delta v = -\frac{\rho v c^2}{v+c}\frac{\delta S}{S}. \tag{91.5}$$

The shock wave moves from left to right with velocity $v_1 > c_1$ in the medium at rest with specified values of p_1 and ρ_1. The motion behind the shock in medium 2 is given by the solution (91.5) throughout the part of the pipe to the left of the point reached by the discontinuity at a given instant. After the shock has passed, all quantities in each cross-section of the pipe remain constant in time and equal to the values they had at the moment of passage: the pressure p_2, the density ρ_2, and the velocity v_1-v_2 (in accordance with the notation used in this chapter, v_2 denotes the gas relative to the moving shock wave; its velocity relative to the pipe walls is then v_1-v_2). With this notation, and again separating the variable parts of these quantities, the equation (91.5) can be written

$$\frac{\delta S}{S} = -\frac{v_1-v_2+c_2}{\rho_2(v_1-v_2)c_2{}^2}[\delta p_2+\rho_2 c_2(\delta v_1-\delta v_2)]. \tag{91.6}$$

All the quantities $\delta v_1, \delta v_2, \delta p_2$ can be expressed in terms of one of them, say δv_1. For this purpose, we write the varied relations (85.1), (85.2) at the discontinuity (for given p_1 and ρ_1):

$$\rho_1\delta v_1 = v_2\delta\rho_2+\rho_2\delta v_2, \qquad 2j(\delta v_1-\delta v_2) = \delta p_2+v_2{}^2\delta\rho_2,$$

where $j=\rho_1 v_1=\rho_2 v_2$ is the unperturbed mass flux; and we use also the relation

$$\delta p_2 = (\mathrm{d}p_2/\mathrm{d}\rho_2)\delta\rho_2,$$

where the derivative is taken along the Hugoniot adiabatic. The calculation gives the following final relation between the change δv_1 in the shock wave velocity relative to the gas at rest in front of it and the change δS in the cross-sectional area of the pipe:

$$-\frac{1}{S}\frac{\delta S}{\delta v_1} = \frac{v_1-v_2+c_2}{v_1 c_2}\left[\frac{1+2v_2/c_2-h}{1+h}\right], \tag{91.7}$$

again with the notation

$$h = -(j/\rho_2)^2\mathrm{d}\rho_2/\mathrm{d}p_2 = j^2\mathrm{d}V_2/\mathrm{d}p_2. \tag{91.8}$$

The coefficient of the square brackets in (91.7) is positive. The sign of $\delta v_1/\delta S$ is therefore determined by that of the expression in the brackets. For all stable shocks it is positive, and so $\delta v_1/\delta S<0$. When either of the conditions (90.12), (90.13) for corrugation instability is satisfied, however, the expression in the brackets becomes negative, and so $\delta v_1/\delta S>0$.

This leads to an intuitive interpretation of the origin of the instability. Figure 62 shows the corrugated surface of the shock wave, moving to the right; the arrows show schematically the direction of the streamlines. As the shock moves, the area δS increases on the parts of the surface that project forwards, and decreases on those that lag behind. When $\delta v_1/\delta S < 0$, this retards the forward projections and accelerates the lagging parts, so that the surface tends to smooth out. When $\delta v_1/\delta S > 0$, on the other hand, the perturbation of the surface is intensified, the forward projections moving further forward, and the lagging parts being left further behind.†

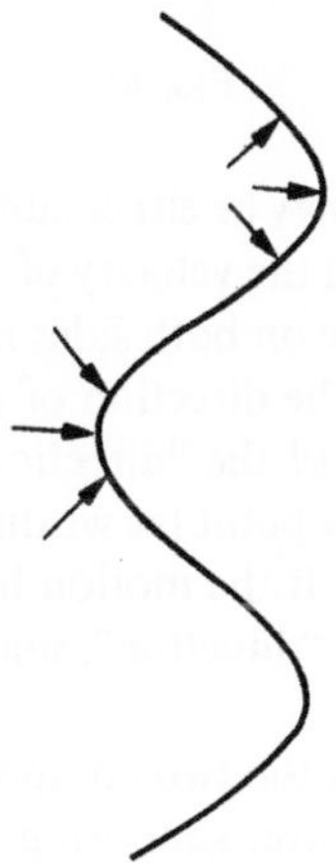

FIG. 62

§92. Oblique shock waves

Let us consider a steady shock wave, and abandon the system of coordinates used hitherto, in which the gas velocity is perpendicular to the shock surface element considered. The streamlines can intersect the surface of such a shock wave at any angle, and in so doing are refracted: the tangential component of the gas velocity is unchanged, while the normal component is, according to (87.4), diminished: $v_{1t} = v_{2t}$, $v_{1n} > v_{2n}$. It is therefore clear that the streamlines approach the shock wave as they pass through it (cf. Fig. 63). Thus the streamlines are always refracted in a definite direction in passing through the shock wave.

We take as the x-axis the direction of the gas velocity $\mathbf{v}_1$ in front of the shock wave; let ϕ be the angle between the surface of discontinuity and the x-axis (Fig. 63). The possible values of ϕ are restricted only by the condition that the normal component of $\mathbf{v}_1$ be greater than the velocity of sound c_1. Since $v_{1n} = v \sin\phi$, it follows that ϕ can have any value in the range between $\frac{1}{2}\pi$ and the Mach angle α_1:

$$\alpha_1 < \phi < \tfrac{1}{2}\pi, \quad \sin\alpha_1 = c_1/v_1 \equiv 1/\mathrm{M}_1.$$

† The expression (91.7) for any (not polytropic) medium and its relationship to the conditions for corrugation instability of shock waves were noted by S. G. Sugak, V. E. Fortov, and A. L. Ni (1981).

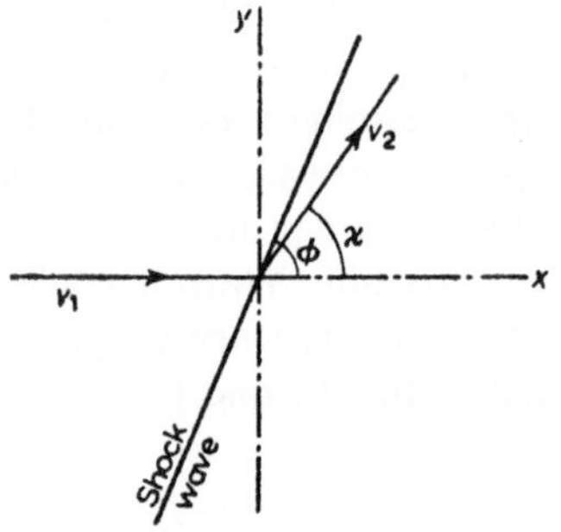

FIG. 63

The motion behind a shock wave may be either subsonic or supersonic (only the normal velocity component need be less than the velocity of sound c_2); the motion in front of it is necessarily supersonic. If the gas flow on both sides is supersonic, every disturbance must be propagated along the surface in the direction of the tangential component of the gas velocity. In this sense we can speak of the "direction" of a shock wave, and distinguish shock waves leaving and reaching any point (as we did for characteristics, the motion near which is always supersonic; see §82). If the motion behind the shock is subsonic, there is strictly no meaning in speaking of its "direction", since disturbances can be propagated in all directions on its surface.

We shall derive a relation between the two components of the gas velocity after it has passed through an oblique shock wave, supposing that we have a polytropic gas. The continuity of the velocity component tangential to the shock means that $v_1 \cos\phi = v_{2x}\cos\phi + v_{2y}\sin\phi$, or

$$\tan\phi = (v_1 - v_{2x})/v_{2y}. \tag{92.1}$$

Next we use formula (89.6), in which v_1 and v_2 denote the velocity components normal to the plane of the shock wave and must be replaced by $v_1 \sin\phi$ and $v_{2x}\sin\phi - v_{2y}\cos\phi$, so that

$$\frac{v_{2x}\sin\phi - v_{2y}\cos\phi}{v_1 \sin\phi} = \frac{\gamma-1}{\gamma+1} + \frac{2c_1^2}{(\gamma+1)v_1^2 \sin^2\phi}. \tag{92.2}$$

We can eliminate the angle ϕ from these two relations. After some simple transformations, we obtain the following formula which determines the relation between v_{2x} and v_{2y} (for given v_1 and c_1):

$$v_{2y}^2 = (v_1 - v_{2x})^2 \frac{2(v_1 - c_1^2/v_1)/(\gamma+1) - (v_1 - v_{2x})}{v_1 - v_{2x} + 2c_1^2/(\gamma+1)v_1}. \tag{92.3}$$

This formula can be more intelligibly written by introducing the critical velocity. According to Bernoulli's equation and the definition of the critical velocity, we have $w_1 + \frac{1}{2}v_1^2 = \frac{1}{2}v_1^2 + c_1^2/(\gamma-1) = (\gamma+1)c_*^2/2(\gamma-1)$ (see §89, Problem 1), whence

$$c_*^2 = [(\gamma-1)v_1^2 + 2c_1^2]/(\gamma+1). \tag{92.4}$$

Using this in (92.3), we obtain

$$v_{2y}^2 = (v_1 - v_{2x})^2 \frac{v_1 v_{2x} - c_*^2}{2v_1^2/(\gamma+1) - v_1 v_{2x} + c_*^2}. \tag{92.5}$$

Equation (92.5) is called the equation of the *shock polar* (A. Busemann 1931). Figure 64 shows a graph of the function $v_{2y}(v_{2x})$; it is a cubic curve, called a strophoid. It crosses the axis of abscissae at the points P and Q, corresponding to $v_{2x} = c_*^2/v_1$ and $v_{2x} = v_1$.† A line (OB in Fig. 64) drawn from the origin at an angle χ to the axis of abscissae gives, by the length of the segment between O and the point where it intersects the shock polar, the gas velocity behind a discontinuity which turns the stream through an angle χ. There are two such intersections (A, B), i.e. two different shock waves correspond to a given value of χ. The direction of the shock wave also can be immediately determined from the shock polar: it is given by the direction of the perpendicular from the origin to the line QB or QA (Fig. 64 shows the angle ϕ for a shock corresponding to the point B). As χ decreases, the point A approaches P, corresponding to a normal shock ($\phi = \frac{1}{2}\pi$) with $v_2 = c_*^2/v_1$. The point B approaches Q; the intensity of the shock (velocity discontinuity) tends to zero, and the angle ϕ tends, as it should, to the Mach angle α_1; the tangent to the shock polar at Q makes an angle $\frac{1}{2}\pi + \alpha_1$ with the axis of abscissae.

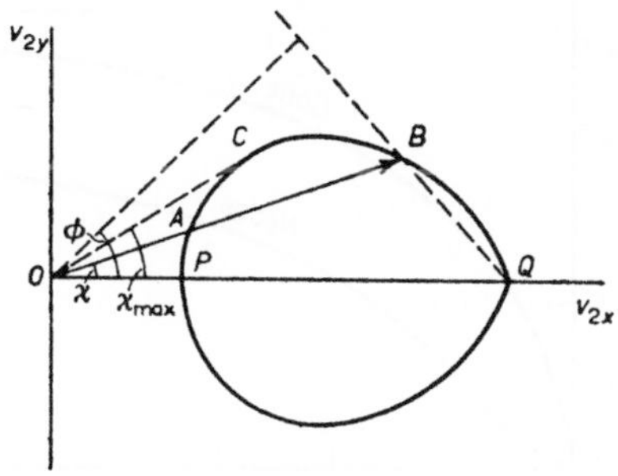

FIG. 64

From the shock polar we can immediately derive the important result that the angle of deviation χ of the stream at the shock wave cannot exceed a certain maximum value χ_{max}, corresponding to the tangent from O to the curve. This quantity is, of course, a function of the Mach number $M_1 = v_1/c_1$, but we shall not give the expression for it, which is very cumbersome. For $M_1 = 1$, $\chi_{max} = 0$; as M_1 increases, χ_{max} increases monotonically, and tends to a finite limit as $M_1 \to \infty$. It is easy to discuss the two limiting cases. If the velocity v_1 is near to c_*, then v_2 is so also, and the angle χ is small; the equation (92.5) of the shock polar can then be written in the approximate form‡

$$\chi^2 = (\gamma+1)(v_1 - v_2)^2(v_1 + v_2 - 2c_*)/2c_*^3, \tag{92.6}$$

† The strophoid actually continues in two branches from the point $v_{2x} = v_1$ (which is a double point) to infinite $|v_{2y}|$; these are not shown in Fig. 64. They have a common vertical asymptote

$$v_{2x} = c_*^2/v_1 + 2v_1/(\gamma+1).$$

The points on these branches have no physical significance; they would give values for v_{2x} and v_{2y} such that $v_{2n}/v_{1n} > 1$, which is impossible.

‡ It is easily seen that equation (92.6) holds also for any (non-polytropic) gas, provided that $(\gamma+1)$ is replaced by $2\alpha_*$ from (102.2).

where we have put $v_{2x} \cong v_2$, $v_{2y} \cong c_*\chi$ in view of the smallness of χ. Hence we easily find†

$$\chi_{max} = \frac{4\sqrt{(\gamma+1)}}{3\sqrt{3}}\left(\frac{v_1}{c_*}-1\right)^{\frac{3}{2}} = \frac{8\sqrt{2}}{3\sqrt{3(\gamma+1)}}(M_1-1)^{\frac{3}{2}} \tag{92.7}$$

In the opposite limiting case $M_1 \to \infty$, the shock polar degenerates to a circle

$$v_{2y}{}^2 = (v_1 - v_{2x})\left(v_{2x} - \frac{\gamma-1}{\gamma+1}v_1\right).$$

It is easy to see that we then have

$$\chi_{max} = \sin^{-1}(1/\gamma). \tag{92.8}$$

Figure 65 shows a graph of χ_{max} as a function of M_1 for air ($\gamma = 1{\cdot}4$); the broken horizontal line gives the limiting value $\chi_{max}(\infty) = 45{\cdot}6°$. The upper curve is a similar graph for flow past a cone (see §113).

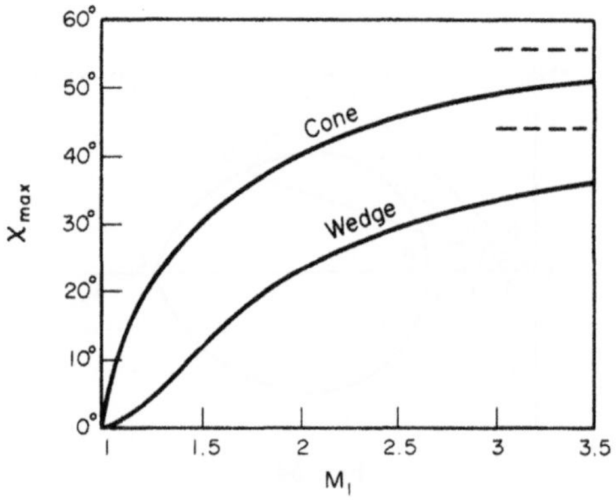

FIG. 65

The circle $v_2 = c_*$ cuts the axis of abscissae between the points P and Q (Fig. 64), and therefore divides the shock polar into two parts corresponding to subsonic and supersonic gas velocities behind the discontinuity. The point where this circle crosses the polar lies to the right of, but very close to, the point C; the whole segment PC therefore corresponds to transitions to subsonic velocities, while CQ (except for a very small segment near C) corresponds to transitions to supersonic velocities.

The pressure and density changes in an oblique shock wave depend only on the velocity components normal to it. The ratios p_2/p_1 and ρ_2/ρ_1 for given M_1 and ϕ are therefore obtained from (89.6) and (89.7) on simply replacing M_1 by $M_1 \sin\phi$:

$$\frac{p_2-p_1}{p_1} = \frac{2\gamma}{\gamma+1}(M_1{}^2\sin^2\phi - 1), \tag{92.9}$$

$$\frac{\rho_2-\rho_1}{\rho_1} = \frac{2(M_1{}^2\sin^2\phi-1)}{(\gamma-1)M_1{}^2\sin^2\phi+2}. \tag{92.10}$$

† It may be noted that this dependence of χ_{max} on $M_1 - 1$ is in agreement with the general similarity law (126.7) for transonic flow.

These increase monotonically when ϕ increases from α_1 (for which $p_2/p_1 = \rho_2/\rho_1 = 1$) to $\frac{1}{2}\pi$, i.e. as we move along the shock polar from Q to P.

We may state for reference the formula giving the angle of deviation χ of the velocity in terms of M_1 and ϕ:

$$\cot\chi = \tan\phi\left[\frac{(\gamma+1)M_1^2}{2(M_1^2\sin^2\phi - 1)} - 1\right], \tag{92.11}$$

and the formula giving the number $M_2 = v_2/c_2$ in terms of M_1 and ϕ:

$$M_2^2 = \frac{2+(\gamma-1)M_1^2}{2\gamma M_1^2\sin^2\phi - (\gamma-1)} + \frac{2M_1^2\cos^2\phi}{2+(\gamma-1)M_1^2\sin^2\phi}; \tag{92.12}$$

when $\phi = \frac{1}{2}\pi$, the latter becomes (89.9).

The two shock waves determined by the shock polar for a given deviation angle χ are said to belong to the *weak* and *strong* families. A shock wave of the strong family (the segment PC of the polar) is strong (the ratio p_2/p_1 is large), makes a large angle ϕ with the direction of the velocity $\mathbf{v}_1$, and converts the flow from supersonic to subsonic. A shock wave of the weak family (the segment QC) is weak, is inclined at a smaller angle to the stream, and almost always leaves the flow supersonic.

As an illustration, Fig. 66 shows the deviation angle χ as a function of the angle ϕ of the discontinuity surface for air ($\gamma = 1{\cdot}4$), for several values of M_1, including the limit $M_1 \to \infty$. The branches shown as continuous curves correspond to shock waves of the weak family, and the broken curves to those of the strong family. The broken line $\chi = \chi_{max}$ is the locus of the points with the maximum deviation angle for each M_1; the continuous line $M_2 = 1$ divides the regions of supersonic and subsonic flow behind the discontinuity. The narrow region between these two lines corresponds to shock waves which belong to

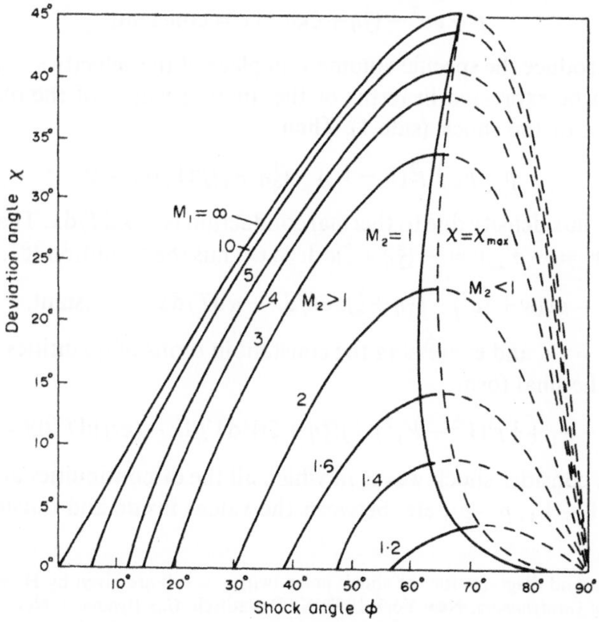

FIG. 66

the weak family but nevertheless convert the flow from supersonic to subsonic. The difference between the values of ϕ on $\chi = \chi_{max}$ and $M_2 = 1$ (for a given M_1) nowhere exceeds 4·5°; that between χ_{max} and the value χ_s on the line $M_2 = 1$ (again for a given M_1) does not exceed 0·5°.†

§93. The thickness of shock waves

Hitherto we have regarded shock waves as geometrical surfaces with zero thickness. We shall now consider the structure of actual surfaces of discontinuity, and we shall see that shock waves in which the discontinuities are small are in reality transition layers with finite thickness, the thickness diminishing as the magnitude of the discontinuities increases. If the discontinuities are not small, the change occurs so sharply that the concept of thickness is meaningless in the macroscopic theory.

To determine the structure and thickness of the transition layer we must take account of the viscosity and thermal conductivity of the gas, which we have hitherto neglected.

The relations (85.1)–(85.3) for a shock wave were obtained from the constancy of the fluxes of mass, momentum and energy. If we consider a surface of discontinuity as a layer with finite thickness, these conditions must be written, not as the equality of the quantities concerned on the two sides of the discontinuity, but as their constancy throughout the thickness of the layer. The first condition, (85.1), is unchanged:

$$\rho v \equiv j = \text{constant}. \tag{93.1}$$

In the other two conditions additional fluxes of momentum and energy, due to internal friction and thermal conduction, must be taken into account.

The momentum flux density (in the x-direction) due to internal friction is given by the component $-\sigma'_{xx}$ of the viscous stress tensor; according to the general expression (15.3) for this tensor, we have $\sigma'_{xx} = (\frac{4}{3}\eta + \zeta)\mathrm{d}v/\mathrm{d}x$. The condition (85.2) then becomes‡

$$p + \rho v^2 - (\tfrac{4}{3}\eta + \zeta)\mathrm{d}v/\mathrm{d}x = \text{constant}.$$

As in §85, we introduce the specific volume V in place of the velocity $v = jV$. The constant on the right can be expressed in terms of the limiting values of the quantities at large distances in front of the shock (side 1). Then

$$p - p_1 + j^2(V - V_1) - (\tfrac{4}{3}\eta + \zeta)j\mathrm{d}V/\mathrm{d}x = 0. \tag{93.2}$$

Next, the energy flux density due to thermal conduction is $-\kappa \mathrm{d}T/\mathrm{d}x$. That due to internal friction is $-\sigma'_{xi}v_i = -\sigma'_{xx}v = -(\frac{4}{3}\eta + \zeta)v\,\mathrm{d}v/\mathrm{d}x$. Thus the condition (85.3) can be written

$$\rho v(w + \tfrac{1}{2}v^2) - (\tfrac{4}{3}\eta + \zeta)v\,\mathrm{d}v/\mathrm{d}x - \kappa\mathrm{d}T/\mathrm{d}x = \text{constant}.$$

Again putting $v = jV$, and expressing the constant in terms of quantities with the suffix 1, we can obtain the final form

$$w - w_1 + \tfrac{1}{2}j^2(V^2 - V_1^{\,2}) - j(\tfrac{4}{3}\eta + \zeta)V\mathrm{d}V/\mathrm{d}x - (\kappa/j)\mathrm{d}T/\mathrm{d}x = 0. \tag{93.3}$$

We shall here consider shock waves in which all the discontinuities are small. Then all the differences $V - V_1$, $p - p_1$, etc. between the values inside and outside the transition

† Detailed graphs and diagrams for the shock polar (with $\gamma = 1{\cdot}4$) are given by H. W. Liepmann and A. Roshko, *Elements of Gasdynamics*, New York 1957; K. Oswatitsch, *Gas Dynamics*, New York 1956.

‡ The positive x-direction is the direction of motion of the gas through the shock wave at rest. If we use a frame of reference in which the gas in front of the shock is at rest, the shock wave itself moves in the negative x-direction.

layer are also small. It is seen from the relations obtained below that $1/\delta$ (where δ is the thickness of the discontinuity) is of the first order in the small quantity $p_2 - p_1$. Thus differentiation with respect to x increases the order of smallness by one; for example, $\mathrm{d}p/\mathrm{d}x$ is a second-order small quantity.

We multiply (93.2) by $\frac{1}{2}(V + V_1)$ and subtract it from (93.3). The result is

$$(w - w_1) - \tfrac{1}{2}(p - p_1)(V + V_1) = (\kappa/j)\,\mathrm{d}T/\mathrm{d}x; \tag{93.4}$$

the third-order term in $(V - V_1)\mathrm{d}V/\mathrm{d}x$ is omitted. We expand the left-hand side of (93.4) in powers of $p - p_1$ and $s - s_1$, taking the pressure and the entropy as the basic independent variables. The terms of first and second order in $p - p_1$ are zero (cf. the derivation of (86.1)), and omitting the higher-order terms gives just $T(s - s_1)$. We write

$$\frac{\mathrm{d}T}{\mathrm{d}x} = \left(\frac{\partial T}{\partial p}\right)_s \frac{\mathrm{d}p}{\mathrm{d}x} + \left(\frac{\partial T}{\partial s}\right)_p \frac{\mathrm{d}s}{\mathrm{d}x}.$$

The $\mathrm{d}s/\mathrm{d}x$ term can be omitted, being a third-order small quantity (see below), and we thus obtain an expression for $s(x)$ in terms of $p(x)$:

$$T(s - s_1) = (\kappa/j)\,(\partial T/\partial p)_s\,\mathrm{d}p/\mathrm{d}x. \tag{93.5}$$

Note that $s - s_1$ in the transition layer is a second-order small quantity, whereas the total discontinuity $s_2 - s_1$ is (as shown in §86) third-order with respect to the pressure discontinuity $p_2 - p_1$. The reason is that, as we shall show below, the pressure in the transition layer varies monotonically from p_1 to p_2, whereas the entropy $s(x)$, which is determined by the derivative $\mathrm{d}p/\mathrm{d}x$, has a maximum within the layer.

An equation for $p(x)$ could be obtained by making a similar expansion of (93.2) and (93.3) and then combining the two. We will, however, use a different and more instructive method, which allows a clearer understanding of the origin of the various terms in the equation.

It has been shown in §79 that a monochromatic weak disturbance of the state of the gas (a sound wave) is damped in the course of propagation, at a rate proportional to the square of the frequency: $\gamma = a\omega^2$. The positive coefficient a can be expressed in terms of the viscosity and the thermal conductivity by (79.6). It was also shown that (for any plane sound wave) this damping can be described by including an extra term in the linearized equation of motion; see (79.9). In this equation, we replace the second time derivative by the second coordinate derivative and change the sign of $\partial p'/\partial x$ (corresponding to wave propagation in the negative x-direction†), and write it as

$$\frac{\partial p'}{\partial t} - c\frac{\partial p'}{\partial x} = ac^3\frac{\partial^2 p'}{\partial x^2}, \tag{93.6}$$

where p' is the variable part of the pressure.

To take account of the slight non-linearity, we have to include a term in $p'\partial p'/\partial x$:

$$\frac{\partial p'}{\partial t} - c\frac{\partial p'}{\partial x} - \alpha_p p'\frac{\partial p'}{\partial x} = ac^3\frac{\partial^2 p'}{\partial x^2}. \tag{93.7}$$

The coefficient α_p in the non-linear term is found by an appropriate expansion of the

† This direction propagation is chosen in accordance with the comment in the last footnote.

equations of fluid dynamics for an ideal (dissipationless) fluid; the result is

$$\alpha_p = \tfrac{1}{2}(c^3/V^2)(\partial^2 V/\partial p^2)_s \tag{93.8}$$

(see the Problem).†

Equation (93.7) describes the propagation of disturbances in a medium with slight dissipation and slight non-linearity. In the case of a weak shock wave, it describes the propagation in a frame of reference where the unperturbed gas (in front of the shock) is at rest. We need to find a solution with a steady (time-independent) profile in which the pressure far from the wave ($x \to \pm\infty$) has limiting values p_2 and p_1; the difference $p_2 - p_1$ is the pressure discontinuity.‡

A steady profile is described by a solution having the form

$$p'(x, t) = p'(x + v_1 t), \tag{93.9}$$

where v_1 is the velocity. Substitution in (93.7) gives

$$\frac{d}{d\xi}\left[(v_1 - c)p' - \tfrac{1}{2}\alpha_p p'^2 - ac^3\frac{dp'}{d\xi}\right] = 0, \qquad \xi = x + v_1 t,$$

whose first integral is

$$ac^3 dp'/d\xi = -\tfrac{1}{2}\alpha_p p'^2 + (v_1 - c)p' + \text{constant}. \tag{93.10}$$

The quadratic trinomial on the right must be zero for the values of p' which correspond to the limiting conditions at infinity, where $dp'/d\xi = 0$. These values are $p_2 - p_1$ and 0, if p' is measured from the unperturbed value p_1 in front of the shock. The trinomial must therefore be representable as

$$-\tfrac{1}{2}\alpha_p[p' - (p_2 - p_1)]p',$$

the constant v_1 being given in terms of p_1 and p_2 by

$$v_1 = c + \tfrac{1}{4}\alpha_p(p_2 - p_1). \tag{93.11}$$

Equation (93.10) takes the following form for the pressure p itself:

$$ac^3 dp/d\xi = -\tfrac{1}{2}\alpha_p(p - p_1)(p - p_2).$$

The solution of this equation satisfying the necessary conditions is

$$p = \tfrac{1}{2}(p_1 + p_2) + \tfrac{1}{2}(p_2 - p_1)\tanh\frac{(p_2 - p_1)(x + v_1 t)}{4ac^3/\alpha_p}.$$

This solves the problem. Returning to the frame of reference in which the shock wave is at

---

† With a new unknown function $u = -p'\alpha_p$, a new independent variable (instead of x) $\zeta = x + ct$, and the notation $\mu = ac^3$, the equation (93.7) can be brought to the form

$$\frac{\partial u}{\partial t} + u\frac{\partial u}{\partial \zeta} = \mu\frac{\partial^2 u}{\partial \zeta^2}, \tag{93.7a}$$

known as *Burgers' equation* (J. M. Burgers 1940).

‡ It will be seen later (§102) that in the absence of dissipation the non-linear effects distort the profile as propagation proceeds, the front becoming gradually steeper. This in turn enhances the dissipative effects, which tend to make the profile less steep (to reduce the gradients of the variable quantities). The mutual compensation of these opposite tendencies makes possible the propagation with a steady profile in a non-linear dissipative medium.

rest, we have as the formula for the pressure variation in it

$$p-\tfrac{1}{2}(p_2+p_1)=\tfrac{1}{2}(p_2-p_1)\tanh(x/\delta), \tag{93.12}$$

where

$$\delta = 8aV^2/(p_2-p_1)(\partial^2 V/\partial p^2)_s. \tag{93.13}$$

Almost the whole change from p_1 to p_2 occurs over a distance of the order of δ, which may be called the *thickness* of the shock wave. We see that this is less for stronger shocks, i.e. greater pressure discontinuities.†

The variation of the entropy across the discontinuity is obtained from (93.5) and (93.12):

$$s-s_1=\frac{\kappa}{16caVT}\left(\frac{\partial T}{\partial p}\right)_s\left(\frac{\partial^2 V}{\partial p^2}\right)_s(p_2-p_1)^2\frac{1}{\cosh^2(x/\delta)}. \tag{93.14}$$

From this we see that the entropy does not vary monotonically, but has a maximum inside the shock, at $x=0$. For $x=\pm\infty$ this formula gives $s=s_1$ in either case; this is because the total entropy change s_2-s_1 is of the third order in p_2-p_1 (cf. (86.1)), whereas $s-s_1$ is of the second order.

Formula (93.12) is quantitatively valid only for sufficiently small differences p_2-p_1. We can, however, use (93.13) qualitatively to determine the order of magnitude of the thickness in cases where the difference p_2-p_1 is of the same order of magnitude as p_1 and p_2 themselves. The velocity of sound in the gas is of the same order as the thermal velocity v of the molecules. The kinematic viscosity is, as we know from the kinetic theory of gases, $\nu \sim lv \sim lc$, where l is the mean free path of the molecules. Hence $a \sim l/c^2$; an estimate of the thermal-conduction term gives the same result. Finally, $(\partial^2 V/\partial p^2)_s \sim V/p^2$, and $pV \sim c^2$. Using these relations in (93.13), we obtain

$$\delta \sim l. \tag{93.15}$$

Thus the thickness of a strong shock is of the same order of magnitude as the mean free path of the gas molecules.‡ In macroscopic gas dynamics, however, where the gas is treated as a continuous medium, the mean free path must be taken as zero. It follows that the methods of gas dynamics cannot strictly be used alone to investigate the internal structure of strong shock waves.

PROBLEMS

PROBLEM 1. Determine the non-linearity coefficient α_p in (93.7) for sound wave propagation in a gas.

SOLUTION. The exact equations of one-dimensional gas flow without dissipation are

$$\frac{\partial v}{\partial t}+v\frac{\partial v}{\partial x}=-\frac{1}{\rho}\frac{\partial p}{\partial x}, \qquad \frac{\partial \rho}{\partial t}+\frac{\partial}{\partial x}(\rho v)=0. \tag{1}$$

We expand these as far as second-order small terms, putting

$$p=p_0+p', \qquad \rho=\rho_0+p'/c^2+\tfrac{1}{2}p'^2(\partial^2\rho/\partial p^2)_s. \tag{2}$$

† For a shock wave propagated in a mixture there is also a contribution to the thickness from diffusion processes in the transition layer, calculated by S. P. D'yakov, *Zhurnal éksperimental'noĭ i teoreticheskoĭ fiziki* **27**, 283, 1954.

Weak shock waves remain stable with respect to transverse modulation (see the eighth footnote to §90) even when their dissipative structure is taken into account (M. D. Spektor, *JETP Letters* **35**, 221, 1983).

‡ A strong shock wave causes a considerable increase in temperature; l denotes the mean free path for some mean temperature of the gas in the shock.

The second-order terms can be simplified by bringing all of them to a form containing the product $p'\partial p'/\partial x$. To do so, we note that, for a wave propagated in the negative x-direction with velocity c, differentiations with respect to t and x/c are equivalent, and $v = -p'/c\rho_0$. Then equations (1) and (2) become

$$\frac{\partial v}{\partial t}+\frac{1}{\rho}\frac{\partial p'}{\partial x}=0, \tag{3}$$

$$\frac{\partial v}{\partial x}+\frac{1}{\rho c^2}\frac{\partial p'}{\partial t}=c\rho\left(\frac{\partial^2 V}{\partial p^2}\right)_s p'\frac{\partial p'}{\partial x}; \tag{4}$$

the suffix zero in the constant equilibrium values has been omitted. We have used also the relation

$$\left(\frac{\partial^2 \rho}{\partial p^2}\right)_s=\frac{2}{\rho c^4}-\rho^2\left(\frac{\partial^2 V}{\partial p^2}\right)_s, \tag{5}$$

where $V = 1/\rho$ is the specific volume. Differentiation of (3) with respect to x and (5) with respect to t, followed by subtraction, gives

$$\left(\frac{1}{c}\frac{\partial}{\partial t}-\frac{\partial}{\partial x}\right)\left(\frac{1}{c}\frac{\partial}{\partial t}+\frac{\partial}{\partial x}\right)p'=c^2\rho^2\left(\frac{\partial^2 V}{\partial p^2}\right)_s\frac{\partial}{\partial x}\left(p'\frac{\partial p'}{\partial x}\right).$$

To the same accuracy, we replace $\partial/\partial x+(1/c)\partial/\partial t$ on the left by $2\partial/\partial x$. Lastly, cancelling $\partial/\partial x$ on each side and comparing the result with (93.7), we get α_p in accordance with (93.8).

An equation for v can be obtained directly from (93.7) without repeating calculations similar to the above. The sum of the first-order terms on the left of (93.7) contains the operator $\partial/\partial t - c\partial/\partial x$, which is to be regarded as a first-order term; it gives zero when applied to $p'(x, t)$ in the linear approximation. We thus obtain an equation for $v(x, t)$ in the required approximation on simply replacing p' in (93.7) according to the linear relation $p' = -\rho c v$:

$$\frac{\partial v}{\partial t}-c\frac{\partial v}{\partial x}+\alpha_v v\frac{\partial v}{\partial x}=ac^3\frac{\partial^2 v}{\partial x^2}, \tag{6}$$

where

$$\alpha_v=\tfrac{1}{2}(c^4/V^3)(\partial^2 V/\partial p^2)_s.$$

This α_v is dimensionless; for a polytropic gas, $\alpha_v = \frac{1}{2}(\gamma+1)$.

Problem 2. Use a non-linear substitution to convert Burgers' equation (93.7a) to a linear thermal-conduction equation (E. Hopf 1950).

Solution. By the substitution

$$u(\zeta, t) = -2\mu\frac{\partial}{\partial\zeta}\log\phi(\zeta, t), \tag{1}$$

equation (93.7a) is brought to the form

$$2\mu\frac{\partial}{\partial\zeta}\left[\frac{1}{\phi}\left(-\frac{\partial\phi}{\partial t}+\mu\frac{\partial^2\phi}{\partial\zeta^2}\right)\right]=0,$$

whence

$$\frac{\partial\phi}{\partial t}-\mu\frac{\partial^2\phi}{\partial\zeta^2}=\phi\frac{\mathrm{d}f(t)}{\mathrm{d}t}, \tag{2}$$

where $\mathrm{d}f/\mathrm{d}t$ is an arbitrary function of t. By making the change $\phi\to\phi e^f$ (which does not affect the function $u(\zeta, t)$ sought), we convert this equation to the required form

$$\partial\phi/\partial t=\mu\partial^2\phi/\partial\zeta^2. \tag{3}$$

The solution with the initial condition $\phi(\zeta, 0) = \phi_0(\zeta)$ is given by (51.3):

$$\phi(\zeta, t)=\frac{1}{2}\frac{1}{\sqrt{(\pi\mu t)}}\int_{-\infty}^{\infty}\phi_0(\zeta')\exp[-(\zeta-\zeta')^2/4\mu t]\,\mathrm{d}\zeta'. \tag{4}$$

The initial function $\phi_0(\zeta)$ is related to the initial value of $u(\zeta, t)$ by

$$\log\phi_0(\xi)=-\frac{1}{2\mu}\int_0^{\zeta}u_0(\zeta)\mathrm{d}\zeta, \tag{5}$$

the lower limit of the integral being chosen arbitrarily.

§94. Shock waves in a relaxing medium

A considerable increase in the thickness of a shock wave may be caused by the presence in the gas of comparatively slow relaxation processes (slow chemical reactions, a slow energy transfer between different degrees of freedom of the molecule, and so on (Ya. B Zel'dovich 1946)).†

Let τ be a time of the order of magnitude of the relaxation time. Both the initial and the final state of the gas must be states of complete equilibrium; it is therefore immediately clear that the total thickness of the shock wave will be of the order of τv_1, the distance traversed by the gas in the time τ. It is also found that, if the shock strength is above a certain limit, its structure becomes more complex; this may be seen as follows.

In Fig. 67 the continuous curve shows the shock adiabatic drawn through a given initial point 1, on the assumption that the final states of the gas are states of complete equilibrium; the slope of the tangent at the point 1 is given by the "equilibrium" velocity of sound, denoted in §81 by c_0. The dashed curve shows the shock adiabatic through the same point 1, on the assumption that the relaxation processes are "frozen" and do not occur. The slope of the tangent to this curve at the point 1 depends on the velocity of sound denoted in §81 by c_∞.

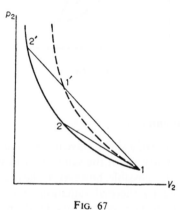

FIG. 67

If the velocity of the shock wave is such that $c_0 < v_1 < c_\infty$, the chord 12 lies as shown in Fig. 67 (the lower chord). In this case we have a simple increase in the shock thickness, all intermediate states between the initial state 1 and the final state 2 being represented in the pV-plane by points on the segment 12. This follows from the fact that (neglecting ordinary viscosity and thermal conduction) all the states through which the gas passes satisfy the equations of conservation of mass, $\rho v = j =$ constant, and of momentum, $p + j^2 V =$ constant (cf. the similar but more detailed discussion in §129).

If, however, $v_1 > c_\infty$, the chord takes the position 11'2'. No point lying between 1 and 1' corresponds to any actual state of the gas; the first real point (after 1) is 1', which corresponds to a state in which the relaxation equilibrium is no different from that in state 1. The compression of the gas from state 1 to state 1' occurs discontinuously, and afterwards (over distances $\sim v_1 \tau$) it is gradually compressed to the final state 2'.

† For example, in diatomic gases, for temperatures behind the shock wave $\sim$ 1000–3000 K, the excitation of intramolecular vibrations is a slow relaxation process. At higher temperatures, the role of such a process falls to thermal dissociation of molecules into their constituent atoms.

If the equilibrium and non-equilibrium shock adiabatics intersect (Fig. 68), there can exist shock waves of a further type: if the shock velocity is such that the chord 12 meets the adiabatics above their intersection, as in Fig. 68, the relaxation is accompanied by a pressure decrease from the value corresponding to the point 1′ to that for point 2 (S.P. D'yakov 1954).†

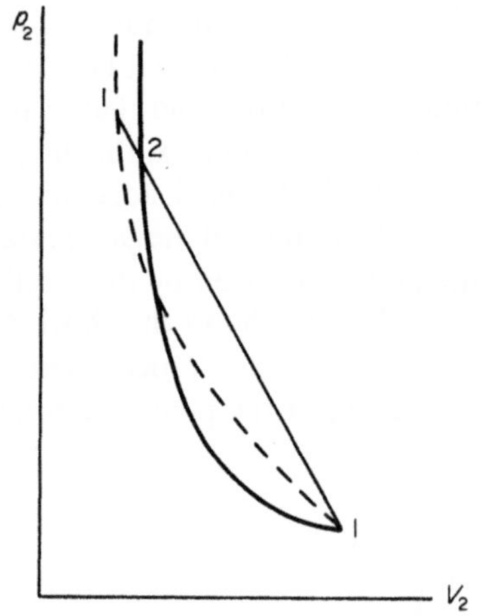

FIG. 68

§95. **The isothermal discontinuity**

The discussion of the structure of a shock wave in §93 involves the assumption that the viscosity and thermal conductivity are of the same order of magnitude, as is usually the case. The case where $\chi \gg \nu$ is also possible, however. If the temperature is sufficiently high, additional heat is transferred by thermal radiation in equilibrium with the matter. Radiation has a much smaller effect on the viscosity (i.e. the momentum transfer), and so ν may be small compared with χ. We shall now see that this inequality leads to a very important difference in the structure of the shock wave.

Neglecting terms which contain the viscosity, we can write equations (93.2) and (93.3), which determine the structure of the transition layer, as

$$p + j^2 V = p_1 + j^2 V_1, \tag{95.1}$$

$$\frac{\kappa}{j}\frac{\mathrm{d}T}{\mathrm{d}x} = w + \tfrac{1}{2}j^2 V^2 - w_1 - \tfrac{1}{2}j^2 V_1^{\,2}. \tag{95.2}$$

The right-hand side of (95.2) is zero only at the boundaries of the layer. Since the temperature behind the shock wave must be higher than that in front of it, it follows that

† Such a case might in principle occur in a dissociating polyatomic gas if in the equilibrium state sufficiently complete dissociation of the molecules into smaller parts occurs behind the shock wave. The dissociation increases the specific-heat ratio γ and therefore reduces the limiting compression in the shock wave, if it is so complete that heating the gas does not require any appreciable expenditure of energy on continuing the dissociation.

we have

$$dT/dx > 0 \tag{95.3}$$

everywhere in the transition layer, i.e. the temperature increases monotonically.

All quantities in the layer are functions of a single variable, the coordinate x, and therefore are functions of one another. Differentiating (95.1) with respect to V, we obtain

$$\left(\frac{\partial p}{\partial T}\right)_V \frac{dT}{dV} + \left(\frac{\partial p}{\partial V}\right)_T + j^2 = 0.$$

The derivative $(\partial p/\partial T)_V$ is always positive in gases. The sign of the derivative dT/dV is therefore the reverse of that of the sum $(\partial p/\partial V)_T + j^2$. In state 1 we have $j^2 > -(\partial p_1/\partial V_1)_s$ (since $v_1 > c_1$), and, since the adiabatic compressibility is always less than the isothermal compressibility, $j^2 > -(\partial p_1/\partial V_1)_T$. On side 1, therefore, $dT_1/dV_1 < 0$. If this derivative remains negative everywhere in the transition layer, then, as the gas is compressed (V decreasing), the temperature increases monotonically, in accordance with (95.3), from side 1 to side 2. In other words, we have a shock wave whose thickness is much increased by the high thermal conductivity (possibly to such an extent that even to call it a shock wave is mere convention).

If, however, the shock is so strong (see (95.7)) that

$$j^2 < -(\partial p_2/\partial V_2)_T, \tag{95.4}$$

then we have in state 2 $dT_2/dV_2 > 0$, so that the function $T(V)$ has a maximum somewhere between V_1 and V_2 (Fig. 69). It is clear that the transition from state 1 to state 2, with V changing continuously, then becomes impossible, since the inequality (95.3) cannot be satisfied everywhere.

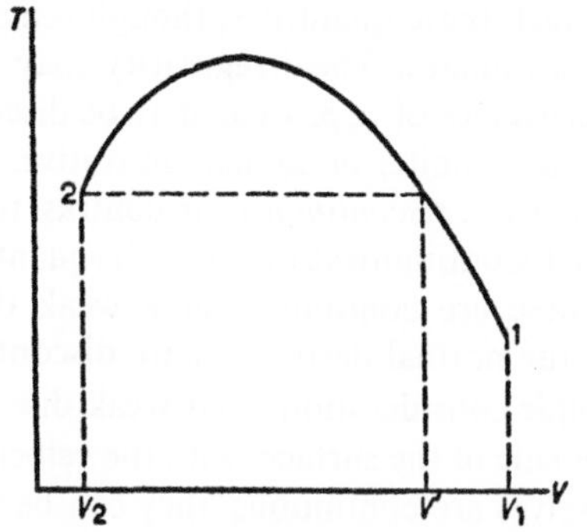

FIG. 69

Consequently, we have the following pattern of transition from the initial state 1 to the final state 2. First comes a region where the gas is gradually compressed from the specific volume V_1 to some V' (the value for which $T(V') = T_2$ for the first time; see Fig. 69); the thickness of this region is determined by the thermal conductivity, and may be considerable. The compression from V' to V_2 then occurs discontinuously, the temperature remaining constant at T_2. This may be called an *isothermal discontinuity*.

Let us determine the variation of the pressure and density in an isothermal discontinuity, assuming that we have a perfect gas. The condition of continuity of momentum flux (95.1), applied to the two sides of the discontinuity, gives $p'+j^2V' = p_2+j^2V_2$. For a perfect gas $V = RT/\mu p$; since $T' = T_2$, we have

$$p' + \frac{j^2RT_2}{\mu p'} = p_2 + \frac{j^2RT_2}{\mu p_2}.$$

This quadratic equation for p' has the solutions $p' = p_2$ (trivial) and

$$p' = j^2RT_2/\mu p_2 = j^2V_2. \tag{95.5}$$

We can express j^2 in the form (85.6), obtaining $p' = (p_2-p_1)V_2/(V_1-V_2)$, and, substituting V_2/V_1 from (89.1), we have for a polytropic gas

$$p' = \tfrac{1}{2}[(\gamma+1)p_1 + (\gamma-1)p_2]. \tag{95.6}$$

Since we must have $p_2 > p'$, we find that an isothermal discontinuity occurs only when the ratio of the pressures p_2 and p_1 satisfies

$$p_2/p_1 > (\gamma+1)/(3-\gamma) \tag{95.7}$$

(Rayleigh 1910). This condition can, of course, be obtained directly from (95.4).

Since, for a given temperature, the gas density is proportional to the pressure, the density ratio in an isothermal discontinuity is equal to the pressure ratio;

$$\rho'/\rho_2 = V_2/V' = p'/p_2, \tag{95.8}$$

and tends to $\frac{1}{2}(\gamma-1)$ as p_2 increases.

§96. **Weak discontinuities**

Besides surface discontinuities, at which the quantities ρ, p, $\mathbf{v}$ etc. are discontinuous, we can also have surfaces at which these quantities, though remaining continuous, are not regular functions of the coordinates. The irregularity may be of various kinds. For example, the first spatial derivatives of ρ, p, $\mathbf{v}$ etc. may be discontinuous on a surface, or these derivatives may become infinite; or higher derivatives may behave in the same manner. We call such surfaces *weak discontinuities*, in contrast to the strong discontinuities (shock waves and tangential discontinuities), in which the quantities ρ, p, $\mathbf{v}$, . . . themselves are discontinuous. Since these are continuous at a weak discontinuity, so are their tangential derivatives; only the normal derivatives are discontinuous.

It is easy to see from simple considerations that weak discontinuities are propagated relative to the gas (on either side of the surface) with the velocity of sound. For, since the functions ρ, p, $\mathbf{v}$, . . . themselves are continuous, they can be "smoothed" by modifying them only near the surface of discontinuity, and only by arbitrarily small amounts, in such a way that the smoothed functions have no singularity. The true distribution of the pressure, say, can thus be represented as a superposition of a perfectly smooth function p_0, free from all singularities, and a very small perturbation p' of this distribution near the surface of discontinuity; and the latter, like any small perturbation, is propagated, relative to the gas, with the velocity of sound.

It must be emphasized that, for a shock wave, the smoothed functions would differ from the true ones by quantities which in general are not small, and the foregoing arguments are therefore invalid. If, however, the discontinuities in the shock wave are sufficiently small, those arguments are again applicable, and such a shock wave is propagated with the velocity of sound, a result which was obtained by another method in §86.

If the flow is steady in a given coordinate system, then the surface of discontinuity is at rest in that system, and the gas flows through it. The gas velocity component normal to the surface must equal the velocity of sound. If we denote by α the angle between the direction of the gas velocity and the tangent plane to the surface, then $v_n = v \sin \alpha = c$, or $\sin \alpha = c/v$, i.e. a surface of weak discontinuity intersects the streamlines at the Mach angle. In other words, a surface of weak discontinuity is one of the characteristic surfaces, a result which is entirely reasonable if we recall the physical significance of the latter: they are surfaces along which small perturbations are propagated (see §82). It is clear that, in steady flow of a gas, weak discontinuities can occur only at velocities not less than that of sound.

Weak discontinuities differ fundamentally from strong ones in the manner of their occurrence. We shall see that shock waves can be formed as a direct result of the gas flow, the boundary conditions being continuous (for instance, the formation of shock waves in a sound wave, §102). In contrast to this, weak discontinuities cannot occur spontaneously; they are always the result of some singularity of the initial or boundary conditions of the flow. These singularities may be of various kinds, like the weak discontinuities themselves. For example, a weak discontinuity may occur on account of the presence of angles on the surface of a body past which the flow takes place; in this case the first spatial derivatives of the velocity are discontinuous. A weak discontinuity is also formed when the curvature of the surface of the body is discontinuous, without there being an angle; in this case the second spatial derivatives of the velocity are discontinuous, and so on. Finally, any singularity in the time variation of the flow results in a non-steady weak discontinuity.

The gas velocity component tangential to the surface of a weak discontinuity is always directed away from the point (e.g. an angle on the surface of a body) from which the perturbation begins which causes the discontinuity; we shall say that the discontinuity starts from this point. This is an example of the fact that, in a supersonic flow, perturbations are propagated downstream.

The presence of viscosity and thermal conduction results in a finite thickness of a weak discontinuity, which is therefore in reality a transition layer, like a shock wave. The thickness of the latter, however, depends only on its strength and is constant in time, whereas the thickness of a weak discontinuity increases with time after its formation. It is easy to determine the qualitative law governing this increase. To do so, we use the fact that the motion of a weak discontinuity follows the same equations as the propagation of any weak sound disturbance. When viscosity and thermal conduction occur, a perturbation which is initially concentrated in a small volume (a wave packet) expands as it moves in the course of time; the manner of this expansion has been determined in §79. We can therefore conclude that the thickness δ of a weak discontinuity is

$$\delta \sim \sqrt{(ac^3 t)}, \tag{96.1}$$

where t is the time from the formation of the discontinuity and a the coefficient of the squared frequency in the sound absorption coefficient (79.6). If the discontinuity is at rest, then the time t must be replaced by l/c, where l is the distance from the point where the discontinuity starts (e.g. for a weak discontinuity starting from an angle on the surface of a body, l is the distance from the vertex of the angle); consequently $\delta \sim \sqrt{(ac^2 l)}$.†

† We must emphasize, however, that the analogy with sound would not suffice for a quantitative determination of the structure of a weak discontinuity. The reason is that, to determine the law of sound damping, the amplitude may be assumed infinitesimal and the linearized equations of motion may be used. For weak discontinuities, as for weak shock waves (§93), the non-linearity of the equations has to be taken into account, since otherwise there would be no discontinuities. An example of such an analysis is given in §99, Problem 6.

To conclude this section, we should make the following remark, analogous to the one at the end of §82. We stated there that, among the various perturbations of the state of a gas in motion, perturbations of entropy (at constant pressure) and vorticity are distinct in their properties. Such perturbations do not move relative to the gas, and are not propagated with the velocity of sound. Hence the surfaces at which the entropy and vorticity† are weakly discontinuous are at rest relative to the gas, and move with it relative to a fixed system of coordinates. Such discontinuities may be called *weak tangential discontinuities*; they pass through streamlines, and are in this respect entirely analogous to the strong tangential discontinuities.

† A weak discontinuity of the vorticity implies a weak discontinuity of the velocity component tangential to the surface of discontinuity; for example, the normal derivatives of the tangential velocity may be discontinuous.

CHAPTER X

ONE-DIMENSIONAL GAS FLOW

§97. Flow of gas through a nozzle

Let us consider steady flow of a gas out of a large vessel through a tube with variable cross-section (a *nozzle*). We shall suppose that the gas flow is uniform over the cross-section at every point in the tube, and that the velocity is along the axis of the tube. For this to be so, the tube must not be too wide, and its cross-sectional area S must vary fairly slowly along its length. Thus all quantities characterizing the flow will be functions only of the coordinate along the axis of the tube. Under these conditions we can apply the relations obtained in §83, which are valid along streamlines, directly to the variation of quantities along the axis.

The mass of gas passing through a cross-section of the tube in unit time (the *discharge*) is $Q = \rho v S$; this must evidently be constant along the tube:

$$Q = S\rho v = \text{constant}. \tag{97.1}$$

The linear dimensions of the vessel are supposed very large in comparison with the diameter of the tube. The velocity of the gas in the vessel may therefore be taken as zero, and accordingly all quantities with the suffix 0 in the formulae of §83 will be the values of those quantities in the vessel.

We have seen that the flux density $j = \rho v$ cannot exceed a certain limiting value j_*. It is therefore clear that the possible values of the total discharge Q have (for a given tube and a given state of the gas in the vessel) an upper limit Q_{max}, which is easily determined. If the value j_* of the flux density were reached anywhere except at the narrowest point of the tube, we should have $j > j_*$ for cross-sections with smaller S, which is impossible. The value $j = j_*$ can therefore be attained only at the narrowest point of the tube; let the cross-sectional area there be S_{min}. Then the upper limit to the total discharge is

$$Q_{max} = \rho_* v_* S_{min} = \sqrt{(\gamma p_0 \rho_0)}[2/(\gamma+1)]^{(1+\gamma)/2(\gamma-1)} S_{min}. \tag{97.2}$$

Let us first consider a nozzle which narrows continually towards its outer end, so that the minimum cross-sectional area is at that end (Fig. 70). By (97.1), the flux density j increases monotonically along the tube. The same is true of the gas velocity v, and the pressure accordingly falls monotonically. The greatest possible value of j is reached if v attains the value c just at the outer end of the tube, i.e. if $v_1 = c_1 = v_*$ (the suffix 1 denotes quantities pertaining to the outer end). At the same time, $p_1 = p_*$.

Let us now follow the change in the manner of outflow of the gas when the external pressure p_e diminishes. When this pressure decreases from p_0, the pressure inside the vessel, to p_*, the pressure p_1 at the outer end of the tube decreases also, and the two pressures p_1 and p_e remain equal; that is, the whole of the pressure drop from p_0 to p_e occurs in the nozzle. The velocity v_1 with which the gas leaves the tube, and the total discharge $Q = j_1 S_{min}$, increase monotonically, however. For $p_e = p_*$ this velocity becomes equal to

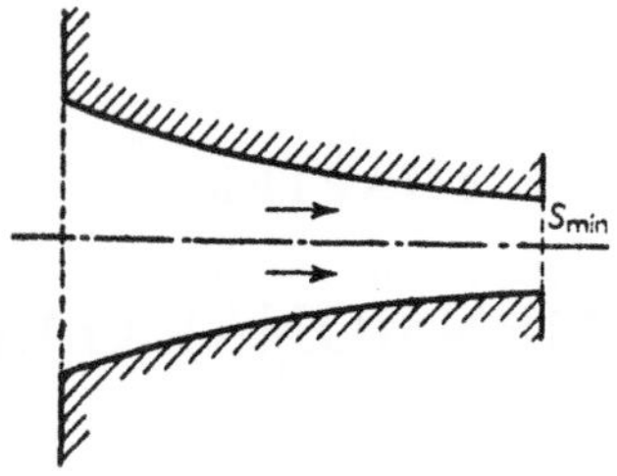

FIG. 70

the local velocity of sound, and the discharge reaches the value Q_{max}. When the external pressure decreases further, the pressure p_1 remains constant at p_*, and the fall of pressure from p_* to p_e occurs outside the tube, in the surrounding medium. In other words, the pressure drop along the tube cannot be greater than from p_0 to p_*, whatever the external pressure. For air ($p_* = 0{\cdot}53p_0$), the maximum pressure drop is $0{\cdot}47p_0$. The velocity at the end of the tube and the discharge also remain constant for $p_e < p_*$. Thus the gas cannot acquire a supersonic velocity in flowing through a nozzle of this kind.

The impossibility of achieving supersonic velocities by flow through a continually narrowing nozzle is due to the fact that a velocity equal to the local velocity of sound can be reached only at the very end of such a tube. It is clear that a supersonic velocity can be attained by means of a nozzle which first narrows and then widens again (Fig. 71). This is called a *de Laval nozzle.*

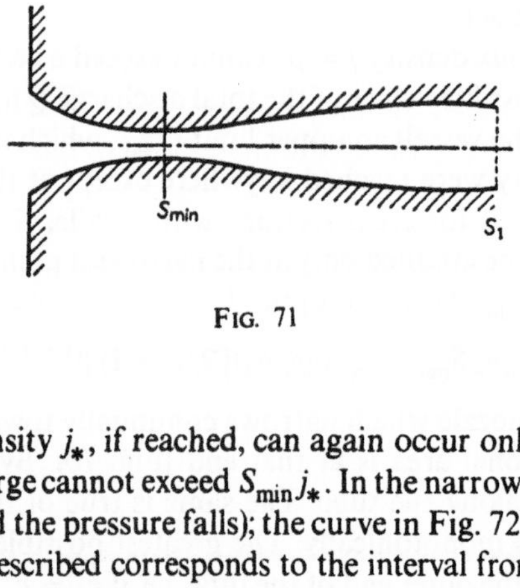

FIG. 71

The maximum flux density j_*, if reached, can again occur only at the narrowest cross-section, so that the discharge cannot exceed $S_{min}j_*$. In the narrowing part of the nozzle, the flux density increases (and the pressure falls); the curve in Fig. 72 shows j as a function† of p, and the variation just described corresponds to the interval from c to b. If the maximum flux density is reached at the cross-section S_{min} (the point b in Fig. 72), the pressure continues to diminish in the widening part of the nozzle, while j begins to decrease also, corresponding to the segment ba of the curve. At the outer end of the tube j takes a definite value, $j_{1\,max} = j_* S_{min}/S_1$, and the pressure has the corresponding value, denoted in Fig.

† According to formulae (83.15–83.17), the dependence is

$$j = \left(\frac{p}{p_0}\right)^{1/\gamma}\left\{\frac{2\gamma}{\gamma-1}p_0\rho_0\left[1-\left(\frac{p}{p_0}\right)^{(\gamma-1)/\gamma}\right]\right\}^{\frac{1}{2}}.$$

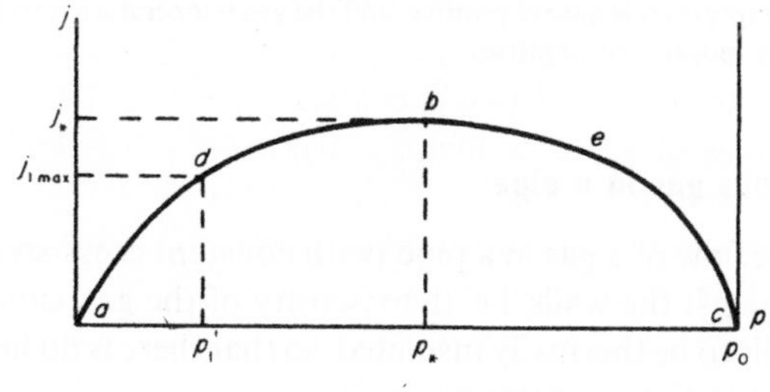

FIG. 72

72 by p_1, at some point d on the curve. If, however, only some point e is reached at the cross-section S_{min}, the pressure increases in the widening part of the nozzle, corresponding to a return down the curve from e towards c. At first sight it might appear that we might pass discontinuously from cb to ab, without going through the point b, by the formation of a shock wave. This, however, is impossible, since the gas "entering" the shock wave cannot have a subsonic velocity.

Bearing in mind these results, let us now investigate the manner of variation in the outflow when the external pressure p_e is gradually increased. For small pressures, from zero to p_1', the pressure p_* and velocity $v_* = c_*$ are reached at the cross-section S_{min}. In the widening part of the nozzle the velocity continues to increase, so that there results a supersonic flow of the gas, and the pressure accordingly continues decreasing, reaching the value p_1' at the outer end of the tube, whatever the pressure p_e. The pressure falls from p_1 to p_e outside the nozzle, in the rarefaction wave which leaves the edge of the tube mouth (see §112).

When p_e exceeds p_1', an oblique shock wave leaves the edge of the tube mouth, compressing the gas from p_1' to p_e (§112). We shall see, however, that a steady shock wave can leave a solid surface only if its intensity is not too great (§111). Hence, when the external pressure increases further, the shock wave soon begins to move into the nozzle, with separation occurring in front of it on the inner surface of the tube. For some value of p_e the shock wave reaches the narrowest cross-section and then disappears; the flow becomes everywhere subsonic, with separation on the walls of the widening part of the nozzle. All these complex phenomena are, of course, three-dimensional.

PROBLEM

A small amount of heat is supplied over a short segment of a tube to a gas in steady flow in the tube. Determine the change in the gas velocity when it passes through this segment. The gas is assumed polytropic.

SOLUTION. Let Sq be the amount of heat supplied per unit time, S being the cross-sectional area of the tube at the segment concerned. The mass flux density $j = \rho v$ and the momentum flux density $p + jv$ are the same on both sides of the heated segment; hence $\Delta p = -j\Delta v$, where Δ denotes the change in a quantity in passing through the segment. The difference in the energy flux density $(w + \frac{1}{2}v^2)j$ is q. Writing $w = \gamma p/(\gamma-1)\rho = \gamma pv/(\gamma-1)j$, we obtain (supposing Δv and Δp small) $vj\Delta v + \gamma(p\Delta v + v\Delta p)/(\gamma-1) = q$. Eliminating Δp, we find $\Delta v = (\gamma-1)q/\rho(c^2 - v^2)$. We see that, in subsonic flow, the supply of heat accelerates the flow ($\Delta v > 0$), while in supersonic flow it retards it.

Writing the gas temperature as $T = \mu p/R\rho = \mu pv/Rj$ (R being the gas constant), we find

$$\Delta T = \frac{\mu}{Rj}(v\Delta p + p\Delta v) = \frac{\mu(\gamma-1)q}{Rj(c^2-v^2)}\left(\frac{c^2}{\gamma} - v^2\right).$$

For supersonic flow, this expression is always positive, and the gas temperature is increased; for subsonic flow, however, ΔT may be either positive or negative.

§98. Flow of a viscous gas in a pipe

Let us consider the flow of a gas in a pipe (with constant cross-section) so long that the friction of the gas against the walls, i.e. the viscosity of the gas, cannot be neglected. We shall suppose the walls to be thermally insulated, so that there is no heat exchange between the gas and the surrounding medium.

For gas velocities of the order of or exceeding the velocity of sound (the only case we shall discuss here), the gas flow in the pipe is, of course, turbulent if the radius of the pipe is not small. The turbulence of the flow is important, as regards our problem, only in one respect: we have seen in §43 that, in turbulent flow, the (mean) velocity is practically the same almost everywhere in the cross-section of the pipe, and falls rapidly to zero very close to the walls. We shall therefore suppose that the gas velocity v is a constant over the cross-section, and define it so that the product $S\rho v$ (S being the cross-sectional area) is equal to the total discharge through the cross-section.

Since the total discharge $S\rho v$ is constant along the pipe, and S is assumed constant, the mass flux density must also be constant:

$$j = \rho v = \text{constant}. \tag{98.1}$$

Next, since the pipe is thermally insulated, the total energy flux carried by the gas through any cross-section must also be constant. This flux is $S\rho v(w+\frac{1}{2}v^2)$, and by (98.1) we have

$$w+\tfrac{1}{2}v^2 = w+\tfrac{1}{2}j^2V^2 = \text{constant}. \tag{98.2}$$

The entropy s of the gas does not, of course, remain constant, but increases as the gas moves along the pipe, because of the internal friction. If x is the coordinate along the pipe, with x increasing downstream, we can write

$$ds/dx > 0. \tag{98.3}$$

We now differentiate (98.2) with respect to x. Since $dw = Tds + Vdp$, we have

$$T\frac{ds}{dx}+V\frac{dp}{dx}+j^2V\frac{dV}{dx} = 0.$$

Next, substituting

$$\frac{dV}{dx} = \left(\frac{\partial V}{\partial p}\right)_s\frac{dp}{dx}+\left(\frac{\partial V}{\partial s}\right)_p\frac{ds}{dx}, \tag{98.4}$$

we obtain

$$\left[T+j^2V\left(\frac{\partial V}{\partial s}\right)_p\right]\frac{ds}{dx} = -V\left[1+j^2\left(\frac{\partial V}{\partial p}\right)_s\right]\frac{dp}{dx}. \tag{98.5}$$

By a well-known formula of thermodynamics, $(\partial V/\partial s)_p = (T/c_p)(\partial V/\partial T)_p$. The coefficient of thermal expansion is positive for gases. We therefore conclude, using (98.3), that the left-hand side of (98.5) is positive. The sign of the derivative dp/dx is therefore that of $-[1+j^2(\partial V/\partial p)_s] = (v/c)^2-1$. We see that

$$dp/dx \lessgtr 0 \quad \text{for} \quad v \lessgtr c. \tag{98.6}$$

Thus, in subsonic flow, the pressure decreases downstream, as for an incompressible fluid. For supersonic flow, however, it increases.

We can similarly determine the sign of the derivative $\mathrm{d}v/\mathrm{d}x$. Since $j = v/V =$ constant, the sign of $\mathrm{d}v/\mathrm{d}x$ is the same as that of $\mathrm{d}V/\mathrm{d}x$. The latter can be expressed in terms of the positve derivative $\mathrm{d}s/\mathrm{d}x$ by means of (98.4) and (98.5). The result is that

$$\mathrm{d}v/\mathrm{d}x \gtrless 0 \quad \text{for} \quad v \lessgtr c, \tag{98.7}$$

i.e. the velocity increases downstream for subsonic flow and decreases for supersonic flow.

Any two thermodynamic quantities for a gas flowing in a pipe are functions of one another, independent of (*inter alia*) the resistance law for the pipe. These functions depend on the constant j as a parameter, and are given by the equation $w + \frac{1}{2}j^2V^2 =$ constant, which is obtained by eliminating the velocity from the equations of conservation of mass and energy for the gas.

Let us ascertain the nature of the curves giving, for example, the entropy as a function of pressure. Rewriting (98.5) in the form

$$\frac{\mathrm{d}s}{\mathrm{d}p} = V\frac{(v/c)^2 - 1}{T + j^2V(\partial V/\partial s)_p},$$

we see that, at the point where $v = c$, the entropy has an extremum. It is easy to see that s has a maximum. For the second derivative of s with respect to p at this point is

$$\left[\frac{\mathrm{d}^2 s}{\mathrm{d}p^2}\right]_{v=c} = -\frac{j^2V(\partial^2V/\partial p^2)_s}{T + j^2V(\partial V/\partial s)_p} < 0;$$

we assume, as usual, that the derivative $(\partial^2V/\partial p^2)_s$ is positive.

The curves giving s as a function of p are therefore as shown in Fig. 73. The region of subsonic velocities lies to the right of the maximum, and that of supersonic velocities to the left. When the parameter j increases, we go to lower curves. For, differentiating equation (98.2) with respect to j for constant p, we have

$$\frac{\mathrm{d}s}{\mathrm{d}j} = -\frac{jV^2}{T + j^2V(\partial V/\partial s)_p} < 0.$$

We can draw an interesting conclusion from the above results. Let the gas velocity at the entrance to the pipe be less than that of sound. The entropy increases downstream, and the

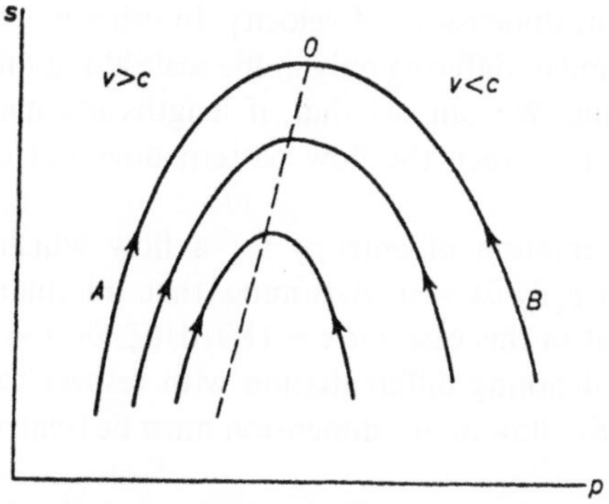

FIG. 73

pressure decreases; this corresponds to a movement along the right-hand branch of the curve $s = s(p)$, from B towards O (Fig. 73). This can, however, continue only until the entropy reaches its maximum value. A further movement along the curve beyond O (i.e. into the region of supersonic velocities) is not possible, since the entropy of the gas would have to decrease as it moved along the pipe. The transition between the branches BO and OA cannot even be effected by a shock wave, since the gas entering a shock wave cannot move with subsonic velocity.

Thus we conclude that, if the gas velocity at the entrance to the pipe is less than that of sound, the flow remains subsonic everywhere in the pipe. The gas velocity becomes equal to the local velocity of sound only at the other end of the pipe, if at all (it does so if the pressure of the external medium into which the gas issues is sufficiently low).

In order that the gas should have supersonic velocities in the pipe, its velocity at the entrance must be supersonic. By the general properties of supersonic flow (the impossibility of propagating disturbances upstream), the flow will then be entirely independent of the conditions at the outlet of the pipe. In particular, the entropy will increase along the pipe in a quite definite manner, and its maximum value will be attained at a definite distance $x = l_k$ from the entrance. If the total length l of the pipe is less than l_k, the flow is supersonic throughout the pipe (corresponding to movement on the branch AO from A towards O). If, on the other hand, $l > l_k$, the flow cannot be supersonic throughout the pipe, nor can there be a smooth transition to subsonic flow, since we can move along the branch OB only in the direction shown by the arrow. In this case, therefore, a shock wave must necessarily be formed, which discontinuously changes the flow from supersonic to subsonic. The pressure is thereby increased, and we pass from the branch AO to BO without going through the point O. The flow is entirely subsonic beyond the discontinuity.

§99. One-dimensional similarity flow

An important class of one-dimensional non-steady gas flows is formed by flows occurring in conditions where there are characteristic velocities but not characteristic lengths. The simplest example of such a flow is given by gas flow in a semi-infinite cylindrical pipe terminated by a piston, when the piston begins to move with constant velocity.

Such a flow is defined by the velocity parameter and by parameters which give, say, the gas pressure and density at the initial instant. We can, however, form no combination of these parameters which has the dimensions of length or time. It therefore follows that the distributions of all quantities can depend on the coordinate x and the time t only through the ratio x/t, which has the dimensions of velocity. In other words, these distributions at various instants will be similar, differing only in the scale along the x-axis, which increases proportionally to the time. We can say that, if lengths are measured in a unit which increases proportionally to t, then the flow pattern does not change. This is called a *similarity flow*.

The equation of conservation of entropy for a flow which depends on only one coordinate, x, is $\partial s/\partial t + v_x \partial s/\partial x = 0$. Assuming that all quantities depend only on $\xi = x/t$, and noticing that in this case $\partial/\partial x = (1/t)\mathrm{d}/\mathrm{d}\xi$, $\partial/\partial t = -(\xi/t)\mathrm{d}/\mathrm{d}\xi$, we obtain $(v_x - \xi)s' = 0$ (the prime denoting differentiation with respect to ξ). Hence $s' = 0$, i.e. s = constant†; thus similarity flow in one dimension must be isentropic as well as adiabatic.

† The assumption that $v_x - \xi = 0$ would contradict the other equations of motion; from (99.3) we should have v_x = constant, contrary to hypothesis.

Likewise, from the y and z components of Euler's equation: $\partial v_y/\partial t + v_x \partial v_y/\partial x = 0$, $\partial v_z/\partial t + v_x \partial v_z/\partial x = 0$, we find that v_y and v_z are constants, which we can take as zero without loss of generality.

Next, the equation of continuity and the x-component of Euler's equation are

$$\frac{\partial \rho}{\partial t} + \rho \frac{\partial v}{\partial x} + v \frac{\partial \rho}{\partial x} = 0, \tag{99.1}$$

$$\frac{\partial v}{\partial t} + v \frac{\partial v}{\partial x} = -\frac{1}{\rho}\frac{\partial p}{\partial x}; \tag{99.2}$$

here and henceforward we write v_x as v simply. In terms of the variable ξ, these equations become

$$(v - \xi)\rho' + \rho v' = 0, \tag{99.3}$$

$$(v - \xi)v' = -p'/\rho = -c^2\rho'/\rho. \tag{99.4}$$

In the second equation we have put $p' = (\partial p/\partial \rho)_s \rho' = c^2 \rho'$, since the entropy is constant.

These equations have, first of all, the trivial solution $v =$ constant, $\rho =$ constant, i.e. a uniform flow with constant velocity. To find a non-trivial solution, we eliminate ρ' and v' from the equations, obtaining $(v - \xi)^2 = c^2$, whence $\xi = v \pm c$. We shall take the plus sign:

$$x/t = v + c; \tag{99.5}$$

this choice of sign means that we take the positive x-axis in a definite direction, selected in a manner shown later. Finally, putting $v - \xi = -c$ in (99.3), we obtain $c\rho' = \rho v'$, or $\rho \mathrm{d}v = c\mathrm{d}\rho$. The velocity of sound is a function of the thermodynamic state of the gas; taking as the fundamental thermodynamic quantities the entropy s and the density ρ, we can represent the velocity of sound as a function $c(\rho)$ of the density, for any given value of the constant entropy. With c understood as such a function, we can write

$$v = \int c\mathrm{d}\rho/\rho = \int \mathrm{d}p/c\rho. \tag{99.6}$$

This formula can also be written

$$v = \int \sqrt{(-\mathrm{d}p\,\mathrm{d}V)}, \tag{99.7}$$

in which the choice of independent variable remains open.

Formulae (99.5) and (99.6) give the required solution of the equations of motion. If the function $c(\rho)$ is known, then the velocity v can be calculated as a function of density from (99.6). Equation (99.5) then determines the density as an implicit function of x/t, and so the dependence of all the other quantities on x/t is determined also.

We can derive some general properties of the solution thus obtained. Differentiating equation (99.5) with respect to x, we have

$$t\frac{\partial \rho}{\partial x}\frac{\mathrm{d}(v+c)}{\mathrm{d}\rho} = 1. \tag{99.8}$$

For the derivative of $v + c$ we have, by (99.6),

$$\frac{\mathrm{d}(v+c)}{\mathrm{d}\rho} = \frac{c}{\rho} + \frac{\mathrm{d}c}{\mathrm{d}\rho} = \frac{1}{\rho}\frac{\mathrm{d}(\rho c)}{\mathrm{d}\rho}.$$

But

$$\rho c = \rho \sqrt{(\partial p/\partial \rho)} = 1/\sqrt{(-\partial V/\partial p)};$$

differentiating, we have

$$\mathrm{d}(\rho c)/\mathrm{d}\rho = c^2 \mathrm{d}(\rho c)/\mathrm{d}p = \tfrac{1}{2}\rho^3 c^5 (\partial^2 V/\partial p^2)_s. \qquad (99.9)$$

Thus

$$\mathrm{d}(v+c)/\mathrm{d}\rho = \tfrac{1}{2}\rho^2 c^5 (\partial^2 V/\partial p^2)_s > 0. \qquad (99.10)$$

It therefore follows from (99.8) that $\partial \rho/\partial x > 0$ for $t > 0$. Since $\partial p/\partial x = c^2 \partial \rho/\partial x$, we conclude that $\partial p/\partial x > 0$ also. Finally, we have $\partial v/\partial x = (c/\rho)\,\partial \rho/\partial x$, so that $\partial v/\partial x > 0$. The inequalities

$$\partial \rho/\partial x > 0, \qquad \partial p/\partial x > 0, \qquad \partial v/\partial x > 0 \qquad (99.11)$$

therefore hold.

The meaning of these inequalities becomes clearer if we follow the variation of quantities, not along the x-axis for given t, but with time for a given gas element as it moves about. This variation is given by the total time derivative; for the density, for example, we have, using the equation of continuity, $\mathrm{d}\rho/\mathrm{d}t = \partial \rho/\partial t + v\,\partial \rho/\partial x = -\rho\,\partial v/\partial x$. By the third inequality (99.11), this quantity is negative, and therefore so is $\mathrm{d}p/\mathrm{d}t$:

$$\mathrm{d}\rho/\mathrm{d}t < 0, \qquad \mathrm{d}p/\mathrm{d}t < 0. \qquad (99.12)$$

Similarly (using Euler's equation (99.2)) we can see that $\mathrm{d}v/\mathrm{d}t < 0$; this, however, does not mean that the magnitude of the velocity diminishes with time, since v may be negative.

The inequalities (99.12) show that the density and pressure of any gas element decrease as it moves. In other words, the gas is continually rarefied as it moves. Such a flow may therefore be called a *non-steady rarefaction wave*.†

A rarefaction wave can be propagated only a finite distance along the x-axis; this is seen from the fact that formula (99.5) would give an infinite velocity for $x \to \pm\infty$, which is impossible.

Let us apply formula (99.5) to a plane bounding the region of space occupied by the rarefaction wave. Here x/t is the velocity of this boundary relative to the fixed coordinate system chosen. Its velocity relative to the gas itself is $(x/t) - v$ and is, by (99.5), equal to the local velocity of sound. This means that the boundaries of a rarefaction wave are weak discontinuities. The similarity flow in different cases is therefore made up of rarefaction waves and regions of constant flow, separated by surfaces of weak discontinuity. There may also, of course, be regions of constant flow separated by shock waves.

The choice of sign in (99.5) is now seen to correspond to the fact that these weak discontinuities are assumed to move in the positive x-direction relative to the gas. The inequalities (99.11) arise from this choice, but the inequalities (99.12), of course, do not depend on the direction of the x-axis.

We are usually concerned, in actual problems, with a rarefaction wave bounded on one side by a region where the gas is at rest. Let this region (I in Fig. 74) be to the right of the rarefaction wave. Region II is the rarefaction wave, and region III contains gas moving

† This flow can occur only as the result of the presence of a singularity in the initial conditions (for example, the piston velocity changes discontinuously at $t = 0$). The opposite flow could occur only by the action of a compressive piston moving a particular manner.

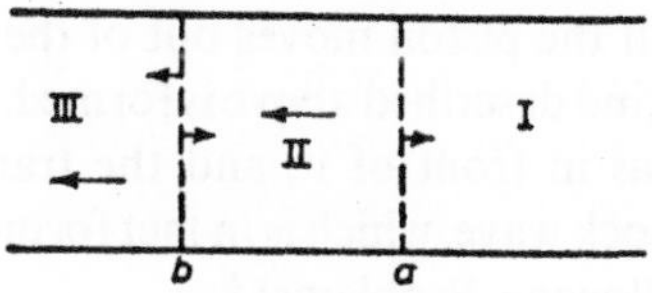

FIG. 74

with constant velocity. The arrows in the figure show the direction of motion of the gas, and of the weak discontinuities bounding the rarefaction wave; the discontinuity a always moves into the gas at rest, but the discontinuity b may move in either direction, depending on the velocity reached in the rarefaction wave (see Problem 2). We may give explicitly the relations between the various quantities in such a rarefaction wave, assuming that we have a polytropic gas. For an adiabatic process $\rho T^{1/(1-\gamma)}$ = constant. Since the velocity of sound is proportional to $\sqrt{T}$, we can write this relation as

$$\rho = \rho_0 (c/c_0)^{2/(\gamma-1)}. \tag{99.13}$$

Substituting this expression in the integral (99.6), we obtain

$$v = \frac{2}{\gamma-1}\int dc = \frac{2}{\gamma-1}(c-c_0);$$

the constant of integration is chosen so that $c = c_0$ for $v = 0$ (we use the suffix 0 to refer to the point where the gas is at rest). We shall express all quantities in terms of v, bearing in mind that, with the above situation of the various regions, the gas velocity is in the negative x-direction, i.e. $v < 0$. Thus

$$c = c_0 - \tfrac{1}{2}(\gamma-1)|v|, \tag{99.14}$$

which determines the local velocity of sound in terms of the gas velocity. Substituting in (99.13), we find the density to be

$$\rho = \rho_0 [1 - \tfrac{1}{2}(\gamma-1)|v|/c_0]^{2/(\gamma-1)}, \tag{99.15}$$

and similarly the pressure is

$$p = p_0 [1 - \tfrac{1}{2}(\gamma-1)|v|/c_0]^{2\gamma/(\gamma-1)}. \tag{99.16}$$

Finally, substituting (99.14) in formula (99.5), we obtain

$$|v| = \frac{2}{\gamma+1}\left(c_0 - \frac{x}{t}\right), \tag{99.17}$$

which gives v as a function of x and t.

The quantity c cannot be negative, by definition. We can therefore draw from (99.14) the important conclusion that the velocity must satisfy the inequality

$$|v| \leqslant 2c_0/(\gamma-1); \tag{99.18}$$

when the velocity reaches this limiting value, the gas density (and also p and c) becomes zero. Thus a gas originally at rest and expanding non-steadily in a rarefaction wave can be accelerated only to velocities not exceeding $2c_0/(\gamma-1)$.

We have already mentioned, at the beginning of this section, a simple example of similarity flow, namely that which occurs in a cylindrical pipe in which a piston begins to move with constant velocity. If the piston moves out of the pipe, it creates a rarefaction, and a rarefaction wave of the kind described above is formed. If, however, the piston moves inwards, it compresses the gas in front of it, and the transition to the original lower pressure can occur only in a shock wave, which is in fact formed in front of a piston moving forward in a pipe (see the following Problems).†

PROBLEMS

PROBLEM 1. A gas is in a semi-infinite cylindrical pipe terminated by a piston. At an initial instant the piston begins to move into the pipe with constant velocity U. Determine the resulting flow, assuming the gas to be polytropic.

SOLUTION. A shock wave is formed in front of the piston, and moves along the pipe. At the initial instant this shock and the piston are coincident, but at subsequent instants the shock is ahead of the piston, and a region of gas lies between them (region 2). In front of the shock wave (region 1), the gas pressure is equal to its initial value p_1, and its velocity relative to the pipe is zero. In region 2, the gas moves with constant velocity, equal to the velocity U of the piston (Fig. 75). The difference in velocity between regions 1 and 2 is therefore also U, and, by formulae (85.7) and (89.1), we can write

$$U = \sqrt{[(p_2-p_1)(V_1-V_2)]}$$
$$= (p_2-p_1)\sqrt{\{2V_1/[(\gamma-1)p_1+(\gamma+1)p_2]\}}.$$

Hence we find the gas pressure p_2 between the piston and the shock wave to be given by

$$\frac{p_2}{p_1} = 1+\frac{\gamma(\gamma+1)U^2}{4c_1^2}+\frac{\gamma U}{c_1}\sqrt{\left[1+\frac{(\gamma+1)^2U^2}{16c_1^2}\right]}.$$

Knowing p_2, we can calculate, from formulae (89.4), the velocity of the shock wave relative to the gas on each side of it. Since gas 1 is at rest, the velocity of the shock relative to it is equal to the rate of propagation of the shock in the pipe. If the x coordinate (along the pipe) is measured from the initial position of the piston (the gas being on the side $x > 0$), we find the position of the shock wave at time t to be

$$x = t\{\tfrac{1}{4}(\gamma+1)U+\sqrt{[\tfrac{1}{16}(\gamma+1)^2U^2+c_1^2]}\},$$

while the position of the piston is $x = Ut$.

PROBLEM 2. The same as Problem 1, but for the case where the piston moves out of the pipe with velocity U.

SOLUTION. The piston adjoins a region of gas (region 1 in Fig. 76a) which moves in the negative x-direction with constant velocity $-U$, equal to the velocity of the piston. Then follows a rarefaction wave (2), in which the

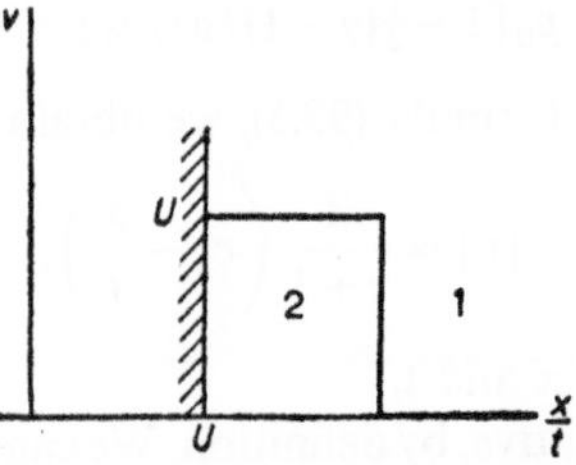

FIG. 75

† We may mention also an analogous similarity flow in three dimensions: the centrally symmetrical gas flow caused by a uniformly expanding sphere. A spherical shock wave, expanding with constant velocity, is formed in front of the sphere. Unlike what happens in the one-dimensional case, the velocity of the gas between the sphere and the shock is not constant; the equation which determines it as a function of the ratio r/t (and therefore the rate of propagation of the shock wave) cannot be integrated analytically.

This problem has been discussed by L. I. Sedov (1945; see his book *Similarity and Dimensional Methods in Mechanics*, London 1959) and by G. I. Taylor, *Proceedings of the Royal Society*, A**186**, 273, 1946.

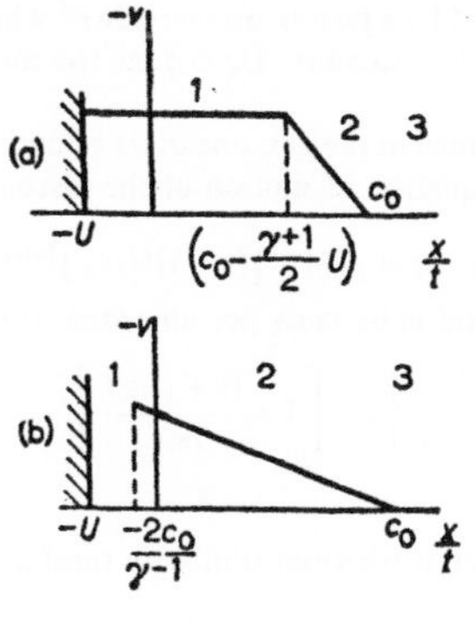

FIG. 76

gas moves in the negative x-direction, its velocity varying linearly from $-U$ to zero according to (99.17). The pressure varies according to (99.16) from $p_1 = p_0[1-\frac{1}{2}(\gamma-1)U/c_0]^{2\gamma/(\gamma-1)}$ in gas 1 to p_0 in the gas 3, which is at rest. The boundary of regions 1 and 2 is given by the condition $v = -U$; according to (99.17), we have $x = [c_0-\frac{1}{2}(\gamma+1)U]t = (c-U)t$, where c is the velocity of sound in gas 1. At the boundary of regions 2 and 3, $v = 0$, whence $x = c_0t$. Both boundaries are weak discontinuities; the second is always propagated to the right (i.e. away from the piston), but the first may be propagated either to the right (as shown in Fig. 76a) or to the left (if the piston velocity $U > 2c_0/(\gamma+1)$).

The flow pattern just described can occur only if $U < 2c_0/(\gamma-1)$. If $U > 2c_0/(\gamma-1)$, a vacuum is formed in front of the piston (the gas cannot follow the piston), which extends from the piston to the point $x = -2c_0t/(\gamma-1)$ (region 1 in Fig. 76b). At this point, $v = -2c_0/(\gamma-1)$; then follow region 2, in which the velocity decreases to zero at the point $x = c_0t$, and region 3, where the gas is at rest.

PROBLEM 3. A gas occupies a semi-infinite cylindrical pipe ($x > 0$) terminated by a valve. At time $t = 0$, the valve is opened, and the gas flows into the external medium, the pressure p_e in which is less than the initial pressure p_0 in the pipe. Determine the resulting flow.

SOLUTION. Let $-v_e$ be the gas velocity which corresponds to the external pressure p_e according to formula (99.16); for $x = 0$ and $t > 0$, we must have $v = -v_e$. If $v_e < 2c_0/(\gamma+1)$, the velocity distribution shown in Fig. 77a results. For $v_e = 2c_0/(\gamma+1)$ (corresponding to a rate of outflow equal to the local velocity of sound at the end of the pipe: this is easily seen by putting $v = c$ in formula (99.14)), the region of constant velocity vanishes and the pattern shown in Fig. 77b is obtained. The quantity $2c_0/(\gamma+1)$ is the greatest possible rate of outflow from the pipe in the conditions stated. If the external pressure p_e is such that

$$p_e < p_0[2/(\gamma+1)]^{2\gamma/(\gamma-1)}, \tag{1}$$

the corresponding velocity v_e exceeds $2c_0/(\gamma+1)$. In reality, the pressure at the pipe outlet would still be equal to the limiting value (the right-hand side of (1)), and the rate of outflow would be $2c_0/(\gamma+1)$; the remaining pressure drop (to p_e) occurs in the external medium.

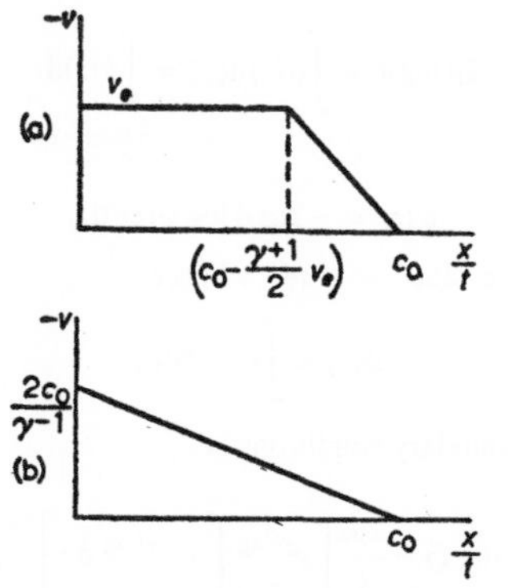

FIG. 77

Problem 4. An infinite pipe is divided by a piston, on one side of which ($x < 0$) there is, at the initial instant, gas at pressure p_0, and on the other side a vacuum. Determine the motion of the piston as the gas expands.

Solution. A rarefaction wave is formed in the gas; one of its boundaries moves to the right with the piston, and the other moves to the left. The equation of motion of the piston is

$$m\mathrm{d}U/\mathrm{d}t = p_0[1-\tfrac{1}{2}(\gamma-1)U/c_0]^{2\gamma/(\gamma-1)},$$

where U is the velocity of the piston and m its mass per unit area. Integrating, we obtain

$$U(t) = \frac{2c_0}{\gamma-1}\left\{1-\left[1+\frac{(\gamma+1)p_0 t}{2mc_0}\right]^{-(\gamma-1)/(\gamma+1)}\right\}.$$

Problem 5. Determine the flow in an isothermal similarity rarefaction wave.

Solution. The isothermal velocity of sound is $c_T = \sqrt{(\partial p/\partial\rho)_T} = \sqrt{(RT/\mu)}$, and for constant temperature $c_T = \text{constant} = c_{T_0}$. According to (99.5) and (99.6), we therefore have

$$v = c_{T_0}\log(\rho/\rho_0) = c_{T_0}\log(p/p_0) = (x/t) - c_{T_0}$$

Problem 6. Using Burgers' equation (§93), determine the structure due to dissipation in a weak discontinuity between a rarefaction wave and a gas at rest.

Solution. Let the gas at rest be to the left of the discontinuity, and the rarefaction wave to the right, so that the discontinuity moves to the left. Neglecting dissipation, we have in the first region $v = 0$. In the second, the flow is described by (99.5) and (99.6) with the sign of c reversed, and v is small near the discontinuity; as far as terms of the first order in v, we have

$$x/t = v - c \cong -c_0 + \left(1+\frac{\rho_0}{c_0}\frac{\mathrm{d}c_0}{\mathrm{d}\rho_0}\right) = -c_0 + \alpha_0 v,$$

where α is defined by (102.2), and the suffix 0, which denotes values for $v = 0$, will be omitted henceforward.
As far as second-order small terms the velocity in the wave propagated to the left obeys equation (6) in §93, Problem 1, or Burgers' equation

$$\frac{\partial u}{\partial t} + u\frac{\partial u}{\partial\zeta} = \mu\frac{\partial^2 u}{\partial\zeta^2},$$

where $\mu = ac^3$ and the unknown $u = \alpha v$ is expressed as a function of t and $\zeta = x + ct$; ζ measures the distance from the weak discontinuity at any instant. We have to find a continuous solution of this equation with the boundary conditions $u = \zeta/t$ for $\zeta \to \infty$, $u = 0$ for $\zeta \to -\infty$, corresponding to flow without dissipation. According to the expression (96.1) for the expansion of a weak discontinuity, t should appear in the solution combined with ζ as $z = \zeta/\sqrt{t}$. Such a solution can satisfy the specified boundary conditions if

$$u(t, \zeta) = (1/\zeta)\psi(\zeta/\sqrt{t}).$$

The function ψ is related to ϕ in §93, Problem 2, by

$$-2\mu\log\phi = \int\psi(z)\mathrm{d}\zeta/\zeta = \int\psi(z)\mathrm{d}z/z$$

so that ψ depends only on z, with

$$\psi(z) = -2\mu z\,\mathrm{d}\log\phi(z)/\mathrm{d}z.$$

Equation (3) in that Problem becomes $2\mu\phi'' = -z\phi'$, whence

$$\phi(z) = \int e^{-z^2/4\mu}\mathrm{d}z.$$

The solution which satisfies the boundary conditions is

$$u(z, \zeta) = \frac{2\mu z}{\zeta}\left[e^{z^2/4\mu}\int_z^\infty e^{-z^2/4\mu}\,\mathrm{d}z\right]^{-1},$$

or finally

$$v(\zeta, t) = \sqrt{\frac{\mu}{\alpha^2 t}}\left[e^{\zeta^2/4\mu t} \int\limits_{\zeta/2\sqrt{(\mu t)}}^{\infty} e^{-z^2}\,dz\right]^{-1}$$

which gives the structure of the weak discontinuity.

§100. Discontinuities in the initial conditions

One of the most important reasons for the occurrence of surfaces of discontinuity in a gas is the possibility of discontinuities in the initial conditions. These conditions (i.e. the initial distributions of velocity, pressure, etc.) may in general be prescribed arbitrarily. In particular, they need not be everywhere continuous, but may be discontinuous on various surfaces. For example, if two masses of gas at different pressures are brought together at some instant, their surface of contact will be a surface of discontinuity of the initial pressure distribution.

It is of importance that the discontinuities of the various quantities in the initial conditions (or, as we shall say, in the *initial discontinuities*) can have any values whatever; no relation between them need exist. We know, however, that certain conditions must hold on stable surfaces of discontinuity in a gas; for instance, the discontinuities of density and pressure in a shock wave are related by the shock adiabatic. It is therefore clear that, if these conditions are not satisfied in the initial discontinuity, it cannot continue to be a discontinuity at subsequent instants. Instead, the initial discontinuity in general splits into several discontinuities, each of which is one of the possible types (shock wave, tangential discontinuity, weak discontinuity); in the course of time, these discontinuities move apart.†

During a short interval of time after the initial instant $t = 0$, the discontinuities formed from the initial discontinuity do not move apart to great distances, and the flow under consideration therefore takes place in a relatively small volume adjoining the surface of initial discontinuity. As usual, it suffices to consider separate portions of this surface, each of which may be regarded as plane. We need therefore consider only a plane surface of discontinuity, which we take as the yz-plane. It is evident from symmetry that the discontinuities formed from the initial discontinuity will also be plane, and perpendicular to the x-axis. The flow pattern will depend on the coordinate x only (and on the time), so that the problem is one-dimensional. There being no characteristic parameters of length and time, we have a similarity problem, and the results obtained in §99 can be used.

The discontinuities formed from the initial discontinuity must evidently move away from their point of formation, i.e. away from the position of the initial discontinuity. It is easy to see that either one shock wave, or one pair of weak discontinuities bounding a rarefaction wave, can move in each direction (the positive and negative x-direction). For, if there were, say, two shock waves formed at the same point at time $t = 0$ and both propagated in the positive x-direction, the leading one would have to move more rapidly than the other. According to the general properties of shock waves, however, the leading shock wave must move, relative to the gas behind it, with a velocity less than the velocity of sound c in that gas, and the following shock must move, relative to the same gas, with a velocity exceeding c (c being a constant in the region between the shock waves), i.e. it must overtake the other. For the same reason, a shock wave and a rarefaction wave cannot move

† A general discussion of this topic has been given by N. E. Kochin (1926).

in the same direction; to see this, it is sufficient to notice that weak discontinuities move with the velocity of sound relative to the gas on each side of them. Finally, two rarefaction waves formed at the same time cannot become separated, since the velocities of their backward fronts are the same.

As well as shock waves and rarefaction waves, a tangential discontinuity must in general be formed from an initial discontinuity. Such a discontinuity must occur if the transverse velocity components v_y, v_z are discontinuous in the initial discontinuity. Since these velocity components do not change in a shock or rarefaction wave, their discontinuities always occur at a tangential discontinuity, which remains at the position of the initial discontinuity; on each side of this discontinuity, v_y and v_z are constant (in reality, of course, the instability of a tangential velocity discontinuity causes its gradual smoothing into a turbulent region).

A tangential discontinuity must occur, however, even if v_y and v_z are continuous at the initial discontinuity (without loss of generality, we can, and shall, assume that they are zero). This is shown as follows. The discontinuities formed from the initial discontinuity must make it possible to go from a given state 1 of the gas on one side of the initial discontinuity to a given state 2 on the other side. The state of the gas is determined by three independent quantities, e.g. p, ρ and $v_x = v$. It is therefore necessary to have three arbitrary parameters in order to go from state 1 to an arbitrary state 2 by some choice of the discontinuities. We know, however, that a shock wave, perpendicular to the stream, propagated in a gas whose thermodynamic state is given, is completely determined by one parameter (§85). The same is true of a rarefaction wave; as we see from formulae (99.14)–(99.16), when the state of the gas entering a rarefaction wave is given, the state of the gas leaving it is completely determined by one parameter. We have seen, moreover, that at most one wave (rarefaction or shock) can move in each direction. We therefore have at our disposal only two parameters, which are not sufficient.

The tangential discontinuity formed at the position of the initial discontinuity furnishes the third parameter required. The pressure is continuous there, but the density (and therefore the temperature and entropy) is not. The tangential discontinuity is stationary with respect to the gas on both sides of it and the arguments about the "overtaking" of two waves propagated in the same direction therefore do not apply to it.

The gases on the two sides of the tangential discontinuity do not mix, since there is no motion of gas through a tangential discontinuity; in all the examples given below, these gases may be different substances.

Figure 78 shows schematically all possible types of break-up of an initial discontinuity. The continuous line shows the variation of the pressure along the x-axis; the variation of the density would be given by a similar line, the only difference being that there would be a further jump at the tangential discontinuity. The vertical lines show the discontinuities formed, and the arrows show their direction of propagation and that of the gas flow. The coordinate system is always that in which the tangential discontinuity is at rest, together with the gas in the regions 3 and 3′ which adjoin it. The pressures, densities and velocities of the gases in the extreme left-hand (1) and right-hand (2) regions are the values of these quantities at time $t = 0$ on each side of the initial discontinuity.

In the first case, which we write $I \rightarrow S_{\leftarrow} T S_{\rightarrow}$ (Fig. 78a), the initial discontinuity I gives two shock waves S, propagated in opposite directions, and a tangential discontinuity T between them. This case occurs when two masses of gas collide with a large relative velocity.

In the case $I \rightarrow S_{\leftarrow} T R_{\rightarrow}$ (Fig. 78b), a shock wave is propagated on one side of the

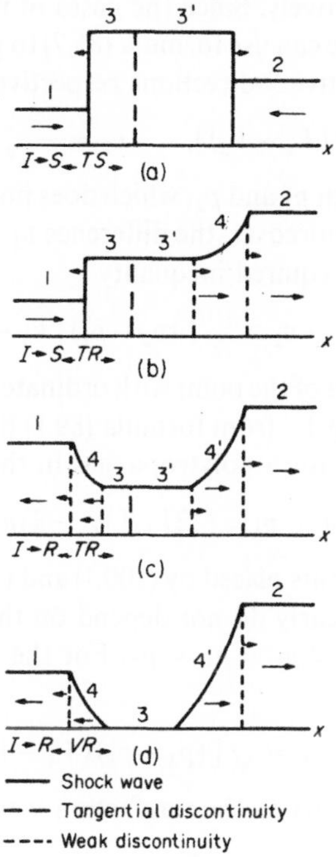

FIG. 78

tangential discontinuity, and a rarefaction wave R on the other side. This case occurs, for instance, if two masses of gas at relative rest ($v_2 - v_1 = 0$) and at different pressures are brought into contact at the initial instant. For, of all the cases shown in Fig. 78, the second is the only one in which gases 1 and 2 are moving in the same direction, and so the equation $v_1 = v_2$ is possible.

In the third case ($I \to R_{\leftarrow}\ TR_{\rightarrow}$, Fig. 78c), a rarefaction wave is propagated on each side of the tangential discontinuity. If gases 1 and 2 separate with a sufficiently great relative velocity $v_2 - v_1$, the pressure may decrease to zero in the rarefaction waves. We then have the pattern shown in Fig. 78d; a vacuum 3 is formed between regions 4 and 4′.

We can derive the analytical conditions which determine the manner in which the initial discontinuity breaks up, as a function of its parameters. We shall suppose in every case that $p_2 > p_1$, and take the positive x-direction from region 1 to region 2 (as in Fig. 78).

Since the gases on the two sides of the initial discontinuity may be of different substances, we shall distinguish them as gases 1 and 2.

(1) $I \to S_{\leftarrow}\ TS_{\rightarrow}$. If $p_3 = p_{3'}$, $v_3 = v_{3'}$, V_3 and $V_{3'}$ are the pressures, velocities and specific volumes in the resulting regions 3 and 3′, then we have $p_3 > p_2 > p_1$, and the volumes V_3 and $V_{3'}$ are the abscissae of the points with ordinate p_3 on the shock adiabatics through

(p_1, V_1) and (p_2, V_2) respectively. Since the gases in regions 3 and 3′ are at rest in the coordinate system chosen, we can use formula (85.7) to give the velocities v_1 and v_2, which are in the positive and negative x-directions respectively:

$$v_1 = \sqrt{[(p_3 - p_1)(V_1 - V_3)]}, \qquad v_2 = -\sqrt{[(p_3 - p_2)(V_2 - V_{3'})]}.$$

The least value of p_3, for given p_1 and p_2, which does not contradict the initial assumption $(p_3 > p_2 > p_1)$ is p_2. Since, moreover, the difference $v_1 - v_2$ is a monotonically increasing function of p_3, we find the required inequality

$$v_1 - v_2 > \sqrt{[(p_2 - p_1)(V_1 - V')]}, \tag{100.1}$$

where V' denotes the abscissa of the point with ordinate p_2 on the shock adiabatic for gas 1 through (p_1, V_1). Calculating V' from formula (89.1) (in which V_2 is replaced by V'), we obtain the condition (100.1) for a polytropic gas in the form

$$v_1 - v_2 > (p_2 - p_1)\sqrt{\{2V_1/[(\gamma_1 - 1)p_1 + (\gamma_1 + 1)p_2]\}}. \tag{100.2}$$

It should be noted that the limits placed by (100.1) and (100.2) on the possible values of the velocity difference $v_1 - v_2$ clearly do not depend on the coordinate system chosen.

(2) $I \to S_{\leftarrow} TR_{\rightarrow}$. Here $p_1 < p_3 = p_{3'} < p_2$. For the gas velocity in region 1 we again have

$$v_1 = \sqrt{[(p_3 - p_1)(V_1 - V_3)]},$$

and the total change in velocity in the rarefaction wave 4 is, by (99.7),

$$v_2 = \int_{p_3}^{p_2} \sqrt{(-\mathrm{d}p\,\mathrm{d}V)}.$$

For given p_1 and p_2, p_3 can lie between them. Replacing p_3 in the difference $v_2 - v_1$ by p_1 and then by p_2, we obtain the condition

$$-\int_{p_1}^{p_2} \sqrt{(-\mathrm{d}p\,\mathrm{d}V)} < v_1 - v_2 < \sqrt{[(p_2 - p_1)(V_1 - V')]}. \tag{100.3}$$

Here V' has the same significance as in the previous case; the upper limit of the difference $v_1 - v_2$ must be calculated for gas 1, and the lower limit for gas 2. For a polytropic gas we have

$$-\frac{2c_2}{\gamma_2 - 1}\left[1 - \left(\frac{p_1}{p_2}\right)^{(\gamma_2 - 1)/2\gamma_2}\right] < v_1 - v_2$$
$$< (p_2 - p_1)\sqrt{\{2V_1/[(\gamma_1 - 1)p_1 + (\gamma_1 + 1)p_2]\}}, \tag{100.4}$$

where $c_2 = \sqrt{(\gamma_2 p_2 V_2)}$ is the velocity of sound in gas 2 in the state (p_2, V_2).

(3) $I \to R_{\leftarrow} TR_{\rightarrow}$. Here $p_2 > p_1 > p_3 = p_{3'} > 0$. By the same method we find the following condition for this case to occur:

$$-\int_0^{p_1} \sqrt{(-\mathrm{d}p\,\mathrm{d}V)} - \int_0^{p_2} \sqrt{(-\mathrm{d}p\,\mathrm{d}V)} < v_1 - v_2 < -\int_{p_1}^{p_2} \sqrt{(-\mathrm{d}p\,\mathrm{d}V)}. \tag{100.5}$$

The first integral in the first member is calculated for gas 1, and the others for gas 2. For a polytropic gas we find

$$-\frac{2c_1}{\gamma_1-1}-\frac{2c_2}{\gamma_2-1}<v_1-v_2<-\frac{2c_2}{\gamma_2-1}\left[1-\left(\frac{p_1}{p_2}\right)^{(\gamma_2-1)/2\gamma_2}\right], \tag{100.6}$$

where $c_1=\sqrt{(\gamma_1 p_1 V_1)}$, $c_2=\sqrt{(\gamma_2 p_2 V_2)}$. If

$$v_1-v_2<-\frac{2c_1}{\gamma_1-1}-\frac{2c_2}{\gamma_2-1}, \tag{100.7}$$

a vacuum is formed between the rarefaction waves ($I \to R_{\leftarrow} VR_{\rightarrow}$).

The problem of a discontinuity in the initial conditions includes that of various collisions between plane surfaces of discontinuity. At the instant of collision, the two planes coincide, and form some initial discontinuity, which then leads to one of the patterns described above. The collision of two shock waves, for instance, results in two other shock waves, which move away from the tangential discontinuity remaining between them: $S_{\rightarrow}S_{\leftarrow} \to S_{\leftarrow}TS_{\rightarrow}$. When one shock wave overtakes another, there are two possibilities: $S_{\rightarrow}S_{\rightarrow} \to S_{\leftarrow}TS_{\rightarrow}$ and $S_{\rightarrow}S_{\rightarrow} \to R_{\leftarrow}TS_{\rightarrow}$. In either case a shock wave continues in the same direction.

The problem of the reflection and transmission of shock waves by a tangential discontinuity (boundary of two media) also comes under this heading. Here two cases are possible: $S_{\rightarrow}T \to S_{\leftarrow}TS_{\rightarrow}$ and $S_{\rightarrow}T \to R_{\leftarrow}TS_{\rightarrow}$. The wave transmitted into the second medium is always a shock (see also the following Problems).†

PROBLEMS

PROBLEM 1. A plane shock wave is reflected from a rigid plane surface. Determine the gas pressure behind the reflected wave (H. Hugoniot 1885).

SOLUTION. When a shock wave is incident on a rigid wall, a reflected shock wave is propagated away from the wall. We denote by the suffixes 1, 2 and 3 respectively quantities pertaining to the undisturbed gas in front of the incident shock, the gas behind this shock (which is also the gas in front of the reflected shock) and the gas behind the reflected shock; see Fig. 79, where the arrows indicate the direction of motion of the shock waves and of the

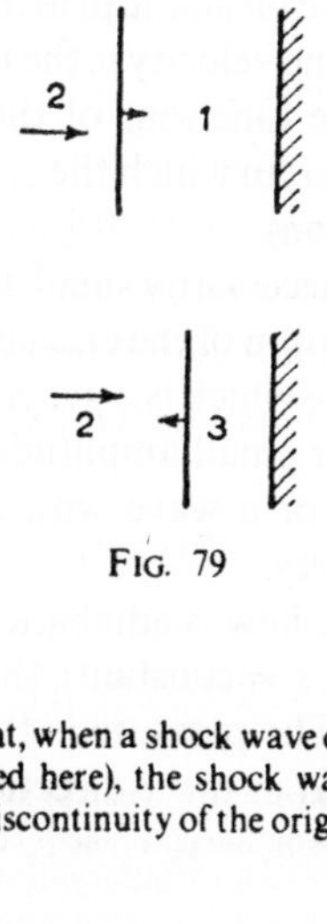

FIG. 79

† For completeness we should mention that, when a shock wave collides with a weak discontinuity (a problem which is not of the similarity type considered here), the shock wave continues to be propagated in the same direction, but behind it there remain a weak discontinuity of the original kind and a weak tangential discontinuity (see the end of §96).

gas itself. The gas in regions 1 and 3, which adjoin the wall, is at rest relative to the wall. The relative velocity of the gases on the two sides of the discontinuity is the same in both the incident and the reflected shock wave, and equal to the velocity of gas 2. Using formula (85.7) for the relative velocity, we therefore have $(p_2-p_1)(V_1-V_2) = (p_3-p_2)(V_2-V_3)$. The equation of the shock adiabatic (89.1) for each shock gives

$$\frac{V_2}{V_1} = \frac{(\gamma+1)p_1+(\gamma-1)p_2}{(\gamma-1)p_1+(\gamma+1)p_2}, \qquad \frac{V_3}{V_2} = \frac{(\gamma+1)p_2+(\gamma-1)p_3}{(\gamma-1)p_2+(\gamma+1)p_3}.$$

We can eliminate the specific volumes from these three equations, and the result is

$$(p_3-p_2)^2[(\gamma+1)p_1+(\gamma-1)p_2] = (p_2-p_1)^2[(\gamma+1)p_3+(\gamma-1)p_2].$$

This is a quadratic equation for p_3, which has the trivial root $p_3 = p_1$; cancelling $p_3 - p_1$, we obtain

$$\frac{p_3}{p_2} = \frac{(3\gamma-1)p_2-(\gamma-1)p_1}{(\gamma-1)p_2-(\gamma+1)p_1},$$

which determines p_3 from p_1 and p_2. In the limiting case of a very strong incident shock, the further compression of the gas in the reflected shock is given by $p_3 = (3\gamma-1)p_2/(\gamma-1)$, $V_3/V_1 = (\gamma-1)/\gamma$, while for a weak shock $p_3 - p_2 = p_2 - p_1$, corresponding to the sound-wave approximation.

PROBLEM 2. Find the condition for a shock wave to be reflected from a plane boundary between two gases.

SOLUTION. Let $p_1 < p_{2'}$, V_1, $V_{2'}$, be the pressures and specific volumes of the two media before the incidence of the shock wave (propagated in gas 2), at their surface of separation, and p_2, V_2 the values behind the shock wave. The condition for the reflected wave to be a shock wave is given by the inequality (100.2), in which we must now put

$$v_1 - v_2 = \sqrt{[(p_2-p_{2'})(V_{2'}-V_2)]}.$$

Expressing all quantities in terms of the ratio of pressures p_2/p_1 and the initial specific volumes V_1, $V_{2'}$, we obtain

$$\frac{V_1}{(\gamma_1+1)p_2/p_1+(\gamma_1-1)} < \frac{V_{2'}}{(\gamma_2+1)p_2/p_1+(\gamma_2-1)}.$$

§101. One-dimensional travelling waves

In discussing sound waves in §64, we assumed the amplitude of oscillations in the wave to be small. The result was that the equations of motion were linear and were easily solved. A particular solution of these equations is any function of $x \pm ct$ (a plane wave), corresponding to a *travelling wave* whose profile moves with velocity c, its shape remaining unchanged; by the *profile* of a wave we mean the distribution of density, velocity, etc., along the direction of propagation. Since the velocity v, the density ρ and the pressure p (and the other quantities) in such a wave are functions of the same quantity $x \pm ct$, they can be expressed as functions of one another, in which the coordinates and time do not explicitly appear ($p = p(\rho)$, $v = v(p)$, and so on).

When the wave amplitude is not necessarily small, these simple relations do not hold. It is found, however, that a general solution of the exact equations of motion can be obtained, in the form of a travelling plane wave which is a generalization of the solution $f(x \pm ct)$ of the approximate equations valid for small amplitudes. To derive this solution, we shall begin from the requirement that, for a wave with any amplitude, the velocity can be expressed as a function of the density.

In the absence of shock waves the flow is adiabatic. If the gas is homogeneous at some initial instant (so that, in particular, s = constant), then s = constant at all times, and we shall assume this in what follows. The pressure is thus a function of the density only.

In a plane sound wave propagated in the x-direction, all quantities depend on x and t only, and for the velocity we have $v_x = v$, $v_y = v_z = 0$. The equation of continuity is

$\partial\rho/\partial t + \partial(\rho v)/\partial x = 0$, and Euler's equation is

$$\frac{\partial v}{\partial t} + v\frac{\partial v}{\partial x} + \frac{1}{\rho}\frac{\partial p}{\partial x} = 0.$$

Using the fact that v is a function of ρ only, we can write these equations as

$$\frac{\partial \rho}{\partial t} + \frac{\mathrm{d}(\rho v)}{\mathrm{d}\rho}\frac{\partial \rho}{\partial x} = 0, \qquad (101.1)$$

$$\frac{\partial v}{\partial t} + \left(v + \frac{1}{\rho}\frac{\mathrm{d}p}{\mathrm{d}v}\right)\frac{\partial v}{\partial x} = 0. \qquad (101.2)$$

Since

$$\frac{\partial\rho/\partial t}{\partial\rho/\partial x} = -\left(\frac{\partial x}{\partial t}\right)_\rho,$$

we have from (101.1)

$$\left(\frac{\partial x}{\partial t}\right)_\rho = \frac{\mathrm{d}(\rho v)}{\mathrm{d}\rho} = v + \rho\frac{\mathrm{d}v}{\mathrm{d}\rho},$$

and similarly from (101.2)

$$\left(\frac{\partial x}{\partial t}\right)_v = v + \frac{1}{\rho}\frac{\mathrm{d}p}{\mathrm{d}v}. \qquad (101.3)$$

Since the value of ρ uniquely determines that of v, the derivatives for constant ρ and constant v are the same, i.e. $(\partial x/\partial t)_\rho = (\partial x/\partial t)_v$, so that $\rho\ \mathrm{d}v/\mathrm{d}\rho = (1/\rho)\mathrm{d}p/\mathrm{d}v = (c^2/\rho)\mathrm{d}\rho/\mathrm{d}v$. Thus $\mathrm{d}v/\mathrm{d}\rho = \pm c/\rho$, whence

$$v = \pm\int\frac{c}{\rho}\mathrm{d}\rho = \pm\int\frac{\mathrm{d}p}{\rho c}. \qquad (101.4)$$

This gives the general relation between the velocity and the density or pressure in the wave.†

Next, we can combine (101.3) and (101.4) to give $(\partial x/\partial t)_v = v + (1/\rho)\mathrm{d}p/\mathrm{d}v = v \pm c(v)$, or, integrating,

$$x = t[v \pm c(v)] + f(v), \qquad (101.5)$$

where $f(v)$ is an arbitrary function of the velocity, and $c(v)$ is given by (101.4).

Formulae (101.4) and (101.5) give the required general solution (B. Riemann 1860). They determine the velocity (and therefore all other quantities) as an implicit function of x and t, i.e. the wave profile at every instant. For any given value of v, we have $x = at + b$, i.e. the point where the velocity has a given value moves with constant velocity; in this sense, the solution obtained is a travelling wave. The two signs in (101.5) correspond to waves propagated (relative to the gas) in the positive and negative x-directions.

The flow described by the solution (101.4) and (101.5) is often called a *simple wave*, and we shall use this expression below. It should be noticed that the similarity flow discussed in

† In a wave with small amplitude we have $\rho = \rho_0 + \rho'$, and (101.4) gives in the first approximation $v = c_0\rho'/\rho_0$ (where $c_0 = c(\rho_0)$), i.e. the usual formula (64.12).

§99 is a particular case of a simple wave, corresponding to $f(v) = 0$ in (101.5).

We can write out explicitly the relations for a simple wave in a polytropic gas; we assume that there is a point in the wave for which $v = 0$, as usually happens in practice. Since formula (101.4) is the same as (99.6), we have by analogy with formulae (99.14)–(99.16)

$$c = c_0 \pm \tfrac{1}{2}(\gamma - 1)v, \tag{101.6}$$

$$\left.\begin{aligned} \rho &= \rho_0(1 \pm \tfrac{1}{2}(\gamma - 1)v/c_0)^{2/(\gamma-1)}, \\ p &= p_0(1 \pm \tfrac{1}{2}(\gamma - 1)v/c_0)^{2\gamma/(\gamma-1)}. \end{aligned}\right\} \tag{101.7}$$

Substituting (101.6) in (101.5), we obtain

$$x = t(\pm c_0 + \tfrac{1}{2}(\gamma + 1)v) + f(v). \tag{101.8}$$

It is sometimes convenient to write this solution in the form

$$v = F[x - (\pm c_0 + \tfrac{1}{2}(\gamma + 1)v)t], \tag{101.9}$$

where F is another arbitrary function.

From formulae (101.6) and (101.7) we again see (as in §99) that the velocity in a direction opposite to that of the propagation of the wave (relative to the gas itself) is of limited magnitude; for a wave propagated in the positive x-direction we have

$$-v \leqslant 2c_0/(\gamma - 1). \tag{101.10}$$

A travelling wave described by formulae (101.4) and (101.5) is essentially different from the one obtained in the limiting case of small amplitudes. The velocity of a point in the wave profile is

$$u = v \pm c; \tag{101.11}$$

it may be conveniently regarded as a superposition of the propagation of a disturbance relative to the gas with the velocity of sound and the movement of the gas itself with velocity v. The velocity u is now a function of the density, and therefore is different for different points in the profile. Thus, in the general case of a plane wave with arbitrary amplitude, there is no definite constant "wave velocity". Since the velocities of different points in the wave profile are different, the profile changes its shape in the course of time.

Let us consider a wave propagated in the positive x-direction, for which $u = v + c$. The derivative of $v + c$ with respect to the density has been calculated in §99; see (99.10). We have seen that $\mathrm{d}u/\mathrm{d}\rho > 0$. The velocity of propagation of a given point in the wave profile therefore increases with the density. If we denote by c_0 the velocity of sound for a density equal to the equilibrium density ρ_0, then in compressions $\rho > \rho_0$ and $c > c_0$, while in rarefactions $\rho < \rho_0$ and $c < c_0$.

The inequality of the velocity of different points in the wave profile causes its shape to change in the course of time: the points of compression move forward and those of rarefaction are left behind (Fig. 80b). Finally, the profile may become such that the function $\rho(x)$ (for given t) is no longer one-valued; three different values of ρ correspond to some x (the dashed line in Fig. 80c).† This is, of course, physically impossible. In reality, discontinuities are formed where ρ is not one-valued, and ρ is consequently one-valued everywhere except at the discontinuities themselves. The wave profile then has the form shown by the continuous line in Fig. 80c. The surfaces of discontinuity are thus formed at points a wavelength apart.

† This change in the wave profile is often referred to as *turn-over*.

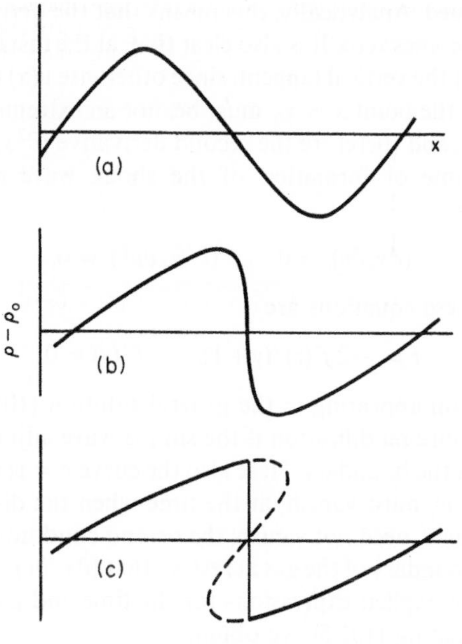

FIG. 80

When the discontinuities are formed, the wave ceases to be a simple wave. The cause of this can be briefly stated thus: when surfaces of discontinuity are present, the wave is reflected from them, and therefore ceases to be a wave travelling in one direction. The assumption on which the whole derivation is based, namely that there is a one-to-one relation between the various quantities, consequently ceases to be valid in general.

The presence of discontinuities (shock waves) results, as was mentioned in §85, in the dissipation of energy. The formation of discontinuities therefore leads to a marked damping of the wave. This is evident from Fig. 80. When the discontinuity is formed, the highest part of the wave profile is cut off. In the course of time, as the profile is bent over, its height becomes less, and the profile is smoothed to one with smaller amplitude, i.e. the wave is damped.

It is clear from the above that discontinuities must ultimately be formed in every simple wave which contains regions where the density decreases in the direction of propagation. The only case where discontinuities do not occur is a wave in which the density everywhere increases monotonically in the direction of propagation (such, for example, is the wave formed when a piston moves out of an infinite pipe filled with gas; see the Problems at the end of this section).

Although the wave is no longer a simple one when a discontinuity has been formed, the time and place of formation of the discontinuity can be determined analytically. We have seen that the occurrence of discontinuities is mathematically due to the fact that, in a simple wave, the quantities p, ρ and v become many-valued functions of x (for given t) at times greater than a certain definite value t_0, whereas for $t < t_0$ they are one-valued functions. The time t_0 is the time of formation of the discontinuity. It is evident from geometrical considerations that, at the instant t_0, the curve giving, say, v as a function of x becomes vertical at some point $x = x_0$, which is the point where the function is subsequently many-valued. Analytically, this means that the derivative $(\partial v/\partial x)_t$ becomes infinite, and $(\partial x/\partial v)_t$ becomes zero. It is also clear that, at the instant t_0, the curve $v = v(x)$ must lie on both sides of the vertical tangent, since otherwise $v(x)$ would already be many-valued. In other words, the point $x = x_0$ must be, not an extremum of the function $x(v)$, but a point of inflexion, and therefore the second derivative $(\partial^2 x/\partial v^2)_t$ must also vanish. Thus the place and time of formation of the shock wave are determined by the simultaneous equations

$$(\partial x/\partial v)_t = 0, \qquad (\partial^2 x/\partial v^2)_t = 0. \tag{101.12}$$

For a polytropic gas these equations are

$$t = -2f'(v)/(\gamma+1), \qquad f''(v) = 0, \tag{101.13}$$

where $f(v)$ is the function appearing in the general solution (101.8).

These conditions require modification if the simple wave adjoins a gas at rest and the shock wave is formed at the boundary. Here also the curve $v = v(x)$ must become vertical, i.e. the derivative $(\partial x/\partial v)_t$ must vanish, at the time when the discontinuity occurs. The second derivative, however, need not vanish; the second condition here is simply that the velocity be zero at the boundary of the gas at rest, so that $(\partial x/\partial v)_t = 0$ for $v = 0$. From this condition we can obtain explicit expressions for the time and place of formation of the discontinuity. Differentiating (101.5), we obtain

$$t = -f'(0)/\alpha_0, \qquad x = \pm c_0 t + f(0), \tag{101.14}$$

where α_0 is the value, for $v = 0$, of the quantity α defined by formula (102.2). For a polytropic gas

$$t = -2f'(0)/(\gamma+1). \tag{101.15}$$

PROBLEMS

PROBLEM 1. A gas is in a semi-infinite cylindrical pipe ($x > 0$) terminated by a piston. At time $t = 0$ the piston begins to move with a uniformly accelerated velocity $U = \pm at$. Determine the resulting flow, assuming the gas to be polytropic.

SOLUTION. If the piston moves out of the pipe ($U = -at$), the result is a simple rarefaction wave, whose forward front is propagated to the right, through gas at rest, with velocity c_0; in the region $x > c_0 t$ the gas is at rest. At the surface of the piston, the gas and the piston must have the same velocity, i.e. we must have $v = -at$ for $x = -\frac{1}{2}at^2$ ($t > 0$). This condition gives for the function $f(v)$ in (101.8)

$$f(-at) = -c_0 t + \tfrac{1}{2}\gamma a t^2.$$

Hence we have

$$x - [c_0 + \tfrac{1}{2}(\gamma+1)v]t = f(v)$$
$$= c_0 v/a + \tfrac{1}{2}\gamma v^2/a,$$

whence

$$-v = [c_0 + \tfrac{1}{2}(\gamma+1)at]/\gamma - \sqrt{\{[c_0 + \tfrac{1}{2}(\gamma+1)at]^2 - 2a\gamma(c_0 t - x)\}}/\gamma. \tag{1}$$

This formula gives the change in velocity over the region between the piston and the forward front $x = c_0 t$ of the wave (Fig. 81a) during the time interval $t = 0$ to $t = 2c_0/(\gamma-1)a$. The gas velocity is everywhere to the left, like that of the piston, and decreases monotonically in magnitude in the positive x-direction; the density and pressure increase monotonically in that direction. For $t > 2c_0/(\gamma-1)a$, the inequality (101.10) does not hold for the piston velocity, and so the gas can no longer follow the piston. A vacuum is then formed in a region adjoining the piston, beyond which the gas velocity decreases from $-2c_0/(\gamma-1)$ to zero according to formula (1).

If the piston moves into the pipe ($U = at$), a simple compression wave is formed; the corresponding solution is obtained by merely changing the sign of a in (1) (Fig. 81b). It is valid, however, only until a shock wave is formed;

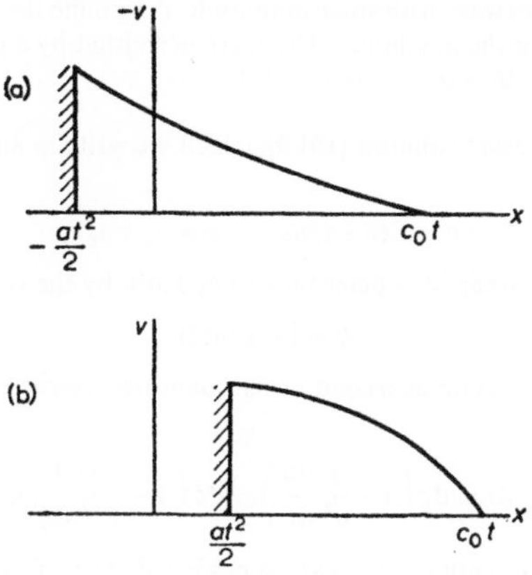

FIG. 81

the time when this happens is determined from formula (101.15), and is

$$t = 2c_0/a(\gamma+1).$$

PROBLEM 2. The same as Problem 1, but for the case where the piston moves in any manner.

SOLUTION. Let the piston begin to move at time $t = 0$ according to the law $x = X(t)$ (with $X(0) = 0$); its velocity is $U = X'(t)$. The boundary condition on the piston ($v = U$ for $x = X$) gives $v = X'(t)$, $f(v) = X(t) - t[c_0 + \frac{1}{2}(\gamma+1)X'(t)]$. If we now regard t as a parameter, these two equations determine the function $f(v)$ in parametric form. Denoting the parameter by τ, we can write the solution as

$$v = X'(\tau), \quad x = X(\tau) + (t-\tau)[c_0 + \tfrac{1}{2}(\gamma+1)X'(\tau)], \tag{2}$$

which determines, in parametric form, the requied function $v(t, x)$ in the simple wave which is caused by the motion of the piston.

PROBLEM 3. Determine the time and place of formation of the shock wave when the piston (Problem 1) moves according to the law $U = at^n$ ($n > 0$).

SOLUTION. If $a < 0$, i.e. the piston moves out of the pipe, a simple rarefaction wave results, in which no shock wave is formed. We therefore assume that $a > 0$, i.e. the piston moves into the pipe, causing a simple compression wave.

When the function $v(x, t)$ is given by the parametric formulae (2), and $X = a\tau^{n+1}/(n+1)$, the time and place of formation of the shock wave are given by the equations

$$\left.\begin{aligned} \left(\frac{\partial x}{\partial \tau}\right)_t &= -c_0 + \tfrac{1}{2}t\tau^{n-1}an(\gamma+1) - \tfrac{1}{2}a\tau^n[\gamma - 1 + n(\gamma+1)] = 0, \\ \left(\frac{\partial^2 x}{\partial \tau^2}\right)_t &= \tfrac{1}{2}t\tau^{n-2}an(n-1)(\gamma+1) - \tfrac{1}{2}an\tau^{n-1}[\gamma - 1 + n(\gamma+1)] = 0, \end{aligned}\right\} \tag{3}$$

where the second equation must be replaced by $\tau = 0$ if we are concerned with the formation of a shock wave at the forward front of the simple wave.

For $n = 1$ we find $\tau = 0$, $t = 2c_0/a(\gamma+1)$, i.e. the shock wave is formed at the forward front at a finite time after the motion begins, in accordance with the results of Problem 1.

For $n < 1$, the derivative $\partial x/\partial \tau$ is of variable sign (and therefore the function $v(x)$ for given t is many-valued) for any $t > 0$. This means that a shock wave is formed at the piston as soon as it begins to move.

For $n > 1$ the shock wave is formed, not at the forward front of the simple wave, but at some intermediate point given by (3). Having determined τ and t from (3), we can then find the place of formation of the discontinuity from (2). The result is

$$t = \left(\frac{2c_0}{a}\right)^{1/n} \frac{1}{\gamma+1}\left[\frac{n+1}{n-1}\gamma + 1\right]^{(n-1)/n},$$

$$x = 2c_0\left(\frac{2c_0}{a}\right)^{1/n}\left[\frac{\gamma}{\gamma+1} + \frac{n-1}{n+1}\right]\frac{1}{(n-1)^{(n-1)/n}[\gamma - 1 + n(\gamma+1)]^{1/n}}.$$

PROBLEM 4. For a plane (sound) wave with small amplitude, determine the time-averaged values of quantities in the approximation quadratic in the amplitude. The wave is emitted by a piston moving in accordance with some law $z = X(t)$, $U = X'(t)$, $X(0) = 0$, $\overline{X} = 0$, $\overline{U} = 0$.†

SOLUTION. We start from the exact solution (101.9), which we write in an equivalent form with a different choice of argument:

$$v = F(t - x/u), \qquad u = c_0 + \alpha_0 v, \tag{4}$$

where $\alpha_0 = \frac{1}{2}(\gamma+1)$, or $v = F(\xi)$, where ξ is determined implicitly by the equation‡

$$\xi = t - x/u(\xi). \tag{5}$$

We shall show that, in a calculation as far as second-order quantities, averaging over t is equivalent to averaging over ξ. For a given x,

$$dt = d\xi\left(1 - \frac{x}{u^2}\frac{du}{d\xi}\right) \cong d\xi\left(1 - \frac{x\alpha_0}{c_0{}^2}\frac{dv}{d\xi}\right);$$

in the denominator u^2, the small quantity $v \ll c_0$ can be neglected. The effect sought is due to cumulative non-linear distortions of the profile, and is found by solving (4) for v. Hence

$$\int_{t_1}^{t_2} v\,dt = \int_{\xi_1}^{\xi_2} \left\{F - \frac{x\alpha_0}{c_0{}^2}F\frac{dF}{d\xi}\right\}d\xi$$

$$= \int_{\xi_1}^{\xi_2} F\,d\xi - \frac{x\alpha_0}{c_0{}^2}[F^2(\xi_2) - F^2(\xi_1)],$$

The second term is always finite and makes no contribution when averaged over a long interval of time. Since also

$$\xi_2 - \xi_1 \cong t_2 - t_1 + (\alpha_0 x/c_0{}^2)(v_2 - v_1)$$

$$\cong t_2 - t_1,$$

we reach the result that $\bar{v}^t = \bar{v}^\xi$, where the index beside the bar shows the variable over which the averaging is done (and will be omitted henceforward); the average over t is therefore independent of x.

For the problem with an oscillating piston, $F(\xi)$ is determined by equation (2), which may be written as

$$v(\tau) = X'(\tau), \quad \tau = \xi + X(\tau)/u(\tau)$$

or, since the oscillation amplitude is small,

$$\tau \cong \xi + (1/c_0)X(\xi), \quad v(\tau) \cong U(\xi) + (1/c_0)X(\xi)dU(\xi)/d\xi.$$

Averaging the last expression gives

$$\bar{v} = (1/c_0)\overline{X\,dU/d\xi} = (1/c_0)\overline{d(XU)/d\xi} - (1/c_0)\overline{U^2}$$

and, since the mean value of a total derivative is zero,

$$\bar{v} = -\overline{U^2}/c_0. \tag{6}$$

To the same accuracy, the time-averaged mass flux density is

$$\overline{\rho v} = \rho_0\bar{v} + \overline{\rho' v} = \rho_0\bar{v} + \rho_0\overline{v^2}/c_0.$$

Using (6) and the relation (in the same approximation) $\overline{v^2} = \overline{U^2}$, we find that $\overline{\rho v} = 0$, as it should be, according to the conservation of mass, in the purely one-dimensional case, where no mass is transferred "sideways". The mean energy flux density is

$$\bar{q} = \overline{\rho w v} = w_0\overline{\rho v} + \rho_0\overline{w' v} = \overline{p' v} = \rho_0 c_0\overline{v^2}$$

(cf. §65); thus $\bar{q} = \rho_0 c_0 \overline{U^2}$.

To calculate $\overline{p'}$ and $\overline{\rho'}$, we have to express p' and ρ' in terms of v as far as v^2 terms. From (101.7), or from (101.4),

† The solution follows that by L. A. Ostrovskiĭ (1968).

‡ For waves with small amplitude, (4) is valid for any (not necessarily polytropic) gas if α_0 is defined by (102.2).

and (101.6) for a non-polytropic gas, we have

$$\rho'/\rho_0 = v/c_0 + (2-\alpha)v^2/2c_0^2, \qquad p' = c^2\rho' + (\alpha-1)\rho_0 v^2,$$

and on averaging†

$$\overline{\rho'} = -\tfrac{1}{2}\alpha\rho_0\overline{U^2}/c_0^2, \qquad \overline{p'} = -\tfrac{1}{2}(2-\alpha)\rho_0\overline{U^2}. \tag{7}$$

Note that $\overline{p'}$ here is not zero even in the quadratic approximation; cf. the end of §65.

§102. Formation of discontinuities in a sound wave

A travelling plane sound wave, being an exact solution of the equations of motion, is also a simple wave. We can use the general results obtained in §101 to derive some properties of small-amplitude sound waves in the second approximation (the first approximation being that which gives the ordinary linear wave equation).

We must notice first of all that a discontinuity must ultimately appear in each wavelength of a sound wave. This leads to a very marked damping of the wave, as shown in §101. It must be remarked, however, that this happens only for a sufficiently strong sound wave; a weak sound wave is damped by the usual effects of viscosity and thermal conduction before the effects of higher order in the amplitude can develop.

The distortion of the wave profile has another effect also. If the wave is purely harmonic at some instant, it ceases to be so at later instants, on account of the change in shape of the profile. The motion, however, remains periodic, with the same period as before. When the wave is expanded in a Fourier series, terms with frequencies $n\omega$ (n being integral and ω being the fundamental frequency) appear, as well as that with frequency ω. Thus the distortion of the profile as the sound wave is propagated may be regarded as the appearance in it of higher harmonics in addition to the fundamental frequency.

The velocity u of points in the wave profile (the wave being propagated in the positive x-direction) is obtained, in the first approximation, by putting in (101.11) $v = 0$, i.e. $u = c_0$, corresponding to the propagation of the wave with no change in its profile. In the next approximation we have

$$u = c_0 + \rho'\partial u/\partial\rho_0 = c_0 + (\partial u/\partial\rho_0)\rho_0 v/c_0,$$

or, using the expression (99.10) for the derivative $\partial u/\partial\rho$,

$$u = c_0 + \alpha_0 v, \tag{102.1}$$

where we have put for brevity‡

$$\alpha = (c^4/2V^3)(\partial^2 V/\partial p^2)_s. \tag{102.2}$$

For a polytropic gas, $\alpha = \frac{1}{2}(\gamma+1)$, and formula (102.1) agrees with the exact formula (see (101.8)) for the velocity u.

In the general case of arbitrary amplitude, the wave is no longer simple after the discontinuities have appeared. A small-amplitude wave, however, is still simple in the second approximation even when discontinuities are present. This can be seen as follows. The changes in velocity, pressure and specific volume in a shock wave are related by $v_1 - v_2 = \sqrt{[(p_2-p_1)(V_1-V_2)]}$. The change in the velocity v over a segment of the x-

† These formulae were derived, with more restrictive assumptions, by A. Eichenwald (1932).

‡ In §93, Problem 1, this quantity was denoted by α_p.

axis in a simple wave is

$$v_2 - v_1 = \int_{p_1}^{p_2} \sqrt{(-\partial V/\partial p)}\,\mathrm{d}p.$$

A simple calculation, using an expansion in series, shows that these two expressions differ only by terms of the third order (it must be borne in mind that the change in entropy at a discontinuity is of the third order of smallness, while in a simple wave the entropy is constant). Hence it follows that, as far as terms of the second order, a sound wave on either side of a discontinuity in it remains simple, and the appropriate boundary condition is satisfied at the discontinuity itself. In higher approximations this is no longer true, on account of the appearance of waves reflected from the surface of discontinuity.

Let us now derive the condition which determines the location of the discontinuities in a travelling sound wave (again in the second approximation). Let u be the velocity of the discontinuity relative to a fixed coordinate system, and v_1, v_2 the velocities of the gases on each side of it. Then the condition that the mass flux be continuous is $\rho_1(v_1 - u) = \rho_2(v_2 - u)$, whence $u = (\rho_1 v_1 - \rho_2 v_2)/(\rho_1 - \rho_2)$. As far as the second-order terms, this is equal to the derivative $\mathrm{d}(\rho v)/\mathrm{d}\rho$ at the point where v is equal to $\frac{1}{2}(v_1 + v_2)$. Since, in a simple wave, $\mathrm{d}(\rho v)/\mathrm{d}\rho = v + c$, we have, by (102.1),

$$u = c_0 + \tfrac{1}{2}\alpha_0(v_1 + v_2). \tag{102.3}$$

From this we can obtain the following simple geometrical condition which determines the position of the shock wave. In Fig. 82 the curve shows the velocity profile corresponding to the simple wave; let *ae* be the discontinuity, and x_s its position. The difference of the shaded areas *abc* and *cde* is the integral

$$\int_{v_1}^{v_2} (x - x_s)\,\mathrm{d}v$$

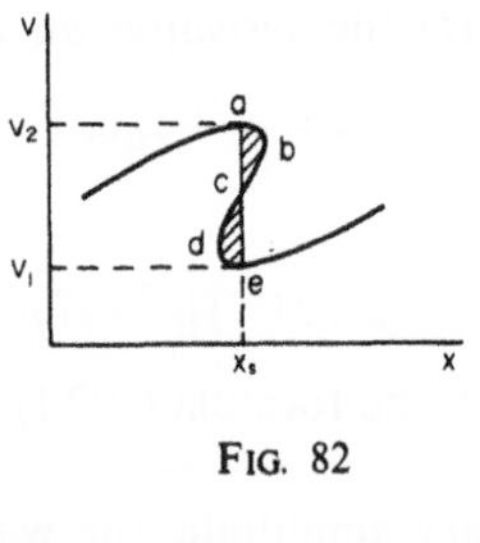

FIG. 82

taken along the curve *abcde*. In the course of time, the wave profile moves; let us calculate the time derivative of the above integral. Since the velocity $\mathrm{d}x/\mathrm{d}t$ of points in the wave profile is given by formula (102.1), and the velocity $\mathrm{d}x_s/\mathrm{d}t$ of the discontinuity by (102.3), we have

$$\frac{\mathrm{d}}{\mathrm{d}t}\int_{v_1}^{v_2} (x - x_s)\,\mathrm{d}v = \alpha\left\{\int_{v_1}^{v_2} v\,\mathrm{d}v - \tfrac{1}{2}(v_1 + v_2)\int_{v_1}^{v_2} \mathrm{d}v\right\} = 0;$$

in differentiating the integral, we must notice that, although the limits of integration v_1 and v_2 also vary with time, $x - x_s$ always vanishes at the limits, and so we need only differentiate the integrand.

Thus the integral $\int (x - x_s)\mathrm{d}v$ remains constant in time. Since it is zero at the instant when the shock wave is formed (the points *a* and *e* then coinciding), it follows that we always have

$$\int\limits_{abcde} (x - x_s)\mathrm{d}v = 0. \tag{102.4}$$

Geometrically this means that the areas *abc* and *cde* are equal, a condition which determines the position of the discontinuity.

The formation of discontinuities in a sound wave is an example of the spontaneous occurrence of shock waves in the absence of any singularity in the external conditions of the flow. It must be emphasized that, although a shock wave can appear spontaneously at a particular instant, it cannot disappear in the same manner. Once formed, a shock wave decays only asymptotically as the time becomes infinite.

Let us consider a single one-dimensional compression pulse, in which a shock wave has already been formed, and ascertain how this shock will finally be damped. In the later stages of its propagation, a sound pulse containing a shock wave will have a triangular velocity profile, the linear profile remaining linear as it changes shape.†

Let the profile be given at some instant (which we take as $t = 0$) by the triangle *ABC* in Fig. 83a; the values of quantities at this instant are denoted by the suffix 1.‡ If the points in this profile moved with the velocities (102.1), we should obtain after time *t* a profile *A′B′C′* (Fig. 83b). In reality, the discontinuity moves to *E*, and the actual profile will be *A′DE*. The

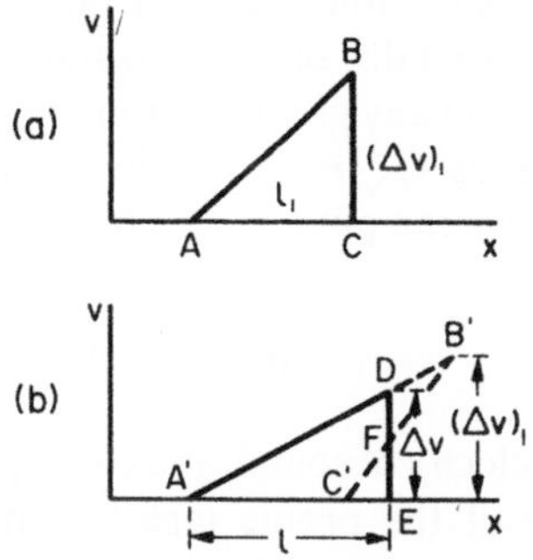

FIG. 83

† Here and later, a distribution profile of the velocity v is mentioned, simply in order not to complicate the formulae. A quantity having greater practical interest is the excess pressure p', which differs from v only by a constant factor, $p' = v/\rho_0 c_0$; similar results are valid for it. The sign of v is the same as that of p', so that $v > 0$ and $v < 0$ correspond to compression and rarefaction respectively. The rate of movement of points in the profile is expressed in terms of p' by

$$u = c_0(1 + \nu_0 p'/p_0), \quad \nu = \alpha p/\rho c^2;$$

for a polytropic gas, $\nu = \frac{1}{2}(\gamma + 1)/\gamma$.

‡ The suffix 0 denoting the equilibrium values will be omitted.

areas $DB'F$ and $C'FE$ are equal, by (102.4), and therefore the area $A'DE$ of the new profile is equal to the area ABC of the original profile. Let l be the length of the sound pulse at time t, and Δv the velocity discontinuity in the shock wave. During a time t, the point B moves a distance $\alpha t(\Delta v)_1$ relative to C; the tangent of the angle $B'AC'$ is therefore $(\Delta v)_1/[l_1 + \alpha t(\Delta v)_1]$, and we obtain the condition of equal areas ABC and $A'DE$ in the form

$$l_1(\Delta v)_1 = l^2(\Delta v)_1/[l_1 + \alpha t(\Delta v)_1)],$$

whence

$$\left.\begin{aligned} l &= l_1\sqrt{[1+\alpha(\Delta v)_1 t/l_1]}, \\ \Delta v &= (\Delta v)_1/\sqrt{[1+\alpha(\Delta v)_1 t/l_1]}. \end{aligned}\right\} \tag{102.5}$$

The total energy of a travelling sound pulse (per unit area of its front) is

$$E = \rho\int v^2\,\mathrm{d}x = E_1/\sqrt{[1+\alpha(\Delta v)_1 t/l]}. \tag{102.6}$$

For $t\to\infty$ the strength of the shock wave and its energy decrease asymptotically as $1/\sqrt{t}$ (or, equivalently, as $1/\sqrt{x}$ with the distance $x = ct$). The pulse length increases as $\sqrt{t}$. Note also that the limiting slope of the profile $\Delta v/l \to 1/\alpha t$ is independent of the shock strength and of the pulse length.

Let us now consider the limiting properties (at large distances from the source) of shock waves formed in cylindrical and spherical sound waves (L. D. Landau 1945). We take first the cylindrical case.

At sufficiently large distances r from the axis, any small section of such a wave may be regarded as plane. The velocity of any point in the wave profile is then given by formula (102.1). If, however, we wish to use this formula to follow the motion of any point in the wave profile over long intervals of time, we must take into account the fact that the amplitude of a cylindrical wave falls off with distance as $1/\sqrt{r}$, even in the first approximation. This means that, at any given point in the profile, v is not constant, as it is for a plane wave, but decreases as $1/\sqrt{r}$. If v_1 is the value of v (for a given point in the profile) at a (large) distance r_1, we can put $v = v_1\sqrt{(r_1/r)}$. Thus the velocity u of points in the wave profile is

$$u = c + \alpha v_1\sqrt{(r_1/r)}. \tag{102.7}$$

The first term is the ordinary velocity of sound, and corresponds to movement of the wave without change in the shape of the profile (apart from the general decrease of the amplitude as $1/\sqrt{r}$, that is, taking as the profile the distribution of $v\sqrt{r}$). The second term results in a distortion of the profile. The amount δr of additional movement of points in the profile during a time $(r-r_1)/c$ is found by integrating over $\mathrm{d}r/c$:

$$\delta r = 2\alpha(v_1/c)\sqrt{r_1}(\sqrt{r}-\sqrt{r_1}). \tag{102.8}$$

The distortion of the profile of a cylindrical wave increases more slowly than for a plane wave, where δx is proportional to the distance x traversed by the wave, but here too it does of course lead ultimately to the formation of discontinuities. Let us consider shock waves formed in a single cylindrical sound pulse which has reached a large distance from the source (the axis).

The cylindrical case is distinguished from the plane case primarily by the fact that a single pulse cannot consist of compression only or rarefaction only; if the sound wave front is followed by a region of compression, this in turn must be followed by a region of rarefaction (see §71).† The point of maximum rarefaction will lag behind all those to the rear of it, and the profile therefore turns over to form a discontinuity. Thus, in a cylindrical sound pulse, two shock waves are formed. In the leading one, the velocity increases abruptly from zero; then follows a region in which the compression gradually decreases into a rarefaction, after which the pressure again increases discontinuously in the second shock. A cylindrical sound pulse is, however, distinctive (in comparison with the plane and spherical cases) also in that it cannot have a backward front; v tends to zero only asymptotically. This has the result that in the rear discontinuity v increases not to zero but only to some negative non-zero value, afterwards tending asymptotically to zero. This leads to a profile of the kind shown in Fig. 84.

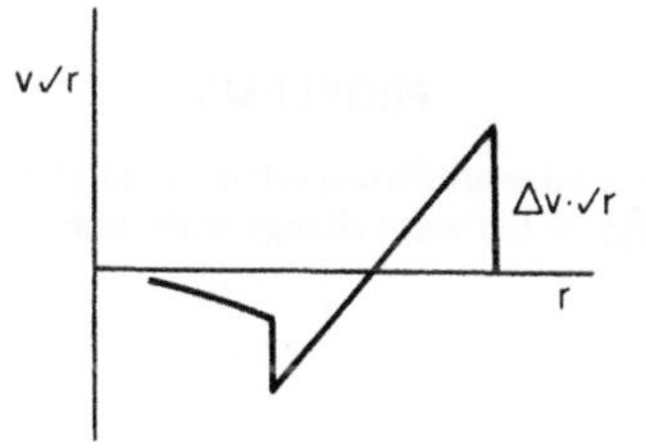

FIG. 84

The manner of the final damping of the shock waves with time (or, equivalently, with the distance r from the axis) can be found in the same way as for the plane case discussed above. It is seen from the previous result that the limiting form corresponds to the time when the displacement δr at the top of the profile becomes large in comparison with the "original" pulse width l_1 (by which is meant, for example, the distance from the leading shock wave to the point where $v = 0$). This displacement on the path from r_1 to $r \ll r_1$ is

$$\delta r \cong (2\alpha/c)(\Delta v)_1 \sqrt{(r_1 r)},$$

where $(\Delta v)_1$ is the "original" discontinuity (at distance r_1) on the leading shock. The "final" slope of the linear part of the profile between the shock waves is then $\cong \sqrt{r_1}(\Delta v)_1/\delta r \cong c/2\alpha\sqrt{r}$. The condition of constant area of the profile gives $l_1\sqrt{r_1}(\Delta v)_1 = l^2 c/\alpha\sqrt{r}$, whence $l \propto r^{1/4}$, instead of $l \propto x^{1/2}$ in the plane case. The limiting decrease of Δv in the leading shock is then given by $l\sqrt{r}\Delta v = $ constant, i.e.

$$\Delta v \propto r^{-3/4} \tag{102.9}$$

Lastly, let us consider the spherical case.‡ The general decrease in amplitude of the outgoing sound wave takes place as $1/r$, where r is now the distance from the centre Repeating the arguments given above for the cylindrical case, we find as the velocity of points in the wave profile

$$u = c + \alpha v_1 r_1/r, \tag{102.10}$$

† This type of configuration will be the one considered. It pertains, in particular, to the application of the results to shock waves formed in supersonic motion of a body with finite size (§122).

‡ For example, a shock wave formed in an explosion and considered at large distances from the source.

and hence the displacement δr of points in the profile on the path from r_1 to r:

$$\delta r = (\alpha v_1 r_1/c)\log(r/r_1). \tag{102.11}$$

We see that the profile distortion in a spherical wave increases with distance only logarithmically, much more slowly than in the plane case or even the cylindrical one.

Spherical propagation of a compression sound wave must be accompanied, as in the cylindrical case, by a following rarefaction (see §70). Here also, two discontinuities must be formed (but a single spherical pulse can have a backward front, in which case v increases discontinuously to zero).† By the same method as before, we find the limiting relations for the increase in the pulse length and the decrease in the strength of the shock wave:

$$l \propto \sqrt{\log(r/a)}, \quad \Delta v \propto 1/r\sqrt{\log(r/a)}, \tag{102.12}$$

where a is constant having the dimensions of length.‡

PROBLEMS

Problem 1. At the initial instant, the wave profile consists of an infinite series of "teeth", as shown in Fig. 85.†† Determine how the profile and energy of the wave change with time.

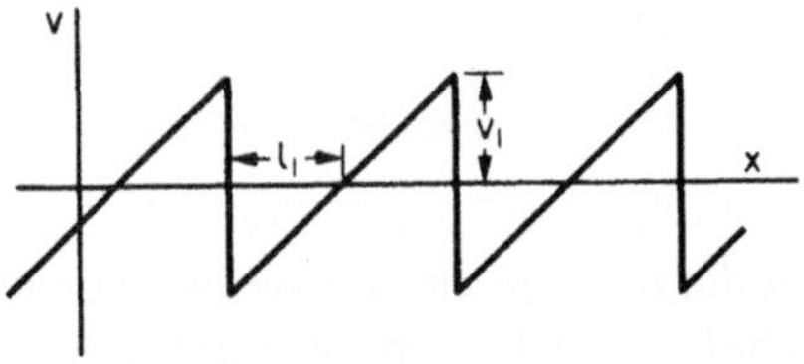

Fig. 85

Solution. It is evident that, at subsequent instants, the wave profile will have the same form, with l_0 unchanged but the height v_t less then v_1. Let us consider one "tooth": at time $t = 0$, the ordinate through the point where $v = v_t$ cuts off a part $v_t l_1/v_1$ of the base of the triangle. During a time t, this point moves forward a distance $\alpha v_t t$. The condition that the base of the triangle be unchanged in length is $v_t l_1/v_1 + \alpha t v_t = l_1$, whence $v_t = v_1/(1+\alpha v_1 t/l_1)$. As $t \to \infty$, the wave amplitude diminishes as $1/t$. The energy is $E = E_0/(1+\alpha v_1 t/l_1)^2$, i.e. it diminishes as $1/t^2$ for $t \to \infty$.

Problem 2. Determine the intensity of the second harmonic formed by the distortion of the profile of a monochromatic spherical wave.

Solution. Writing the wave in the form $rv = A\cos(kr - \omega t)$, we can allow for the distortion, in the first approximation, by adding δr to r on the right-hand side of this equation, and expanding in powers of δr. This gives, by (102.11),

$$rv = A\cos(kr-\omega t) - (\alpha k/2c)\,A^2\log(r/r_1)\sin 2(kr-\omega t);$$

† Since in practice a gas always exhibits ordinary sound absorption due to thermal conduction and viscosity, the slowness of the distortion in a spherical wave may have the result that it is absorbed before discontinuities can be formed.

‡ This constant is not in general equal to r_1. The reason is that the argument of the logarithm has to be dimensionless, and therefore, when $r \gg r_1$, we cannot simply neglect $\log r_1$ in (102.11). The determination of the coefficient of r in the large logarithm requires a more exact allowance for the original form of the profile.

†† This is the asymptotic form of the profile for any periodic wave.

here r_1 must be taken as a distance at which the wave can still be regarded, with sufficient accuracy, as strictly monochromatic. The second term in this formula is the second harmonic in the spectral resolution of the wave. Its total (time-averaged) intensity I_2 is

$$I_2 = (\alpha^2 k^2/8\pi c^3 \rho_0)\log^2(r/r_1) I_1^2,$$

where $I_1 = 2\pi c\rho A^2$ is the intensity of the first harmonic.

§103. Characteristics

The definition of characteristics, given in §82, as lines along which small disturbances are propagated (in the approximation of geometrical acoustics) has general validity, and is not restricted to the plane steady supersonic flow discussed in §82.

For one-dimensional non-steady flow, we can introduce the characteristics as lines in the xt-plane whose slope dx/dt is equal to the velocity of propagation of small disturbances relative to a fixed coordinate system. Disturbances propagated relative to the gas with the velocity of sound, in the positive or negative x-direction, move relative to the fixed coordinate system with velocity $v \pm c$. The differential equations of the two families of characteristics, which we shall call C_+ and C_-, are accordingly

$$(dx/dt)_+ = v + c, \qquad (dx/dt)_- = v - c. \tag{103.1}$$

Disturbances transmitted with the gas are propagated in the xt-plane along characteristics belonging to a third family C_0, for which

$$(dx/dt)_0 = v. \tag{103.2}$$

These are just the "streamlines" in the xt-plane; cf. the end of §82.† It should be emphasized that, for characteristics to exist, it is no longer necessary for the gas flow to be supersonic. The "directional" propagation of disturbances, as evidenced by the characteristics, is here simply due to the causal relation between the motions at successive instants.

As an example, let us consider the characteristics of a simple wave. For a wave propagated in the positive x-direction we have, by (101.5), $x = t(v+c) + f(v)$. Differentiating this relation, we have

$$dx = (v+c)\, dt + [t + tc'(v) + f'(v)]\, dv.$$

Along a characteristic C_+, we have $dx = (v+c)\, dt$; comparing the two equations, we find that along such a characteristic $[t + tc'(v) + f'(v)]\, dv = 0$. The expression in brackets cannot vanish identically, and therefore $dv = 0$, i.e. v = constant. Thus we conclude that, along any characteristic C_+, the velocity is constant, and therefore so are all other quantities. The same property holds for the characteristics C_- in a wave propagated to the left. We shall see in §104 that this is no accident, but is a mathematical consequence of the nature of simple waves.

From this property of the characteristics C_+ for a simple wave, we can in turn conclude that they are a family of straight lines in the xt-plane; the velocity is constant along the lines $x = t[v + c(v)] + f(v)$ (101.5). In particular, for a similarity rarefaction wave (a simple wave with $f(v) = 0$), these lines form a pencil through the origin in the xt-plane. For this reason, a similarity simple wave is said to be *centred.*

Figure 86 shows the family of characteristics C_+ for the simple rarefaction wave formed when a piston moves out of a pipe with acceleration. It is a family of diverging straight

† The same equations (103.1) and (103.2) determine the characteristics for non-steady spherically symmetrical flow, if x is replaced by the radial coordinate r (the characteristics now being lines in the rt-plane).

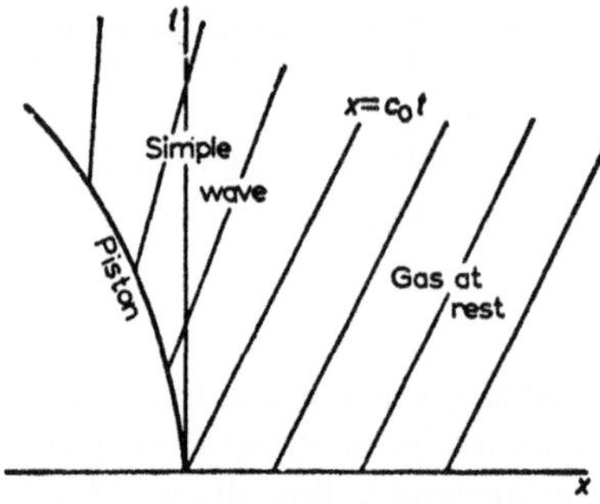

FIG. 86

lines, which begin from the curve $x = X(t)$ giving the motion of the piston. To the right of the characteristic $x = c_0 t$ lies a region of gas at rest, where the characteristics become parallel.

Figure 87 is a similar diagram for the simple compression wave formed when a piston moves into a pipe with acceleration. In this case the characteristics are converging straight lines, which eventually intersect. Since every characteristic has a constant value of v, their intersection shows that the function $v(x, t)$ is many-valued, which is physically meaningless. This is the geometrical interpretation of the result obtained in §101: a simple compression wave cannot exist indefinitely, and a shock wave must be formed in it. The geometrical interpretation of the conditions (101.12), which determine the time and place of formation of the shock wave, is as follows. The intersecting family of rectilinear characteristics has an envelope, which, for a certain least value of t, has a cusp; this gives the instant at which many-valuedness first occurs. If the equations of the characteristics are given in the parametric form $x = x(v)$, $t = t(v)$, the position of the cusp is given by equations (101.12).†

We shall now indicate briefly how the physical definition, given above, of the characteristics as lines along which disturbances are propagated corresponds to the

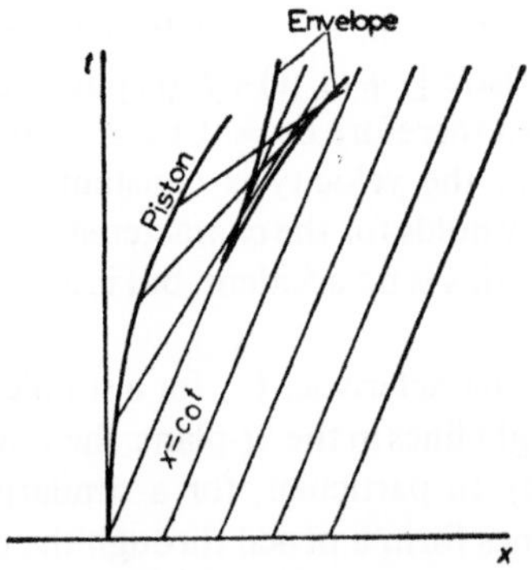

FIG. 87

† The whole of the region between the two branches of the envelope is occupied by three sets of characteristics, in accordance with the three-valuedness caused by the turn-over of the wave profile. The particular case where the shock wave occurs at the boundary of the gas at rest corresponds to that where one branch of the envelope is part of the characteristic $x = c_0 t$.

mathematical sense of the word in the theory of partial differential equations. Let us consider a partial differential equation having the form

$$A\frac{\partial^2 \phi}{\partial x^2}+2B\frac{\partial^2 \phi}{\partial x\,\partial t}+C\frac{\partial^2 \phi}{\partial t^2}+D=0, \tag{103.3}$$

which is linear in the second derivatives; the coefficients A, B, C, D can be any functions, both of the independent variables x, t and of the unknown function ϕ and its first derivatives.† Equation (103.3) is of the elliptic type if $B^2-AC<0$ everywhere, and of the hyperbolic type if $B^2-AC>0$. In the latter case, the equation

$$A\,\mathrm{d}t^2-2B\,\mathrm{d}x\,\mathrm{d}t+C\,\mathrm{d}x^2=0, \tag{103.4}$$

or

$$\mathrm{d}x/\mathrm{d}t=[B\pm\sqrt{(B^2-AC)}]/C, \tag{103.5}$$

determines two families of curves in the xt-plane, the *characteristics* (for a given solution $\phi(x,t)$ of equation (103.3)). We may point out that, if the coefficients A, B, C are functions only of x and t, then the characteristics are independent of the particular solution ϕ.

Let a given flow correspond to some solution $\phi=\phi_0(x,t)$ of equation (103.3), and let a small perturbation ϕ_1 be applied to it. We assume that this perturbation satisfies the conditions for geometrical acoustics to be valid: it does not greatly affect the flow (ϕ_1 and its first derivatives are small), but varies considerably over short distances (the second derivatives of ϕ_1 are relatively large). Putting in equation (103.3) $\phi=\phi_0+\phi_1$, we then obtain for ϕ_1 the equation

$$A\frac{\partial^2 \phi_1}{\partial x^2}+2B\frac{\partial^2 \phi_1}{\partial x\,\partial t}+C\frac{\partial^2 \phi_1}{\partial t^2}=0,$$

with $\phi=\phi_0$ in the coefficients A, B, C. Following the method used in changing from wave optics to geometrical optics, we write $\phi_1=ae^{i\psi}$, where the function ψ (the *eikonal*) is large, and obtain for ψ the equation

$$A\left(\frac{\partial \psi}{\partial x}\right)^2+2B\frac{\partial \psi}{\partial x}\frac{\partial \psi}{\partial t}+C\left(\frac{\partial \psi}{\partial t}\right)^2=0. \tag{103.6}$$

The equation of ray propagation in geometrical acoustics is obtained by equating $\mathrm{d}x/\mathrm{d}t$ to the group velocity: $\mathrm{d}x/\mathrm{d}t=\mathrm{d}\omega/\mathrm{d}k$, where $k=\partial\psi/\partial x$, $\omega=-\partial\psi/\partial t$. Differentiating the relation $Ak^2-2Bk\omega+C\omega^2=0$, we obtain $\mathrm{d}x/\mathrm{d}t=(B\omega-Ak)/(C\omega-Bk)$, and, eliminating k/ω by the same relation, we again arrive at equation (103.5).

PROBLEM

Find the equation of the second family of characteristics in a centred simple wave in a polytropic gas.

SOLUTION. In a centred simple wave propagated into gas at rest to the right of it, we have $x/t=v+c=c_0+\frac{1}{2}(\gamma+1)v$. The characteristics C_+ form the pencil $x=\text{constant}\times t$. The characteristics C_-, on the other hand, are determined by the equation

$$\frac{\mathrm{d}x}{\mathrm{d}t}=v-c=\frac{3-\gamma}{\gamma+1}\frac{x}{t}-\frac{4}{\gamma+1}c_0.$$

† The velocity potential satisfies an equation of this form in one-dimensional non-steady flow.

integrating, we find

$$= -\frac{2}{\gamma-1}c_0 t+\frac{\gamma+1}{\gamma-1}c_0 t_0\left(\frac{t}{t_0}\right)^{(3-\gamma)/(\gamma+1)}$$

where the constant of integration has been chosen so that the characteristic C_- passes through the point $x = c_0 t_0, t = t_0$ on the characteristic C_+ ($x = c_0 t$) which is the boundary between the simple wave and the region at rest.

The "streamlines" in the xt-plane are given by the equation

$$\frac{dx}{dt} = v = \frac{2}{\gamma+1}\left(\frac{x}{t}-c_0\right),$$

whence, for the characteristic C_0,

$$x = -\frac{2}{\gamma-1}c_0 t+\frac{\gamma+1}{\gamma-1}c_0 t_0\left(\frac{t}{t_0}\right)^{2/(\gamma+1)}$$

§104. Riemann invariants

An arbitrary small disturbance is in general propagated along all three characteristics (C_+, C_-, C_0) leaving a given point in the xt-plane. However, an arbitrary disturbance can be separated into parts each of which is propagated along only one characteristic.

Let us first consider isentropic gas flow. We write the equation of continuity and Euler's equation in the form

$$\frac{\partial p}{\partial t}+v\frac{\partial p}{\partial x}+\rho c^2\frac{\partial v}{\partial x} = 0,$$

$$\frac{\partial v}{\partial t}+v\frac{\partial v}{\partial x}+\frac{1}{\rho}\frac{\partial p}{\partial x} = 0;$$

in the equation of continuity we have replaced the derivatives of the density by those of the pressure, using the formulae

$$\frac{\partial \rho}{\partial t} = \left(\frac{\partial \rho}{\partial p}\right)_s\frac{\partial p}{\partial t} = \frac{1}{c^2}\frac{\partial p}{\partial t}, \qquad \frac{\partial \rho}{\partial x} = \frac{1}{c^2}\frac{\partial p}{\partial x}.$$

Dividing the first equation by $\pm\rho c$ and adding it to the second, we obtain

$$\frac{\partial v}{\partial t}\pm\frac{1}{\rho c}\frac{\partial p}{\partial t}+\left(\frac{\partial v}{\partial x}\pm\frac{1}{\rho c}\frac{\partial p}{\partial x}\right)(v\pm c) = 0. \tag{104.1}$$

We now introduce as new unknown functions

$$J_+ = v+\int dp/\rho c, \qquad J_- = v-\int dp/\rho c, \tag{104.2}$$

which are called *Riemann invariants*. It should be remembered that, in isentropic flow, ρ and c are definite functions of p, and the integrals on the right-hand sides are therefore definite functions. For a polytropic gas

$$J_+ = v+2c/(\gamma-1), \qquad J_- = v-2c/(\gamma-1). \tag{104.3}$$

In terms of these quantities, the equations of motion take the simple form

$$\left[\frac{\partial}{\partial t}+(v+c)\frac{\partial}{\partial x}\right]J_+ = 0, \qquad \left[\frac{\partial}{\partial t}+(v-c)\frac{\partial}{\partial x}\right]J_- = 0. \tag{104.4}$$

The differential operators acting on J_+ and J_- are just the operators of differentiation along the characteristics C_+ and C_- in the xt-plane. Thus we see that J_+ and J_- remain constant along each characteristic C_+ or C_- respectively. We can also say that small perturbations of J_+ are propagated only along the characteristics C_+, and those of J_- only along C_-.

In the general case of anisentropic flow, the equations (104.1) cannot be written in the form (104.4), since $\mathrm{d}p/\rho c$ is not a perfect differential. These equations, however, still permit the separation of perturbations propagated along characteristics of only one family. Such perturbations are those of the form $\delta v \pm \delta p/\rho c$, where δv and δp are arbitrary small perturbations of the velocity and pressure. Their propagation is described by the linearized equations

$$\left[\frac{\partial}{\partial t} + (v \pm c)\frac{\partial}{\partial x}\right]\left(\delta v \pm \frac{\delta p}{\rho c}\right) = 0. \tag{104.5}$$

In order to obtain a complete system of equations of motion, these must be supplemented by the adiabatic equation

$$\left[\frac{\partial}{\partial t} + v\frac{\partial}{\partial x}\right]\delta s = 0, \tag{104.6}$$

which shows that perturbations δs are propagated along the characteristics C_0. An arbitrary small perturbation can always be separated into independent parts of the three kinds mentioned.

A comparison with formula (101.4) shows that the Riemann invariants (104.2) are the quantities which, in simple waves, are constant throughout the region of the flow at all times: J_- is constant in a simple wave propagated to the right, and J_+ in one travelling to the left. Mathematically, this is the fundamental property of simple waves, from which follows, in particular, the property mentioned in §103: one family of characteristics consists of straight lines. For example, let the wave be propagated to the right. Each characteristic C_+ has a constant value of J_+ and, furthermore, a constant value of J_-, which value is the same everywhere. Since both J_+ and J_- are constant, it follows that v and p are constant (and therefore so are all the other quantities), and we obtain the property of the characteristics C_+ deduced in §103, which in turn shows that they are straight lines.

If the flow in two adjoining regions of the xt-plane is described by two analytically different solutions of the equations of motion, then the boundary between the regions is a characteristic. For this boundary is a discontinuity in the derivatives of some quantity, i.e. it is a weak discontinuity, and therefore must necessarily coincide with some characteristic.

The following property of simple waves is of great importance in the theory of one-dimensional isentropic flow. The flow in a region adjoining a region of constant flow (in which v = constant, p = constant) must be a simple wave.

This statement is very easily proved. Let the region 1 in the xt-plane be bounded on the right by a region (2) of constant flow (Fig. 88). Both invariants J_+ and J_- are evidently constant in the latter region, and both families of characteristics are straight lines. The boundary between the two regions is a characteristic C_+, and the lines C_+ in one region do not enter the other region. The characteristics C_- pass continuously from one region to the other, and carry the constant value of J_- into region 1 from region 2. Thus J_- is constant throughout region 1 also, so that the flow in the latter is a simple wave.

The ability of characteristics to transmit constant values of certain quantities throws

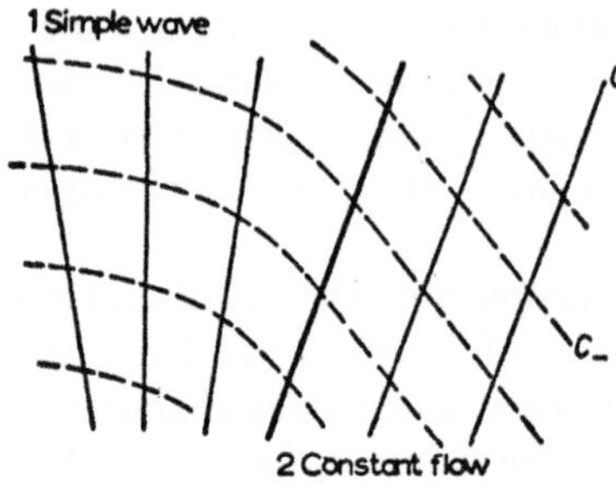

FIG. 88

some light on the general problem of initial and boundary conditions for the equations of fluid dynamics. In particular cases of physical interest, there is usually no doubt about the choice of these conditions, which is dictated by physical considerations. In more complex cases, however, mathematical considerations based on the general properties of characteristics may be useful.

We shall discuss specifically a one-dimensional isentropic gas flow. Mathematically, a problem of gas dynamics usually amounts to the determination of two unknown functions (for instance, v and p) in a region of the xt-plane lying between two given curves (OA and OB in Fig. 89a), on which the boundary conditions are known. The problem is to find how many quantities can take given values on these curves. In this respect it is very important to know how each curve is situated relative to the directions (shown by arrows in Fig. 89) of the two characteristics C_+ and C_- leaving† each point of it. Two cases can occur: either both characteristics lie on the same side of the curve, or they do not. In Fig. 89a, the curve OA belongs to the first case and the curve OB to the second. It is clear that, for a complete determination of the unknown functions in the region AOB, the values of two quantities must be given on the curve OA (e.g. the two invariants J_+ and J_-), and those of only one quantity on OB. For the values of the second quantity are transmitted to the curve OB from the curve OA by the characteristics of the corresponding family, and therefore cannot be given arbitrarily.‡ Similarly, Figs. 89b and c show cases where one and two quantities respectively are given on each bounding curve.

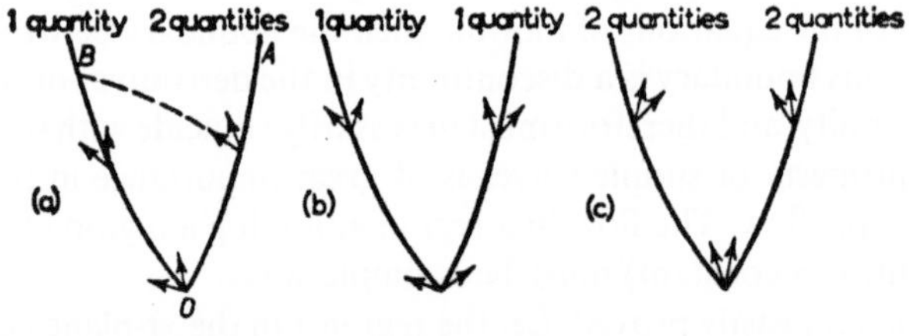

FIG. 89

† In the xt-plane, the characteristics leaving a given point are those which go in the direction of t increasing.

‡ An example of this case may be given as an illustration: the gas flow when a piston moves into or out of an infinite pipe. Here we are concerned with finding a solution of the equations of gas dynamics in the region of the xt-plane lying between two lines, the positive x-axis and the line $x = X(t)$ which gives the movement of the piston (Figs. 86, 87). On the first line the values of two quantities are given (the initial conditions $v = 0, p = p_0$ for $t = 0$), and on the second line those of one quantity ($v = u$, where $u(t)$ is the velocity of the piston).

It should also be mentioned that, if the bounding curve coincides with a characteristic, two independent quantities cannot be specified on it, since their values are related by the condition that the corresponding Riemann invariant be constant.

The problem of specifying boundary conditions for the general case of anisentropic flow can be discussed in a similar manner.

We have everywhere above spoken of the characteristics of one-dimensional flow as lines in the *xt*-plane. The characteristics can, however, also be defined in the plane of any two variables describing the flow. For example, we can consider the characteristics in the *vc*-plane. For isentropic flow, the equations of these characteristics are given simply by $J_+ = \text{constant}$, $J_- = \text{constant}$, with various constants on the right; we call these characteristics Γ_+ and Γ_-. For a polytropic gas these are, by (104.3), two families of parallel lines (Fig. 90).

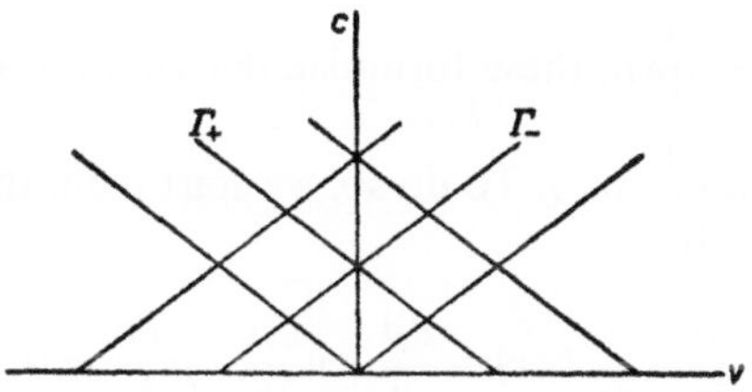

FIG. 90

It should be noted that these characteristics are entirely determined by the properties of the gas, and do not depend on any particular solution of the equations of motion. This is because the equation of isentropic flow in the variables *v*, *c* is (as we shall see in §105) a linear second-order partial differential equation with coefficients which depend only on the independent variables.

The characteristics in the *xt* and *vc* planes are transformations of one another involving the particular solution of the equations of motion. The transformation need not be one-to-one, however. In particular, only one characteristic in the *vc*-plane corresponds to a given simple wave, and all the characteristics in the *xt*-plane are transformed into it. For a wave travelling to the right (e.g.), it is one of the characteristics Γ_-; the characteristics C_- are transformed into the line Γ_-, and the characteristics C_+ into its various points.

§105. Arbitrary one-dimensional gas flow

Let us now consider the general problem of arbitrary one-dimensional isentropic gas flow (without shock waves). We shall first show that this problem can be reduced to the solution of a linear differential equation.

Any one-dimensional flow (i.e. a flow depending on only one spatial coordinate) must be a potential flow, since any function $v(x, t)$ can be written as a derivative: $v(x, t) = \partial\phi(x, t)/\partial x$. We can therefore use, as a first integral of Euler's equation, Bernoulli's equation (9.3): $\partial\phi/\partial t + \frac{1}{2}v^2 + w = 0$. From this, we find the differential

$$\begin{aligned} d\phi &= \frac{\partial\phi}{\partial x}\,dx + \frac{\partial\phi}{\partial t}\,dt \\ &= v\,dx - (\tfrac{1}{2}v^2 + w)\,dt. \end{aligned}$$

Here the independent variables are x and t; we now change to the independent variables v and w. To do so, we use Legendre's transformation; putting

$$d\phi = d(xv) - x\,dv - d[t(w+\tfrac{1}{2}v^2)] + t\,d(w+\tfrac{1}{2}v^2)$$

and replacing ϕ by a new auxiliary function

$$\chi = \phi - xv + t(w+\tfrac{1}{2}v^2),$$

we obtain

$$d\chi = -x\,dv + t\,d(w+\tfrac{1}{2}v^2) = t\,dw + (vt-x)\,dv,$$

where χ is regarded as a function of v and w. Comparing this relation with the equation $d\chi = (\partial\chi/\partial w)\,dw + (\partial\chi/\partial v)\,dv$, we have $t = \partial\chi/\partial w$, $vt - x = \partial\chi/\partial v$, or

$$t = \partial\chi/\partial w, \qquad x = v\,\partial\chi/\partial w - \partial\chi/\partial v. \tag{105.1}$$

If the function $\chi(v, w)$ is known, these formulae determine v and w as functions of the coordinate x and the time t.

We now derive an equation for χ. To do so, we start from the equation of continuity, which has not yet been used:

$$\frac{\partial\rho}{\partial t} + \frac{\partial}{\partial x}(\rho v) \equiv \frac{\partial\rho}{\partial t} + v\frac{\partial\rho}{\partial x} + \rho\frac{\partial v}{\partial x} = 0.$$

We transform this equation to one in terms of the variables v, w. Writing the partial derivatives as Jacobians, we have

$$\frac{\partial(\rho, x)}{\partial(t, x)} + v\frac{\partial(t, \rho)}{\partial(t, x)} + \rho\frac{\partial(t, v)}{\partial(t, x)} = 0,$$

or, multiplying by $\partial(t, x)/\partial(w, v)$,

$$\frac{\partial(\rho, x)}{\partial(w, v)} + v\frac{\partial(t, \rho)}{\partial(w, v)} + \rho\frac{\partial(t, v)}{\partial(w, v)} = 0.$$

To expand these Jacobians we must use the following result. According to the equation of state of the gas, ρ is a function of any two other independent thermodynamic quantities; for example, we may regard ρ as a function of w and s. If s = constant, we have simply $\rho = \rho(w)$, and the density is independent of v. Expanding the Jacobians, we therefore have

$$\frac{d\rho}{dw}\frac{\partial x}{\partial v} - v\frac{d\rho}{dw}\frac{\partial t}{\partial v} + \rho\frac{\partial t}{\partial w} = 0.$$

Substituting here the expressions (105.1) for t and x, we obtain

$$\frac{1}{\rho}\frac{d\rho}{dw}\left(\frac{\partial\chi}{\partial w} - \frac{\partial^2\chi}{\partial v^2}\right) + \frac{\partial^2\chi}{\partial w^2} = 0.$$

If s = constant, we have $dw = dp/\rho$, whence $d\rho/dw = (d\rho/dp)(dp/dw) = \rho/c^2$. We finally have for χ the equation

$$c^2\frac{\partial^2\chi}{\partial w^2} - \frac{\partial^2\chi}{\partial v^2} + \frac{\partial\chi}{\partial w} = 0; \tag{105.2}$$

here the velocity of sound c is to be regarded as a function of w. The problem of integrating the non-linear equations of motion has thus been reduced to that of solving a linear equation.

Let us apply this result to the case of a polytropic gas. We have $c^2 = (\gamma-1)w$, and the fundamental equation (105.2) becomes

$$(\gamma-1)w\frac{\partial^2\chi}{\partial w^2}-\frac{\partial^2\chi}{\partial v^2}+\frac{\partial\chi}{\partial w}=0. \tag{105.3}$$

This equation has an elementary general integral if $(3-\gamma)/(\gamma-1)$ is an even integer:

$$(3-\gamma)/(\gamma-1)=2n, \quad \text{or} \quad \gamma=(3+2n)/(2n+1), \quad n=0,1,2,\ldots. \tag{105.4}$$

This condition is satisfied by monatomic ($\gamma=\frac{5}{3}, n=1$) and diatomic ($\gamma=\frac{7}{5}, n=2$) gases. Expressing γ in terms of n, we can rewrite (105.3) as

$$\frac{2}{2n+1}w\frac{\partial^2\chi}{\partial w^2}-\frac{\partial^2\chi}{\partial v^2}+\frac{\partial\chi}{\partial w}=0. \tag{105.5}$$

We denote by χ_n a function which satisfies this equation for a given n. For the function χ_0 we have

$$2w\frac{\partial^2\chi_0}{\partial w^2}-\frac{\partial^2\chi_0}{\partial v^2}+\frac{\partial\chi_0}{\partial w}=0.$$

Introducing in place of w the variable $u=\sqrt{(2w)}$, we obtain

$$\frac{\partial^2\chi_0}{\partial u^2}-\frac{\partial^2\chi_0}{\partial v^2}=0.$$

This is just the ordinary wave equation, whose general solution is

$$\chi_0=f_1(u+v)+f_2(u-v),$$

f_1 and f_2 being arbitrary functions. Thus

$$\chi_0=f_1[\sqrt{(2w)}+v]+f_2[\sqrt{(2w)}-v]. \tag{105.6}$$

We shall now show that, if the function χ_n is known, the function χ_{n+1} can be obtained by differentiation. For, differentiating equation (105.5) with respect to w, we easily find on rearrangement

$$\frac{2}{2n+1}w\frac{\partial^2}{\partial w^2}\left(\frac{\partial\chi_n}{\partial w}\right)+\frac{2n+3}{2n+1}\frac{\partial}{\partial w}\left(\frac{\partial\chi_n}{\partial w}\right)-\frac{\partial^2}{\partial v^2}\left(\frac{\partial\chi_n}{\partial w}\right)=0.$$

Putting $v=v'\sqrt{[(2n+1)/(2n+3)]}$, we have for $\partial\chi_n/\partial w$ the equation

$$\frac{2}{2n+3}w\frac{\partial^2}{\partial w^2}\left(\frac{\partial\chi_n}{\partial w}\right)+\frac{\partial}{\partial w}\left(\frac{\partial\chi_n}{\partial w}\right)-\frac{\partial^2}{\partial v'^2}\left(\frac{\partial\chi_n}{\partial w}\right)=0,$$

which is equation (105.5) for the function $\chi_{n+1}(w,v')$. Thus we conclude that

$$\chi_{n+1}(w,v')=\frac{\partial}{\partial w}\chi_n(w,v)=\frac{\partial}{\partial w}\chi_n\left(w,v'\sqrt{\frac{2n+1}{2n+3}}\right). \tag{105.7}$$

Using this formula n times and taking χ_0 from (105.6), we find that the general solution of equation (105.5) is

$$\chi = \frac{\partial^n}{\partial w^n}\{f_1[\sqrt{[2(2n+1)w]}+v]+f_2[\sqrt{[2(2n+1)w]}-v]\},$$

or

$$\chi = \frac{\partial^{n-1}}{\partial w^{n-1}}\left\{\frac{F_1[\sqrt{[2(2n+1)w]}+v]+F_2[\sqrt{[2(2n+1)w]}-v]}{\sqrt{w}}\right\}, \quad (105.8)$$

where F_1 and F_2 are again two arbitrary functions.

If we express w in terms of the velocity of sound by $w = c^2/(\gamma-1) = \frac{1}{2}(2n+1)c^2$, the solution (105.8) becomes

$$\chi = \left(\frac{\partial}{c\partial c}\right)^{n-1}\left\{\frac{1}{c}F_1\left(c+\frac{v}{2n+1}\right)+\frac{1}{c}F_2\left(c-\frac{v}{2n+1}\right)\right\}. \quad (105.9)$$

The expressions $c \pm v/(2n+1) = c \pm \frac{1}{2}(\gamma-1)v$ which are the arguments of the arbitrary functions are just the Riemann invariants (104.3), which are constant along the characteristics.

In applications it is often necessary to calculate the values of the function $\chi(v, c)$ on a characteristic. The following formula† is useful for this purpose:

$$\left(\frac{\partial}{c\partial c}\right)^{n-1}\left\{\frac{1}{c}F\left(c\pm\frac{v}{2n+1}\right)\right\} = \frac{1}{2^{n-1}}\left(\frac{\partial}{\partial c}\right)^{n-1}\frac{F(2c+a)}{c^n}, \quad (105.10)$$

with $\pm v/(2n+1) = c+a$ (a being an arbitrary constant).

Let us now ascertain the relation between the general solution just found and the solution of the equations of gas dynamics which describes a simple wave. The latter is distinguished by the property that in it v is a definite function of w: $v = v(w)$, and therefore the Jacobian $\Delta = \partial(v, w)/\partial(x, t)$ vanishes identically. In transforming to the variables v and w, however, we divided the equation of motion by this Jacobian, and the solution for which $\Delta \equiv 0$ is therefore "lost". Thus a simple wave cannot be directly obtained from the general integral of the equations of motion, but is a special integral of these equations.

To understand the nature of this special integral, we must observe that it can be obtained from the general integral by a certain passage to a limit, which is closely related to the physical significance of the characteristics as the paths of propagation of small disturbances. Let us suppose that the region of the vw-plane in which the function $\chi(v, w)$ is

---

† It is most simply derived by using Cauchy's theorem in the theory of functions of a complex variable. For an arbitrary function $F(c+u)$ we have

$$\left(\frac{\partial}{c\partial c}\right)^{n-1}\frac{F(c+u)}{c} = 2^{n-1}\left(\frac{\partial}{\partial c^2}\right)^{n-1}\frac{F(c+u)}{c}$$
$$= 2^{n-1}\frac{(n-1)!}{2\pi i}\oint\frac{F(\sqrt{z}+u)}{\sqrt{z}(z-c^2)^n}dz,$$

where the integral is taken along a contour in the complex z-plane which encloses the point $z = c^2$. Putting now $u = c+a$ and substituting in the integral $\sqrt{z} = 2\zeta - c$, we obtain

$$\frac{1}{2^{n-1}}\frac{(n-1)!}{2\pi i}\oint\frac{F(2\zeta+a)}{\zeta^n(\zeta-c)^n}d\zeta,$$

where the contour of integration encloses the point $\zeta = c$; again applying Cauchy's theorem, we have the result (105.10).

not zero becomes a very narrow strip along a characteristic. The derivatives of χ in the direction transverse to the characteristic then take a very wide range of values, since χ diminishes very rapidly in that direction. Such solutions $\chi(v, w)$ of the equations of motion must exist. For, regarded as a perturbation in the vw-plane, they satisfy the conditions of geometrical acoustics, and are therefore non-zero along characteristics, as such perturbations must be.

It is clear from the foregoing that, for such a function χ, the time $t = \partial\chi/\partial w$ will take ar arbitrarily large range of values. The derivative of χ along the characteristic, however, is finite. Along a characteristic (for instance, a characteristic Γ_-) we have

$$\frac{dJ_-}{dv} = 1 - \frac{1}{\rho c}\frac{dp}{dw}\frac{dw}{dv} = 1 - \frac{1}{c}\frac{dw}{dv} = 0.$$

The derivative of χ with respect to v along a characteristic, which we denote by $-f(v)$, is therefore

$$\frac{d\chi}{dv} = \frac{\partial\chi}{\partial v} + \frac{\partial\chi}{\partial w}\frac{\partial w}{\partial v} = \frac{\partial\chi}{\partial v} + c\frac{\partial\chi}{\partial w} = -f(v).$$

Expressing the partial derivatives of χ in terms of x and t by (105.1), we obtain the relation $x = (v+c)t + f(v)$, i.e. the equation (101.5) for a simple wave. The relation (101.4), which gives the relation between v and c in a simple wave, is necessarily satisfied, since J_- is constant along a characteristic Γ_-.

We have shown in §104 that, if the solution of the equations of motion reduces to constant flow in some part of the xt-plane, then there must be a simple wave in the adjoining regions. The motion described by the general solution (105.8) must therefore be separated from a region of constant flow (in particular, a region of gas at rest) by a simple wave. The boundary between the simple wave and the general solution, like any boundary between two analytically different solutions, is a characteristic. In solving particular problems, the value of the function $\chi(w, v)$ on this boundary characteristic must be determined.

The joining condition at the boundary between the simple wave and the general solution is obtained by substituting the expressions (105.1) for x and t in the equation of the simple wave $x = (v \pm c)t + f(v)$; this gives

$$\frac{\partial\chi}{\partial v} \pm c\frac{\partial\chi}{\partial w} + f(v) = 0.$$

Moreover, in a simple wave (and therefore on the boundary characteristic), we have $dv = \pm dp/\rho c = \pm dw/c$, or $\pm c = dw/dv$. Substituting this in the above condition, we obtain

$$\frac{\partial\chi}{\partial v} + \frac{\partial\chi}{\partial w}\frac{dw}{dv} + f(v) = \frac{d\chi}{dv} + f(v) = 0,$$

or, finally,

$$\chi = -\int f(v)\,dv, \qquad (105.11)$$

which determines the required boundary value of χ. In particular, if the simple wave has a centre at the origin, i.e. if $f(v) \equiv 0$, then $\chi =$ constant; since the function χ is defined only to within an additive constant, we can without loss of generality take $\chi = 0$ on the boundary characteristic.

PROBLEMS

PROBLEM 1. Determine the resulting flow when a centred rarefaction wave is reflected from a solid wall.

SOLUTION. Let the rarefaction wave be formed at the point $x = 0$ at time $t = 0$, and propagated in the positive x-direction; it reaches the wall after a time $t = l/c_0$, where l is the distance to the wall. Figure 91 shows the characteristics for the reflection of the wave. In regions 1 and 1′ the gas is at rest; in region 3 it moves with a constant velocity $v = -U$.† Region 2 is the incident rarefaction wave (with rectilinear characteristics C_+), and region 5 is the reflected wave (with rectilinear characteristics C_-). Region 4 is the "region of interaction", in which the solution is required; the linear characteristics become curved on entering this region. The solution is entirely determined by the boundary conditions on the segments ab and ac. On ab (i.e. on the wall) we must have $v = 0$ for $x = l$; by (105.1), we hence obtain the condition $\partial\chi/\partial v = -l$ for $v = 0$. The boundary ac with the rarefaction wave is part of a characteristic C_-, and we therefore have $c - \frac{1}{2}(\gamma-1)v = c - v/(2n+1) = \text{constant}$; since, at the point a, $v = 0$ and $c = c_0$, the constant is c_0. On this boundary χ must be zero, so that we have the condition $\chi = 0$ for $c - v/(2n+1) = c_0$. It is easily seen that a function of the form (105.9) which satisfies these conditions is

$$\chi = \frac{l(2n+1)}{2^n n!}\left(\frac{\partial}{c\,\partial c}\right)^{n-1}\left\{\frac{1}{c}\left[\left(c - \frac{v}{2n+1}\right)^2 - c_0^2\right]^n\right\}, \tag{1}$$

and this gives the required solution.

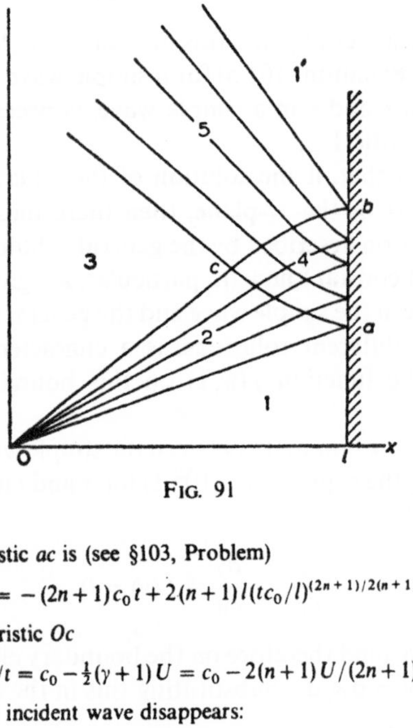

FIG. 91

The equation of the characteristic ac is (see §103, Problem)

$$x = -(2n+1)c_0 t + 2(n+1)l(tc_0/l)^{(2n+1)/2(n+1)}.$$

Its intersection with the characteristic Oc

$$x/t = c_0 - \tfrac{1}{2}(\gamma+1)U = c_0 - 2(n+1)U/(2n+1)$$

determines the time at which the incident wave disappears:

$$t_c = \frac{l(2n+1)^{n+1}c_0^n}{[(2n+1)c_0 - U]^{n+1}}.$$

In Fig. 91 it is assumed that $U < 2c_0/(\gamma+1)$; in the opposite case, the characteristic Oc is in the negative x-direction (Fig. 92). The interaction of the incident and reflected waves then lasts for an infinite time (not, as in Fig. 91, for a finite time).

The function (1) also describes the interaction between two equal centred rarefaction waves which leave the points $x = 0$ and $x = 2l$ at time $t = 0$ and are propagated towards each other; this is evident from symmetry (Fig. 93).

† If the rarefaction wave is due to a piston which begins to move out of a pipe at a constant velocity, then U is the velocity of the piston.

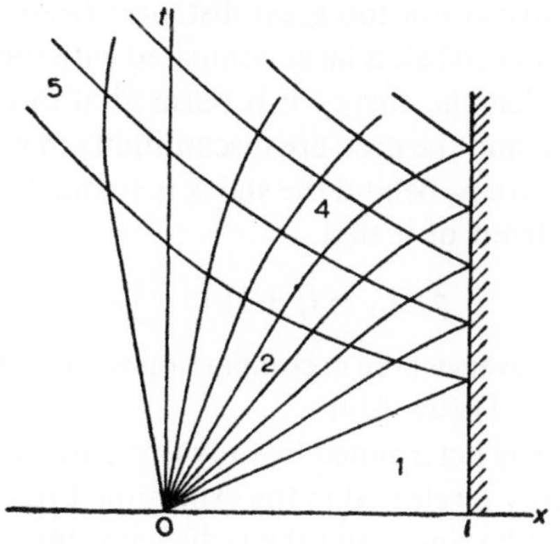

FIG. 92

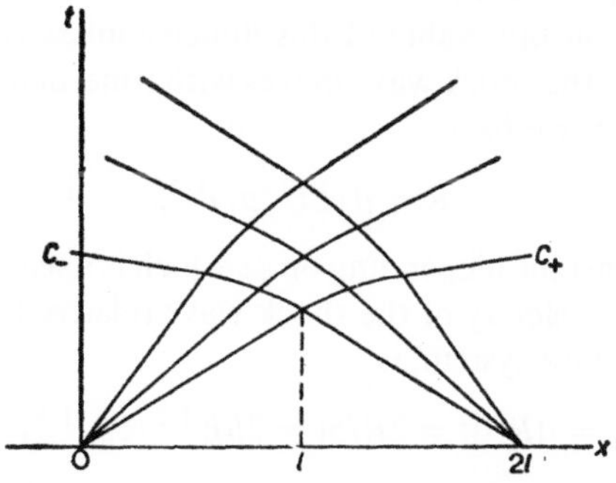

FIG. 93

PROBLEM 2. Derive the equation analogous to (105.3) for one-dimensional isothermal flow of a perfect gas.

SOLUTION. For isothermal flow, the heat function w in Bernoulli's equation is replaced by

$$\mu = \int \mathrm{d}p/\rho = c_T^2 \int \mathrm{d}\rho/\rho = c_T^2 \log \rho,$$

where $c_T^2 = (\partial p/\partial \rho)_T$ is the square of the isothermal velocity of sound. For a perfect gas $c_T =$ constant. Taking the quantity μ (instead of w) as an independent variable, we obtain, by the same method as in the text, the following linear equation with constant coefficients:

$$c_T^2 \frac{\partial^2 \chi}{\partial \mu^2} + \frac{\partial \chi}{\partial \mu} - \frac{\partial^2 \chi}{\partial v^2} = 0.$$

§106. A strong explosion

Let us consider the propagation of a strong spherical shock wave resulting from a strong explosion, that is, the instantaneous release of a large amount of energy E in a small volume. The gas in which the wave is propagated will be assumed to be polytropic.†

† The solution given below was found independently by L. I. Sedov (1946) and J. von Neumann (1947). The problem was treated less fully by G. I. Taylor (1941, published 1950), who did not derive an analytical solution.

We shall consider the wave at not too great distances from the source, where it is still strong. These distances are nevertheless large compared with the dimensions of the source: this enables us to assume that the energy E is released at one point, the origin.

The shock wave is strong, and the pressure discontinuity in it is therefore very large. We shall suppose that the pressure p_2 behind the shock is so much larger than the pressure p_1 of the undisturbed gas in front of it that

$$p_2/p_1 \gg (\gamma+1)/(\gamma-1).$$

This enables us to neglect p_1 everywhere in comparison with p_2; the density ratio ρ_2/ρ_1 has its limiting value $(\gamma+1)/(\gamma-1)$ (see §89).

Thus the gas flow pattern is determined by only two parameters: the initial gas density ρ_1, and the amount of energy E released in the explosion. From these parameters and the two independent variables (the time t and the radial coordinate r), we can form only one dimensionless combination, which we write as $r(\rho_1/Et^2)^{1/5}$. Consequently, we have a certain type of similarity flow.

We can say, first of all, that the position of the shock wave itself at every instant must correspond to a certain constant value of this dimensionless combination. This gives at once the manner in which the shock wave moves with time; denoting by R the distance of the shock from the origin, we have

$$R = \beta\,(Et^2/\rho_1)^{1/5}, \tag{106.1}$$

where β is a numerical constant (depending on γ), which is itself determined by solving the equations of motion. The velocity of the shock wave relative to the undisturbed gas, i.e. relative to a fixed coordinate system, is

$$u_1 = \mathrm{d}R/\mathrm{d}t = 2R/5t = 2\beta E^{1/5}/5\rho_1{}^{1/5}t^{3/5}. \tag{106.2}$$

Thus in this problem the movement of the shock wave can be determined (to within a constant factor) by simple dimensional arguments.

The gas pressure p_2, the density ρ_2 and the velocity $v_2 = u_2 - u_1$ (relative to a fixed coordinate system) at the back of the shock can be expressed in terms of u_1 by means of the formulae derived in §89. According to (89.10) and (89.11),†

$$v_2 = 2u_1/(\gamma+1), \qquad \rho_2 = \rho_1(\gamma+1)/(\gamma-1), \qquad p_2 = 2\rho_1 u_1{}^2/(\gamma+1). \tag{106.3}$$

The density is constant in time, while v_2 and p_2 decrease as $t^{-3/5}$ and $t^{-6/5}$ respectively. We may also note that the pressure p_2 due to the shock increases with the total energy of the explosion as $E^{2/5}$.

Let us next determine the gas flow throughout the region behind the shock. Instead of the gas velocity v, the density ρ and the squared velocity of sound $c^2 = \gamma p/\rho$ (which replaces the pressure p as a variable), we introduce dimensionless variables V, G, Z defined by‡

$$v = 2rV/5t, \qquad \rho = \rho_1 G, \qquad c^2 = 4r^2 Z/25t^2. \tag{106.4}$$

They can be functions only of a single independent dimensionless "similarity" variable, which we define as

$$\xi = r/R(t) = (r/\beta)(\rho_1/Et^2)^{1/5}. \tag{106.5}$$

† We here denote by u_1 and u_2 the velocities of the shock wave, relative to the gas, given by formulae (89.11).

‡ The symbol V in §§106 and 107 should not be confused with the specific volume used elsewhere.

In accordance with (106.3), their values at the discontinuity surface ($\xi = 1$) must be

$$V(1) = 2/(\gamma+1), \qquad G(1) = (\gamma+1)/(\gamma-1), \qquad Z(1) = 2\gamma(\gamma-1)/(\gamma+1)^2. \quad (106.6)$$

The equations of centrally-symmetrical adiabatic gas flow are

$$\left.\begin{aligned} &\frac{\partial v}{\partial t} + v\frac{\partial v}{\partial r} = -\frac{1}{\rho}\frac{\partial p}{\partial r}, \qquad \frac{\partial \rho}{\partial t} + \frac{\partial(\rho v)}{\partial r} + \frac{2\rho v}{r} = 0, \\ &\left(\frac{\partial}{\partial t} + v\frac{\partial}{\partial r}\right)\log\frac{p}{\rho^\gamma} = 0. \end{aligned}\right\} \quad (106.7)$$

The last equation is the equation of conservation of entropy, with the expression (83.12) for the entropy of a polytropic gas substituted. After substituting (106.4), we obtain a set of ordinary differential equations for the functions V, G and Z. The integration of these equations is facilitated by the fact that one integral can be obtained immediately, using the following arguments.

The fact that we have neglected the pressure p_1 of the undisturbed gas means that we neglect the original energy of the gas in comparison with the energy E which it acquires as a result of the explosion. It is therefore clear that the total energy of the gas within the sphere bounded by the shock wave is constant and equal to E. Furthermore, since we have a similarity flow, it is evident that the energy of the gas inside any sphere of a smaller radius, which increases with time in such a way that ξ = any constant (not only 1), must remain constant; the radial velocity of points on this sphere is $v_n = 2r/5t$ (cf. (106.2)).

It is easy to write down the equation which expresses the constancy of this energy. On the one hand, an amount of energy $dt.\, 4\pi r^2 \rho v(w + \frac{1}{2}v^2)$ leaves the sphere (whose area is $4\pi r^2$) in time dt. On the other hand, the volume of the sphere is increased in that time by $dt.v_n.4\pi r^2$, and the energy of the gas in this extra volume is $dt.4\pi r^2 \rho v_n(\varepsilon + \frac{1}{2}v^2)$. Equating the two expressions, substituting ε and w from (83.10) and (83.11), and introducing the dimensionless functions by (106.4), we obtain

$$Z = \frac{\gamma(\gamma-1)(1-V)V^2}{2(\gamma V - 1)}, \quad (106.8)$$

which is the required integral. It automatically satisfies the boundary conditions (106.6).

When the integral (106.8) is known, the integration of the equations is elementary though laborious. The second and third equations (106.7) give

$$\left.\begin{aligned} &\frac{dV}{d\log\xi} - (1-V)\frac{d\log G}{d\log\xi} = -3V, \\ &\frac{d\log Z}{d\log\xi} - (\gamma-1)\frac{d\log G}{d\log\xi} = -\frac{5-2V}{1-V}. \end{aligned}\right\} \quad (106.9)$$

From these two equations we can express the derivatives $dV/d\log\xi$ and $d\log G/dV$, by means of (106.8), as functions of V only, and then an integration with the boundary conditions (106.6) gives

$$\xi^5 = [\tfrac{1}{2}(\gamma+1)V]^{-2}\left\{\frac{\gamma+1}{7-\gamma}[5-(3\gamma-1)V]\right\}^{\nu_1}\left[\frac{\gamma+1}{\gamma-1}(\gamma V-1)\right]^{\nu_2}$$

$$G = \frac{\gamma+1}{\gamma-1}\left[\frac{\gamma+1}{\gamma-1}(\gamma V-1)\right]^{\nu_3}\left\{\frac{\gamma+1}{7-\gamma}[5-(3\gamma-1)V]\right\}^{\nu_4}\left[\frac{\gamma+1}{\gamma-1}(1-V)\right]^{\nu_5},$$

$$\nu_1 = -\frac{13\gamma^2-7\gamma+12}{(3\gamma-1)(2\gamma+1)}, \quad \nu_2 = \frac{5(\gamma-1)}{2\gamma+1}, \quad \nu_3 = \frac{3}{2\gamma+1},$$

$$\nu_4 = -\frac{\nu_1}{2-\gamma}, \quad \nu_5 = -\frac{2}{2-\gamma}. \tag{106.10}$$

Formulae (106.8) and (106.10) give the complete solution of the problem. The constant β in the definition of the independent variable ξ is determined by the condition

$$E = \int_0^R \rho\,[\tfrac{1}{2}v^2 + c^2/\gamma(\gamma-1)]\,4\pi r^2\,dr,$$

which states that the total energy of the gas is equal to the energy E of the explosion. In terms of the dimensionless quantities, this condition becomes

$$\beta^5\frac{16\pi}{25}\int_0^1 G[\tfrac{1}{2}V^2 + Z/\gamma(\gamma-1)]\,\xi^4\,d\xi = 1. \tag{106.11}$$

For air ($\gamma = 7/5$), $\beta = 1{\cdot}033$.

It is easily seen from (106.10) that, as $\xi \to 0$, V tends to a constant limit and G to zero:

$$V - 1/\gamma \propto \xi^{5/\nu_2}, \quad G \propto \xi^{5\nu_3/\nu_2}.$$

Hence it follows that v/v_2 and ρ/ρ_2 as functions of $r/R = \xi$ tend to zero as $\xi \to 0$:

$$v/v_2 \propto r/R, \quad \rho/\rho_2 \propto (r/R)^{3/(\gamma-1)}; \tag{106.12}$$

the pressure ratio p/p_2 tends to a constant limit, and the temperature ratio accordingly becomes infinite.†

Figure 94 shows the quantities v/v_2, p/p_2 and ρ/ρ_2 as functions of r/R for air ($\gamma = 1{\cdot}4$). The very rapid decrease of the density into the sphere is noticeable: almost all the gas is in a relatively thin layer behind the shock wave. This is, of course, due to the fact that the gas on the surface of greatest radius (R) has a density six times the normal density.‡

§107. **An imploding spherical shock wave**

There are a number of instructive features in the problem of a strong shock wave converging to a centre.†† We shall not be concerned with the specific mechanism whereby

† These statements relate to values $\gamma < 7$, the function $V(\xi)$ varying from $V(1) = 2/(\gamma+1)$ to $V(0) = 1/\gamma$. For actual gases whose thermodynamic functions could be approximated by the expressions for a polytropic gas, the inequality is certainly satisfied; the upper limit of γ in practice is 5/3 for a monatomic gas. For formal completeness, however, it may be noted that when $\gamma > 7$ the function $V(\xi)$ varies from $2/(\gamma+1)$ for $\xi = 1$ to the limit of unity reached at a value $\xi_0 < 1$ which depends on γ; at this point, G is zero, and an expanding spherical vacuum is formed.

‡ The results of calculations for other values of γ are given by L. I. Sedov, *Similarity and Dimensional Methods in Mechanics*, Chapter IV, §11, London 1959. The corresponding problem with cylindrical symmetry is also discussed.

†† This was discussed independently by G. Guderley (1942) and by L. D. Landau and K. P. Stanyukovich (1944, published 1955).

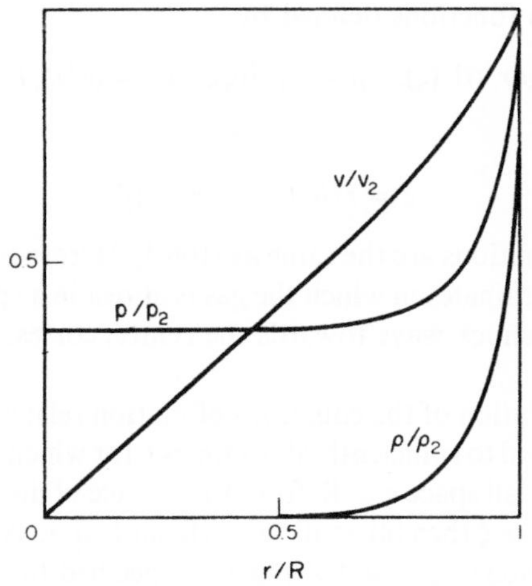

FIG. 94

such a shock wave is formed; it is sufficient to suppose it produced by some kind of "spherical piston" which gives the gas an initial impetus, and the shock becomes stronger as the centre is approached.

We will consider the gas flow at the stage when the radius R of the spherical surface of discontinuity is already much less than its initial value, the radius R_0 of the "piston". At this stage the flow is largely (to an extent described below) independent of the specific initial conditions. The shock wave will be assumed so strong that the pressure p_1 of the gas in front of it is (as in §106) negligible in comparison with the pressure p_2 behind it. The total energy of the gas in the (variable) region $r \sim R \ll R_0$ considered is by no means constant; it will be shown later to decrease in the course of time.

The spatial scale of the flow in question can only be determined by the time-dependent radius $R(t)$ of the shock wave, and the scale of the velocity by the derivative $\mathrm{d}R\ \mathrm{d}t$. Under these conditions, it is reasonable to suppose that there is similarity flow with the independent variable $\xi = r/R(t)$. The function $R(t)$, however, cannot be determined by dimensional arguments alone.

Let the time when the shock wave is focused (i.e. when $R = 0$) be taken as $t = 0$. Then the times before focusing correspond to $t < 0$. We shall seek the function $R(t)$ in the form

$$R(t) = A(-t)^{\alpha}, \tag{107.1}$$

with an initially unknown *similarity index* α. It is found that this index is determined by the condition for a solution of the equations of motion to exist (in the region $r \ll R_0$) with the necessary boundary conditions. This also determines the dimensions of the constant parameter A, whose value, however, remains indeterminate and can in principle be found only by solving the gas flow problem as a whole, that is, by joining the similarity solution to the solution at distances $r \sim R_0$, which depends on the specific initial conditions. This parameter alone governs the flow for $R \ll R_0$ in relation to the way in which the shock wave is initially formed.

We shall now show how the problem thus formulated is solved. As in §106, we use dimensionless unknown functions defined by

$$v = (\alpha r/t)V(\xi), \quad \rho = \rho_1 G(\xi), \quad c^2 = (\alpha^2 r^2/t^2)Z(\xi), \tag{107.2}$$

where

$$\xi = r/R(t) = r/A(-t)^\alpha; \tag{107.3}$$

when $\alpha = 2/5$, these definitions are the same as (106.4). Here, v is the radial velocity of the gas, relative to fixed coordinates in which the gas is at rest in a sphere with radius $r = R_0$; the gas moves with the shock wave towards the centre, corresponding to $v < 0$, so that $V(\xi) > 0$.

In fact, the desired solution of the equations of motion relates only to the region $r \sim R$ behind the shock wave and to sufficiently short times t, for which $R \ll R_0$. But formally the solution obtained covers all space $r \geqslant R$, from the surface of discontinuity to infinity, and all times $t \leqslant 0$; the variable ξ then takes all values from 1 to ∞. Accordingly, the boundary conditions for the functions G, V and Z must be specified for $\xi = 1$ and $\xi = \infty$.

The value $\xi = 1$ corresponds to the surface of the shock wave; the boundary conditions there are the same as (106.6).

To establish the conditions at infinity (for ξ), we note that when $t = 0$ (the shock is focused) all the quantities v, ρ and c^2 must remain finite at any finite distance from the centre. When $t = 0$ and $r \neq 0$, we have $\xi = \infty$. If the functions $v(r, t)$ and $c^2(r, t)$ then remain finite, it follows that $V(\xi)$ and $Z(\xi)$ must tend to zero:

$$V(\infty) = 0, \quad Z(\infty) = 0. \tag{107.4}$$

Substitution of (107.2) and (107.3) brings the equations (106.7) to the form

$$\left.\begin{aligned} (1-V)\frac{\mathrm{d}V}{\mathrm{d}\log\xi} - \frac{Z}{\gamma}\frac{\mathrm{d}\log G}{\mathrm{d}\log\xi} - \frac{1}{\gamma}\frac{\mathrm{d}Z}{\mathrm{d}\log\xi} &= \frac{2}{\gamma}Z - V\left(\frac{1}{\alpha} - V\right), \\ \frac{\mathrm{d}V}{\mathrm{d}\log\xi} - (1-V)\frac{\mathrm{d}\log G}{\mathrm{d}\log\xi} &= -3V, \\ (\gamma-1)Z\frac{\mathrm{d}\log G}{\mathrm{d}\log\xi} - \frac{\mathrm{d}Z}{\mathrm{d}\log\xi} &= \frac{2Z(1/\alpha - V)}{1-V}; \end{aligned}\right\} \tag{107.5}$$

compare (106.9) for the last two equations. The independent variable ξ appears here only as the differential d log ξ; the constant log A disappears from the equations entirely, as stated above.

The coefficients of the derivatives in equations (107.5) and the right-hand sides involve only V and Z, not G.† By solving these equations for the derivatives, we can express the latter in terms of the two functions V and Z. This gives

$$\frac{\mathrm{d}\log\xi}{\mathrm{d}V} = -\frac{Z-(1-V)^2}{(3V-\kappa)Z - V(1-V)(1/\alpha - V)}, \tag{107.6}$$

$$(1-V)\frac{\mathrm{d}\log G}{\mathrm{d}\xi} = 3V - \frac{(3V-\kappa)Z - V(1-V)(1/\alpha - V)}{Z-(1-V)^2}, \tag{107.7}$$

where $\kappa = 2(1-\alpha)/\alpha\gamma$. As a third equation, we write the result of dividing $\mathrm{d}Z/\mathrm{d}\log\xi$ by

† This is the advantage of using v, ρ and c^2 as the fundamental variables instead of v, p and p.

$dV/d\log\xi$:

$$\frac{dZ}{dV} = \frac{Z}{1-V}\left\{\frac{[Z-(1-V)^2][2/\alpha-(3\gamma-1)V]}{(3V-\kappa)Z-V(1-V)(1/\alpha-V)}+\gamma-1\right\}. \tag{107.8}$$

If the required solution of (107.8) has been found, i.e. the functional $Z(V)$, the solution of equations (107.6) and (107.7) to find $\xi(V)$ and then $G(\xi)$ is reduced to quadratures.

The whole problem thus reduces to first solving equation (107.8). The integral curve in the VZ-plane must start from a point Y with coordinates $V(1)$, $Z(1)$, the image of the shock wave in the VZ-plane. This point specifies the solution of (107.8) for a given α: the integral curve of a first-order equation is uniquely defined by one (non-singular) point on it. Let us next ascertain the condition which establishes the value of α giving the "correct" integral curve.

This condition follows from an obvious physical requirement, that the dependence of all quantities on ξ must be single-valued: to each value of ξ there must correspond unique values of V, G and Z. This means that throughout the range of variation of ξ ($1 \leqslant \xi \leqslant \infty$, i.e. $0 \leqslant \log\xi \leqslant \infty$) the functions $\xi(V)$, $\xi(G)$ and $\xi(Z)$ cannot have extrema. Thus the derivatives $d\log\xi/dV$ etc. can nowhere be zero. In Fig. 95, curve 1 is the parabola

$$Z = (1-V)^2. \tag{107.9}$$

It is easily seen that Y lies above this parabola.† The integral curve corresponding to the solution of the problem stated must reach the origin in accordance with the limiting condition (107.4), and must therefore cross the parabola (107.9). According to

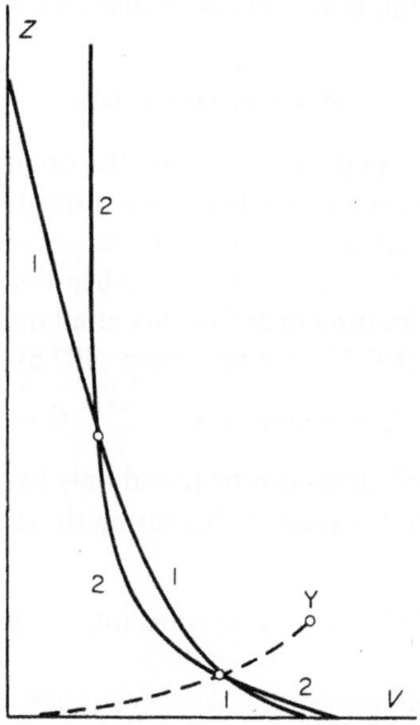

FIG. 95

† A result which simply expresses the fact that the gas velocity at the back of the discontinuity surface is less than the velocity of sound in the gas.

(107.6)–(107.8), all the derivatives mentioned are expressed by fractions with $Z-(1-V)^2$ in the numerator. If these do not become zero, then at the point where the integral curve intersects the parabola (107.9) we must also have

$$(3V-\kappa)Z = V(1-V)(1/\alpha - V). \qquad (107.10)$$

That is, the integral curve must pass through the point where the parabola (107.9) meets the curve (107.10) (curve 2 in Fig. 95); this point is a singularity of (107.8), since $dZ/dV = 0/0$. The same condition determines the value of the similarity index α; two values given by numerical calculation are

$$\alpha = 0{\cdot}6884 \text{ for } \gamma = 5/3, \qquad \alpha = 0{\cdot}7172 \text{ for } \gamma = 7/5. \qquad (107.11)$$

After passing through the singular point, the integral curve goes to the origin O corresponding to the limiting values (107.4). To elucidate the mathematical situation, we will briefly describe the distribution of the integral curves of equation (107.8) in the VZ-plane (for the "correct" value of α), without going through the calculations.†

The curves (107.9) and (107.10) intersect, in general, at two points as shown by the circles in Fig. 95 (in addition to the unimportant point $V = 1$, $Z = 0$ on the abscissa axis). The equation also has a singular point c where the curve (107.10) intersects the straight line $(3\gamma-1)V = 2/\alpha$ (on which the second factor in the numerator of (107.8) is zero). The point a through which the "correct" integral curve passes is a saddle point; b and c are nodes. The origin is also a node singularity. Near it, equation (107.8) becomes

$$\frac{dZ}{dV} = \frac{2Z}{V+\kappa Z}.$$

An elementary integration of this homogeneous equation shows that, as $V \to 0$, $Z(V)$ tends to zero faster than V:

$$Z \simeq \text{constant} \times V^2. \qquad (107.12)$$

There is thus an infinity of integral curves leaving the origin, with different values of the constant in (107.12). All of them go to either b or c, except one which goes to the saddle point a (one of the two separatrices, the only integral curves through the saddle point).‡

The origin corresponds to $\xi = \infty$, i.e. the time when the shock wave is focused at the centre. Let us determine the limiting distributions of all quantities with respect to radial distances at this time. With (107.12), we find from (107.6) – (107.7) that

$$V = \text{constant} \times \xi^{-1/\alpha}, \ Z = \text{constant} \times \xi^{-2/\alpha}, \ G = \text{constant as } \xi \to \infty; \quad (107.13)$$

the values of the constant coefficients can be found only by a numerical determination of the integral curve over its whole extent. Substituting these expressions in the definitions (107.2), we find††

$$|v| \propto c \propto r^{-(1/\alpha-1)}, \qquad \rho = \text{constant}, \qquad p \propto r^{-2(1/\alpha-1)}. \qquad (107.14)$$

† The procedure is to use the general methods of qualitative differential equation theory. The types of singular points for a first-order equation are classified as described by V. V. Stepanov, *Differential Equations* (*Kurs differentsial'nykh uravnenii*), Chapter II; G. Birkhoff and G.-C. Rota, *Ordinary Differential Equations*, 2nd ed., Waltham (Mass.), 1969.

‡ The picture described is found to be valid only for $\gamma < \gamma_1 = 1{\cdot}87 \ldots$. When $\gamma = \gamma_1$ and α has the "correct" value, the points a and b coincide; when $\gamma > \gamma_1$, the distribution of integral curves changes and a fuller investigation is needed. However, in actual cases, $\gamma \leqslant 5/3$; see the penultimate footnote to §106.

†† The limiting value of ρ/ρ_1 at the time of focusing is 20·1 for $\gamma = 7/5$ and 9·55 for $\gamma = 5/3$.

These relations could also be derived directly from dimensional arguments (when the dimensions of A are known). We have two available parameters, ρ_1 and A, and one variable, r; from these, only one combination with the dimensions of velocity can be formed, $A^{1/\alpha}r^{1-1/\alpha}$. The quantity with the dimensions of density must be ρ_1 itself.

Let us also find the time dependence of the total energy of the gas in the region of similarity flow. The radial size of this region is of the order of the radius R of the shock wave, and decreases with R. Let us arbitrarily take as the boundary of the similarity region a particular value $r/R = \xi_1$. The total energy of the gas in a spherical shell between radii R and $\xi_1 R$ is given, in dimensionless variables, by

$$E_{\text{sim}} = \frac{\alpha^2 \rho_1 R^5}{t^2} \int_1^{\xi_1} G[\tfrac{1}{2}V^2 + Z/\gamma(\gamma-1)].4\pi\xi^2\, d\xi;$$

cf. (106.11). The integral is a constant.† Hence we find

$$E_{\text{sim}} \propto R^{5-2/\alpha} \propto (-t)^{5\alpha-2} \tag{107.15}$$

For any actual value of γ, the exponent is positive. Although the shock wave itself becomes stronger as it approaches the centre, the volume of the region of similarity flow decreases, and this reduces the total energy contained in it.

After focusing at the centre, a "reflected" shock wave is formed, which (when $t > 0$) expands to meet the gas moving towards the centre. This is again a similarity flow, with the same index α, so that the expansion occurs according to $R \propto t^\alpha$. We shall not give here a more detailed analysis of this stage.‡

The problem thus provides an example of similarity flow, but one in which the similarity index (i.e. the form of the similarity variable ξ) cannot be determined from dimensional arguments; it is found only by solving the equations of motion themselves, using the conditions imposed by the physical formulation of the problem. Mathematically, it is characteristic that these conditions are formulated as requiring that the integral curve of a first-order differential equation should pass through a singular point of the equation. The similarity index is in general irrational.††

§108. Shallow-water theory

There is a remarkable analogy between gas flow and the flow in a gravitational field of an incompressible fluid with a free surface, when the depth of the fluid is small (compared with the characteristic dimensions of the problem, such as the dimensions of the irregularities on the bottom of the vessel). In this case the vertical component of the fluid velocity may be neglected in comparison with its velocity parallel to the surface, and the latter may be regarded as constant throughout the depth of the fluid. In this (*hydraulic*) approximation, the fluid can be regarded as a "two-dimensional" medium having a definite

† The integral diverges as $\xi_1 \to \infty$. This is because the similarity regime does not apply at distances $r \gg R$.

‡ But simply mention that the reflection of the shock wave is accompanied by a further compression of the gas, reaching 145 for $\gamma = 7/5$ and 32·7 for $\gamma = 5/3$.

†† Another example of this kind of similarity flow is the propagation of a shock wave formed by a short sharp impact on a gas-filled half-space (C. F. von Weizsäcker 1954, Ya. B. Zel'dovich 1956). The problem is also described in chapter XII of the book by Zel'dovich and Raizer cited in §86, and in G. I. Barenblatt's *Similarity, Self-Similarity, and Intermediate Asymptotics*, New York 1979

velocity $\mathbf{v}$ at each point and also characterized at each point by a quantity h, the depth of the fluid.

The corresponding general equations of motion differ from those obtained in §12 only in that the changes in quantities during the motion need not be assumed small, as they were in §12 in discussing long gravity waves with small amplitude. Consequently, the second-order velocity terms in Euler's equation must be retained. In particular, for one-dimensional flow in a channel, depending only on one coordinate x (and on the time), the equations are

$$\left.\begin{aligned}\frac{\partial h}{\partial t}+\frac{\partial(vh)}{\partial x}&=0,\\ \frac{\partial v}{\partial t}+v\frac{\partial v}{\partial x}&=-g\frac{\partial h}{\partial x};\end{aligned}\right\}\tag{108.1}$$

the depth h is here assumed constant across the channel.

Long gravity waves are, in a general sense, small perturbations of the flow now under consideration. The results of §12 show that such perturbations are propagated relative to the fluid with a finite velocity, namely

$$c=\sqrt{(gh)}.\tag{108.2}$$

This velocity here plays the part of the velocity of sound in gas dynamics. Just as in §82, we can conclude that, if the fluid moves with velocities $v<c$ (*streaming flow*), the effect of the perturbations is propagated both upstream and downstream. If the fluid moves with velocities $v>c$ (*shooting flow*), however, the effect of the perturbations is propagated only into certain regions downstream.

The pressure p (reckoned from the atmospheric pressure at the free surface) varies with depth in the fluid according to the hydrostatic law $p=\rho g(h-z)$, where z is the height above the bottom. It is useful to note that, if we introduce the quantities

$$\bar{\rho}=\rho h,\qquad \bar{p}=\int_0^h p\,\mathrm{d}z=\tfrac{1}{2}\rho g h^2=g\bar{\rho}^2/2\rho,\tag{108.3}$$

then equations (108.1) become

$$\frac{\partial\bar{\rho}}{\partial t}+\frac{\partial(v\bar{\rho})}{\partial x}=0,\qquad \frac{\partial v}{\partial t}+v\frac{\partial v}{\partial x}=-\frac{1}{\bar{\rho}}\frac{\partial\bar{p}}{\partial x},\tag{108.4}$$

which are formally identical with the equations of adiabatic flow of a polytropic gas with $\gamma=2$ ($\bar{p}\propto\bar{\rho}^2$). This enables us to apply immediately to shallow-water theory all the results of gas dynamics for flow in the absence of shock waves. If shock waves are present, however, the results of shallow-water theory differ from those of perfect-gas dynamics.

A "shock wave" in a fluid in a channel is a discontinuity in the fluid height h, and therefore in the fluid velocity v (what is called a *hydraulic jump*). The relations between the values of the quantities on the two sides of the discontinuity can be obtained from the conditions of continuity of the fluxes of mass and momentum. The mass flux density (per unit width of the channel) is $j=\rho vh$. The momentum flux density is obtained by integrating $p+\rho v^2$ over the depth of the fluid, and is

$$\int_0^h (p+\rho v^2)\,dz = \tfrac{1}{2}\rho g h^2 + \rho v^2 h.$$

The conditions of continuity therefore give two equations:

$$\left.\begin{aligned} v_1 h_1 &= v_2 h_2, \\ v_1^2 h_1 + \tfrac{1}{2} g h_1^2 &= v_2^2 h_2 + \tfrac{1}{2} g h_2^2. \end{aligned}\right\} \quad (108.5)$$

These give the relations between the four quantities v_1, v_2, h_1, h_2, two of which can be specified arbitrarily. Expressing the velocities v_1 and v_2 in terms of the heights h_1 and h_2, we obtain

$$v_1^2 = \tfrac{1}{2} g h_2 (h_1 + h_2)/h_1, \qquad v_2^2 = \tfrac{1}{2} g h_1 (h_1 + h_2)/h_2. \quad (108.6)$$

The energy fluxes on the two sides of the discontinuity are not the same, and their difference is the amount of energy dissipated in the discontinuity per unit time. The energy flux density in the channel is

$$q = \int_0^h \left(\frac{p}{\rho} + \tfrac{1}{2} v^2\right) \rho v\,dz = \tfrac{1}{2} j (gh + v^2).$$

Using (108.6), we find the difference to be

$$q_1 - q_2 = gj(h_1^2 + h_2^2)(h_2 - h_1)/4h_1 h_2.$$

Let the fluid move through the discontinuity from side 1 to side 2. Then the fact that energy is dissipated means that $q_1 - q_2 > 0$, and we conclude that

$$h_2 > h_1, \quad (108.7)$$

i.e. the fluid moves from the smaller to the greater height. We then can deduce from (108.6) that

$$v_1 > c_1 = \sqrt{(gh_1)}, \qquad v_2 < c_2 = \sqrt{(gh_2)}, \quad (108.8)$$

is complete analogy to the results for shock waves in gas dynamics. The inequalities (108.8) could also be derived as the necessary conditions for the discontinuity to be stable, as in §88.

PROBLEM

Find the stability condition for a tangential discontinuity in shallow water, i.e. a line such that the liquid on either side is moving with different velocities (S. V. Bezdenkov and O. P. Pogutse 1983).

SOLUTION. Because of the analogy mentioned above between the hydrodynamics of shallow water and polytropic gas dynamics, the problem is equivalent to that of the stability of a tangential discontinuity in a gas (§84, Problem 1). There is a difference, however, because in the shallow-water case we have to consider perturbations depending only on the coordinates in the plane of the liquid layer (parallel and perpendicular to the velocity v), not on the depth coordinate† z; the shallow-water approximation corresponds to perturbations with wavelength $\lambda \gg h$. The velocity v_k found in §84, Problem 1, is therefore now the limit of instability: the discontinuity is stable for $v > v_k$, where v is the velocity change there. Since the density and the depth are the same on either side of the discontinuity, the velocity of sound is the same on either side, $c_1 = c_2 = \sqrt{(gh)}$, and the discontinuity is therefore stable if $v > 2\sqrt{(2gh)}$.

† Corresponding to y in §84, Problem 1.

CHAPTER XI

THE INTERSECTION OF SURFACES OF DISCONTINUITY

§109. **Rarefaction waves**

The line of intersection of two shock waves is, mathematically, a singular line of two functions describing the gas flow. The vertex of an acute angle on the surface of a body past which the gas flows is always such a singular line. It is found that the gas flow near the singular line can be investigated in a general manner (L. Prandtl and T. Meyer 1908).

In considering the region near a small segment of the singular line, we may regard the latter as a straight line, which we take as the z-axis in a system of cylindrical polar coordinates r, ϕ, z. Near the singular line, all quantities depend considerably on the angle ϕ, but their dependence on the coordinate r is only slight, and for sufficiently small r it can be neglected. The dependence on the coordinate z is also unimportant; the change in the flow pattern over a small segment of the singular line may be neglected.

Thus we have to investigate a steady flow in which all quantities are functions of ϕ only. The equation of conservation of entropy, $\mathbf{v}\cdot\mathbf{grad}\, s = 0$, gives $v_\phi \mathrm{d}s/\mathrm{d}\phi = 0$, whence $s = \text{constant}$,† i.e. the flow is isentropic. In Euler's equation we can therefore replace $\mathbf{grad}\, p/\rho$ by $\mathbf{grad}\, w$: $(\mathbf{v}\cdot\mathbf{grad})\mathbf{v} = -\mathbf{grad}\, w$. In cylindrical polar coordinates, we have three equations:

$$\frac{v_\phi}{r}\frac{\mathrm{d}v_r}{\mathrm{d}\phi} - \frac{v_\phi^2}{r} = 0, \qquad \frac{v_\phi}{r}\frac{\mathrm{d}v_\phi}{\mathrm{d}\phi} + \frac{v_r v_\phi}{r} = -\frac{1}{r}\frac{\mathrm{d}w}{\mathrm{d}\phi}, \qquad v_\phi \frac{\mathrm{d}v_z}{\mathrm{d}\phi} = 0.$$

From the last of these we have $v_z = \text{constant}$, and without loss of generality we can put $v_z = 0$, regarding the flow as two-dimensional; this is simply a matter of suitably defining the velocity of the coordinate system along the z-axis. The first two equations can be written

$$v_\phi = \mathrm{d}v_r/\mathrm{d}\phi, \tag{109.1}$$

$$v_\phi\left(\frac{\mathrm{d}v_\phi}{\mathrm{d}\phi} + v_r\right) = -\frac{1}{\rho}\frac{\mathrm{d}p}{\mathrm{d}\phi} = -\frac{\mathrm{d}w}{\mathrm{d}\phi}. \tag{109.2}$$

Substituting (109.1) in (109.2), we have

$$v_\phi \frac{\mathrm{d}v_\phi}{\mathrm{d}\phi} + v_r \frac{\mathrm{d}v_r}{\mathrm{d}\phi} = -\frac{\mathrm{d}w}{\mathrm{d}\phi},$$

† If $v_\phi = 0$, we easily deduce from the equations of motion given below that $v_r = 0$, $v_z \neq 0$. Such a flow would correspond to the intersection of surfaces of tangential discontinuity (with a discontinuous velocity v_z), and is of no interest, since such discontinuities are unstable.

or integrating,

$$w+\tfrac{1}{2}(v_\phi^2+v_r^2) = \text{constant}. \tag{109.3}$$

We may notice that equation (109.1) implies that **curl v** = 0, i.e. we have potential flow, as a result of which Bernoulli's equation (109.3) holds.

Next, the equation of continuity, div(ρ**v**) = 0, gives

$$\rho v_r + \frac{d}{d\phi}(\rho v_\phi) = \rho\left(v_r + \frac{dv_\phi}{d\phi}\right) + v_\phi \frac{d\rho}{d\phi} = 0. \tag{109.4}$$

Using (109.2), we obtain

$$\left(\frac{dv_\phi}{d\phi}+v_r\right)\left(1-v_\phi^2\frac{d\rho}{dp}\right) = 0.$$

The derivative $dp/d\rho$, or more correctly $(\partial p/\partial \rho)_s$, is just the square of the velocity of sound. Thus

$$\left(\frac{dv_\phi}{d\phi}+v_r\right)\left(1-\frac{v_\phi^2}{c^2}\right) = 0. \tag{109.5}$$

This equation can be satisfied in either of two ways. Firstly, we may have $dv_\phi/d\phi + v_r = 0$. Then, from (109.2), p = constant and ρ = constant, and from (109.3) we find that $v^2 = v_r^2 + v_\phi^2$ = constant, i.e. the velocity is constant in magnitude. It is easy to see that in this case the velocity is constant in direction also. The angle χ between the velocity and some given direction in the plane of the motion is (Fig. 96)

$$\chi = \phi + \tan^{-1}(v_\phi/v_r). \tag{109.6}$$

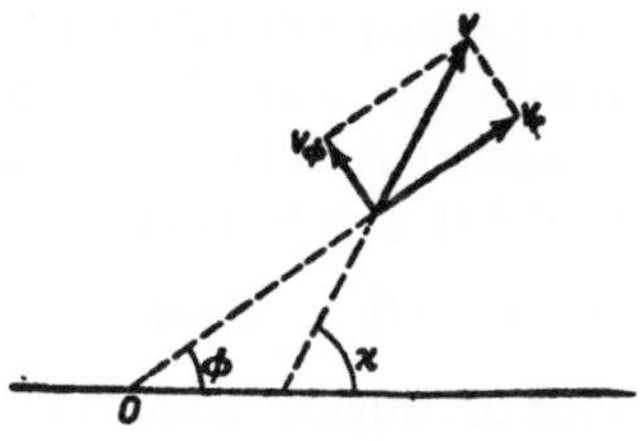

FIG. 96

Differentiating this expression with respect to ϕ and using formulae (109.1) and (109.2), we easily obtain

$$d\chi/d\phi = -(v_r/\rho v_\phi v^2)dp/d\phi. \tag{109.7}$$

Since p = constant, it follows that χ = constant. Thus, if the first factor in (109.5) is zero, we have the trivial solution of a uniform flow.

Secondly, equation (109.5) can be satisfied by putting $1 - v_\phi^2/c^2 = 0$, i.e. $v_\phi = \pm c$. The radial velocity is given by (109.3). Denoting the constant in that equation by w_0, we find that

$$v_\phi = \pm c, \qquad v_r = \pm\sqrt{[2(w_0 - w) - c^2]}.$$

In this solution, the velocity component v_ϕ perpendicular to the radius vector is equal to the local velocity of sound at every point. The total velocity $v = \sqrt{(v_\phi^2 + v_r^2)}$ therefore exceeds that of sound. Both the magnitude and the direction of the velocity are different at different points. Since the velocity of sound cannot vanish, it is clear that the function $v_\phi(\phi)$, which is continuous, must everywhere be $+c$, or else everywhere $-c$. By measuring the angle ϕ in the appropriate direction, we can take $v_\phi = c$. We shall see below that the choice of the sign of v_r follows from physical considerations, and that the plus sign must be taken. Thus

$$v_\phi = c, \qquad v_r = \sqrt{[2(w_0 - w) - c^2]}. \tag{109.8}$$

From the equation of continuity (109.4) we have $\mathrm{d}\phi = -\mathrm{d}(\rho v_\phi)/\rho v_r$. Substituting (109.8) and integrating, we have

$$\phi = -\int \frac{\mathrm{d}(\rho c)}{\rho\sqrt{[2(w_0 - w) - c^2]}}. \tag{109.9}$$

If the equation of state of the gas and the adiabatic equation are known (we recall that s is constant), this formula can be used to determine all quantities as functions of the angle ϕ. Thus formulae (109.8) and (109.9) completely determine the gas flow.

Let us now study in more detail the solution which we have obtained. First of all, we notice that the straight lines ϕ = constant intersect the streamlines at every point at the Mach angle (whose sine is $v_\phi/v = c/v$), i.e. they are characteristics. Thus one family of characteristics (in the xy-plane) is a pencil of straight lines through the singular point, and has an important property in this case: all quantities are constant along each characteristic. In this respect the solution concerned plays the same part in the theory of steady two-dimensional flow as does the similarity flow discussed in §99 in the theory of non-steady one-dimensional flow. We shall return to this point in §115.

It is seen from (109.9) that $(\rho c)' < 0$, the prime denoting differentiation with respect to ϕ. Putting $(\rho c)' = \rho' \mathrm{d}(\rho c)/\mathrm{d}\rho$ and noticing that the derivative $\mathrm{d}(\rho c)/\mathrm{d}\rho$ is positive (see (99.9)), we find that $\rho' < 0$, and therefore so are the derivatives $p' = c^2\rho'$ and $w' = p'/\rho$. Next, from the fact that w' is negative it follows that the velocity $v = \sqrt{[2(w_0 - w)]}$ increases with ϕ. Finally, from (109.7), $\chi' > 0$. Thus we have

$$\mathrm{d}p/\mathrm{d}\phi < 0, \qquad \mathrm{d}\rho/\mathrm{d}\phi < 0, \qquad \mathrm{d}v/\mathrm{d}\phi > 0, \qquad \mathrm{d}\chi/\mathrm{d}\phi > 0. \tag{109.10}$$

In other words, when we go round the singular point in the direction of flow, the density and pressure decrease, while the magnitude of the velocity increases and its direction rotates in the direction of flow.

The flow just described is often called a *rarefaction wave*, and we shall use this name in what follows.

It is easy to see that a rarefaction wave cannot exist throughout the region surrounding the singular line. For, since v increases monotonically with ϕ, a complete circuit round the origin (i.e. a change of ϕ by 2π) would give a value for v different from the initial one, which is impossible. For this reason, the actual pattern of flow round the singular line must be composed of a series of sectors separated by planes ϕ = constant which are surfaces of discontinuity. In each of these regions we have either a rarefaction wave or a flow with constant velocity. The number and nature of these regions for various particular cases will be established in the following sections. Here we shall simply mention that the boundary between a rarefaction wave and a uniform flow must be a weak discontinuity: it cannot be a tangential discontinuity (of v_r), since the normal velocity component $v_\phi = c$ does not vanish on it. Nor can it be a shock wave, since the normal velocity component v_ϕ must be greater than the velocity of sound on one side of such a discontinuity and smaller on the other side, whereas in our problem we always have $v_\phi = c$ on one side of the boundary.

An important conclusion can be drawn from the foregoing. Disturbances which cause weak discontinuities leave the singular line (the z-axis) and are propagated away from it. This means that the weak discontinuities bounding the rarefaction wave must be ones which leave this line, i.e. the velocity component v_r tangential to the weak discontinuity must be positive. This justifies the choice of the sign of v_r made in (109.8).

Let us now apply these formulae to a polytropic gas. In such a gas $w = c^2/(\gamma-1)$, while the equation of the Poisson adiabatic can be written

$$\rho c^{-2/(\gamma-1)} = \text{constant}, \qquad p c^{-2\gamma/(\gamma-1)} = \text{constant}; \tag{109.11}$$

cf. (99.13). Using these formulae, we can put the integral (109.9) in the form

$$\phi = -\sqrt{\frac{\gamma+1}{\gamma-1}}\int\frac{dc}{\sqrt{(c_*^2-c^2)}},$$

where c_* is the critical velocity (see (83.14)). Hence

$$\phi = \sqrt{\frac{\gamma+1}{\gamma-1}}\cos^{-1}\frac{c}{c_*}+\text{constant},$$

or, if we measure ϕ in such a way that the constant is zero,

$$v_\phi = c = c_*\cos\sqrt{[(\gamma-1)/(\gamma+1)]}\,\phi. \tag{109.12}$$

According to formula (109.8) we therefore have

$$v_r = \sqrt{\frac{\gamma+1}{\gamma-1}}\,c_*\sin\sqrt{\frac{\gamma-1}{\gamma+1}}\,\phi. \tag{109.13}$$

Next, using the Poisson adiabatic equation in the form (109.11), we can find the pressure as a function of the angle ϕ:

$$p = p_*\cos^{2\gamma/(\gamma-1)}\sqrt{\frac{\gamma-1}{\gamma+1}}\,\phi. \tag{109.14}$$

Finally, we have for the angle χ (109.6)

$$\chi = \phi+\tan^{-1}\left(\sqrt{\frac{\gamma-1}{\gamma+1}}\cot\sqrt{\frac{\gamma-1}{\gamma+1}}\,\phi\right), \tag{109.15}$$

the angles χ and ϕ being measured from the same initial line.

Since we must have $v_r > 0, c > 0$, the angle ϕ in these formulae can vary only between 0 and $\phi_{\max}$, where

$$\phi_{\max} = \tfrac{1}{2}\pi\sqrt{[(\gamma+1)/(\gamma-1)]}. \tag{109.16}$$

This means that the rarefaction wave can occupy a sector whose angle does not exceed $\phi_{\max}$; for a diatomic gas (air, for example), this angle is 219·3°. When ϕ varies from 0 to $\phi_{\max}$, the angle χ varies from $\frac{1}{2}\pi$ to $\phi_{\max}$. Thus the direction of the velocity in the rarefaction wave can turn through an angle not exceeding $\phi_{\max}-\frac{1}{2}\pi$ (= 129·3° for air).

For $\phi = \phi_{\max}$ the pressure is zero. In other words, if the rarefaction wave occupies the maximum angle, the weak discontinuity on one side is a boundary with a vacuum, and is, of course, a streamline; we have $v_\phi = c = 0$, $v_r = v = \sqrt{[(\gamma+1)/(\gamma-1)]}c_* = v_{max}$, i.e. the velocity is radial and attains its limiting value v_{max} (see §83).

Figure 97 shows graphs of p/p_*, c_*/v and $\chi_1 = \chi - \frac{1}{2}\pi$ as functions of the angle ϕ for air ($\gamma = 1{\cdot}4$).

It is useful to note the form of the curve in the $v_x v_y$-plane defined by formulae (109.12) and (109.13) (called the *velocity hodograph*). It is an arc of an epicycloid between circles with radii $v = c_*$ and $v = v_{max}$ (Fig. 98).

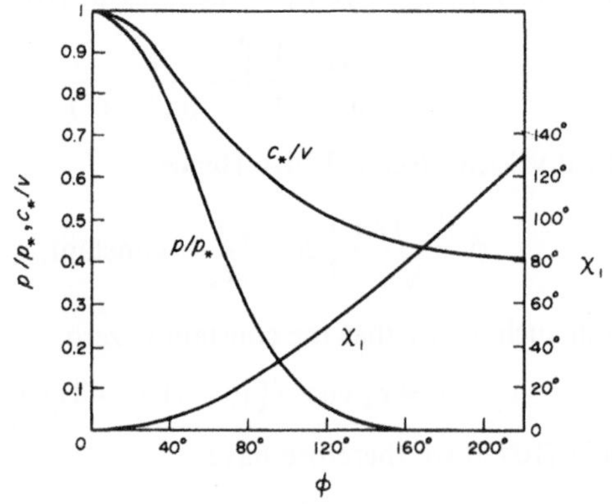

FIG. 97

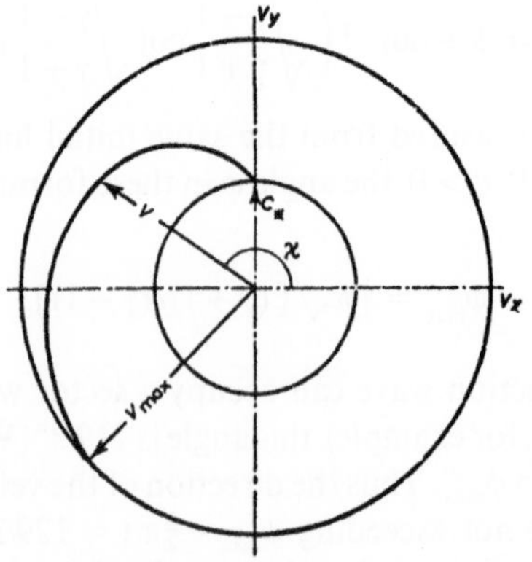

FIG. 98

PROBLEMS

PROBLEM 1. Determine the form of the streamlines in a rarefaction wave.

SOLUTION. The equation of the streamlines for two-dimensional flow is, in polar coordinates, $dr/v_r = r d\phi/v_\phi$. Substituting (109.12) and (109.13) and integrating, we obtain

$$r = r_0 \cos^{-(\gamma+1)/(\gamma-1)} \sqrt{[(\gamma-1)/(\gamma+1)]}\phi.$$

These streamlines form a family of similar curves concave toward the origin, which is the centre of similarity.

PROBLEM 2. Determine the maximum possible angle between the weak discontinuities bounding a rarefaction wave, for given values v_1, c_1 of the gas velocity and the velocity of sound at one discontinuity.

SOLUTION. The angle ϕ corresponding to the first discontinuity is, by (109.12),

$$\phi_1 = \sqrt{\frac{\gamma+1}{\gamma-1}} \cos^{-1}\frac{c_1}{c_*}.$$

The value of ϕ_2 is ϕ_{max}, so that the angle required is

$$\phi_2 - \phi_1 = \sqrt{\frac{\gamma+1}{\gamma-1}} \sin^{-1}\frac{c_1}{c_*}.$$

The critical velocity c_* is given in terms of v_1 and c_1 by Bernoulli's equation:

$$w_1 + \tfrac{1}{2}v_1^2 = \frac{c_1^2}{\gamma-1} + \tfrac{1}{2}v_1^2 = \frac{\gamma+1}{2(\gamma-1)}c_*^2.$$

The maximum possible angle through which the gas velocity can turn in a rarefaction wave is accordingly, by (109.15), the difference $\chi_{max} = \chi(\phi_1) - \chi(\phi_2)$:

$$\chi_{max} = \sqrt{\frac{\gamma+1}{\gamma-1}} \sin^{-1}\frac{c_1}{c_*} - \sin^{-1}\frac{c_1}{v_1}.$$

As a function of v_1/c_1, χ_{max} is greatest for $v_1/c_1 = 1$:

$$\chi_{max} = \tfrac{1}{2}\pi\left(\sqrt{\frac{\gamma+1}{\gamma-1}} - 1\right).$$

For $v_1/c_1 \to \infty$, χ_{max} tends to zero:

$$\chi_{max} = \frac{2}{\gamma-1}\frac{c_1}{v_1}.$$

§110. Classification of intersections of surfaces of discontinuity

Shock waves can intersect along a line. In considering the flow near a small segment of this line, we can assume that it is a straight line, and that the surfaces of discontinuity are planes. It is therefore sufficient to discuss the intersection of plane shock waves.

The line of intersection of two discontinuities is, mathematically, a singular line, as has already been mentioned at the beginning of §109. The flow pattern near this line consists of a number of sectors, in each of which we have either uniform flow or a rarefaction wave of the kind described in §109. It is possible to give a general classification of the possible types of intersection of surfaces of discontinuity.†

First of all, we must make the following remark. If the gas flow on both sides of a shock wave is supersonic, then (as mentioned at the beginning of §92) we can speak of the "direction" of the shock wave, and accordingly distinguish shock waves leaving the line of

† This is due to L. D. Landau (1944), with supplementary points (relating to the interaction of shock waves with tangential and weak discontinuities) added by S. P. D'yakov (1954).

intersection from those reaching it. In the former case, the tangential velocity component is directed away from the line of intersection, and we can say that the disturbances which cause the discontinuity leave this line. In the latter case, the disturbances leave a point not on the line of intersection.

If the flow on one side of the shock wave is subsonic, then disturbances are propagated in both directions along its surface, and the "direction" of the shock has, strictly, no meaning. In the arguments given below, however, what is important is that disturbances leaving the point of intersection can be propagated along such a discontinuity. In this sense, such shock waves play the same part in the following discussion as the purely supersonic shocks which leave the intersection, and we shall include both kinds in the term "shocks which leave the intersection".

Figures 99–104 show the flow patterns in a plane perpendicular to the line of intersection. We can assume, without loss of generality, that the flow occurs in this plane. The velocity component parallel to the line of intersection (which lies in all the planes of discontinuity) must be the same in all regions round the line of intersection, and can therefore be made to vanish by an appropriate choice of the coordinate system.

Let us first mention some configurations that are certainly impossible. It is easy to see that there can be no intersection of shock waves in which no shock reaches the intersection. For instance, in the intersection of two shock waves leaving the intersection, shown in Fig. 99a, the streamlines of the flow incident from the left would deviate in opposite directions, whereas the velocity should be constant throughout region 2, and this difficulty cannot be overcome by adding any further discontinuities in region 2.† Similarly, we can see that the intersection of a shock wave and a rarefaction wave both leaving the intersection, shown in Fig. 99b, is impossible; although the velocity in region 2 can be constant in direction, the pressure cannot be constant, since it increases in a shock wave but decreases in a rarefaction wave.

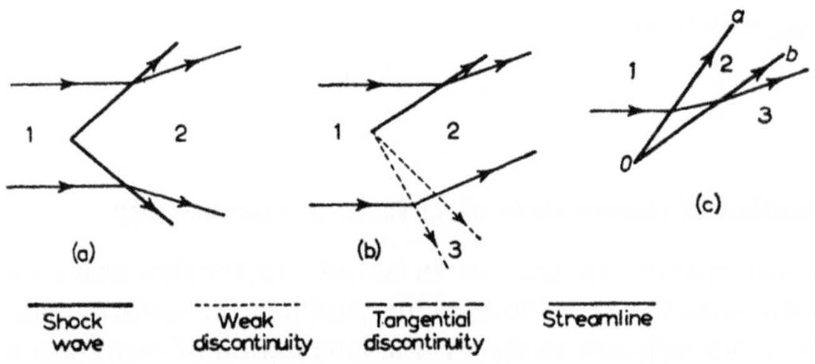

FIG. 99

Next, since the intersection cannot affect shock waves reaching it, the simultaneous intersection (along a common line) of more than two such waves, which are due to other causes, would be an improbable coincidence. Thus only one or two shock waves can reach the intersection.

† In order not to encumber the discussion with repetitive arguments, we shall not give similar considerations for cases where there are regions of subsonic flow and the shock leaving the intersection is actually a shock wave bounded by a subsonic region.

The following fact is very important. The gas flowing past a point of intersection can pass through only one shock or rarefaction wave leaving this point. For example, let the gas pass through two successive shock waves leaving the point O, as shown in Fig. 99c. Since the normal velocity component v_{2n} behind the shock Oa is $v_{2n} < c_2$, the velocity component in region 2 normal to the shock Ob must also be less than c_2, in contradiction to a fundamental property of shock waves. Similarly, we can see that the gas cannot pass through two successive rarefaction waves, or a shock wave and a rarefaction wave, leaving the point O.

These arguments evidently cannot be extended to shock waves reaching the point of intersection.

We can now proceed to enumerate the possible types of intersection. Figure 100 shows an intersection involving one shock wave Oa reaching it and two shock waves Ob, Oc leaving it. This case may be regarded as the splitting of one shock wave into two.† It is easy to see that, besides the two shock waves leaving, there must be formed a tangential discontinuity Od lying between them, which separates the gas flowing through Ob from that flowing through Oc.‡ For the shock Oa is due to other causes, and is therefore completely defined. This means that the thermodynamic quantities (p and ρ, say) and the velocity $\mathbf{v}$ have given values in regions 1 and 2. There remain at our disposal, therefore, only two quantities (the angles giving the directions of the discontinuities Ob and Oc) with which to satisfy, in general, four conditions (the constancy of p, ρ and two velocity components) in the region 3–4, which would have to be satisfied in the absence of the tangential discontinuity Od. The addition of the latter, however, reduces the number of conditions to two (the constancy of the pressure and of the direction of the velocity).

An arbitrary shock wave, however, cannot divide in this manner. A shock wave reaching the intersection is defined by two parameters (for a given thermodynamic state of gas 1), say the Mach number M_1 of the incident stream and the ratio of pressures p_1/p_2. It can divide in two only in a certain region in the plane of these two parameters.††

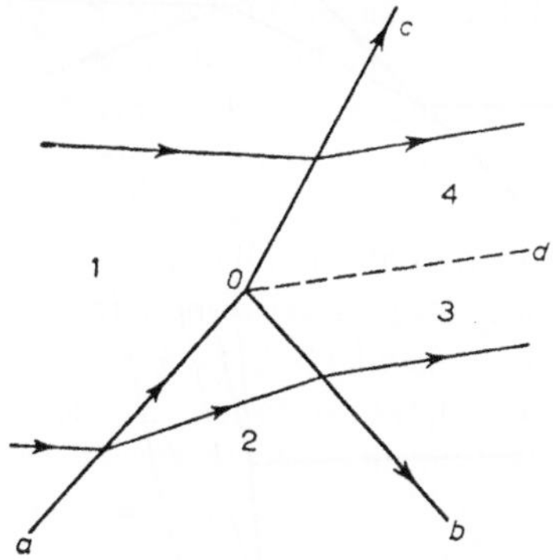

FIG. 100

† It should be noticed that a shock wave cannot divide into a shock and a rarefaction wave; it is easily seen that the changes in the pressure and the direction of the velocities in the two waves leaving cannot be reconciled.

‡ As usual, the tangential discontinuity in reality becomes a turbulent region.

†† The determination of this region involves very laborious algebraic or numerical calculations. Again, the "direction" of the shock waves is important. Cases with two shocks reaching the intersection and one leaving it would constitute an intersection of two discontinuities due to other causes, and therefore reaching the point of intersection with given values of all parameters. Their fusion into one shock is possible only when these arbitrary parameters are related in a certain way, and this would be an improbable coincidence.

Intersections involving two shock waves reaching them can be regarded as "collisions" of two shocks due to other causes. Here two essentially different cases are possible, as shown in Fig. 101.

In the first case, the collision of two shock waves results in two other shock waves leaving the point of intersection. If all the necessary conditions are to be fulfilled, a tangential discontinuity must again be formed, and it must lie between the two resulting shock waves.

In the second case, instead of two shock waves, there are formed one shock wave and one rarefaction wave.

Two colliding shock waves are defined by three parameters (for instance, M_1 and the ratios p_1/p_2, p_1/p_3). The types of intersection just described are possible only for certain ranges of values of these parameters. If the values of the parameters do not lie in these ranges, the collision of the shock waves must be preceded by their breaking up.

Let us next consider the types of intersection that can occur when a shock wave meets a tangential discontinuity.

Figure 102a shows the reflection of a shock wave from the boundary between gas in motion and gas at rest. Region 5 contains gas at rest, separated from the gas in motion by a tangential discontinuity. In the two regions 1 and 4 adjoining it, the pressure must be the same and equal to p_5. Since the pressure increases in a shock wave, it is clear that the shock must be reflected from the tangential discontinuity as a rarefaction wave 3, which reduces the pressure to its initial value. The tangential discontinuity has a kink at the point of intersection.

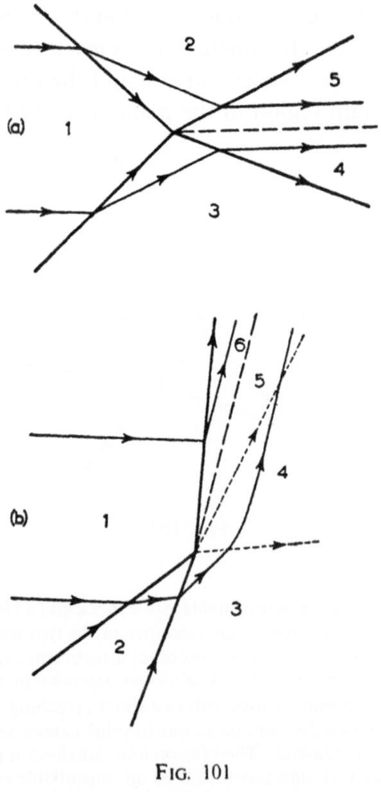

FIG. 101

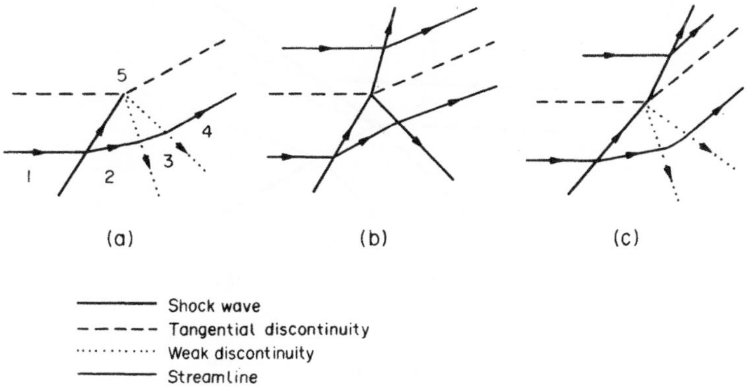

FIG. 102

The intersection of a shock wave with a tangential discontinuity having a non-zero but subsonic flow velocity on the other side is impossible. Neither a shock wave nor a rarefaction wave can penetrate into a subsonic region, and in this region there can thus be only a trivial flow with constant velocity, so that the tangential discontinuity cannot have a kink. The shock wave cannot be reflected as a rarefaction wave, since this would necessarily give the tangential discontinuity a kink; reflection as a shock wave is also impossible, since it would then be impossible to satisfy the condition of equal pressures at the tangential discontinuity.

If the flow on either side of a tangential discontinuity is supersonic, there are two possible configurations. In one (Fig. 102b), in addition to the incident shock wave, reflected and refracted shocks are formed; the tangential discontinuity has a kink. In the other (Fig. 102c), a reflected rarefaction wave is formed, and a refracted shock wave transmitted into the other gas. Both can occur only in certain ranges of the parameters of the incident shock wave and the tangential discontinuity.†

The interaction of two tangential discontinuities may yield a configuration with no shock wave reaching it and two leaving it; in the absence of the tangential discontinuities, this is impossible, as shown above. In Fig. 103, the gas in region 1 is at rest; the configuration is, however, evidently possible only if there is supersonic flow in regions 2 and 5.

We may briefly discuss the intersection of a shock wave with a weak discontinuity arriving from an external source. Here two cases can occur, according as the flow behind the shock wave is supersonic or subsonic. In the former case (Fig. 104a), the weak discontinuity is refracted at the shock wave into the space behind the latter; the shock itself is not refracted at the intersection, but has a singularity of a higher order, like that at a weak discontinuity. Moreover, the entropy change in the shock wave must cause behind it a weak tangential discontinuity, at which the derivatives of the entropy are discontinuous.

If, however, the flow becomes subsonic behind the shock wave, the weak discontinuity

† These two configurations are a kind of generalization of the cases shown in Figs. 100 and 101b.

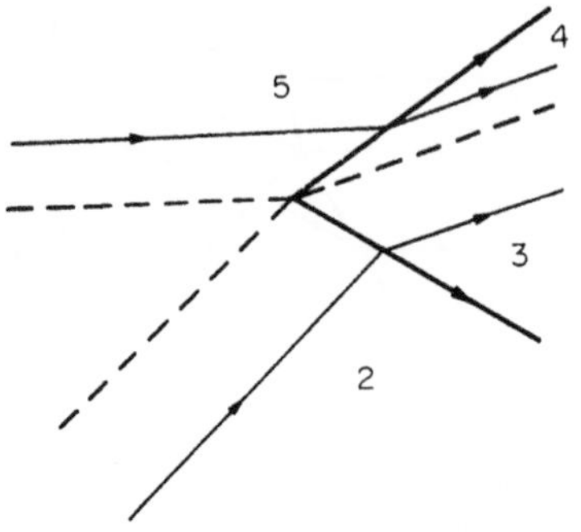

FIG. 103

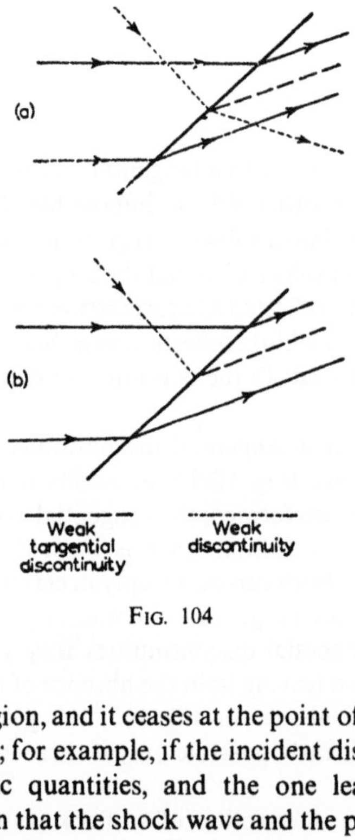

FIG. 104

cannot penetrate into this region, and it ceases at the point of intersection (Fig. 104b). The latter is now a singular point; for example, if the incident discontinuity relates to the first derivatives of hydrodynamic quantities, and the one leaving is a weak tangential discontinuity, it can be shown that the shock wave and the pressure distribution near the intersection have logarithmic singularities. Furthermore, as in the previous case, a weak tangential discontinuity of the entropy must occur behind the shock wave.†

The foregoing discussion of the interaction of shock waves with weak discontinuities applies also to that with weak tangential discontinuities. If the flow behind the shock wave is supersonic, a weak discontinuity and a weak tangential discontinuity are formed there; if it is subsonic, only a refracted weak tangential discontinuity occurs.

† A detailed quantitative analysis of the intersection of shock waves with weak discontinuities is given by S. P. D'yakov, *Soviet Physics JETP* **6**, 729, 739, 1958.

Finally, let us refer also to the interaction between weak and tangential discontinuities. If the flow on either side of the tangential discontinuity is supersonic, then reflected and refracted weak discontinuities are formed, in addition to the incident one; if there is subsonic flow beyond the tangential discontinuity, the weak discontinuity does not penetrate there, and undergoes "total internal reflection".

§111. The intersection of shock waves with a solid surface

An important part in the phenomenon of steady interaction of shock waves with the surface of a body is played by their intersection with the boundary layer. This interaction is very complex, and a detailed discussion is outside the scope of this book; we shall give here only some general results.†

The pressure is discontinuous in a shock wave, and increases in the direction of motion of the gas. Hence, if the shock wave intersects the surface, there must be a finite increment of pressure over a very short distance near the place of intersection, i.e. there must be a very large positive pressure gradient. We know, however, that such a rapid increase in pressure cannot occur near a solid wall (see the end of §40); it would cause separation, and the pattern of flow round the body is changed in such a way that the shock wave moves away to a sufficient distance from the surface. An exception occurs only when the shock wave is weak. It is clear from the proof given at the end of §40 that the impossibility of a positive pressure discontinuity at the boundary layer is a consequence of the assumption that this discontinuity is large: it must exceed a certain limit depending on the value of R, which diminishes when R increases.

Thus we reach the following important conclusions. The steady intersection of strong shock waves with a solid surface is impossible. A solid surface can intersect only weak shock waves, and the limiting intensity decreases with increasing R. The maximum permissible intensity of the shock wave also depends on whether the boundary layer is laminar or turbulent. If the boundary layer is turbulent, the onset of separation is retarded (§45). In a turbulent boundary layer, therefore, stronger shock waves can leave the surface of the body than in a laminar boundary layer.

It should be emphasized that these arguments rely on the fact that the boundary layer exists in front of the shock wave (i.e upstream of it). The results obtained therefore do not relate to shock waves which leave the leading edge of the body; the latter can occur, for instance in flow past an acute-angled wedge, a case which is discussed in detail in §112. In the latter case the gas reaches the vertex of the angle from outside, i.e. from a region in which there is no boundary layer. It is therefore clear that the present arguments do not deny that shock waves can occur which leave the vertex of such an angle.

In subsonic flow, separation can occur only when the pressure in the main stream increases downstream along the surface. In supersonic flow, however, it is found that separation can occur even when the pressure decreases downstream. Such a phenomenon can occur by the combination of a weak shock wave with a separation, the pressure increase necessary for separation taking place in the shock wave; the pressure may either increase or decrease downstream in the region in front of the shock wave.

All the above discussion relates only to a steady intersection, with the shock wave and the

† The boundary layer necessarily contains a subsonic part adjoining the surface, into which the shock wave cannot penetrate. In speaking of the intersection, we ignore this fact, which does not affect the following discussion.

body at relative rest. Let us now consider non-steady intersections, when a moving shock wave is incident on a solid body, so that the line of intersection moves on the surface. Such an intersection is accompanied by reflection of the shock wave: besides the incident wave, a reflected wave leaving the body is formed.

We shall examine the phenomenon in a system of coordinates which moves with the line of intersection; in this system the shock waves are steady. The simplest type of reflection occurs when the reflected wave leaves the line of intersection itself; this is called *regular* reflection (Fig. 105). If the angle of incidence α_1 and the intensity of the incident shock are given, the flow in region 2 is uniquely determined. The gas velocity in the reflected shock must be turned through an angle such that it is again parallel to the surface. When this angle is given, the position and intensity of the reflected shock are obtained from the equation of the shock polar. For a given angle, the shock polar determines two different shock waves, those of the weak and strong families (§92). Experimental results show that in fact the reflected shock always belongs to the weak family, and we shall assume this in what follows. It should be pointed out that, when the intensity of the incident shock tends to zero, the intensity of the reflected shock then tends to zero also, and the angle of reflection α_2 tends to the angle of incidence α_1, as we should expect in accordance with the acoustic approximation. In the limit $\alpha_1 \to 0$, the reflected shock of the weak family passes continuously into the shock obtained when a shock wave is incident "frontally" (§100, Problem 1).

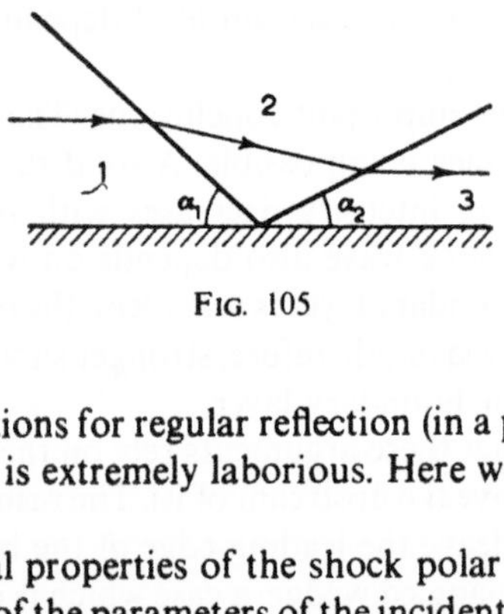

FIG. 105

The mathematical calculations for regular reflection (in a perfect gas) offer no difficulty in principle, but the algebra is extremely laborious. Here we shall give only some of the results.†

It is clear from the general properties of the shock polar that regular reflection is not possible for arbitrary values of the parameters of the incident shock (the angle of incidence α_1 and the ratio p_2/p_1). For a given ratio p_2/p_1 there is a maximum possible angle α_{1k}, and for $\alpha_1 > \alpha_{1k}$ regular reflection is impossible. As $p_2/p_1 \to \infty$, the maximum angle tends to a value which depends on γ ($= 40°$ for air). As $p_2/p_1 \to 1$, α_{1k} tends to $90°$, i. e. regular reflection is possible for any angle of incidence. Figure 106 shows α_{1k} as a function of p_1/p_2 for $\gamma = 7/5$ and $5/3$.

The angle of reflection α_2 is not in general the same as the angle of incidence. There is a value α_* of the angle of incidence such that, if $\alpha_1 < \alpha_*$, the angle of reflection $\alpha_2 < \alpha_1$; if $\alpha_1 > \alpha_*$, on the other hand, $\alpha_2 > \alpha_1$. The value of α_* is $\frac{1}{2}\cos^{-1}\frac{1}{2}(\gamma - 1)$ ($= 39{\cdot}2°$ for air); it does not depend on the intensity of the incident shock.

† A more detailed account of the reflection of shock waves is given by R. Courant and K. O. Friedrichs, *Supersonic Flow and Shock Waves*, New York 1948, Chapter IV, by R. von Mises, *Mathematical Theory of Compressible Fluid Flow*, New York 1958, §23, and by W. Bleakney and A. H. Taub, *Reviews of Modern Physics* 21, 584, 1949.

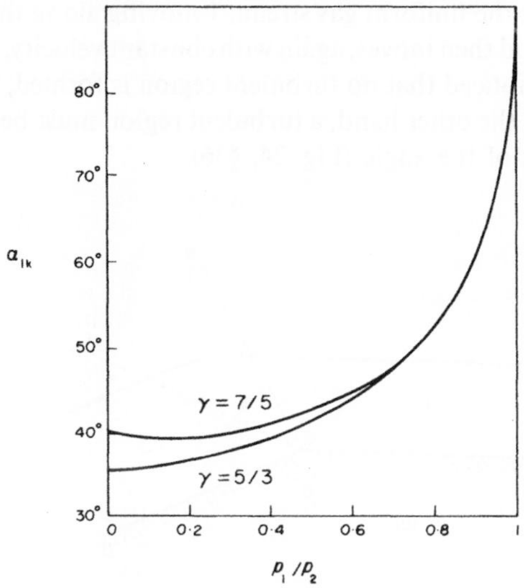

FIG. 106

For $\alpha_1 > \alpha_{1k}$ regular reflection is impossible, and the incident shock wave must break up at a distance from the surface, so that we have the pattern shown in Fig. 107, with three shock waves, and a tangential discontinuity leaving the point where the incident shock wave divides. This is called *Mach reflection.*

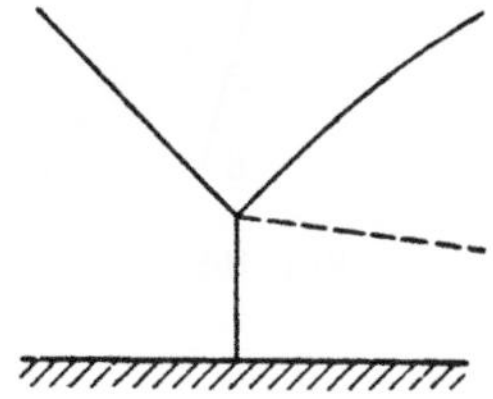

FIG. 107

§112. Supersonic flow round an angle

In investigating the flow near the vertex of an angle on the surface, it is again sufficient to consider small portions of the vertex and suppose it straight, the angle being formed by two intersecting planes. We shall speak of flow outside an angle if the angle is greater than π, and of flow inside an angle if it is less than π.

Subsonic flow past an angle is not essentially different from the flow of an incompressible fluid. Supersonic flow, however, is entirely different; an important property of it is the occurrence of discontinuities leaving the vertex of the angle.

Let us first consider the possible flow patterns when a supersonic gas stream reaches the vertex along one of the sides of the angle. In accordance with the general properties of supersonic flow, the stream remains uniform up to the vertex. The turning of the stream into the direction parallel to the other side of the angle occurs in a rarefaction wave leaving the vertex, and the flow pattern consists of three regions separated by weak discontinuities

(Oa and Ob in Fig. 108): the uniform gas stream 1 moving along the side AO is turned in the rarefaction wave 2 and then moves, again with constant velocity, along the other side of the angle. It should be noticed that no turbulent region is formed; in a similar flow of an incompressible flüid, on the other hand, a turbulent region must be formed, with a line of separation at the vertex of the angle (Fig. 24, §36).

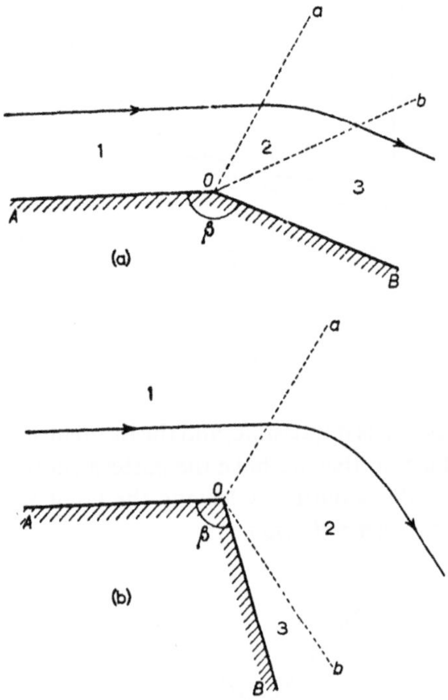

FIG. 108

Let v_1 be the velocity of the incident stream (1 in Fig. 108), and c_1 the velocity of sound in it. The position of the weak discontinuity Oa is determined immediately from the Mach number $M_1 = v_1/c_1$ by the condition that it intersect the streamlines at the Mach angle. The changes in velocity and pressure in the rarefaction wave are determined by formulae (109.12)–(109.15); all that is needed is the direction from which the angle ϕ in these formulae is to be measured. The straight line $\phi = 0$ corresponds to $v = c = c_*$; for $M_1 > 1$, there is in fact no such line, since $v/c > 1$ everywhere. However, if the rarefaction wave is imagined to be formally extended into the region to the left of Oa, we can use formula (109.12), and we find that the discontinuity Oa must correspond to a value of ϕ given by

$$\phi_1 = \sqrt{\frac{\gamma+1}{\gamma-1}}\cos^{-1}\frac{c_1}{c_*},$$

and that ϕ must increase from Oa to Ob. The position of the discontinuity Ob is determined by the fact that the direction of the velocity becomes parallel to the side OB of the angle.

The angle through which the stream turns in the rarefaction wave cannot exceed the value χ_{max} determined in §109, Problem 2. If the angle β round which the flow occurs is less than $\pi - \chi_{max}$, the rarefaction wave cannot turn the stream through the necessary angle, and we have the flow pattern shown in Fig. 108b. The rarefaction in the wave 2 then proceeds to zero pressure (reached on the line *Ob*), so that the rarefaction wave is separated from the wall by a vacuum (region 3).

The flow pattern described above is not the only possible one, however. Figures 109 and 110 show patterns in which a region of gas at rest adjoins the second side of the angle, this region being separated from the moving gas by a tangential discontinuity; as usual, the latter becomes a turbulent region, so that the case considered corresponds to the presence of separation.† The stream is turned through a certain angle in a rarefaction wave (Fig. 109) or in a shock wave (Fig. 110). The latter case, however, is possible only if the shock wave is not too strong (in accordance with the general considerations given in §111).

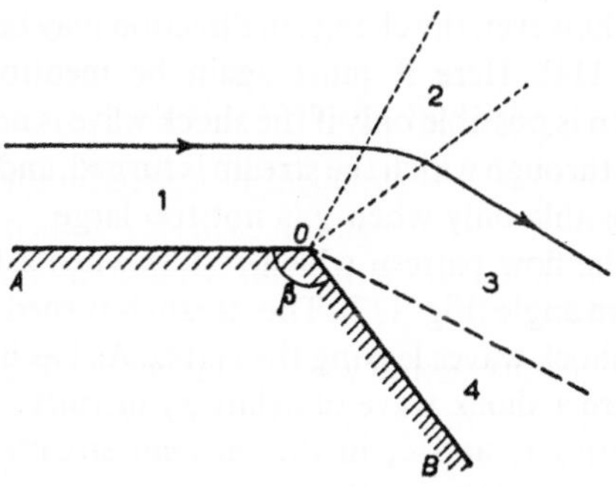

FIG. 109

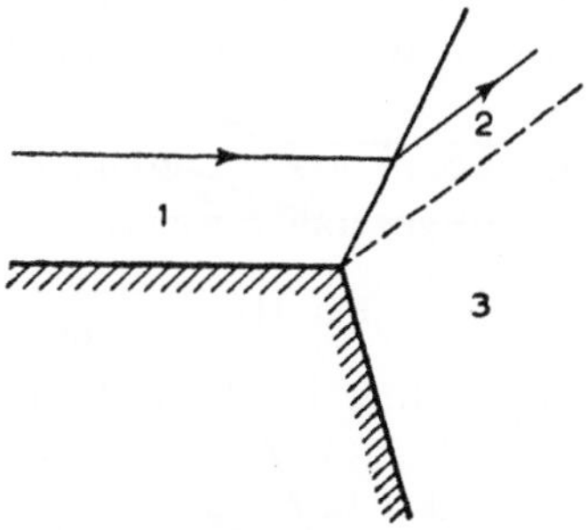

FIG. 110

Which of these flow patterns will occur in any particular case depends in general on the conditions far from the angle. For instance, when gas flows out of a nozzle (the vertex of the angle being here the edge of the outlet), the relation between the pressure p_1 of the outgoing gas and the pressure p_e of the external medium is of importance. If $p_e < p_1$, the flow is of the type shown in Fig. 109; the position and angle of the rarefaction wave are then determined by the condition that the pressure in regions 3 and 4 be equal to p_e. The smaller

† According to experimental results, the compressibility of the gas somewhat diminishes the angle of the turbulent region resulting from the tangential discontinuity.

p_e, the greater the angle through which the stream must be turned. If, however, the angle β (Fig. 109) is large, the gas pressure cannot reach the required value p_e; the direction of the velocity becomes parallel to the side OB of the angle before the pressure falls to p_e. The flow near the edge of the outlet will then be as shown in Fig. 107. The pressure near the outer side OB of the outlet is entirely determined by the angle β, and does not depend on the pressure p_e; the final decrease of the pressure to p_e occurs only at a distance from the outlet.

If $p_e > p_1$, on the other hand, the flow round the edge of the outlet is of the type shown in Fig. 110, with a shock wave which leaves the edge and raises the pressure from p_1 to p_e. This is possible, however, only if the difference between p_e and p_1 is not too large, i.e. the shock wave is not too strong; otherwise there is separation at the inner surface of the nozzle, and the shock wave moves into the nozzle, in the manner described in §97.

Next, let us consider flow inside an angle. In the subsonic case such a flow is accompanied by separation at a point ahead of the vertex (see the end of §40). For a supersonic incident flow, however, the change in direction may be effected by a shock wave leaving the vertex (Fig. 111). Here it must again be mentioned that such a simple separationless flow pattern is possible only if the shock wave is not too strong. Its intensity increases with the angle χ through which the stream is turned, and we can therefore say that separationless flow is possible only when χ is not too large.

Let us now consider the flow pattern which results when a free supersonic stream is incident on the vertex of an angle (Fig. 112). The stream is turned into directions parallel to the sides of the angle by shock waves leaving the vertex. As has been shown in §111, this is the exceptional case where a shock wave of arbitrary intensity can leave a solid surface.

If we know the velocities $\mathbf{v}_1$ and c_1 in the incident stream 1, we can determine the positions of the shock waves and the gas flow in the regions behind them. The direction of the velocity $\mathbf{v}_2$ must be parallel to the side OA of the angle: $v_{2y}/v_{2x} = \tan\chi$. Thus v_2 and the

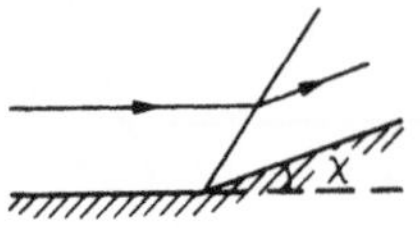

FIG. 111

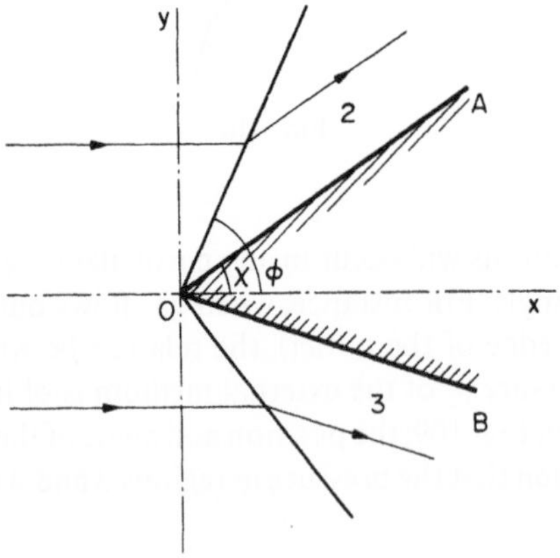

FIG. 112

angle ϕ giving the position of the shock wave can be determined immediately from the shock polar, using a chord through the origin at the known angle χ to the axis of abscissae (Fig. 64), as explained in §92. We have seen that, for a given χ, the shock polar gives two different shock waves, with different values of ϕ. One of these (corresponding to the point B in Fig. 64) is the weaker, and in general leaves the flow supersonic; the other, stronger, shock renders the flow subsonic. In the present case of flow past an angle on a finite solid surface, we must always take the former, i.e. the weak shock. It should be borne in mind that this choice is really decided by the conditions of the flow far from the angle. In flow past a very acute angle (χ small), the resulting shock wave must obviously be very weak. It is natural to suppose that, as the angle increases, the intensity of the shock increases monotonically; this corresponds to a movement along the arc QC of the shock polar (Fig. 64), from Q towards C.†

We have also seen in §92 that the angle through which the velocity vector is turned in a shock wave cannot exceed a certain value χ_{max}, which depends on M_1. The flow pattern described above is therefore impossible if either of the sides of the angle makes an angle greater than χ_{max} with the direction of the incident stream. In this case the gas flow near the angle must be subsonic; this is achieved by the appearance of a shock wave somewhere in front of the angle (see §122). Since χ_{max} increases monotonically with M_1, we can also say that, for a given value of the angle χ, M_1 for the incident stream must be greater than a certain value $M_{1,min}$.

Finally, it may be mentioned that, if the sides of the angle are situated, relative to the incident stream, as shown in Fig. 113, then a shock wave is of course formed on only one side of the angle; the stream is turned on the other side by a rarefaction wave.

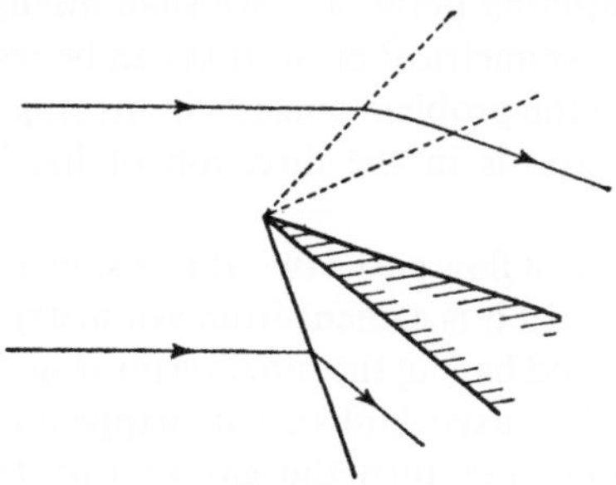

FIG. 113

PROBLEMS

PROBLEM 1. Determine the position and intensity of the shock wave in flow past a very small angle ($\chi \ll 1$) for Mach numbers such that $M_1\chi \ll 1$.

SOLUTION. When $\chi \ll 1$, the shock polar gives two values: close to $\frac{1}{2}\pi$ (near P in Fig. 64) and close to the Mach angle α_1 (near Q). The relevant shock wave in the weak family corresponds to the latter. From (92.11), when $\chi \ll 1$,

$$M_1{}^2 \sin^2\phi - 1 \cong \chi \cdot \tfrac{1}{2}(\gamma+1)M_1{}^2 \tan\alpha_1$$
$$= \chi \cdot \tfrac{1}{2}(\gamma+1)M_1{}^2/\sqrt{(M_1{}^2-1)}.$$

† Cf., however, the first footnote to §113. The purely formal problem of flow past a wedge formed by the intersection of two infinite planes is of no physical interest.

Substitution of this in (92.9) gives

$$(p_2-p_1)/p_1 = \gamma M_1{}^2\chi/\sqrt{(M_1{}^2-1)}.$$

The angle ϕ is sought in the form $\phi = \alpha_1+\varepsilon$, where $\varepsilon \ll \alpha_1$; the same formula gives

$$\phi-\alpha_1 = \tfrac{1}{4}(\gamma+1)M_1{}^2\chi/\sqrt{(M_1{}^2-1)}.$$

When $M_1 \gg 1$, the angle $\alpha_1 \cong 1/M_1$, and for the above expressions to be valid we must have $M_1\chi \ll 1$.

PROBLEM 2. The same as Problem 1, but for a Mach number so great that $M_1\chi \gg 1$.

SOLUTION. In this case, ϕ and χ have the same order of magnitude. From (92.11), $\phi = \tfrac{1}{2}(\gamma+1)\chi$. The pressure ratio is, from (92.9),

$$p_2/p_1 = 2\gamma M_1{}^2\phi^2/(\gamma+1) = \tfrac{1}{2}\gamma(\gamma+1)M_1{}^2\chi^2.$$

The value of M_2 behind the shock is, from (92.12),

$$M_2 = \frac{1}{\chi}\sqrt{\frac{2}{\gamma(\gamma-1)}},$$

and thus remains large in comparison with unity but not in comparison with $1/\chi$. In the same approximation,

$$\rho_2/\rho_1 = (\gamma+1)/(\gamma-1), \qquad v_2/v_1 = 1;$$

the difference $v_1-v_2 \sim v_1\chi^2$. The decrease in the Mach number is therefore actually due only to the increase in the velocity of sound: $M_2/M_1 = c_1/c_2$.

§113. Flow past a conical obstacle

The problem of steady supersonic flow near a pointed projection on the surface of a body is three-dimensional, and is very much more complicated than that of flow past an angle with a line vertex. A problem that has been completely solved is that of axially symmetrical flow past a projecting point, and we shall discuss this case.

Near its vertex, an axially symmetrical projection can be regarded as a right cone with circular cross-section, and so the problem consists in investigating the flow of a uniform stream past a cone whose axis is in the direction of incidence. The flow pattern is qualitatively as follows.

As in the analogous problem of flow past a two-dimensional angle, a shock wave must be formed (A. Busemann 1929), and it is evident from symmetry that this shock is a conical surface coaxial with the cone and having the same vertex (Fig. 114 shows the cross-section of the cone by a plane through its axis). Unlike what happens in the two-dimensional case, however, the shock wave does not turn the gas velocity through the whole angle χ necessary for the gas to flow along the surface of the cone (2χ being the vertical angle of the cone). After passing through the surface of discontinuity, the streamlines are curved, and asymptotically approach the generators of the cone. This curvature is accompanied by a continuous increase in density (besides the increase which occurs at the shock itself) and by a corresponding decrease in the velocity.

The change in the magnitude and direction of the velocity at the shock wave itself is determined by the shock polar; here again, the solution which occurs corresponds to the "weak" branch of the polar.† Accordingly, for each value of the incident stream Mach number $M_1 = v_1/c_1$, there is a definite limiting value of the half-angle χ_{max} of the cone, beyond which such flow becomes impossible and the shock wave detaches itself from the

† This may not be so, however, for certain "exotic" shapes of the obstacle. For example, it appears that the "strong" family shock may occur in flow past a cone at the forward end of a broad blunt body.

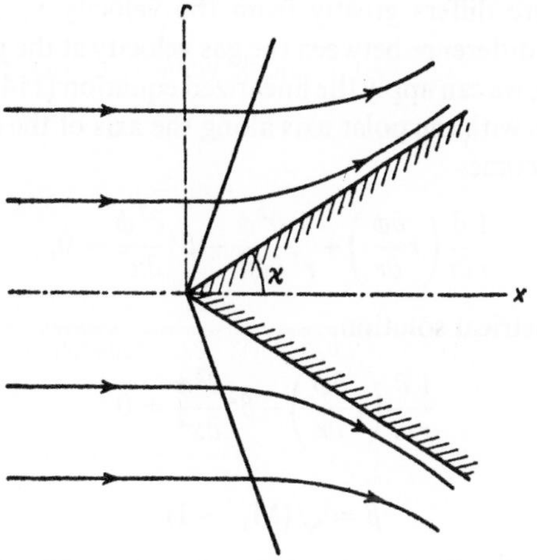

FIG. 114

cone vertex. Since there is a further rotation of the flow beyond the shock, the values of χ_{max} for flow past a cone exceed (for the same M_1) those in the two-dimensional case of flow past a wedge. Immediately behind the shock wave, the gas flow is usually supersonic, but may be subsonic when χ is close to χ_{max}. The supersonic flow may become subsonic as the cone surface is approached, in which case the velocity passes through that of sound on a certain conical surface.

The conical shock wave intersects all streamlines in the incident flow at the same angle, and is therefore of constant intensity. Hence it follows (see §114) that we have isentropic potential flow behind the shock wave also.

From the symmetry of the problem and its similarity properties (there are no characteristic constant lengths in the conditions imposed), it is evident that the distribution of all quantities (velocity, pressure) in the flow behind the shock wave will depend only on the angle θ which the radius vector from the vertex of the cone to the point considered makes with the axis of the cone (the x-axis in Fig. 114). Accordingly, the equations of motion are ordinary differential equations; the boundary conditions on these equations at the shock wave are determined by the equation of the shock polar, while those at the surface of the cone are that the velocity should be parallel to the generators. These equations, however, cannot be integrated analytically, and have to be solved numerically. We refer the reader elsewhere† for the results of the calculations, and merely give the curve (Fig. 65, §92) which shows the maximum possible angle χ_{max} as a function of M_1. We may also mention that, as $M_1 \to 1$, the angle χ_{max} tends to zero:

$$\chi_{max} = \text{constant} \times \sqrt{[(M_1 - 1)/(\gamma + 1)]}, \tag{113.1}$$

as may be deduced from the general law of transonic similarity (126.11); the constant is independent both of M_1 and of the gas involved.

An analytical solution of the problem of flow past a cone is possible only in the limit of small vertical angles (T. von Kármán and N. B. Moore 1932). It is evident that in this case

† See G. I. Taylor and J. W. Maccoll, *Proceedings of the Royal Society* A**139**, 278, 1933; J. W. Maccoll, *ibid.* **159**, 459, 1937; and N. E. Kochin, I. A. Kibel' and N. V. Roze, *Theoretical Hydromechanics* (*Teoreticheskaya gidromekhanika*), Part 2, §27, Moscow 1963.

the gas velocity nowhere differs greatly from the velocity $\mathbf{v}_1$ of the incident stream. Denoting by $\mathbf{v}$ the small difference between the gas velocity at the point considered and $\mathbf{v}_1$, and using its potential ϕ, we can apply the linearized equation (114.4); if we take cylindrical polar coordinates x, r, ω with the polar axis along the axis of the cone (ω being the polar angle), this equation becomes

$$\frac{1}{r}\frac{\partial}{\partial r}\left(r\frac{\partial\phi}{\partial r}\right)+\frac{1}{r^2}\frac{\partial^2\phi}{\partial\omega^2}-\beta^2\frac{\partial^2\phi}{\partial x^2}=0, \tag{113.2}$$

or, for an axially symmetrical solution,

$$\frac{1}{r}\frac{\partial}{\partial r}\left(r\frac{\partial\phi}{\partial r}\right)-\beta^2\frac{\partial^2\phi}{\partial x^2}=0, \tag{113.3}$$

where

$$\beta=\sqrt{(\mathrm{M}_1{}^2-1)}. \tag{113.4}$$

In order that the velocity distribution should be a function of θ only, the potential must be of the form $\phi = xf(\xi)$, where $\xi = r/x = \tan\theta$. Substituting this, we obtain for the function $f(\xi)$ the equation

$$\xi(1-\beta^2\xi^2)f''+f'=0,$$

of which the solution is elementary. The trivial solution $f=$ constant corresponds to a uniform flow; the other solution is

$$f=\text{constant}\times[\sqrt{(1-\beta^2\xi^2)}-\cosh^{-1}(1/\beta\xi)].$$

The boundary condition on the surface of the cone (i.e. for $\xi=\tan\chi\cong\chi$) is

$$v_r/(v_1+v_x)\cong(1/v_1)\partial\phi/\partial r=\chi \tag{113.5}$$

or $f'=v_1\chi$. Hence the constant is $v_1\chi^2$, and we have the following expression for the potential in the region $x>\beta r$:†

$$\phi=v_1\chi^2[\sqrt{(x^2-\beta^2r^2)}-x\cosh^{-1}(x/\beta r)]. \tag{113.6}$$

It should be noticed that ϕ has a logarithmic singularity for $r\to 0$.

We can now find the velocity components:

$$\left.\begin{aligned} v_x &= -v_1\chi^2\cosh^{-1}(x/\beta r),\\ v_r &= (v_1\chi^2/r)\sqrt{(x^2-\beta^2r^2)}. \end{aligned}\right\} \tag{113.7}$$

The pressure on the surface of the cone is calculated from formula (114.5); since ϕ has a logarithmic singularity for $r\to 0$, the velocity v_r on the surface of the cone (i.e. for small r) is large compared with v_x, and therefore we need retain only the term in $v_r{}^2$ in the formula for the pressure. The result is

$$p-p_1=\rho_1v_1{}^2\chi^2[\log(2/\beta\chi)-\tfrac{1}{2}]. \tag{113.8}$$

All these formulae, which have been derived by means of a linearized theory, cease to be valid for large M_1, comparable with $1/\chi$ (see §127).

---

† In this approximation, the cone $x=\beta r$ is a surface of weak discontinuity. In the next approximation, a shock wave occurs whose strength (relative to the pressure discontinuity) is proportional to χ^4; the vertical semi-angle exceeds the Mach angle by an amount that is likewise proportional to χ^4.

CHAPTER XII

TWO-DIMENSIONAL GAS FLOW

§114. Potential flow of a gas

In what follows we shall meet with many important cases where the flow of a gas can be regarded as potential flow almost everywhere. Here we shall derive the general equations of potential flow and discuss the question of their validity.†

After passing through a shock wave, potential flow of a gas usually becomes rotational flow. An exception, however, is formed by cases where a steady potential flow passes through a shock wave whose intensity is constant over its area; such, for example, is the case where a uniform stream passes through a shock wave intersecting every streamline at the same angle.‡ The flow behind the shock wave is then potential flow also. To prove this, we use Euler's equation in the form

$$\tfrac{1}{2}\,\mathbf{grad}\,v^2 - \mathbf{v}\times\mathbf{curl}\,\mathbf{v} = -(1/\rho)\,\mathbf{grad}\,p$$

(cf. (2.10)), or

$$\mathbf{grad}\,(w+\tfrac{1}{2}v^2) - \mathbf{v}\times\mathbf{curl}\,\mathbf{v} = T\,\mathbf{grad}\,s,$$

where we have used the thermodynamic relation $\mathrm{d}w = T\mathrm{d}s + \mathrm{d}p/\rho$. In potential flow, however, $w+\frac{1}{2}v^2$ = constant in front of the shock wave, and this quantity is continuous at the shock; it is therefore constant everywhere behind the shock wave, so that

$$\mathbf{v}\times\mathbf{curl}\,\mathbf{v} = -T\,\mathbf{grad}\,s. \qquad (114.1)$$

The potential flow in front of the shock wave is isentropic. In the general case of an arbitrary shock wave, for which the discontinuity of entropy varies over its surface, $\mathbf{grad}\,s \neq 0$ in the region behind the shock, and $\mathbf{curl}\,\mathbf{v}$ is therefore also not zero. If, however, the shock wave is of constant intensity, then the discontinuity of entropy in it is constant, so that the flow behind the shock is also isentropic, i.e. $\mathbf{grad}\,s = 0$. From this it follows that either $\mathbf{curl}\,\mathbf{v} = 0$ or the vectors $\mathbf{v}$ and $\mathbf{curl}\,\mathbf{v}$ are everywhere parallel. The latter, however, is impossible; at the shock wave, $\mathbf{v}$ always has a non-zero normal component, but the normal component of $\mathbf{curl}\,\mathbf{v}$ is always zero (since it is given by the tangential derivatives of the tangential velocity components, which are continuous).

Another important case where potential flow continues despite the shock wave is that of a weak shock. We have seen (§86) that in such a shock wave the discontinuity of entropy is of the third order relative to the discontinuity of pressure or velocity. We therefore see from (114.1) that $\mathbf{curl}\,\mathbf{v}$ behind the shock is also of the third order. This enables us to

† In §114, the flow is not yet assumed to be two-dimensional.

‡ We have already met with this situation in connection with supersonic flow past a wedge or cone (§§112, 113).

assume that we have potential flow behind the shock wave, the error being of a higher order of smallness.

We shall now derive the general equation for the velocity potential in an arbitrary steady potential flow of a gas. To do so, we eliminate the density from the equation of continuity $\operatorname{div}(\rho\mathbf{v}) \equiv \rho \operatorname{div}\mathbf{v} + \mathbf{v}\cdot\mathbf{grad}\,\rho = 0$, using Euler's equation

$$(\mathbf{v}\cdot\mathbf{grad})\mathbf{v} = -(1/\rho)\,\mathbf{grad}\,p = -(c^2/\rho)\,\mathbf{grad}\,\rho$$

and obtaining

$$c^2 \operatorname{div}\mathbf{v} - \mathbf{v}\cdot(\mathbf{v}\cdot\mathbf{grad})\mathbf{v} = 0.$$

Introducing the velocity potential by $\mathbf{v} = \mathbf{grad}\,\phi$ and expanding in components, we obtain the equation

$$(c^2-\phi_x^{\,2})\phi_{xx} + (c^2-\phi_y^{\,2})\phi_{yy} + (c^2-\phi_z^{\,2})\phi_{zz} - \\ -2(\phi_x\phi_y\phi_{xy} + \phi_y\phi_z\phi_{yz} + \phi_z\phi_x\phi_{zx}) = 0, \tag{114.2}$$

where the suffixes here denote partial derivatives. In particular, for two-dimensional flow we have

$$(c^2-\phi_x^{\,2})\phi_{xx} + (c^2-\phi_y^{\,2})\phi_{yy} - 2\phi_x\phi_y\phi_{xy} = 0. \tag{114.3}$$

In these equations, the velocity of sound must itself be expressed in terms of the velocity; this can in principle be done by means of Bernoulli's equation, $w + \frac{1}{2}v^2 = \text{constant}$, and the isentropic equation, $s = \text{constant}$. For a polytropic gas, c as a function of v is given by formula (83.18).

Equation (114.2) is much simplified if the gas velocity nowhere differs greatly in magnitude or direction from that of the stream incident from infinity.† This implies that the shock waves (if any) are weak, and so the potential flow is not destroyed.

We denote by $\mathbf{v}'$ the small difference between the gas velocity $\mathbf{v}$ at a given point and that of the main stream, $\mathbf{v}_1$. The potential ϕ is replaced by that of the velocity $\mathbf{v}'$: $\mathbf{v}' = \mathbf{grad}\,\phi'$. The equation for this potential is obtained from (114.2) by substituting $\phi = \phi' + xv_1$; we take the x-axis in the direction of the vector $\mathbf{v}_1$. We then regard ϕ' as a small quantity, and omit all terms of order higher than the first, obtaining the following linear equation:

$$(1-\mathrm{M}_1^{\,2})\frac{\partial^2\phi'}{\partial x^2} + \frac{\partial^2\phi'}{\partial y^2} + \frac{\partial^2\phi'}{\partial z^2} = 0, \tag{114.4}$$

where $\mathrm{M}_1 = v_1/c_1$; the velocity of sound is, of course, given its value at infinity.

The pressure at any point is determined in terms of the velocity in the same approximation, by a formula which can be obtained as follows. We regard p as a function of w (for given s), and use the fact that $(\partial w/\partial p)_s = 1/\rho$, writing $p - p_1 \cong (\partial p/\partial w)_s(w-w_1)$ $= \rho_1(w-w_1)$. From Bernoulli's equation we have

$$w - w_1 = -\tfrac{1}{2}[(\mathbf{v}_1+\mathbf{v})^2 - \mathbf{v}_1^{\,2}] \cong -\tfrac{1}{2}(v_y^{\,2}+v_z^{\,2}) - v_1 v_x,$$

so that

$$p - p_1 = -\rho_1 v_1 v_x - \tfrac{1}{2}\rho_1(v_y^{\,2}+v_z^{\,2}). \tag{114.5}$$

† One such case was discussed in §113 (flow past a narrow cone), and others will be found in connection with gas flow past arbitrary thin bodies.

In this expression the term in the squared transverse velocity must in general be retained, since, in the region near the x-axis (and, in particular, on the surface of the body itself), the derivatives $\partial\phi'/\partial y$, $\partial\phi'/\partial z$ may be large compared with $\partial\phi'/\partial x$.

Equation (114.4), however, is not valid if the number M_1 is very close to unity (*transonic* flow), so that the coefficient of the first term is small. It is clear that, in this case, terms of higher order in the x-derivatives of ϕ must be retained. To derive the corresponding equation, we return to the original equation (114.2); when the terms which are certainly small are neglected, this becomes

$$\left(1-\frac{\phi_x^2}{c^2}\right)\phi_{xx}+\phi_{yy}+\phi_{zz}=0. \qquad (114.6)$$

In the present case, the velocity $v_x \cong v$, and the velocity of sound c is close to the critical velocity c_*. We can therefore put $c-c_*=(v-c_*)\,(\mathrm{d}c/\mathrm{d}v)_{v=c_*}$, or $c-v=(c_*-v)[1-(\mathrm{d}c/\mathrm{d}v)_{v=c_*}]$. Using the fact that, for $v=c=c_*$, we have by (83.4) $\mathrm{d}\rho/\mathrm{d}v=-\rho/c$, we put (for $v=c_*$)

$$\frac{\mathrm{d}c}{\mathrm{d}v}=\frac{\mathrm{d}c}{\mathrm{d}\rho}\frac{\mathrm{d}\rho}{\rho v}=-\frac{\rho}{c}\frac{\mathrm{d}c}{\mathrm{d}\rho},$$

so that

$$c-v=[(c_*-v)/c]\,\mathrm{d}(\rho c)/\mathrm{d}\rho=\alpha_*(c_*-v). \qquad (114.7)$$

We have here used the expression (99.9) for the derivative $\mathrm{d}(\rho c)/\mathrm{d}\rho$, while α_* denotes the value of α (102.2) for $v=c_*$; for a polytropic gas, α is constant, so that $\alpha_*=\alpha=\frac{1}{2}(\gamma+1)$. To the same accuracy, this equation can be written as

$$v/c-1=\alpha_*(v/c_*-1). \qquad (114.8)$$

This gives the general relation between the Mach numbers M and M_* in transonic flow.

Using this formula, we can put

$$1-\frac{v_x^2}{c^2}\cong 1-\frac{v^2}{c^2}\cong 2\left(1-\frac{v}{c}\right)\cong 2\alpha_*\left(1-\frac{v}{c_*}\right).$$

Finally, we introduce a new potential by the substitution $\phi\to c_*(x+\phi)$, so that

$$\frac{\partial\phi}{\partial x}=\frac{v_x}{c_*}-1,\qquad \frac{\partial\phi}{\partial y}=\frac{v_y}{c_*},\qquad \frac{\partial\phi}{\partial z}=\frac{v_z}{c_*}. \qquad (114.9)$$

Substituting these formulae in (114.6), we obtain the following final equation for the velocity potential in a transonic flow (with the velocity everywhere almost parallel to the x-axis):

$$2\alpha_*\frac{\partial\phi}{\partial x}\frac{\partial^2\phi}{\partial x^2}=\frac{\partial^2\phi}{\partial y^2}+\frac{\partial^2\phi}{\partial z^2}. \qquad (114.10)$$

The properties of the gas appear here only through the constant α_*. We shall see later that this constant governs the entire dependence of the properties of transonic flow on the nature of the gas.

The linearized equation (114.4) becomes invalid also in another limiting case, that of very large values of M_1; however, the appearance of strong shock waves has the result that potential flow cannot actually occur for such values of M_1 (see §127).

§115. Steady simple waves

Let us determine the general form of those solutions of the equations of steady two-dimensional supersonic gas flow which describe flows in which there is a uniform plane-parallel stream at infinity, which then turns through an angle as it flows round a curved profile. We have already met a particular case of such a solution in discussing the flow near an angle; the flow considered was essentially a plane-parallel one along one side of the angle, which turned at the vertex of the angle. In this particular solution all quantities (the two velocity components, the pressure and the density) were functions of only one variable, the angle ϕ. Each of these quantities could therefore be expressed as a function of any other. Since this solution must be a particular case of the required general solution, it is natural to seek the latter on the assumption that each of the quantities p, ρ, v_x, v_y (the plane of the motion being taken as the xy-plane) can be expressed as a function of any other. This assumption is, of course, a very considerable restriction on the solution of the equations of motion, and the solution thus obtained is not the general integral of those equations. In the general case, each of the quantities p, ρ, v_x, v_y, which are functions of the two coordinates x, y, can be expressed as a function of any two of them.

Since we have a uniform stream at infinity, in which all quantities, and in particular the entropy s, are constants, and since in steady flow of an ideal fluid the entropy is constant along the streamlines, it is clear that s = constant in all space if there are no shock waves in the gas, as we shall assume.

Euler's equations and the equation of continuity are

$$v_x\frac{\partial v_x}{\partial x}+v_y\frac{\partial v_x}{\partial y} = -\frac{1}{\rho}\frac{\partial p}{\partial x}, \qquad v_x\frac{\partial v_y}{\partial x}+v_y\frac{\partial v_y}{\partial y} = -\frac{1}{\rho}\frac{\partial p}{\partial y};$$

$$\frac{\partial}{\partial x}(\rho v_x)+\frac{\partial}{\partial y}(\rho v_y) = 0.$$

Writing the partial derivatives as Jacobians, we can convert these equations to the form

$$v_x\frac{\partial(v_x, y)}{\partial(x, y)}-v_y\frac{\partial(v_x, x)}{\partial(x, y)} = -\frac{1}{\rho}\frac{\partial(p, y)}{\partial(x, y)},$$

$$v_x\frac{\partial(v_y, y)}{\partial(x, y)}-v_y\frac{\partial(v_y, x)}{\partial(x, y)} = \frac{1}{\rho}\frac{\partial(p, x)}{\partial(x, y)};$$

$$\frac{\partial(\rho v_x, y)}{\partial(x, y)}-\frac{\partial(\rho v_y, x)}{\partial(x, y)} = 0.$$

We now take x and p as independent variables. In order to effect this transformation, we need only multiply the above equations by $\partial(x, y)/\partial(x, p)$, obtaining the same equations except that $\partial(x, p)$ replaces $\partial(x, y)$ in the denominator of each Jacobian. We now expand the Jacobians, bearing in mind that all the quantities ρ, v_x, v_y are assumed to be functions of p but not of x, so that their partial derivatives with respect to x are zero. We then obtain

$$\left(v_y - v_x\frac{\partial y}{\partial x}\right)\frac{dv_x}{dp} = \frac{1}{\rho}\frac{\partial y}{\partial x}, \qquad \left(v_y - v_x\frac{\partial y}{\partial x}\right)\frac{dv_y}{dp} = -\frac{1}{\rho},$$

$$\left(v_y - v_x\frac{\partial y}{\partial x}\right)\frac{d\rho}{dp}+\rho\left(\frac{dv_y}{dp}-\frac{\partial y}{\partial x}\frac{dv_x}{dp}\right) = 0.$$

Here $\partial y/\partial x$ denotes $(\partial y/\partial x)_p$. All the quantities in these equations except $\partial y/\partial x$ are functions of p only, by hypothesis, and x does not appear explicitly. We can therefore conclude, first of all, that $\partial y/\partial x$ also is a function of p only: $(\partial y/\partial x)_p = f_1(p)$, whence

$$y = xf_1(p) + f_2(p), \tag{115.1}$$

where $f_2(p)$ is an arbitrary function of the pressure.

No further calculations are necessary if we use the particular solution, already known, for a rarefaction wave in flow past an angle (§§109, 112). It will be recalled that, in this solution, all quantities (including the pressure) are constants along any straight line (characteristic) through the vertex of the angle. This particular solution evidently corresponds to the case where the arbitrary function $f_2(p)$ in the general expression (115.1) is identically zero. The function $f_1(p)$ is determined by the formulae obtained in §109.

Equation (115.1) for various constant p gives a family of straight lines in the xy-plane. These lines intersect the streamlines at every point at the Mach angle. This is seen immediately from the fact that the lines $y = xf_1(p)$ in the particular solution with $f_2 \equiv 0$ have this property. Thus one of the families of characteristics (those leaving the surface of the body) consists, in the general case, of straight lines along which all quantities remain constant; these lines, however, are no longer concurrent.

The properties of the flow described above are, mathematically, entirely analogous to those of one-dimensional simple waves, in which one family of characteristics is a family of straight lines in the xt-plane (see §§101, 103, 104). Hence the class of flows under consideration occupies the same place in the theory of steady (supersonic) two-dimensional flow as do simple waves in non-steady one-dimensional flow. On account of this analogy, such flows are also called *simple waves*; in particular, the rarefaction wave which corresponds to the case $f_2 \equiv 0$ is called a *centred simple wave.*

As in the non-steady case, one of the most important properties of steady simple waves is that the flow in any region of the xy-plane bounded by a region of uniform flow is a simple wave (cf. §104).

We shall now show how the simple wave corresponding to flow round a given profile can be constructed. Figure 115 shows the profile in question; to the left of the point O it is straight, but to the right it begins to curve. In supersonic flow the effect of the curvature is, of course, propagated only downstream of the characteristic OA which leaves the point O. Hence the flow to the left of this characteristic is uniform; we denote by the suffix 1 quantities pertaining to this region. All the characteristics there are parallel and at an angle to the x-axis which is equal to the Mach angle $\alpha_1 = \sin^{-1}(c_1/v_1)$.

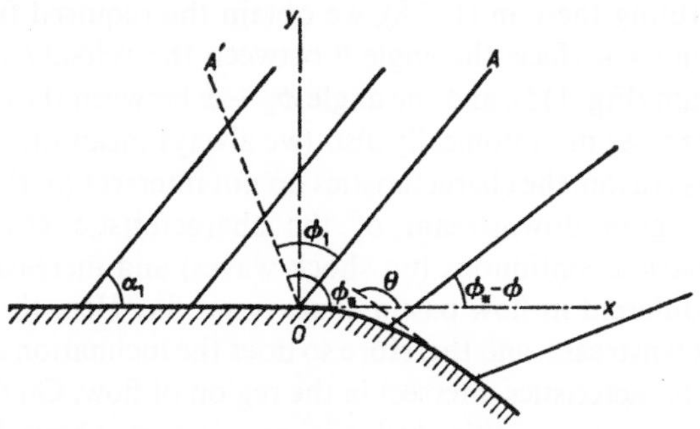

FIG. 115

In formulae (109.12)–(109.15), the angle ϕ of the characteristics is measured from the line on which $v = c = c_*$. This means (cf. §112) that the characteristic OA must have a value of ϕ given by

$$\phi_1 = \sqrt{\frac{\gamma+1}{\gamma-1}}\cos^{-1}\frac{c_1}{c_*},$$

and the angle ϕ is to be measured from OA' (Fig. 115). The angle between the characteristics and the x-axis is then $\phi_* - \phi$, where $\phi_* = \alpha_1 + \phi_1$. According to formulae (109.12)–(109.15), the velocity and pressure are given in terms of ϕ by

$$v_x = v\cos\theta, \qquad v_y = v\sin\theta, \tag{115.2}$$

$$v^2 = c_*{}^2\left[1+\frac{2}{\gamma-1}\sin^2\sqrt{\frac{\gamma-1}{\gamma+1}}\phi\right], \tag{115.3}$$

$$\theta = \phi_* - \phi - \tan^{-1}\left(\sqrt{\frac{\gamma-1}{\gamma+1}}\cot\sqrt{\frac{\gamma-1}{\gamma+1}}\phi\right), \tag{115.4}$$

$$p = p_*\cos^{2\gamma/(\gamma-1)}\sqrt{\frac{\gamma-1}{\gamma+1}}\phi. \tag{115.5}$$

The equation of the characteristics can be written

$$y = x\tan(\phi_* - \phi) + F(\phi). \tag{115.6}$$

The arbitrary function $F(\phi)$ is determined as follows when the form of the profile is given. Let the latter be $Y = Y(X)$, where X and Y are the coordinates of points on it. At the surface, the gas velocity is tangential, i.e.

$$\tan\theta = \mathrm{d}Y/\mathrm{d}X. \tag{115.7}$$

The equation of the line through the point (X, Y) at an angle $\phi_* - \phi$ to the x-axis is

$$y - Y = (x - X)\tan(\phi_* - \phi).$$

This equation is the same as (115.6) if we put

$$F(\phi) = Y - X\tan(\phi_* - \phi). \tag{115.8}$$

Starting from the given equation $Y = Y(X)$ and equation (115.7), we express the form of the profile in parametric equations $X = X(\theta)$, $Y = Y(\theta)$, the parameter being the inclination θ of the tangent. Substituting θ in terms of ϕ from (115.4), we obtain X and Y as functions of ϕ; finally, substituting these in (115.8), we obtain the required function $F(\phi)$.

In flow past a convex surface, the angle θ between the velocity vector and the x-axis decreases downstream (Fig. 115), and the angle $\phi_* - \phi$ between the characteristic and the x-axis therefore decreases monotonically also (we always mean the characteristic leaving the surface). For this reason, the characteristics do not intersect (in the region of flow, that is). Thus, in the region downstream of the characteristic OA (which is a weak discontinuity), we have a continuous (no shock waves) and increasingly rarefied flow.

The situation is different in flow past a concave profile. Here the inclination θ of the tangent increases downstream, and therefore so does the inclination of the characteristics. Consequently, the characteristics intersect in the region of flow. On different non-parallel characteristics, however, all quantities (velocity, pressure, etc.) have different values. Thus all these quantities become many-valued at points where characteristics intersect, which is physically impossible. We have already met a similar phenomenon in connection with a non-steady one-dimensional simple compression wave (§101). As in that case, it signifies that in reality a shock wave is formed. The position of the discontinuity cannot be completely determined from the solution under consideration, since this was derived on the assumption that there are no discontinuities. The only result that can be obtained is the place where the shock wave begins (the point O in Fig. 116, where the shock is shown by the continuous line OB). It is the point of intersection of characteristics whose streamline lies nearest to the surface of the body. On streamlines passing below O (i.e. nearer to the surface) the solution is everywhere single-valued; its many-valuedness begins at O. The equations for the coordinates x_0, y_0 of this point can be obtained in the same way as the corresponding equations which determine the time and place of formation of the discontinuity in a one-dimensional non-steady simple wave. If we regard the inclination of the characteristics as a function of the coordinates (x, y) of points through which they pass, then this function becomes many-valued when x and y exceed certain values x_0, y_0. In §101 the situation was the same in relation to the function $v(x, t)$, and so we need not repeat the arguments used there, but can write down immediately the equations

$$(\partial y/\partial \phi)_x = 0, \qquad (\partial^2 y/\partial \phi^2)_x = 0, \tag{115.9}$$

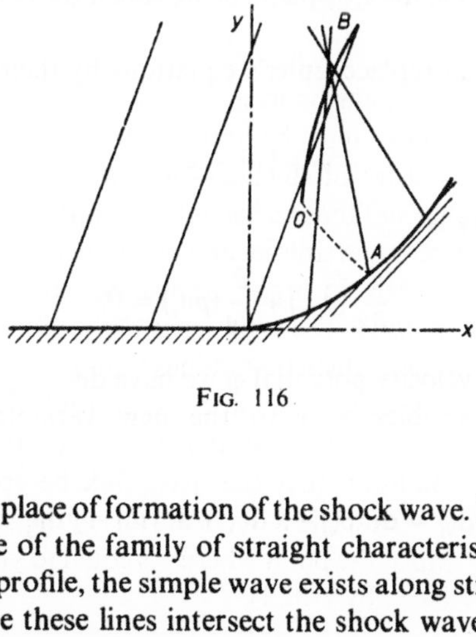

FIG. 116

which now determine the place of formation of the shock wave. Mathematically, this point is a cusp on the envelope of the family of straight characteristics (cf. §103).

In flow past a concave profile, the simple wave exists along streamlines passing above O as far as the points where these lines intersect the shock wave. The streamlines passing below O do not intersect the shock wave at all, but we cannot conclude from this that the solution in question is valid at all points on these streamlines. The reason is that the shock wave has a perturbing effect even on the gas which flows along these streamlines, and so alters the flow from what it would be in the absence of the shock wave. By a property of supersonic flow, however, these perturbations reach only the gas downstream of the characteristic OA (of the second family) which leaves the point where the shock wave begins. Thus the solution under consideration is valid everywhere to the left of AOB. The line OA itself is a weak discontinuity. We see that there cannot be a continuous (no shock waves) simple compression wave everywhere in flow past a concave surface, which would correspond to the simple rarefaction wave in flow past a convex surface.

The shock wave formed in flow past a concave profile is an example of a shock which "begins" at a point inside the stream, away from the solid walls. The point where the shock begins has some general properties, which may be noted here. At the point itself the intensity of the shock wave is zero, and near the point it is small. In a weak shock wave, however, the discontinuities of entropy and vorticity are of the third order of smallness, and so the change in the flow on passing through the shock differs from a continuous potential isentropic change only by quantities of the third order. Hence it follows that, in the weak discontinuities which leave the point where the shock wave begins, only the third derivatives of the various quantities can be discontinuous. There will in general be two such discontinuities: a weak discontinuity coinciding with the characteristic, and a weak tangential discontinuity coinciding with the streamline (see the end of §96).

§116. Chaplygin's equation: the general problem of steady two-dimensional gas flow

Having dealt with steady simple waves, let us now consider the general problem of an arbitrary steady two-dimensional potential flow. We assume that the flow is isentropic and contains no shock waves.

It is possible to reduce this problem to the solution of a single linear partial differential equation (S. A. Chaplygin 1902). This is achieved by means of a transformation to new independent variables, the velocity components v_x, v_y; this transformation is often called the *hodograph transformation*, the v_xv_y-plane being called the *hodograph plane* and the xy-plane the *physical plane*.

For potential flow we can replace Euler's equations by their first integral, Bernoulli's equation:

$$w+\tfrac{1}{2}v^2 = w_0. \tag{116.1}$$

The equation of continuity is

$$\frac{\partial}{\partial x}(\rho v_x)+\frac{\partial}{\partial y}(\rho v_y) = 0. \tag{116.2}$$

For the differential of the velocity potential ϕ we have $d\phi = v_x\,dx + v_y\,dy$. We transform from the independent variables x, y to the new variables v_x, v_y by Legendre's transformation, putting

$$d\phi = d(xv_x) - x\,dv_x + d(yv_y) - y\,dv_y,$$

introducing the function

$$\Phi = -\phi + xv_x + yv_y, \tag{116.3}$$

and obtaining

$$d\Phi = x\,dv_x + y\,dv_y,$$

where Φ is regarded as a function of v_x and v_y. Hence

$$x = \partial\Phi/\partial v_x, \qquad y = \partial\Phi/\partial v_y. \tag{116.4}$$

It is more convenient, however, to use, not the Cartesian components of the velocity, but its magnitude v and the angle θ which it makes with the x-axis:

$$v_x = v\cos\theta, \qquad v_y = v\sin\theta. \tag{116.5}$$

The appropriate transformation of the derivatives gives, instead of (116.4),

$$x = \cos\theta\frac{\partial\Phi}{\partial v} - \frac{\sin\theta}{v}\frac{\partial\Phi}{\partial\theta}, \qquad y = \sin\theta\frac{\partial\Phi}{\partial v} + \frac{\cos\theta}{v}\frac{\partial\Phi}{\partial\theta}. \tag{116.6}$$

The relation between the potential ϕ and the function Φ is given by the simple formula

$$\phi = -\Phi + v\,\partial\Phi/\partial v. \tag{116.7}$$

Finally, in order to obtain the equation which determines the function $\Phi(v, \theta)$, we must transform the equation of continuity (116.2) to the new variables. Writing the derivatives as Jacobians:

$$\frac{\partial(\rho v_x, y)}{\partial(x, y)} - \frac{\partial(\rho v_y, x)}{\partial(x, y)} = 0,$$

multiplying by $\partial(x, y)/\partial(v, \theta)$ and substituting (116.5), we have

$$\frac{\partial(\rho v\cos\theta, y)}{\partial(v, \theta)} - \frac{\partial(\rho v\sin\theta, x)}{\partial(v, \theta)} = 0.$$

To expand these Jacobians, we must substitute (116.6) for x and y. Furthermore, since the entropy s is a given constant, if we express the density as a function of s and w and substitute $w = w_0 - \frac{1}{2}v^2$ we find that the density can be written as a function of v only: $\rho = \rho(v)$. We therefore obtain, after a simple calculation, the equation

$$\frac{\mathrm{d}(\rho v)}{\mathrm{d}v}\left(\frac{\partial\Phi}{\partial v} + \frac{1}{v}\frac{\partial^2\Phi}{\partial\theta^2}\right) + \rho v\frac{\partial^2\Phi}{\partial v^2} = 0.$$

According to (83.5),

$$\frac{\mathrm{d}(\rho v)}{\mathrm{d}v} = \rho\left(1 - \frac{v^2}{c^2}\right),$$

and so we have finally *Chaplygin's equation* for the function $\Phi(v, \theta)$:

$$\frac{\partial^2\Phi}{\partial\theta^2} + \frac{v^2}{1 - v^2/c^2}\frac{\partial^2\Phi}{\partial v^2} + v\frac{\partial\Phi}{\partial v} = 0. \tag{116.8}$$

Here the velocity of sound is a known function $c(v)$, determined by the equation of state of the gas together with Bernoulli's equation.

The equation (116.8), together with the relations (116.6), is equivalent to the equations of motion. Thus the problem of solving the non-linear equations of motion is reduced to the solution of a linear equation for the function $\Phi(v, \theta)$. It is true that the boundary conditions on this equation are non-linear. These conditions are as follows. At the surface of the body, the gas velocity must be tangential. Expressing the equation of the surface in the parametric form $X = X(\theta)$, $Y = Y(\theta)$ (as in §115), and substituting X and Y in place of x and y in (116.6), we obtain two equations, which must be satisfied for all values of θ; this is not possible for every function $\Phi(v, \theta)$. The boundary condition is, in fact, that these two equations are compatible for all θ, i.e. one of them must be deducible from the other.

The satisfying of the boundary conditions, however, does not ensure that the resulting solution of Chaplygin's equation determines a flow that is actually possible everywhere in the physical plane. The following condition must also be met: the Jacobian $\Delta \equiv \partial(x, y)/\partial(\theta, v)$ must nowhere be zero, except in the trivial case when all its four component derivatives vanish. It is easy to see that, unless this condition holds, the solution becomes complex when we pass through the line (called the *limiting line*) in the xy-plane given by the equation $\Delta = 0$.† For, let $\Delta = 0$ on the line $v = v_0(\theta)$, and suppose that $(\partial y/\partial\theta)_v \neq 0$. Then we have

$$-\Delta\left(\frac{\partial\theta}{\partial y}\right)_v = \frac{\partial(x, y)}{\partial(v, \theta)}\frac{\partial(v, \theta)}{\partial(v, y)} = \frac{\partial(x, y)}{\partial(v, y)} = \left(\frac{\partial x}{\partial v}\right)_y = 0.$$

Hence we see that, near the limiting line, v is determined as a function of x (for given y) by

$$x - x_0 = \tfrac{1}{2}(\partial^2 x/\partial v^2)_y (v - v_0)^2,$$

and v becomes complex on one side or the other of the limiting line.‡

It is easy to see that a limiting line can occur only in regions of supersonic flow. A direct calculation, using the relations (116.6) and equation (116.8), gives

$$\Delta = \frac{1}{v}\left[\left(\frac{\partial^2\Phi}{\partial\theta\,\partial v} - \frac{1}{v}\frac{\partial\Phi}{\partial\theta}\right)^2 + \frac{v^2}{1 - v^2/c^2}\left(\frac{\partial^2\Phi}{\partial v^2}\right)^2\right]. \tag{116.9}$$

It is clear that, for $v \leqslant c$, $\Delta > 0$, and Δ can become zero only if $v > c$.

The appearance of limiting lines in the solution of Chaplygin's equation indicates that, under the given conditions, a continuous flow throughout the region is impossible, and shock waves must occur. It should be emphasized, however, that the position of these shocks is not the same as that of the limiting lines.

In §115 we discussed the particular case of steady two-dimensional supersonic flow (a simple wave), which is characterized by the fact that the velocity in it is a function only of its direction: $v = v(\theta)$. This solution cannot be obtained from Chaplygin's equation, since $1/\Delta \equiv 0$, and the solution is lost when the equation of continuity is multiplied by the Jacobian Δ in the transformation to the hodograph plane. The situation is exactly analogous to that found in the theory of non-steady one-dimensional flow. The remarks made in §105 concerning the relation between the simple wave and the general integral of equation (105.2) are wholly applicable to the relation between the steady simple wave and the general integral of Chaplygin's equation.

The fact that the Jacobian Δ is positive in subsonic flow gives a rule for finding the direction in which the velocity vector turns along the flow (A. A. Nikol'skiĭ and G. I. Taganov 1946). We have identically

$$\frac{1}{\Delta} \equiv \frac{\partial(\theta, v)}{\partial(x, y)} = \frac{\partial(\theta, v)}{\partial(x, v)}\frac{\partial(x, v)}{\partial(x, y)},$$

or

$$\frac{1}{\Delta} = \left(\frac{\partial\theta}{\partial x}\right)_v\left(\frac{\partial v}{\partial y}\right)_x. \tag{116.10}$$

In a subsonic flow $\Delta > 0$, and we see that the derivatives $(\partial\theta/\partial x)_v$ and $(\partial v/\partial y)_x$ have the same sign. This has a simple geometrical significance: if we move along a line v = constant

† There is no objection to a passage through points where Δ becomes infinite. If $1/\Delta = 0$ on some line, this merely means that the correspondence between the xy and $v\theta$ planes is no longer one-to-one: in going round the xy-plane, we cover some part of the $v\theta$-plane two or three times.

‡ This result clearly remains valid even if $(\partial^2 x/\partial v^2)_y$ vanishes with Δ but $(\partial x/\partial v)_y$ again changes sign for $v = v_0$, i.e. the difference $x - x_0$ is proportional to a higher even power of $v - v_0$.

$\equiv v_0$, with the region $v < v_0$ to the right, the angle θ increases monotonically, i.e. the velocity vector turns always counterclockwise. This result holds, in particular, for the line of transition between subsonic and supersonic flow, on which $v = c = c_*$.

In conclusion, we may give Chaplygin's equation for a polytropic gas, writing c explicitly in terms of v:

$$\frac{\partial^2\Phi}{\partial\theta^2} + v^2 \frac{1-(\gamma-1)v^2/(\gamma+1)c_*^{\,2}}{1-v^2/c_*^{\,2}} \frac{\partial^2\Phi}{\partial v^2} + v\frac{\partial\Phi}{\partial v} = 0. \qquad (116.11)$$

This equation has a family of particular integrals expressible in terms of hypergeometric functions.†

§117. Characteristics in steady two-dimensional flow

Some general properties of characteristics in steady (supersonic) two-dimensional flow have already been discussed in §82. We shall now derive the equations which give the characteristics in terms of a given solution of the equations of motion.

In steady two-dimensional supersonic flow there are, in general, three families of characteristics. All small disturbances, except those of entropy and vorticity, are propagated along two of these families (which we call the characteristics C_+ and C_-); disturbances of entropy and vorticity are propagated along characteristics (C_0) of the third family, which coincide with the streamlines. For a given flow, the streamlines are known, and the problem is to determine the characteristics belonging to the first two families.

The directions of the characteristics C_+ and C_- passing through each point in the plane lie on opposite sides of the streamline through that point, and make with it an angle equal to the local value of the Mach angle α (Fig. 51, §82). We denote by m_0 the slope of the streamline at a given point, and by m_+, m_- the slopes of the characteristics C_+, C_-. Then we have

$$\frac{m_+ - m_0}{1+m_0 m_+} = \tan\alpha, \qquad \frac{m_- - m_0}{1+m_0 m_-} = -\tan\alpha,$$

whence

$$m_\pm = \frac{m_0 \pm \tan\alpha}{1 \pm m_0\tan\alpha};$$

the upper signs everywhere relate to C_+ and the lower to C_-. Substituting $m_0 = v_y/v_x$, $\tan\alpha = c/\sqrt{(v^2-c^2)}$ and simplifying, we obtain the following expression for the slopes of the characteristics:

$$m_\pm \equiv \left(\frac{\mathrm{d}y}{\mathrm{d}x}\right)_\pm = \frac{v_x v_y \pm c\sqrt{(v^2-c^2)}}{v_x^{\,2}-c^2}. \qquad (117.1)$$

If the velocity distribution is known, this is a differential equation which determines the characteristics C_+ and C_-.‡

† See, for instance, L. I. Sedov, *Two-dimensional Problems in Hydrodynamics and Aerodynamics*, Chapter X, New York 1965; R. von Mises, *Mathematical Theory of Compressible Fluid Flow*, §20, New York 1958.

‡ Equation (117.1) also determines the characteristics for steady axially symmetrical flow if v_y and y are replaced by v_r and r, where r is the cylindrical polar coordinate (the distance from the axis of symmetry, which is the x-axis); it is clear that the derivation is unchanged if we consider, instead of the xy-plane, an xr-plane through the axis of symmetry.

Besides the characteristics in the *xy*-plane, we may consider those in the hodograph plane, which are especially useful in the discussion of isentropic potential flow; we shall take this case in what follows. Mathematically, these are the characteristics of Chaplygin's equation (116.8), which is of hyperbolic type for $v > c$. Following the general method familiar in mathematical physics (see §103), we form from the coefficients the equation of the characteristics:

$$\mathrm{d}v^2 + \mathrm{d}\theta^2\, v^2/(1 - v^2/c^2) = 0,$$

or

$$\left(\frac{\mathrm{d}\theta}{\mathrm{d}v}\right)_{\pm} = \pm\frac{1}{v}\sqrt{\left(\frac{v^2}{c^2} - 1\right)}. \tag{117.2}$$

The characteristics given by this equation do not depend on the particular solution of Chaplygin's equation considered, because the coefficients in that equation are independent of Φ. The characteristics in the hodograph plane are a transformation of the characteristics C_+ and C_- in the physical plane, and we call them respectively the characteristics Γ_+ and Γ_-, in accordance with the signs in (117.2).

The integration of equation (117.2) gives relations of the form $J_+(v, \theta) = \text{constant}$, $J_-(v, \theta) = \text{constant}$. The functions J_+ and J_- are quantities which remain constant along the characteristics C_+ and C_- (i.e. Riemann invariants). For a polytropic gas, equation (117.2) can be integrated explicitly. There is, however, no need to go through the calculations, since the result can be seen from formulae (115.3) and (115.4). For, according to the general properties of simple waves (see §104), the dependence of v on θ for a simple wave is given by the condition that one of the Riemann invariants be constant in all space. The arbitrary constant in formulae (115.3) and (115.4) is ϕ_*; eliminating the parameter ϕ from these formulae, we obtain

$$J_{\pm} = \theta \pm \left\{\sin^{-1}\sqrt{\left[\tfrac{1}{2}(\gamma+1)\left(1 - \frac{c_*{}^2}{v^2}\right)\right]} - \right.$$
$$\left. - \sqrt{\frac{\gamma+1}{\gamma-1}}\sin^{-1}\sqrt{\left[\tfrac{1}{2}(\gamma-1)\left(\frac{v^2}{c_*{}^2} - 1\right)\right]}\right\}. \tag{117.3}$$

The characteristics in the hodograph plane are a family of epicycloids, occupying the space between two circles with radii $v = c_*$ and $v = \sqrt{[(\gamma+1)/(\gamma-1)]}c_*$ (Fig. 117).

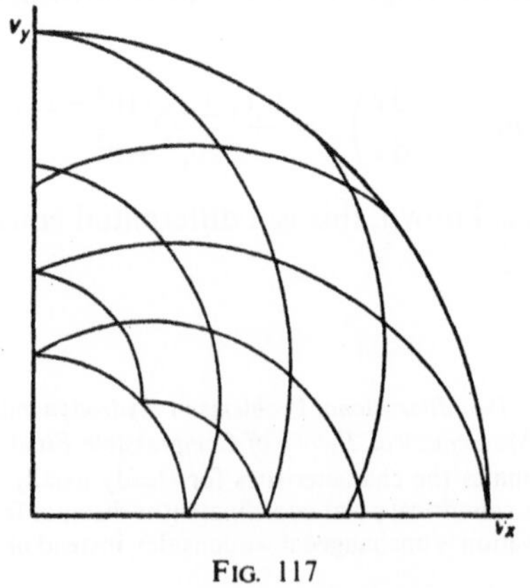

FIG. 117

For isentropic potential flow, the characteristics Γ_+, Γ_- have the following important property: the families Γ_+, Γ_- are orthogonal to the families C_-, C_+ respectively (it is assumed that the coordinate axes of x and y are mapped parallel to those of v_x and v_y).†

To prove this, we start from equation (114.3) for two-dimensional potential flow, which has the form

$$A\frac{\partial^2\phi}{\partial x^2}+2B\frac{\partial^2\phi}{\partial x\,\partial y}+C\frac{\partial^2\phi}{\partial y^2}=0, \tag{117.4}$$

with no free term. The slopes $m_\pm$ of the characteristics $C_\pm$ are the roots of the quadratic

$$Am^2-2Bm+C=0.$$

Let us consider the expression $\mathrm{d}v_x{}^+\,\mathrm{d}x^-+\mathrm{d}v_y{}^+\,\mathrm{d}y^-$, in which the velocity differentials are taken along the characteristics Γ_+, and the coordinate differentials along C_-. We have, identically,

$$\mathrm{d}v_x{}^+\,\mathrm{d}x^-+\mathrm{d}v_y{}^+\,\mathrm{d}y^-$$

$$=\frac{\partial^2\phi}{\partial x^2}\mathrm{d}x^+\,\mathrm{d}x^-+\frac{\partial^2\phi}{\partial x\,\partial y}(\mathrm{d}x^+\,\mathrm{d}y^-+\mathrm{d}x^-\,\mathrm{d}y^+)+\frac{\partial^2\phi}{\partial y^2}\mathrm{d}y^+\,\mathrm{d}y^-.$$

Dividing by $\mathrm{d}x^+\,\mathrm{d}x^-$, we obtain as the coefficients of $\partial^2\phi/\partial x\,\partial y$ and $\partial^2\phi/\partial y^2$ respectively $m_++m_-=2B/A$ and $m_+m_-=C/A$. It is then clear that the expression is zero, by (117.4). Thus

$$\mathrm{d}v_x{}^+\,\mathrm{d}x^-+\mathrm{d}v_y{}^+\,\mathrm{d}y^-=\mathbf{dv}^+\cdot\mathbf{dr}^-=0.$$

Similarly, $\mathbf{dv}^-\cdot\mathbf{dr}^+=0$. These equations are equivalent to the result stated.

§118. The Euler–Tricomi equation. Transonic flow

The investigation of the properties resulting from the transition between subsonic and supersonic flow is of fundamental interest. Steady flows in which this transition occurs are called *mixed* or *transonic* flows, and the surface where the transition occurs is called the *transitional* or *sonic* surface.

Chaplygin's equation is particularly useful in investigating the flow near the transition, since it is much simplified there. At the boundary where the transition occurs $v=c=c_*$, and near it (in the transonic region) the differences $v-c$ and $v-c_*$ are small; they are related by (114.8):

$$(v/c)-1=\alpha_*[(v/c_*)-1].$$

Let us effect the corresponding simplification in Chaplygin's equation. The third term in equation (116.8) is small compared with the second, which contains $1-v^2/c^2$ in the denominator. In the second term we put approximately

$$\frac{v^2}{1-v^2/c^2}=\frac{c_*{}^2}{2(1-v/c)}=\frac{c_*}{2\alpha_*(1-v/c_*)}.$$

Finally, replacing the velocity v by a new variable

$$\eta=(2\alpha_*)^{\frac{1}{3}}(v-c_*)/c_*, \tag{118.1}$$

† This does not apply to the characteristics of axially symmetrical flow in the xr-plane.

we obtain the required equation in the form

$$\frac{\partial^2\Phi}{\partial\eta^2} - \eta\frac{\partial^2\Phi}{\partial\theta^2} = 0. \tag{118.2}$$

An equation of this form is called in mathematical physics the *Euler–Tricomi equation.*† In the half-plane $\eta > 0$ it is hyperbolic, but in $\eta < 0$ it is elliptic. We shall discuss here some mathematical properties of this equation which are important in connection with various physical problems.

The characteristics of equation (118.2) are given by the equation $\eta d\eta^2 - d\theta^2 = 0$, which has the general integral

$$\theta \pm \tfrac{2}{3}\eta^{3/2} = C, \tag{118.3}$$

where C is an arbitrary constant. This equation represents two families of curves in the $\eta\theta$-plane, which are branches of semi-cubical parabolae in the right half-plane with cusps on the θ-axis (Fig. 118).

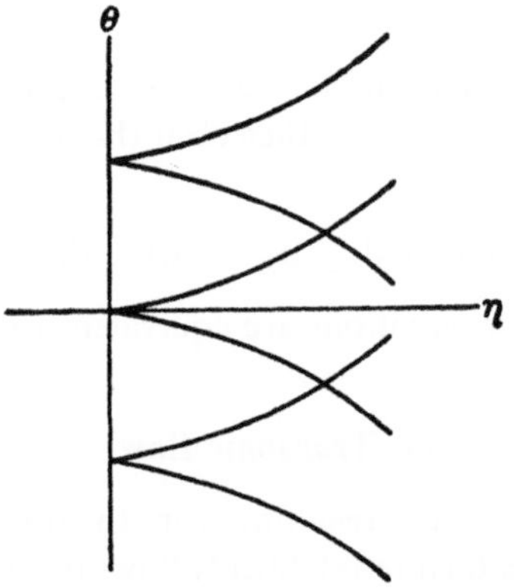

FIG. 118

In investigating the flow in a small region‡ of space, where the direction of the gas velocity varies only slightly, we can always take the direction of the x-axis such that the angle θ measured from it is small throughout the region considered. The equations (116.6) which determine the coordinates x, y from the function $\Phi(\eta, \theta)$ are then much simplified also:†† $x = (2\alpha_*)^{1/3}\partial\Phi/\partial\eta$, $y = \partial\Phi/\partial\theta$. In order to avoid the appearance of the factor $(2\alpha_*)^{1/3}$, we shall replace the coordinate x in §§118–121 by $x(2\alpha_*)^{-1/3}$, and call the latter quantity x. Then

$$x = \partial\Phi/\partial\eta, \qquad y = \partial\Phi/\partial\theta. \tag{118.4}$$

† The application of this equation to the problem here considered is due to F. I. Frankl' (1945).

‡ This phrase must not be taken literally, of course. The region concerned may be the neighbourhood of the point at infinity, i.e. the region at large distances from the body.

†† We omit a factor $1/c_*$ on the right-hand sides; this simply means that Φ is replaced by $c_*\Phi$, which does not affect equation (118.2) and is therefore always permissible.

It is useful to note that, since it is so simply related to Φ, the function $y(\eta, \theta)$ (but not $x(\eta, \theta)$) also satisfies the Euler–Tricomi equation. Using this fact, we can write the Jacobian of the transformation from the physical plane to the hodograph plane as

$$\Delta = \frac{\partial(x, y)}{\partial(\theta, \eta)} = \Phi_{\eta\theta}{}^2 - \Phi_{\eta\eta}\Phi_{\theta\theta} = \left(\frac{\partial y}{\partial \eta}\right)^2 - \eta\left(\frac{\partial y}{\partial \theta}\right)^2. \tag{118.5}$$

As has already been mentioned, the Euler–Tricomi equation has usually to be applied to investigate the properties of the solution near the origin in the $\eta\theta$-plane. In cases of physical interest, the origin is a singular point of the solution. For this reason especial significance attaches to the family of particular integrals of the Euler–Tricomi equation which possess certain properties of homogeneity. These solutions are homogeneous in the variables θ^2 and η^3; such solutions must exist, since the transformation $\theta^2 \to a\theta^2, \eta^3 \to a\eta^3$ leaves the equation (118.2) unchanged. We shall seek these solutions in the form $\Phi = \theta^{2k} f(\xi)$, $\xi = 1 - 4\eta^3/9\theta^2$, where k is a constant, the degree of homogeneity of the function Φ with respect to the transformation mentioned. We have taken the variable ξ so that it vanishes on the characteristics which pass through the point $\eta = \theta = 0$. Making the above substitution, we obtain for the function $f(\xi)$ the equation

$$\xi(1-\xi)f'' + [\tfrac{5}{6} - 2k - \xi(\tfrac{3}{2} - 2k)]f' - k(k - \tfrac{1}{2})f = 0.$$

This is a hypergeometric equation. Using the well-known expressions for the two independent integrals of that equation, we find the required solution (for $2k + \frac{1}{6}$ not integral):

$$\Phi_k = \theta^{2k}\left[AF\left(-k, \quad -k+\tfrac{1}{2}; \quad -2k+\tfrac{5}{6}; \quad 1 - \frac{4\eta^3}{9\theta^2}\right) + \right.$$
$$\left. + B\left(1 - \frac{4\eta^3}{9\theta^2}\right)^{2k+1/6} F\left(k+\tfrac{1}{6}, \quad k+\tfrac{2}{3}; \quad 2k+\tfrac{7}{6}; \quad 1 - \frac{4\eta^3}{9\theta^2}\right)\right]. \tag{118.6}$$

Using the relations between hypergeometric functions of arguments $z, 1/z, 1-z, 1/(1-z)$ and $z/(1-z)$, we can also put this solution in five other forms, all of which are needed in various problems.† We shall give two of these:

$$\Phi_k = \theta^{2k}\left[AF\left(-k, \quad -k+\tfrac{1}{2}; \quad \tfrac{2}{3}; \quad \frac{4\eta^3}{9\theta^2}\right) + \right.$$
$$\left. + B\frac{\eta}{\theta^{2/3}} F\left(-k+\tfrac{1}{3}, \quad -k+\tfrac{5}{6}; \quad \tfrac{4}{3}; \quad \frac{4\eta^3}{9\theta^2}\right)\right], \tag{118.7}$$

$$\Phi_k = \eta^{3k}\left[AF\left(-k, \quad -k+\tfrac{1}{3}; \quad \tfrac{1}{2}; \quad \frac{9\theta^2}{4\eta^3}\right) + \right.$$
$$\left. + B\frac{\theta}{\eta^{3/2}} F\left(-k+\tfrac{1}{2}, \quad -k+\tfrac{5}{6}; \quad \tfrac{3}{2}; \quad \frac{9\theta^2}{4\eta^3}\right)\right]; \tag{118.8}$$

the constants A and B in formulae (118.6)–(118.8) are not the same, of course. These expressions yield at once the following important property of the functions Φ_k, which is not evident from (118.6): the lines $\eta = 0$ and $\theta = 0$ are not singular lines (it is seen from

† The relevant formulae are given, for example, in *QM*, Mathematical Appendices, e.

(118.7) that, near $\eta = 0$, Φ_k can be expanded in integral powers of η, and from (118.8) the same is true of θ). It is seen from the expression (118.6) that the characteristics, on the other hand, are singular lines of the general (i.e. containing the two constants A and B) homogeneous integral Φ_k of the Euler–Tricomi equation: if $2k+\frac{1}{6}$ is not an integer, the factor $(9\theta^2-4\eta^3)^{2k+1/6}$ has branch points, while if $2k+\frac{1}{6}$ is an integer, one term of (118.6) is meaningless† (or degenerates to the other term if $2k+\frac{1}{6}=0$), and must be replaced by the second independent solution of the hypergeometric equation, which in this case has a logarithmic singularity.

The following relations hold between the integrals Φ_k with different values of k:

$$\Phi_k = \Phi_{-k-1/6}\,(9\theta^2-4\eta^3)^{2k+1/6}, \tag{118.9}$$

$$\Phi_{k-1/2} = \partial\Phi_k/\partial\theta. \tag{118.10}$$

The first of these follows immediately from (118.6), and the second from the fact that $\partial\Phi_k/\partial\theta$ satisfies the Euler–Tricomi equation, and its degree of homogeneity is that of $\Phi_{k-1/2}$. In these formulae Φ_k means, of course, the general expression, with two arbitrary constants.

In investigating the solution near the point $\eta=\theta=0$, we have to follow its variation along a contour round this point. For example, let the function Φ_k (118.6) represent the solution at the point A near the characteristic $\theta=\frac{2}{3}\eta^{3/2}$ (Fig. 119), and suppose that we require the form of the solution near the characteristic $\theta=-\frac{2}{3}\eta^{3/2}$ (at the point B). The passage from A to B involves crossing the axis of abscissae, and $\theta=0$ is a singular line of the hypergeometric functions in the expression (118.6), so that their argument is infinite there. In order to go from A to B, therefore, it is necessary to transform the hypergeometric functions into functions of the reciprocal argument $9\theta^2/(9\theta^2-4\eta^3)$, for which $\theta=0$ is not a singularity, and then change the sign of θ, finally returning to the original argument by repeating the transformation. In this way we obtain the following transformation formulae for the functions which appear in (118.6):

$$\left.\begin{aligned} F_1 &\to \frac{F_1}{2\sin(2k+\frac{1}{6})\pi} + F_2\cdot 2^{-4k-1/3}\frac{\Gamma(-2k-\frac{1}{6})\Gamma(-2k+\frac{5}{6})}{\Gamma(-2k)\Gamma(-2k+\frac{2}{3})},\\ F_2 &\to \frac{-F_2}{2\sin(2k+\frac{1}{6})\pi} + F_1\cdot 2^{4k+1/3}\frac{\Gamma(2k+\frac{1}{6})\Gamma(2k+\frac{7}{6})}{\Gamma(2k+1)\Gamma(2k+\frac{1}{3})}, \end{aligned}\right\} \tag{118.11}$$

where F_1 and F_2 signify

$$\left.\begin{aligned} F_1 &= |\theta|^{2k}F\left(-k,\quad -k+\tfrac{1}{2};\quad -2k+\tfrac{5}{6};\quad 1-\frac{4\eta^3}{9\theta^2}\right),\\ F_2 &= |\theta|^{2k}\left|1-\frac{4\eta^3}{9\theta^2}\right|^{2k+1/6}F\left(k+\tfrac{1}{6},\quad k+\tfrac{2}{3};\quad 2k+\tfrac{7}{6};\quad 1-\frac{4\eta^3}{9\theta^2}\right), \end{aligned}\right\} \tag{118.12}$$

in which the moduli of θ and $1-4\eta^3/9\theta^2$ are taken in the coefficients of the hypergeometric functions.

We can similarly obtain transformation formulae for the passage from A' to B' (Fig. 119) round the origin in the opposite direction. The calculations are more involved, since we have to pass through three singularities of the hypergeometric function (one with $\theta=0$

† We recall that the series $F(\alpha, \beta; \gamma; z)$ is meaningless for $\gamma = 0, -1, -2, \ldots$.

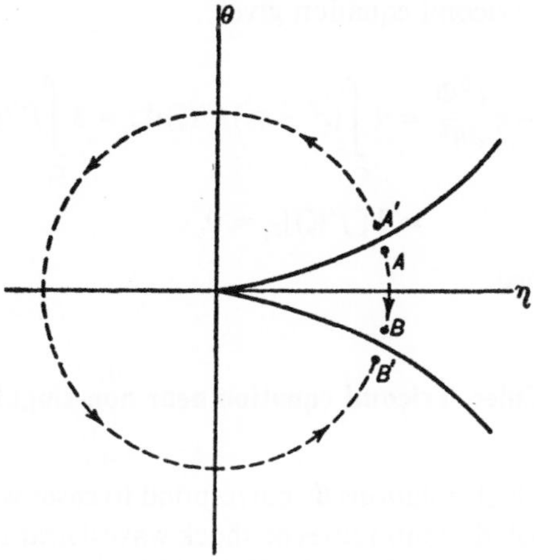

FIG. 119

and two with $\eta = 0$; we recall that the singularities of a hypergeometric function with argument z are $z = 1$ and $z = \infty$). The final formulae are

$$\left.\begin{aligned} F_1 &\to -\frac{\sin(4k-\frac{1}{6})\pi}{\sin(2k+\frac{1}{6})\pi}F_1 + F_2\cdot 2^{-4k+2/3}\cos(2k+\tfrac{1}{6})\pi\frac{\Gamma(-2k-\frac{1}{6})\Gamma(-2k+\frac{5}{6})}{\Gamma(-2k)\Gamma(-2k+\frac{2}{3})},\\ F_2 &\to \frac{\sin(4k-\frac{1}{6})\pi}{\sin(2k+\frac{1}{6})\pi}F_2 + F_1\cdot 2^{4k+4/3}\cos(2k+\tfrac{1}{6})\pi\frac{\Gamma(2k+\frac{1}{6})\Gamma(2k+\frac{7}{6})}{\Gamma(2k+1)\Gamma(2k+\frac{1}{3})}. \end{aligned}\right\} \quad (118.13)$$

As well as this family of homogeneous solutions there are, of course, other families of particular integrals of the Euler–Tricomi equation. We may mention here a family which results from a Fourier expansion in terms of θ. If we seek Φ in the form

$$\Phi_\nu = g_\nu(\eta)e^{\pm i\nu\theta}, \quad (118.14)$$

where ν is an arbitrary constant, we obtain for the function g_ν the equation $g_\nu'' + \nu^2\eta g_\nu = 0$. This is the equation for the Airy function; its general integral is

$$g_\nu(\eta) = \sqrt{\eta}Z_{\frac{1}{3}}(\tfrac{2}{3}\nu\eta^{3/2}), \quad (118.15)$$

where $Z_{\frac{1}{3}}$ is an arbitrary linear combination of Bessel functions of order $\frac{1}{3}$.

Finally, it is useful to bear in mind that the general integral of the Euler–Tricomi equation may be written

$$\Phi = \int\limits_{C_z} f(\zeta)\,dz, \qquad \zeta = z^3 - 3\eta z + 3\theta, \quad (118.16)$$

where $f(\zeta)$ is an arbitrary function and the integration in the complex z-plane is taken along any contour C_z at whose ends the derivative $f'(\zeta)$ has equal values. For a direct substitution of (118.16) in the Euler–Tricomi equation gives

$$\frac{\partial^2\Phi}{\partial\eta^2} - \eta\frac{\partial^2\Phi}{\partial\theta^2} = 9\int\limits_{C_z}(z^2-\eta^2)f''(\zeta)\,\mathrm{d}z = 3\int\limits_{C_\zeta}f''(\zeta)\,\mathrm{d}\zeta$$

$$= 3\,[f'(\zeta)]_{C_\zeta} = 0,$$

i.e. the equation is satisfied.

§119. Solutions of the Euler–Tricomi equation near non-singular points of the sonic surface

Let us now ascertain which solutions Φ_k correspond to cases where the gas flow has no physical singularities (weak discontinuities or shock waves) near the transition. To do this it is more convenient to start, not from the Euler–Tricomi equation itself, but from the equation for the velocity potential in the physical plane. This equation has been derived in §114; for a two-dimensional flow, equation (114.10) becomes, with the substitution $x \to x(2\alpha_*)^{1/3}$,

$$\frac{\partial\phi}{\partial x}\frac{\partial^2\phi}{\partial x^2} = \frac{\partial^2\phi}{\partial y^2}. \tag{119.1}$$

We recall that the potential ϕ in this equation is defined so that its derivatives with respect to the coordinates give the velocity according to the equations

$$\partial\phi/\partial x = \eta, \qquad \partial\phi/\partial y = \theta. \tag{119.2}$$

We may also note that the Euler–Tricomi equation can be obtained directly from equation (119.1) by changing to the independent variables θ, η by Legendre's transformation, with $\Phi = -\phi + x\eta + y\theta$, or

$$\phi = -\Phi + \eta\partial\Phi/\partial\eta + \theta\,\partial\Phi/\partial\theta. \tag{119.3}$$

Taking the origin in the xy-plane at the point on the transition or *sonic line* whose neighbourhood we are investigating, we expand ϕ in powers of x and y. In the general case, the first term of an expansion which satisfies equation (119.1) is

$$\phi = xy/a. \tag{119.4}$$

Here $\theta = x/a$, $\eta = y/a$, so that

$$\Phi = a\theta\eta. \tag{119.5}$$

It is clear from the degree of homogeneity of this function that it corresponds to one of the functions $\Phi_{5/6}$; this is the second term of the expression (118.7), in which the hypergeometric function with $k = 5/6$ reduces to 1 simply: $\eta\theta F(-\frac{1}{2}, 0; \frac{4}{3}; 4\eta^3/9\theta^2) = \eta\theta$.

If we wish to find the equation of the sonic line in the physical plane, the first term of the expansion does not suffice. The next term is of degree 1, i.e. it corresponds to one of the functions Φ_1, namely the first term in the expression (118.7), which reduces to a polynomial for $k = 1$:

$$\theta^2 F(-1, \ -\tfrac{1}{2}; \ \tfrac{2}{3}; \ 4\eta^3/9\theta^2) = \theta^2 + \tfrac{1}{3}\eta^3.$$

Thus the first two terms of the expansion of Φ are

$$\Phi = a\eta\theta + b(\theta^2 + \tfrac{1}{3}\eta^3). \tag{119.6}$$

Hence

$$\left.\begin{aligned} x &= a\theta + b\eta^2, \\ y &= a\eta + 2b\theta. \end{aligned}\right\} \tag{119.7}$$

The sonic line ($\eta = 0$) is the straight line $y = 2bx/a$.

To find the equation of the characteristics in the physical plane we need only the first term of the expansion. Substituting $\theta = x/a$, $\eta = y/a$ in the equation of the hodograph characteristics $\theta = \pm\frac{2}{3}\eta^{3/2}$, we obtain $x = \pm\frac{2}{3}y^{3/2}/\sqrt{a}$, i.e. again two branches of a semicubical parabola with a cusp on the sonic line (the thick line in Fig. 120). This property of the characteristics is evident also from the following simple argument. At points on the sonic line, the Mach angle is $\frac{1}{2}\pi$. This means that the tangents to the characteristics of the two families coincide, so that there is a cusp (Fig. 120). The streamlines intersect the sonic line perpendicularly to the characteristics, and do not have singularities there.

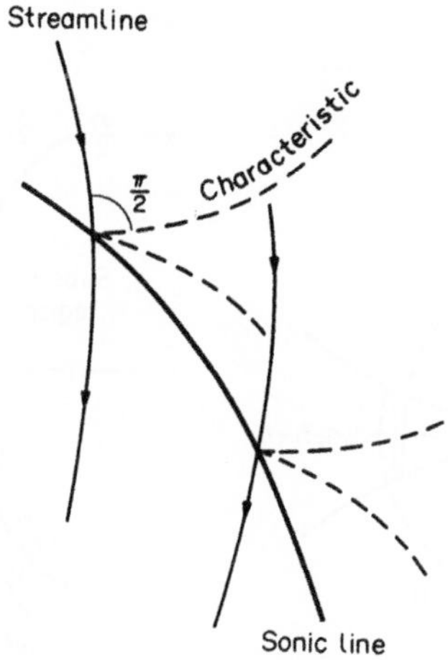

FIG. 120

The solution (119.6) is not applicable in the exceptional case where the streamline is perpendicular to the sonic line at the point considered.† Near such a point the flow is evidently symmetrical about the x-axis. This case requires special consideration, which has been given by F. I. Frankl' and S. V. Fal'kovich (1945).

The symmetry of the flow means that, when the sign of y is changed, the velocity v_y changes sign and v_x remains unchanged. That is, the potential ϕ must be an even function of y, and the potential Φ an even function of θ. The first terms in the expansion of ϕ in this case therefore have the form

$$\phi = \tfrac{1}{2}ax^2 + \tfrac{1}{2}a^2xy^2 + \tfrac{1}{24}a^3y^4; \tag{119.8}$$

† This would correspond to the case $a = 0$ in (119.6); the solution then ceases to hold, because the Jacobian Δ vanishes on the line $\eta = 0$.

the relative order of smallness of x and y is not known *a priori*, so that all three terms may be of the same order. Hence we find the following formulae for the transformation from the physical plane to the hodograph plane:

$$\left.\begin{aligned} \eta &= ax + \tfrac{1}{2}a^2y^2, \\ \theta &= a^2xy + \tfrac{1}{6}a^3y^3. \end{aligned}\right\} \qquad (119.9)$$

Without explicitly solving these equations for x and y, we can easily see that the degree of the function $y(\theta, \eta)$ is $\frac{1}{6}$. Hence the corresponding function Φ has $k = \frac{1}{6} + \frac{1}{2} = \frac{2}{3}$, i.e. it is a particular case of the general integral $\Phi_{2/3}$.

Eliminating x from equations (119.9), we obtain a cubic equation for the function $y(\theta, \eta)$:

$$(ay)^3 - 3\eta ay + 3\theta = 0. \qquad (119.10)$$

For $9\theta^2 - 4\eta^3 > 0$, i.e. throughout the region to the left of the hodograph characteristics which pass through the point $\eta = \theta = 0$ (including the whole of the subsonic region $\eta < 0$; Fig. 121), this equation has only one real root, which must be the function $y(\theta, \eta)$. In the region to the right of the characteristics, all three roots are real, and we must take the one which is the continuation of the real root in the region to the left.

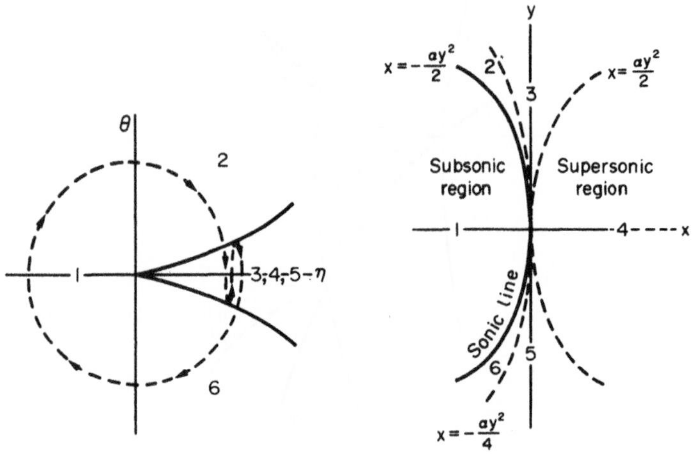

FIG. 121

The characteristics in the physical plane (which pass through the origin) are obtained by substituting the expressions (119.9) in the equation $4\eta^3 = 9\theta^2$. This gives two parabolae:

$$\left.\begin{aligned} &\text{the characteristics 23 and 56:} \quad x = -\tfrac{1}{4}ay^2, \\ &\text{the characteristics 34 and 45:} \quad x = \tfrac{1}{2}ay^2. \end{aligned}\right\} \qquad (119.11)$$

The numbers show which two regions in the physical plane are separated by the characteristic in question. The sonic line ($\eta = 0$ in the hodograph plane) is the parabola $x = -\frac{1}{2}ay^2$ in the physical plane (the thick line in Fig. 121). We may notice the following property of the point where the sonic line intersects the axis of symmetry: four branches of characteristics leave this point, whereas only two leave any other point on the sonic line.

Figure 121 shows by corresponding numbers the regions of the hodograph plane which correspond to the various regions of the physical plane. This correspondence is not one-to-one;† when we go completely round the origin in the physical plane, the region between the two characteristics in the hodograph plane is covered three times, as shown by the dashed line in Fig. 121, which is twice reflected from the characteristics.

Since the function $y(\theta, \eta)$ itself satisfies the Euler–Tricomi equation, it must be obtainable from the general integral $\Phi_{1/6}$. Near the characteristic 23 in the physical plane, it is

$$y = \frac{1}{a}\left(\frac{3}{2}\theta\right)^{1/3} F\left(-\tfrac{1}{6},\ \tfrac{1}{3};\ \tfrac{1}{2};\ 1-\frac{4\eta^3}{9\theta^2}\right); \tag{119.12}$$

the first term in (118.6) has no singularity on this characteristic. Continuing this analytically to the neighbourhood of the characteristic 56 (by a path through the subsonic region 1, i.e. by means of formulae (118.13)), we obtain the same function there. Near the characteristics 34 and 45, however, $y(\theta, \eta)$ is given by linear combinations of that function and

$$\theta^{1/3}\sqrt{\left(\frac{4\eta^3}{9\theta^2}-1\right)}F\left(\tfrac{1}{3},\ \tfrac{5}{6};\ \tfrac{3}{2};\ 1-\frac{4\eta^3}{9\theta^2}\right), \tag{119.13}$$

i.e. the second term of (118.6). These combinations are obtained by analytical continuation, using formulae (118.11); here it must be borne in mind that the square root in (119.13) changes sign at each reflection from a hodograph characteristic.

Mathematically, these results show that the functions $\Phi_{1/6}$ are linear combinations of the roots of the cubic equation

$$f^3 - 3\eta f + 3\theta = 0, \tag{119.14}$$

i.e. they are algebraic functions.‡ As well as $\Phi_{1/6}$, all the Φ_k with

$$k = \tfrac{1}{6} \pm \tfrac{1}{2}n, \qquad n = 0, 1, 2, \ldots \tag{119.15}$$

reduce to algebraic functions; they are obtained from $\Phi_{1/6}$, according to formulae (118.9) and (118.10), by successive differentiation, a remark due to F. I. Frankl' (1947).

The functions Φ_k with

$$k = \pm\tfrac{1}{2}n, \qquad k = \tfrac{1}{3} \pm \tfrac{1}{2}n, \tag{119.16}$$

in which the hypergeometric function reduces to a polynomial,†† also reduce to algebraic functions; e.g. for $k = \frac{1}{2}n$ we have the first term of the expression (118.6), and for $k = -\frac{1}{2}n$ the second term.

These three families of algebraic functions Φ_k include, in particular, all the functions which can be potentials Φ corresponding to flows having no singularity in the physical plane. In such flows, all the terms in the expansion of Φ near an asymmetric point on the sonic line (the first two terms of which are given by formula (119.6)) must have either $k = \frac{5}{6} + \frac{1}{2}n$ or $k = 1 + \frac{1}{2}n$. The expansion of Φ near a symmetric point, however, which begins with a term with $k = \frac{2}{3}$, can also contain functions with $k = \frac{2}{3} + \frac{1}{2}n$.

---

† In accordance with the fact that $\Delta = \infty$ on the characteristic $x = \frac{1}{2}ay^2$ in the physical plane; see the first footnote to §116.

‡ It is not convenient in practice to use the explicit forms of these functions, which are obtained from (119.14) by Cardan's formula.

†† Here it must be recalled that $F(\alpha, \beta; \gamma; z)$ reduces to a polynomial if α (or β) is such that $\alpha = -n$ or $\gamma - \alpha = -n$.

§120. Flow at the velocity of sound

The simplified form of Chaplygin's equation (i.e. the Euler–Tricomi equation) must, in principle, be used to investigate the basic qualitative properties of steady two-dimensional flow past bodies, resulting from the existence of transonic regions. These include, in the first place, problems concerning the formation of shock waves. In the transonic region, the shock wave is weak, and this is the reason why the Euler–Tricomi equation is applicable under these conditions. We have seen in §§86 and 114 that in a weak shock the changes in the entropy and vorticity are higher-order small quantities; in the first approximation, therefore, we can assume isentropic potential flow behind the discontinuity also.

We shall discuss here a problem of theoretical importance, that of the nature of steady two-dimensional flow past a body when the velocity of the incident stream is exactly equal to the velocity of sound. We shall see, in particular, that a shock wave must extend from the surface of the body to infinity. From this we can draw the important conclusion that the shock wave must first appear for a Mach number M_∞ which is certainly less than unity.

For, let us consider two-dimensional flow past a body ("wing") with infinite span and arbitrary (not necessarily symmetrical) cross-section. Here we are interested in the flow pattern at distances from the body which are large compared with its dimension. For convenience we shall first describe the results in a qualitative manner, and afterwards give a quantitative calculation.

In Fig. 122, AB and $A'B'$ are sonic lines, so that the subsonic region lies to the left of them (upstream); the arrow shows the direction of the main stream, which we shall take as the x-axis, with the origin anywhere near the body. At a certain distance from the sonic line we have shock waves leaving the body (EF and $E'F'$ in Fig. 122). It is found that the characteristics leaving the body (between the sonic line and the shock wave) can be divided into two groups. The characteristics in the first group meet the sonic line and end there (that is to say, they are reflected from it as characteristics which reach the body; Fig. 122 shows one such characteristic). The characteristics in the second group end at the shock wave. The two groups are separated by *limiting characteristics*, the only ones which go to infinity and meet neither the sonic line nor the shock wave (CD and $C'D'$ in Fig. 122). Since disturbances (caused, for instance, by a change in the shape of the body) which are

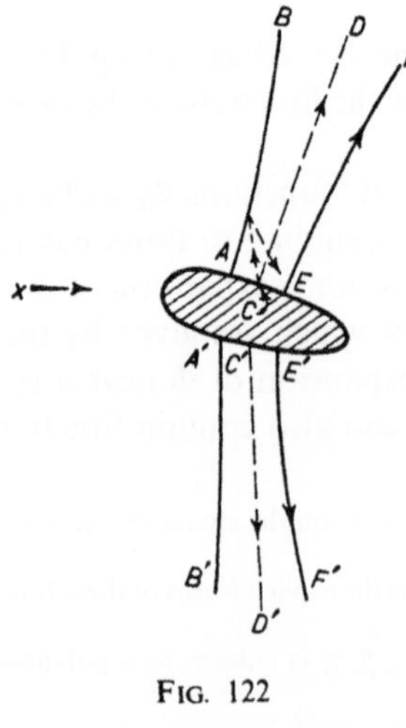

FIG. 122

propagated from the body along characteristics of the first group reach the boundary of the subsonic region, it is clear that the part of the supersonic region which lies between the sonic line and the limiting characteristic affects the subsonic region, but the flow to the right of the limiting characteristics has no effect on the flow to the left: the flow to the left is not affected by a disturbance of the flow to the right (such as a change in the profile to the right of C or C'). The flow behind the shock wave has, as we know, no effect on the flow in front of it. Thus the whole flow can be divided into three parts (to the left of $DCC'D'$, between $DCC'D'$ and $FEE'F'$, and to the right of $FEE'F'$), such that the flow in the second part has no effect on that in the first, and the flow in the third part has no effect on that in the second.

We shall now give a quantitative account (and verification) of the flow pattern just described.

The origin in the hodograph plane ($\theta = \eta = 0$) corresponds to an infinitely distant region of the physical plane, and the hodograph characteristics leaving the origin correspond to the limiting characteristics CD and $C'D'$. Figure 123 shows the neighbourhood of the origin, the letters corresponding to those in Fig. 122. The shock wave corresponds not to one line but to two lines in the hodograph plane (corresponding to the gas flow on the two sides of the discontinuity); the regions between these lines (shaded in Fig. 123) do not correspond to any part of the physical plane.

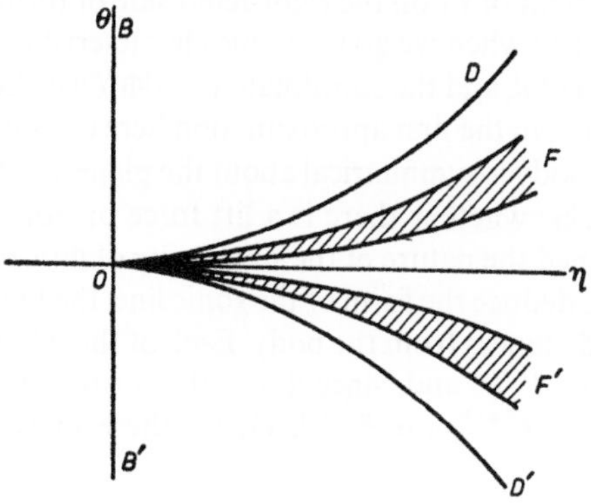

FIG. 123

We must ascertain, first of all, which of the general integrals Φ_k corresponds to this case. If $\Phi(\theta, \eta)$ is of degree k, then the functions $x = \partial\Phi/\partial\eta$ and $y = \partial\Phi/\partial\theta$ are homogeneous and of degree $k - \frac{1}{3}$ and $k - \frac{1}{2}$ respectively. As θ and η tend to zero we must, in general, reach infinity in the physical plane (x and y tend to infinity). It is evident that, for this to be so, we must have $k < \frac{1}{3}$. The limiting characteristics in the physical plane, however, need not lie entirely at infinity, i.e. $y = \pm\infty$ need not hold everywhere on the curve $9\theta^2 = 4\eta^3$. In that case (for $2k + \frac{1}{6} < \frac{5}{6}$), the second term in the brackets in (118.6) must be zero. Thus the function $\Phi(\theta, \eta)$ must be given by the first term of (118.6):

$$\Phi = A\theta^{2k} F\left(-k, \quad -k+\tfrac{1}{2}; \quad -2k+\tfrac{5}{6}; \quad 1-\frac{4\eta^3}{9\theta^2}\right). \tag{120.1}$$

The function $y(\theta, \eta)$ (which also satisfies the Euler–Tricomi equation) has the same form, but with $k - \frac{1}{2}$ instead of k.

If the expression (120.1) is valid near (e.g.) the upper characteristic ($\theta = +\frac{2}{3}\eta^{3/2}$), however, it will not be valid near the lower characteristic also ($\theta = -\frac{2}{3}\eta^{3/2}$) for an arbitrary $k < \frac{1}{3}$. We must therefore require also that the form (120.1) of the function $\Phi(\theta, \eta)$ is maintained on going round the origin in the hodograph plane from one characteristic to the other through the half-plane $\eta < 0$ (the path $A'B'$ in Fig. 119). This path corresponds to a passage in the physical plane from distant points on one of the limiting characteristics to distant points on the other, along a path which passes through the subsonic region and therefore nowhere intersects the shock wave, at which the flow is discontinuous. The transformation of the hypergeometric function in (120.1) in going along such a path is given by the first formula (118.13), and we must require that the coefficient of F_2 in this formula be zero. This condition is fulfilled (for $k < \frac{1}{3}$) when $k = \frac{1}{6} - \frac{1}{2}n$ ($n = 0, 1, 2, \ldots$). Of these values, only one can be taken, namely

$$k = -\tfrac{1}{3}; \tag{120.2}$$

it can be shown that all values of k with $n > 1$ give a mapping of the hodograph plane into the physical plane which is not one-to-one (in going once round the former we go more than once round the latter), and so the physical flow is many-valued, which is of course impossible. The value $k = \frac{1}{6}$, on the other hand, gives a solution in which we do not go to infinity in every direction in the physical plane when θ and η tend to zero; such a solution is, evidently, likewise physically impossible.

For $k = -\frac{1}{3}$ the coefficient of F_1 on the right-hand side of formula (118.13) is unity, i.e. the function Φ is unchanged when we go from one characteristic to the other. This means that Φ is an even function of θ, and the coordinate $y = \partial\Phi/\partial\theta$ is therefore an odd function. Physically, this means that, in the first approximation here considered, the flow pattern at large distances from the body is symmetrical about the plane $y = 0$, whatever the shape of the body, and in particular whether there is a lift force or not.

Thus we have determined the nature of the singularity of $\Phi(\eta, \theta)$ at the point $\eta = \theta = 0$. From this we can at once deduce the form of the sonic line, the limiting characteristics and the shock wave at great distances from the body. Each of these lines must correspond to a definite value of the ratio θ^2/η^3 and, since Φ has the form $\theta^{-2/3} f(\eta^3/\theta^2)$, we find from formulae (118.4) that $x \propto \theta^{-4/3}$, $y \propto \theta^{-5/3}$. Hence these lines are given by equations having the form

$$x = \text{constant} \times y^{4/5}, \tag{120.3}$$

with various values of the constant. Along these lines, θ and η decrease according to

$$\theta = \text{constant} \times y^{-3/5}, \qquad \eta = \text{constant} \times y^{-2/5}. \tag{120.4}$$

These results are due to F. I. Frankl' (1947) and K. G. Guderley (1948).†

In what follows we shall, for definiteness, write the formulae with the signs appropriate to the upper half-plane ($y > 0$).

We shall show how the coefficients in these formulae may be calculated. The value $k = -\frac{1}{3}$ is one of those for which the Φ_k reduce to algebraic functions (see §119). The particular

† Similar results can be obtained for axially symmetrical flow (with $M_\infty = 1$). In cylindrical polar coordinates x, r, the form of the sonic surface, the limiting characteristic and the shock wave, and the velocity variations, are given (far from the body) by $x = \text{constant} \times r^{4/7}$, $v_x \propto r^{-6/7}$, $v_r \propto r^{-9/7}$. See K. G. Guderley, *The Theory of Transonic Flow*, Oxford 1962; S. V. Fal'kovich and I. A. Chernov, *Journal of Applied Mathematics and Mechanics* **28**, 342, 1965.

integral which determines Φ in the present case can be written as $\Phi = \frac{1}{2}a_1\,\partial f/\partial\theta$, where a_1 is an arbitrary positive constant, and f is that root of the cubic equation

$$f^3 - 3\eta f + 3\theta = 0 \tag{120.5}$$

which is the one real root for $9\theta^2 - 4\eta^3 > 0$. Hence

$$\Phi = \tfrac{1}{2}a_1\,\partial f/\partial\theta = -a_1/2(f^2-\eta), \tag{120.6}$$

and we have for the coordinates

$$\left.\begin{aligned} x &= \partial\Phi/\partial\eta = \tfrac{1}{2}a_1(f^2+\eta)/(f^2-\eta)^3, \\ y &= \partial\Phi/\partial\theta = -a_1 f/(f^2-\eta)^3. \end{aligned}\right\} \tag{120.7}$$

These formulae can be put in a convenient parametric form by using as a parameter $s = f^2/(f^2-\eta)$. Then

$$\left.\begin{aligned} x/y^{4/5} &= a_1{}^{1/5}(2s-1)/2s^{2/5}, \\ \eta y^{2/5} &= a_1{}^{2/5}s^{1/5}(s-1), \\ \theta y^{3/5} &= \tfrac{1}{3}a_1{}^{3/5}s^{4/5}(3-2s), \end{aligned}\right\} \tag{120.8}$$

which give, in parametric form, η and θ as functions of the coordinates. The parameter s takes positive values from zero upwards ($s = 0$ corresponding to $x = -\infty$, i.e. to the stream incident from infinity). In particular, the value $s = \frac{1}{2}$ corresponds to $x = 0$, i.e. it gives the velocity distribution for large y in a plane perpendicular to the x-axis and passing near the body. The value $s = 1$ corresponds to the sonic line ($\eta = 0$), and $s = \frac{4}{3}$, as is easily seen, to the limiting characteristic. The value of the constant a_1 depends on the actual shape of the body, and can be determined only from an exact solution of the problem in all space.

Formulae (120.8) relate only to the region in front of the shock wave. The necessity for the shock to appear can be seen as follows. A simple calculation from formula (118.5) gives for the Jacobian Δ the expression $a_1{}^2(4f^2-\eta)/(f^2-\eta)^3$. It is easy to see that $\Delta > 0$ (and does not vanish) on the characteristics and everywhere to the left of them, corresponding to the region upstream of the limiting characteristics in the physical plane. To the right of the characteristics, however, Δ becomes zero, and so a shock wave must appear in this region.

The boundary conditions at the shock wave which must be satisfied by the solution of the Euler–Tricomi equation are as follows. Let θ_1, η_1 and θ_2, η_2 be the values of θ and η on the two sides of the discontinuity. First of all, they must correspond to the same curve in the physical plane, i.e.

$$x(\theta_1, \eta_1) = x(\theta_2, \eta_2), \qquad y(\theta_1, \eta_1) = y(\theta_2, \eta_2). \tag{120.9}$$

Next, the condition that the velocity component tangential to the discontinuity be continuous (i.e. that the derivative of the potential ϕ along the discontinuity be continuous) is equivalent to the condition that the potential itself be continuous:

$$\phi(\theta_1, \eta_1) = \phi(\theta_2, \eta_2); \tag{120.10}$$

the potential ϕ is determined from the function Φ by (119.3). Finally, another condition can be obtained from the limiting form of the equation (92.6) of the shock polar, which gives a relation between the velocity components on the two sides of the discontinuity. Replacing the angle χ in (92.6) by $\theta_2-\theta_1$, and introducing η_1, η_2 in place of v_1, v_2, we obtain the relation

$$2(\theta_2-\theta_1)^2=(\eta_2-\eta_1)^2(\eta_2+\eta_1). \tag{120.11}$$

In the present case, the solution of the Euler–Tricomi equation behind the shock wave (the region between OF and OF' in the hodograph plane, Fig. 123) has the same form (120.5), (120.6), but of course with a different constant coefficient (which we call $-a_2$) in place of a_1. The four simultaneous equations (120.9)–(120.11) determine the ratio a_2/a_1 and relate the quantities η_1, θ_1, η_2, θ_2. The solution of these equations is fairly complicated; it gives the following results. The shock wave corresponds to the value $s=\frac{1}{6}(5\sqrt{3}+8)=2{\cdot}78$ of the parameter s in formulae (120.8), which give the form of the shock and the velocity distribution on the forward side of the discontinuity. In the region behind (downstream of) the shock, the coefficient $-a_2$ is negative, and $f^2/(f^2-\eta)$ takes negative values. Using as the parameter the positive quantity $s=f^2/(\eta-f^2)$, we have instead of (120.8) the formulae

$$\left.\begin{aligned} x/y^{4/5}&=a_2{}^{1/5}(2s+1)/2s^{2/5}, \qquad \eta y^{2/5}=a_2{}^{2/5}s^{1/5}(s+1),\\ \theta y^{3/5}&=-\tfrac{1}{3}a_2{}^{3/5}s^{4/5}(2s+3), \end{aligned}\right\} \tag{120.12}$$

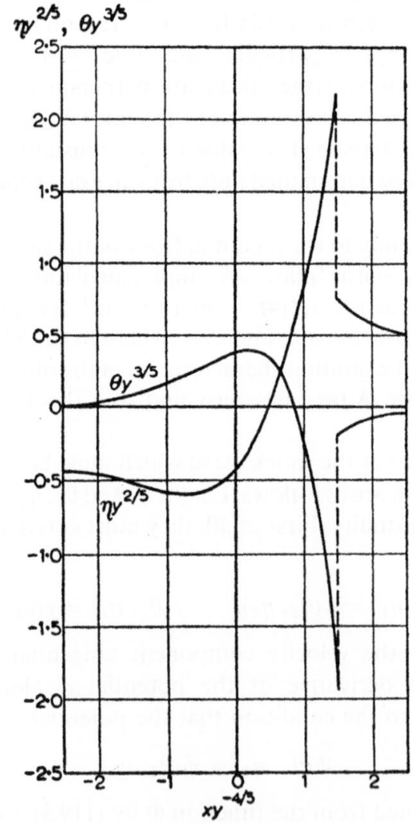

FIG. 124

where

$$a_2/a_1 = (9\sqrt{3}+1)/(9\sqrt{3}-1) = 1{\cdot}14,$$

and s takes values from $\frac{1}{6}(5\sqrt{3}-8) = 0{\cdot}11$ on the shock wave to zero at an infinite distance downstream.

Figure 124 shows graphs of $\eta y^{2/5}$ and $\theta y^{3/5}$ as functions of $xy^{-4/5}$, calculated from formulae (120.8) and (120.12) (the constant a_1 being arbitrarily taken as unity).

§121. The reflection of a weak discontinuity from the sonic line

Let us consider by means of the Euler–Tricomi equation the reflection of a weak discontinuity from the sonic line.

We shall assume that the weak discontinuity incident on the sonic line (reaching the point of intersection) is of the ordinary type, formed (say) by flow past an acute angle, i.e. the first spatial derivatives of the velocity are discontinuous in it. It is reflected from the sonic line as another discontinuity, the nature of which, however, is unknown a priori and must be determined by investigating the flow near the point of intersection. We take this point as the origin in the xy-plane, and the x-axis in the direction of the gas velocity there, so that it corresponds to the origin in the hodograph plane also.

Weak discontinuities coincide with characteristics, as we know. Let the characteristic Oa in the hodograph plane (Fig. 125a) correspond to the incident discontinuity. Since the coordinates x, y are continuous at the discontinuity, the first derivatives Φ_η, Φ_θ must be continuous also. The second derivatives of Φ, on the other hand, can be expressed in terms of the first spatial derivatives of the velocity, and therefore must be discontinuous. Denoting the discontinuities of quantities by placing them in brackets, we therefore have

$$\text{on } Oa \quad [\Phi_\eta] = [\Phi_\theta] = 0; \quad [\Phi_{\theta\theta}], \quad [\Phi_{\theta\eta}], \quad [\Phi_{\eta\eta}] \neq 0. \tag{121.1}$$

The functions Φ themselves in the regions 1 and 2 on each side of the characteristic Oa must not have singularities on the characteristic. Such a solution can be constructed from the second term in (118.6) with $k = 11/12$, which is proportional to the square of the difference $1 - 4\eta^3/9\theta^2$ (the other independent solution $\Phi_{11/12}$ has a singularity on the characteristic; see below). The first derivatives of this function vanish on the characteristic, and the second derivatives are finite. Furthermore, Φ can include those particular solutions of the Euler–Tricomi equation which do not give singularities of the flow in the physical plane. The solution of this kind which is of the lowest degree in θ and η is $\eta\theta$ (§119). Thus we seek Φ near the characteristic Oa and on either side of it in the forms:

$$\left.\begin{aligned} \Phi_{a1} &= -A\eta\theta - B\xi^2\theta^{11/6}F(\tfrac{13}{12}, \tfrac{19}{12}; 3; \xi), \\ \Phi_{a2} &= -A\eta\theta - C\xi^2\theta^{11/6}F(\tfrac{13}{12}, \tfrac{19}{12}; 3; \xi), \end{aligned}\right\} \tag{121.2}$$

where the suffixes $a1$ and $a2$ denote regions 1 and 2 near the characteristic and on each side of it; A, B, C are constants, and $\xi \equiv 1 - 4\eta^3/9\theta^2$; on the characteristics, $\xi = 0$.

We shall see that there are two cases, depending on the sign of the product AB, the weak discontinuity being reflected as either a logarithmic weak discontinuity or a weak shock wave.

Reflection as a weak discontinuity

Let us take the first case (L. D. Landau and E. M. Lifshitz 1954). A second characteristic in the hodograph plane (Ob in Fig. 125a) corresponds to the weak discontinuity reflected

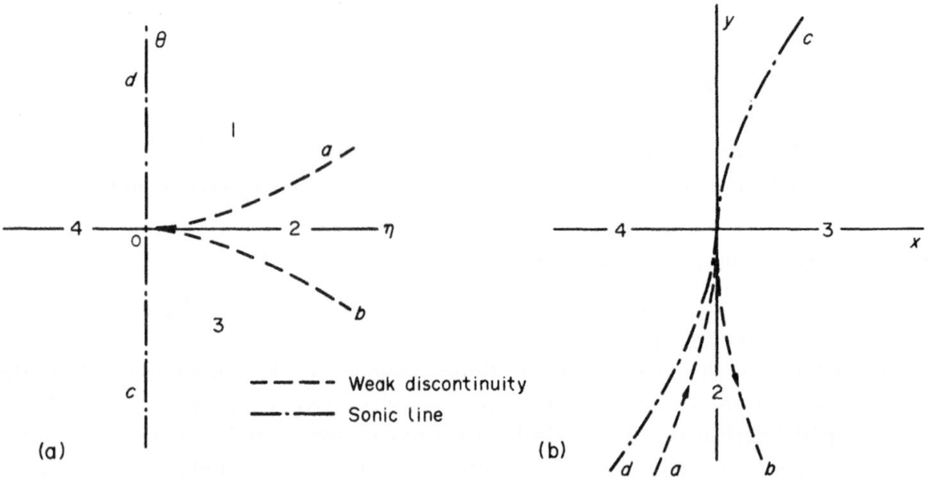

FIG. 125

from the sonic line. The form of the function Φ near this characteristic is obtained by analytical continuation of the functions (121.2), using (118.11)–(118.13). For $k = 11/12$, however, the function F_1 is meaningless, and therefore we cannot use these formulae directly. Instead, we must first put $k = (11/12)+\varepsilon$, and then let ε tend to zero. Logarithmic terms then appear, in accordance with the general theory of hypergeometric functions.

The calculation with (118.13) gives the following expression for the function Φ in region 3 near the characteristic Ob (we retain terms up to the second order in ξ):

$$\Phi_{b3} = -A\theta\eta + (B/\pi)(-\theta)^{11/6}\{\xi^2 \log|\xi| + c_0 + c_1\xi + c_2\xi^2\}, \tag{121.3}$$

where c_0, c_1, c_2 are numerical constants.†

A similar transformation (using formula (118.11)) of the function Φ_{a2} from the neighbourhood of the characteristic Oa to that of Ob gives an expression for Φ_{b2} similar to (121.3), with $\frac{1}{2}C$ in place of B. The coordinates x and y of points on the characteristic in the physical plane are calculated as the derivatives (118.4) for $\xi = 0$. Starting from (121.3), we find

$$\left.\begin{aligned} x &= -A\theta - (12^{1/3}Bc_1/\pi)(-\theta)^{7/6}, \\ y &= -A(-\tfrac{3}{2}\theta)^{2/3} - (B/\pi)(\tfrac{11}{6}c_0 + 2c_1)(-\theta)^{5/6}, \end{aligned}\right\} \tag{121.4}$$

and differentiation of Φ_{b2} gives the same with $\frac{1}{2}C$ instead of B. The condition that the coordinates x and y be continuous at the characteristic Ob therefore gives

$$C = 2B. \tag{121.5}$$

Next, for this type of reflection to occur, there must be no limiting lines in the hodograph plane (and therefore no non-physical regions in that plane), i.e. the Jacobian Δ must not vanish anywhere. Near the characteristic Oa, Δ can be calculated from the

† Their values are $c_0 = -2^9.3^4/385 = -108$, $c_1 = 288/7 = 41{\cdot}1$, $c_2 = 4{\cdot}86$.

functions (121.2), and is seen to be positive; the leading term in Δ is $\cong A^2$. Near the characteristic Ob, a calculation using (121.3) gives

$$\Delta \cong A - 16(3/2)^{1/6} AB\eta^{1/4} \log|\xi|. \qquad (121.6)$$

As we approach the characteristic, the logarithm tends to $-\infty$, and the second term is the leading one. The condition $\Delta > 0$ therefore gives $AB > 0$, i.e. A and B must have the same sign.

Finally, to determine the form of the sonic line, we need an expression for Φ near the axis $\eta = 0$. An expression valid near the upper half is obtained by simply transforming the hypergeometric function in Φ (121.2) into hypergeometric functions of argument $1 - \xi = 4\eta^3/9\theta^2$, which vanishes for $\eta = 0$.† On retaining only terms of the lowest degrees in η, we obtain

$$\Phi_d = -A\eta\theta - \frac{2\Gamma(1/3)}{\Gamma(23/12)\Gamma(17/12)} B\theta^{11/6} = -A\eta\theta - 6{\cdot}25B\theta^{11/6}. \qquad (121.7)$$

An analytical continuation into the region near the lower half of the axis gives

$$\Phi_c = -A\eta\theta - 6{\cdot}25\sqrt{3}B\theta^{11/6}; \qquad (121.8)$$

the calculations are similar to those used in deriving the transformation formulae (118.13).

We can now determine the form of all the lines under consideration. On the characteristics we have, omitting terms of higher order, $x = -A\theta$, $y = -A\eta$. We arbitrarily suppose that the upper characteristic ($\theta > 0$) corresponds to the weak discontinuity reaching the intersection. Since the gas velocity is in the positive x-direction, this discontinuity is the one which reaches the intersection if it lies in the half-plane $x < 0$. Hence it follows that the constant A, and therefore the constant B also, must be positive. The equation of the line of discontinuity in the physical plane is

$$-y = (\tfrac{3}{2})^{2/3} A^{1/3}(-x)^{2/3} = 1{\cdot}31 A^{1/3}(-x)^{2/3}. \qquad (121.9)$$

The reflected discontinuity, which corresponds to the lower characteristic, is given by the equation‡

$$-y = 1{\cdot}31 A^{1/3} x^{2/3} \qquad (121.10)$$

(Fig. 125b, in which the lines and regions are marked in correspondence with those in Fig. 125a).

The equation of the sonic line is obtained from the functions (121.7) and (121.8). Effecting the differentiation with respect to η and θ, and then putting $\eta = 0$, we obtain from (121.7) the equation of the part for which $\theta > 0$: $x = -A\theta$, $y = -\frac{11}{6} . 6{\cdot}25B\theta^{5/6}$, whence

$$y = -11{\cdot}4BA^{-5/6}(-x)^{5/6}. \qquad (121.11)$$

This is the lower part of the sonic line in Fig. 125b. Similarly, we obtain from (121.8) the equation of the upper part of this line:

$$y = 11{\cdot}4\sqrt{3}BA^{+5/6}x^{5/6}. \qquad (121.12)$$

Thus both discontinuities and both branches of the sonic line have a common tangent (the

† The transformation is given, for example, in *QM*, Mathematical Appendices, formula (e.7).

‡ When the first correction terms (the second terms in (121.4)) are taken into account, the equation of the reflected discontinuity is

$$-y = 1{\cdot}31 A^{\frac{1}{3}} x^{\frac{2}{3}} - 10{\cdot}5 BA^{-5/6} x^{5/6}. \qquad (121.10a)$$

y-axis) at the point of intersection O. Near this point the two branches of the sonic line are on opposite sides of the y-axis.

On the discontinuity which reaches O, the spatial derivatives of the velocity are discontinuous; as a characteristic quantity we may consider the discontinuity of the derivative $(\partial\eta/\partial x)_y$. Using the fact that

$$\left(\frac{\partial\eta}{\partial x}\right)_y = \frac{\partial(\eta, y)}{\partial(x, y)} = \frac{\partial(\eta, y)}{\partial(\eta, \theta)} \Big/ \frac{\partial(x, y)}{\partial(\eta, \theta)} = -\frac{1}{\Delta}\frac{\partial^2\Phi}{\partial\theta^2}$$

and formulae (121.2), (121.5), we obtain

$$[(\partial\eta/\partial x)_y]_1^2 = 8(\tfrac{3}{2})^{1/6}(B/A^2)\eta^{-1/4} = 8{\cdot}56BA^{-7/4}(-y)^{-1/4}. \tag{121.13}$$

Thus this discontinuity increases as $(-y)^{-1/4}$ as we approach the point of intersection.

On the reflected weak discontinuity, the derivatives of the velocity are not discontinuous, but the velocity distribution has a curious logarithmic singularity. Calculating the coordinates x and y as functions of η, θ from (121.3) (keeping only the first term in the braces), we can put the dependence of η on x for given y near the reflected discontinuity in the parametric form

$$\left.\begin{aligned} \eta &= \frac{|y|}{A} + \frac{x - x_0}{2\sqrt{(A|y|)}} - \frac{1}{6A}|y|\zeta, \\ x - x_0 &= \frac{1}{3\sqrt{A}}|y|^{3/2}\zeta - 5{\cdot}7\frac{B|y|^{7/4}}{\pi A^{7/4}}\zeta\log|\zeta|, \end{aligned}\right\} \tag{121.14}$$

where ζ is the parameter and $x_0 = x_0(y)$ is the equation of the discontinuity in the physical plane.

REFLECTION AS A SHOCK WAVE

Let us now consider the other case, that of reflection of a weak discontinuity from the sonic line as a shock wave (L. P. Gor'kov and L. P. Pitaevskiĭ 1962).†

This case occurs if $AB < 0$. From (121.6), we see that here there are two limiting lines which are exponentially close to the characteristic Ob: the Jacobian Δ is zero for

$$|\xi| \cong \frac{2}{|\theta|}|\theta + \tfrac{2}{3}\eta^{3/2}|e^{-\Theta}, \quad \Theta = \frac{A\pi(2/3)^{1/6}}{16|B|\eta^{1/4}}. \tag{121.15}$$

It is evident from the start that the boundaries of the non-physical region in the hodograph plane (Ob_2 and Ob_3 in Fig. 126a) will also be exponentially close to the characteristic, and the shock wave is therefore exponentially weak.

Neglecting the exponentially small values of ξ on Ob_2 and Ob_3, we find for the coordinates x and y on them the same expressions as on either side of the characteristic Ob in the previous case. The continuity condition for the coordinates at the shock wave therefore always gives the previous relation (121.5). Accordingly, we have the same expression (121.13) for the change in the velocity derivative at the incident discontinuity. Again assuming that the latter corresponds to the upper characteristic Oa in the hodograph plane, we have as before $A > 0$, so that now $B < 0$. It is seen from (121.13), therefore, that the physical criterion for the two cases of weak-discontinuity reflection is the sign of the change in the velocity derivative at the incident discontinuity.

† The possibility in principle of such reflection was noted by K. G. Guderley (1948).

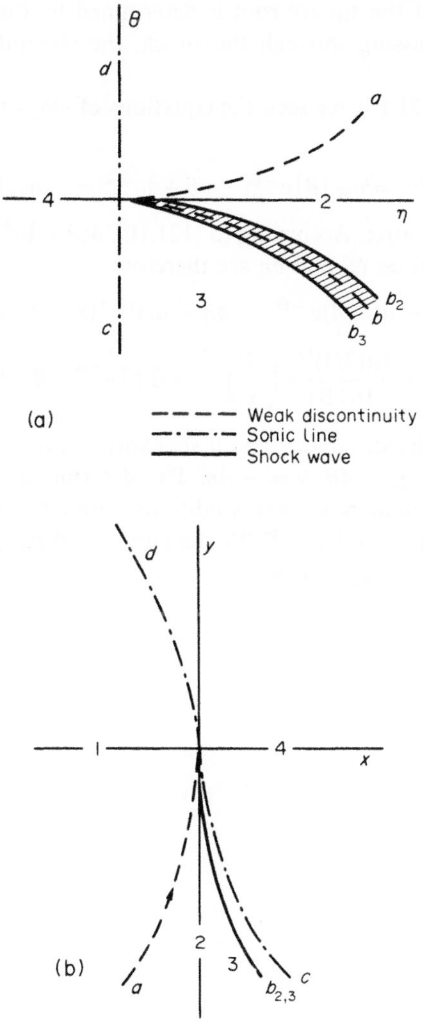

FIG. 126

The equations (121.9) and (121.10) for the incident weak discontinuity and reflected shock wave lines remain the same, if exponentially small corrections are neglected. However, since the sign of B is different, the configuration of these lines in the physical plane is changed, as shown in Fig. 126b.

To determine the strength of the shock wave, i.e. the changes $\delta\theta$ and $\delta\eta$ there, we have to use the complete boundary conditions to be satisfied at the shock by the solution of the Euler–Tricomi equation. These have already been formulated as (120.9)–(120.11). The last of these, the equation of the shock polar, becomes $(\delta\theta)^2 = \eta(\delta\eta)^2$, where $\delta\theta = \theta_{b2} - \theta_{b3}$, $\delta\eta = \eta_{b2} - \eta_{b3}$ are exponentially small discontinuities at the shock; the suffixes $b2$ and $b3$ relate to the lines Ob_2 and Ob_3 in the hodograph plane, that is, to the front and back of the shock in the physical plane respectively. Hence

$$\delta\theta = \sqrt{\eta}\delta\eta; \tag{121.16}$$

the choice of the sign of the square root is determined by the fact that, when the gas velocity decreases on passing through the shock, the streamlines must approach the surface of discontinuity.

In accordance with (121.15), we seek the equations of Ob_2 and Ob_3 in the hodograph plane as

$$\theta + \tfrac{2}{3}\eta^{3/2} = a_{b2}\,|\theta|\,e^{-\Theta}, \qquad \theta + \tfrac{2}{3}\eta^{3/2} = -a_{b3}\,|\theta|\,e^{-\Theta},$$

where a_{b2} and a_{b3} are positive. According to (121.16), $\delta(\theta + \frac{2}{3}\eta^{3/2}) = \delta\theta + \sqrt{\eta}\,d\eta = 2\,\delta\theta$. The required discontinuities $\delta\theta$ and $\delta\eta$ are therefore

$$\left.\begin{aligned} \delta\theta = a(x/A)e^{-\Theta}, \quad & \delta\eta = a(\tfrac{2}{3})^{1/3}(x/A)^{2/3}e^{-\Theta}, \\ \Theta = \frac{A\pi(2/3)^{1/3}}{16|B|}\left(\frac{A}{x}\right)^{1/6} & = 0{\cdot}17A^{7/6}/|B|x^{1/6}, \end{aligned}\right\} \tag{121.17}$$

where $a = \frac{1}{2}(a_{b2} + a_{b3})$; the variables η and θ are expressed in terms of the coordinates in the physical plane by $x \cong -A\theta$, $y = -A\eta$. The determination of a has to take into account also all the remaining boundary conditions, with terms both linear and quadratic in the exponentially small quantity $e^{-\Theta}$. We shall not give these quite lengthy calculations, but simply the result: $a_{b2} = a_{b3} = a = 5{\cdot}2$.

CHAPTER XIII

FLOW PAST FINITE BODIES

§122. The formation of shock waves in supersonic flow past bodies

Simple arguments show that, in supersonic flow past an arbitrary body, a shock wave must be formed in front of the body. For the disturbances in the supersonic flow caused by the presence of the body are propagated only downstream. Hence a uniform supersonic stream incident on the body would be unperturbed as far as the leading end of the body. The normal component of the gas velocity would then be non-zero at the surface there, in contradiction to the necessary boundary condition. The resolution of this difficulty can only be the occurrence of a shock wave, as a result of which the gas flow between it and the leading end of the body becomes subsonic.

Thus a shock wave is formed in front of the body when the incident flow is supersonic; it is called the *bow wave*. When the leading end of the body is blunt, the bow wave does not touch the body. In front of the shock wave, the flow is uniform; behind it, the flow is modified and bends round the body (Fig. 127a). The surface of the shock wave extends to infinity, and at great distances from the body, where the shock is weak, it intersects the incident streamlines at an angle approaching the Mach angle. A characteristic feature of flow past a blunt-ended body is the existence of a subsonic flow region behind the shock wave at the most forward part of its surface; this region extends to the body itself, and thus lies between the discontinuity surface, the body, and a lateral sonic surface (the broken curves in Fig. 127a).

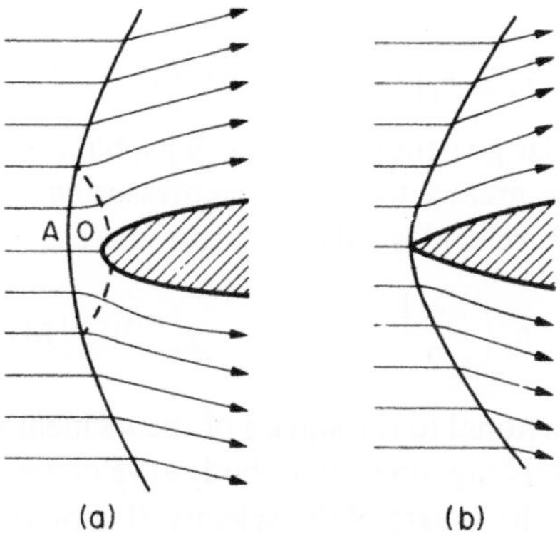

FIG. 127

The shock wave can touch the body only when the leading end of the latter is pointed. The surface of discontinuity then has a point at the same place (Fig. 127b); in asymmetric flow, part of this surface may be a weak discontinuity.

For a body with a given shape, however, this type of flow pattern is possible only for velocities exceeding a certain limit; at lower velocities, the shock wave is detached from the leading end of the body (see §§113), even if the latter is pointed.

Let us consider axially symmetrical supersonic flow past a solid of revolution and determine the gas pressure at the rounded leading end of the body (the stagnation point O in Fig. 127a). It is evident from symmetry that the streamline which terminates at O intersects the shock wave at right angles, so that the velocity component at A normal to the surface of discontinuity is the same as the total velocity. The values of quantities in the incident stream will be denoted, as usual, by the suffix 1, and the values behind the shock wave at the point A by the suffix 2. The latter are determined from formulae (89.6) and (89.7):

$$p_2 = p_1[2\gamma M_1^2 - (\gamma - 1)]/(\gamma + 1),$$

$$v_2 = c_1 \frac{2 + (\gamma - 1)M_1^2}{(\gamma + 1)M_1}, \qquad \rho_2 = \rho_1 \frac{(\gamma + 1)M_1^2}{2 + (\gamma - 1)M_1^2}.$$

The pressure p_0 at the point O (where the gas velocity $v = 0$) can now be obtained by means of the formulae which give the variation of quantities along a streamline. We have (see §83, Problem)

$$p_0 = p_2\left[1 + \frac{\gamma - 1}{2}\frac{v_2^2}{c_2^2}\right]^{\gamma/(\gamma - 1)},$$

and a simple calculation gives

$$p_0 = p_1\left(\frac{\gamma + 1}{2}\right)^{(\gamma + 1)/(\gamma - 1)} \frac{M_1^2}{[\gamma - (\gamma - 1)/2M_1^2]^{1/(\gamma - 1)}}. \qquad (122.1)$$

This determines the pressure at the leading end for a supersonic incident flow ($M_1 > 1$).

For comparison, we give the formula for the pressure at the stagnation point obtained for a continuous adiabatic retardation of the gas, with no shock wave (as would be true for a subsonic incident flow):

$$p_0 = p_1[1 + \tfrac{1}{2}(\gamma - 1)M_1^2]^{\gamma/(\gamma - 1)}. \qquad (122.2)$$

For $M_1 = 1$, the two formulae give the same value of p_0, but for $M_1 > 1$ the pressure given by formula (122.2) is always greater than the true pressure p_0 given by formula (122.1).†

In the limit of very large velocities ($M_1 \gg 1$), formula (122.1) gives

$$p_0 = p_1\left(\frac{\gamma + 1}{2}\right)^{(\gamma + 1)/(\gamma - 1)} \gamma^{-1/(\gamma - 1)} M_1^2, \qquad (122.3)$$

i.e. the pressure p_0 is proportional to the square of the incident velocity. From this result we can conclude that the total drag force on the body at velocities large compared with that of sound is proportional to the square of the velocity. It should be noticed that this is the

† This statement is true generally, and does not depend on the assumption of a polytropic or even a perfect gas in (122.1), (122.2). For, when a shock wave is present, the entropy s_0 of the gas at O is greater than s_1, whereas if the shock wave were absent s_0 would be equal to s_1. The heat function is in either case $w_0 = w_1 + \frac{1}{2}v_1^2$, since the quantity $w + \frac{1}{2}v^2$ is unchanged when a streamline intersects a normal compression discontinuity. From the thermodynamic identity $dw = T\,ds + dp/\rho$, it follows that the derivative $(\partial p/\partial s)_w = -\rho T < 0$, i.e. an increase in entropy when w remains constant involves a decrease in pressure, whence the result follows.

same as the law governing the drag force at velocities small compared with that of sound but yet so large that the Reynolds number is large (see §45).

Besides the fact that shock waves must be formed, we can also say that in supersonic flow past a finite body there must be two successive shock waves at large distances from the body (L. Landau 1945). For the disturbances caused by the body at large distances are small, and can therefore be regarded as a cylindrical sound wave outgoing from the x-axis (which passes through the body parallel to the direction of flow); considering the flow, as usual, in a coordinate system where the body is at rest, we have a wave in which the time is represented by x/v_1, and the rate of propagation by $v_1/\sqrt{(M_1^2-1)}$ (see §123). We can therefore apply immediately the results obtained in §102 for a cylindrical wave at large distances from the source. We thus arrive at the following pattern of shock waves far from the body: in the first shock, the pressure increases discontinuously, so that behind it there is a condensation; then follows a region where the pressure gradually decreases into a rarefaction, after which the pressure again increases discontinuously in the second shock. The intensity of the leading shock decreases as $r^{-3/4}$ with increasing distance from the x-axis, and the distance between the two shocks increases as $r^{1/4}$.†

Let us now examine the appearance and development of the shock waves as the number M_1 gradually increases. A supersonic region first appears for some value of M_1 less than unity, as a region adjoining the surface of the body. At least one shock wave occurs in this region, usually at the edge of the supersonic region.

As M_1 increases, the supersonic region expands, and the length of the shock wave increases. This is the shock wave whose existence for $M_1 = 1$ has been demonstrated (for the two-dimensional case) in §120; it follows also that the shock wave must first appear for $M_1 < 1$.

As soon as M_1 exceeds unity, another shock wave appears, the bow wave, which intersects the whole of the infinitely wide incident stream of gas. For M_1 exactly unity, the flow in front of the body is entirely subsonic. For $M_1 > 1$ but arbitrarily close to unity, therefore, the supersonic part of the incident stream, and consequently the bow wave, are arbitrarily far in front of the body. As M_1 increases further, the bow wave gradually approaches the body.

The shock wave in the local supersonic region must intersect the sonic line in some way. We shall discuss the two-dimensional case. The nature of this intersection is not yet fully understood. If the shock terminates at the intersection, its strength falls to zero there, and the flow is transonic everywhere in the plane near the intersection point. The flow pattern in such a case must be given by the appropriate solution of the Euler–Tricomi equation. In addition to the usual conditions that the solution be single-valued in the physical plane and the boundary conditions at the shock wave, the following conditions must also be satisfied: (1) if the flow is supersonic on both sides of the shock (as when only the shock terminates at the intersection, being "supported" by the sonic line), then the shock wave must be one that reaches the intersection; (2) characteristics in the supersonic region which reach the intersection cannot have any flow singularities, since these could arise only from the intersection itself and would therefore have to be carried away from the intersection

† For shock waves formed in axially symmetrical flow past narrow pointed bodies, the quantitative coefficients in these relationships can be determined; see the second footnote to §123.

point. The existence of a solution of the Euler–Tricomi equation satisfying these requirements seems to be not yet proved.†

Another possible configuration of the shock wave and the sonic line in the local supersonic region is for only the sonic line to terminate at the intersection (Fig. 128b); the shock wave strength need not be zero there, and so the flow near it is transonic only on one side of the shock. The shock wave itself may have one end "supported" by a solid surface, and the other end (or both ends) starting within the supersonic stream (cf. the end of §115).

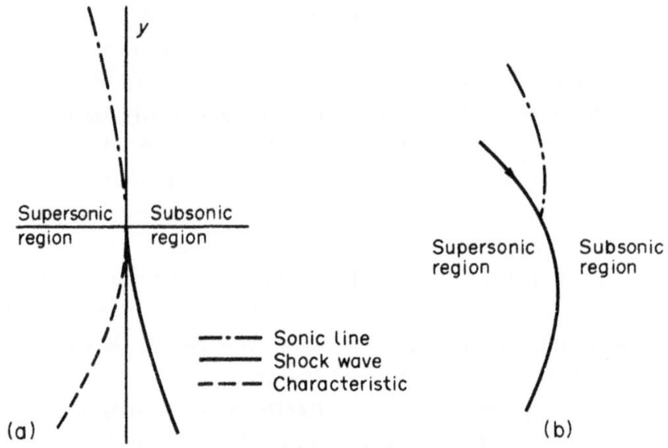

FIG. 128

§123. Supersonic flow past a pointed body

The shape which a body must have in order to be streamlined in supersonic flow, i.e. to be subject to as small a drag force as possible, is quite different from the corresponding shape for subsonic flow. We may recall that, in the subsonic case, streamlined bodies are those which are elongated, rounded in front, and pointed behind. In supersonic flow past such a body, however, a strong shock wave would be formed in front of it, leading to a considerable increase in the drag. In the supersonic case, therefore, a long streamlined body must be pointed at both ends, and the angle of the point must be small; if the body is inclined to the direction of flow, the angle between them (angle of attack) must also be small.

In steady supersonic flow past a body of this shape, the gas velocity is nowhere very different in magnitude or direction from the incident velocity, even near the body, and the shock waves formed are weak; the intensity of the bow wave decreases with the angle at the front of the body. Far from the body, the gas flow consists of outgoing sound waves. The main part of the drag can be regarded as due to the conversion of kinetic energy of the

† Germain has found several types of solution of the Euler–Tricomi equation which might represent the intersection of a shock wave with the sonic line, but they have not been at all fully investigated. Some of them do not satisfy condition (1) above. Figure 128a shows a case which might correspond to the termination of a shock wave forming the boundary of the local supersonic region: at the point of intersection, the shock wave and the sonic line both terminate and have a common tangent, and lie on opposite sides of it (the gas moves from left to right). The fulfilment of condition (2) has not been tested, however. Only the possible range of k has been determined ($\frac{3}{4} < k < \frac{11}{12}$), but it is not known whether one can satisfy the condition for the coordinates to be continuous at the shock wave in the physical plane. See P. Germain, *Progress in Aeronautical Sciences* 5, 143, 1964.

moving body into the energy of the sound waves which it emits. This drag, which occurs only in supersonic flow, is called *wave drag*;† it can be calculated in a general form valid for any cross-section of the body (T. von Kármán and N. B. Moore 1932).

The nature of the flow just described makes it possible to use the linearized equation (114.4) for the potential:

$$\frac{\partial^2 \phi}{\partial y^2} + \frac{\partial^2 \phi}{\partial z^2} - \beta^2 \frac{\partial^2 \phi}{\partial x^2} = 0, \tag{123.1}$$

where we have introduced for brevity the positive constant

$$\beta^2 = (v_1{}^2 - c_1{}^2)/c_1{}^2; \tag{123.2}$$

the x-axis is in the direction of the flow, the suffix 1 denotes quantities pertaining to the incident stream, and $1/\beta$ is just the tangent of the Mach angle.

Equation (123.1) is formally identical with the two-dimensional wave equation with x/v_1 representing the time and v_1/β the velocity of the waves. This is no accident; the physical significance is that the gas flow far from the body consists, as already mentioned, of outgoing sound waves emitted by the body. If the gas at infinity is regarded as being at rest, and the body as being in motion, the cross-section of the body at a given point in space will vary with time, and the distance to which a disturbance is propagated at time t (i.e. the distance to the Mach cone) will increase as $v_1 t/\beta$. Thus we shall have a two-dimensional emission of sound (propagated with velocity v_1/β) by the variable profile.

Using this "sonic analogy" as a guide, we can immediately write down the required expression for the velocity potential of the gas, using formula (74.15) for the potential of cylindrical sound waves emitted from a source (at distances large compared with the dimension of the source) and replacing ct by x/β.

Let $S(x)$ be the area of the cross-section of the body in a plane perpendicular to the direction of flow (the x-axis), and l the length of the body in that direction; we take the origin at the leading end of the body. Then

$$\phi(x, r) = -\frac{v_1}{2\pi} \int_0^{x-\beta r} \frac{S'(\xi)d\xi}{\sqrt{[(x-\xi)^2 - \beta^2 r^2]}}; \tag{123.3}$$

the lower limit is taken as zero, since $S(x) \equiv 0$ for $x < 0$ (and for $x > l$).

Thus we have completely determined the gas flow at distances r from the axis which are large compared with the thickness of the body.‡ Disturbances leaving the body in a supersonic flow are, of course, propagated only into the region behind the cone $x - \beta r = 0$, whose vertex is at the leading end of the body; in front of this cone we have simply $\phi = 0$ (uniform flow). Between the cones $x - \beta r = 0$ and $x - \beta r = l$, the potential is determined by formula (123.3); behind the latter cone (whose vertex is at the trailing end of the body)

---

† The total drag is obtained by adding to the wave drag the forces due to friction and to separation at the trailing end of the body.

‡ For axially symmetrical flow past a solid of revolution, (123.3) is valid for all r up to the surface of the body. In particular, it again gives (113.6) for flow past a narrow cone.

On the other hand, if this linear-approximation solution is considered at a great distance from the body, a correction can be applied to it for the non-linear distortion of the profile, as was done in §102 for a cylindrical sound wave. This gives the strength of the shock wave at large distances from a narrow pointed solid of revolution, including the dependence on M_1, i.e. the coefficient in the law of damping ($\propto r^{-3/4}$) described in §122. See G. B. Whitham, *Linear and Nonlinear Waves*, New York 1974, §9.3.

the upper limit of the integral in (123.3) is the constant l. Both these cones are weak discontinuities, in the approximation considered; in reality, they are weak shock waves.

The drag force acting on the body is just the x-component of the momentum carried away by the sound waves per unit time. We take a cylindrical surface with large radius r and axis along the x-axis. The x-component of the momentum flux density through this surface is $\Pi_{xr} = \rho v_r(v_x + v_1) \cong \rho_1 (\partial\phi/\partial r)(v_1 + \partial\phi/\partial x)$. On integration over the whole surface, the first term gives zero, since the integral of ρv_r is the total mass flux through the surface, which is zero. Thus

$$F_x = -2\pi r \int_{-\infty}^{\infty} \Pi_{xr}\,dx = -2\pi r\rho_1 \int_{-\infty}^{\infty} \frac{\partial\phi}{\partial r}\frac{\partial\phi}{\partial x}\,dx. \qquad (123.4)$$

At large distances (in the wave region), the derivatives of the potential can be calculated as in §74 (see formula (74.17)), and we have

$$\frac{\partial\phi}{\partial r} = -\beta\frac{\partial\phi}{\partial x} = \frac{v_1}{2\pi}\sqrt{\frac{\beta}{2r}} \int_0^{x-\beta r} \frac{S''(\xi)\,d\xi}{\sqrt{(x-\xi-\beta r)}}.$$

This expression is substituted in (123.4), and the squared integral is written as a double integral; putting for brevity $x - \beta r = X$, we obtain

$$F_x = \frac{\rho_1 v_1^2}{4\pi} \int_{-\infty}^{\infty}\int_0^X\int_0^X \frac{S''(\xi_1)S''(\xi_2)\,d\xi_1\,d\xi_2\,dX}{\sqrt{[(X-\xi_1)(X-\xi_2)]}}.$$

The integration over X can be effected; after changing the order of integration, the integral is from the greater of ξ_1 and ξ_2 to infinity. We first take as the upper limit a large but finite quantity L, which later tends to infinity. Thus

$$F_x = -\frac{\rho_1 v_1^2}{2\pi} \int_0^l\int_0^{\xi_2} S''(\xi_1)S''(\xi_2)[\log(\xi_2-\xi_1) - \log 4L]\,d\xi_1\,d\xi_2.$$

The integral of the term containing the constant factor log $4L$ is zero, since not only the area $S(x)$ but also its derivative $S'(x)$ vanishes at the pointed ends of the body. We therefore have

$$F_x = -\frac{\rho_1 v_1^2}{2\pi} \int_0^l\int_0^{\xi_2} S''(\xi_1)S''(\xi_2)\log(\xi_2-\xi_1)\,d\xi_1\,d\xi_2,$$

or

$$F_x = -\frac{\rho_1 v_1^2}{4\pi} \int_0^l\int_0^l S''(\xi_1)S''(\xi_2)\log|\xi_2-\xi_1|\,d\xi_1\,d\xi_2. \qquad (123.5)$$

This is the required formula for the wave drag on a narrow pointed body.† The order of magnitude of the integral is $(S/l^2)^2 l^2$, where S is some mean cross-sectional area of the

† The lift (for a body not axially symmetrical or a non-zero angle of attack) is zero in the approximation here considered.

body. Hence $F_x \sim \rho_1 v_1^2 S^2/l^2$. The drag coefficient for an elongated body may be conventionally defined, in terms of the square of the length, as

$$C_x = F_x/\tfrac{1}{2}\rho_1 v_1^2 l^2. \tag{123.6}$$

Then, in this case,

$$C_x \sim S^2/l^4; \tag{123.7}$$

it is proportional to the square of the cross-sectional area.

We may point out the formal analogy between formula (123.5) and formula (47.4) for the induced drag on a thin wing; the function $\Gamma(z)$ in (47.4) is here replaced by $v_1 S'(x)$. On account of this analogy we can use, to calculate the integral in (123.5), the method described at the end of §47.

It should also be noticed that the wave drag given by formula (123.5) is unchanged if the direction of flow is reversed: the integral is independent of the direction in which the body extends. This property of the drag force is characteristic of the linearized theory.†

Finally, let us briefly discuss the range of applicability of this formula. This subject may be approached as follows. The amplitude of oscillation of the gas particles in the sound waves emitted by the body is of the order of magnitude of the thickness of the body, which we denote by δ. The velocity of the oscillations is accordingly of the order of the ratio $\delta:(l/v_1)$ of the amplitude δ to the period l/v_1 of the wave. The linear approximation for the propagation of sound waves (i.e. the linearized equation for the potential), however, always requires that the gas velocity be small compared with the velocity of sound, i.e. we must have $v_1/\beta \gg v_1\delta/l$, or, what is in practice the same,

$$\mathrm{M}_1 \ll l/\delta. \tag{123.8}$$

Thus the theory given above becomes inapplicable for values of M_1 comparable with the ratio of length to thickness of the body.

It is also inapplicable, of course, in the opposite limiting case where M_1 is close to unity and the linearization of the equations is invalid.

PROBLEM

Determine the form of the elongated solid of revolution which experiences the smallest drag for a given volume V and length l.

SOLUTION. On account of the analogy mentioned in the text, we introduce a variable θ such that $x = \frac{1}{2}l(1-\cos\theta)$ $(0 \leqslant \theta \leqslant \pi$; the origin of x is at the leading end of the body); and write the function $f(x) = S'(x)$ as

$$f = -l\sum_{n=2}^{\infty} A_n \sin n\theta;$$

the condition $S = 0$ for $x = 0$ and l means that only terms with $n \geqslant 2$ can appear in the sum. The drag coefficient is then

$$C_x = \tfrac{1}{4}\pi \sum_{n=2}^{\infty} n A_n^2.$$

The area $S(x)$ and the total volume V of the body are calculated from the function $f(x)$ as

$$S = \int_0^x f(x)\mathrm{d}x, \qquad V = \int_0^l S(x)\mathrm{d}x.$$

† It also holds in the theory of the wave drag on thin wings given in §125.

A simple calculation gives $V = \pi l^3 A_2/16$, i.e. the volume is determined by the coefficient A_2 alone. The minimum F_x is therefore reached if $A_n = 0$ for $n \geqslant 3$. The result is

$$C_{x,\min} = (128/\pi)(V/l^3)^2 = (9\pi/2)(S_{\max}/l^2)^2.$$

The cross-sectional area of the body is $S = \frac{1}{4}l^2 A_2 \sin^3\theta$, and the radius as a function of x is therefore $R(x) = \sqrt{2(8/\pi)(V/3l^4)^{1/2}}[x(l-x)]^{3/4}$. The body is symmetrical about the plane $x = \frac{1}{2}l$.†

§124. Subsonic flow past a thin wing

Let us consider subsonic flow of a gas past a thin streamlined wing. As for an incompressible fluid, a wing which is streamlined for subsonic flow must be thin, pointed at the trailing edge, and rounded at the leading edge, and the angle of attack must be small. We take the direction of flow as the x-axis and the direction of the span as the z-axis.

The gas velocity nowhere‡ differs greatly from the velocity $\mathbf{v}_1$ of the incident stream, so that we can use the linearized equation (114.4) for the potential:

$$(1-\mathrm{M}_1{}^2)\frac{\partial^2\phi}{\partial x^2}+\frac{\partial^2\phi}{\partial y^2}+\frac{\partial^2\phi}{\partial z^2}=0. \tag{124.1}$$

At the surface of the wing (which we call C), the velocity must be tangential; introducing a unit vector $\mathbf{n}$ along the normal to the surface, we can write this condition as

$$\left(v_1+\frac{\partial\phi}{\partial x}\right)n_x+\frac{\partial\phi}{\partial y}n_y+\frac{\partial\phi}{\partial z}n_z=0.$$

Since the wing is flattened and the angle of attack is small, the normal $\mathbf{n}$ is almost parallel to the y-axis, so that $|n_y|$ is almost unity, while n_x and n_z are small. We can therefore neglect the second-order terms $n_x\partial\phi/\partial x$ and $n_z\partial\phi/\partial z$, and replace n_y by ± 1 ($+1$ on the upper surface of the wing and -1 on the lower surface). Thus the boundary condition on equation (124.1) is

$$v_1 n_x \pm \partial\phi/\partial y = 0. \tag{124.2}$$

Since the wing is assumed thin, $\partial\phi/\partial y$ on its surface can be taken as the limiting value for $y \to 0$.

The solution of equation (124.1) with the condition (124.2) can easily be reduced to the solution of a problem of incompressible flow. To do so, we use instead of the coordinates x, y, z the variables

$$x' = x, \qquad y' = y\sqrt{(1-\mathrm{M}_1{}^2)}, \qquad z' = z\sqrt{(1-\mathrm{M}_1{}^2)}. \tag{124.3}$$

In these variables, equation (124.1) becomes

$$\frac{\partial^2\phi}{\partial x'^2}+\frac{\partial^2\phi}{\partial y'^2}+\frac{\partial^2\phi}{\partial z'^2}=0, \tag{124.4}$$

i.e. Laplace's equation. The surface of the body is replaced by another, C', obtained by leaving unchanged the profiles of cross-sections by planes parallel to the xy-plane, but reducing in the ratio $\sqrt{(1-\mathrm{M}_1{}^2)}$ all dimensions in the direction of the span (the z-direction).

† Although $R(x)$ vanishes at the ends of the body, $R'(x)$ becomes infinite, i.e. the body is not pointed; the approximation underlying the method is therefore not strictly applicable near the ends.

‡ Except in a small region near the leading edge of the wing, where there is a stagnation line.

The boundary condition (124.2) then becomes

$$v_1 n_x \pm \frac{\partial \phi}{\partial y'} \sqrt{(1-M_1^2)} = 0,$$

and it can be reduced to the previous form by introducing in place of ϕ a new potential ϕ':

$$\phi' = \phi\sqrt{(1-M_1^2)}. \tag{124.5}$$

We then have for ϕ' Laplace's equation with the boundary condition

$$v_1 n_x \pm \partial\phi'/\partial y' = 0, \tag{124.6}$$

which must be satisfied for $y' = 0$.

Equation (124.4) with the boundary condition (124.6), is, however, the equation which must be satisfied by the velocity potential of an incompressible fluid flowing past the surface C'. Thus the problem of determining the velocity distribution in compressible flow past a wing with surface C is equivalent to that of finding the velocity distribution in incompressible flow past a wing with surface C'.

Next, let us consider the lift force F_y acting on the wing. First of all, we note that the derivation of Zhukovskiĭ's formula (38.4) given in §38 is entirely valid for a compressible fluid, since the variable density ρ of the fluid can be replaced in that approximation by a constant ρ_1. Thus

$$F_y = -\rho_1 v_1 \int \Gamma \, dz, \tag{124.7}$$

where the integration is taken along the span l_z of the wing. From the relation (124.5) and the equality of the transverse profiles of the wings C and C' it follows that the velocity circulation Γ in compressible flow past the wing C is related to the circulation Γ' in incompressible flow past the wing C' by

$$\Gamma' = \Gamma\sqrt{(1-M_1^2)}. \tag{124.8}$$

Substituting this in (124.7) and changing to an integration over z', we obtain

$$F_y = -\rho_1 v_1 \int \Gamma' \, dz'/(1-M_1^2).$$

The numerator is the lift force on the wing C' in an incompressible fluid. Denoting it by F'_y, we have

$$F_y = F'_y/(1-M_1^2). \tag{124.9}$$

Introducing the lift coefficients

$$C_y = F_y/\tfrac{1}{2}\rho_1 v_1^2 l_x l_z, \qquad C'_y = F'_y/\tfrac{1}{2}\rho_1 v_1^2 l_x l_z'$$

(where l_x, l_z and l_x, $l_z' = l_z\sqrt{(1-M_1^2)}$ are the lengths of the wings C and C' in the x and z directions), we can rewrite this equation as

$$C_y = C'_y/\sqrt{(1-M_1^2)}. \tag{124.10}$$

For wings with large span (and constant profile), the lift coefficient in an incompressible fluid is proportional to the angle of attack, and does not depend on the length or width of the wing:

$$C'_y = \text{constant} \times \alpha, \tag{124.11}$$

where the constant depends only on the shape of the profile (see §46). In this case, therefore, (124.10) can be replaced by

$$C_y = C_y^{(0)}/\sqrt{(1-\mathrm{M}_1^2)}, \tag{124.12}$$

where C_y and $C_y^{(0)}$ are the lift coefficients for the same wing in compressible and incompressible fluids respectively. Thus we have the rule that the lift force acting on a long wing in a compressible fluid is $1/\sqrt{(1-\mathrm{M}_1^2)}$ times that on the same wing (at the same angle of attack) in an incompressible fluid (L. Prandtl 1922, H. Glauert 1928).

Similar relations can be obtained for the drag force. Together with Zhukovskii's formula for the lift force, formula (47.4) for the induced drag on a wing is also entirely applicable to compressible flow. Effecting the same transformations (124.3) and (124.8), we obtain

$$F_x = F'_x/(1-\mathrm{M}_1^2), \tag{124.13}$$

where F'_x is the drag on the wing C' in an incompressible fluid. When the span increases, the induced drag tends to a constant limit (§47). For sufficiently long wings we can therefore replace F'_x by $F_x^{(0)}$ (the drag in an incompressible fluid for the wing C). Then the drag coefficient is

$$C_x = C_x^{(0)}/(1-\mathrm{M}_1^2). \tag{124.14}$$

Comparing this with (124.12), we see that the ratio C_y^2/C_x is the same for compressible and incompressible fluids.

All the results given here are, of course, invalid for values of M_1 close to unity, since the linearized theory then becomes inapplicable.

§125. Supersonic flow past a wing

If a wing is streamlined in a supersonic stream, it must be pointed at both ends, like the thin bodies discussed in §123.

Here we shall consider only the flow past a thin wing with very large span, the profile being constant along the span. Regarding the span as infinite, we have a two-dimensional gas flow (in the xy-plane). Instead of equation (123.1), we now have for the potential the equation

$$\frac{\partial^2\phi}{\partial y^2} - \beta^2\frac{\partial^2\phi}{\partial x^2} = 0, \tag{125.1}$$

with the boundary condition

$$[\partial\phi/\partial y]_{y\to\pm 0} = \mp v_1 n_x, \tag{125.2}$$

where the signs $\mp$ on the right relate to the upper and lower surfaces of the wing respectively. Equation (125.1) is a one-dimensional wave equation, and its general solution has the form $\phi = f_1(x-\beta y) + f_2(x+\beta y)$. The fact that disturbances which affect the flow start from the body means that above the wing ($y > 0$) we must have $f_2 \equiv 0$, so that $\phi = f_1(x-\beta y)$, and below the wing ($y < 0$) $\phi = f_2(x+\beta y)$. For definiteness, we shall consider the region above the wing, where $\phi = f(x-\beta y)$. The function f is determined from the condition (125.2) by putting $n_x \cong -\zeta_2'(x)$, where $y = \zeta_2(x)$ is the equation of the upper part of the wing profile (Fig. 129a). We have $[\partial\phi/\partial y]_{y\to +0} = -\beta f'(x) = v_1\zeta_2'(x)$, whence $f = -v_1\zeta_2(x)/\beta$. Thus the velocity distribution for $y > 0$ is given by the potential

$$\phi(x, y) = -(v_1/\beta)\zeta_2(x - \beta y). \tag{125.3}$$

Similarly we obtain, for $y < 0$, $\phi = (v_1/\beta)\zeta_1(x + \beta y)$, where $y = \zeta_1(x)$ is the equation of the lower part of the profile. It should be noticed that the potential, and therefore the other quantities, are constant along the straight lines $x \pm \beta y$ = constant (the characteristics), in accordance with the results of §115, of which the solution just obtained is a particular case.

The flow pattern is qualitatively as follows. Weak discontinuities (aAa' and bBb' in Fig. 129b) leave the pointed leading and trailing edges.† In the regions in front of the discontinuity aAa' and behind bBb' the flow is uniform, but between them it is turned so as to go round the surface of the wing; the flow here is a simple wave, and in the present linearized approximation the characteristics are all parallel and inclined at the Mach angle of the incident stream.

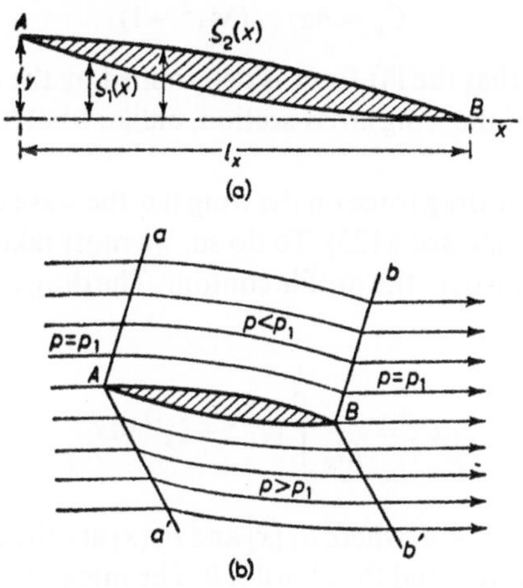

FIG. 129

The pressure distribution is given by the formula $p - p_1 = -\rho_1 v_1 \partial\phi/\partial x$; the term in v_y^2 in the general formula (114.5) can here be omitted, since v_x and v_y are of the same order of magnitude. Substituting (125.3) and introducing the *pressure coefficient* C_p, we obtain in the upper half-plane $C_p = (p - p_1)/\frac{1}{2}\rho_1 v_1^2 = 2\zeta_2'(x - \beta y)/\beta$. In particular, the pressure coefficient on the upper surface of the wing is

$$C_{p2} = 2\zeta_2'(x)/\beta. \tag{125.4}$$

Similarly, we find for the lower surface

$$C_{p1} = -2\zeta_1'(x)/\beta. \tag{125.5}$$

† This statement is valid only in the approximation used here. In reality we have not weak discontinuities but weak shock waves or narrow centred rarefaction waves, depending on the direction in which the velocity is turned by them. For the profile shown in Fig. 129b, for example, Aa and Bb' are rarefaction waves, while Aa' and Bb are shock waves.

The streamline leaving the trailing edge (B in Fig. 129b) is actually a tangential discontinuity of the velocity (which in practice becomes a narrow turbulent wake).

It should be noted that the pressure at any point on the wing profile depends only on the slope of the profile contour at that point.

Since the angle between the profile contour and the x-axis is always small, the vertical component of the pressure force can be taken, with sufficient accuracy, as the pressure itself. The resultant lift force on the wing is equal to the difference of the pressures on the lower and upper surfaces. The lift coefficient is therefore

$$C_y = \frac{1}{l_x}\int_0^{l_x} (C_{p1} - C_{p2})dx = \frac{4l_y}{\beta l_x};$$

see Fig. 129a for the definition of l_x, l_y. We define the angle of attack α as the angle between the chord AB through the ends of the profile (Fig. 129a) and the x-axis: $\alpha \cong l_y/l_x$, and obtain the following simple formula:

$$C_y = 4\alpha/\sqrt{(M_1{}^2 - 1)} \tag{125.6}$$

(J. Ackeret 1925). We see that the lift force is determined by the angle of attack, and does not depend on the form of the wing cross-section, unlike what happens for subsonic flow (see formula (48.7)).

Let us next determine the drag force on the wing (i.e. the wave drag, which is of the same nature as that on thin bodies; see §123). To do so, we must take the x-component of the pressure force and integrate over the profile contour. The drag coefficient is then found to be

$$C_x = \frac{2}{\beta l_x}\int_0^{l_x} (\zeta_1'^2 + \zeta_2'^2)dx. \tag{125.7}$$

We put $\zeta_1' = \theta_1 - \alpha$, $\zeta_2' = \theta_2 - \alpha$, where $\theta_1(x)$ and $\theta_2(x)$ are the angles between the upper and lower parts of the contour and the chord AB. The integrals of θ_1 and θ_2 are evidently zero, and the result is therefore

$$C_x = [4\alpha^2 + 2(\overline{\theta_1{}^2} + \overline{\theta_2{}^2})]/\sqrt{(M_1{}^2 - 1)}; \tag{125.8}$$

the bar denotes an average with respect to x. For a given angle of attack, the drag coefficient is seen to be least for a wing in the form of a flat plate (for which $\theta_1 = \theta_2 = 0$). In this case $C_x = \alpha C_y$. If we apply formula (125.8) to a rough surface, we find that the roughness may result in a considerable increase in the drag, even if the height of the irregularities is small.† For the drag is independent of the height of the irregularities if the mean slope of the surface, i.e. the mean ratio of the height of the irregularities to the distance between them, remains constant.

Finally, we may make the following remark. Here, as everywhere, when we speak of a wing we imply that its edges are perpendicular to the flow. The generalization to the case of any angle γ between the direction of flow and the edge (the *angle of yaw*) is quite obvious. It is clear that the forces on an infinite wing with constant cross-section depend only on the component of the incident velocity normal to its edges; in an ideal fluid, the velocity component parallel to the edges does not result in a force. The forces acting on a wing at an

† But nevertheless greater than the thickness of the boundary layer.

angle of yaw other than $\frac{1}{2}\pi$ in a stream with Mach number M_1 are therefore the same as those on the same wing for $\gamma = \frac{1}{2}\pi$ in a stream with Mach number $M_1 \sin\gamma$. In particular, if $M_1 > 1$ but $M_1 \sin\gamma < 1$, the wave drag, which is peculiar to supersonic flow, will not occur.

§126. The law of transonic similarity

The theory of supersonic and subsonic flow past thin bodies developed in §§123–125 is not applicable to transonic flow, when the linearized equation for the potential becomes invalid. In this case the flow pattern in all space is given by the non-linear equation (114.10):

$$2\alpha_* \frac{\partial\phi}{\partial x}\frac{\partial^2\phi}{\partial x^2} = \frac{\partial^2\phi}{\partial y^2} + \frac{\partial^2\phi}{\partial z^2} \tag{126.1}$$

(or, for two-dimensional flow, by the equivalent Euler–Tricomi equation). The solution of these equations for particular cases is very difficult, however. The similarity rules which can be established for such flows, without finding any particular solution, are therefore of great interest.

Let us first consider two-dimensional flow, and let

$$Y = \delta f(x/l) \tag{126.2}$$

be the equation which gives the shape of the thin contour past which the flow takes place, l being its length (in the direction of flow) and δ some characteristic thickness ($\delta \ll l$). By varying the two parameters l and δ, we obtain a family of similar contours.

The equation of motion is

$$2\alpha_* \frac{\partial\phi}{\partial x}\frac{\partial^2\phi}{\partial x^2} = \frac{\partial^2\phi}{\partial y^2}, \tag{126.3}$$

with the following boundary conditions. At infinity, the velocity equals the velocity $\mathbf{v}_1$ of the undisturbed stream, i.e.

$$\frac{\partial\phi}{\partial y} = 0, \qquad \frac{\partial\phi}{\partial x} = M_{1*} - 1 = (M_1 - 1)/\alpha_*; \tag{126.4}$$

see the definition of the potential ϕ, (114.9). On the profile, the velocity must be tangential:

$$v_y/v_x \cong \partial\phi/\partial y = \mathrm{d}Y/\mathrm{d}X = (\delta/l)f'(x/l); \tag{126.5}$$

since the profile is thin, this condition can be imposed at $y = 0$.

We introduce dimensionless variables thus:

$$x = l\bar{x}, \qquad y = l\bar{y}/(\theta\alpha_*)^{1/3}, \qquad \phi = (l\theta^{2/3}/\alpha_*{}^{1/3})\bar{\phi}(\bar{x}, \bar{y}); \tag{126.6}$$

here $\theta = \delta/l$ gives the angular thickness of the wing or angle of attack. Then

$$2\frac{\partial\bar{\phi}}{\partial\bar{x}}\frac{\partial^2\bar{\phi}}{\partial\bar{x}^2} = \frac{\partial^2\bar{\phi}}{\partial\bar{y}^2},$$

with the following boundary conditions:

$$\partial\bar{\phi}/\partial\bar{x} = K, \qquad \partial\bar{\phi}/\partial\bar{y} = 0 \text{ at infinity,}$$
$$\partial\bar{\phi}/\partial\bar{y} = f'(\bar{x}) \text{ at } \bar{y} = 0,$$

where

$$K = (M_1 - 1)/(\alpha_* \theta)^{2/3}. \tag{126.7}$$

These conditions contain only one parameter, K. Thus we have obtained the required similarity law: two-dimensional transonic flows with the same value of K are similar, as is shown by formulae (126.6) (S. V. Fal'kovich 1947).

It should be noticed that the expression (126.7) involves only a single parameter α_* which characterizes the properties of the gas itself. The similarity law therefore determines also the similarity with respect to a change in the gas.

In the approximation here considered, the pressure is given by the formula $p - p_1 \cong -\rho_1 v_1 (v_x - v_1)$. A calculation using the expressions (126.6) shows that the pressure coefficient on the profile has the form

$$C_p = \frac{p - p_1}{\frac{1}{2}\rho_1 {v_1}^2} = \frac{\theta^{2/3}}{{\alpha_*}^{1/3}} P\left(K, \frac{x}{l}\right).$$

The drag and lift coefficients are given by integrals along the contour of the profile:

$$C_x = \frac{1}{l}\oint C_p \frac{dY}{dx} dx,$$

$$C_y = \frac{1}{l}\oint C_p dx,$$

and therefore have the form†

$$C_x = \frac{\theta^{5/3}}{{\alpha_*}^{1/3}} f_x(K), \qquad C_y = \frac{\theta^{2/3}}{{\alpha_*}^{1/3}} f_y(K). \tag{126.8}$$

In an entirely similar manner, we can obtain the similarity law for a three-dimensional thin body whose shape is given by equations having the form

$$Y = \delta f_1(x/l), \qquad Z = \delta f_2(x/l), \tag{126.9}$$

with the two parameters δ and l ($\delta \ll l$). There is an important difference from the two-dimensional case, because the potential has a logarithmic singularity for $y \to 0$, $z \to 0$ (see, for instance, the formulae for flow past a narrow cone in §113). Hence the boundary condition at the x-axis must determine, not the derivatives $\partial\phi/\partial y$, $\partial\phi/\partial z$ themselves, but the products $y\,\partial\phi/\partial y = Y dY/dx$, $z\,\partial\phi/\partial z = Z\,dZ/dx$, which remain finite. It is easy to see that in this case the similarity transformation is (again with $\theta = \delta/l$)

$$x = l\bar{x}, \qquad y = (l/\theta{\alpha_*}^{\frac{1}{2}})\bar{y}, \qquad z = (l/\theta{\alpha_*}^{\frac{1}{2}})\bar{z}, \qquad \phi = l\theta^2\bar{\phi}, \tag{126.10}$$

the similarity parameter being

$$K = (M_1 - 1)/\theta^2\alpha_* \tag{126.11}$$

† The range of validity of these formulae is given by the condition $|M_1 - 1| \ll 1$. The linearized theory, however, corresponds to large K, i.e. $|M_1 - 1| \gg \theta^{2/3}$. In the range $1 \gg M_1 - 1 \gg \theta^{2/3}$, formulae (126.8) must therefore become the formulae (125.6)–(125.8) given by the linearized theory. This means that, for large K, the functions f_x and f_y must be proportional to $K^{-1/2}$.

(T. von Kármán 1947). The pressure coefficient at the surface of the body is found to have the form $C_p = \theta^2 P(K, x/l)$, and the drag coefficient is accordingly†

$$C_x = \theta^4 f(K). \tag{126.12}$$

All these formulae hold, of course, for both small positive and small negative values of $M_1 - 1$. If $M_1 = 1$ exactly, the similarity parameter $K = 0$, and the functions in formulae (126.8) and (126.12) reduce to constants, so that these formulae completely determine C_x and C_y as functions of θ and α_*, which represents the properties of the gas.

§127. The law of hypersonic similarity

The linearized theory is invalid for supersonic flow past thin pointed bodies for very large values of the Mach number M_1 (*hypersonic flow*), as has already been mentioned at the end of §114. A simple similarity rule which can be established for this case is therefore of interest.

The shock waves formed in such flow are at a small angle to the direction of flow, of the order of the ratio $\theta = \delta/l$ of thickness to length of the body. These shocks are in general curved and also strong; the velocity discontinuity in them is relatively small, but the pressure discontinuity (and therefore the entropy discontinuity) is large. The gas flow is therefore not in general potential flow.

We shall assume that the Mach number M_1 is of the order of $1/\theta$ or greater. A shock wave reduces the local value of M, but the latter always remains of the order of $1/\theta$ (see §112, Problem 2), so that M is large everywhere.

We use the "sonic analogy" mentioned in §123: a three-dimensional problem of steady flow past a thin body with variable cross-section $S(x)$ is equivalent to a two-dimensional problem of non-steady emission of sound waves by a contour whose area varies with time according to the law $S(v_1 t)$; the velocity of sound is represented by $v_1/\sqrt{(M_1^2 - 1)}$, or, for large M_1, by c_1 simply. It should be emphasized that the only condition necessary for the two problems to be equivalent is that the ratio δ/l be small; this enables us to regard small annular regions of the surface of the body as cylindrical. For large M_1, however, the rate of propagation of the emitted waves is comparable with the velocity of the gas particles in the waves (cf. the end of §123), and the problem therefore has to be solved on the basis of the exact (non-linearized) equations.

The velocity perturbation is small (in comparison with the velocity v_1 of the incident stream) in any supersonic flow past a narrow pointed body. In hypersonic flow, the perturbation of the longitudinal velocity is also small in comparison with the transverse velocities which occur:

$$v_y \sim v_x \sim v_1 \theta, \qquad v_x - v_1 \sim v_1 \theta^2. \tag{127.1}$$

The pressure and density changes, however, are not small:

$$(p - p_1)/p_1 \sim M_1^2 \theta^2, \qquad (\rho_2 - \rho_1)/\rho_1 \sim 1, \tag{127.2}$$

and the pressure change can even be indefinitely large when $M_1 \theta \gg 1$; cf. §112, Problem 2.

The sonic analogy applies, however, only to the two-dimensional problem of flow in the yz-plane, which is perpendicular to the incident stream. In this two-dimensional problem,

† In the range $1 \gg M_1 - 1 \gg \theta^2$, we must obtain the formula (123.7) given by the linearized theory, according to which $C_x \propto \theta^4$; this means that the function $f(K)$ tends to a constant as K increases.

the linear velocity of the source is of the order of $v_1\theta$; the only other independent parameters of the problem are the velocity of sound c_1, the dimension δ of the source, and the density ρ_1.† From these we can form only one dimensionless combination,

$$K = \mathrm{M}_1\theta, \tag{127.3}$$

which is the similarity parameter.‡ The scales of length for the coordinates y, z and of time must be taken to have the appropriate dimensions, and be formed from the same parameters, e.g. δ and $\delta/v_1\theta = l/v_1$. The natural parameter for the coordinate x is the length l of the body. We can then say that

$$v_y = v_1\theta v_y', \qquad v_z = v_1\theta v_z', \qquad p = \rho_1 v_1^2\theta^2 p', \qquad \rho = \rho_1\rho', \tag{127.4}$$

where v_y', v_z', p' and ρ' are functions of the dimensionless variables x/l, y/δ, z/δ and the parameter K; from (127.1) and (127.2), these functions are of the order of unity.††

The drag force F_x is calculated as the integral

$$F_x = \oint p\,dy\,dz,$$

taken over the whole surface of the body; according to the boundary condition $v_n = 0$, the term $v_x(\mathbf{v}\cdot\mathbf{n})$ in the momentum flux density is zero at the surface of the body, $\mathbf{n}$ being the normal to the surface. Changing to dimensionless variables in accordance with (127.4), we get the drag coefficient C_x, defined by (123.6), as

$$C_x = 2\theta^4\oint p'\,dy'\,dz'$$

The remaining integral is a function of the dimensionless parameter K. Thus

$$C_x = \theta^4 f(K). \tag{127.5}$$

The same similarity law obviously occurs in the two-dimensional case of flow past a thin wing with infinite span. The drag and lift coefficients then have the form

$$C_x = \theta^3 f_x(K), \qquad C_y = \theta^2 f_y(K). \tag{127.6}$$

In applying the relations (127.5) and (127.6), it must be remembered that the similarity of the flows implies that the shape, size and orientation of the bodies relative to the incident stream are obtained from one another by merely changing the scale δ along the y and z axes and l along the x-axis. This means, in particular, that if the angle of attack α is not zero the ratio α/θ must be the same for similar configurations.

---

† We are considering, of course, not only the equations of motion of the gas, but also the boundary conditions on them at the surface of the body and the conditions which must be satisfied at the shock waves which are formed. We take the case of a polytropic gas, so that the gas-dynamic properties depend only on the dimensionless parameter γ; the similarity rule obtained below, however, does not determine the dependence of the flow on this parameter.

In flow with $\mathrm{M}_1 \gg 1$, there is considerable heating of the gas, and there may be a considerable consequent change in its thermodynamics properties. The quantitative signficance of the formulae for a polytropic gas (the specific heat being assumed constant) is therefore in practice limited, at hypersonic velocities.

‡ If M_1 is not supposed large, we obtain a similarity rule with parameter $K = \theta\sqrt{(\mathrm{M}_1^2 - 1)}$. This is of no interest, however, since for small M_1 the linearized theory determines all quantities as functions of this parameter.

†† The similarity law for hypersonic flow was formulated by H. S. Tsien (1946). Its relationship to the sonic analogy extended to the non-linear problem was noted by W. D. Hayes (1947). In the specialist literature, this is called the "piston analogy".

As $K \to \infty$, the functions of this parameter in (127.5) and (127.6) tend to constant limits. This is a consequence of the existence of a limiting flow regime as $M_1 \to \infty$, whose properties are independent of M_1 over a considerable region (S. V. Vallander 1947; K. Oswatitsch 1951). By this we mean a region between the forward strongest part of the bow wave and the surface of the body in the flow not too far from its forward part; it is this region, where the pressure is greatest, which determines the forces acting on the body. If the flow is described by the "reduced" velocity v/v_1, pressure $p/\rho_1 v_1^2$ and density ρ/ρ_1 as functions of the dimensionless coordinates, the pattern of flow past a body having a given shape is in this region independent of M_1 in the limit. The reason is that, when expressed in terms of these variables, not only the equations of fluid dynamics and the boundary conditions on the surface of the body but also all conditions at the shock wave surface are independent of M_1. The term "considerable region" is used because the quantities neglected in the latter conditions are of relative order $1/M_1^2 \sin^2\phi$, where ϕ is the angle between $\mathbf{v}_1$ and the surface of discontinuity; at large distances, where the shock wave is weak, this angle tends to the Mach angle $\sin^{-1}(1/M_1) \cong 1/M_1$, and so the expansion parameter is no longer small: $1/M_1^2 \sin^2\phi \sim 1$.†

PROBLEM

Determine the lift force on a flat wing with infinite span inclined at a small angle of attack α to the direction of flow, for $M_1 \gtrsim 1/\alpha$ (R. D. Linnell 1949).

SOLUTION. The flow pattern is as shown in Fig. 130: a shock wave and a rarefaction wave leave each of the two edges of the plate, and the stream is turned in them through an angle α in opposite directions.

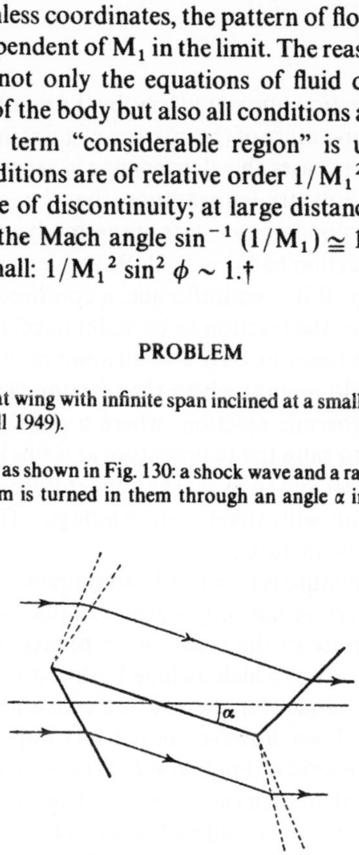

FIG. 130

According to the sonic analogy, the problem of steady flow past such a plate is equivalent to that of non-steady one-dimensional gas flow on each side of a piston moving with uniform velocity αv_1. In front of the piston a shock wave is formed, and behind it a rarefaction wave (see §99, Problems 1 and 2). Using the results there obtained, we find the required lift force as the difference of the pressures on the two sides of the plate. The lift coefficient is

$$C_y = \alpha^2 \left\{ \frac{2}{\gamma K^2} + \frac{\gamma+1}{2} + \sqrt{\left[\frac{4}{K^2} + \left(\frac{\gamma+1}{2} \right)^2 \right]} \right\} - \frac{2\alpha^2}{\gamma K^2} \left[1 - \frac{\gamma-1}{2} K \right]^{2\gamma/(\gamma-1)}$$

where $K = \alpha M_1$. For $K \geq 2/(\gamma-1)$, a vacuum is formed under the plate, and the second term must be omitted. In the range $1 \ll M_1 \ll 1/\alpha$, this formula becomes $C_y = 4\alpha/M_1$, as given by the linearized theory, in accordance with the fact that both procedures are applicable in that range.

† The detailed proof is given by G. G. Chernyi, *Introduction to Hypersonic Flow*, New York 1961, chapter I, §4.

CHAPTER XIV

FLUID DYNAMICS OF COMBUSTION

§128. **Slow combustion**

The speed of a chemical reaction (measured, say, by the number of molecules reacting in unit time) depends on the temperature of the mixture of gases in which it occurs, increasing with the temperature. In many cases this dependence is very marked.† The speed of the reaction may be so small at ordinary temperatures that the reaction hardly occurs, even though the gas mixture corresponding to a state of thermodynamic (chemical) equilibrium would be one in which the reaction had occurred. When the temperature rises sufficiently, the reaction proceeds rapidly. If it is endothermic, a continuous supply of heat from an external source is necessary for the reaction to be maintained; if the temperature is merely raised at the beginning of the reaction, only a small amount of matter reacts, and thereby reduces the gas temperature to a point where the reaction ceases. The situation is quite different for a strongly exothermic reaction, where a considerable quantity of heat is evolved. Here it is sufficient to raise the temperature at a single point; the reaction which begins at that point evolves heat and so raises the temperature of the surrounding gas, and the reaction, once having begun, will extend to the whole gas. This is called *slow combustion* or simply *combustion* of a gas mixture.‡

The combustion of a gas mixture is necessarily accompanied by motion of the gas. The process of combustion is therefore not only a chemical phenomenon but also one of gas dynamics. In general, the nature of the combustion process has to be determined by a solution of simultaneous equations which include both those of chemical kinetics for the reaction and those of gas dynamics for the mixture concerned.

The situation is much simplified, however, in the very important case (the one usually encountered) where the characteristic dimension l of the problem is large (in a sense to be defined later). We shall see that, in such cases, the problems of gas dynamics and chemical kinetics can be, to a certain extent, considered separately.

The region of burnt gas (i.e. the region where the reaction is over and the gas is a mixture of combustion products) is separated from the gas where combustion has not yet begun by a transition layer, where the reaction is in progress (the *combustion zone* or *flame*); in the course of time, this layer moves forward, with a velocity which may be called the velocity of propagation of combustion in the gas. The magnitude of this velocity depends on the amount of heat transfer from the combustion zone to the cold gas mixture. The main mechanism of heat transfer is ordinary conduction (V. A. Mikhel'son 1890).

---

† The reaction rate usually depends exponentially on the temperature, being nearly proportional to a factor of the form $e^{-U/T}$, where U is a constant for any given reaction and is called the *activation energy*. The greater U, the more strongly the reaction rate depends on the temperature.

‡ It should be borne in mind that, in a mixture capable of combustion, the spontaneous propagation of the combustion may be impossible in certain circumstances. This limitation is due to heat losses resulting from such factors as conduction through the walls of a pipe in which combustion occurs, radiation losses, etc. For this reason combustion is not possible in pipes with very small radius, for example.

We denote by δ the order of magnitude of the width of the combustion zone. It is determined by the mean distance over which heat evolved in the reaction is propagated during the time τ for which the reaction lasts (at the point concerned). The time τ is characteristic of the reaction, and depends only on the thermodynamic state of the gas undergoing combustion (and not on the parameter l). If χ is the thermometric conductivity of the gas, we have (see (51.6))†

$$\delta \sim \sqrt{(\chi\tau)}. \tag{128.1}$$

Let us now make more precise the above assumption: we shall suppose that the characteristic dimension is large compared with the width of the combustion zone ($l \gg \delta$) When this condition holds, the problem of gas dynamics can be considered separately. In determining the gas flow, we can neglect the width of the combustion zone, regarding it as a surface which separates the combustion products from the unburnt gas. On this surface (the *flame front*) the state of the gas changes discontinuously, i.e. it is a surface of discontinuity.

The velocity v_1 of this discontinuity relative to the gas itself (in a direction normal to the front) is called the *normal velocity* of the flame. In a time τ, the combustion is propagated through a distance of the order of δ, and so the flame velocity is‡

$$v_1 \sim \delta/\tau \sim \sqrt{(\chi/\tau)}. \tag{128.2}$$

The ordinary thermometric conductivity of the gas is of the order of the mean free path of the molecules multiplied by their thermal velocity or, what is the same thing, the mean free time τ_{fr} multiplied by the square of this velocity. Since the thermal velocity of the molecules is of the same order as the velocity of sound, we have $v_1/c \sim \sqrt{(\chi/\tau c^2)} \sim \sqrt{(\tau_{fr}/\tau)}$. Not every collision between molecules results in a chemical reaction between them; on the contrary, only a very small fraction of colliding molecules react. This means that $\tau_{fr} \ll \tau$, and therefore $v_1 \ll c$. Thus the flame velocity is, in this case, small compared with the velocity of sound.††

On the surface of discontinuity which replaces the combustion zone, the fluxes of mass, momentum and energy must be continuous, as at any discontinuity. The first of these conditions, as usual, determines the ratio of the components, normal to the surface, of the gas velocities relative to the discontinuity: $\rho_1 v_1 = \rho_2 v_2$, or

$$v_1/v_2 = V_1/V_2, \tag{128.3}$$

where V_1 and V_2 are the specific volumes of the unburnt gas and the combustion products. According to the general results obtained in §84 for arbitrary discontinuities, the tangential velocity component must be continuous if the normal component is discontinuous. The streamlines are therefore refracted at the discontinuity.

---

† To avoid misunderstanding, it should be mentioned that, when τ depends markedly on the temperature, a fairly large coefficient should appear in formula (128.1) if τ is the value for the temperature of the combustion products. The important fact for our purposes, however, is that δ does not depend on l.

‡ As an example, it may be mentioned that the flame velocity in a mixture of methane (6 per cent) and air is only 5 cm/sec, whereas in detonating mixture ($2H_2 + O_2$) it is 1000 cm/sec; the widths of the combustion zones in these two cases are about 5×10^{-2} cm and 5×10^{-4} cm respectively.

†† The diffusion of the components of the burning mixture also has a certain effect on the propagation of combustion; this does not alter the orders of magnitude of the flame velocity and width. We are, however, always considering the combustion of already mixed gases, not cases where the reactants are separated and combustion occurs only as they diffuse into one another.

On account of the smallness of the normal velocity of the flame relative to that of sound, the condition of continuity of the momentum flux reduces to the continuity of pressure, and that for the energy flux reduces to the continuity of the heat function:

$$p_1 = p_2, \qquad w_1 = w_2. \tag{128.4}$$

In using these conditions, it must be remembered that the gases on the two sides of the discontinuity under consideration are chemically different, and so the thermodynamic quantities are not the same functions of one another.

For a polytropic gas we have $w_1 = w_{01} + c_{p1}T_1$, $w_2 = w_{02} + c_{p2}T_2$; the constant terms cannot be put equal to zero as for a single gas (by an appropriate choice of the zero of energy), since w_{01} and w_{02} are different. We put $w_{01} - w_{02} = q$; this is just the heat evolved (per unit mass) in the reaction, if the reaction occurs at a temperature of absolute zero. Then we obtain the following relations between the thermodynamic quantities for the unburnt gas (1) and the burnt gas (2):

$$p_1 = p_2, \qquad T_2 = \frac{q}{c_{p2}} + \frac{c_{p1}}{c_{p2}}T_1, \qquad V_2 = V_1\frac{\gamma_1(\gamma_2-1)}{\gamma_2(\gamma_1-1)}\left(\frac{q}{c_{p1}T_1}+1\right). \tag{128.5}$$

Since the flame has a definite normal velocity, independent of the gas velocities themselves, the flame front has a definite form for steady combustion in a moving gas. An example is the combustion of gas leaving the end of a tube (a burner outlet). If v is the gas velocity averaged over the cross-section of the tube, it is evident that $v_1 S_1 = vS$, where S is the cross-sectional area of the tube and S_1 the total surface area of the flame front.

There arises the question of the limits of stability of this flow regime under small perturbations, that is, the conditions for it actually to occur. Since the gas velocity is much less than that of sound, in examining the stability of the flame front we can regard the gas as an incompressible ideal fluid, the normal velocity of the flame front being taken as a given constant. Such an investigation leads to the result that the flame front is unstable (L. D. Landau 1944; see Problem 1). In this form the investigation is valid only for large Reynolds numbers $v_1 l/\nu_1$ and $v_2 l/\nu_2$. When the viscosity of the gas is taken into account, however, it cannot here result in very large critical Reynolds numbers.

Such instability would have to give rise to spontaneous turbulence of the flame. The experimental results, on the other hand, show that this does not occur, at least up to very large Reynolds numbers. This is because of the presence, in actual flows, of various factors of fluid dynamics and thermal diffusion which stabilize the flame. An account of these complex topics is outside the scope of the present book; here we shall give only some brief comments on certain possible causes of stabilization.

The influence of curvature of the front on the combustion rate may be important in this respect. If only thermal conduction is considered, v_1 increases on the parts of the front that are concave towards the original combustible mixture, because of the better heat transfer to unburnt mixture within the concavity, and decreases on the convex parts; this effect tends to straighten the front, i.e. has a stabilizing action. A change in the diffusion regime, however, can be shown by similar arguments to have a destabilizing action. Thus the overall sign of the effect depends on the relation between the thermometric conductivity and the diffusion coefficient (I. P. Drozdov and Ya. B. Zel'dovich 1943). For a phenomenological description of the influence of front curvature on the combustion rate v_1, we can include in it a term proportional to the curvature (G. H. Markstein 1951); when the sign of this term is appropriately chosen, its inclusion in the boundary conditions eliminates the instability of short-wavelength perturbations.† The development of perturbations that are unstable in the linear approximation can be stabilized at a certain steady-amplitude limit as a result of non-linear effects (R. E. Petersen and N. W. Emmons 1956; Ya. B. Zel'dovich 1966); this mechanism may give rise to a "cellular" structure of the flame.‡

A flame propagated in a mixture of combustible gases results in a motion of the surrounding gas up to a considerable distance. The fact that a motion of the gas must accompany combustion is evident from the fact that, because of the difference between the velocities v_1 and v_2, the combustion products must move with velocity $v_1 - v_2$ relative to the unburnt gas. In some cases this motion results in the formation of shock waves. These shocks bear no direct relation to the process of combustion, and their occurrence is due to the necessity of satisfying the boundary conditions. Let us consider, for example, combustion propagated from the closed end of a pipe. In Fig. 131, *ab* is the combustion zone. The gas in regions 1 and 3 is the original unburnt mixture, while that in region 2 consists of combustion products. The velocity v_1 with which the combustion zone moves relative to the gas 1 in front of it is determined by the properties of the reaction and the conditions of heat transfer, and must be regarded as given. The velocity v_2 with which the flame moves relative to gas 2 is then determined at once by the condition (128.3). At the closed end of the pipe, the gas velocity must vanish, and so the gas in region 2 will be at rest. Gas 1, therefore, must move relative to the pipe with a constant velocity $v_2 - v_1$. In the forward part of the pipe, far from the flame, the gas is again at rest. This condition can be satisfied only by the presence of a shock wave (*cd* in Fig. 131), in which the gas velocity is discontinuous in such a way that gas 3 is at rest. From the given discontinuity of velocity we can find the discontinuities in the other quantities and the velocity of propagation of the shock itself. Thus we see that the flame front acts as a piston on the gas in front of it. The shock wave moves faster than the flame, so that the mass of gas set in motion increases in the course of time.††

For sufficiently large Reynolds numbers, the gas flow which accompanies combustion in a pipe becomes turbulent, and this in turn affects the flame which causes the motion. There are still many unsolved problems of turbulent combustion, and they will not be discussed here.

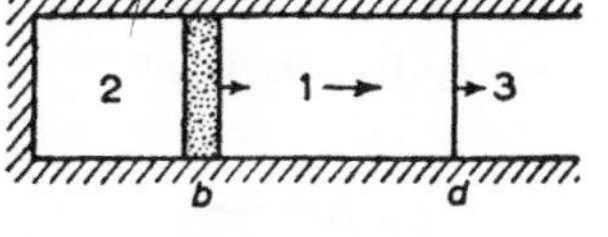

FIG. 131

† In the notation of Problem 1, the expression for v_1 including this effect is to be written as $v_1 = v_1^{(0)}(1 - \mu\partial^2\zeta/\partial y^2)$, where $v_1^{(0)}$ is the combustion rate with a plane front, and μ is an empirical constant (having the dimensions of length) which is positive when there is stabilization.

‡ A detailed account of these topics has been given by Ya. B. Zel'dovich, G. I. Barenblatt, V. B. Librovich, and G. M. Makhviladze, *Mathematical Theory of Combustion and Explosion* (*Matematicheskaya teoriya goreniya i vzryva*), Moscow 1980, chapters 4 and 6.

†† In actual cases, the combustion front in a pipe is usually convex towards the original gas mixture ahead of it. This gives rise to a specific mechanism of flame stabilization with respect to small-scale perturbations. The propagation of combustion along the normal to the front causes a "stretching" of the front, and perturbations arising anywhere on it are carried to the pipe walls, where they disappear; the steady shape of the front is maintained by the gas flow ahead of it. See Ya. B. Zel'dovich, A. G. Istratov, N. I. Kidin and V. B. Librovich, *Combustion Science and Technology* **24**, 1, 1980.

PROBLEMS

PROBLEM 1. Investigate the stability of a plane flame front in slow combustion, with respect to infinitesimal disturbances.

SOLUTION. We take the plane of discontinuity (flame front) as the yz-plane in coordinates where it is at rest, with the unperturbed gas velocity in the positive x-direction. On the flow with constant velocities v_1, v_2 (on the two sides of the discontinuity) we superpose a perturbation periodic in the y-direction and in time. From the equations of motion

$$\operatorname{div} \mathbf{v}' = 0, \qquad \partial \mathbf{v}'/\partial t + (\mathbf{v}\cdot\mathbf{grad})\mathbf{v}' = -(1/\rho)\mathbf{grad}\, p' \tag{1}$$

($\mathbf{v}$, ρ being either $\mathbf{v}_1$, ρ_1 or $\mathbf{v}_2$, ρ_2), we obtain as in §29 the equation

$$\triangle p' = 0. \tag{2}$$

On the surface of discontinuity (i.e. for $x \cong 0$) the following conditions must be satisfied: the equation of continuity of pressure

$$p'_1 = p'_2, \tag{3}$$

the condition of continuity of the velocity component tangential to the surface

$$v'_{1y} + v_1 \partial\zeta/\partial y = v'_{2y} + v_2 \partial\zeta/\partial y \tag{4}$$

(where $\zeta(y, t)$ is the small displacement of the surface of discontinuity along the x-axis due to the disturbance), and the condition that the gas velocity normal to the surface of discontinuity be unchanged,

$$v'_{1x} - \partial\zeta/\partial t = v'_{2x} - \partial\zeta/\partial t = 0. \tag{5}$$

In the region $x < 0$ (the unburnt gas 1), the solution of equations (1) and (2) can be written

$$\left.\begin{aligned} v'_{1x} &= Ae^{iky+kx-i\omega t}, \qquad v'_{1y} = iAe^{iky+kx-i\omega t},\\ p'_1 &= A\rho_1\left(\frac{i\omega}{k} - v_1\right)e^{iky+kx-i\omega t}. \end{aligned}\right\} \tag{6}$$

In the region $x > 0$ (the combustion products, gas 2), besides the solution having the form constant $\times\, e^{iky-kx-i\omega t}$, we must take into account another particular solution of equations (1) and (2), in which the dependence on y and t is given by the same factor $e^{iky-i\omega t}$. This solution is obtained by putting $p' = 0$; then the right-hand side of Euler's equation is zero, and the resulting homogeneous equation has a solution in which v'_x and v'_y are proportional to $e^{iky-i\omega t+i\omega x/v}$. The reason why this solution need be taken into consideration only in gas 2, and not in gas 1, is that our ultimate purpose is to determine whether frequencies ω can exist having positive imaginary parts; for such ω, however, the factor $e^{i\omega x/v}$ increases without limit with $|x|$ for $x < 0$, and so such a solution is not possible in region 1. Again choosing appropriate values of the constant coefficients, we seek a solution for $x > 0$ in the form

$$\left.\begin{aligned} v'_{2x} &= Be^{iky-kx-i\omega t} + Ce^{iky-i\omega t+i\omega x/v_2},\\ v'_{2y} &= -iBe^{iky-kx-i\omega t} - (\omega/kv_2)Ce^{iky-i\omega t+i\omega x/v_2},\\ p'_2 &= -B\rho_2[v_2 + (i\omega/k)]e^{iky-kx-i\omega t}. \end{aligned}\right\} \tag{7}$$

Putting also

$$\zeta = De^{iky-i\omega t}, \tag{8}$$

and substituting these expressions in the conditions (3)–(5), we obtain four homogeneous equations for the coefficients A, B, C, D.† A simple calculation (using the fact that $j \equiv \rho_1 v_1 = \rho_2 v_2$) gives the following condition for these equations to be compatible:

$$\Omega^2(v_1 + v_2) + 2\Omega k v_1 v_2 + k^2 v_1 v_2 (v_1 - v_2) = 0, \tag{9}$$

where $\Omega = -i\omega$. If $v_1 > v_2$, this equation has either two negative real roots or two complex conjugate roots with $\operatorname{re}\Omega < 0$; in this case, the flow is stable. If $v_1 < v_2$ (and accordingly $\rho_1 > \rho_2$), both roots of (9) are real, and one is positive:

$$\Omega = kv_1 \frac{\mu}{1+\mu}\left[\sqrt{\left(1 + \mu - \frac{1}{\mu}\right)} - 1\right],$$

† The flow described by (6) is a potential flow; that described by (7) has $\mathbf{curl}\, \mathbf{v}'_2 \neq 0$. Thus the movement of the combustion products behind the perturbed front is rotational.

where $\mu = \rho_1/\rho_2$, so that the flow is unstable. This case occurs for a combustion front, since the combustion product density ρ_2 is always less than the density ρ_1 of the unburnt gas, because of the considerable heating.

It should be noted that im $\Omega = 0$. This means that the disturbances are not propagated along the front, but are amplified as stationary waves. Instability occurs for disturbances with any wavelength; the growth rate increases with k, but it should be remembered that the analysis in which the front is treated as a geometrical surface is valid only for disturbances whose wavelengths are much greater than δ ($k\delta \ll 1$). For given k, the growth rate increases with μ.

PROBLEM 2. Combustion occurs on the surface of a liquid, the reaction taking place in vapour evaporating from the surface.† Determine the stability condition in this case, taking into account the effect of the gravitational field and capillary forces (L. D. Landau 1944).

SOLUTION. Let us consider the combustion zone in vapour near the liquid surface as a surface of discontinuity, but now let this surface have a surface tension α. The calculations are entirely similar to those of Problem 1, the only difference being that, instead of the boundary condition (3), we now have $p'_1 - p'_2 = -\alpha\partial^2\zeta/\partial y^2 + (\rho_1 - \rho_2)g\zeta$; medium 1 is the liquid and medium 2 the burnt gas. The conditions (4) and (5) are unchanged. In place of equation (9) we obtain

$$\Omega^2(v_1+v_2) + 2\Omega k v_1 v_2 + \left[k^2(v_1 - v_2) + \frac{gk(\rho_1-\rho_2)+\alpha k^3}{j}\right]v_1 v_2 = 0.$$

The stability condition in this case is that the roots of this equation should have negative real parts, i.e. the free term must be positive for all k. This requirement gives the stability condition $j^4 < 4\alpha g\rho_1{}^2\rho_2{}^2/(\rho_1-\rho_2)$. Since the density of the gaseous combustion products is small compared with that of the liquid ($\rho_1 \gg \rho_2$), the condition becomes in practice

$$j^4 < 4\alpha g\rho_1\rho_2{}^2.$$

PROBLEM 3. Determine the temperature distribution in the gas in front of a plane flame.

SOLUTION. In a system of coordinates moving with the front the temperature distribution is steady, and the gas moves with velocity $-v_1$. The equation of thermal conduction,

$$\mathbf{v}\cdot\mathbf{grad}\, T = -v_1\, \mathrm{d}T/\mathrm{d}x = \chi\, \mathrm{d}^2T/\mathrm{d}x^2,$$

has the solution $T = T_0 e^{-v_1 x/\chi}$, where T_0 is the temperature on the flame front, the temperature far from the front being taken as zero.

§129. Detonation

In the type of combustion (slow combustion) described above, the propagation through the gas is due to the heating which results from the direct transfer of heat from the burning gas to that which is still unburnt. Another entirely different mechanism of propagation of combustion, involving shock waves, is also possible. The shock wave heats the gas as it passes; the gas temperature behind the shock is higher than in front of it. If the shock wave is sufficiently strong, the rise in temperature which it causes may be sufficient for combustion to begin. The shock wave will then "ignite" the gas mixture as it moves, i.e. the combustion will be propagated with the velocity of the shock, or much faster than ordinary combustion. This mechanism of propagation of combustion is called *detonation*.

When the shock wave passes some point in the gas, the reaction begins at that point, and continues until all the gas there is burnt, i.e. for a time τ which characterizes the kinetics of the reaction concerned.‡ It is therefore clear that the shock wave will be followed by a layer moving with it in which combustion is occurring, and the width of this layer is equal to the

† The reaction takes place in the vapour without involving any extraneous substances (such as atmospheric oxygen), i.e. it is a spontaneous decomposition reaction.

‡ This time is, however, itself dependent on the strength of the shock, decreasing rapidly as the shock becomes stronger, on account of the increase in the reaction rate with rising temperature.

speed of propagation of the shock multiplied by the time τ. It is of importance that the width does not depend on the dimensions of any bodies that are present. When the characteristic dimensions of the problem are sufficiently large, therefore, we can regard the shock wave and the combustion zone following it as a single surface of discontinuity which separates the burnt and unburnt gases. We call such a surface a *detonation wave.*

At a detonation wave the flux densities of mass, energy and momentum must be continuous, and the relations (85.1)–(85.10) derived previously, which follow from these continuity conditions alone, remain valid. In particular, the equation

$$w_1 - w_2 + \tfrac{1}{2}(V_1 + V_2)(p_2 - p_1) = 0 \qquad (129.1)$$

holds; the suffix 1 always pertains to the unburnt gas and the suffix 2 to the combustion products. The curve of p_2 as a function of V_2 given by this equation is called the *detonation adiabatic*. Unlike the shock adiabatic considered earlier, this curve does not pass through the given initial point (p_1, V_1). The fact that the shock adiabatic passes through this point is due to the fact that w_1 and w_2 are the same functions of p_1, V_1 and p_2, V_2 respectively, whereas this does not now hold, on account of the chemical difference between the two gases. In Fig. 132 the continuous line shows the detonation adiabatic. The ordinary shock adiabatic for the unburnt gas mixture is drawn (dashed) through the point (p_1, V_1). The detonation adiabatic always lies above the shock adiabatic, because a high temperature is reached in combustion, and the gas pressure is therefore greater than it would be in the unburnt gas for the same specific volume.

The previous formula (85.6) holds for the mass flux density:

$$j^2 = (p_2 - p_1)/(V_1 - V_2), \qquad (129.2)$$

so that graphically j^2 is again the slope of the chord from the point (p_1, V_1) to any point (p_2, V_2) on the detonation adiabatic (for instance, the chord *ac* in Fig. 132). It is seen at once from the diagram that j^2 cannot be less than the slope of the tangent *aO*. The flux j is just the mass of gas which is ignited per unit time per unit area of the surface of the detonation wave; we see that, in a detonation, this quantity cannot be less than a certain limiting value j_{min} (which depends on the initial state of the unburnt gas).

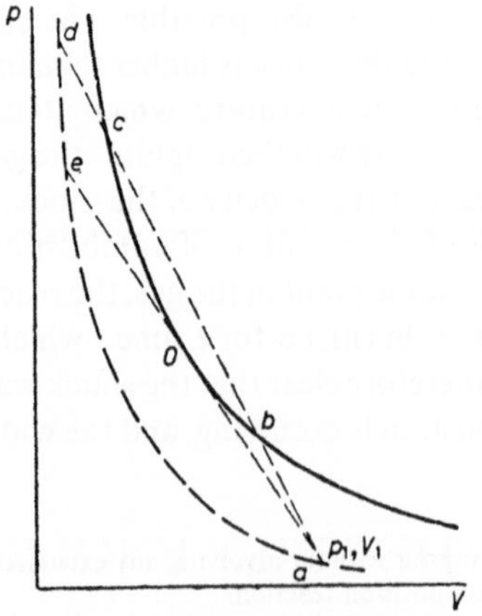

FIG. 132

Formula (129.2) is a consequence only of the conditions of continuity of the fluxes of mass and momentum. It therefore holds (for a given initial state of the gas) not only for the final state of the combustion products, but also for all intermediate states, in which only part of the reaction energy has been evolved.† In other words, the pressure p and specific volume V of the gas in any state obey the linear relation

$$p = p_1 + j^2(V_1 - V), \tag{129.3}$$

which is shown graphically by the chord *ad* (V. A. Mikhel'son 1890).

Let us now use a procedure developed by Ya. B. Zel'dovich (1940) to investigate the variation of the state of the gas through the layer of finite width which a detonation wave actually is. The forward front of the detonation wave is a true shock wave in the unburnt gas 1. In it, the gas is compressed and heated to a state represented by the point d (Fig. 132) on the shock adiabatic of gas 1. The chemical reaction begins in the compressed gas, and as the reaction proceeds the state of the gas is represented by a point which moves down the chord *da*; heat is evolved, the gas expands, and its pressure decreases. This continues until combustion is complete and the whole heat of the reaction has been evolved. The corresponding point is c, which lies on the detonation adiabatic representing the final state of the combustion products. The lower point b at which the chord *ad* intersects the detonation adiabatic cannot be reached for a gas in which combustion is caused by compression and heating in a shock wave.‡

Thus we conclude that the detonation is represented, not by the whole of the detonation adiabatic, but only by the upper part, lying above the point O where this adiabatic touches the straight line aO drawn from the initial point a.

It has been shown in §87 that, at the point where $\mathrm{d}(j^2)/\mathrm{d}p_2 = 0$, i.e. where the shock adiabatic touches the chord 12, the velocity v_2 is equal to the corresponding velocity of sound c_2. This result has been obtained only from the conservation laws for the surface of discontinuity, and is therefore entirely applicable to the detonation wave also. On the ordinary shock adiabatic for a single gas there are no points with $\mathrm{d}(j^2)/\mathrm{d}p_2 = 0$, as has been shown in §87. On the detonation adiabatic, however, there is such a point, namely the point O. At the same time as $v_2 = c_2$, we have at such a point the inequality (87.10) $\mathrm{d}(v_2/c_2)/\mathrm{d}p_2 < 0$, and therefore $v_2 < c_2$ at higher values of p_2, above O. Since detonation corresponds to the upper part only of the adiabatic, above the point O, we conclude that

$$v_2 \leqslant c_2, \tag{129.4}$$

i.e. a detonation wave moves relative to the gas just behind it with a velocity equal to or less than that of sound; the equality $v_2 = c_2$ holds for a detonation corresponding to the point O (called the *Chapman–Jouguet point*).††

The velocity of the detonation wave relative to gas 1 is always supersonic (even for the point O):

$$v_1 > c_1. \tag{129.5}$$

---

† Here it is assumed that diffusion and viscosity may be neglected in the combustion zone, so that mass and momentum transfer take place only by fluid flow.

‡ For completeness, it should also be mentioned that a discontinuous transition from state c to state b in another shock wave is also impossible, since the gas would have to cross such a shock from high pressure to low pressure.

†† It should be recalled that the velocities v_1, v_2 always signify the velocities normal to the surface of discontinuity.

This is most simply seen directly from Fig. 132. The velocity of sound c_1 is given graphically by the slope of the tangent to the shock adiabatic for gas 1 (dashed curve) at the point *a*. The velocity v_1, on the other hand, is given by the slope of the chord *ac*. Since all the chords concerned are steeper than the tangent, we always have $v_1 > c_1$. Moving with supersonic velocity, the detonation wave, like a shock wave, does not affect the state of the gas in front of it. The velocity v_1 with which the detonation wave moves relative to the unburnt gas at rest is the velocity of propagation of the detonation.

Since $v_1/V_1 = v_2/V_2 \equiv j$, and $V_1 > V_2$, it follows that $v_1 > v_2$. The difference $v_1 - v_2$ is evidently the velocity of the combustion products relative to the unburnt gas. This difference is positive, i.e. the combustion products move in the direction of propagation of the detonation wave.

We may note also the following. In §87 it was also shown that $ds_2/d(j^2) > 0$. At the point where j^2 has a minimum, s_2 therefore also has a minimum. This point is *O*, and we conclude that it corresponds to the least value of the entropy s_2 on the detonation adiabatic. The entropy s_2 also has an extremum at *O* if we consider the change in state along the line *ae* (since the slopes of the curve and the tangent at *O* are the same). This extremum, however, is a maximum (V. A. Mikhel'son). For a displacement from *e* to *O* corresponds to the change of state as the combustion reaction occurs in the compressed gas, and this is accompanied by the evolution of heat and an increase in entropy; a passage from *O* to *a*, however, would correspond to the endothermic conversion of the combustion products into the original gases, with a decrease in entropy.

If the detonation is caused by a shock wave which is produced by some external source and is then incident on the gas, any point on the upper part of the detonation adiabatic may correspond to the detonation. It is of particular interest, however, to consider a detonation which is due to the combustion process itself. We shall see in §130 that, in a number of important cases, such a detonation must correspond to the Chapman–Jouguet point, so that the velocity of the detonation wave relative to the combustion products just behind it is exactly equal to the velocity of sound, while the velocity $v_1 = jV_1$ relative to the unburnt gas has its least possible value.†

Let us now derive the relations between the various quantities in a detonation wave in a polytropic gas. Substituting in the general equation (129.1) the heat function in the form

$$w = w_0 + c_p T = w_0 + \gamma p V/(\gamma - 1),$$

we obtain

$$\frac{\gamma_2+1}{\gamma_2-1}p_2V_2 - \frac{\gamma_1+1}{\gamma_1-1}p_1V_1 - V_1p_2 + V_2p_1 = 2q, \tag{129.6}$$

where $q = w_{01} - w_{02}$ again denotes the heat of the reaction, reduced to the absolute zero of temperature. The curve $p_2(V_2)$ given by this equation is a rectangular hyperbola. For $p_2/p_1 \to \infty$, the ratio of densities tends to a finite limit $\rho_2/\rho_1 = V_1/V_2 = (\gamma_2+1)/(\gamma_2-1)$; this is the greatest compression that can be achieved in a detonation wave.

The formulae are much simplified in the important case of strong detonation waves, which are obtained when the heat evolved in the reaction is large compared with the

† This result was put forward as a hypothesis by D. L. Chapman (1899) and E. Jouguet (1905); its theoretical justification is due to Ya. B. Zel'dovich (1940), and independently to J. von Neumann (1942) and W. Döring (1943).

internal heat energy of the original gas, i.e. $q \gg c_{v1}T_1$. In this case we can neglect the terms containing p_1 in (129.6), obtaining

$$p_2\left(\frac{\gamma_2+1}{\gamma_2-1}V_2 - V_1\right) = 2q. \qquad (129.7)$$

Let us consider in more detail a detonation corresponding to the Chapman–Jouguet point, which is of particular interest, from the above discussion. At this point $j^2 = c_2{}^2/V_2{}^2 = \gamma_2 p_2/V_2$. From this relation and (129.2) we can express p_2 and V_2 in the form

$$p_2 = (p_1 + j^2V_1)/(\gamma_2+1), \qquad V_2 = \gamma_2(p_1+j^2V_1)/j^2(\gamma_2+1). \qquad (129.8)$$

Substituting these expressions in equation (129.6) and replacing j by v_1/V_1, we have after a simple reduction the following biquadratic equation for the velocity v_1:

$$v_1{}^4 - 2v_1{}^2[(\gamma_2{}^2-1)q + (\gamma_2{}^2-\gamma_1)c_{v1}T_1] + \gamma_2{}^2(\gamma_1-1)^2c_{v1}{}^2T_1{}^2 = 0,$$

where the temperature has been introduced by $T = pV/(c_p - c_v) = pV/c_v(\gamma-1)$. Hence†

$$v_1 = \sqrt{\{\tfrac{1}{2}(\gamma_2-1)[(\gamma_2+1)q + (\gamma_1+\gamma_2)c_{v1}T_1]\}} + $$
$$+ \sqrt{\{\tfrac{1}{2}(\gamma_2+1)[(\gamma_2-1)q + (\gamma_2-\gamma_1)c_{v1}T_1]\}}. \qquad (129.9)$$

This formula determines the velocity of propagation of the detonation in terms of the temperature T_1 of the original gas mixture.

We can rewrite formulae (129.8) in the form

$$\frac{p_2}{p_1} = \frac{v_1{}^2 + (\gamma_1-1)c_{v1}T_1}{(\gamma_2+1)(\gamma_1-1)c_{v1}T_1}, \qquad \frac{V_2}{V_1} = \frac{\gamma_2[v_1{}^2 + (\gamma_1-1)c_{v1}T_1]}{(\gamma_2+1)v_1{}^2}. \qquad (129.10)$$

Together with (129.9), they determine the ratios of pressure and density between the combustion products and the unburnt gas at temperature T_1.

The velocity v_2 is calculated as $v_2 = V_2v_1/V_1$, using formulae (129.9) and (129.10). The result is

$$v_2 = \sqrt{\{\tfrac{1}{2}(\gamma_2-1)[(\gamma_2+1)q + (\gamma_1+\gamma_2)c_{v1}T_1]\}} + $$
$$+ \frac{\gamma_2-1}{\gamma_2+1}\sqrt{\{\tfrac{1}{2}(\gamma_2+1)[(\gamma_2-1)q + (\gamma_2-\gamma_1)c_{v1}T_1]\}}. \qquad (129.11)$$

The difference $v_1 - v_2$, i.e. the velocity of the combustion products relative to the unburnt gas, is

$$v_1 - v_2 = \sqrt{\{2[(\gamma_2-1)q + (\gamma_2-\gamma_1)c_{v1}T_1]/(\gamma_2+1)\}}. \qquad (129.12)$$

The temperature of the combustion products is calculated from the formula

$$c_{v2}T_2 = v_2{}^2/\gamma_2(\gamma_2-1) \qquad (129.13)$$

(since $v_2 = c_2$).

---

† If $x^4 - 2px^2 + q = 0$, then

$$x = \sqrt{[p \pm \sqrt{(p^2-q)}]} = \sqrt{[\tfrac{1}{2}(p+\sqrt{q})]} \pm \sqrt{[\tfrac{1}{2}(p-\sqrt{q})]}.$$

The two signs in this case correspond to the fact that two tangents can be drawn from the point a to the detonation adiabatic: one upwards, as shown in Fig. 132, and the other downwards. The upward tangent, in which we are interested, has the steeper slope, and we accordingly take the plus sign.

All these somewhat complex formulae are much simplified for strong detonation waves. In this case the velocities are given by the simple formulae

$$v_1 = \sqrt{[2(\gamma_2^2-1)q]}, \qquad v_1 - v_2 = v_1/(\gamma_2+1). \tag{129.14}$$

The thermodynamic state of the combustion products is given by the formulae

$$\left. \begin{aligned} V_2/V_1 &= \gamma_2/(\gamma_2+1), \qquad T_2 = 2\gamma_2 q/c_{v2}(\gamma_2+1), \\ \frac{p_2}{p_1} &= \frac{2(\gamma_2-1)}{\gamma_1-1}\frac{q}{c_{v1}T_1} = \frac{\gamma_1 v_1^2}{(\gamma_2+1)c_1^2}. \end{aligned} \right\} \tag{129.15}$$

If we compare formulae (129.15) with the corresponding formulae (128.5) for slow combustion, we notice that, in the limiting case $q \gg c_{v1}T_1$, the ratio of the temperatures of the combustion products after detonation and slow combustion is $T_{2,\text{det}}/T_{2,\text{com}} = 2\gamma_2^2/(\gamma_2+1)$. This ratio always exceeds unity (since $\gamma_2 > 1$).

PROBLEM

Determine the thermodynamic quantities for the gas immediately behind the shock wave which is the forward front of a strong detonation wave corresponding to the Chapman–Jouguet point.

SOLUTION. Immediately behind the shock wave we have unburnt gas, and its state is represented by the point e where the tangent aO produced (Fig. 132) intersects the shock adiabatic of gas 1, shown dashed. Denoting the coordinates of this point by (p_1', V_1'), we have, firstly, by equation (89.1) for the shock adiabatic of gas 1,

$$\frac{V_1'}{V_1} = \frac{(\gamma_1+1)p_1 + (\gamma_1-1)p_1'}{(\gamma_1-1)p_1 + (\gamma_1+1)p_1'}$$

and, secondly, $(p_1'-p_1)/(V_1-V_1') = j^2 = v_1^2/V_1^2$. Taking v_1 from (129.14), we obtain

$$p_1' = p_1\frac{4(\gamma_2^2-1)}{\gamma_1^2-1}\frac{q}{c_{v1}T_1}, \qquad V_1' = V_1\frac{\gamma_1-1}{\gamma_1+1},$$

$$T_1' = \frac{q}{c_{v1}}\frac{4(\gamma_2^2-1)}{(\gamma_1+1)^2}.$$

The ratio of the pressure p_1' to the pressure p_2 behind the detonation wave is

$$p_1'/p_2 = 2(\gamma_2+1)/(\gamma_1+1).$$

§130. The propagation of a detonation wave

Let us now consider some actual cases of the propagation of detonation waves in a gas initially at rest. We take first the case of detonation in a gas in a pipe closed at one end ($x = 0$). The boundary conditions in this case are that the gas velocity is zero both in front of the detonation wave (which does not affect the state of the gas in front of it) and at the closed end of the pipe. Since the gas acquires a non-zero velocity when the detonation wave passes, the velocity must diminish in the region between the detonation wave and the closed end of the pipe. In order to determine the resulting flow pattern, we notice that in this case there is no length parameter which might characterize the conditions of flow along the pipe (the x-direction). We have seen in §99 that, in such cases, the gas velocity can change either in a shock wave (separating two regions where the velocity is constant) or in a similarity rarefaction wave.

Let us first assume that the detonation wave does not correspond to the Chapman–Jouguet point on the adiabatic. Then its velocity of propagation relative to the gas behind it is $v_2 < c_2$. It is easy to see that, in this case, neither a shock wave nor a weak discontinuity (the forward front of a rarefaction wave) can follow the detonation wave. For the former would have to move, relative to the gas in front of it, with a velocity exceeding c_2, and the latter with a velocity equal to c_2, and either would overtake the detonation wave. Thus, on the above assumption, the velocity of the gas moving behind the detonation wave cannot decrease, i.e. the boundary condition at $x = 0$ cannot be satisfied.

This condition can be satisfied only for a detonation wave corresponding to the Chapman–Jouguet point. Then $v_2 = c_2$, and a rarefaction wave can follow the detonation wave. It is formed at $x = 0$ when the detonation begins, and its forward front coincides with the detonation wave.

Thus we reach the important result that a detonation wave propagated in a pipe, with the gas ignited at the closed end, must correspond to the Chapman–Jouguet point. It moves relative to the gas just behind it with a velocity equal to the local velocity of sound. The detonation wave adjoins a rarefaction wave, in which the gas velocity (relative to the pipe) falls monotonically to zero. The point where the velocity becomes zero is a weak discontinuity. Behind this discontinuity the gas is at rest (Fig. 133a).

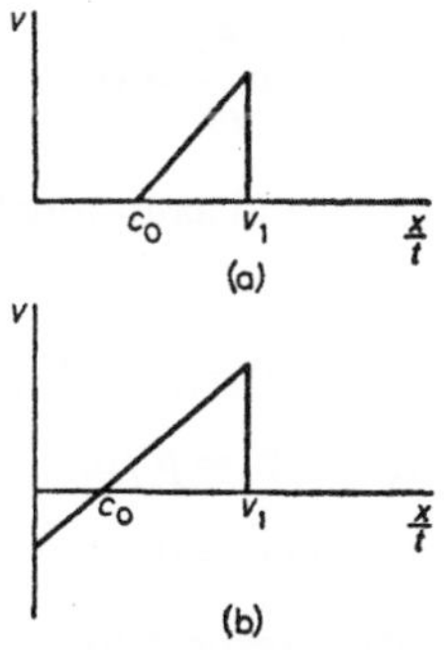

FIG. 133

Let us now consider a detonation wave propagated from the open end of a pipe. The pressure of the gas in front of the detonation wave must be equal to the original pressure, which clearly equals the external pressure. It is evident that, in this case also, the velocity must decrease somewhere behind the detonation wave. If the gas velocity were constant between the end of the pipe and the detonation wave, it would follow that gas was being sucked into the open end of the pipe from outside; this would be impossible, since the gas pressure in the pipe would be greater than the external pressure on account of the pressure increase in a detonation wave. For the same reasons as in the previous case, the detonation wave must correspond to the Chapman–Jouguet point. The resulting flow pattern is shown schematically in Fig. 133b. Immediately behind the detonation wave is a similarity rarefaction wave, in which the velocity decreases monotonically towards the end of the pipe, changing sign at some point. This means that, in the end section of the pipe, the gas moves towards the open end and flows out of it; the velocity with which it leaves the pipe equals the local velocity of sound, and its pressure exceeds the external pressure. We have seen in §97 that such a flow is possible.†

† We everywhere ignore the heat losses that may accompany the propagation of a detonation wave. As in the case of slow combustion, these losses may prevent the propagation. For detonation in a pipe, the losses are due mainly to heat loss through the pipe walls and the retardation of the gas by friction.

Let us next consider the important case of a spherically symmetrical outgoing detonation wave whose centre is the point where the gas is first ignited (Ya. B. Zel'dovich 1942). Since the gas must be at rest both in front of the detonation wave and near the centre, the gas velocity must decrease from the detonation wave towards the centre. As with the flow in a pipe, there are no characteristic parameters having the dimensions of length. The result must therefore be a similarity flow, the coordinate x being replaced by the distance r from the centre. Thus all quantities are functions only of the ratio r/t.†

For centrally symmetrical flow ($v_r = v(r, t)$, $v_\phi = v_\theta = 0$), the equations of motion are as follows. The equation of continuity is

$$\frac{\partial \rho}{\partial t} + \frac{\partial (v\rho)}{\partial r} + \frac{2v\rho}{r} = 0;$$

Euler's equation is

$$\frac{\partial v}{\partial t} + v\frac{\partial v}{\partial r} = -\frac{1}{\rho}\frac{\partial p}{\partial r},$$

and the equation of conservation of entropy is

$$\frac{\partial s}{\partial t} + v\frac{\partial s}{\partial r} = 0.$$

Introducing the variable $\xi = r/t$ (> 0) and assuming that all quantities are functions of ξ only, we obtain

$$(\xi - v)\rho'/\rho = v' + 2v/\xi, \tag{130.1}$$

$$(\xi - v)v' = p'/\rho, \tag{130.2}$$

$$(\xi - v)s' = 0, \tag{130.3}$$

the prime denoting differentiation with respect to ξ. We cannot have $v = \xi$, since this contradicts the first equation. From the third equation, therefore, $s' = 0$, i.e. $s =$ constant. We can therefore write $p' = c^2\rho'$, and equation (130.2) becomes

$$(\xi - v)v' = c^2\rho'/\rho. \tag{130.4}$$

Substituting ρ'/ρ from (130.1), we obtain the relation

$$\left[\frac{(\xi - v)^2}{c^2} - 1\right]v' = \frac{2v}{\xi}. \tag{130.5}$$

Equations (130.4) and (130.5) cannot be integrated analytically, but the properties of their solutions can be investigated.

The region where the gas flow is of the type considered is bounded, as we shall see below, by two spheres, of which the outer is the surface of the detonation wave itself, and the inner is the surface of a weak discontinuity, where the velocity is zero.

Let us first examine the properties of the solution near the point where v is zero. It is easy to see that, where $v = 0$, $\xi = c$ also:

$$v = 0, \qquad \xi = c. \tag{130.6}$$

† The dimensionless similarity variable in this problem may be taken as $r/t\sqrt{q}$, where the constant parameter q is the heat of reaction per unit mass.

For, when v tends to zero, $\log v \to -\infty$; hence, when ξ decreases to the value corresponding to the inner boundary of the region in question, the derivative $d \log v/d\xi$ must tend to $+\infty$. From (130.5), however, we have for $v = 0$

$$d \log v/d\xi = 2c^2/\xi(\xi^2 - c^2).$$

This expression can tend to $+\infty$ only if $\xi \to c$.

At the origin, the radial velocity must vanish, by symmetry. Thus there is a region of gas at rest round the origin; this is the region inside the sphere $\xi = c_0$, where c_0 is the velocity of sound for the point where $v = 0$.

Let us ascertain the properties of the function $v(\xi)$ near the point (130.6). From (130.5) we have

$$v\frac{d\xi}{dv} = \tfrac{1}{2}\xi\left[\frac{(\xi - v)^2}{c^2} - 1\right].$$

As fas as quantities of the first order (such as v, $\xi - c_0$ and $c - c_0$), we have after a simple calculation $v\,d(\xi - c_0)/dv = (\xi - c_0) - (v + c - c_0)$. According to (102.1) we have $v + c - c_0 = \alpha_0 v$, where α_0 is a positive constant, the value of (102.2) for $v = 0$, and we obtain the following linear first-order differential equation for $\xi - c_0$ as a function of v:

$$v\,d(\xi - c_0)/dv - (\xi - c_0) = -\alpha_0 v.$$

The solution of this equation is

$$\xi - c_0 = \alpha_0 v \log(\text{constant}/v). \tag{130.7}$$

This implicitly determines the function $v(\xi)$ near the point where $v = 0$.

We see that the inner boundary is a surface of weak discontinuity: the velocity tends continuously to zero. The curve of $v(\xi)$ has a horizontal tangent at this point ($dv/d\xi = 0$). The weak discontinuity involved is very unusual: the first derivative is continuous, but all higher derivatives are infinite (as is easily seen from (130.7)). The ratio r/t for $v = 0$ is clearly just the velocity of motion of the boundary relative to the gas; according to (130.6), it is equal to the local velocity of sound, as it should be for a weak discontinuity.

We have also for small v, by (130.7),

$$\begin{aligned}\xi - v - c &= (\xi - c_0) - (v + c - c_0)\\ &= \alpha_0 v[\log(\text{constant}/v) - 1].\end{aligned}$$

For small v, this quantity is positive: $\xi - v - c > 0$. We shall show that the difference $(\xi - v) - c$ cannot change sign anywhere in the region of the flow considered. Let us consider a point, if there is one, where

$$\xi - v = c, \qquad v \neq 0. \tag{130.8}$$

We see from (130.5) that the derivative v' must be infinite at this point, i.e.

$$d\xi/dv = 0. \tag{130.9}$$

The second derivative $d^2\xi/dv^2$ is shown by a simple calculation (using the conditions (130.8) and (130.9)) to be $d^2\xi/dv^2 = -\alpha_0\xi/c_0 v$, which is not zero. This means that ξ as a function of v has a maximum at the point in question. Thus the function $v(\xi)$ exists only for ξ less than the value corresponding to the conditions (130.8), and this value is the other boundary of the region considered. Since $\xi - v - c$ can vanish only at the boundary of the region, and $\xi - v - c > 0$ for small v, we conclude that

$$\xi - v > c \tag{130.10}$$

everywhere in the region.

It is now easy to see that the outer boundary of the region of the flow considered must in fact be at the point where the conditions (130.8) hold. To see this, we notice that the difference $r/t - v$, where r is the coordinate of the boundary, is just the velocity of the boundary relative to the gas behind it. A surface on which $r/t - v > c$, however, cannot be the surface of a detonation wave (where we must have $r/t - v \leqslant c$). We therefore conclude that the outer boundary of the region considered can only be the point where (130.8) holds. On this boundary v falls discontinuously to zero, and the velocity of the boundary relative to the gas just behind it is equal to the local velocity of sound. This means that the detonation wave must correspond to the Chapman–Jouguet point on the detonation adiabatic.†

We thus have the following flow pattern for spherical propagation of a detonation. The detonation wave, like that in a pipe, must correspond to the Chapman–Jouguet point. Immediately behind it is a spherical similarity rarefaction wave, in which the gas velocity decreases to zero. The decrease is monotonic, since, by (130.5), the derivative $dv/d\xi$ can vanish only if $v = 0$ also. The gas pressure and density also decrease monotonically, since by (130.4) and (130.10) the derivative p' always has the same sign as v'. The curve giving v as a function of r/t has a vertical tangent at the outer boundary (by (130.9)) and a horizontal tangent at the inner boundary (Fig. 134). The inner boundary is a weak discontinuity, near which the dependence of v on r/t is given by equation (130.7). The gas within the sphere bounded by the weak discontinuity is at rest. The total mass of gas at rest is, however, very small (cf. the remarks at the end of §106).

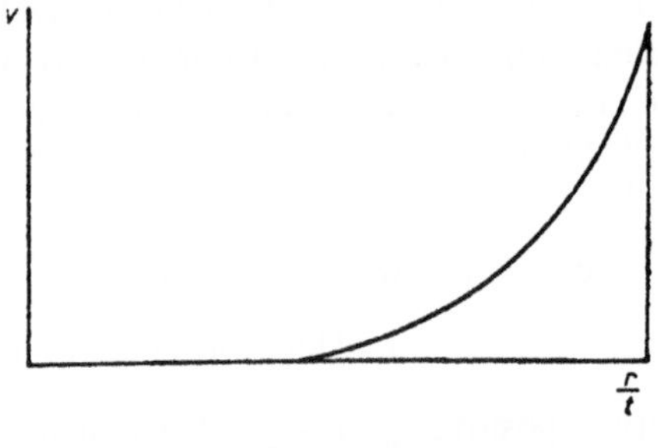

FIG. 134

Thus, in all the typical cases of spontaneous one-dimensional spherical propagation of detonation which we have considered, the boundary conditions in the region behind the detonation wave give a unique velocity for the latter, which corresponds to the Chapman–Jouguet point (the whole of the detonation adiabatic below this point being excluded by the arguments of §129). The achievement, in a pipe with constant cross-section, of a detonation corresponding to the part of the adiabatic above the

† We may notice for completeness that v = constant is not a solution of the equations of centrally symmetrical flow. Hence the detonation wave cannot be followed by a region of constant velocity.

Chapman–Jouguet point would require an artificial compression of the combustion products by a piston moving with a supersonic velocity (see Problem 3). Such detonation waves are said to be *over-compressed.*

It should be emphasized, however, that these conclusions are not universally valid, and there are cases of propagation of a detonation where an over-compressed detonation wave occurs spontaneously. In particular, an over-compressed detonation wave is formed when an ordinary detonation wave goes from a wide pipe into a narrow one. This phenomenon occurs because, when a detonation wave reaches a narrowing of the pipe, it is partly reflected, and the pressure of the combustion products moving from the wide part to the narrow part is considerably increased (cf. Problem 4) (B. V. Aĭvazov and Ya. B. Zel'dovich 1947).†

A general remark should be made regarding the theory given in §§129 and 130. The structure of the detonation wave has been assumed steady, and uniform over its area; it is one-dimensional, in the sense that the propagation of all quantities in the combustion zone is assumed to depend only on the transverse coordinate. However, the experimental results now available indicate that this picture is highly idealized and could serve only to give an averaged description of the process; the picture actually observed is usually quite different. The structure is in fact far from steady, and basically three-dimensional; the wave area has a complex small-scale structure which varies rapidly with time. This is due to the instability which arises in particular from the strong (exponential) temperature dependence of the reaction rate: even a slight change in temperature when the shock front undergoes distortion has a large effect on the progress of the reaction, and this instability becomes stronger with increasing ratio of the reaction activation energy to the gas temperature behind the shock. The non-uniform and non-steady form of the structure of the detonation wave is particularly clear in conditions close to the limit of propagation in a pipe: the combustible mixture is ignited almost entirely at isolated, eccentrically located and spirally moving, highly deformed parts of the shock front (*spinning detonation*). The study of the possible mechanisms of all these complex phenomena is outside the scope of the present book.‡

PROBLEMS

PROBLEM 1. Determine the gas flow when a detonation wave is propagated from the closed end of a pipe.

SOLUTION. The velocity v_1 of the detonation wave relative to the gas at rest in front of it, and its velocity v_2 relative to the burnt gas just behind it, are given in terms of the temperature T_1 by formulae (129.11), (129.12); v_1 is also the velocity of the wave relative to the pipe, so that its coordinate is $x = v_1 t$. The velocity (relative to the pipe) of the combustion products at the detonation wave is $v_1 - v_2$. The velocity v_2 equals the local velocity of sound. Since the velocity of sound is related to the gas velocity v in a similarity rarefaction wave by $c = c_0 + \frac{1}{2}(\gamma - 1)v$, we have $v_2 = c_0 + \frac{1}{2}(\gamma_2 - 1)(v_1 - v_2)$, whence $c_0 = \frac{1}{2}(\gamma_2 + 1)v_2 - \frac{1}{2}(\gamma_2 - 1)v_1$. For a strong detonation wave we have, by (129.14), simply $c_0 = \frac{1}{2}v_1$. The quantity c_0 is the velocity of the backward boundary of the rarefaction wave. The velocity varies linearly between the two boundaries (Fig. 133a).

PROBLEM 2. The same as Problem 1, but for a pipe with an open end.

† Over-compression arises also in the propagation of an ingoing cylindrical or spherical detonation wave; see Ya. B. Zel'dovich, *Soviet Physics JETP* **9**, 550, 1959.

‡ Some relevant monographs and review articles are: K. I. Shchelkin and Ya. K. Troshin, *Gasdynamics of Combustion*, Baltimore 1965; R. I. Soloukhin, *Shock Waves and Detonations in Gases*, Baltimore 1966; R. I. Soloukhin, *Soviet Physics Uspekhi* **6**, 523, 1964; A. K. Oppenheim and R. I. Soloukhin, *Annual Review of Fluid Mechanics* **5**, 31, 1973.

SOLUTION. The velocities v_1 and v_2 are determined as in the previous case, and so c_0 is the same also. The rarefaction wave, however, now extends, not to the point where $v = 0$, but to the end of the pipe ($x = 0$, Fig. 133b). We see from the formula $x/t = v + c$ (99.5) that the gas leaves the open end of the pipe with a velocity $v = -c$ equal to the local velocity of sound. Putting $-v = c = c_0 + \frac{1}{2}(\gamma_2 - 1)v$, we therefore find the velocity of outflow to be $[-v]_{x=0} = 2c_0/(\gamma_2 + 1)$. For a strong detonation wave this velocity is $v_1/(\gamma_2 + 1)$, equal in magnitude to that of the gas just behind it.

PROBLEM 3. The same as Problem 1, but for a detonation wave propagated in a pipe whose end is closed by a piston which begins to move forward with a constant velocity U.

SOLUTION. If $U < v_1$, the velocity distribution in the gas has the form shown in Fig. 135a. The gas velocity decreases from $v_1 - v_2$ at $x/t = v_1$ to U at $x/t = c_0 + \frac{1}{2}(\gamma + 1)U$, with the same value of c_0 as before. Then follows a region in which the gas moves with constant velocity U.

If $U > v_1$, however, the detonation wave cannot correspond to the Chapman–Jouguet point (since the piston would overtake it). In this case we have an over-compressed detonation wave, corresponding to a point on the adiabatic above the Chapman–Jouguet point. It is determined by the fact that the discontinuity of velocity in the detonation wave must equal the velocity of the piston: $v_1 - v_2 = U$. Throughout the region between the detonation wave and the piston, the gas moves with constant velocity U (Fig. 135b).

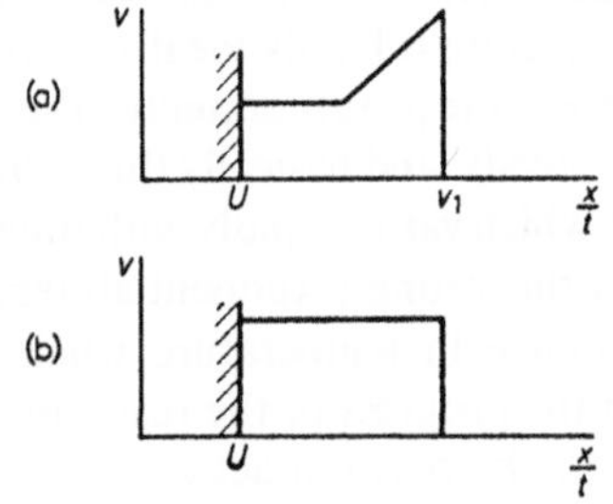

FIG. 135

PROBLEM 4. Determine the pressure at a perfectly rigid wall when a strong plane detonation wave normally incident is reflected from it (K. P. Stanyukovich 1946).

SOLUTION. When a detonation wave is incident on a wall, a reflected shock wave is formed and propagated in the opposite direction, through the combustion products. The calculations are entirely similar to those in §100, Problem 1. With the same notation, we obtain the three relations

$$p_2(V_1 - V_2) = (p_3 - p_2)(V_2 - V_3), \qquad V_2/V_1 = \gamma_2/(\gamma_2 + 1),$$

$$\frac{V_3}{V_2} = \frac{(\gamma_2 + 1)p_2 + (\gamma_2 - 1)p_3}{(\gamma_2 - 1)p_2 + (\gamma_2 + 1)p_3};$$

here we have neglected p_1 in comparison with p_2, but p_2 and p_3 are of the same order of magnitude. Eliminating the volumes, we obtain a quadratic equation for p_3, and must take the root which is greater than p_2:

$$\frac{p_3}{p_2} = \frac{5\gamma_2 + 1 + \sqrt{(17\gamma_2^2 + 2\gamma_2 + 1)}}{4\gamma_2}.$$

It should be noted that this quantity is almost independent of γ_2, varying from 2·6 to 2·3 as γ_2 varies from 1 to ∞.

§131. The relation between the different modes of combustion

It has been shown in §129 that detonation corresponds to points on the upper part of the detonation adiabatic for the combustion process concerned. Since the equation of this adiabatic is a consequence only of the conservation laws for mass, momentum and energy (applied to the initial and final states of the burning gas), it is clear that the points representing the state of the reaction products must lie on the same curve for any other mode of combustion in which the combustion zone can be regarded as a surface of discontinuity of some kind. Let us now ascertain the physical significance of the remainder of the curve.

We draw through the point (p_1, V_1) (point 1 in Fig. 136) vertical and horizontal lines $1A$ and $1A'$, and the two tangents $1O$ and $1O'$ to the adiabatic. The points A, A', O, O' where these lines intersect or touch the curve divide the adiabatic into five parts. The part lying above O corresponds to detonation, as we have said. We shall now consider the other parts of the curve.

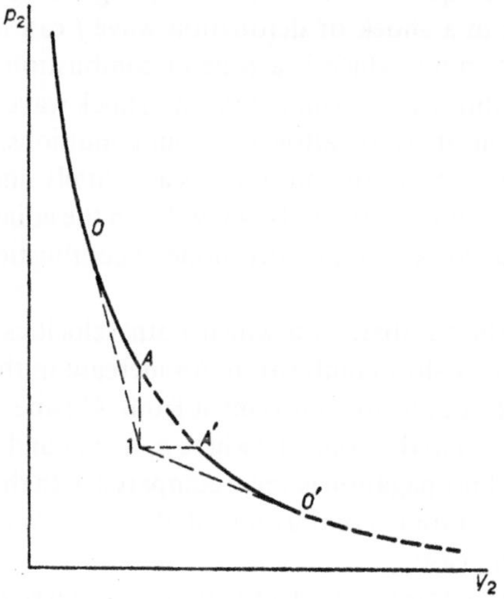

FIG. 136

First of all, it is easy to see that the section AA' has no physical significance. For we have on this section $p_2 > p_1$, $V_2 > V_1$, and so the mass flux j is imaginary; cf. (129.2).

At the points of contact O and O', the derivative $\mathrm{d}(j^2)/\mathrm{d}p_2$ is zero; it has been shown in §129 (referring to §87) that at such points we have $v_2/c_2 = 1$ and $\mathrm{d}(v_2/c_2)/\mathrm{d}p_2 < 0$. Hence it follows that above the points of contact $v_2/c_2 < 1$, and below them $v_2/c_2 > 1$. The relation between v_1 and c_1 is always easily found by considering the slopes of the corresponding chords and tangents, as was done in §129 for the part above O. The result is that the following inequalities hold on the various sections of the adiabatic:

$$\left.\begin{array}{lll} \text{above } O & v_1 > c_1, & v_2 < c_2; \\ \text{on } AO & v_1 > c_1, & v_2 > c_2; \\ \text{on } A'O' & v_1 < c_1, & v_2 < c_2; \\ \text{below } O' & v_1 < c_1, & v_2 > c_2. \end{array}\right\} \qquad (131.1)$$

At O and O', $v_2 = c_2$. As we approach A, the flux j, and therefore the velocities v_1, v_2, tend to infinity. As we approach A', however, j and the velocities v_1, v_2 tend to zero.

In §88 we have defined the condition for shock waves to be evolutionary, as a necessary condition for their existence. We have seen that the condition depends on the relation between the number of parameters determining the perturbation and the number of boundary conditions which the perturbation must satisfy at the surface of discontinuity.

All these considerations can also be applied to the surfaces of discontinuity here considered. In particular, the calculation made in §88 of the number of parameters of the perturbation for each case (131.1), shown in Fig. 57, remains valid. For a detonation (the adiabatic above O) the number of boundary conditions is the same as for an ordinary shock wave, and the condition for an evolutionary shock remains as before. In the absence of detonation (below O) the situation is different, because of the change in the number of boundary conditions. Here, the velocity of propagation is determined only by the properties of the chemical reaction and by the conditions of heat transfer from the combustion zone to the cold gas in front of it. This means that the mass flux j through the combustion zone is a given quantity (more precisely, a given function of the state of the unburnt gas 1), whereas in a shock or detonation wave j can have any value. Hence it follows that, on a discontinuity which is a zone of combustion without detonation, the number of boundary conditions is one more than at a shock wave: the condition that j have a given value is added. Thus there are altogether four conditions, and we now conclude in the same manner as in §87 that the discontinuity is absolutely unstable only in the case $v_1 < c_1, v_2 > c_2$, which corresponds to points below O' on the adiabatic. Consequently, this part of the curve does not correspond to any mode of combustion that can be realized in practice.

The section $A'O'$ of the adiabatic, on which both velocities v_1 and v_2 are subsonic, corresponds to the ordinary slow combustion. An increase in the rate of propagation of combustion, i.e. in j, corresponds to a movement from A' (where $j = 0$) towards O'. The formulae (128.5) correspond to the point A' (where $p_1 = p_2$), and are valid if j is sufficiently small, viz. if the velocity of propagation is small compared with that of sound. The point O' corresponds to the "most rapid" combustion of this type. We shall give the formulae pertaining to this limiting case.

The point O', like O, is a point of contact between the curve and the tangent from the point 1. Hence the formulae relating to O' can be obtained immediately from formulae (129.8)–(129.11) for O by appropriately changing the signs (see the footnote to (129.9)). In formulae (129.9) and (129.11) for v_1 and v_2 we change the sign of the second radical, and the sign of the expression (129.12) for $v_1 - v_2$ is therefore changed also. Formulae (129.10) are unchanged if v_1 is taken to have its new value. All these formulae are much simplified if the heat of reaction is large ($q \gg c_{v1} T_1$). We then obtain

$$\left.\begin{aligned} v_1 &= \gamma_2 p_1 V_1/\sqrt{[2(\gamma_2^2-1)q]}, \qquad v_2 = \sqrt{[2(\gamma_2-1)q/(\gamma_2+1)]}, \\ p_2/p_1 &= 1/(\gamma_2+1), \qquad c_{v2} T_2 = 2q/\gamma_2(\gamma_2+1). \end{aligned}\right\} \tag{131.2}$$

The following remark must be made here. We have seen that, in slow combustion in a closed pipe, a shock wave must be formed in front of the combustion zone. For large velocities of propagation of combustion, this shock wave is strong, and it may considerably affect the state of the gas which enters the combustion zone. It is therefore, strictly speaking, useless to investigate the change in the manner of combustion with increasing velocity, the state p_1, V_1 of the unburnt gas remaining unchanged. In order to reach the point O' we must create conditions of combustion in which no shock wave is formed. This can be done, for instance, in combustion in a pipe open at both ends, with a continuous removal of combustion products at the rear end. The rate of removal must be such that the combustion zone remains at rest, and so no shock wave is formed.†

† Ordinary slow combustion in a tube may spontaneously change into detonation. This is preceded by a spontaneous acceleration of the flame, the detonation wave being formed ahead of it. The possible mechanisms of these processes are discussed in the works cited in the last footnote and in the penultimate footnote to §128.

The part AO of the adiabatic corresponds to combustion without detonation, propagated at supersonic speed. This can in principle occur in the presence of very high heat transfer, e.g. by radiation, which gives combustion rates j exceeding the value corresponding to O'.

In conclusion, we may call attention to the following general differences (besides those contained in the inequalities (131.1)) between the modes of combustion corresponding to the upper and lower parts of the adiabatic. Above A we have $p_2 > p_1$, $V_2 < V_1$, $v_2 < v_1$. That is, the reaction products have a pressure and density greater than that of the original gas, and move behind the combustion front with velocity $v_1 - v_2$. In the region below A, however, the inequalities are reversed: $p_2 < p_1$, $V_2 > V_1$, $v_2 > v_1$, and the combustion products are less dense than the original gas.

§132. Condensation discontinuities

There is a formal similarity between detonation waves and what are called *condensation discontinuities*; these occur, for instance, in the flow of a gas containing supersaturated water vapour.† The discontinuities are the result of a sudden condensation of vapour occurring very rapidly in a narrow region, which can be regarded as a surface of discontinuity separating the original gas from a gas containing condensed vapour (a *fog*). It should be emphasized that condensation discontinuities are a distinct physical phenomenon, and do not result from the compression of gas in an ordinary shock wave; the latter effect cannot lead to condensation, since the increase of pressure in the shock wave has less effect on the degree of supersaturation than the increase of temperature.

Like combustion, the condensation of a vapour is an exothermic process. The heat of reaction q is represented by the heat evolved per unit mass of gas by the condensation of the vapour.‡ The condensation adiabatic which gives p_2 as a function of V_2 for a given state p_1, V_1 of the original uncondensed gas has the same form as the combustion adiabatic shown in Fig. 136. The relations between the velocities of propagation of the discontinuity v_1, v_2 and the velocities of sound c_1, c_2 for the various parts of the condensation adiabatic are given by the inequalities (131.1). However, not all the four cases enumerated in (131.1) can actually occur.

First of all, the question arises whether condensation discontinuities are evolutionary. In this respect their properties are entirely similar to those of combustion zones. We have seen (§131) that the difference in stability between combustion zones and ordinary shock waves is due to the existence of a further condition (that the flux j have a given value) which must be satisfied at the surface. In the case of condensation discontinuities there is again a further condition: the thermodynamic state of the gas 1 in front of the discontinuity must be one for which rapid condensation of the vapour begins. This condition gives a relation between the pressure and temperature of gas 1. We therefore conclude immediately that the whole of the adiabatic below O', for which $v_1 < c_1$, $v_2 > c_2$, is excluded, since it does not correspond to stable discontinuities.

---

† The theoretical analysis was begun by K. Oswatitsch (1942) and S. Z. Belen'kii (1945).

‡ The heat q is not, strictly speaking, the usual latent heat of condensation, since the process occurring in the condensation zone includes not only the isothermal condensation of the vapour, but also a general change in the gas temperature. However, if the degree of supersaturation is not too small (a condition usually satisfied), the difference is unimportant.

It is easy to see that discontinuities corresponding to the part above O, for which $v_1 > c_1$, $v_2 < c_2$, also cannot occur in practice. Such a discontinuity would move with supersonic velocity relative to the gas in front of it, and so its presence would have no effect on the state of that gas. Consequently, the discontinuity would have to be formed along a surface determined by the conditions of flow, namely the surface on which the necessary conditions for the onset of rapid condensation would be fulfilled in continuous flow. The velocity of the discontinuity relative to the gas behind it, on the other hand, would be subsonic in this case. The equations of subsonic flow, however, in general have no solution for which all quantities take prescribed values on a given surface.†

Thus only two types of condensation discontinuity are possible: (1) supersonic discontinuities (the section AO of the adiabatic) for which

$$v_1 > c_1, \qquad v_2 > c_2, \qquad p_2 > p_1, \qquad V_2 < V_1 \tag{132.1}$$

and the condensation involves a compression, (2) subsonic discontinuities (the section $A'O'$ of the adiabatic), for which

$$v_1 < c_1, \qquad v_2 < c_2, \qquad p_2 < p_1, \qquad V_2 > V_1 \tag{132.2}$$

and the condensation involves a rarefaction.

The value of the flux j increases monotonically along the section $A'O'$ from A' (where $j = 0$) to O', and decreases monotonically along AO from A (where $j = \infty$) to O. The range of values of j (and therefore the range of values of the velocity $v_1 = jV_1$) between those corresponding to O and O' is "forbidden", and cannot occur in a condensation discontinuity. The total mass of condensed vapour is usually very small compared with the mass of the original gas. We can therefore regard both gases 1 and 2 as perfect gases; for the same reason, the specific heats of the two gases may be supposed equal. Then the value of v_1 at the point O is determined by formula (129.9), and its value at O' by the same formula with the sign of the second radical changed; putting $\gamma_1 = \gamma_2 \equiv \gamma$ and $c_1^2 = \gamma(\gamma-1)c_v T_1$, we find the forbidden range of values of v_1 to be

$$\begin{aligned} &\sqrt{[c_1^2 + \tfrac{1}{2}(\gamma^2-1)q]} - \sqrt{[\tfrac{1}{2}(\gamma^2-1)q]} < v_1 \\ &< \sqrt{[c_1^2 + \tfrac{1}{2}(\gamma^2-1)q]} + \sqrt{[\tfrac{1}{2}(\gamma^2-1)q]}. \end{aligned} \tag{132.3}$$

PROBLEM

Determine the limiting values of the ratio of pressures p_2/p_1 in a condensation discontinuity, assuming that $q/c_1^2 \ll 1$.

SOLUTION. On the section $A'O'$ of the condensation adiabatic (Fig. 136), the ratio p_2/p_1 increases monotonically from O' to A', taking values in the range

$$1 - \gamma\sqrt{[2(\gamma-1)q/(\gamma+1)c_1^2]} \leqslant p_2/p_1 \leqslant 1.$$

On the section AO, this ratio increases from A to O, taking values in the range

$$1 + \gamma(\gamma-1)q/c_1^2 \leqslant p_2/p_1 \leqslant 1 + \gamma\sqrt{[2(\gamma-1)q/(\gamma+1)c_1^2]}.$$

† Similar arguments hold in the case where the total velocity $\mathbf{v}_2$ (of which $v_2 < c_2$ is the component normal to the discontinuity) is supersonic.

To avoid misunderstanding, it should be mentioned that a condensation discontinuity with $v_1 > c_1$, $v_2 < c_2$ may actually (for certain conditions of vapour content and shape of the surface past which the flow occurs) be simulated by a true condensation discontinuity with $v_1 > c_1$, $v_2 > c_2$, closely followed by a shock wave which renders the flow subsonic.

CHAPTER XV

RELATIVISTIC FLUID DYNAMICS

§133. The energy-momentum tensor

The necessity of allowing for relativistic effects in fluid dynamics may not only be due to a large velocity of the macroscopic flow (comparable with that of light). The equations of fluid dynamics are considerably modified also when this velocity is not large but those of the microscopic motion of the fluid particles are large.

To derive the relativistic equations of fluid dynamics, we must first establish the form of the energy-momentum 4-tensor T^{ik} for a fluid in motion.† We recall that $T^{00} = T_{00}$ is the energy density, $T^{0\alpha}/c = -T_{0\alpha}/c$ the momentum component density; $T^{\alpha\beta} = T_{\alpha\beta}$ form the momentum flux density tensor; the energy flux density $cT^{0\alpha}$ differs from the momentum density only by the factor c^2.

The momentum flux through an element $\mathrm{d}\mathbf{f}$ of the surface of a body‡ is just the force acting on this element. Hence $T^{\alpha\beta}\,\mathrm{d}f_\beta$ is the α-component of the force acting on the surface element. Let us consider some volume element in a frame of reference in which it is at rest (the *local proper frame* or *local rest frame*; quantities in this frame are said to have *proper values*). In this frame, Pascal's law is valid, that is, the pressure exerted by a given portion of the fluid is the same in all directions and is everywhere perpendicular to the surface on which it acts. We can therefore write $T^{\alpha\beta}\,\mathrm{d}f_\beta = p\,\mathrm{d}f_\alpha$, whence $T_{\alpha\beta} = p\delta_{\alpha\beta}$.

The components $T^{0\alpha}$ which give the momentum density are zero in the local proper frame. The component T^{00} is the proper internal energy density of the fluid, which in this chapter will be denoted by e.

In the local rest frame, then, the energy-momentum tensor has the form

$$T^{ik} = \begin{pmatrix} e & 0 & 0 & 0 \\ 0 & p & 0 & 0 \\ 0 & 0 & p & 0 \\ 0 & 0 & 0 & p \end{pmatrix} \tag{133.1}$$

It is now easy to find an expression for T^{ik} in any frame of reference. To do so, we use the fluid 4-velocity u^i. In the local rest frame, $u^0 = 1$, $u^\alpha = 0$. The expression for T^{ik} which gives (133.1) with these values of u^i is

$$T^{ik} = wu^iu^k - pg^{ik}, \tag{133.2}$$

† This section largely reproduces *Fields*, §35; it is given here so as to make the discussion complete in itself. The notation is the same as there. Latin suffixes and indices take the values 0, 1, 2, 3; $x^0 = ct$ is the time coordinate (in this chapter, c is the velocity of light). The letters $\alpha, \beta, \gamma, \ldots$ from the beginning of the Greek alphabet take the values 1, 2, 3, corresponding to the spatial coordinates. The Galilean metric (special theory of relativity) corresponds to the metric tensor with components $g_{00} = 1$, $g_{11} = g_{22} = g_{33} = -1$.

‡ For the three-dimensional vector $\mathrm{d}\mathbf{f}$ (and the velocity vector $\mathbf{v}$ below) in Cartesian coordinates, there is no need to distinguish contravariant and covariant components, and these will be written with suffixes. The same applies to the three-dimensional unit tensor $\delta_{\alpha\beta}$.

where $w = e + p$ is the heat function per unit volume. This is the required expression for the energy-momentum tensor.†

The components T^{ik} in three-dimensional form are

$$\left.\begin{aligned} T^{\alpha\beta} &= \frac{wv_\alpha v_\beta}{c^2(1-v^2/c^2)} + p\delta_{\alpha\beta}, \\ T^{0\alpha} &= \frac{wv_\alpha}{c(1-v^2/c^2)}, \qquad T^{00} = \frac{w}{1-v^2/c^2} - p = \frac{e+pv^2/c^2}{1-v^2/c^2}. \end{aligned}\right\} \tag{133.3}$$

The non-relativistic case is that of small velocities ($v \ll c$) and small velocities of the internal (microscopic) motion of the fluid particles. In passing to the limit it must be borne in mind that the relativistic internal energy e includes the rest energy nmc^2 of the fluid particles (m being the rest mass of one particle). It must also be remembered that the particle number density n is referred to unit proper volume; in non-relativistic expressions, however, the energy density is referred to unit volume in the laboratory frame, in which the fluid element concerned is in motion. We must therefore put, in the limit, $mn \to \rho\sqrt{(1-v^2/c^2)} \cong \rho - \rho v^2/2c^2$, where ρ is the ordinary non-relativistic mass density. Both the non-relativistic energy density $\rho\varepsilon$ and the pressure are small compared with ρc^2.

We thus find that the limiting value of T_{00} is $\rho c^2 + \rho\varepsilon + \frac{1}{2}\rho v^2$, i.e. it is ρc^2 together with the non-relativistic energy density. The corresponding limiting form of the tensor $T_{\alpha\beta}$ is $\rho v_\alpha v_\beta + p\delta_{\alpha\beta}$, i.e. it coincides, as it should, with the usual expression for the momentum flux density, which we have denoted in §7 by $\Pi_{\alpha\beta}$.

The simple relation between the momentum density and the energy flux density (divided by c^2) no longer holds in the non-relativistic limit, because the non-relativistic energy does not include the rest energy. The components $T^{0\alpha}/c$ form a three-dimensional vector, approximately equal to $\rho\mathbf{v} + (\rho\varepsilon + p + \frac{1}{2}\rho v^2)\mathbf{v}/c^2$. Hence we see that the limiting value of the momentum density is just $\rho\mathbf{v}$, as it should be; for the energy flux density we have, omitting the term $\rho c^2\mathbf{v}$, the expression $(\rho\varepsilon + p + \frac{1}{2}\rho v^2)\mathbf{v}$, in agreement with the result obtained in §6.

§134. The equations of relativistic fluid dynamics

The equations of motion are contained in

$$\partial T_i^k/\partial x^k = 0, \tag{134.1}$$

which expresses the laws of conservation of energy and momentum for the physical system to which the tensor T^{ik} pertains. Using the expression (133.2) for T^{ik}, we obtain the equations of fluid motion; it is necessary, however, to use also the law of conservation of numbers of particles, which is not contained in (134.1). The energy-momentum tensor (133.2) does not take account of any dissipative processes (including viscosity and thermal conduction), and therefore the equations relate to an ideal fluid.

To derive an equation for the conservation of particle numbers (the equation of continuity), we use the particle flux 4-vector n^i. Its time component is the number density of

† In all formulae in this chapter, the thermodynamic quantities have their proper values. Quantities such as e and w (and the entropy density σ below) are taken per unit volume in the local rest frame.

particles, and the three spatial components form the three-dimensional particle flux vector. It is evident that the vector n^i must be proportional to the 4-velocity u^i, so that

$$n^i = nu^i, \tag{134.2}$$

where n is a scalar; it is clear from the definition of n that this scalar is just the proper number density of particles.† The equation of continuity is obtained by simply equating to zero the 4-divergence of the flux vector:

$$\partial(nu^i)/\partial x^i = 0. \tag{134.3}$$

Let us now return to equation (134.1). Differentiating the expression (133.2), we obtain

$$\frac{\partial T_i^k}{\partial x^k} = u_i \frac{\partial (wu^k)}{\partial x^k} + wu^k \frac{\partial u_i}{\partial x^k} + \frac{\partial p}{\partial x^i} = 0. \tag{134.4}$$

We multiply this equation by u^i, i.e. project it on the direction of the 4-velocity. Since $u_i u^i = -1$, $u_i \partial u^i/\partial x^k = 0$, we find

$$\frac{\partial (wu^k)}{\partial x^k} - u^k \frac{\partial p}{\partial x^k} = 0. \tag{134.5}$$

With the identity $wu^k = nu^k(w/n)$ and the equation of continuity (134.3), we obtain

$$nu^k \left[\frac{\partial}{\partial x^k}\left(\frac{w}{n}\right) - \frac{1}{n}\frac{\partial p}{\partial x^k} \right] = 0.$$

By the thermodynamic relation

$$\mathrm{d}(w/n) = T\mathrm{d}(\sigma/n) + (1/n)\,\mathrm{d}p \tag{134.6}$$

(where T is the temperature and σ the entropy per unit proper volume),‡ the expression in square brackets is $T\partial(\sigma/n)/\partial x^k = 0$. Without the factor nT, we have

$$u^k \partial(\sigma/n)/\partial x^k \equiv \mathrm{d}(\sigma/n)/\mathrm{d}s = 0, \tag{134.7}$$

which states that the flow is adiabatic; $\mathrm{d}/\mathrm{d}s$ denotes differentiation along the world line of the fluid element concerned.

By the equation of continuity (134.3), equation (134.7) can also be written as the vanishing of the 4-divergence of the entropy flux σu^i:

$$\partial(\sigma u^i)/\partial x^i = 0. \tag{134.8}$$

† At very high temperatures, new particles may be formed in the substance, so that the total number of particles of all kinds is changed. In such cases n must be taken as a conserved macroscopic quantity describing the number of particles. For example, in the case of electron pair formation, n may be taken as the number of electrons that would remain if all pairs were annihilated. A convenient definition of n is the number density of baryons (that of antibaryons, if any, being taken as negative). However, ultra-relativistic fluid dynamics may also apply to problems in which there is no conserved macroscopic characteristic of the number of particles in the system, the latter being governed by the conditions of thermodynamic equilibrium; for example, problems involving the multiple formation of particles in collisions between fast nucleons. The derivation of the equations of fluid dynamics in such cases is discussed in Problem 2.

‡ Such a relation has to be written for a particular quantity of matter, not for a particular volume, which might contain a variable number of particles. In (134.6) it is written for the heat function per particle, and $1/n$ is the volume per particle.

We now project equation (134.1) on a direction perpendicular to u^i. This produces the combination†

$$\frac{\partial T_i^k}{\partial x^k} - u_i u^k \frac{\partial T_k^l}{\partial x^l} = 0;$$

the expression on the left gives zero identically on scalar multiplication by u^i. A simple calculation leads to the equation

$$wu^k \frac{\partial u_i}{\partial x^k} = \frac{\partial p}{\partial x^i} - u_i u^k \frac{\partial p}{\partial x^k}. \tag{134.9}$$

The three spatial components of this equation are the relativistic generalization of Euler's equation; the time component is implied by the other three.

Equation (134.9) can be put in a different form for the case of isentropic flow (similarly to the change from (2.3) to (2.9) with the non-relativistic Euler's equation). When σ/n = constant, (134.6) gives

$$\frac{\partial p}{\partial x^i} = n \frac{\partial}{\partial x^i} \frac{w}{n},$$

and (134.9) becomes

$$u^k \frac{\partial}{\partial x^k}\left(\frac{w}{n} u_i\right) = \frac{\partial}{\partial x^i} \frac{w}{n}. \tag{134.10}$$

If the flow is also steady, with all quantities independent of time, the spatial components of (134.10) give

$$\gamma(\mathbf{v}\cdot\mathbf{grad})(\gamma w\mathbf{v}/n) + c^2\,\mathbf{grad}\,(w/n) = 0.$$

Scalar multiplication by $\mathbf{v}$ readily leads to the result $(\mathbf{v}\cdot\mathbf{grad})(\gamma w/n) = 0$. It follows that along any streamline

$$\gamma w/n = \text{constant}. \tag{134.11}$$

This is the relativistic generalization of Bernoulli's equation.‡

Without assuming that the isentropic flow is steady, we can easily see that the equations (134.10) have solutions in the form

$$wu_i/n = -\partial\phi/\partial x^i, \tag{134.12}$$

where ϕ is a function of coordinates and time; these are the relativistic analogue of potential flow in non-relativistic fluid dynamics (I. M. Khalatnikov 1954). To verify this, we note that, from the symmetry of the derivatives $\partial^2\phi/\partial x^i \partial x^k$ in i and k,

$$\frac{\partial}{\partial x^k}(wu_i/n) = \frac{\partial}{\partial x^i}(wu_k/n);$$

† For the 4-velocity components we have (see *Fields*, §4)

$$u^i = (\gamma, \gamma\mathbf{v}/c),\ u_i = (\gamma, -\gamma\mathbf{v}/c),$$

where for brevity, in this chapter only, $\gamma = 1/\sqrt{(1-v^2/c^2)}$.

‡ For $v \ll c$, $w/n = mc^2 + mw_{\text{non-r}}$, where $w_{\text{non-r}}$ is the non-relativistic heat function per unit mass, denoted in §5 by w, and (134.11) becomes (5.3).

scalar multiplication by u^k and expansion of the derivative on the right does in fact bring us back to (134.10). The spatial and time components of (134.12) give

$$\gamma w \mathbf{v}/nc = \mathbf{grad}\,\phi, \qquad c\gamma w/n + \partial\phi/\partial t = 0.$$

The first of these becomes, in the non-relativistic limit, the ordinary condition for potential flow; the second becomes equation (9.3), with the change $\phi/cm \to \phi$.

Let us consider the propagation of sound in a substance having a relativistic equation of state (i.e. one in which the pressure is comparable with the internal energy density, including the rest energy). The equations of fluid dynamics for the sound waves can be linearized; it is convenient to start from the equations of motion in the original form (134.1), and not the equivalent form (134.8), (134.9). Substituting the expressions (133.3) for the components of the energy-momentum tensor and retaining only quantities of the same order of smallness as the wave amplitude, we obtain the equations

$$\partial e'/\partial t = -w\,\mathrm{div}\,\mathbf{v}, \qquad (w/c^2)\partial\mathbf{v}/\partial t = -\mathbf{grad}\,p', \tag{134.13}$$

where the prime denotes the variable parts of quantities. Eliminating $\mathbf{v}$, we find $\partial^2 e'/\partial t^2 = c^2 \triangle p'$. Finally, putting $e' = (\partial e/\partial p)_{\mathrm{ad}}\,p'$, we obtain the wave equation for p', with the velocity of sound†

$$u = c\sqrt{(\partial p/\partial e)_{\mathrm{ad}}}; \tag{134.14}$$

the suffix ad signifies that the derivative is taken for an adiabatic process, i.e. for constant σ/n. This formula differs from the corresponding non-relativistic expression in that the mass density is replaced by e/c^2. With the ultra-relativistic equation of state $p = \frac{1}{3}e$, the velocity of sound is $u = c/\sqrt{3}$.

Finally, let us discuss briefly the equations of fluid dynamics in the presence of significant gravitational fields, i.e. in the general theory of relativity. They are obtained from equations (134.8) and (134.9) by simply replacing the ordinary derivatives by the covariant ones:‡

$$wu^k u_{i;k} = -\partial p/\partial x^i - u_i u^k \partial p/\partial x^k, \qquad (\sigma u^i)_{;i} = 0. \tag{134.15}$$

From these equations we can derive the condition of mechanical equilibrium in a gravitational field. In equilibrium, the field is static; we can take a frame of reference in which the substance is at rest ($u^\alpha = 0$, $u^0 = 1/\sqrt{g_{00}}$), all quantities are independent of time, and the mixed components of the metric tensor are zero ($g_{0\alpha} = 0$). The spatial components of equation (134.15) then give

$$w\Gamma_{\alpha 0}{}^0 u^0 u_0 = \tfrac{1}{2}(w/g_{00})\partial g_{00}/\partial x^\alpha = -\partial p/\partial x^\alpha,$$

or

$$\frac{1}{w}\frac{\partial p}{\partial x^\alpha} = -\frac{1}{2}\frac{\partial}{\partial x^\alpha}\log g_{00}. \tag{134.16}$$

† Denoted in this chapter by u.

‡ In general, these equations are quite complicated. They are written in full, in terms of the three-dimensional space metric tensor ($\gamma_{\alpha\beta}$ in *Fields*, §84) by R. A. Nelson, *General Relativity and Gravitation* **13**, 569, 1981. The equations of fluid dynamics in the first approximation beyond the Newtonian are given by S. Chandrasekhar, *Astrophysical Journal* **142**, 1488, 1965, and by C. W. Misner, K. S. Thorne and J. A. Wheeler, *Gravitation*, San Francisco 1973, §39.11.

This is the required equation of equilibrium. In the non-relativistic limit $w = \rho c^2$, $g_{00} = 1 + 2\phi/c^2$ (ϕ being the Newtonian gravitational potential), and equation (134.16) becomes **grad** $p = -\rho$ **grad** ϕ, i.e. the usual equation of hydrostatics.

PROBLEMS

PROBLEM 1. Find the solution of the equations of relativistic fluid dynamics which describes a one-dimensional non-steady simple wave.

SOLUTION. In a simple wave, all quantities can be expressed as functions of any one of them (see §101). Writing the equations of motion in the form

$$\frac{\partial T_{00}}{c\,\partial t} - \frac{\partial T_{01}}{\partial x} = 0, \qquad \frac{\partial T_{01}}{c\,\partial t} - \frac{\partial T_{11}}{\partial x} = 0, \tag{1}$$

and supposing T_{00}, T_{01}, T_{11} to be functions of one another, we obtain $dT_{00}dT_{11} = (dT_{01})^2$. Here we must substitute $T_{00} = eu_0{}^2 + pu_1{}^2$, $T_{01} = wu_0u_1$, $T_{11} = eu_1{}^2 + pu_0{}^2$, using the fact that $u_1{}^2 - u_0{}^2 = -1$; it is convenient to introduce a parameter η such that $u_0 = \cosh\eta$, $u_1 = \sinh\eta$. The result is

$$\tanh^{-1}(v/c) = \pm(1/c)\int (u/w)\,de, \tag{2}$$

where u is the velocity of sound. Next, from (1) we find $\partial x/\partial t = c\,dT_{01}/dT_{00}$, and a calculation of the derivative gives

$$x = \frac{t(v \pm u)}{1 \pm uv/c^2} + f(v). \tag{3}$$

Formulae (2) and (3) give the required solution.

PROBLEM 2. Derive the equations of fluid dynamics for an ultra-relativistic medium with an indeterminate number of particles that is itself determined by the conditions of thermodynamic equilibrium.

SOLUTION. The condition of thermodynamic equilibrium which determines the number of particles in such a medium is that all the chemical potentials be zero. Then $e - T\sigma + p = 0$, i.e. $w = T\sigma$, and the thermodynamic expression for the differential of the heat function with a fixed (unit) volume and zero chemical potentials is $dw = Td\sigma + dp$; these two formulae give $dp = \sigma dT$.† Equation (134.5) (in which we have not yet used the continuity equation) gives the adiabatic equation in the form (134.8). Equation (134.9) becomes

$$u^k\partial(Tu_i)/\partial x^k = \partial T/\partial x^i.$$

§135. Shock waves in relativistic fluid dynamics

The theory of shock waves in relativistic fluid dynamics is constructed analogously to the non-relativistic theory (A. H. Taub 1948).

As in §85, we consider the surface of discontinuity in coordinates where it is at rest and the gas moves at right angles to it in the $x^1 \equiv x$-direction from side 1 to side 2. The conditions of continuity for the particle number, momentum and energy flux densities are

$$[n^x] = [nu^x] = 0, \qquad [T^{xx}] = [w(u^x)^2 + p] = 0,$$

$$c[T^{0x}] = c[wu^0u^x] = 0,$$

† With the ultra-relativistic equation of state $p = \frac{1}{3}e$, the above expressions readily lead to $e \propto T^4$, $\sigma \propto T^3$, the same as for black-body radiation (*SP* 1, §63), as we should expect.

or, after substitution of the 4-velocity components,

$$v_1\gamma_1/V_1 = v_2\gamma_2/V_2 \equiv j, \tag{135.1}$$

$$w_1 v_1^2\gamma_1^2/c^2 + p_1 = w_2 v_2^2 \gamma_2^2/c^2 + p_2, \tag{135.2}$$

$$w_1 v_1 \gamma_1^2 = w_2 v_2 \gamma_2^2, \tag{135.3}$$

where $\gamma_1 = 1/\sqrt{(1-v_1^2/c^2)}$, $\gamma_2 = 1/\sqrt{(1-v_2^2/c^2)}$; $V_1 = 1/n_1$ and $V_2 = 1/n_2$ are the volumes per particle.†

From (135.1) and (135.2),

$$j^2 = (p_2 - p_1)c^2/(w_1 V_1^2 - w_2 V_2^2). \tag{135.4}$$

We next use (135.1) to rewrite (135.3) as

$$w_1^2 V_1^2 \gamma_1^2 = w_2^2 V_2^2 \gamma_2^2.$$

We express γ_1^2 and γ_2^2 in terms of j^2 from (135.1) and then substitute j^2 from (135.4); simple algebra gives as the relativistic equation of the shock adiabatic (the *Taub adiabatic*)

$$w_1^2 V_1^2 - w_2^2 V_2^2 + (p_2 - p_1)(w_1 V_1^2 + w_2 V_2^2) = 0. \tag{135.5}$$

We shall also give expressions for the gas velocities on either side of the discontinuity; these can be derived by elementary algebra from (135.2) and (135.3):‡

$$\frac{v_1}{c} = \sqrt{\left[\frac{(p_2-p_1)(e_2+p_1)}{(e_2-e_1)(e_1+p_2)}\right]}, \quad \frac{v_2}{c} = \sqrt{\left[\frac{(p_2-p_1)(e_1+p_2)}{(e_2-e_1)(e_2+p_1)}\right]}. \tag{135.6}$$

The relative velocity of the gases on either side of the discontinuity is, according to the relativistic rule of velocity addition,

$$v_{12} = \frac{v_1 - v_2}{1 - v_1 v_2/c^2} = c\sqrt{\frac{(p_2-p_1)(e_2-e_1)}{(e_1+p_2)(e_2+p_1)}}. \tag{135.7}$$

In the non-relativistic limit, if we put $e \cong mc^2 n = mc^2/V$ and neglect p in comparison with e, equations (135.4), (135.6) and (135.7) become (85.4), (85.6) and (85.7), with the changed definitions of j and V (see the last footnote but one).†† The ultra-relativistic equation of state $p = \frac{1}{3}e$ and (135.6) give

$$\frac{v_1}{c} = \sqrt{\left[\frac{3e_2+e_1}{3(3e_1+e_2)}\right]}, \quad \frac{v_2}{c} = \sqrt{\left[\frac{3e_1+e_2}{3(3e_2+e_1)}\right]}; \tag{135.8}$$

$v_1 v_2 = \frac{1}{3}c^2$. As the shock wave becomes stronger ($e_2 \to \infty$), v_1 tends to the velocity of light and v_2 to $\frac{1}{3}c$.

As in Chapter IX we plotted the shock adiabatic in the Vp-plane, the natural variables for representing the relativistic shock adiabatic are wV^2 and pc^2; in these coordinates, j^2 gives the slope of the chord from the initial point 1 on the adiabatic to any other point 2.

---

† In the non-relativistic limit, the particle number flux defined by (135.1) differs by a factor $1/m$ from the mass flux density denoted by j in §85. The volumes V as defined here and in §85 differ by a factor m also.

‡ In the calculations, it is convenient to substitute $v/c = \tanh\phi$, $\gamma = \cosh\phi$.

†† In the passage to the limit from the adiabatic equation (135.5) to the non-relativistic equation (85.10), this approximation is insufficient; we have to put $w = nmc^2 + nm\varepsilon + p$ (where ε is the non-relativistic internal energy per unit mass), and take the limit as $c \to \infty$ after dividing (135.5) by c^2.

Weak relativistic shock waves can be treated in exactly the same way as for the non-relativistic case in §86 (I. M. Khalatnikov 1954). Without repeating the calculations, we shall give the result for the entropy discontinuity, which is a third-order small quantity relative to the pressure discontinuity:

$$\sigma_2 - \sigma_1 = \frac{1}{12}\left[\frac{1}{wV^2T}\left(\frac{\partial^2(wV^2)}{\partial p^2}\right)_{\text{ad}}\right]_1 (p_2 - p_1)^3. \tag{135.9}$$

Since we must have $\sigma_2 > \sigma_1$, we see that the shock wave is a compression wave if

$$(\partial^2(wV^2)/\partial p^2)_{\sigma V} > 0. \tag{135.10}$$

This is the relativistic generalization of the condition (86.2) in non-relativistic fluid dynamics.† When $p_2 > p_1$, we get from (135.4) and (135.5)

$$w_2 V_2^{\,2} < w_1 V_1^{\,2}, \qquad w_2 V_2 > w_1 V_1;$$

hence, in turn, it follows that we always have $V_2 < V_1$: the volume V must decrease even more rapidly than wV increases. The velocities v_1 and v_2 for a weak shock are, of course, equal to that of sound in the first approximation: since the entropy change is third-order, the expressions (135.6) become the derivative (134.14) when $p_2 \to p_1, e_2 \to e_1$.‡ Arguments entirely similar to those in §86 show that in the next approximation $v_1 > u_1$, $v_2 < u_2$.

Thus the direction of variation of quantities in a weak relativistic shock wave satisfies, subject to the condition (135.10), the same inequalities as in the non-relativistic case. The generalization to shock waves of any strength can be made in exactly the same way as in §87.††

At the same time, it should be emphasized that the inequalities $v_1 > u_1$ and $v_2 < u_2$ are valid for relativistic (as well as non-relativistic) shock waves, whatever the thermodynamic conditions, because of the need for the shock to be evolutionary. In the derivation of these inequalities (§88) only the sign of the sound disturbance propagation velocities $u \pm v$ in the moving fluid relative to the surface of discontinuity at rest was important. According to the relativistic rule of velocity addition, these are given by $(u \pm v)/(1 \pm uv/c^2)$, and the signs depend only on the numerators, so that the reasoning of §88 remains valid.

§136. Relativistic equations for flow with viscosity and thermal conduction

The finding of the relativistic equations of fluid dynamics in the presence of dissipative processes (viscosity and thermal conduction) amounts to determining the form of the additional terms in the energy-momentum tensor and in the particle flux density vector. Denoting these terms by τ_{ik} and v_i respectively, we write

$$T_{ik} = -pg_{ik} + wu_i u_k + \tau_{ik}, \tag{136.1}$$

† Using the thermodynamic relation for the heat function per particle, $\mathrm{d}(wV) = V\mathrm{d}p$ (when σV = constant), we find that the condition (135.10) is equivalent to

$$(\partial^2 V/\partial p^2)_{\text{ad}} > (3/w)|(\partial V/\partial p)_{\text{ad}}|.$$

In the non-relativistic limit, the right-hand side is replaced by zero.

‡ The expression (135.4) becomes the derivative $-c^2[\mathrm{d}p/\mathrm{d}(wV^2)]_1$. With the thermodynamic relations $\mathrm{d}(eV) = -p\mathrm{d}V$, $\mathrm{d}(wV) = V\mathrm{d}p$ (when σV = constant), we easily see that this derivative multiplied by $V_1^{\,2}$ is equal to $u_1^{\,2}/(1-u_1^{\,2})$, as it should be.

†† See K. S. Thorne, *Astrophysical Journal* **179**, 897, 1973.

$$n_i = nu_i + \nu_i. \tag{136.2}$$

The equations of motion are again contained in $\partial T_i^k/\partial x^k = 0$, $\partial n^i/\partial x^i = 0$.

First of all, however, we must discuss more closely the concept of the velocity u^i itself. In relativistic mechanics, an energy flux necessarily involves a mass flux. Hence, when there is (e.g.) a heat flux, the definition of the velocity in terms of the mass flux density (as in non-relativistic fluid dynamics) has no direct meaning. We now define the velocity by the condition that, in the proper frame of any given fluid element, the momentum of the element be zero and its energy expressible in terms of the other thermodynamic quantities by the same formulae as when dissipative processes are absent. This means that, in the proper frame, the components τ_{00} and $\tau_{0\alpha}$ of the tensor τ_{ik} are zero; since, in this frame, $u^\alpha = 0$ also, we have (in any frame) the tensor equation

$$\tau_{ik} u^k = 0. \tag{136.3}$$

A similar relation,

$$\nu_i u^i = 0, \tag{136.4}$$

must hold for the vector ν_i, since the component n^0 of the particle flux 4-vector n^i in the proper frame must, by definition, equal the particle number density n.

The required form of the tensor τ_{ik} and the vector ν_i can be established from the requirements of the law of increase of entropy. This law must be contained in the equations of motion (in the same way as the condition of constant entropy for an ideal fluid was obtained in §134 from these equations). By simple transformations, using the equation of continuity, we easily obtain the equation

$$u^i \frac{\partial T_i^k}{\partial x^k} = T\frac{\partial}{\partial x^i}(\sigma u^i) + \mu\frac{\partial \nu^i}{\partial x^i} + u^i\frac{\partial \tau_i^k}{\partial x^k},$$

where $\mu = (w - T\sigma)/n$ is the relativistic chemical potential, and we have used the thermodynamic relation

$$\mathrm{d}\mu = (1/n)\,\mathrm{d}p - (\sigma/n)\,\mathrm{d}T \tag{136.5}$$

for its differential. Finally, using the relation (136.3), we can rewrite this equation as

$$\frac{\partial}{\partial x^i}\left(\sigma u^i - \frac{\mu}{T}\nu^i\right) = -\nu^i\frac{\partial}{\partial x^i}\left(\frac{\mu}{T}\right) + \frac{\tau_i^k}{T}\frac{\partial u^i}{\partial x^k}. \tag{136.6}$$

The expression on the left must be the 4-divergence of the entropy flux, and that on the right the increase in entropy owing to dissipative processes. Thus the entropy flux density 4-vector is

$$\sigma^i = \sigma u^i - (\mu/T)\nu^i, \tag{136.7}$$

and τ_{ik} and ν^i must be linear functions of the gradients of velocity and thermodynamic quantities, such as to make the right-hand side of equation (136.6) necessarily positive. This condition, together with (136.3) and (136.4), uniquely determines the form of the symmetrical 4-tensor τ_{ik} and the 4-vector ν_i:

$$\tau_{ik} = -c\eta\left(\frac{\partial u_i}{\partial x^k} + \frac{\partial u_k}{\partial x^i} - u_k u^l\frac{\partial u_i}{\partial x^l} - u_i u^l\frac{\partial u_k}{\partial x^l}\right) - c(\zeta - \tfrac{2}{3}\eta)\frac{\partial u^l}{\partial x^l}(g_{ik} - u_i u_k), \tag{136.8}$$

$$\nu_i = -\frac{\kappa}{c}\left(\frac{nT}{w}\right)^2\left[\frac{\partial}{\partial x^i}\left(\frac{\mu}{T}\right) - u_i u^k\frac{\partial}{\partial x^k}\left(\frac{\mu}{T}\right)\right]. \tag{136.9}$$

Here η and ζ are the two viscosity coefficients, and κ the thermal conductivity, taken in accordance with their non-relativistic definitions. In the non-relativistic limit, the $\tau_{\alpha\beta}$ reduce to the components of the three-dimensional viscous stress tensor $\sigma'_{\alpha\beta}$ (15.3).

Pure thermal conduction corresponds to an energy flux with no particle flux. The condition of zero particle flux is $nu^\alpha + v^\alpha = 0$. The spatial components $u^\alpha = -v^\alpha/n$ of the 4-velocity are of the first order in the gradients; since the expressions (136.8) and (136.9) are written only as far as this order, the 4-velocity component u^0 must be taken as unity: $u_0{}^2 = 1 + u_\alpha u^\alpha = 1 + v_\alpha v^\alpha/n^2 \cong 1$. To the same accuracy, we must omit the second term in the square brackets in (136.9). The energy flux density is then

$$cT^{0\alpha} = -cT_\alpha^{\ 0} = -cwu_\alpha u^0 = cwv_\alpha/n$$

$$= \frac{\kappa nT^2}{w^2}\frac{\partial}{\partial x^\alpha}\left(\frac{\mu}{T}\right).$$

Using the thermodynamic relation (136.5) in the form

$$\mathrm{d}(\mu/T) = -(w/nT^2)\mathrm{d}T + (1/nT)\mathrm{d}p,$$

we find the energy flux

$$-\kappa[\mathbf{grad}\, T - (T/w)\,\mathbf{grad}\, p]. \tag{136.10}$$

It is seen that, in the relativistic case, the thermal conduction heat flux is proportional, not just to the temperature gradient, but to a certain combination of the temperature and pressure gradients; in the non-relativistic limit, $w \cong nmc^2$, and the **grad** p term is to be omitted.

CHAPTER XVI

DYNAMICS OF SUPERFLUIDS

§137. Principal properties of superfluids

At temperatures close to absolute zero, quantum effects begin to be of primary importance in the properties of fluids, which are then called *quantum fluids*. In reality, only helium remains a fluid down to absolute zero; all other liquids solidify well before quantum effects become noticeable in them. There are, however, two helium isotopes, He^4 and He^3, whose atoms obey different statistics. The He^4 nucleus has no spin, and the spin of the whole atom is therefore also zero; these atoms obey Bose–Einstein statistics. The He^3 atoms have a spin $\frac{1}{2}$ due to the nucleus, and obey Fermi–Dirac statistics. This difference has a fundamental effect on the properties of the corresponding quantum liquids, which are called *Bose liquids* and *Fermi liquids* respectively. Only the former will be discussed in this chapter.

At 2·19 K, liquid He^4 has a second-order phase transition at the *λ-point*.† Below this point, liquid helium (there called helium II) has several remarkable properties, the most significant of which is *superfluidity* (discovered by P. L. Kapitza in 1938), the property of flowing without viscosity in narrow capillaries or gaps.

The theory of superfluidity was developed by L. D. Landau (1941). The microscopic theory is given in *SP* 2, Chapter III. Here we shall consider only macroscopic superfluid dynamics, which can be derived from the principles of the microscopic theory.‡

The basis of the dynamics of helium II is the following fundamental result of the microscopic theory. At temperatures other than zero, helium II behaves as if it were a mixture of two different liquids. One of these is a superfluid, and moves with zero viscosity along a solid surface. The other is a normal viscous fluid. It is of great importance that no friction occurs between these two parts of the liquid in their relative motion, i.e. no momentum is transferred from one to the other.

It should, however, be most decidedly emphasized that regarding the liquid as a mixture of normal and superfluid parts is no more than a convenient description of the phenomena which occur in a quantum fluid. Like any description of quantum phenomena in classical terms, it falls short of adequacy. In reality, we ought to say that a quantum fluid, such as helium II, can execute two motions at once, each of which involves its own effective mass (the sum of the two effective masses being equal to the actual total mass of the fluid). One of these motions is normal, i.e. has the same properties as the motion of an ordinary viscous fluid, but the other is the motion of a superfluid. The two motions occur without any

† Such points form a curve in the phase diagram of helium in the pT-plane. The temperature 2·19 K is the intersection of this curve with the liquid–vapour equilibrium curve.

‡ The He^3 Fermi liquid also becomes superfluid, but at much lower temperatures, $\sim 10^{-3}$ K. The dynamics of this superfluid is more complex, on account of the complexity of the order parameter which describes its state (cf. *SP* 2, §54).

transfer of momentum from one to the other. We can, in a certain sense, speak of the superfluid and normal parts of the fluid mass, but this does not mean that the fluid can actually be separated into two such parts.†

With careful note taken of these reservations concerning the true nature of the phenomena in helium II, we can use the terms *superfluid part* and *normal part* of the fluid to give a convenient concise description of these phenomena. We shall, however, prefer to use the more exact terms *superfluid flow* and *normal flow*, without associating them with the components of a "mixture of two parts" of the fluid.

The concept of two kinds of flow enables us to give a simple explanation of the main observed dynamical properties of helium II. The absence of viscosity when helium II flows in a narrow passage is the result of frictionless superfluid flow in the passage; we can say that the normal part remains in the vessel, flowing much more slowly through the passage at a velocity in accordance with its viscosity and the passage width. The measurement of the viscosity of helium II from the damping of torsional oscillations of a disk immersed in it, on the other hand, gives non-zero values; the rotation of the disk causes a normal flow near it, which brings the disk to rest by virtue of the viscosity pertaining to that flow. Thus, in experiments on flow through a capillary or gap, the superfluid flow is observed, whereas in experiments on the rotation of a disk in helium II the normal flow is observed.

Besides the absence of viscosity, the superfluid flow has two other important properties: it does not involve heat transfer, and it is always potential flow. Both these properties also follow from the microscopic theory, according to which the normal flow is actually the flow of an "excitation gas" (we may recall that the collective thermal motion of the atoms in a quantum fluid can be regarded as a system of excitations, which behave like quasi-particles moving in the volume occupied by the fluid and have definite momenta and energies).

The entropy of helium II is determined by the statistical distribution of the elementary excitations. In any flow, therefore, in which the excitation gas is at rest, there is no macroscopic transfer of entropy. This means that the superfluid flow involves no entropy transfer, and therefore no heat transfer. Hence it follows that a superfluid flow of helium II is thermodynamically reversible.

The transfer of heat by the normal flow is the only mechanism of heat transfer in helium II. It is therefore of the nature of convection, and is fundamentally different from ordinary thermal conduction. Any difference of temperature in helium II causes internal flow, both normal and superfluid; the two flows may balance as regards mass transfer, so that no macroscopic mass transfer occurs in the fluid.

In what follows we shall denote by $\mathbf{v}_s$ and $\mathbf{v}_n$ the velocities of the superfluid and normal flow respectively. The heat-transfer mechanism described above means that the entropy flux density is the product $\mathbf{v}_n \rho s$ of the velocity $\mathbf{v}_n$ and the entropy per unit volume (s being the entropy per unit mass). The heat flux density is obtained by multiplying the entropy flux density by T, i.e. it is

$$\mathbf{q} = \rho T s \mathbf{v}_n. \tag{137.1}$$

† Independently of Landau's work, the qualitative idea of macroscopically describing helium II by separating its density into two parts and using two velocity distributions was put forward by L. Tisza (1940), and it enabled him to predict also the existence of two kinds of sound waves in helium II (see §141 below). However, the basic microscopic ideas were incorrect, and Tisza's papers do not contain a consistent theory of superfluidity or superfluid dynamics.

The potential flow of the superfluid part corresponds to the equation

$$\mathbf{curl}\, \mathbf{v}_s = 0, \tag{137.2}$$

which must hold at any instant throughout the volume of the fluid. This property is the macroscopic expression of the property of the helium II energy spectrum which underlies the microscopic theory of superfluidity: the elementary excitations which have long wavelengths (i.e. small momenta and energies) are sound quanta or *phonons.* Hence the macroscopic superfluid dynamics can include only sound vibrations, a result which follows from the condition (137.2).†

Since it is potential flow, a steady superfluid flow exerts no force on a solid body (d'Alembert's paradox; see §11). The normal flow, on the other hand, exerts a drag force. If the flow is such that the superfluid and normal mass transfers balance, we have a very unusual flow: a force acts on a body immersed in helium II, but there is no net mass transfer.

PROBLEM

A small temperature difference ΔT is set up between the ends of a capillary containing helium II. Determine the heat flux along the capillary.

SOLUTION. According to (138.3), the pressure difference between the two ends of the capillary is $\Delta p = \rho s \Delta T$. This difference produces in the capillary a normal flow whose velocity averaged over the cross-section is $\bar{v}_n = R^2 \Delta p / 8\eta l$, where R and l are the radius and length of the capillary, η the viscosity for normal flow (cf. (17.10)). The total heat flux is $T\rho s \bar{v}_n \pi R^2 = T\pi R^4 \rho^2 s^2 \Delta T / 8\eta l$. There is a superfluid flow in the opposite direction, whose velocity is determined by the condition that there be no net mass transfer: $v_s = -\bar{v}_n \rho_n / \rho_s$.

§138. The thermo-mechanical effect

The *thermo-mechanical effect* in helium II is as follows: when helium flows out of a vessel through a narrow capillary, a rise in temperature occurs in the vessel, and a cooling where the helium flows out of the capillary into another vessel.‡ This phenomenon has the natural explanation that the flow into a capillary is mainly superfluid, and therefore transfers no heat, so that the heat remaining in the vessel is distributed over a smaller quantity of helium II. In flow out of a capillary the opposite effect is seen.

It is easy to find the quantity of heat Q absorbed when unit mass of helium enters a vessel through a capillary. The incoming fluid transfers no entropy. If the helium in the vessel were to remain at its initial temperature T, an amount of heat Ts would be needed, to compensate the decrease in entropy per unit mass due to the addition of unit mass of helium with zero entropy. This means that, when unit mass of helium enters a vessel containing helium at temperature T, an amount of heat

$$Q = Ts \tag{138.1}$$

is absorbed. Conversely when unit mass of helium leaves a vessel containing helium at temperature T, an amount of heat Ts is evolved.

† A more complete microscopic proof is given in *SP* 2, §26.

‡ A very slight thermo-mechanical effect must, strictly speaking, occur for any fluid; the anomaly in helium II is the magnitude of the effect. The effect in ordinary fluids is an irreversible phenomenon similar to the thermo-electric Peltier effect (and is actually observed in rarefied gases; see *PK*, §14, Problem 1). Such an effect occurs in helium II also, but is masked by another considerably larger effect described below, which occurs only in helium II and is not an irreversible phenomenon like the Peltier effect.

Let us now consider two vessels containing helium II at temperatures T_1 and T_2, connected by a narrow capillary. Since the superfluid can flow freely along the capillary, mechanical equilibrium is rapidly established. The superfluid, however, does not transfer heat, and so thermal equilibrium (in which the temperature of the helium in the two vessels is the same) is established considerably more slowly.

The condition of mechanical equilibrium is easily written down by using the fact that this equilibrium is established for constant entropies s_1, s_2 of the helium in the two vessels. If ε_1, ε_2 are the internal energies per unit mass of helium at temperatures T_1, T_2, the condition of mechanical equilibrium (minimum energy) effected by superfluid flow is $(\partial\varepsilon_1/\partial N)_{s_1} = (\partial\varepsilon_2/\partial N)_{s_2}$, where N is the number of atoms in unit mass of helium. The derivative $(\partial\varepsilon/\partial N)_s$ is the chemical potential μ. We therefore obtain the equilibrium condition

$$\mu(p_1, T_1) = \mu(p_2, T_2), \tag{138.2}$$

where p_1 and p_2 are the pressures in the two vessels.

In what follows we shall understand by the chemical potential μ not the usual thermodynamic potential (Gibbs free energy) per particle (atom), but the thermodynamic potential per unit mass of helium. These differ only by a constant factor, the mass of a helium atom.

If the pressures p_1, p_2 are small, then, expanding in powers of the pressures and recalling that $(\partial\mu/\partial p)_T$ is the specific volume (which depends only slightly on the temperature), we obtain

$$\frac{\Delta p}{\rho} = \mu(0, T_1) - \mu(0, T_2) = \int_{T_1}^{T_2} s\,dT,$$

where $\Delta p = p_2 - p_1$. If the temperature difference $\Delta T = T_2 - T_1$ is also small, then, expanding in powers of ΔT and recalling that $(\partial\mu/\partial T)_p = -s$, we obtain

$$\Delta p/\Delta T = \rho s \tag{138.3}$$

(H. London 1939). Since $s > 0$, $\Delta p/\Delta T > 0$.

§139. The equations of superfluid dynamics

We shall now derive a complete system of equations describing macroscopically (phenomenologically) the flow of helium II. From the above discussion, we are concerned with equations of motion which involve at every point two velocities $\mathbf{v}_s$ and $\mathbf{v}_n$, and not one as in ordinary fluid dynamics. It is found that the required system of equations can be uniquely determined simply from the requirements imposed by Galileo's relativity principle and by the necessary conservation laws (using also the properties of the flow expressed by equations (137.1) and (137.2)).

It should be borne in mind that helium II actually ceases to be superfluid at high velocities. This phenomenon of *critical velocities* means that the equations of superfluid dynamics for helium are physically significant only when the velocities v_s and v_n are not too large.† Nevertheless, we shall first derive these equations without making any assumptions concerning the velocities $\mathbf{v}_s$ and $\mathbf{v}_n$, since, if higher powers of the velocities are neglected, the equations cannot be consistently derived from the conservation laws. The transition to the physically significant case of small velocities will be made in the final equations.

We denote by $\mathbf{j}$ the mass flux density; this quantity is also the momentum of unit volume (cf. the footnote to §49). We write

$$\mathbf{j} = \rho_s \mathbf{v}_s + \rho_n \mathbf{v}_n \tag{139.1}$$

as a sum of the fluxes pertaining to the superfluid and normal flows. The coefficients ρ_s and ρ_n may be called the superfluid and normal densities. Their sum is the actual density ρ of helium II:

$$\rho = \rho_s + \rho_n. \tag{139.2}$$

The quantities ρ_s and ρ_n are, of course, functions of the temperature; ρ_n vanishes at absolute zero, where helium II becomes wholly superfluid,‡ while ρ_s vanishes at the λ-point, where the liquid becomes wholly normal.

The density ρ and the flux $\mathbf{j}$ must satisfy the equation of continuity

$$\partial\rho/\partial t + \operatorname{div} \mathbf{j} = 0, \tag{139.3}$$

which expresses the law of conservation of mass. The law of conservation of momentum gives an equation

$$\frac{\partial j_i}{\partial t} + \frac{\partial \Pi_{ik}}{\partial x_k} = 0, \tag{139.4}$$

where Π_{ik} is the momentum flux density tensor.

We shall not at present consider dissipative processes. Then the flow is reversible, and the entropy of the fluid is also conserved. Since the entropy flux is $\rho s \mathbf{v}_n$, we can write the law of conservation of entropy as

$$\partial(\rho s)/\partial t + \operatorname{div}(\rho s \mathbf{v}_n) = 0. \tag{139.5}$$

Equations (139.3)–(139.5) must be supplemented by an equation which gives the time derivative of the velocity $\mathbf{v}_s$. This equation must be such that we have potential flow at all times; this means that the derivative of $\mathbf{v}_s$ must be the gradient of a scalar. We can write the equation as

$$\frac{\partial \mathbf{v}_s}{\partial t} + \mathbf{grad}\left(\tfrac{1}{2}v_s^2 + \mu\right) = 0, \tag{139.6}$$

where μ is some scalar.

Equations (139.4) and (139.6) become significant, of course, only when we obtain values for the still undefined quantities Π_{ik} and μ. To do so, we must use the law of conservation of

† The existence of a limiting velocity for superfluid flow follows from the microscopic theory: the specific form of the elementary excitation energy spectrum in helium II means that the Landau condition is not satisfied at high velocities (see *SP* 2, §23). The observed critical velocities, however, are well below this limit, and depend on the particular conditions of the flow, e.g. they are higher for flow in narrow passages than in large volumes. Physically, these phenomena involve the formation of quantized vortices; similar (but straight) vortex filaments are formed when liquid helium rotates in a cylindrical vessel (see *SP* 2, §29). Such effects will not be considered in the present chapter.

‡ If the helium II contains an admixture (of the isotope He^3), then ρ_n is not zero even at 0 K.

energy and arguments based on Galileo's relativity principle. The equations (139.3)–(139.6) must imply the law of conservation of energy, which is expressed by an equation of the form

$$\partial E/\partial t + \operatorname{div} \mathbf{Q} = 0, \tag{139.7}$$

where E is the energy in unit volume of the fluid and $\mathbf{Q}$ the energy flux density. Galileo's relativity principle enables us to determine all quantities as functions of one velocity ($\mathbf{v}_s$) and the given relative velocity $\mathbf{v}_n - \mathbf{v}_s$ of the two simultaneous flows.

We use both the original coordinate system K and a system K_0 in which the velocity of the superfluid flow of a given fluid element is zero. The system K_0 moves relative to K with a velocity equal to the superfluid velocity $\mathbf{v}_s$ in the original system. The values of all quantities in K are related to their values in K_0 (which we distinguish by the suffix 0) by the following transformation formulae of mechanics:†

$$\left.\begin{aligned} \mathbf{j} &= \rho\mathbf{v}_s + \mathbf{j}_0, \\ E &= \tfrac{1}{2}\rho v_s^2 + \mathbf{j}_0\cdot\mathbf{v}_s + E_0, \\ \mathbf{Q} &= (\tfrac{1}{2}\rho v_s^2 + \mathbf{j}_0\cdot\mathbf{v}_s + E_0)\mathbf{v}_s + \tfrac{1}{2}v_s^2\mathbf{j}_0 + \mathbf{\Pi}_0\cdot\mathbf{v}_s + \mathbf{Q}_0, \\ \Pi_{ik} &= \rho v_{si}v_{sk} + v_{si}j_{0k} + v_{sk}j_{0i} + \Pi_{0ik}. \end{aligned}\right\} \tag{139.8}$$

Here $\mathbf{\Pi}_0\cdot\mathbf{v}_s$ denotes the vector whose components are $\Pi_{0ik}v_{sk}$.

In the system K_0, the fluid element considered executes only one motion, a normal flow with velocity $\mathbf{v}_n - \mathbf{v}_s$. Hence the quantities $\mathbf{j}_0$, E_0, $\mathbf{Q}_0$ and Π_{0ik} can depend only on the difference $\mathbf{v}_n - \mathbf{v}_s$, and not on $\mathbf{v}_n$ and $\mathbf{v}_s$ separately; in particular, the vectors $\mathbf{j}_0$ and $\mathbf{Q}_0$ must be parallel to the vector $\mathbf{v}_n - \mathbf{v}_s$. Thus formulae (139.8) give the dependence of the quantities concerned on $\mathbf{v}_s$ for given $\mathbf{v}_n - \mathbf{v}_s$.

The energy E_0, as a function of ρ, s and the momentum $\mathbf{j}_0$ per unit volume, satisfies the thermodynamic relation

$$\mathrm{d}E_0 = \mu\,\mathrm{d}\rho + T\,\mathrm{d}(\rho s) + (\mathbf{v}_n - \mathbf{v}_s)\cdot\mathrm{d}\mathbf{j}_0, \tag{139.9}$$

where μ is the chemical potential (thermodynamic potential per unit mass). The first two terms correspond to the usual thermodynamic relation for a fluid at rest with constant volume (in this case unity), and the last term shows that the derivative of the energy with respect to the momentum is the velocity.

The momentum $\mathbf{j}_0$ (the mass flux density in K_0) is evidently just $\mathbf{j}_0 = \rho_n(\mathbf{v}_n - \mathbf{v}_s)$; the first of (139.8) is then the same as (139.1).

The subsequent calculations are as follows in general outline. In the equation of conservation of energy (139.7) we substitute E and $\mathbf{Q}$ from (139.8), calculating the derivative $\partial E_0/\partial t$ in terms of ρ, ρs and $\mathbf{j}_0$ by means of the identity (139.9). We then eliminate

† These formulae are a direct consequence of Galileo's relativity principle, and therefore hold for any system. They can be derived by considering, for instance, an ordinary fluid. The momentum flux density tensor in ordinary fluid dynamics is $\Pi_{ik} = \rho v_i v_k + p\delta_{ik}$. The fluid velocity $\mathbf{v}$ in the system K is related to the velocity $\mathbf{v}_0$ in K_0 by $\mathbf{v} = \mathbf{v}_0 + \mathbf{u}$, where $\mathbf{u}$ is the relative velocity of the two systems. Substituting in Π_{ik}, we have

$$\Pi_{ik} = p\delta_{ik} + \rho v_{0i}v_{0k} + \rho v_{0i}u_k + \rho u_i v_{0k} + \rho u_i u_k.$$

Putting $\Pi_{0ik} = p\delta_{ik} + \rho v_{0i}v_{0k}$ and $\mathbf{j}_0 = \rho\mathbf{v}_0$, we obtain the transformation formula for the tensor Π_{ik} given in (139.8). The remaining formulae are obtained similarly.

all the time derivatives ($\dot{\rho}$, $\dot{\mathbf{v}}_s$, etc.) by means of the equations (139.3)–(139.6). The fairly laborious calculations lead, after considerable cancelling of terms, to the result

$$-\Pi_{0ik}\frac{\partial v_{si}}{\partial x_k}+w_i\frac{\partial}{\partial x_k}\Pi_{0ik}+p\,\mathrm{div}\,\mathbf{v}_s-\mathbf{w}\cdot\mathbf{grad}\,p+\rho_n\mathbf{w}\cdot(\mathbf{w}\cdot\mathbf{grad})\mathbf{v}_n+$$

$$+\mathrm{div}[\mathbf{w}(T\rho s+\rho_n\mu)]+(\rho_n-\rho s)\mathbf{w}\cdot\mathbf{grad}(\phi-\mu)=\mathrm{div}\,\mathbf{Q}_0;$$

the scalar in (139.6) is here temporarily denoted by ϕ in place of μ, and for brevity $\mathbf{w} = \mathbf{v}_n - \mathbf{v}_s$; also,

$$p = -E_0 + T\rho s + \mu\rho + \rho_n(\mathbf{v}_n - \mathbf{v}_s)^2 \tag{139.10}$$

(the significance of this quantity will appear later). This equation of energy conservation must be satisfied identically, with $\mathbf{Q}_0, \Pi_0$ and ϕ depending only on the thermodynamic variables and on the velocity $\mathbf{w}$, not on any of their gradients (since dissipative processes are not being considered). These conditions uniquely determine the choice of expressions for $\mathbf{Q}_0, \Pi_0$ and ϕ.

First, we must put $\phi = \mu$; that is, the scalar in (139.6) is the same as the chemical potential of the fluid defined by (139.9) (and has therefore been denoted by the same letter). The other quantities are

$$\mathbf{Q}_0 = (T\rho s + \rho_n\mu)\mathbf{w} + \rho_n w^2\mathbf{w},$$

$$\Pi_{0ik} = p\delta_{ik} + \rho_n w_i w_k.$$

Substituting these expressions in formulae (139.8), we obtain the following final expressions for the energy flux density and the momentum flux density tensor:

$$\mathbf{Q} = (\mu + \tfrac{1}{2}v_s^2)\mathbf{j} + T\rho s\mathbf{v}_n + \rho_n\mathbf{v}_n(\mathbf{v}_n^2 - \mathbf{v}_n\cdot\mathbf{v}_s), \tag{139.11}$$

$$\Pi_{ik} = \rho_n v_{ni}v_{nk} + \rho_s v_{si}v_{sk} + p\delta_{ik}. \tag{139.12}$$

The expression (139.12) is in form a natural generalization of the formula $\Pi_{ik} = \rho v_i v_k + p\delta_{ik}$ in ordinary fluid dynamics. The quantity p defined by (139.10) may naturally be regarded as the fluid pressure; in a fluid entirely at rest, (139.10) is of course the same as is usually defined, since $\Phi = \mu\rho$ becomes the ordinary thermodynamic potential per unit volume of fluid.†

Equations (139.3)–(139.6), with $\mathbf{j}$ and Π_{ik} defined by (139.1) and (139.12), form the required complete equations of fluid dynamics. They are very complicated, mainly because the quantities ρ_s, ρ_n, μ and s in them are functions not only of the thermodynamic variables p and T but also of $w^2 = (\mathbf{v}_n - \mathbf{v}_s)^2$, the squared relative velocity of the two flows. The latter is a scalar invariant under Galilean transformations of the reference frame and under rotation of the fluid as a whole; it is specific to a superfluid, need not be zero in thermodynamic equilibrium, and must appear in the equation of state of the fluid along with p and T.

---

† The usual thermodynamic definition of the pressure as the mean force acting on unit area relates to a medium at rest. In ordinary fluid dynamics, however, there is no ambiguity in the definition of pressure (dissipative processes being neglected), since we can always take a coordinate system in which the fluid volume element concerned is at rest. In superfluid dynamics, however, we can eliminate only one of the two simultaneous motions by a suitable choice of the coordinate system, and so the usual definition of pressure cannot be applied.

The expression (139.10) corresponds to defining the pressure as $p = -\partial(E_0 V)/\partial V$, the derivative of the total energy of the fluid with given values of its total mass ρV, total entropy $\rho s V$ and total momentum $\rho\mathbf{w}V$ of the relative motion.

The equations are, however, much simplified in the physically interesting case where the velocities are small (compared with that of second sound; see §141).

In this case, firstly, we can neglect the dependence of ρ_s and ρ_n on $\mathbf{w}$; the expression (139.1) for $\mathbf{j}$ is then essentially the first terms in an expansion in powers of $\mathbf{v}_n$ and $\mathbf{v}_s$. A similar expansion is needed for the other thermodynamic quantities which appear in the equations.

Differentiating (139.10) and using (139.9), we obtain the following expression for the differential of the chemical potential:

$$d\mu = -s\,dT + \frac{1}{\rho}dp - (\rho_n/\rho)\mathbf{w}\cdot d\mathbf{w}. \tag{139.13}$$

This shows that the first two terms in the expansion of μ in powers of $\mathbf{w}$ are

$$\mu(p, T, \mathbf{w}) \cong \mu(p, T) - \tfrac{1}{2}(\rho_n/\rho)w^2, \tag{139.14}$$

where the right-hand side involves the ordinary chemical potential $\mu(p, T)$ and density $\rho(p, T)$ of the fluid at rest. Differentiating this with respect to temperature and pressure, we find the corresponding expansions for the entropy and the density:

$$\left.\begin{aligned} s(p, T, \mathbf{w}) &\cong s(p, T) + \tfrac{1}{2}w^2 \frac{\partial}{\partial T}(\rho_n/\rho), \\ \rho(p, T, \mathbf{w}) &\cong \rho(p, T) + \tfrac{1}{2}\rho^2 w^2 \frac{\partial}{\partial p}(\rho_n/\rho). \end{aligned}\right\} \tag{139.15}$$

These are to be substituted in the equations of fluid dynamics, which will then be valid as far as terms of the second order in the velocities. The inclusion in $\mathbf{j}$ of the w^2 dependence of ρ_s and ρ_n would give third-order terms.†

The inclusion in the dynamical equations of terms which take account of dissipative processes in superfluids will be dealt with in §140. Here, we shall formulate the boundary conditions on these equations.

Firstly, the perpendicular component of the mass flux $\mathbf{j}$ must vanish at any solid surface at rest. To determine the conditions on $\mathbf{v}_n$, we must recall that the normal flow is actually a flow of a thermal excitation gas. In flow along a solid surface, the excitation quanta interact with the surface, and this must be described macroscopically as the "adhesion" of the normal fluid to the surface, as in ordinary viscous fluids. In other words, the tangential component of the velocity $\mathbf{v}_n$ must be zero at a solid surface.

The component of $\mathbf{v}_n$ perpendicular to the surface need not vanish, since the excitation quanta can be absorbed or emitted by the surface, corresponding simply to heat transfer between the fluid and the surface. The boundary condition requires only that the heat flux perpendicular to the surface be continuous. The temperature itself has a discontinuity at

† Note that the equations of fluid dynamics with ρ_s regarded as a given function of p and T may become unsuitable near the λ-point. The reason is that, as this point (or any second-order phase transition) is approached, there is an unlimited increase in the relaxation time to establish the equilibrium value of the order parameter and in the correlation length of its fluctuations; but in superfluid He^4 the order parameter is the condensate wave function, whose squared modulus determines ρ_s; see *SP* 2, §§26, 28, and *PK* §103 regarding relaxation in superfluids. The fluid dynamics equations with a given $\rho_s(p, T)$ are valid only so long as the characteristic distances and times of the flow are much larger than the correlation length and the relaxation time respectively. Otherwise, the complete equations must include also those which determine ρ_s; see V. L. Ginzburg and A. A. Sobyanin, *Soviet Physics Uspekhi* **19**, 773, 1977; *Journal of Low Temperature Physics* **49**, 507, 1982.

the boundary which is proportional to the heat flux: $\Delta T = Kq$, with a proportionality coefficient which depends on the properties of both the fluid and the solid. The occurrence of this discontinuity is due to the peculiar nature of heat transfer in helium II. All the resistance to heat transfer between the solid and the fluid is in the fluid adjoining the surface, since the convective propagation of heat in the fluid meets with almost no resistance. Consequently, the whole of the temperature drop which causes the heat transfer occurs at the surface itself.

An interesting property of these boundary conditions is that the heat exchange between the solid surface and the moving fluid results in tangential forces on the surface. If the x-axis is perpendicular to the surface, and the y-axis tangential, the tangential force per unit area is equal to the component Π_{xy} of the momentum flux tensor. Since we must have $j_x = \rho_n v_{nx} + \rho_s v_{sx} = 0$ on the surface, we find for this surface the non-zero expression $\Pi_{xy} = \rho_s v_{sx} v_{sy} + \rho_n v_{nx} v_{ny} = \rho_n v_{nx}(v_{ny} - v_{sy})$. We can write this in terms of the heat flux $\mathbf{q} = \rho s T \mathbf{v}_n$ as

$$\Pi_{xy} = (\rho_n/\rho s T) q_x (v_{ny} - v_{sy}), \tag{139.16}$$

where q_x is the heat flux from the solid surface to the fluid, which is continuous at the surface.

In the absence of heat transfer between the solid surface and the fluid, the component of $\mathbf{v}_n$ perpendicular to the surface is also zero. The boundary conditions $j_x = 0$ and $\mathbf{v}_n = 0$ (with the x-axis perpendicular to the surface) are equivalent to $v_{sx} = 0$ and $\mathbf{v}_n = 0$. In this case, therefore, we obtain the usual boundary conditions for an ideal fluid for $\mathbf{v}_s$, and those for a viscous fluid for $\mathbf{v}_n$.

Finally, let us say a few words about the dynamics of liquid He^4 containing some other substance (in practice, the isotope He^3). In addition to the equations expressing the conservation of mass, momentum, entropy and the potential flow, the complete equations for the fluid dynamics of the mixture must include one which expresses the conservation of each of the two substances separately. This is

$$\partial(\rho c)/\partial t + \operatorname{div} \mathbf{i} = 0,$$

where c is the mass content of He^3 in the mixture and $\mathbf{i}$ its mass flux density. However, the requirements imposed by the conservation laws and by Galilean invariance are sufficient to establish the form of all the equations only if an expression for $\mathbf{i}$ is known. This is given by the statement that the He^3 impurity takes part only in the normal flow, i.e. $\mathbf{i} = \rho c \mathbf{v}_n$.†

§140. **Dissipative processes in superfluids**

To take account of dissipative processes in the equations of superfluid dynamics we must, as with ordinary fluids, include additional terms linear in the spatial derivatives of the velocities and the temperature. The form of these terms can be established with certainty from the requirements imposed by the law of increase of entropy and Onsager's principle of the symmetry of the kinetic coefficients (I. M. Khalatnikov 1952).

As before, ρ and $\mathbf{j}$ are the mass and momentum of unit volume of the fluid. The

---

† The full derivation of the equations of fluid dynamics for mixtures is given by I. M. Khalatnikov, *Theory of Superfluidity* (*Teoriya sverkhtekuchesti*), Moscow 1971, chapter XIII. These equations cease to be valid at very low temperatures, when the elementary excitations due to the impurity atoms show quantum degeneracy.

continuity condition is again (139.3). In equations (139.4), (139.6) and (139.7), we must add terms on the right:

$$\frac{\partial j_i}{\partial t}+\frac{\partial \Pi_{ik}}{\partial x_k}=-\frac{\partial \Pi'_{ik}}{\partial x_k}, \tag{140.1}$$

$$\partial \mathbf{v}_s/\partial t+\mathbf{grad}\,(\tfrac{1}{2}v_s^2+\mu)=-\mathbf{grad}\,\phi', \tag{140.2}$$

$$\partial E/\partial t+\operatorname{div}\mathbf{Q}=-\operatorname{div}\mathbf{Q}'. \tag{140.3}$$

The entropy equation no longer has the form of the conservation equation (139.5); on the contrary, Π', ϕ' and $\mathbf{Q}'$ must be determined so as to ensure that the entropy increases. To do so, we again substitute in the energy conservation equation (140.3) the derivative $\partial E_0/\partial t$ expressed by means of (139.9), and then eliminate the derivatives $\dot\rho$, $\partial\mathbf{j}/\partial t$, $\dot{\mathbf{v}}_s$ by means of (139.3), (140.1) and (140.2). Here it is assumed that $\mathbf{Q}$ and Π are given by the known equations (139.11) and (139.12); consequently, all terms cancel except those involving the entropy or the dissipative quantities Π', $\mathbf{Q}'$ and ϕ'. The result is

$$T\left\{\frac{\partial(\rho s)}{\partial t}+\operatorname{div}(\rho s\mathbf{v}_n)\right\}=-\operatorname{div}\{\mathbf{Q}'+\rho_s\mathbf{w}\phi'-(\Pi'\mathbf{v}_n)+\phi'\operatorname{div}(\rho_s\mathbf{w})-\Pi'_{ik}\,\partial v_{ni}/\partial x_k, \tag{140.4}$$

where again $\mathbf{w}=\mathbf{v}_n-\mathbf{v}_s$.

The expressions linear in the gradients for Π', ϕ' and $\mathbf{Q}'$ which ensure the increase of the entropy are†

$$\Pi'_{ik}=-\eta\left(\frac{\partial v_{ni}}{\partial x_k}+\frac{\partial v_{nk}}{\partial x_i}-\tfrac{2}{3}\delta_{ik}\operatorname{div}\mathbf{v}_n\right)-\delta_{ik}\zeta_1\operatorname{div}(\rho_s\mathbf{w})-\delta_{ik}\zeta_2\operatorname{div}\mathbf{v}_n, \tag{140.5}$$

$$\phi'=\zeta_3\operatorname{div}(\rho_s\mathbf{w})+\zeta_4\operatorname{div}\mathbf{v}_n, \tag{140.6}$$

$$\mathbf{Q}'=-\phi'\rho_s\mathbf{w}+(\Pi'\cdot\mathbf{v}_n)-\kappa\,\mathbf{grad}\,T; \tag{140.7}$$

Π'_{ik} includes a combination of derivatives of $\mathbf{v}_n$ with zero trace, as in ordinary fluid dynamics. From Onsager's principle,

$$\zeta_1=\zeta_4, \tag{140.8}$$

leaving five independent kinetic coefficients.‡

Finally, substituting the expressions (140.5)–(140.7) in (140.4), we have after some simple algebra

$$T\left\{\frac{\partial(\rho s)}{\partial t}+\operatorname{div}\left(\rho s\mathbf{v}_n-\frac{\kappa}{T}\mathbf{grad}\,T\right)\right\}=2R, \tag{140.9}$$

---

† Here we also use the condition that rotation of the normal part of the fluid as a whole, with $\mathbf{v}_n=\boldsymbol{\Omega}\times\mathbf{r}$, should not cause any dissipation; cf. §15.

‡ We shall not give here in full the arguments entirely similar to those in §59, for example, but merely note that ζ_1 is the coefficient of $\operatorname{div}(\rho_s\mathbf{w})$ in Π', and on the right of (140.4) this term appears multiplied by $\operatorname{div}\mathbf{v}_n$; conversely, ζ_4 is the coefficient of $\operatorname{div}\mathbf{v}_n$ in ϕ', and on the right of (140.4) this is multiplied by $\operatorname{div}(\rho_s\mathbf{w})$.

where

$$2R = \tfrac{1}{2}\eta\left(\frac{\partial v_{ni}}{\partial x_k} + \frac{\partial v_{nk}}{\partial x_i} - \tfrac{2}{3}\delta_{ik}\,\mathrm{div}\,\mathbf{v}_n\right)^2 +$$

$$+ 2\zeta_1\,\mathrm{div}\,\mathbf{v}_n\,\mathrm{div}\,\rho_s\mathbf{w} + \zeta_2(\mathrm{div}\,\mathbf{v}_n)^2 + \zeta_3\,(\mathrm{div}\,\rho_s\mathbf{w})^2 + (\kappa/T)(\mathbf{grad}\,T)^2; \qquad (140.10)$$

R is called the *dissipative function*. This is analogous to the general equation (49.5) of heat transfer in ordinary fluid dynamics.† The right-hand side determines the rate of increase of the entropy of the fluid and must be positive definite. It follows that all the coefficients $\eta, \zeta_1, \zeta_2, \zeta_3, \kappa$ are positive, and also $\zeta_1{}^2 \leqslant \zeta_2\zeta_3$. The first viscosity η, related to the normal flow, is analogous to the viscosity of an ordinary fluid; κ is formally analogous to the thermal conductivity of an ordinary fluid. The second viscosity coefficients here are three in number ($\zeta_1, \zeta_2, \zeta_3$), instead of one as in ordinary fluid dynamics.

One further remark is, however, necessary regarding these results. The energy dissipated in the fluid is, of course, invariant under a Galilean transformation of the reference frame. The derivatives of the velocity satisfy this condition, of course, but in a superfluid the velocity difference $\mathbf{w} = \mathbf{v}_n - \mathbf{v}_s$ is also a Galilean invariant. The dissipative fluxes in a superfluid may therefore depend not only on the gradients of the velocities and thermodynamic quantities, but also on $\mathbf{w}$ itself. As already noted in §139, this difference must in practice be regarded as small, and in that sense the expressions (140.5) and (140.6) contain not all the terms that are in principle possible, but only the largest terms.‡

PROBLEM

Separate the equations for normal and superfluid flow in an incompressible superfluid (assuming not only the total density ρ but also ρ_s and ρ_n separately to be constant).

SOLUTION. The dissipative terms in the entropy equation are second-order small quantities and can be omitted here; then s = constant also, and from equations (139.3) and (139.5) we have $\mathrm{div}\,\mathbf{v}_s = \mathrm{div}\,\mathbf{v}_n = 0$. In the momentum flux density tensor, we retain the term linear in the velocity gradients which arises from the viscosity of the normal flow:

$$\Pi'_{ik} = -\eta\left(\frac{\partial v_{ni}}{\partial x_k} + \frac{\partial v_{nk}}{\partial x_i}\right).$$

Substitution of this, with Π_{ik} from (139.12), gives

$$\rho_s\partial\mathbf{v}_s/\partial t + \rho_n\partial\mathbf{v}_n/\partial t + \rho_s(\mathbf{v}_s\cdot\mathbf{grad})\mathbf{v}_s + \rho_n(\mathbf{v}_n\cdot\mathbf{grad})\mathbf{v}_n$$

$$= -\mathbf{grad}\,p + \eta\,\mathrm{div}\,\mathbf{v}_n,$$

† The comments at the end of §49 about the definition of the entropy in a state close to thermodynamic equilibrium remain valid here.

‡ If we ignore this condition, the variety of admissible terms in the dissipative fluxes becomes much greater (and the kinetic coefficients themselves are then, in general, functions of w); for example, ϕ' contains terms in $\mathbf{w}\cdot\mathbf{grad}\,T$ and $w_1 w_k \partial v_{ni}/\partial x_k$. The total number of unknown kinetic coefficients which describe the dissipation in helium II is then 13 (A. Clark 1963). See S. J. Putterman, *Superfluid Hydrodynamics*, Amsterdam 1974, Appendix VI.

It may be noted here that the terms in $\mathrm{div}\,(\rho_s\mathbf{w})$ have been included in (140.5) and (140.6) because this combination of derivatives arises naturally in the exact equations. To the accuracy used, it would be correct to write $\rho_s\,\mathrm{div}\,\mathbf{w}$.

or

$$\rho_n \partial \mathbf{v}_n/\partial t + \rho_n(\mathbf{v}_n \cdot \mathbf{grad})\mathbf{v}_n + \rho_s \,\mathbf{grad}\, \tfrac{1}{2}v_s^2 + \rho_s \,\mathbf{grad}\, \partial\phi/\partial t$$
$$= -\mathbf{grad}\, p + \eta \,\mathrm{div}\, \mathbf{v}_n,$$

with the superfluid flow potential according to $\mathbf{v}_s = \mathbf{grad}\, \phi_s$, using the fact that $(\mathbf{v}_s \cdot \mathbf{grad})\mathbf{v}_s = \mathbf{grad}\, \frac{1}{2}v_s^2$. Since $\mathrm{div}\, \mathbf{v}_s = 0$, the potential ϕ_s satisfies Laplace's equation, $\triangle \phi_s = 0$. We use as auxiliary quantities the "pressures" p_n and p_s of the normal and superfluid flows: $p = p_0 + p_n + p_s$, where p_0 is the pressure at infinity, and p_s is given by the usual formula for an ideal fluid:

$$p_s = -\rho_s \partial\phi_s/\partial t - \tfrac{1}{2}\rho_s v_s^2.$$

The equation for the velocity $\mathbf{v}_n$ then becomes

$$\frac{\partial \mathbf{v}_n}{\partial t} + (\mathbf{v}_n \cdot \mathbf{grad})\mathbf{v}_n = -\frac{1}{\rho_n}\mathbf{grad}\, p_n + \frac{\eta}{\rho_n}\triangle \mathbf{v}_n.$$

This equation is formally identical with the Navier–Stokes equation for a fluid with density ρ_n and viscosity η.

Thus the problem of the flow of incompressible helium II reduces to two problems of ordinary fluid dynamics, one for an ideal fluid and the other for a viscous fluid. The superfluid flow is determined by Laplace's equation with a boundary condition on the normal derivative $\partial\phi_s/\partial n$, as in the ordinary problem of potential flow of an ideal fluid past a body. The normal flow is determined by the Navier–Stokes equation, with the same boundary condition on $\mathbf{v}_n$ (in the absence of heat exchange between the surface and the fluid) as in ordinary flow of a viscous fluid. The pressure distribution is then determined by $p_0 + p_n + p_s$.

To determine the temperature distribution, we write in (139.6), with μ from (139.14), $\mathbf{v}_s = \mathbf{grad}\, \phi_s$ and, integrating, we obtain $\mu(p, T) + \frac{1}{2}v_s^2 - \frac{1}{2}(\rho_n/\rho)(\mathbf{v}_n - \mathbf{v}_s)^2 + \partial\phi_s/\partial t = \text{constant}$. The changes of temperature and pressure in an incompressible fluid are small, and we have as far as terms of the first order $\mu - \mu_0 = -s(T - T_0) + (p - p_0)/\rho$, where T_0 and p_0 are the temperature and pressure at infinity. Substituting this expression in the above integral, and using p_n and p_s, we obtain

$$T - T_0 = \frac{\rho_n}{\rho s}\left[\frac{p_n}{\rho_n} - \frac{p_s}{\rho_s} - \tfrac{1}{2}(\mathbf{v}_n - \mathbf{v}_s)^2\right].$$

§141. The propagation of sound in superfluids

Let us apply the equations of fluid dynamics for helium II to the propagation of sound in it. As usual, the velocities in the sound wave are supposed small, and the density, pressure and entropy almost equal to their constant equilibrium values. Then we can linearize the equations, neglecting the terms quadratic in the velocity in (139.12)–(139.14), and regard the entropy ρs as constant in the term $\mathrm{div}\,(\rho s \mathbf{v}_n)$ in (139.5) (since the term already contains the small quantity $\mathbf{v}_n$). Thus the equations of fluid dynamics become

$$\partial\rho/\partial t + \mathrm{div}\, \mathbf{j} = 0, \tag{141.1}$$

$$\partial(\rho s)/\partial t + \rho s \,\mathrm{div}\, \mathbf{v}_n = 0, \tag{141.2}$$

$$\partial \mathbf{j}/\partial t + \mathbf{grad}\, p = 0, \tag{141.3}$$

$$\partial \mathbf{v}_s/\partial t + \mathbf{grad}\, \mu = 0. \tag{141.4}$$

Differentiating (141.1) with respect to time and substituting (141.3), we obtain

$$\partial^2\rho/\partial t^2 = \triangle p. \tag{141.5}$$

By the thermodynamic relation $d\mu = -s\,dT + dp/\rho$, we have $\mathbf{grad}\, p = \rho s \,\mathbf{grad}\, T$

$+\rho\,\mathbf{grad}\,\mu$. Substituting $\mathbf{grad}\,p$ from (141.3) and $\mathbf{grad}\,\mu$ from (141.4), we obtain

$$\rho_n \frac{\partial}{\partial t}(\mathbf{v}_n - \mathbf{v}_s) + \rho s\,\mathbf{grad}\,T = 0.$$

We take the divergence of this equation, substituting for $\operatorname{div}(\mathbf{v}_s - \mathbf{v}_n)$ the expression $(\rho/s\rho_s)\partial s/\partial t$, which follows from the equation

$$\begin{aligned}\frac{\partial s}{\partial t} &= \frac{1}{\rho}\frac{\partial(\rho s)}{\partial t} - \frac{s}{\rho}\frac{\partial \rho}{\partial t} \\ &= -s\operatorname{div}\mathbf{v}_n + (s/\rho)\operatorname{div}\mathbf{j} \\ &= (s\rho_s/\rho)\operatorname{div}(\mathbf{v}_s - \mathbf{v}_n).\end{aligned}$$

The result is

$$\partial^2 s/\partial t^2 = (\rho_s s^2/\rho_n)\triangle T. \qquad (141.6)$$

Equations (141.5) and (141.6) determine the propagation of sound in a superfluid. Since there are two equations, we see that there are two velocities of propagation of sound.

We write s, p, ρ and T as $s = s_0 + s'$, $p = p_0 + p'$, etc., where the primed letters are the small changes in the corresponding quantities in the sound wave, and those with the suffix zero (which we omit, for brevity) their constant equilibrium values. Then we can write

$$\rho' = \frac{\partial \rho}{\partial p}p' + \frac{\partial \rho}{\partial T}T', \qquad s' = \frac{\partial s}{\partial p}p' + \frac{\partial s}{\partial T}T',$$

and the equations (141.5) and (141.6) become

$$\frac{\partial \rho}{\partial p}\frac{\partial^2 p'}{\partial t^2} - \triangle p' + \frac{\partial \rho}{\partial T}\frac{\partial^2 T'}{\partial t^2} = 0,$$

$$\frac{\partial s}{\partial p}\frac{\partial^2 p'}{\partial t^2} + \frac{\partial s}{\partial T}\frac{\partial^2 T'}{\partial t^2} - \frac{\rho_s s^2}{\rho_n}\triangle T' = 0.$$

We seek a solution of these equations in the form of a plane wave, in which p' and T' are proportional to a factor $e^{-i\omega(t - x/u)}$ (the velocity of sound being here denoted by u). The condition of compatibility of the two equations is

$$u^4\frac{\partial(s,\rho)}{\partial(T,p)} - u^2\left(\frac{\partial s}{\partial T} + \frac{\rho_s s^2}{\rho_n}\frac{\partial \rho}{\partial p}\right) + \frac{\rho_s s^2}{\rho_n} = 0,$$

where $\partial(s,\rho)/\partial(T,p)$ denotes the Jacobian of the transformation from s, ρ to T, p. By a simple transformation, using the thermodynamic relations, this equation can be reduced to

$$u^4 - u^2\left[\left(\frac{\partial p}{\partial \rho}\right)_s + \frac{\rho_s T s^2}{\rho_n c_v}\right] + \frac{\rho_s T s^2}{\rho_n c_v}\left(\frac{\partial p}{\partial \rho}\right)_T = 0, \qquad (141.7)$$

c_v being the specific heat per unit mass. This quadratic equation in u^2 gives the two velocities of propagation of sound in helium II. For $\rho_s = 0$, one root is zero, and we obtain, as we should expect, only the ordinary velocity of sound $u = \sqrt{(\partial p/\partial \rho)_s}$.

The specific heats c_p and c_v of helium II are actually very nearly the same at all temperatures far from the λ-point (since the coefficient of thermal expansion is small). By a well-known thermodynamic formula, the isothermal and adiabatic compressibilities are then very nearly the same also: $(\partial p/\partial \rho)_T = (\partial p/\partial \rho)_s c_v/c_p \cong (\partial p/\partial \rho)_s$. Denoting the common value of c_p and c_v by c, and that of $(\partial p/\partial \rho)_T$ and $(\partial p/\partial \rho)_s$ by $\partial p/\partial \rho$, we obtain from equation (141.7) the following expressions for the velocities of sound:

$$u_1 = \sqrt{(\partial p/\partial \rho)}, \qquad u_2 = \sqrt{(Ts^2 \rho_s/c\rho_n)}. \tag{141.8}$$

One of these, u_1, is almost constant, while the other, u_2, depends markedly on temperature, vanishing with ρ_s at the λ-point.†

Near the λ-point, however, the thermal expansion coefficient is not small, and the difference between c_p and c_v cannot be neglected. To derive an expression for u_2 in this case, we must omit the second term in the square brackets in (141.7), which contains ρ_s, and the u^4 term, which in this case is small, since u_2 tends to zero. Also, we have to put $\rho_n \cong \rho$. The result is

$$u_2 = \sqrt{(Ts^2 \rho_s/c_p \rho)}. \tag{141.9}$$

For u_1, we get (141.8), where $\partial p/\partial \rho$ is to be taken as $(\partial p/\partial \rho)_s$, i.e. the ordinary formula for the velocity of sound.

Regarding the derivation of (141.9), it is to be noted that this is valid only at sufficiently low frequencies, the upper limit of frequency decreasing as the λ-point is approached. The reason is that, as already mentioned in the penultimate footnote to §139, the order parameter relaxation time τ increases indefinitely near the λ-point; but formula (141.9), which does not take account of sound dispersion and absorption, is valid only when $\omega\tau \ll 1$. For u_1 there is an additional damping near the λ-point because of order parameter relaxation, in accordance with the general statements in §81.

At very low temperatures, where nearly all the elementary excitations in the fluid are phonons, the quantities ρ_n, c and s are related by‡ $c = 3s$, $\rho_n = cT\rho/3u_1^2$, and $\rho_s \cong \rho$. Substituting these expressions in formula (141.8) for u_2, we find $u_2 = u_1/\sqrt{3}$. Thus, as the temperature tends to zero, the velocities u_1 and u_2 tend to finite limits, and their ratio tends to $\sqrt{3}$.

In order to elucidate more clearly the physical nature of the two kinds of sound wave in helium II, let us consider a plane sound wave (E. M. Lifshitz 1944). In such a wave, the velocities $\mathbf{v}_n$, $\mathbf{v}_s$ and the variable parts T', p' of the temperature and pressure are proportional to one another. We introduce proportionality coefficients by

$$\mathbf{v}_n = a\mathbf{v}_s, \qquad p' = bv_s, \qquad T' = cv_s. \tag{141.10}$$

A simple calculation, using equations (141.1)–(141.6), and working to the necessary accuracy, gives

$$\left.\begin{aligned} a_1 &= 1 + \frac{\beta\rho}{\rho_s s}\frac{u_1^2 u_2^2}{(u_1^2 - u_2^2)}, \qquad b_1 = \rho u_1, \qquad c_1 = \frac{\beta T u_1^3}{c(u_1^2 - u_2^2)}, \\ &\text{and} \\ a_2 &= -\frac{\rho_s}{\rho_n} + \frac{\beta\rho}{\rho_n s}\frac{u_1^2 u_2^2}{(u_1^2 - u_2^2)}, \qquad b_2 = \frac{\beta\rho u_1^2 u_2^3}{s(u_1^2 - u_2^2)}, \qquad c_2 = -u_2/s. \end{aligned}\right\} \tag{141.11}$$

---

† The problem of sound propagation in mixtures of liquid He^4 and He^3 is discussed in chapter XIII of Khalatnikov's book cited at the end of §139.

‡ These formulae are easily derived from those for the thermodynamic quantities of helium II (*SP* 2, §§22, 23).

Here $\beta = -(1/\rho)\partial\rho/\partial T$ is the coefficient of thermal expansion; since it is small, the quantities which involve β are small in comparison with those which do not.

We see that, in a sound wave of the first type, $\mathbf{v}_n \approx \mathbf{v}_s$, i.e. to a first approximation the fluid in any given volume element oscillates as a whole in such a wave, the normal and superfluid parts moving together. This type of wave clearly corresponds to an ordinary sound wave in an ordinary fluid.

In a wave of the second type, however, we have $\mathbf{v}_n \cong -\rho_s \mathbf{v}_s/\rho_n$, i.e. the total flux density $\mathbf{j} = \rho_s\mathbf{v}_s + \rho_n\mathbf{v}_n \cong 0$. Thus, in a *second-sound* wave the superfluid and normal parts move in opposition, the centre of mass of any given volume element remaining at rest to a first approximation, and the total mass flux being zero. Such a wave is evidently peculiar to superfluids.

There is another important difference between the two types of wave, which is seen from formulae (141.11). In a sound wave of the ordinary type, the amplitude of the pressure oscillations is relatively large, while that of the temperature oscillations is small. In a second-sound wave, however, the relative amplitude of the temperature oscillations is large compared with that of the pressure oscillations. In this sense we can say that second-sound waves are undamped temperature waves.†

In an approximation in which the thermal expansion is neglected, second-sound waves are purely temperature oscillations (with $\mathbf{j} = 0$), while first-sound waves are pressure oscillations (with $\mathbf{v}_s = \mathbf{v}_n$). Accordingly, their equations of motion are completely separable: in equation (141.6), we write $s' = cT'/T$, obtaining

$$\partial^2 T'/\partial t^2 = u_2^{\ 2}\triangle T', \tag{141.12}$$

and in equation (141.5) we write $\rho' = p'\partial\rho/\partial p$, obtaining

$$\partial^2 p'/\partial t^2 = u_1^{\ 2}\triangle p'. \tag{141.13}$$

These properties of sound waves in helium II are closely related to the various methods of generating them (E. M. Lifshitz 1944). The usual mechanical methods of generating sound by means of oscillating solid bodies are extremely unsuitable for producing second sound, since the resulting sound intensity is negligibly small in comparison with that of the ordinary sound emitted at the same time. In helium II, however, there are also specific means of sound generation. These include the use of solid surfaces with a periodically varying temperature; the emitted second-sound intensity is in that case much greater than that of first sound, as is to be expected in view of the above-mentioned difference in the nature of the temperature oscillations in these waves (see Problems 1 and 2).

When a second-sound wave with high amplitude is propagated, its profile is gradually deformed by the non-linearity, and this ultimately causes discontinuities, as in the case of ordinary sound in ordinary fluids (cf. §§101, 102). Let us consider these effects for a one-dimensional travelling second-sound wave (I. M. Khalatnikov 1952).

In a one-dimensional travelling wave, all quantities (ρ, p, T, v_s, v_n) can be expressed as functions of one parameter, which may be taken, for example, as any one of these (§101). The velocity U of a point in the wave profile is dx/dt taken for a particular value of this parameter. The coordinate and time derivatives of each quantity are related by $\partial/\partial t = -U\partial/\partial x$.

† They have, of course, no connection with the damped temperature waves in an ordinary thermally conducting medium (§52).

Instead of v_s and v_n, it will be more convenient to use $v = j/\rho$ and $w = v_n - v_s$; we choose coordinates in which v is zero at the relevant point in the profile. The dynamical equations (139.3)–(139.6) (with Π, μ, ρ, s from (139.12)–(139.15)) give

$$-U\frac{\partial\rho}{\partial p}p' - U\rho^2\frac{\partial}{\partial p}\left(\frac{\rho_n}{\rho}\right)ww' + \rho v' = 0, \qquad (141.14)$$

$$p' + 2(\rho_s\rho_n/\rho)ww' - U\rho v' = 0, \qquad (141.15)$$

$$\left[-\rho U\frac{\partial s}{\partial T} + w\frac{\partial}{\partial T}(\rho_s s)\right]T' + sw\frac{\partial\rho_s}{\partial p}p' + \left[\rho_s s - Uw\frac{\partial\rho_n}{\partial T}\right]w' = 0, \qquad (141.16)$$

$$\left[-\rho s + Uw\frac{\partial\rho_n}{\partial T}\right]T' + \left[1 + Uwp\frac{\partial}{\partial p}\left(\frac{\rho_n}{\rho}\right)\right]p' + [\rho_n U - (\rho_n\rho_s/\rho)w]w' - [U\rho + w\rho_n]v' = 0. \qquad (141.17)$$

Here, all terms above the second order of smallness are omitted, and also all terms containing the thermal-expansion coefficient; the prime everywhere denotes differentiation with respect to the parameter.†

In the second-sound wave, the relative amplitude of the oscillations of p and v is much less than those of T and w; we can therefore also omit the terms in wp' and wv'. To determine U, it is sufficient to consider (141.16) and the difference of (141.15) and (141.17). The compatibility condition for the two linear equations thus obtained for T' and w' gives the quadratic

$$\rho_n U^2\frac{\partial s}{\partial T} - Uw\left[\frac{4\rho_s\rho_n}{\rho}\frac{\partial s}{\partial T} - 2s\frac{\partial\rho_n}{\partial T}\right] - \rho_s s^2 = 0,$$

whence

$$U = u_2 + w\left(\frac{2\rho_s}{\rho} - \frac{sT}{\rho_n c}\frac{\partial\rho_n}{\partial T}\right).$$

Here, u_2 is the local value of the second-sound velocity, varying from point to point in the wave profile together with the difference δT between the temperature and its equilibrium value. Expanding u_2 in powers of δT, we find

$$u_2 = u_{20} + (\partial u_2/\partial T)\delta T = u_{20} + (\partial u_2/\partial T)\rho_n u_2 w/\rho s,$$

where u_{20} is the equilibrium value of u_2. The final result is

$$U = u_{20} + w\frac{\rho_s sT}{\rho c}\frac{\partial}{\partial T}\log\frac{u_{20}{}^3 c}{T}. \qquad (141.18)$$

When the wave profile distortion is sufficiently great, discontinuities are formed (cf. §102), in this case temperature discontinuities. Their velocity of propagation is the mean of the velocities U on either side of the discontinuity:

$$c_{20} + \tfrac{1}{2}(w_1 + w_2)(\rho_s sT/\rho c)\partial \log(u_{20}{}^3 c/T)/\partial T \qquad (141.19)$$

where w_1 and w_2 are the values of w on either side.

The coefficient of w in (141.18) may be either positive or negative. The points where w is greater may accordingly either lead or lag behind those where it is less, and the

† Not the variable part of oscillatory quantities, as it did earlier in this section.

discontinuity correspondingly arises at either the forward or the backward front of the wave, unlike ordinary sound, where the shock wave is always formed at the forward front.

PROBLEMS

PROBLEM 1. Determine the ratio of intensities of the first and second sound emitted by a plane oscillating in a direction perpendicular to itself.

SOLUTION. We seek the velocities v_s (along the x-axis, which is perpendicular to the plane) in the first and second sound waves in the forms

$$v_{s1} = A_1 \cos\omega(t - x/u_1), \qquad v_{s2} = A_2 \cos\omega(t - x/u_2)$$

respectively. At the surface of the oscillating plane, the velocities v_s and v_n must be equal to the velocity of the plane, which we denote by $v_0 \cos\omega t$. This gives the equations $A_1 + A_2 = v_0$, $a_1 A_1 + a_2 A_2 = v_0$, where the coefficients a_1 and a_2 are given by (141.11). The (time) average energy density in a sound wave in helium II is $\rho_s \overline{v_s^2} + \rho_n \overline{v_n^2} = \frac{1}{2}A^2(\rho_s + \rho_n a^2)$; the energy flux (intensity) is obtained by multiplying by the corresponding velocity of sound u. The ratio of the intensities of the second and first sound waves is found to be

$$\frac{I_2}{I_1} = \frac{A_2^2(\rho_s + \rho_n a_2^2)u_2}{A_1^2(\rho_s + \rho_n a_1^2)u_1} \cong \frac{\beta^2 T u_2^3}{c u_1}.$$

Here we have assumed that $u_2 \ll u_1$, which is valid down to very low temperatures. The ratio is always small.

PROBLEM 2. The same as Problem 1, but for a surface whose temperature varies periodically.

SOLUTION. It is sufficient to use the boundary condition $j = 0$, which must hold at a fixed surface. This gives $\rho_s(A_1 + A_2) + \rho_n(a_1 A_1 + a_2 A_2) = 0$, whence

$$|A_2/A_1| = (\rho_n a_1 + \rho_s)/(\rho_n a_2 + \rho_s) \cong s/\beta u_2^2.$$

The ratio of intensities is found to be $I_2/I_1 = c/T\beta^2 u_1 u_2$. This is very large.

PROBLEM 3. Determine the velocity of sound propagated along a capillary whose diameter is much less than the viscous penetration depth $\delta \sim \sqrt{(\eta/\rho_n\omega)}$ (K. R. Atkins 1959).†

SOLUTION. Under the conditions stated, it can be assumed that the normal flow in the capillary is completely stopped by friction against the walls ($\mathbf{v}_n = 0$). The linearized equations (141.1), (141.2), and (141.4) become‡

$$\dot{\rho}' + \rho_s \operatorname{div} \mathbf{v}_s = 0, \qquad \dot{\mathbf{v}}_s + \mathbf{grad}\,\mu' = \dot{\mathbf{v}}_s - s\,\mathbf{grad}\,T' + \frac{1}{\rho}\mathbf{grad}\,p' = 0,$$

$$\partial(s\rho)/\partial t = \rho\dot{s}' + s\dot{\rho}' = 0,$$

where the prime denotes the variable parts of quantities in the wave. Again neglecting the thermal expansion of the fluid, we find from the third equation $p's/u_1^2 = -T'\rho c/T$. Now, eliminating v_s from the first two equations, we obtain the wave equation $\ddot{p}' - u^2 \triangle p' = 0$, in which the velocity of propagation u is given by

$$u^2 = (\rho_s/s)u_1^2 + (\rho_n/\rho)u_2^2.$$

PROBLEM 4. Find the absorption coefficients for first and second sound in helium II.

SOLUTION. The calculation is similar to that in §79 for sound in ordinary fluids; instead of (79.1), we use (140.10). Neglecting all terms which involve the thermal-expansion coefficient β (in (141.10), (141.11) and elsewhere), we find the absorption coefficients

$$\gamma_1 = \frac{\omega^2}{2\rho u_1^3}(\tfrac{4}{3}\eta + \zeta_2),$$

$$\gamma_2 = \frac{\omega^2 \rho_s}{2\rho\rho_n u_2^3}(\tfrac{4}{3}\eta + \zeta_2 + \rho^2\zeta_3 - 2\rho\zeta_1 + \rho_n\kappa/\rho_s c).$$

---

† This is usually called *fourth sound*. The name *third sound* is given to waves propagated in a film of helium II on a solid surface; they depend considerably on the van der Waals interaction between the fluid in the film and the solid.

‡ The momentum conservation equation (141.3) is to be omitted: it does not apply in these conditions, where an external force has to be exerted on the capillary in order to keep it at rest.

INDEX

Printed and bound by CPI Group (UK) Ltd, Croydon, CR0 4YY
03/10/2024
01040334-0014